208

Springer Texts in Statistics

Advisors:
Stephen Fienberg Ingram Olkin

Springer

New York
Berlin
Heidelberg
Barcelona
Budapest
Hong Kong
London
Milan
Paris
Santa Clara
Singapore
Tokyo

Springer Texts in Statistics

Continued at end of book

J.D. Jobson

Applied Multivariate
Data Analysis

Volume II: Categorical and Multivariate Methods

With 85 illustrations in 108 parts

With a diskette

Springer

J.D. Jobson
Faculty of Business
University of Alberta
Edmonton, Alberta T6G 2R6
Canada

Mathematics Subject Classification: 62-07, 62J05, 62J10, 62K10, 62K15

Library of Congress Cataloging-in-Publication Data
Jobson, J.D.
    Applied multivariate data analysis.
    (Springer texts in statistics)
    System requirements for disk: IBM PC.
    Includes bibliographical references and indexes.
    Contents: v. 1. Regression and experimental
design   v. 2. Categorical and multivariate methods.
    1. Multivariate analysis. I. Title. II. Series.
QA278.J58   1991   519.5   91-221219
ISBN 0-387-97660-4 (v 1: Springer-Verlag : New York :
Berlin : Heidelberg : alk. paper)
ISBN 3-540-97660-4 (v. 1 : Springer-Verlag : Berlin :
Heidelberg : New York : alk. paper)
ISBN 0-387-97804-6 (v. 2 : Springer-Verlag : New York :
Berlin : Heidelberg : alk. paper)
ISBN 3-540-97804-6 (v. 2 : Springer-Verlag : Berlin :
Heidelberg : New York : alk. paper)

Printed on acid-free paper.

Production managed by Francine McNeill; manufacturing supervised by Vincent Scelta.
Photocomposed pages prepared from the author's TₑX file.
Printed and bound by Hamilton Printing Company, Rennselaer, NY.
Printed in the United States of America.

9 8 7 6 5 4 3

ISBN 0-387-97804-6 Springer-Verlag New York Berlin Heidelberg
ISBN 3-540-97804-6 Springer-Verlag Berlin Heidelberg New York

To Leslie and Heather

Some Quotations From Early Statisticians

All sciences of observation follow the same course. One begins by observing a phenomenon, then studies all associated circumstances, and finally, if the results of observation *can be expressed numerically* [Quetelet's italics], estimates the intensity of the causes that have concurred in its formation. This course has been followed in studying purely material phenomena in physics and astronomy; it will likely also be the course followed in the study of phenomena dealing with moral behavior and the intelligence of man. Statistics begins with the gathering of numbers; these numbers, collected on a large scale with care and prudence, have revealed interesting facts and have led to the conjecture of laws ruling the moral and intellectual world, much like those that govern the material world. It is the whole of these laws that appears to me to constitute *social physics*, a science which, while still in its infancy, becomes incontestably more important each day and will eventually rank among those sciences most beneficial to man. (Quetelet, 1837)

The investigation of causal relations between economic phenomena presents many problems of peculiar difficulty, and offers many opportunities for fallacious conclusions. Since the statistician can seldom or never make experiments for himself, he has to accept the data of daily experience, and discuss as best he can the relations of a whole group of changes; he cannot, like the physicist, narrow down the issue to the effect of one variation at a time. The problems of statistics are in this sense far more complex than the problems of physics. (Yule, 1897)

Some people hate the very name of statistics, but I find them full of beauty and interest. Whenever they are not brutalized, but delicately handled by the higher methods, and are warily interpreted, their power of dealing with complicated phenomena is extraordinary. They are the only tools by which an opening can be cut through the formidable thicket of difficulties that bars the path of those who pursue the Science of man. (Galton, 1908)

Preface

A Second Course in Statistics

The past decade has seen a tremendous increase in the use of statistical data analysis and in the availability of both computers and statistical software. Business and government professionals, as well as academic researchers, are now regularly employing techniques that go far beyond the standard two-semester, introductory course in statistics. Even though for this group of users short courses in various specialized topics are often available, there is a need to improve the statistics training of future users of statistics while they are still at colleges and universities. In addition, there is a need for a survey reference text for the many practitioners who cannot obtain specialized courses.

With the exception of the statistics major, most university students do not have sufficient time in their programs to enroll in a variety of specialized one-semester courses, such as data analysis, linear models, experimental design, multivariate methods, contingency tables, logistic regression, and so on. There is a need for a second survey course that covers a wide variety of these techniques in an integrated fashion. It is also important that this second course combine an overview of theory with an opportunity to practice, including the use of statistical software and the interpretation of results obtained from real data.

Topics

This two-volume survey is designed to provide a second two-semester course in statistics. The first volume outlines univariate data analysis and provides an extensive overview of regression models. The first volume also surveys the methods of analysis of variance and experimental design including their relationship to the regression model. The second volume begins with a survey of techniques for analyzing multidimensional contingency tables and then outlines the traditional topics of multivariate methods. It also includes discussions of logistic regression, cluster analysis, multidimensional scaling and correspondence analysis, which are not always included in surveys of multivariate methods. In each volume an appendix is provided to review the basic concepts of linear and matrix algebra. The appendix also includes a series of exercises in linear algebra for student practice.

Mathematics Background

The text assumes a background equivalent to one semester each of linear algebra and calculus, as well as the standard two-semester introductory course in statistics. Calculus is almost never used in the text other than in the theoretical questions at the end of each chapter. The one semester of calculus is an indication of the ideal mathematics comfort level. The linear algebra background is needed primarily to understand the presentation of formulae. Competence with linear algebra however is required to complete many of the theoretical questions at the end of each chapter. These background prerequisites would seem to be a practical compromise given the wide variety of potential users.

Examples and Exercises

In addition to an overview of theory, the text also includes a large number of examples based on actual research data. Not only are numerical results given for the examples but interpretations for the results are also discussed. The text also provides data analysis exercises and theoretical questions for student practice. The data analysis exercises are based on real data which is also provided with the text. The student is therefore able to improve by "working out" on the favorite local software. The theoretical questions can be used to raise the theoretical level of the course or can be omitted without any loss of the applied aspects of the course. The theoretical questions provide useful training for those who plan to take additional courses in statistics.

Use as a Text

The two volumes can be used independently for two separate courses. Volume I can be used for a course in regression and design, and Volume II can be used for a course in categorical and multivariate methods. A quick review of multiple regression and analysis of variance may be required if the second volume is to be used without the first. If the entire text is to be used in two semesters some material in each chapter can be omitted. A number of sections can be left for the student to read or for the student's future reference. Large portions of most chapters and/or entire topics can be omitted without affecting the understanding of other topics discussed later in the text. A course in applied multivariate data analysis for graduate students in a particular field of specialization can be derived from the text by concentrating on a particular selection of topics.

This two-volume survey should be useful for a second course in statistics for most college juniors or seniors. Also, for the undergraduate statistics major, this text provides a useful second course, which can be combined with other specialized courses in time series, stochastic processes, sampling theory, nonparametric statistics and mathematical statistics. Because the text includes the topics normally found in traditional second courses, such as regression analysis or multivariate methods, this course provides a broader substitute by also including other topics such as data analysis, multidimensional contingency tables, logistic regression, correspondence analysis and multidimensional scaling. The set of theoretical questions in the book can provide useful practice for statistics majors who have already been exposed to mathematical statistics.

For graduate students in business and the social and biological sciences, this survey of applied multivariate data analysis is a useful first year graduate course, which could then be followed by other more specialized courses, such as econometrics, structural equation models, time series analysis or stochastic processes. By obtaining this background early in the graduate program the student is then well prepared to read the research literature in the chosen discipline and at a later stage to analyze research data. This course is also useful if taken concurrently with a course in the research methodology of the chosen discipline. I have found the first year of the Ph.D. program to be the ideal time for this course, since later in their programs Ph.D. students are too often preoccupied with their own area of specialization and research tasks.

Author's Motivation and Use of Text

The author's motivation for writing this text was to provide a two-semester overview of applied multivariate data analysis for beginning Ph.D. students in the Faculty of Business at the University of Alberta. The quantitative background assumed for the business Ph.D. student using this text is equivalent to what is required in most undergraduate business programs in North America — one semester each of linear algebra and calculus and a two-semester introduction to statistics. Many entering Ph.D. students have more mathematics background but do not usually have more statistics background. A selection of topics from the text has also been used for an elective course in applied multivariate data analysis for second year MBA students. For the MBA elective course much less emphasis is placed on the underlying theory.

Because of the many different fields of interest within business Ph.D. programs — Accounting, Finance, Marketing, Organization Analysis and Industrial Relations — the topical needs, interests and level of mathematical sophistication of the graduate students differ greatly. Some will pursue a strong statistics minor, whereas others will take very little statistics training beyond this course.

In my Ph.D. class the wide variety of needs are handled simultaneously by assigning portfolios of theoretical questions to the statistics minor student, while the less theoretically oriented students are assigned a paper. The paper topic may involve a discussion of the application of one or more of the statistical techniques to a particular field or an overview of techniques not discussed in the text. A small number of classes are devoted exclusively to the discussion of the theory questions. For the theory classes only the "theory folk" need attend. All students are required to complete data analysis exercises and to provide written discussions of the results. For the data analysis exercises great emphasis is placed on the quality of the interpretation of the results. Graduate students often have greater difficulty with the interpretation of results than with the understanding of the principles.

Quotations

The quotations by Quetelet (1837) and Yule (1897) were obtained from pages 193 and 348, respectively, of *The History of Statistics: The Measurement of Uncertainty Before: 1900*, by Stephen Stigler, published by Harvard University Press, Cambridge, MA, 1986.

The quotation by Galton (1908) was obtained from *An Introduction to Mathematical Statistics and its Applications*, Second Edition, by Richard J. Larsen and Morris L. Marx, published by Prentice–Hall, 1986.

Acknowledgments

The production of this text has benefited greatly from the input and assistance of many individuals. The Faculty of Business at the University of Alberta has born most of the cost of production. Steve Beveridge, John Brown, Royston Greenwood, Jim Newton and John Waterhouse were helpful in making funds available. John Brown and John Waterhouse also provided much encouragement during the development and early implementation of this text in our Ph.D. program.

The bulk of the typing has been done by two very able typists, Anna Fujita and Shelley Hey. Both individuals have been tireless in providing error-free typing through the uncountable number of drafts. Shelley Hey's patience with the excessive number of "just one more little change" encounters and her willingness to meet all panic deadlines is greatly appreciated. The numerous graphs and figures could not have been carried out without the capable assistance of Anna Fujita.

The examples and data analysis exercises have been generated from data sets provided to me by my colleagues and former students. Colleagues who have allowed me to use their research data or have suggested examples include Alice and Masao Nakamura, Rodney Schneck, Ken Lemke, Chris Vaughn, John Waterhouse and Chris Janssen. Graduate students who have gifted me their data include Rebecca Johnson, Nancy Button, Nancy Keown, Pamela Norton, Diane Ewanishan, Clarke Carson, Caroline Pinkert-Rust, Frank Kobe and Cam Morrell. I am also grateful to G.C. McDonald for the use of the air pollution data and to SPSS for the bank employee salary data.

Mary Allen, Subash Bagui, Alice Nakamura, Caroline Pinkert-Rust and Tom Scott read parts of the manuscript and were extremely helpful in improving the overall presentation. I am also grateful to the graduate students who read the final drafts of the text as part of the course requirements. This class was extremely helpful in locating errors and in suggesting improvements. This class of patient students included Greg Berry, John Buffington, Alessandro Butti, Carla Carnaghan, Alain Duncan, Pietro Gottardo, Mark Harcourt, Ujwal Kayande, Vitor Marciano, Michael Mauws, Shaun McQuitty, Andrew Parsons, Vaughan Radcliffe, Peter Roberts, Ashish Sinha, Conor Vibert and Wayne Yu.

The staff at Springer-Verlag including the statistics editor, Dr. Martin Gilchrist, and the production editor, Karen Phillips, have been instrumental in converting the manuscript to book form.

I am also grateful to the University of Alberta who provided my undergraduate training in mathematics, 1959–1963; to Iowa State University who provided my graduate training in statistics, 1968–1971; and to the University of North Carolina who provided a statistics theory update during my study leave in 1980–81. I am extremely fortunate to have been exposed to so many great teachers.

Last and most importantly I am indebted to my wife, Leone, who cheerfully attended to most of my domestic chores while I was preoccupied with this task. At times I think she was the only one who believed that the project would ever be completed. Perhaps that was wishful thinking, as the book was often spread out all over the house. A vote of thanks is also due my two daughters, Leslie and Heather, who cheerfully accepted DAD's excessive devotion to "the book."

Contents

An Alternative Test Statistic-F, Interpretation of the
Discriminant Analysis Solution, Interpretation Using
Correlations, Graphical Approach to Group Characterization, Comparison of Correlation Coefficients
and Discriminant Function Coefficients, Effect of Correlation Structure on Discriminant Analysis, Discriminant Analysis and Canonical Correlation, Discriminant Analysis and Dimension Reduction

Contents of Volume I

6

Contingency Tables

This chapter begins with an introduction for Volume II and then presents a survey of the techniques available for analyzing contingency tables. The introduction consists of a discussion of data matrices measurement scales and an outline of techniques presented in Volume II. The discussion of contingency tables begins in the second section with a review of bivariate analysis for two categorical random variables and includes a discussion of inference techniques for two-dimensional tables. The discussion of two-dimensional tables also includes an introduction to the use of loglinear models. The third section presents a discussion of the application of loglinear models to multidimensional tables based on the maximum likelihood approach to estimation. The logit model is also introduced as a special case of the log-linear model. The last section of the chapter outlines the weighted least squares approach to modeling categorical data. The weighted least squares approach affords a greater variety of models than the maximum likelihood method.

6.1 Multivariate Data Analysis, Data Matrices and Measurement Scales

The past decade has seen tremendous growth in the availability of both computer hardware and statistical software. As a result, the use of multivariate statistical techniques has increased to include most fields of scientific research and many areas of business and public management. In both research and management domains there is increasing recognition of the need to analyze data in a manner that takes into account the interrelationships among variables. *Multivariate data analysis* refers to a wide assortment of such descriptive and inferential techniques. In contrast to univariate statistics, we are concerned with the jointness of the measurements. Multivariate analysis is concerned with the relationships among the measurements across a sample of individuals, items or objects.

6.1.1 DATA MATRICES

The raw input to multivariate statistics procedures is usually an $n \times p$ (n rows by p columns) rectangular array of real numbers called a *data matrix*. The data matrix summarizes observations made on n objects. Each of the n objects is characterized with respect to p variables. The values attained by the variables may represent the measurement of a quantity or a numerical code for a classification scheme. The term object may mean an individual or a unit, whereas the term *variable* is synonomous with attribute, characteristic, response or item. The data matrix is denoted by the $n \times p$ matrix $\mathbf{X}$, and the column vectors of the matrix are denoted by $\mathbf{x}_1, \mathbf{x}_2, \ldots, \mathbf{x}_p$ for the p variables. The elements of $\mathbf{X}$ are denoted by x_{ij}, $i = 1, 2, \ldots, n;\ j = 1, 2, \ldots, p$.

Data Matrix

Variables

	$\mathbf{x}_1$	$\mathbf{x}_2$	$\mathbf{x}_3$	$\cdots$	$\mathbf{x}_p$	Objects
	x_{11}	x_{12}	x_{13}	$\cdots$	x_{1p}	1
$\mathbf{X}=$	x_{21}	x_{22}	x_{23}	$\cdots$	x_{2p}	2
	x_{31}	x_{32}	x_{33}	$\cdots$	x_{3p}	3
	$\vdots$					$\vdots$
	x_{n1}	x_{n2}	x_{n3}		x_{np}	n

The following four examples of data matrices are designed to show the variety of data types that can be encountered.

Example 1. The bus driver absentee records for a large city transit system were sampled in four different months of a calendar year. The purpose of the study was to determine a model to predict absenteeism. For each absentee record, the variables month, day, bus garage, shift type, scheduled off days, seniority, sex and time lost were recorded. Table 6.1 shows the obervations for 10 records.

Example 2. The top 500 companies in Canada ranked by sales dollars in 1985 were compared using information on percent change in sales, net income, rank on net income, percent change in net income, percent return on equity, value of total assets, rank on total assets, ratio of current assets to current liabilities (current ratio) and number of employees. Table 6.2 contains the data for the top ten companies. In this study the researcher was interested in the properties of the distributions of various quantities.

Example 3. A sample of police officers were asked to respond to questions regarding the amount of stress they encounter in performing their regular duties. The officers also responded to questions seeking personal information such as age, education, rank and years of experience. The purpose of the analysis was to identify the dimensions of stress.

TABLE 6.1. Sample From Bus Driver Absenteeism Survey

Month	Day	Garage	Shift	Days Off	Seniority	Sex	Time Lost
1	1	5	3	6	5	0	7.5
1	2	5	13	1	9	1	7.5
4	6	3	9	2	8	0	7.5
2	3	3	7	3	7	1	7.5
3	5	3	7	1	8	0	2.5
1	1	4	3	1	10	0	4.2
1	7	1	5	6	5	0	7.5
2	6	5	13	1	2	0	7.5
3	7	5	10	4	5	0	7.5
4	3	1	9	2	6	1	7.5

The data in Table 6.3 are a sample of responses obtained for 18 stress items and the personal variables age, education, rank and years of experience.

The 18 stress variables are measures of stress due to 1. insufficient resources, 2. unclear job responsibilities, 3. personality conflicts, 4. investigation where there is serious injury or fatality, 5. dealing with obnoxious or intoxicated people, 6. having to use firearms, 7. notifying relatives about death or serious injury, 8. tolerating verbal abuse in public, 9. unsuccessful attempts to solve a series of offences, 10. lack of availability of ambulances, doctors, and so on, 11. poor presentation of a case by the prosecutor resulting in dismissal of the charge, 12. heavy workload, 13. not getting along with unit commander, 14. many frivolous complaints lodged against members of the public, 15. engaging in high-speed chases, 16. becoming involved in physical violence with an offender, 17. investigating domestic quarrels, 18. having to break up fights or quarrels in bars and cocktail lounges.

Example 4. Real estate sales data pertaining to a sample of three bedroom homes sold in a calendar year in a particular area within a city were collected. The variables recorded were list price, sales price, square feet, number of rooms, number of bedrooms, garage capacity, bathroom capacity, extras, chattels, age, month sold, days to sell, listing broker, selling broker and lot type. Table 6.4 shows a sample of 12 observations, one for each month. The purpose of the study was to determine factors that influence selling price.

The four examples outlined above illustrate the variety of data matrices that may be encountered in practice. Before discussing techniques of multivariate analysis it will be useful to outline a system of classification for variables. We shall see later that the variable types influence the method of analysis that can be performed on the data. The next section outlines some terminology that is commonly applied to classify variables.

TABLE 6.2. Sample of Canadian Companies 1985 Financial Data

Rank	Sales (Millions)	% Change in Sales	Net Income (Millions)	% Change in Net Income	Rank on Net Income	% Return Equity	Total Assets (Millions)	Rank on Total Assets	Current Ratio	No. of Employees (Thousands)
1	18,993	713	16.5	−19.2	2	40.5	4355	448	17	1.5
2	15,040	247	2.8	−34.5	13	5.5	21446	123	2	1.1
3	13,353	199	10.2	−33.6	17	19.2	2973	30	29	1.3
4	13,257	1051	25.4	11.8	1	16.6	20583	108	3	1.3
5	8,880	101	7.6	13.6	39	13.4	2616	57	32	1.3
6	8,667	684	1.0	28.3	3	14.9	9202	15	5	2.2
7	7,834	246	10.7	−24.9	14	6.4	9591	70	4	2.3
8	7,040	124	12.2	−29.9	30	34.2	1580	12	63	1.4
9	6,931	67	8.0	9.8	56	14.4	1530	33	68	1.2
10	6,070	146	5.9	−7.6	27	5.9	5799	7	13	2.3

TABLE 6.3. Sample of Responses to Police Survey

Det. No.	Det. Exp.	Tot. Exp.	Age	Rank	Educ.	Stress Variables																	
						1	2	3	4	5	6	7	8	9	10	11	12	13	14	15	16	17	18
1	4	4	1	1	2	4	3	3	4	3	4	6	6	3	4	6	3	3	6	8	8	8	6
9	5	3	2	1	2	12	3	2	8	6	5	5	9	4	6	5	15	5	3	10	4	10	4
5	5	3	3	1	2	12	8	15	1	3	2	8	4	12	10	8	6	4	8	2	2	2	1
4	5	4	2	1	2	8	4	5	1	9	0	2	9	4	4	15	9	3	10	3	6	4	5
2	4	2	1	1	2	5	5	6	15	12	3	4	3	10	5	10	8	4	10	8	8	8	12
3	6	2	3	2	1	12	2	3	6	4	10	9	15	15	5	10	20	4	5	10	8	15	15
4	3	1	2	1	3	6	3	6	4	9	3	6	2	2	3	12	2	3	2	2	4	9	4
8	5	4	2	1	2	6	8	5	2	6	3	1	4	8	10	15	6	8	4	1	3	6	2
6	5	2	2	1	2	6	6	4	3	4	4	4	4	4	4	4	6	0	2	8	8	2	8
5	5	4	2	1	2	2	3	0	6	4	0	6	4	3	0	2	0	0	2	4	6	4	4
1	6	4	3	2	1	12	2	1	3	6	1	2	6	4	3	2	2	0	2	6	9	4	6
8	4	1	2	1	3	6	8	5	2	9	0	6	3	4	4	10	4	2	0	2	8	3	4
3	4	4	2	1	2	6	6	2	2	6	1	2	6	1	2	8	6	1	1	1	4	1	6
2	6	2	3	2	2	2	0	6	3	4	1	1	6	1	3	9	3	0	6	6	1	2	6
1	4	4	2	1	1	2	6	2	6	6	5	6	6	4	3	4	2	0	8	6	8	10	8
4	5	4	2	1	2	3	3	2	4	3	2	6	3	3	3	8	4	4	3	2	2	2	4
7	6	5	4	2	1	8	4	4	8	9	5	6	8	5	4	8	4	0	6	6	8	9	8
6	3	2	1	1	2	2	4	3	6	6	4	4	6	4	5	4	4	0	3	3	4	6	6

TABLE 6.4. Sample of Real Estate Sales Data

	List Price	Selling Price	Square Feet	Rooms	Bed-rooms	Bath	Garages	Extras	Chattels	Age	Days to Sell	Lot Type	List Broker	Selling Broker
1	79.0	75.4	1365	7	4	6	2	3	2	8	31	0	1	3
2	85.0	79.8	1170	6	3	4	2	2	1	7	48	0	5	5
3	103.8	101.5	1302	6	3	6	0	2	3	5	30	0	5	5
4	83.0	80.0	1120	5	3	6	2	1	2	5	28	0	3	1
5	109.8	107.0	1225	6	3	6	2	1	0	6	2	0	4	1
6	93.8	90.0	1160	6	3	6	2	3	2	6	44	0	2	5
7	95.8	89.8	1270	6	3	6	0	2	0	7	51	1	5	4
8	110.8	102.0	1260	6	3	6	2	3	3	7	45	0	2	2
9	97.2	94.0	1219	6	3	4	2	1	2	7	43	0	5	5
10	91.8	85.0	1170	6	3	4	0	1	3	8	25	1	5	5
11	86.0	84.0	1080	6	3	4	0	1	0	10	59	0	1	5
12	105.0	98.0	1345	6	3	6	2	1	1	7	16	0	1	1

6.1.2 MEASUREMENT SCALES

Variables can be classified as being quantitative or qualitative. A *quantitative* variable is one in which the variates differ in magnitude, for example, income, age, weight and GNP. A *qualitative* variable is one in which the variates differ in kind rather than in magnitude, for example, marital status, sex, nationality and hair colour.

Quantitative Scales

Obtaining values for a quantitative variable involves measurement along a scale and a unit of measure. A unit of measure may be *infinitely divisible* (eg., kilometres, metres, centimetres, millimetres) or *indivisible* (eg., family size). When the units of measure are infinitely divisible the variable is said to be *continuous*. In the case of an indivisible unit of measure the variable is said to be *discrete*. A continuous variable (theoretically) can always be measured in finer units; hence, actual measures obtained for such a variable are always approximate in that they are rounded.

Analysis with discrete variables often results in summary measures or parameters taking on values that are not consistent with the scale of measurement (eg., 1.7 children per household). Some variables which are *intrinsically continuous* are difficult to measure and hence are often measured on a discrete scale. For example, the stress variable discussed in Example 3 is an intrinsically continuous variable.

Scales of measurement can also be classified on the basis of the relations among the elements composing the scale. A *ratio scale* is the most versatile scale of measurement in that it has the following properties: (a) Any two values along the scale may be expressed meaningfully as a ratio, (b) the distance between items on the scale is meaningful and (c) the elements along the scale can be ordered from low to high (eg., weight is usually measured on a ratio scale).

An *interval scale*, unlike a ratio scale, does not have a fixed origin; for example, elevation and temperature are measured relative to a fixed point (sea level or freezing point of water). The ratio between 20°C and 10°C is not preserved when these temperatures are converted to Fahrenheit. An interval scale has only properties (b) and (c) above.

An *ordinal scale* is one in which only property (c) is satisfied; for example, the grades A, B, C, D, can be ordered from highest to lowest, but we cannot say that the difference between A and B is equivalent to the difference between B and C, nor can we say that the ratio A/C is equivalent to the ratio B/D.

Qualitative Scales

The fourth type of scale, *nominal*, corresponds to qualitative data. An example would be the variable marital status which has the categories

married, single, divorced, widowed and separated. The five categories can
be assigned coded values such as 1, 2, 3, 4, or 5. Although these coded
values are numerical, they must not be treated as quantitative. None of the
three properties listed above can be applied to the coded data.

On occasion, quantitative variables are treated in an analysis as if they
were nominal. In general, we use the term *categorical* to denote a variable
that is used as if it were nominal. The variable age for example can be
divided into six levels and coded 1, 2, 3, 4, 5, and 6.

Measurement Scales and Analysis

We shall see throughout the remainder of this text that the scale of measurement used to measure a variable will influence the type of analysis
used. The body of statistical techniques that are specially designed for ordinal data are often outlined in texts on nonparametric statistics. Variables
that are measured on ordinal scales can often be handled using techniques
designed for nominal data or interval data. The categories on the ordinal
scale can be treated as the categories of a nominal scale by ignoring the
fact that they can be ordered.

The variables in the data matrix represent the attempt by a researcher
to operationalize various dimensions that are believed to be important in
the research study. For dimensions such as intelligence, stress and job satisfaction, appropriate dimensions are difficult to define and measure. If
there are no appropriate units of measure, dimensions are sometimes oper-
ationalized by using other variables as *surrogates* for direct measurement.
The surrogate variable is usually an accessible and dependable correlate of
the dimension in question; for example, a surrogate variable can be measured and is believed to be strongly correlated with the required dimension.
Because surrogate variables are not in general perfectly correlated with the
required dimension, a number of them are often used to measure the same
dimension. The effectiveness with which a variable operationalizes a dimension is also called its validity. The measurement of validity in practice is
usually complex and inadequate.

6.1.3   DATA COLLECTION AND STATISTICAL INFERENCE

Having decided upon the variables to be measured, an *experimental design* must be formulated which outlines how the data are to be obtained.
The techniques for this are usually found under the theory and practice
of *survey sampling* and the theory and practice of experimental design. In
addition, texts on research methodology also discuss the issues of designs
for obtaining the data. One characteristic of the quality of a research design
is the reliability of the data that are obtained. The *reliability* of the design
refers to the consistency of the data when the same cases are measured at

some other time or by other equivalent variables, or when other samples of cases are used from the same population.

Probability Samples and Random Samples

The majority of multivariate techniques, generally employed to analyze data matrices, assume that the objects selected for the data matrix represent a *random sample* from some well-defined *population* of objects. A random sample is a special case of a *probability sample*. In a probability sampling process the probability of occurrence for all possible samples is known. In some cases the sample may not be a probability sample in that the probability that any particular object will be chosen for the sample cannot be determined. *Haphazard samples* such as volunteers, *representative samples* as judged by an expert and *quota samples* where the objective is to meet certain quotas are examples of *nonprobability samples* that are frequently used. On occasion the data set may represent the entire population (*a census*).

It is important to remember that without probability sampling, probability statements cannot be made about the outcomes from the multivariate analysis procedures. Since many research data sets are not obtained from probability samples, it is important to note that inference results should be stated as being conditional on the assumption of a probability sample.

In addition to the simple random sample there are alternative probability sampling methods that are commonly used. Cluster sampling, stratified sampling, systematic sampling and multiphase sampling are examples of more sophisticated methods which are usually used to reduce cost and improve reliability. Whenever simple random sampling is not used, adjustments have to be made to the standard inference procedures. Probability samples that are not simple random samples are called *complex samples*. Although modifications to some multivariate techniques have been developed for complex samples, they will not be discussed here. Random sampling is discussed in Chapter 1 of Volume I.

Exploratory and Confirmatory Analysis

The statistical techniques outlined in this text include both exploratory analysis and confirmatory analysis. In *exploratory analysis*, the objective is to describe the behavior of the variables in the data matrix, and to search for patterns and relationships that are not attributable to chance. Exploratory analysis includes analyses devoted to data reduction and matrix approximation. *Data reduction* techniques attempt to replace the existing columns or rows of the data matrix by a much smaller number of new values that are representative of the original data. Data reduction and matrix approximation are essentially the same process. In *confirmatory analysis*, certain hypotheses or models that have been prespecified are to be tested to determine whether the data supports the model. The quality of the model is often measured using a *goodness of fit* criterion. In large data sets the use of goodness of fit criteria often results in the model being overfitted; that is a less complex model than the fitted one is sufficient to explain the variation. The use of *cross validation* techniques to further confirm the model is recommended. Cross validation involves checking the fitted model on a second data matrix that comes from the same population but was not used to estimate the model.

6.1.4   An Outline of the Techniques to be Studied

There is no widely accepted system for classifying multivariate techniques, nor is there a standard or accepted order in which the subject is presented. One useful classification is according to the number and types of variables being used, and also according to whether the focus is a comparison of means or a study of the nature of the covariance structure. Some multivariate techniques are concerned with *data analysis* and data reduction, whereas others are concerned with *models* relating various *parameters*. The presentation of topics in the two volumes of this text is governed by the following:

1. What topics can be assumed to be known from a typical introductory course in statistical inference?

2. How many variables in the data matrix are involved in the analysis?

3. What types of variables are involved in the analysis?

4. Is the technique a data reduction procedure?

For the most part the techniques to be studied are designed for continuous and/or *categorical data*. Quantitative variables, with discrete scales or ordinal scales, will sometimes be treated as if they have continuous scales, and in other cases they may be treated as categorical. For the purpose of outlining the techniques, variables are classified as either quantitative or categorical. Occasionally ordinal data techniques will be introduced to present alternative but similar procedures.

The topics in this text are split into two volumes. Volume I is primarily devoted to procedures for linear models. In addition to the linear regression model, this volume also includes univariate data analysis, bivariate data analysis, analysis of variance and partial correlation. Volume II is designed to provide an overview of techniques for categorical data analysis and multivariate methods. The second volume also includes the topics of logistic regression, cluster analysis, multidimensional scaling and correspondence analysis.

Topics in Volume II

The topics in Volume II can be classified using the categories exploratory or confirmatory and can also be classified according to the types of measurement scales used for the variables involved. The topics in the first three chapters (6, 7 and 8) are primarily confirmatory in the sense that the techniques are usually concerned with making inferences about models and distribution parameters. In Chapter 7 however there is some discussion of data analysis techniques for multivariate samples. The last two chapters (9 and 10) are for the most part exploratory and are generally concerned with data reduction and data matrix approximation. In Chapter 9 there is some discussion of inference with respect to the factor analysis model hoever the emphasis in the chapter is data reduction. In Chapter 10 the topics presented are solely concerned with data reduction, matrix approximation and exploratory analysis.

The techniques presented in Chapter 6 are intended for multidimensional contingency tables and hence would be classified as categorical. In Chapter 7 the techniques presented are designed for studying relationships among variables assumed to be distributed as multivariate normal and hence must be continuous. In Chapter 8 the models studied are concerned with relationships between categorical and continuous variables. In particular the concern is whether the relationships among the continuous variables are the same for all categories defined by the categorical variables. In Chapter 9 the topics of principal components and factor analysis are primarily designed for data matrices of continuous variables, whereas correspondence analysis is designed for categorical data. In Chapter 10, cluster analysis is presented for both types of data, whereas multidimensional scaling is concerned with the determination of continuous scales based on ordinal or interval input data.

6.2   Two-Dimensional Contingency Tables

This section presents a discussion of bivariate distributions for categorical random variables and includes an outline of various commonly used sampling models. For inference purposes a sample of n observations is simultaneously cross-classified with respect to the two categorical random variables. The resulting joint frequencies are summarized in a two-dimensional contingency table. The section also surveys procedures for making inferences regarding the relationship between the two variables.

TABLE 6.5. Joint Density for X and Y

		\multicolumn{6}{c}{Y}
		1	2	3	...	c	Total
X	1	f_{11}	f_{12}	f_{12}	...	f_{1c}	$f_{1.}$
	2	f_{21}	f_{22}	f_{23}	...	f_{2c}	$f_{2.}$
	3	f_{31}	f_{32}	f_{33}	...	f_{3c}	$f_{3.}$
	⋮	⋮	⋮	⋮	...	⋮	⋮
	r	f_{r1}	f_{r2}	f_{r3}	...	f_{rc}	$f_{r.}$
	Total	$f_{.1}$	$f_{.2}$	$f_{.3}$	...	$f_{.c}$	1.00

6.2.1 BIVARIATE DISTRIBUTIONS FOR CATEGORICAL DATA

Joint Density Table

The *joint distribution* for a pair of categorical random variables can be illustrated in a two-dimensional table such as Table 6.5. The random variable X is assumed to have a range of values consisting of r categories, whereas the variable Y is assumed to have c categories. The *cell density* or *joint density* for cell (i, j) is denoted by f_{ij}, $i = 1, 2, \ldots, r$; $j = 1, 2, \ldots, c$; where it is understood that the first subscript refers to the row and the second subscript to the column. The *marginal densities* are denoted by $f_{i.}$ and $f_{.j}$ for the row and column variables respectively. The *conditional densities* for the rows given column j will be denoted by $f_{i.}(i \mid j)$ and for the columns given row i by $f_{.j}(j \mid i)$.

Independence

The random variables X and Y are *independent* if the joint density f_{ij} can be expressed as the product of the corresponding marginal densities $f_{i.}$ and $f_{.j}$ for every cell (i, j). Independence can also be defined in terms of the conditional densities and the marginal densities. X and Y are independent if the conditional density for each row is equal to the marginal density for Y or equivalently if the conditional density for each column is equal to the marginal density for X.

Example

The example presented in Table 6.6 illustrates a joint density function for two random variables X and Y. The observations were obtained from a large population of taxpayers in a large number of municipalities. The

TABLE 6.6. Population Joint Density Age versus Opinion on Crime Situation

| Age Level | Opinion Regarding Crime Situation | | | | |
	Not Serious	Slightly Serious	Moderately Serious	Very Serious	Totals
Under 30	0.015	0.076	0.121	0.055	0.267
30 to 39	0.017	0.117	0.111	0.037	0.282
40 to 49	0.012	0.074	0.104	0.032	0.222
50 to 59	0.007	0.034	0.072	0.020	0.133
60 and over	0.001	0.027	0.038	0.030	0.096
Totals	0.052	0.328	0.446	0.174	1.000

TABLE 6.7. Conditional Density for Opinion at Various Levels

i	Level	j	Not Serious 1	Slightly Serious 2	Moderately Serious 3	Very Serious 4	Totals 5
1	Under 30	$f._j(j \mid 1)$	0.056	0.285	0.453	0.206	1.000
2	30 to 39	$f._j(j \mid 2)$	0.060	0.415	0.394	0.131	1.000
3	40 to 49	$f._j(j \mid 3)$	0.054	0.333	0.469	0.144	1.000
4	50 to 59	$f._j(j \mid 4)$	0.053	0.256	0.541	0.150	1.000
5	60 & over	$f._j(j \mid 5)$	0.010	0.281	0.396	0.313	1.000
	Marginal density for opinion	$f._j$	0.052	0.328	0.446	0.174	1.000

densities in the table are assumed to be the population densities. Each taxpayer was asked to respond to a question regarding the seriousness of the crime situation in the neighborhood. The taxpayers were also asked to give their ages.

The column totals and row totals in Table 6.6 provide the marginal densities for crime opinion and age. A comparison of the joint densities f_{ij} with the products of the corresponding marginals $f_i.$ and $f._j$ suggests that the two variables are not independent. The departure from independence is more easily observed from a comparison of the conditional densities $f._j(j \mid i)$ for opinion at each age level. These conditional densities are shown in Table 6.7.

TABLE 6.8. Age versus Opinion on Crime Situation

| | | Row and Column Proportions* | | | | |
| | | Opinion Regarding Crime Situation | | | | |
	Age Level	Not Serious 1	Slightly Serious 2	Moderately Serious 3	Very Serious 4	Row Total Proportion 5
1	Under 30	0.056/0.288	0.285/0.232	0.453/0.271	0.206/0.318	0.267
2	30 to 39	0.060/0.322	0.415/0.357	0.394/0.248	0.131/0.210	0.281
3	40 to 49	0.054/0.237	0.333/0.225	0.469/0.234	0.144/0.185	0.223
4	50 to 59	0.053/0.136	0.256/0.104	0.541/0.162	0.150/0.113	0.133
5	60 & over	0.010/0.017	0.281/0.082	0.396/0.085	0.313/0.174	0.096
	Column total proportion	0.052	0.328	0.446	0.174	1.000

*The number on the left in each cell is the row proportion and the number on the right is the column proportion.

Row and Column Proportions

A comparison of the conditional densities $f_{\cdot j}(j \mid i)$ for opinion on crime given age in Table 6.7, to the marginal density for opinion on crime $f_{\cdot j}$ in Table 6.7, reveals that the opinion very serious $(j = 4)$ is more common among the 60 and over $(i = 5)$ and the under 30 $(i = 1)$ levels than among the middle three levels $(i = 2, 3, 4)$. We can also see that for the 30–39 age group the most likely choice is slightly serious, whereas for the age groups 40 to 49 and 50 to 59 the most likely choice is moderately serious. The variation in the behavior of the conditional densities over the five age categories suggests an interaction between the rows and columns. The conditional densities $f_{\cdot j}(j \mid i)$ are often referred to as *row proportions*, and the marginal density $f_{\cdot j}$ is called the *column total proportion*. In a similar fashion the *column proportions* $f_{i\cdot}(i \mid j)$ can be compared to the *row total proportions* $f_{i\cdot}$ as shown in Table 6.8. The row proportions and column total proportions are also shown in Table 6.8.

Row and Column Profiles

The row and column proportions are also commonly referred to as *row* and *column profiles*. The term *profile* is often used in connection with graphical displays of relationships in a contingency table.

Figures 6.1 and 6.2 display the row and column profiles for the crime opinion table. The figures also contain a plot of the marginal densities (broken line) for the column and row densities respectively. A comparison of the marginal density to the profile can be used to determine the nature

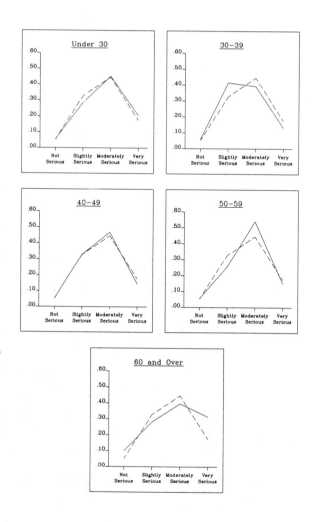

FIGURE 6.1. Row Profiles for Crime Opinion Data

of any departures from independence. For the row profiles in Figure 6.1 we can see that the 60 and over age group has the greatest departure from independence, while the 50–59 and 30–39 age group profiles also display some differences from the column marginal densities. Figure 6.2 shows the column profiles compared to the row marginal densities. The greatest departure from independence occurs in the very serious category. Profile plots are also useful in correspondence analysis to be discussed in Chapter 9.

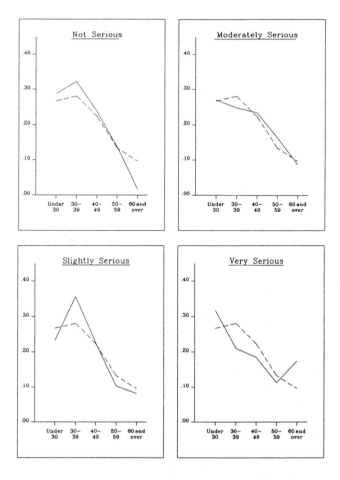

FIGURE 6.2. Column Profiles for Crime Opinion Data

Odds Ratios

The joint distribution can also be studied by examining the *odds ratios*.
The ratio f_{ij}/f_{is} measures the odds of being in column j relative to column s given row i. The ratio f_{tj}/f_{ts} measures the odds for column j versus column s given row t. The odds ratio is the ratio of the two sets of odds and is given by

$$\frac{f_{ij}}{f_{is}} \Big/ \frac{f_{tj}}{f_{ts}} = \frac{f_{ij}f_{ts}}{f_{is}f_{tj}}.$$

The odds ratio is necessarily 1 if the independence assumption holds. Under independence, the odds of being in column j relative to column s do not depend on the row.

TABLE 6.9. Two-Dimensional Contingency Table

		Y Categories					
		1	2	3	...	c	Total
	1	n_{11}	n_{12}	n_{13}	...	n_{1c}	$n_{1.}$
	2	n_{21}	n_{22}	n_{23}	...	n_{2c}	$n_{2.}$
X Categories	3	n_{31}	n_{32}	n_{33}	...	n_{3c}	$n_{3.}$
	$\vdots$	$\vdots$	$\vdots$	$\vdots$	...	$\vdots$	$\vdots$
	r	n_{r1}	n_{r2}	n_{r3}	...	n_{rc}	$n_{r.}$
	Total	$n_{.1}$	$n_{.2}$	$n_{.3}$	...	$n_{.c}$	n

Example

For the crime opinion distribution in Table 6.6, the odds for the category very serious relative to the category not serious in the under 30 age group is $0.055/0.015 = 3.67$, whereas for the 60 and over age group this ratio is $0.030/0.001 = 30.0$. The ratio of the two odds ratios is therefore $3.67/30.0 = 0.12$ which indicates that the odds of very serious relative to not serious are much higher for the 60 and over age group than for the under 30 age group. Since this odds ratio is not 1, crime opinion and age are not independent.

6.2.2 STATISTICAL INFERENCE IN TWO-DIMENSIONAL TABLES

The Two-Dimensional Contingency Table

A *two-dimensional contingency table* is produced when a sample of n observations is simultaneously cross-classified with respect to two categorical random variables. The notation for a two-dimensional contingency table is shown in Table 6.9. The contingency table is similar to the joint density table shown in Table 6.5, except that the joint densities f_{ij} are replaced by the *observed frequencies* or *cell frequencies* n_{ij}, $i = 1, 2, \ldots, r$; $j = 1, 2, \ldots, c$. A contingency table with r rows and c columns is called an $r \times c$ *contingency table*.

The contingency table provides a summary of the *sample joint frequency distribution*. Dividing the sample frequencies by n yields a table of *sample joint densities*. The row and column totals for the contingency table represent the *sample marginal frequency distributions* for the two random variables.

TABLE 6.10. Observed Frequencies: Driver Injury Level versus Seatbelt Usage

Seatbelt Usage	Driver Injury Level				Total
	None	Minimal	Minor	Major/Fatal	
Yes	12,813	647	359	42	13,861
No	65,963	4,000	2,642	303	72,908
Total	78,776	4,647	3,001	345	86,769

Example

An example of a two-dimensional contingency table is given in Table 6.10. Prior to the enactment of seatbelt legislation in the province of Alberta, a study was carried out to determine the usefulness of seatbelts for the prevention of injury. A sample of 86,769 automobile accident reports were studied. For each accident report, the injury level for the driver was classified into one of four categories, none, minimal, minor and major/fatal. Each driver was also classified as to seatbelt usage, yes or no. Table 6.10 displays the 2×4 contingency table produced from this sample. This contingency table will be used to illustrate the inference techniques presented throughout this section.

Sampling Models for Contingency Tables

There are a variety of sampling models that can be used to describe the process that yielded the $(r \times c)$ contingency table of n observations. The most common models are the *multinomial, hypergeometric, Poisson* and *product multinomial*. The most obvious extension of the simple random sample assumed for quantitative bivariate analysis is the multinomial distribution.

Multinomial

For the *multinomial distribution*, a random sample of n observations is selected from an infinite population. The observations are then classified into one of the rc cells of the table. The joint density for the sample cell frequencies is given by

$$f(n_{11}, n_{12}, \ldots, n_{rc}) = \frac{n!}{\prod\limits_{i=1}^{r} \prod\limits_{j=1}^{c} n_{ij}!} \prod_{i=1}^{r} \prod_{j=1}^{c} f_{ij}^{n_{ij}}$$

where $\sum_{i=1}^{r}\sum_{j=1}^{c}n_{ij} = n$. The means, variances, and covariances for the n_{ij} are given by

$$E[n_{ij}] = nf_{ij}, \qquad V[n_{ij}] = nf_{ij}(1 - f_{ij}) \quad \begin{array}{l} i = 1, 2, \ldots, r; \\ j = 1, 2, \ldots, c; \end{array}$$

$$\text{Cov}(n_{ij}, n_{k\ell}) = -nf_{ij}f_{k\ell} \quad \begin{array}{l} i \neq k, \; j \neq \ell \\ i, k = 1, 2, \ldots, r; \\ j, \ell = 1, 2, \ldots, c. \end{array}$$

The maximum likelihood estimators of the cell density parameters f_{ij} are the corresponding sample proportions n_{ij}/n. A useful property of the multinomial is that sums of multinomial random variables are also multinomial. The parameters are also summed to get the corresponding parameters for the distribution of the sums. A special case of the multinomial is the *binomial* where $c = 2$ and $r = 1$. In this case there are only two possible cells.

Hypergeometric

If the population is finite with known population cell frequencies N_{ij}, $i = 1, 2, \ldots, r$, $j = 1, 2, \ldots, c$, the density of the cell frequencies n_{ij} obtained from a random sample of n observations is given by the *hypergeometric density*

$$f(n_{11}, n_{12}, \ldots, n_{rc}) = \prod_{i=1}^{r}\prod_{j=1}^{c} \frac{N_{ij}!}{n_{ij}!(N_{ij} - n_{ij})!} \Big/ \frac{N!}{n!(N - n)!}.$$

The means, variances, and covariances are given by

$$E[n_{ij}] = nf_{ij},$$
$$V[n_{ij}] = \left(\frac{N - n}{N - 1}\right) nf_{ij}(1 - f_{ij}),$$
$$\text{Cov}(n_{ij}, n_{k\ell}) = -\left(\frac{N - n}{N - 1}\right) nf_{ij}f_{k\ell}$$

where $f_{ij} = N_{ij}/N$, $i = 1, 2, \ldots, r$; $j = 1, 2, \ldots, c$. In the case of large finite populations the hypergeometric can be approximated by the multinomial, provided each N_{ij} is large.

Poisson

In the multinomial and hypergeometric densities the total sample size n is fixed. An alternative assumption is to allow n to be a random variable as well. A useful distribution in this case is the *Poisson distribution*. The distributions of the cell frequencies n_{ij} are assumed to be mutually independently distributed as Poisson with parameters $F_{ij} = E[n_{ij}]$. In this case the total sample size n also has a Poisson distribution with parameter $F_{..} = E[n] = \sum_{i=1}^{r}\sum_{j=1}^{c} F_{ij}$. The variance $V[n_{ij}]$ is also given by F_{ij}. The joint density in this case is given by

$$f(n_{11}, n_{12}, \ldots, n_{rc}) = \prod_{i=1}^{r}\prod_{j=1}^{c} F_{ij}^{n_{ij}} e^{-F_{ij}} / n_{ij}!.$$

Since the cell frequencies are assumed to be mutually independent, $\text{Cov}(n_{ij}, n_{k\ell}) = 0$, $i \neq k$, $j \neq \ell$, $i, k = 1, 2, \ldots, r$; $j, \ell = 1, 2, \ldots, c$. The maximum likelihood estimators of the parameters F_{ij} are the sample frequencies n_{ij}. A useful property of the mutually independent Poisson assumption in contingency tables is that the conditional distribution of the n_{ij}, given a fixed n, is a multinomial distribution.

Product Multinomial

The *product multinomial distribution* arises from the joint distribution of two or more independent multinomial distributions. In the two-dimensional contingency table, the *row sample sizes* or *row marginals*, $n_{i.}$, $i = 1, 2, \ldots, r$, may be fixed. In this case the density for the cell frequencies in each row is given by the multinomial. Each row of the table is referred to as a *subpopulation*. The joint density for all r rows is given by the product of the individual row densities and hence the term product multinomial. The product multinomial density for an $r \times c$ contingency table is given by the product of the r multinomial densities corresponding to the rows and hence

$$f(n_{11}, n_{12}, \ldots, n_{rc}) = \prod_{i=1}^{r}\left[\frac{n_{i.}!}{\prod_{j=1}^{c} n_{ij}!} \prod_{j=1}^{c}\left[\frac{f_{ij}}{f_{i.}}\right]^{n_{ij}} \right].$$

The product multinomial can therefore be derived from the multinomial by conditioning on the row sample sizes $n_{i.}$. A product multinomial can also be obtained by fixing the *column marginals* or *column sample sizes*, $n_{.j}$, instead of the row sample sizes.

Example

To characterize the differences among the sampling models, consider the collection of questionnaire returns from a population of taxpayers. Each questionnaire provides information regarding two categorical random variables, X and Y say, state of residence and income category. The entire collection of returned questionnaires, N, is assumed to be a population. The responses generate a two-dimensional table with N_{ij} in the cell (i, j).

A random selection of n questionnaires from this population can be assumed to yield a multinomial or hypergeometric distribution depending on the magnitudes of the population quantities N_{ij}. If the N_{ij} are relatively large, then the multinomial distribution can be assumed.

Suppose that over a fixed time period of two days questionnaires are drawn randomly from the population. The sample size n is not fixed. The number of questionnaires obtained in each cell can be assumed to be Poisson independent of the other cells. The conditional distribution of the cell frequencies given the total sample n in this case is then multinomial.

A third possible sampling scheme first divides the population of questionnaires with respect to the categories of one of the random variables, say state of residence. Random samples of predetermined sizes are then selected from each subpopulation or state. The distribution in this case is a product multinomial or product hypergeometric. This distribution can be obtained from the multinomial or hypergeometric scheme by conditioning on the subpopulation sample sizes.

Test of Independence

A common test of independence between the two categorical random variables X and Y in contingency tables is the *Pearson χ^2 test*. The random sample is assumed to have been drawn from a multinomial population. If X and Y are independent, we would expect that the sample densities n_{ij}/n should be similar to the product of the sample marginal densities $(n_{i.}/n)(n_{.j}/n)$, and hence the estimated *expected frequencies* under independence are $(n_{i.}n_{.j})/n$. If this hypothesis of independence is true, in large samples the Pearson statistic

$$G^2 = \sum_{i=1}^{r}\sum_{j=1}^{c} \frac{(n_{ij} - n_{i.}n_{.j}/n)^2}{n_{i.}n_{.j}/n}$$

has a χ^2 distribution with $(r-1)(c-1)$ degrees of freedom. Large values of G^2 reflect large differences between n_{ij}/n and the product $(n_{i.}/n)(n_{.j})$, and therefore the independence hypothesis is rejected if G^2 is too large.

The Pearson χ^2 statistic is based on the assumption of a multinomial population with rc cells. In large samples, the sample proportions $n_{i.}/n$ and $n_{.j}/n$ are assumed to be normally distributed. The Pearson χ^2 is obtained from the distribution of the sums of squares of standardized normal random variables.

An alternative χ^2 statistic to G^2 is obtainable using the *likelihood ratio approach*. Again assuming a multinomial population, in large samples the statistic

$$H^2 = 2\sum_{i=1}^{r}\sum_{j=1}^{c} n_{ij} \ln\left(\frac{n_{ij}n}{n_{i.}n_{.j}}\right)$$

has a χ^2 distribution with $(r-1)(c-1)$ degrees of freedom if the hypothesis of independence holds. In large samples the two χ^2 statistics are usually quite comparable.

TABLE 6.11. Driver Injury versus Seatbelt Usage Level

Observed and Expected Frequencies under Independence
Driver Injury Level

Seatbelt Usage		None	Minimal	Minor	Major/Fatal	Total
Yes	Observed Frequency	12,813	647	359	42	13,861
	Expected Frequency	12,584.2	742.3	479.4	55.1	
No	Observed Frequency	65,963	4,000	2,642	303	72,908
	Expected Frequency	66,191.8	3,904.7	2,521.6	289.9	
Totals		78,776	4,647	3,001	345	86,769

Example

For the example relating seatbelt usage to driver injury level in automobile accidents, the value of the Pearson χ^2 for the independence hypothesis is $G^2 = 59.224$. This chi square statistic has three degrees of freedom and is significant at the 0.000 level. The likelihood ratio statistic has the value $H^2 = 42.9690$ and a corresponding p-value of 0.000. Generally, the two goodness of fit statistics have similar values. There appears to be a relationship between seatbelt usage and driver injury level.

Table 6.11 compares the observed frequencies to the expected frequencies under independence. A comparison of the numbers in each cell shows the nature of the departure from independence. The frequency of injury for all types of injury for drivers wearing seatbelts is less than expected under independence.

Sampling Model Assumptions

Our test for independence in contingency tables outlined above assumed that the data were obtained as a random sample of size n from a multinomial population. The population units were divided among rc cells with the probability of a unit occuring in cell (i, j) being denoted by f_{ij} where $\sum_{i=1}^{r}\sum_{j=1}^{c}f_{ij} = 1$. The marginal densities for the rows are given by $f_{i\cdot}$, where $f_{i\cdot} = \sum_{j=1}^{c}f_{ij}$, $i = 1, 2, \ldots, r$. Similarly the marginal densities for the columns are given by $f_{\cdot j}$, where $f_{\cdot j} = \sum_{i=1}^{r}f_{ij}$, $j = 1, 2, \ldots, c$. The sample estimates of f_{ij}, $f_{i\cdot}$ and $f_{\cdot j}$ are given by n_{ij}/n, $n_{i\cdot}/n$ and $n_{\cdot j}/n$ respectively. These estimators are the maximum likelihood estimators under the multinomial sampling assumption.

Poisson Distribution

As discussed earlier in this section, two commonly used alternative sampling models lead to the same maximum likelihood estimators as the multinomial. If there is no restriction placed on the total sample size, the cell frequencies n_{ij} may be viewed as random variables with expectation $E[n_{ij}] = F_{ij}$. If each of the n_{ij} are assumed to have an independent Poisson distribution, the maximum likelihood estimators for $E[n_{ij}]$ are given by $(n_i.n._j)/n$. For the Poisson assumption the total sample size n is not fixed. If the sampling is carried out for a fixed period of time and then stopped, the total sample size n acquired up to that point is also a random variable. The conditional distribution of the n_{ij}, given fixed n in this case, is a multinomial distribution; hence, the above procedures can be applied to the Poisson sampling. The independence hypothesis in the Poisson case implies that the true cell means $E[n_{ij}]$ satisfy the independence hypothesis given by

$$F_{ij} = E[n_{ij}] = \frac{E[n_i.]E[n._j]}{E[n]E[n]} \ E[n] = \frac{E[n_i.]E[n._j]}{E[n]}.$$

Product Multinomial Distribution

A second alternative to the multinomial population is called the product multinomial. In the product multinomial, additional restrictions are placed on the sample. Either the row totals $n_i.$ or the column totals $n._j$ are fixed in advance. In this case the sample is restricted to contain a specific number of observations from each category of one of the variables. The maximum likelihood estimators of the unrestricted marginals, either $f_i.$ or $f._j$, are given by $n_i./n$ or $n._j/n$ respectively. The expected cell frequencies under independence are estimated by $(n_i.n._j)/n$ as in the two previous cases. In this case the test is often referred to as a test of homogeneity of row or column proportions.

If the marginals $n_i.$ are fixed, then we are sampling independently from the r row subpopulations. In this case, the independence hypothesis $f_{ij} = f_i.f._j$ is written in the alternative form $f_{ij}/f_i. = f._j$, which states that the conditional densities for each level of j in each row i are equivalent to the marginal densities at each level of j. The estimated expected cell proportions under this model are obtained by rewriting $n_{ij} = (n_i.n._j)/n$ in the form $n_{ij}/n_i. = n._j/n$. The estimated expected row proportions $n_{ij}/n_i.$ for each level of j in each row are expected to be homogeneous over the r rows. Similarly in the case of fixed column marginals the estimated expected column proportions should be homogeneous over the columns.

All three sampling models for the $r \times c$ contingency table yield the same estimates for the expected frequencies under independence. The likelihood ratio χ^2 statistic H^2 given above is therefore identical for all three sampling models.

Standardized Residuals

A useful way of comparing the observed and expected frequencies is to determine the *standardized residuals* for each cell. The components of the Pearson χ^2 statistic provide information about which cells make the largest contribution to χ^2. The square roots of each of the terms in the Pearson χ^2 statistic are commonly called the standardized residuals. For cell (i, j) the standardized residual is given by

$$r_{1ij} = \left(n_{ij} - n_{i.}n_{.j}/n\right) \Big/ \sqrt{(n_{i.}n_{.j})/n}.$$

An alternative method of standardizing the residuals is to use the *Freeman-Tukey deviations* given by

$$r_{2ij} = \sqrt{n_{ij}} + \sqrt{n_{ij}+1} - \sqrt{(4n_{i.}n_{.j}/n)+1}\ .$$

For the observed and expected frequencies in Table 6.11, the standardized residuals are given by 2.0, –3.5, -5.5 and –1.8 for yes and –0.9, 1.5, 2.4 and 0.8 for no. The Freeman–Tukey deviations are 2.0, –3.6, –5.9 and –1.8 for yes and –0.9, 1.5, 2.4 and 0.8 for no. The two methods of determining residuals produce very similar results in both cases.

Correspondence Analysis

An alternative approach to the study of relationships in a two-dimensional contingency table is based on a *singular value decomposition* of the two matrices of row and column proportions or profiles. In this method the departures from independence for the row profiles are characterized in terms of two orthogonal dimensions determined from the singular value decomposition of the row profile deviations from independence. The row categories can then be plotted on the two-dimensional graph to show the departure from independence. A similar plot can be obtained for the column profile deviations from independence. The two pairs of dimensions (one pair for the row profiles, one pair for the column profiles) can be viewed as a scaling of the row and column categories. Thus two scales are derived for each of the two sets of profiles. The scales are determined in such a way that the amount of variation explained among profile deviations is maximized.

This topic will be discussed more extensively in Chapter 9.

6.2.3 MEASURES OF ASSOCIATION

If the independence model does not hold it is sometimes of interest to measure the degree of association between the row and column categories.

A useful approach to the measurement of association uses the concept of the *proportional reduction in error* achieved through knowledge of one variable when predicting the other variable.

Goodman and Kruskal's Lambda

For an individual drawn at random from the population, if no information is available about which row or column the individual belongs to, the best prediction for the cell location would be the one corresponding to the largest row and column population marginal densities. We denote the *maximum row and column marginal densities* by $f_{m\cdot}$ and $f_{\cdot m}$ respectively. Clearly, the probabilities of making an error in each case are $(1 - f_{m\cdot})$ and $(1 - f_{\cdot m})$ respectively.

If the column category for the individual is known, the best prediction for the unknown row category is the one corresponding to the largest density in the given column. Given column j we denote the maximum density by f_{mj}. The probability of making an error in this column is therefore $f_{\cdot j} - f_{mj}$ and over all columns the probability of making an error is $1 - \sum_{j=1}^{c} f_{mj}$.

The difference between the two error probabilities is given by

$$\left(\sum_{j=1}^{c} f_{mj} - f_{m\cdot} \right).$$

After dividing this expression by the probability of error in predicting the row with no information, *Goodman and Kruskal's Lambda* is obtained. This ratio is given by

$$\lambda_{m\cdot} = \left(\sum_{j=1}^{c} f_{mj} - f_{m\cdot} \right) \Big/ (1 - f_{m\cdot}).$$

The measure of association $\lambda_{m\cdot}$ therefore denotes the *proportional reduction in error* for row predictions when the column is known. Similarly, the proportional reduction in error for column predictions given the row is defined by

$$\lambda_{\cdot m} = \left(\sum_{i=1}^{r} f_{im} - f_{\cdot m} \right) \Big/ (1 - f_{\cdot m}).$$

The above two measures of association are said to be *asymmetric* in that one of the two variables is being predicted using the other. A *symmetric measure of association* that combines the logic used above is given by

$$\lambda = \left\{ \left[\frac{1}{2} \sum_{j=1}^{c} f_{mj} + \frac{1}{2} \sum_{i=1}^{r} f_{im} \right] - \frac{1}{2} [f_{\cdot m} - f_{m\cdot}] \right\} \Big/ \left[1 - \frac{1}{2} [f_{\cdot m} + f_{m\cdot}] \right].$$

Example

For the crime opinion population described in Table 6.6, the lambda measures of association are given by

$$\lambda_{m\cdot} = \frac{(0.121 + 0.117 + 0.104 + 0.072 + 0.038) - 0.446}{1 - 0.446} = 0.01$$

$$\lambda_{\cdot m} = \frac{(0.017 + 0.117 + 0.121 + 0.055) - 0.282}{1 - 0.282} = 0.039$$

$$\lambda = \frac{\frac{1}{2}(0.452) + \frac{1}{2}(0.310) - \frac{1}{2}(0.282) - \frac{1}{2}(0.446)}{1 - \frac{1}{2}(0.282) - \frac{1}{2}(0.446)} = 0.027.$$

All three measures of predictive association indicate that the association is extremely weak.

Inference for Lambda

A contingency table can be used to make inferences about the population measure of association defined above. A sample estimator for the Goodman-Kruskal measure of association can be used for this purpose. Replacing the true densities $f_{i\cdot}$ and $f_{\cdot j}$ by sample densities derived from the cell frequencies, n_{ij}, the estimators for the Goodman-Kruskal measures can be obtained. To measure the predictability of the column given the row, the estimator is given by

$$\hat{\lambda}_{\cdot m} = \left[\sum_{i=1}^{r} n_{im} - n_{\cdot m}\right]/[n - n_{\cdot m}],$$

where n_{im} denotes the largest value of n_{ij} in row i, and $n_{\cdot m}$ denotes the largest value of the marginal totals $n_{\cdot j}$.

In large samples the statistic $z = (\hat{\lambda}_{\cdot m} - \lambda_{\cdot m})/\hat{\sigma}_{\lambda_{\cdot m}}$ has a standard normal distribution. The estimator $\hat{\sigma}_{\lambda_{\cdot m}}$ is given by

$$\hat{\sigma}^2_{\lambda_{\cdot m}} = \left(n - \sum_{i=1}^{r} n_{im}\right)\left(\sum n_{im} + n_{\cdot m} - 2n^*_{\cdot m}\right)/(n - n_{\cdot m})^3,$$

where $n^*_{\cdot m}$ denotes the sum of the n_{ij} in the same column as $n_{\cdot m}$. This statistic cannot be applied if $\hat{\lambda}_{\cdot m} = 0$ or $\hat{\lambda}_{\cdot m} = 1$. The hypotheses $H_0: \lambda_{\cdot m} = 0$ and $H_0: \lambda_{\cdot m} = 1$ are rejected unless $\hat{\lambda}_{\cdot m} = 0$ or $\hat{\lambda}_{\cdot m} = 1$ respectively.

In a similar fashion the measure of predictability for the row given the columns $\hat{\lambda}_{m\cdot}$ is given by

$$\hat{\lambda}_{m\cdot} = \left[\sum_{j=1}^{c} n_{mj} - n_{m\cdot}\right]/[n - n_{m\cdot}];$$

where n_{mj} denotes the largest value of n_{ij} in column j and $n_{m\cdot}$ denotes the largest value of the marginal totals $n_{i\cdot}$.

Example

For the sample data presented in Table 6.10, the two measures of association $\hat{\lambda}_{.m}$ and $\hat{\lambda}_{m.}$ are zero. The no-seatbelt category dominates the seatbelt category for every injury level and the no-injury category dominates the other injury categories for both seatbelts and no seatbelts.

6.2.4 MODELS FOR TWO-DIMENSIONAL TABLES

Up to this point in our study of the joint distribution for qualitative bivariate random variables, the departure from independence was characterized by examining the behavior of row and column proportions and/or by the measurement of association between rows and columns. Although such techniques are usually sufficient for characterizing the behavior in two-dimensional tables, higher dimensional tables are more easily studied using models that relate cell frequencies. Before discussing multidimensional tables it will be useful to introduce cell frequency models for two-dimensional tables.

The bivariate distribution can also be characterized in terms of probability models relating *cell probabilities* or densities. In the previous section the *independence model* was evaluated using a χ^2 test. This model is given by

$$f_{ij} = f_{i.}f_{.j}, \qquad i = 1, 2, \ldots, r; \ j = 1, 2, \ldots, c.$$

In addition to the independence model, there are simpler models that could also be used.

Equal Cell Probability Model

The simplest model for a two dimensional table is the *equal cell probability model*

$$f_{ij} = 1/rc, \qquad i = 1, 2, \ldots, r; \ j = 1, 2, \ldots, c$$

implying that all rc possible events are equally likely.

Constant Row or Column Densities

Models that assume constant densities across rows or constant densities down columns are also possible and are given respectively by

$$f_{ij} = (1/c)f_{i.} \text{ and hence constant column densities } f_{.j} = 1/c$$

and

$$f_{ij} = (1/r)f_{.j} \text{ and hence constant row densities } f_{i.} = 1/r.$$

In the *constant column density model* the marginal density in each column is $1/c$ whereas for the *constant row density model* the marginal density in each row is $1/r$. In each case the rows and columns are independent and, in addition, one of the marginals is constant. In the constant column density model the column conditionals are given by $f_{\cdot j}(j \mid i) = 1/c$. Similarly, for the constant row density model $f_{i\cdot}(i \mid j) = 1/r$. An examination of the joint densities in Table 6.6 reveals that none of these simple probability models would adequately describe the behavior of the cell densities.

The Independence Model as a Composite of Three Simple Models

The independence model may be viewed as a less restrictive model than the constant row density and constant column density models, in that the marginal densities are no longer constant. Under independence the conditional densities are given by

$$f_{i\cdot}(i \mid j) = f_{i\cdot}$$

and

$$f_{\cdot j}(j \mid i) = f_{\cdot j}.$$

The conditional densities are therefore equal to the marginal densities. In the independence model the row conditionals are the same for each column and the column conditionals are the same for each row. Unlike the constant row or column density models described above under independence the marginal densities are not constant.

The independence model introduced in Section 6.2.1 may also be written in terms of the three simpler models by writing it in the form

$$\begin{aligned}
f_{ij} &= [1/rc][rf_{i\cdot}][cf_{\cdot j}] \\
&= [1/rc][f_{i\cdot}/(1/r)][f_{\cdot j}/(1/c)], \quad i = 1, 2, \ldots, r, \\
&\qquad\qquad\qquad\qquad\qquad\qquad\quad j = 1, 2, \ldots, c.
\end{aligned}$$

The first term of the product represents the density for cell (i, j) under the constant cell density model. The second term represents the ratio of the marginal density for row i to the marginal density under the constant row marginal model. The third term represents the ratio of the marginal density for column j to the marginal density under the constant column marginal model. In comparison to the three simple models, the independence model may be viewed as the product of an *average effect* $[1/rc]$, a *row effect* $[rf_{i\cdot}]$, and a *column effect* $[cf_{\cdot j}]$.

Example

For the crime opinion distribution introduced in Table 6.6 the average effect is $1/rc = 0.050$. The row effects are 1.337, 1.404, 1.110, 0.663 and 0.481 respectively, and the column effects are 0.210, 1.308, 1.786 and 0.695. The row effects indicate that larger than average proportions of the population fall into the younger age categories, whereas the column effects indicate that larger than average proportions of the population prefer the two moderate

TABLE 6.12. Interactions Between Age and Crime
Opinion

| | Opinion Regarding Crime Situation | | | |
Age Level	Not Serious	Slightly Serious	Moderately Serious	Very Serious
Under 30	1.08	0.87	1.01	1.19
30 to 39	1.14	1.27	0.88	0.75
40 to 49	1.07	1.01	1.05	0.83
50 to 59	1.03	0.78	1.22	0.85
60 & over	0.18	0.85	0.89	1.81

views of the crime situation. Under the independence model, therefore,
the row effects and column effects account for all the variation in the cell
densities.

The Saturated Model

If the independence model does not hold, the joint density may be expressed
as a product of the above effects and a residual. The model is given by

$$f_{ij} = [1/rc][rf_{i\cdot}][cf_{\cdot j}][f_{ij}/f_{i\cdot}f_{\cdot j}].$$

The fourth term

$$f_{ij}/f_{i\cdot}f_{\cdot j}$$

is the added residual term. This residual term guarantees that the equation
holds for all joint densities. This model is called a *saturated model* because
it fits the table perfectly. This last term is called the *interaction term* because it measures the interaction between rows and columns. The degree of
departure of the interaction term from the value 1 indicates the magnitude
and direction of the departure from independence.

Example

Table 6.12 presents the magnitudes of the interactions for the crime opinion
example. A quick perusal of the table reveals the pattern of departures
from independence. The largest and smallest values of the interaction term
are in the highest age category. Relatively few people in the 60 and over
category view the crime situation as not serious, whereas a relatively large
number view the situation as very serious. For the age category 30-39, fewer
individuals view the crime situation as very serious than would be expected
by chance under independence.

Loglinear Characterization for Cell Densities

The above method of modeling the joint distribution employed ratios to column and row marginals that were determined from arithmetic means of row and column densities. Given that we are attempting to model densities that are proportions, the use of *geometric means* rather than arithmetic means may be preferable. The reader may recall that geometric means are usually preferable to arithmetic means when averaging ratios.

The geometric mean of the rc cell densities is given by $\tilde{f}_{..}$ where

$$\ln \tilde{f}_{..} = \frac{1}{rc} \sum_{j=1}^{c} \sum_{i=1}^{r} \ln f_{ij}.$$

The geometric means of the densities in row i and column j are given respectively by $\tilde{f}_{i.}$ and $\tilde{f}_{.j}$ where

$$\ln \tilde{f}_{i.} = \frac{1}{c} \sum_{j=1}^{c} \ln f_{ij}$$

$$\ln \tilde{f}_{.j} = \frac{1}{r} \sum_{i=1}^{r} \ln f_{ij}.$$

A Loglinear Model for Independence

Under the independence assumption we may write a *loglinear model* as

$$\ln f_{ij} = \ln \tilde{f}_{..} + [\ln \tilde{f}_{i.} - \ln \tilde{f}_{..}] + [\ln \tilde{f}_{.j} - \ln \tilde{f}_{..}]$$

and hence

$$f_{ij} = \tilde{f}_{..} \left[\frac{\tilde{f}_{i.}}{\tilde{f}_{..}} \right] \left[\frac{\tilde{f}_{.j}}{\tilde{f}_{..}} \right]$$

if independence holds. Once again this product of three terms may be viewed as an average term multiplied by a row effect and a column effect. The row effect is obtained from the ratio of the geometric mean of the densities in row i, $\tilde{f}_{i.}$, to the overall geometric mean of cell densities, $\tilde{f}_{..}$. Similarly the column effect is obtained from the ratio of the geometric mean of the densities in column j, $\tilde{f}_{.j}$, to the overall geometric mean of all cell densities, $\tilde{f}_{..}$. The row and column effects therefore are now ratios of geometric means of cell densities.

Parameters for the Loglinear Model

From the definitions of the terms in the loglinear model, the terms may be viewed as means. New parameters μ, $\mu_{1(i)}$ and $\mu_{2(j)}$ may be defined as

$$\mu \quad = \frac{1}{rc} \sum_{i=1}^{r} \sum_{j=1}^{c} \ln f_{ij} \quad = \ln \tilde{f}_{..}$$

$$\mu_{1(i)} = \frac{1}{c} \sum_{j=1}^{c} \ln f_{ij} - \ln \tilde{f}_{..} \quad = [\ln \tilde{f}_{i.} - \ln \tilde{f}_{..}]$$

$$\mu_{2(j)} = \frac{1}{r} \sum_{i=1}^{r} \ln f_{ij} - \ln \tilde{f}_{..} \quad = [\ln \tilde{f}_{.j} - \ln \tilde{f}_{..}].$$

The parameter μ is therefore the logarithm of the overall geometric mean of the densities. The parameter $\mu_{1(i)}$ represents the logarithm of the ratio of the geometric mean of the densities in row i to the overall geometric mean. Similarly, the parameter $\mu_{2(j)}$ is the logarithm of the ratio of the geometric mean of the densities in column j to the overall geometric mean. The parameters $\mu_{1(i)}$ and $\mu_{2(j)}$ have the properties $\sum_{i=1}^{r}\mu_{1(i)} = 0$ and $\sum_{j=1}^{c}\mu_{2(j)} = 0$. These parameters therefore are similar to the effect parameters used in analysis of variance.

Under independence the model then becomes

$$\ln f_{ij} = \mu + \mu_{1(i)} + \mu_{2(j)}.$$

This is commonly called the loglinear model for independence in a two-dimensional table. The form of this model is similar to the models used in analysis of variance.

The Loglinear Model with Interaction

If the independence model does not hold, an interaction term can be determined that represents the departure from independence. The density can be expressed in the form

$$f_{ij} = \tilde{f}_{..}.[\tilde{f}_{i.}/\tilde{f}_{..}][\tilde{f}_{.j}/\tilde{f}_{..}]\left[\frac{f_{ij}\tilde{f}_{..}}{\tilde{f}_{i.}\tilde{f}_{.j}}\right].$$

As in the previous models the interaction term $(f_{ij}\tilde{f}_{..})/(\tilde{f}_{i.}\tilde{f}_{.j})$ measures the ratio of the true density to the density suggested by the independence model.

If independence does not hold, the loglinear form of the model can be written as

$$\ln f_{ij} = \mu + \mu_{1(i)} + \mu_{2(j)} + \mu_{12(ij)}$$

where the interaction parameter $\mu_{12(ij)}$ is given by

$$\mu_{12(ij)} = \ln f_{ij} + \ln \tilde{f}_{..} - \ln \tilde{f}_{i.} - \ln \tilde{f}_{.j}.$$

TABLE 6.13. Age versus Opinion on Crime Situation, Logarithms of Cell Densities

Age Level	Opinion Regarding Crime Situation				Row Totals	Row Average
	Not Serious	Slightly Serious	Moderately Serious	Very Serious		
Under 30	−4.200	−2.577	−2.112	−2.900	−11.789	−2.947
30 to 39	−4.075	−2.146	−2.198	−3.297	−11.716	−2.929
40 to 49	−4.423	−2.604	−2.263	−3.442	−12.732	−3.183
50 to 59	−4.962	−3.381	−2.631	−3.912	−14.886	−3.722
60 & over	−6.908	−3.612	−3.270	−3.507	−17.297	−4.324
Column totals	−24.568	−14.320	−12.474	−17.058	−68.420	
Column averages	−4.914	−2.864	−2.495	−3.412		−3.421

The interaction parameters have the properties that $\sum_{i=1}^{r}\mu_{12(ij)} = 0$ and $\sum_{j=1}^{c}\mu_{12(ij)} = 0$. This model is commonly referred to as the *saturated loglinear model* since it describes the densities precisely without any restrictions.

Example

For the crime opinion population discussed above, the logarithms of the cell densities are shown in Table 6.13. The logarithms of the various geometric means can be obtained from the row and column averages in this table. The logarithm of the geometric mean, $\ln \tilde{f}_{..}$, can be obtained from the average of $\ln f_{ij}$ over the entire table. These averages are also shown in Table 6.13.

The logarithms of the various geometric means are given by the row and column averages shown in Table 6.13.

$$\ln \tilde{f}_{..} = -3.421, \quad \ln \tilde{f}_{.1} = -4.914, \quad \ln \tilde{f}_{.2} = -2.864,$$

$$\ln \tilde{f}_{.3} = -2.495, \quad \ln \tilde{f}_{.4} = -3.412,$$

$$\ln \tilde{f}_{1.} = -2.947, \quad \ln \tilde{f}_{2.} = -2.929, \quad \ln \tilde{f}_{3.} = -3.183,$$

$$\ln \tilde{f}_{4.} = -3.722, \quad \ln \tilde{f}_{5.} = -4.324.$$

The loglinear model parameter for the overall mean is given by $\mu = \ln \tilde{f}_{..} = -3.421$. The row and column effects are obtained by subtracting $\ln \tilde{f}_{..}$ from the row and column averages and are given by

$$\mu_{2(1)} = -1.493, \quad \mu_{2(2)} = 0.557, \quad \mu_{2(3)} = 0.926, \quad \mu_{2(4)} = 0.009$$

$$\mu_{1(1)} = 0.474, \quad \mu_{1(2)} = 0.492, \quad \mu_{1(3)} = 0.238,$$

$$\mu_{1(4)} = -0.301, \quad \mu_{1(5)} = -0.903.$$

The reader should note that both the row effects and column effects separately sum to zero. These effects therefore can be compared to zero in order to provide interpretation.

From the column parameters $\mu_{2(j)}$ determined above, we can conclude that there are relatively more individuals in the columns corresponding to slightly serious ($\mu_{2(2)}$) and moderately serious ($\mu_{2(3)}$), whereas for the opinion not serious ($\mu_{2(1)}$) there are relatively few indivduals. The row parameters $\mu_{1(i)}$ indicate that there are relatively more individuals in the first three age categories and relatively fewer in the last two age categories.

Under independence the logarithm of the cell densities, $\ln f_{ij}$, as shown in Table 6.13 should be equal to the overall average μ plus the corresponding row and column effects $\mu_{1(i)}$ and $\mu_{2(j)}$ as determined above. The difference between $\ln f_{ij}$ and the sum of the three parameters represents the departure from independence.

The departure from independence can be shown using the interaction parameters $\mu_{12(ij)}$, which are obtained by subtracting the overall mean μ and the row and column effects $\mu_{1(i)}$ and $\mu_{2(j)}$ from the logarithm of each cell frequency as given in Table 6.13. These interaction terms are

$$
\begin{array}{lll}
\mu_{12(11)} = 0.253, & \mu_{12(12)} = -0.193, & \mu_{12(13)} = -0.098, \\
\mu_{12(14)} = 0.038 & \mu_{12(21)} = 0.355, & \mu_{12(22)} = 0.230, \\
\mu_{12(23)} = -0.200, & \mu_{12(24)} = -0.385 & \mu_{12(31)} = 0.287, \\
\mu_{12(32)} = 0.011, & \mu_{12(33)} = -0.021, & \mu_{12(34)} = -0.277 \\
\mu_{12(41)} = 0.277, & \mu_{12(42)} = -0.220, & \mu_{12(43)} = 0.161, \\
\mu_{12(44)} = -0.220 & \mu_{12(51)} = -1.173, & \mu_{12(52)} = 0.172, \\
\mu_{12(53)} = 0.157, & \mu_{12(54)} = 0.844.
\end{array}
$$

The reader should note that the interaction parameters necessarily sum to zero across each row and down each column. Under independence each of the above interaction parameters would be zero. By comparing the parameter values $\mu_{12(ij)}$ to zero, we can determine that for the opinion not serious, the oldest age category is underrepresented [see $\mu_{12(51)}$] while the remaining age categories are overrepresented. For the opinion very serious, the oldest age category is overrepresented [see $\mu_{12(54)}$] and for ages 50 to 59 this opinion is relatively scarce [see $\mu_{12(44)}$]. Thus in comparison to other age groups individuals in the oldest age category are more likely to choose the opinion very serious and less likely to choose the opinion not serious.

Matrix Notation for Loglinear Model

The loglinear model outlined above can also be shown in matrix notation in a similar fashion to analysis of variance models. The model is given by

$$\ln \mathbf{f} = \mathbf{X}\boldsymbol{\beta} + \boldsymbol{\varepsilon},$$

where

$$\ln \mathbf{f} = \begin{bmatrix} \ln f_{11} \\ \ln f_{21} \\ \vdots \\ \ln f_{r1} \\ \ln f_{12} \\ \ln f_{22} \\ \vdots \\ \ln f_{r2} \\ \vdots \\ \ln f_{1c} \\ \vdots \\ \ln f_{rc} \end{bmatrix}, \quad \mathbf{X} = \begin{bmatrix} 1 & 1 & 0 & \cdots & 0 & 1 & 0 & \cdots & 0 \\ 1 & 1 & 0 & \cdots & 0 & 0 & 1 & & 0 \\ \vdots & \vdots & \vdots & & \vdots & \vdots & \vdots & & \vdots \\ 1 & 1 & 0 & \cdots & 0 & 0 & 0 & \cdots & 1 \\ 1 & 0 & 1 & \cdots & 0 & 1 & 0 & \cdots & 0 \\ 1 & 0 & 1 & & 0 & 0 & 1 & & 0 \\ \vdots & & \vdots & & \vdots & \vdots & \vdots & & \vdots \\ 1 & 0 & 1 & \cdots & 0 & 0 & 0 & \cdots & 1 \\ \vdots & \vdots & & & & \vdots & & & \vdots \\ 1 & 0 & & \cdots & 1 & 1 & 0 & \cdots & 0 \\ 1 & 0 & & \cdots & 1 & 0 & 1 & & \\ \vdots & \vdots & & & \vdots & \vdots & \vdots & & \vdots \\ 0 & 0 & & \cdots & 1 & 0 & 0 & \cdots & 1 \end{bmatrix},$$

$$\boldsymbol{\beta} = \begin{bmatrix} \mu \\ \mu_{1(1)} \\ \mu_{1(2)} \\ \vdots \\ \mu_{1(r)} \\ \mu_{2(1)} \\ \mu_{2(2)} \\ \vdots \\ \mu_{2(c)} \end{bmatrix}, \quad \boldsymbol{\varepsilon} = \begin{bmatrix} \varepsilon_{11} \\ \varepsilon_{21} \\ \vdots \\ \varepsilon_{r1} \\ \varepsilon_{12} \\ \varepsilon_{22} \\ \vdots \\ \varepsilon_{r2} \\ \vdots \\ \varepsilon_{1c} \\ \varepsilon_{2c} \\ \vdots \\ \varepsilon_{rc} \end{bmatrix}.$$

This matrix notation is useful for multidimensional tables and is sometimes used when employing statistical software.

The loglinear model is a useful approach for characterizing variation in a contingency table. It is very useful for the multidimensional tables which will be studied in Section 6.3.

6.2.5 STATISTICAL INFERENCE FOR LOGLINEAR MODELS

Given a two-dimensional contingency table, the loglinear model parameters can be estimated using the sample cell frequencies. Inferences regarding the model parameters can be made using both the Pearson and likelihood ratio χ^2 statistics. This section outlines the approach using the seatbelt data in Table 6.10.

For the loglinear model introduced above, the independence model is written

$$\ln f_{ij} = \mu + \mu_{1(i)} + \mu_{2(j)}, \quad i = 1, 2, \ldots, r; \ j = 1, 2, \ldots, c,$$

where

$$\mu = \frac{1}{rc} \sum_{i=1}^{r} \sum_{j=1}^{c} \ln f_{ij}$$

$$\mu_{1(i)} = \frac{1}{c} \sum_{j=1}^{c} \ln f_{ij} - \mu$$

$$\mu_{2(j)} = \frac{1}{r} \sum_{i=1}^{r} \ln f_{ij} - \mu.$$

If the independence model does not hold, the interactions, $\mu_{12(ij)}$, can be determined from the loglinear model residuals

$$\mu_{12(ij)} = \ln f_{ij} - \mu - \mu_{1(i)} - \mu_{2(j)}, \quad i = 1, 2, \ldots, r; \ j = 1, 2, \ldots, c,$$

hence

$$\ln f_{ij} = \mu + \mu_{1(i)} + \mu_{2(j)} + \mu_{12(ij)}, \quad i = 1, 2, \ldots, r; \ j = 1, 2, \ldots, c,$$

which must fit the data perfectly.

The Loglinear Model Defined in Terms of Cell Frequencies

When fitting loglinear models to contingency tables, it is more common to express the loglinear model in terms of the cell frequencies rather than the cell densities. We shall denote the expected cell frequencies for a sample of size n by $F_{ij} = nf_{ij}, \ i = 1, 2, \ldots, r; \ j = 1, 2, \ldots, c$. The loglinear model is then written

$$\ln F_{ij} = \mu + \mu_{1(i)} + \mu_{2(j)} + \mu_{12(ij)}, \quad i = 1, 2, \ldots, r; \ j = 1, 2, \ldots, c,$$

where

$$\mu = \frac{1}{rc} \sum_{i=1}^{r} \sum_{j=1}^{c} \ln F_{ij}$$

$$\mu_{1(i)} \quad = \quad \frac{1}{c}\sum_{j=1}^{c} \ln F_{ij} - \mu$$

$$\mu_{2(j)} \quad = \quad \frac{1}{r}\sum_{i=1}^{r} \ln F_{ij} - \mu$$

$$\mu_{12(ij)} \quad = \quad \ln F_{ij} - \mu - \mu_{1(i)} - \mu_{2(j)}.$$

Multiplicative Form of the Loglinear Model

Using the antilogarithms of the parameters the loglinear model can be expressed in *multiplicative form* as

$$F_{ij} = \beta_0 \beta_{1(i)} \beta_{2(j)} \beta_{12(ij)},$$

where $\beta_0 = e^{\mu}$, $\beta_{1(i)} = e^{\mu_{1(i)}}$, $\beta_{2(j)} = e^{\mu_{2(j)}}$ and $\beta_{12(ij)} = e^{\mu_{12(ij)}}$.

Estimation for the Loglinear Model

Estimated theoretical frequencies E_{ij} are determined from the observed frequencies n_{ij}. The parameter estimates are then determined by replacing the F_{ij} by the E_{ij} in the above definitions. These estimators are maximum likelihood estimators under the sampling model assumptions introduced above. The parameter estimates are given by

$$\hat{\mu} \quad = \quad \frac{1}{rc}\sum_{i=1}^{r}\sum_{j=1}^{c} \ln E_{ij}$$

$$\hat{\mu}_{1(i)} \quad = \quad \frac{1}{c}\sum_{j=1}^{c} \ln E_{ij} - \hat{\mu}$$

$$\hat{\mu}_{2(j)} \quad = \quad \frac{1}{r}\sum_{i=1}^{r} \ln E_{ij} - \hat{\mu}.$$

For the independence model $E_{ij} = n_i.n_{.j}/n$, while for the saturated model the F_{ij} are estimated using $E_{ij} = n_{ij}$.

Computer Software

The BMDP software will be used to fit the loglinear model in this section.

Example

The logarithms of the expected frequencies for the independence model for the driver injury data of Table 6.10 are shown in Table 6.14. The parameter estimates from this table are given by

$$\hat{\mu} = 7.3880, \quad \hat{\mu}_{1(1)} = -0.8301, \quad \hat{\mu}_{1(2)} = 0.8301, \quad \hat{\mu}_{2(1)} = 2.8823,$$

$$\hat{\mu}_{2(2)} = 0.0518, \quad \hat{\mu}_{2(3)} = -0.3854, \quad \hat{\mu}_{2(4)} = -2.5487.$$

TABLE 6.14. Logarithms of Expected Frequencies Driver Injury Level versus Seatbelt Usage

| Seatbelt | Driver Injury Level | | | | |
Usage	None	Minimal	Minor	Major/Fatal	Total
Yes	9.4402	6.6098	6.1725	4.0091	26.2316
No	11.1003	8.2699	7.8326	5.6695	32.8723
Total	20.5405	14.8797	14.0051	9.6786	59.1039

TABLE 6.15. Logarithms of Observed Frequencies Driver Injury versus Seatbelt Usage

| Seatbelt | Driver Injury Level | | | | |
Usage	None	Minimal	Minor	Major/Fatal	Total
Yes	9.4582	6.4723	5.8833	3.7377	6.3879
No	11.0968	8.2940	7.8793	5.7137	8.2460
Total	10.2775	7.3832	6.8813	4.7257	7.3169

The reader should note how the row effects and column effects separately sum to zero. Under the independence hypothesis, these parameter estimates indicate that the frequencies in the seatbelt row are relatively low and in the no seatbelt row they are consequently very high. The column or injury level parameters indicate that the no injury category has a relatively large frequency and the major/fatal category has a relatively low frequency. We have already seen from the χ^2 test of independence in Section 6.2.2 that the independence model does not fit the data in the table. The values for both the Pearson and likelihood ratio χ^2 statistics were significant at the 0.000 level. Therefore the saturated model must be used to describe the variation in the table.

The logarithms of observed frequencies are presented in Table 6.15. The parameter estimates for the saturated model are obtained by using the observed frequencies n_{ij} for the estimated expected frequencies E_{ij} in the estimation equations above. The interaction parameters are given by

$$\hat{\mu}_{12(ij)} = \ln n_{ij} - \hat{\mu} - \hat{\mu}_{1(i)} - \hat{\mu}_{2(j)}, \quad i = 1, 2, \ldots, r; \; j = 1, 2, \ldots, c.$$

The reader should note that these parameter estimates are different than those obtained for the independence model. In this case E_{ij} has the value n_{ij}, whereas for the independence model $E_{ij} = n_i . n_{.j}/n$.

For the observed frequencies in Table 6.15 the parameter estimates are

$$\hat{\mu} = 7.3169, \qquad \hat{\mu}_{1(1)} = -0.9290, \qquad \hat{\mu}_{1(2)} = 0.9290$$
$$\hat{\mu}_{2(1)} = 2.9606, \qquad \hat{\mu}_{2(2)} = 0.0663, \qquad \hat{\mu}_{2(3)} = -0.4356,$$
$$\hat{\mu}_{2(4)} = -2.5913, \qquad \hat{\mu}_{12(11)} = 0.1097, \qquad \hat{\mu}_{12(12)} = 0.0182,$$
$$\hat{\mu}_{12(13)} = -0.0689, \qquad \hat{\mu}_{12(14)} = -0.0590, \qquad \hat{\mu}_{12(21)} = -0.1097,$$
$$\hat{\mu}_{12(22)} = -0.0182, \qquad \hat{\mu}_{12(23)} = 0.0689, \qquad \hat{\mu}_{12(24)} = 0.0590.$$

The reader should note how the row effects and column effects sum to zero and how the interaction effects separately sum to zero in each row and column. The estimates of the interaction parameters indicate that for individuals who wore seatbelts, the frequency of minimal, minor and major/fatal injuries is lower than for those who did not wear seatbelts. An individual in a car accident who wears a seatbelt therefore is more likely to be in the no injury category and less likely to be in the minimal or major/fatal category than for individuals who do not wear seatbelts.

The antilogarithms of these parameter estimates can be used to express the parameters in terms of the geometric mean and ratios of geometric means. The antilogarithms are given by

$$e^{\hat{\mu}} = 1505.530, \qquad e^{\hat{\mu}_{1(1)}} = 0.395, \qquad e^{\hat{\mu}_{1(2)}} = 2.532,$$
$$e^{\hat{\mu}_{2(1)}} = 19.310, \qquad e^{\hat{\mu}_{2(2)}} = 1.0685, \qquad e^{\hat{\mu}_{2(3)}} = 0.647,$$
$$e^{\hat{\mu}_{2(4)}} = 0.075, \qquad e^{\hat{\mu}_{12(11)}} = 1.116, \qquad e^{\hat{\mu}_{12(12)}} = 1.018,$$
$$e^{\hat{\mu}_{12(13)}} = 0.933, \qquad e^{\hat{\mu}_{12(14)}} = 0.943, \qquad e^{\hat{\mu}_{12(21)}} = 0.896,$$
$$e^{\hat{\mu}_{12(22)}} = 0.982, \qquad e^{\hat{\mu}_{12(23)}} = 1.071, \qquad e^{\hat{\mu}_{12(24)}} = 1.061.$$

Each of the estimated expected frequencies can be written as a product of the appropriate multiplicative parameters

$$E_{ij} = n_{ij} = e^{\hat{\mu}} \cdot e^{\hat{\mu}_{1(i)}} \cdot e^{\hat{\mu}_{2(j)}} \cdot e^{\hat{\mu}_{12(ij)}}.$$

From the antilogarithms of the parameter estimates we can draw conclusions about the cell frequencies. The geometric mean of the cell frequencies is 1505.530. The ratio of the geometric mean of the cell frequencies relative to the overall geometric mean for seatbelt users is 0.395 whereas for seatbelt nonusers this ratio is the inverse of 0.395 = 2.532. Thus a relatively large proportion of the drivers did not wear seatbelts.

The ratios to the overall geometric mean of the geometric means for the injury categories suggest that a large proportion of the frequency did not sustain any injury whereas only a very small proportion obtained a major/fatal injury. For the minimal category the geometric mean was slightly larger than the overall geometric mean, whereas for the minor category the geometric mean was a little below the overall average.

The interaction terms suggest that seatbelt users are less likely to sustain major/fatal or minor injury than seatbelt nonusers. Similarly the minimal injury category and the no injury category are more likely for seatbelt users than for seatbelt nonusers.

Standardized Estimates of Loglinear Parameters

In order to compare the magnitudes of the various parameter estimates, *standardized estimates* should be obtained by dividing the estimates by the standard errors. For the parameter estimates $\hat{\mu}$, $\hat{\mu}_{1(i)}$, $\hat{\mu}_{2(j)}$ and $\hat{\mu}_{12(ij)}$ the asymptotic variances are given by

$$\text{Var}(\hat{\mu}) = \left(\frac{1}{rc}\right)^2 \sum_{i=1}^{r}\sum_{j=1}^{c}\left(\frac{1}{n_{ij}}\right) - \frac{1}{n},$$

$$\text{Var}(\hat{\mu}_{1(i)}) = \left(\frac{1}{rc}\right)^2 \sum_{i=1}^{r}\sum_{j=1}^{c}\left(\frac{1}{n_{ij}}\right) + \left(\frac{r-1}{rc^2}\right)\sum_{j=1}^{c}\left(\frac{1}{n_{ij}}\right),$$

$$\text{Var}(\hat{\mu}_{2(j)}) = \left(\frac{1}{rc}\right)^2 \sum_{i=1}^{r}\sum_{j=1}^{c}\left(\frac{1}{n_{ij}}\right) + \left(\frac{c-1}{cr^2}\right)\sum_{i=1}^{r}\left(\frac{1}{n_{ij}}\right),$$

$$\text{Var}(\hat{\mu}_{12(ij)}) = \left(\frac{1}{rc}\right)^2 \sum_{i=1}^{r}\sum_{j=1}^{c}\left(\frac{1}{n_{ij}}\right) + \left(\frac{r-1}{rc^2}\right)\sum_{j=1}^{c}\left(\frac{1}{n_{ij}}\right)$$
$$+ \left(\frac{c-1}{cr^2}\right)\sum_{i=1}^{r}\left(\frac{1}{n_{ij}}\right) + \frac{(c-1)(r-1)}{rc}\left(\frac{1}{n_{ij}}\right).$$

The square roots of these quantities provide the standard errors.

For the driver injury data of Table 6.10, the parameter estimates, standard errors and standardized estimates are shown in Table 6.16. In large samples these standardized estimates can be treated as standard normal deviates for the purpose of judging significance. From the standardized estimates we can conclude that the interactions corresponding to the minimal and major/fatal categories are not significant.

A Loglinear Representation for Some Simpler Models

In Section 6.2.4 several simple models for the two-dimensional contingency table were introduced. The simplest model is given by

$$f_{ij} = 1/rc \quad i = 1, 2, \ldots, r; \ j = 1, 2, \ldots, c,$$

which indicates that the cell probabilities are equal. For a sample of size n in this case we would expect

$$F_{ij} = n/rc,$$

and hence we would estimate the F_{ij} by $E_{ij} = n/rc$.

A model which assumes constant column probabilities is given by

$$f_{ij} = f_{i\cdot}/c;$$

hence the cell densities in any row are uniform over the c columns. Similarly the constant row probability model is given by

$$f_{ij} = f_{\cdot j}/r.$$

TABLE 6.16. Standardized Parameter Estimates for Saturated Model

Parameter	Estimate	Standard Error	Standardized Estimate
$\mu_{1(1)}$	−0.9290	0.0224	−41.443
$\mu_{1(2)}$	0.9290	0.0224	41.443
$\mu_{2(1)}$	2.9606	0.0227	130.562
$\mu_{2(2)}$	0.0663	0.0046	2.458
$\mu_{2(3)}$	−0.4356	0.1772	−14.537
$\mu_{2(4)}$	−2.5913	0.0624	−41.539
$\mu_{12(11)}$	0.1097	0.0227	4.839
$\mu_{12(12)}$	0.0182	0.0079	0.675
$\mu_{12(13)}$	−0.0689	0.1020	−2.300
$\mu_{12(14)}$	−0.0590	0.0624	−0.945
$\mu_{12(21)}$	−0.1097	0.0227	−4.839
$\mu_{12(22)}$	−0.0182	0.0079	−0.675
$\mu_{12(23)}$	0.0689	0.1020	2.300
$\mu_{12(24)}$	0.0590	0.0624	0.945

For a sample of size n, the expected frequencies for these two models are given by $F_{ij} = F_{i.}/c$ and $F_{ij} = F_{.j}/r$ respectively. The expected cell frequencies F_{ij} for the two models are estimated using the sample cell frequencies n_{ij} and are denoted by $E_{ij} = n_{.j}/r$ for the constant row probability model and by $E_{ij} = n_{i.}/c$ for the constant column probability model.

The loglinear models for the three simple models would be given respectively by

$$\ln F_{ij} = \mu,$$
$$\ln F_{ij} = \mu + \mu_{1(i)},$$
$$\ln F_{ij} = \mu + \mu_{2(j)}.$$

Inference Procedures for the Three Simple Models

For each model the appropriate E_{ij} values are used to determine the parameter estimates using the following equations

$$\hat{\mu} = \frac{1}{rc}\sum_{i=1}^{r}\sum_{j=1}^{c}\ln E_{ij}$$

$$\hat{\mu}_{1(i)} = \frac{1}{c}\sum_{j=1}^{c}\ln E_{ij} - \hat{\mu}$$

$$\hat{\mu}_{2(j)} = \frac{1}{r}\sum_{i=1}^{r}\ln E_{ij} - \hat{\mu}.$$

TABLE 6.17. Comparison of Fitted Frequencies for the Simple Models

Seatbelt Usage	Model	Driver Injury Level				
		None	Minimal	Minor	Major/Fatal	Totals
Yes	Actual	12,813	647	359	42	13,861
	Constant	10,846.1	10,846.1	10,846.1	10,846.1	43,384.4
	Constant Row Prob.	39,388.0	2,323.5	1,500.5	172.5	43,384.5
	Constant Column Prob.	3,465.2	3,465.2	3,465.2	3,465.2	13,861
No	Actual	65,963	4,000	2,642	303	72,908
	Constant	10,846.1	10,846.1	10,846.1	10,846.1	43,384.4
	Constant Row Prob.	39,388.0	2,323.5	1,500.5	172.5	43,384.5
	Constant Column Prob.	18,227	18,227	18,227	18,227	72,907.9
Total	Actual	78,776	4,647	3,001	345	86,769
	Constant Row Prob.	78,775.9	4,647.0	3,001.0	345.0	
	Constant Column Prob.	21,692.2	21,692.2	21,692.2	21,692.2	

The two χ^2 statistics, G^2 and H^2, can also be used to test null hypotheses for these models. The two statistics are given by

$$G^2 = \sum_{i=1}^{r}\sum_{j=1}^{c} \frac{(n_{ij} - E_{ij})^2}{E_{ij}}$$

and

$$H^2 = 2\sum_{i=1}^{r}\sum_{j=1}^{c} n_{ij} \ln\left[\frac{n_{ij}}{E_{ij}}\right].$$

These χ^2 statistics have $(rc - 1 - k)$ degrees of freedom, where k equals the number of parameters estimated from the sample. For the constant cell probability model, $k = 0$, whereas for the constant column and row probability models, $k = (r - 1)$ and $k = (c - 1)$ respectively. For the independence model $k = (r - 1) + (c - 1)$.

Example

For the seatbelt data in Table 6.10, the parameter estimates for the three simple models are $\hat{\mu} = 9.2916$ for the constant cell probability model, $\hat{\mu} = 7.3169$, $\hat{\mu}_{1(1)} = -0.929$, and $\hat{\mu}_{1(2)} = 0.929$ for the constant column probability model, and $\hat{\mu} = 7.6990$, $\hat{\mu}_{2(1)} = 2.882$, $\hat{\mu}_{2(2)} = 0.052$,

$\hat{\mu}_{2(3)} = -0.385$ and $\hat{\mu}_{2(4)} = -2.549$ for the constant row probability model. The estimated frequencies using the three fitted models are summarized in Table 6.17. In the case of the constant row probability model, the column totals are preserved, whereas in the constant column probability model the row totals are preserved. By comparing the actual frequencies to the fitted frequencies one can observe the goodness of fit for the three models. In each case we can see from Table 6.17 that the models do not fit the table.

6.2.6 AN ADDITIVE CHARACTERIZATION FOR CELL DENSITIES

To this point the models introduced for contingency table densities were designed to characterize departures from the independence model. An alternative approach can be obtained using an *additive model* with deviations from the grand mean and from row and column means as measures of row and column effects.

Each cell density f_{ij} can be expressed as a linear model comparable to analysis of variance models. The density f_{ij} can be written as

$$\begin{aligned} f_{ij} &= f_{..} + (f_{i.} - f_{..}) + (f_{.j} - f_{..}) + (f_{ij} - f_{i.} - f_{.j} + f_{..}) \\ &= \mu + \alpha_i + \rho_j + \varepsilon_{ij}, \end{aligned}$$

where $\mu = f_{..}$, $\alpha_i = (f_{i.} - f_{..})$, $\rho_j = (f_{.j} - f_{..})$ and $\varepsilon_{ij} = (f_{ij} - f_{i.} - f_{.j} + f_{..})$. The grand mean is denoted by μ, the row effect by α_i and the column effect by ρ_j. The term ε_{ij} is an interaction term.

As in the case of analysis of variance, matrix notation can be used to represent the model for all rc cells.

$$\mathbf{f} = \mathbf{X}\boldsymbol{\beta} + \boldsymbol{\varepsilon}$$

where the $\mathbf{f}(rc \times 1)$, $\mathbf{X}[rc \times (r+c+1)]$, $\boldsymbol{\beta}[(r+c+1) \times 1]$, $\boldsymbol{\varepsilon}(rc \times 1)$ are given by

$$\mathbf{f} = \begin{bmatrix} f_{11} \\ f_{21} \\ \vdots \\ f_{r1} \\ f_{12} \\ f_{22} \\ \vdots \\ f_{r2} \\ \vdots \\ f_{1c} \\ f_{2c} \\ \vdots \\ f_{rc} \end{bmatrix}, \quad \boldsymbol{\varepsilon} = \begin{bmatrix} \varepsilon_{11} \\ \varepsilon_{21} \\ \vdots \\ \varepsilon_{r1} \\ \varepsilon_{12} \\ \varepsilon_{22} \\ \vdots \\ \varepsilon_{r2} \\ \vdots \\ \varepsilon_{1c} \\ \varepsilon_{2c} \\ \vdots \\ \varepsilon_{rc} \end{bmatrix},$$

$$\mathbf{X} = \begin{bmatrix} 1 & 1 & 0 & \cdots & 0 & 1 & 0 & \cdots & & 0 \\ 1 & 1 & 0 & \cdots & 0 & 0 & 1 & & & 0 \\ \vdots & \vdots & \vdots & & \vdots & \vdots & & 0 & & 0 \\ 1 & 1 & 0 & \cdots & 0 & 0 & 0 & ..0 & & 1 \\ 1 & 0 & 1 & \cdots & 0 & 1 & 0 & \cdots & & 0 \\ 1 & 0 & 1 & & & 0 & 0 & 1 & & \\ \vdots & & & & & \vdots & & 0 & & 0 \\ 1 & 0 & 1 & \cdots & 0 & 0 & 0 & ..0 & & 1 \\ \vdots & \vdots & & & \vdots & & & & & \vdots \\ 1 & 0 & 0 & ..0 & 1 & 1 & 0 & 0.. & & 0 \\ 1 & 0 & 0 & ..0 & 1 & 0 & 1 & 0.. & & \\ \vdots & \vdots & & & \vdots & \vdots & & 0 & & 0 \\ 1 & 0 & 0 & \cdots & 0 & 1 & 0 & ..0 & & 1 \end{bmatrix}$$

and

$$\boldsymbol{\beta} = [\mu \ \alpha_1 \ \alpha_2 \ \cdots \ \alpha_r \ \rho_1 \ \rho_2 \ \cdots \ \rho_c].$$

This matrix notation is similar to the notation introduced for the log-linear model above. The additive model introduced here will be used in the weighted least squares approach to be outlined in Section 6.3. In Section 6.3 the theoretical cell densities will be replaced by row densities p_{ij} where $\sum_{j=1}^{c} p_{ij} = 1$.

Example

For the example given by Table 6.6, the values of the parameters are given by

$$\mu = 1/rc = 1/20 = 0.05,$$
$$\alpha_1 = 0.01675, \quad \alpha_2 = 0.0205, \quad \alpha_3 = 0.0055,$$
$$\alpha_4 = -0.01675, \quad \alpha_5 = -0.026,$$
$$\rho_1 = -0.0396, \quad \rho_2 = 0.0156, \quad \rho_3 = 0.0392, \quad \rho_4 = -0.0152.$$

The interaction parameters are summarized in Table 6.18. As can be seen from the table, the row and column totals for the interactions are zero. In comparison to the previous analyses of this table, the additive interaction effects provide much the same information regarding the departure from independence.

6.2.7 Two-Dimensional Contingency Tables in a Multivariate Setting

In this section, we assume that we are primarily interested in the relationship between two categorical variables X and Y, but that a third categorical variable Z is also related to both X and Y. Assume that the variable Z has ℓ categories. The three-way cross-classification using X, Y and Z

TABLE 6.18. Additive Model Interaction Effects

| | Opinion Regarding Crime Situation | | | |
Age Level	Not Serious	Slightly Serious	Moderately Serious	Very Serious
Under 30	−0.01215	−0.00635	0.01505	0.00345
30 to 39	−0.01390	0.03090	0.00130	−0.01830
40 to 49	−0.00390	0.00290	0.00930	−0.00830
50 to 59	0.01335	−0.01485	−0.00045	0.00195
60 & over	0.01660	−0.01260	−0.02520	0.02120

yields a three-dimensional contingency table with $rc\ell$ cells. Given a sample of n observations, allocation to the $rc\ell$ cells yields cell frequencies n_{ijk}, $i = 1, 2, \ldots, r;\ \ j = 1, 2, \ldots, c;\ \ k = 1, 2, \ldots, \ell$. The important questions are: How does the variable Z affect the relationship between X and Y? When can the variable Z be ignored?

One possible approach to the problem is to carry out an analysis relating X and Y at each level of Z. Thus a total of ℓ contingency tables must be analyzed. If ℓ is large, the number of analyses will also be large, and in some cases the cell frequencies for each level of Z may be quite small. In the next section the loglinear model approach will be used to relate X and Y and control simultaneously for Z. If the sample size is small, the table can be *collapsed* over the categories of Z; however, the collapsing of contingency tables can lead to unusual results, as demonstrated below.

Simpson's Paradox

The importance of controlling for other variables is best illustrated by an example that demonstrates how the relationship between X and Y can be opposite depending on whether Z is controlled or not controlled. Such reversals or contradictions are commonly referred to as *Simpson's paradox*. This paradox is illustrated by the following example.

Example

Consider the following fictitious study relating age, smoking and heart disease. The study shows the cell frequencies in Table 6.19.

In the age 65 and over category, the probability of heart disease for smokers is 95%, and for nonsmokers it is 90%. In the age 40–64 category, the probability of heart disease for smokers is 50% and for nonsmokers is 5%. Thus, in each age category smokers are more likely to have heart disease than nonsmokers.

Collapsing the table over age yields Table 6.20. According to the collapsed table, the probability of heart disease for smokers is 54% whereas for nonsmokers it is 89%. Simpson's Paradox has occurred in that the effect

TABLE 6.19. Contingency Tables Showing Relationships Between Age, Heart Disease and Smoking

| | Age 65 & over | | | Age 40–64 | |
	Smoker	Nonsmoker		Smoker	Nonsmoker
Heart Disease	950	9000	Heart Disease	5000	5
No Heart Disease	50	1000	No Heart Disease	5000	95
Proportion with Heart Disease	0.95	0.90	Proportion with Heart Disease	0.50	0.05

TABLE 6.20. Contingency Table Showing Relationship Between Heart Disease and Smoking After Collapsing on Age

	Smoker	Nonsmoker
Heart Disease	5950	9005
No Heart Disease	5050	1095
Proportion with Heart Disease	0.54	0.89

of smoking on the incidence of heart disease seems to have been reversed from what was obtained when the variable age was controlled. If one were to look at the data by collapsing on heart disease, the reason for the paradox becomes clear. For individuals 40–46, 99% are smokers, whereas for those aged 65 and over, only 9% are smokers. The incidence of heart disease increases with age whereas the tendency to smoke decreases with age, ignoring the effect of age, therefore, produces the paradoxical results.

Example

A second example of Simpson's Paradox is provided by the following fictitious tables of admission statistics for two university faculties. Table 6.21 shows the proportion of male and female applicants admitted to the Faculties of Engineering and Business. The table demonstrates that the overall proportion of female applicants admitted is lower than the overall proportion of male applicants admitted, although for each of the faculties taken separately the reverse is true. Once again collapsing on the faculty variable hides the fact that the number of applications by sex varies considerably

TABLE 6.21. Relationships Between Sex
and Admission Status

	Females	Males
Faculty of Business		
No. of Applicants	160	60
No. of Admissions	40	12
Proportion Admitted	0.25	0.20
Faculty of Engineering		
No. of Applicants	40	140
No. of Admissions	26	84
Proportion Admitted	0.65	0.60
Combined Faculties		
No. of Applicants	200	200
No. of Admissions	66	96
Proportion Admitted	0.33	0.48

between the two faculties. Because the admission rates also vary between the two faculties a lower overall rate of admission for females occurs.

Other examples of Simpson's Paradox commonly occur in practice. Examples showing employment inequity can show opposite results depending on what other variables are included. The effects of collapsing contingency tables is discussed further in Section 6.3. To avoid collapsing tables, multidimensional models can be used to describe the variation in the tables controlling for other effects.

6.2.8   OTHER SOURCES OF INFORMATION

More extensive discussions of statistical techniques for two-dimensional tables are available in Everitt (1977), Reynolds (1977) and Upton (1978). Extensive discussions on testing for independence and the measurement of association can be found in both Reynolds (1977) and Upton (1978). Sampling models and loglinear models for two-dimensional tables are discussed in Bishop, Fienberg and Holland (1975), Andersen (1980) and Freeman (1987).

6.3   Multidimensional Contingency Tables

This section is concerned with statistical inference in *multidimensional contingency tables*. The section begins with an outline of techniques for analyzing a three-dimensional contingency table. These techniques are then extended to higher-dimensional tables. The loglinear model introduced in

Section 6.2 is used to describe the interactions in the multidimensional table. The parameters of the loglinear model are estimated using maximum likelihood estimators that are usually functions of the observed cell frequencies.

6.3.1 THE THREE-DIMENSIONAL CONTINGENCY TABLE

The *three-dimensional contingency table* arises from the cross-classification of the categories associated with three qualitative random variables. Geometrically the table may be viewed as having rows, columns and layers. The subscripts for the rows, columns and layers will be denoted by i, j and k respectively. The number of rows, columns and layers will be denoted by r, c and ℓ respectively. The probability density for cell (i, j, k) will be denoted by f_{ijk} and the theoretical cell frequency by $F_{ijk} = n f_{ijk}$ for a total table frequency of n. The allocation of a sample of size n to the total of $rc\ell$ cells yields cell frequencies n_{ijk}. Table 6.22 shows the n_{ijk} for a sample of size n.

Various marginal totals will be denoted using dots to indicate which subscripts have been summed. For the three possible two-dimensional tables, the cell frequencies are denoted by the marginals $n_{ij\cdot}$, $n_{i\cdot k}$ and $n_{\cdot jk}$. For each of the three variables the univariate marginals are given by $n_{i\cdot\cdot}$, $n_{\cdot j\cdot}$ and $n_{\cdot\cdot k}$.

Example

An example of a three-dimensional table is presented in Table 6.23. For the auto accident data described in Section 6.2, the three-way table shows the relationships between extent of injury, seatbelt usage and driver condition.

Figure 6.3 shows row profiles relating the three injury levels to seatbelt usage (ignoring driver condition) in panel (a), to driver condition (ignoring seatbelt usage) in panel (b) and finally to the four categories of seatbelt usage crossed with driver condition in panel (c). For simplicity the no injury category has been omitted. From the profile plots we can see that the driver condition effect is stronger than the seatbelt effect and that there is some interaction between the two effects. In the case of interaction we can see that the seatbelt effect is more pronounced for drinking drivers than for non-drinking drivers.

Models for Three-Way Tables

The models introduced in Section 6.2 for two-dimensional contingency tables can be viewed as special cases of the set of all possible models for the three-dimensional table. We begin here with the independence model for the three-way table. The independence model requires that the joint density f_{ijk} in cell (i, j, k) be equal to the product of the three univariate marginal densities $f_{ijk} = f_{i\cdot\cdot} f_{\cdot j\cdot} f_{\cdot\cdot k}$. The theoretical frequency for a total

TABLE 6.22. A Three-Dimensional Contingency Table

Layers	Rows	Columns			
		1	2	...	c
1	1	n_{111}	n_{121}	...	n_{1c1}
	2	n_{211}	n_{221}	...	n_{2c1}
	$\vdots$	$\vdots$	$\vdots$		$\vdots$
	r	n_{r11}	n_{r21}	...	n_{rc1}
2	1	n_{112}	n_{122}	...	n_{1c2}
	2	n_{212}	n_{222}	...	n_{2c2}
	$\vdots$	$\vdots$	$\vdots$		$\vdots$
	r	n_{r12}	n_{r22}	...	n_{rc2}
	$\vdots$	$\vdots$	$\vdots$		$\vdots$
ℓ	1	$n_{11\ell}$	$n_{12\ell}$	...	$n_{1c\ell}$
	2	$n_{21\ell}$	$n_{22\ell}$	...	$n_{2c\ell}$
	$\vdots$	$\vdots$	$\vdots$		$\vdots$
	r	$n_{r1\ell}$	$n_{r2\ell}$	...	$n_{rc\ell}$

frequency of n is given by

$$F_{ijk} = nf_{ijk} = F_{i..}F_{.j.}F_{..k}/n^2$$

where $F_{i..} = nf_{i..}$, $F_{.j.} = nf_{.j.}$ and $F_{..k} = nf_{..k}$.

Inference for the Independence Model

Given a sample of size n, the maximum likelihood estimators of the expected cell frequencies under the independence assumption are given by

$$E_{ijk} = n_{i..}n_{.j.}n_{..k}/n^2, \qquad \begin{array}{l} i = 1, 2, \ldots, r; \\ j = 1, 2, \ldots, c; \\ k = 1, 2, \ldots, \ell. \end{array}$$

As in the case of the two-dimensional contingency table, the fitted cell frequencies depend only on the row, column and layer marginals. Using the estimated expected frequencies E_{ijk}, the χ^2 tests of goodness of fit for the independence model are carried out using

$$G^2 = \sum_{i=1}^{r} \sum_{j=1}^{c} \sum_{k=1}^{\ell} \frac{(E_{ijk} - n_{ijk})^2}{E_{ijk}}$$

TABLE 6.23. Frequency Table – Driver Injury Level versus Seatbelt
Usage and Driver Condition

Driver Condition	Seatbelt Usage	Driver Injury Level				Total
		None	Minimal	Minor	Major/Fatal	
Normal	Yes	12500	604	344	38	13486
		(11817.8)	(697.1)	(450.2)	(51.8)	
	No	61971	3519	2272	237	67999
		(62161.0)	(3666.9)	(2368.0)	(272.2)	
	Totals	74471	4123	2616	275	81485
Been Drinking	Yes	313	43	15	4	375
		(766.3)	(45.2)	(29.2)	(3.4)	
	No	3992	481	370	66	4909
		(4030.9)	(283.0)	(153.6)	(17.7)	
	Totals	4305	524	385	70	5284
Total both conditions		78776	4647	3001	345	86769

and

$$H^2 = 2\sum_{i=1}^{r} \sum_{j=1}^{c} \sum_{k=1}^{\ell} n_{ijk}(\ln n_{ijk} - \ln E_{ijk}),$$

both of which are asymptotically χ^2 with $(rc\ell - r - \ell - c + 2)$ degrees of
freedom if the independence hypothesis holds.

Example

The χ^2 test of independence for Table 6.23 yields 1057.47 and 939.90 for
the Pearson and likelihood ratio statistics respectively. Both of these χ^2
statistics have 10 degrees of freedom and are significant at the 0.000 level.
The expected frequencies under the independence model are shown in Table 6.23 in brackets. A comparison of the observed and expected frequencies
in the table permits us to conclude the following:

1. For seatbelt users who appeared normal, the number of accidents
   resulting in no injury was larger than expected, whereas the number
   who sustained any injury was smaller than could be expected under
   independence.

2. For seatbelt users who had been drinking, the number of accidents
   resulting in no injury was less than half the number that could be
   expected under independence. Also, in the minor injury category,
   there were fewer cases than expected.

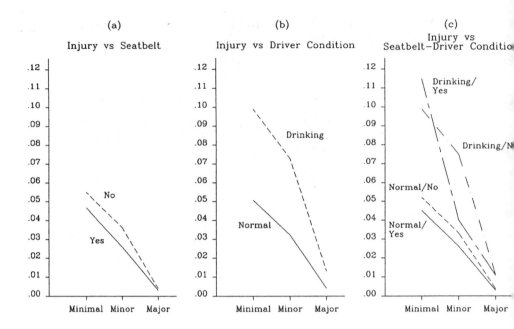

FIGURE 6.3. Profiles Relating Injury Level to Seatbelt Usage and Driver Condition

3. For non-users of seatbelts who appeared normal, the number of accidents in all injury categories was less than could be expected under independence.

4. For non-users of seatbelts who had been drinking, the number of accidents resulting in no injury was less than expected. For the three injury categories, the number of accidents was much larger than could be expected under independence.

Drivers who wore seatbelts and appeared normal sustained fewer injuries than expected, whereas drivers who did not wear seatbelts and had been drinking suffered more injuries than expected under independence. For the remaining two categories, the difference between the observed and expected frequencies seems less obvious. A loglinear model representation for this table will be used below to provide a more systematic approach for identifying the interactions among the three variables. Before attempting to model the variation in the table, a discussion of various model types is required.

Other Models for Three-Way Tables

For the remainder of this section the sampling model assumed is either multinomial or independent Poisson. As in the case of the two-dimensional table, the two distributions are equivalent if the sample size n is fixed. Because the product multinomial places additional restrictions on some marginals, additional requirements must be adhered to in order to obtain maximum likelihood estimates. These requirements will be outlined later in this section.

If the independence model does not fit the three-dimensional table it is often of value to determine if a less restrictive model can be used. A system of models that permits various levels of dependence among the three variables is outlined next.

Partial Independence

Since there are three variables in the table, it is possible to have two variables related to each other that are both independent of a third variable. This model is called the *partial independence model* and is given by

$$f_{ijk} = (f_{ij\cdot})(f_{\cdot\cdot k}).$$

In this case, the third variable with subscript k is independent of the remaining two variables with subscripts i and j. The theoretical frequency is given by

$$F_{ijk} = F_{ij\cdot}F_{\cdot\cdot k}/n$$

and is estimated by

$$E_{ijk} = n_{ij\cdot}n_{\cdot\cdot k}/n.$$

The two-dimensional marginals $n_{ij\cdot}$ are being fitted since $E_{ij\cdot} = n_{ij\cdot}$. The χ^2 goodness of fit statistic in this case has $(rc-1)(\ell-1)$ degrees of freedom.

An example of a partial independence relationship would exist if in Table 6.23 seatbelt usage were independent of both driver condition and driver injury level, but at the same time driver condition and injury level were related.

Conditional Independence

A *conditional independence model* permits two variables to be independent after controlling for a third variable. An example of such a model is provided by

$$f_{ijk} = f_{i\cdot k}f_{\cdot jk}/f_{\cdot\cdot k}$$

where the variables with subscripts i and j are independent at every level of the variable with subscript k. The theoretical frequency is given by

$$F_{ijk} = F_{i\cdot k}F_{\cdot jk}/F_{\cdot\cdot k}$$

and the maximum likelihood estimator is given by

$$E_{ijk} = n_{i\cdot k}n_{\cdot jk}/n_{\cdot\cdot k}.$$

For this model the two-dimensional marginals $n_{i \cdot k}$ and $n_{\cdot jk}$ are being fitted since $E_{i \cdot k} = n_{i \cdot k}$ and $E_{\cdot jk} = n_{\cdot jk}$. The χ^2 goodness of fit statistic has $\ell(r-1)(c-1)$ degrees of freedom.

An example of a conditional independence model in Table 6.23 would occur if, for each of the two driver conditions, driver injury level is independent of seatbelt usage. In this case driver injury level is related to seatbelt usage through the variable driver condition, but if driver condition is held fixed then seatbelt usage and driver injury level are independent. In other words, any relationship between driver injury level and seatbelt usage is due to the relation between driver condition and each of the other two variables. This result is similar to obtaining a zero first-order partial correlation coefficient with three quantitative variables.

No Three-Way Interaction

The next step in moving to less restrictive models is to assume that each pair of variables is related, but that the relation between any pair of variables does not depend on the level of the third. This model is usually referred to as the *no three-way interaction model*. It is not possible to give an expression for f_{ijk} or for F_{ijk} that would permit us to determine the estimators E_{ijk} directly. For this model the E_{ijk} are obtained by a procedure known as *iterative proportional fitting*.

Since the model to be fitted assumes that all possible pairs are related but that there is no three-way interaction, we need only fit a model that preserves the three sets of two-dimensional marginal totals $n_{ij \cdot}$, $n_{\cdot jk}$ and $n_{i \cdot k}$. The steps for iterative proportional fitting proceed as follows:

Step 1. Compute the observed marginal totals $n_{ij \cdot}$, $n_{\cdot jk}$, $n_{i \cdot k}$.

Step 2. Assign the initial value 1 to every estimated cell frequency, that is, $E_{ijk}^{(0)} = 1$, for all i, j, k.

Step 3. Compute new estimates of the E_{ijk} so that they sum to the marginal totals $n_{ij \cdot}$ using

$$E_{ijk}^{(1)} = E_{ijk}^{(0)} \left[\frac{n_{ij \cdot}}{E_{ij \cdot}^{(0)}} \right] \quad \text{for all } i, j, k.$$

Step 4. Compute new estimates of the E_{ijk} so that they sum to the marginal totals $n_{i \cdot k}$ using

$$E_{ijk}^{(2)} = E_{ijk}^{(1)} \left[\frac{n_{i \cdot k}}{E_{i \cdot k}^{(1)}} \right] \quad \text{for all } i, j, k.$$

Step 5. Compute new estimates of the E_{ijk} so that they sum to the marginal totals $n_{.jk}$ using

$$E_{ijk}^{(3)} = E_{ijk}^{(2)} \left[\frac{n_{.jk}}{E_{.jk}^{(2)}} \right] \quad \text{for all } i, j, k.$$

Step 6 and subsequent steps. Repeat the cycle given by Steps 3, 4 and 5 until the changes in the E_{ijk} are smaller than some preassigned number.

For the fitted model the three two-dimensional marginals $E_{ij.}$, $E_{.jk}$ and $E_{i.k}$ will be very close to their observed counterparts $n_{ij.}$, $n_{.jk}$ and $n_{i.k}$. The number of degrees of freedom for a χ^2 goodness of fit test would in this case be $(r-1)(k-1)(c-1)$.

A no three-way interaction model implies that the interaction between any pair does not depend on the third variable. For the data in Table 6.23 a no three-way interaction model would imply that the interaction between seatbelt usage and driver injury level does not depend on driver condition. Similarly, the interaction between driver injury level and driver condition does not depend on seatbelt usage, and the interaction between seatbelt usage and driver condition does not depend on driver injury level.

Saturated Model

As in the case of the two-way contingency table, the most general model for the three-way contingency table is the saturated model that fits the data perfectly. The *saturated model* for the three-way table includes a three-way interaction that allows the two-way interaction between any pair to vary at each level of the third variable. This model will be discussed further with the introduction of the loglinear model for three-way tables below.

Loglinear Models for Three-Way Tables

We begin our discussion of the loglinear model for three-way tables by extending the definitions of the μ parameters introduced in Section 6.2 for the two-way table. The saturated model for the three-way table is given by

$$\ln F_{ijk} = \mu + \mu_{1(i)} + \mu_{2(j)} + \mu_{3(k)} + \mu_{12(ij)} + \mu_{13(ik)} + \mu_{23(jk)} + \mu_{123(ijk)},$$

$$i = 1, 2, \ldots, r; \quad j = 1, 2, \ldots, c; \quad k = 1, 2, \ldots, \ell;$$

where $F_{ijk} = $ true frequency in cell (i, j, k) and

$$\mu = \frac{1}{rc\ell} \sum_{i=1}^{r} \sum_{j=1}^{c} \sum_{k=1}^{\ell} \ln F_{ijk},$$

$$\mu_{1(i)} = \frac{1}{c\ell} \sum_{j=1}^{c} \sum_{k=1}^{\ell} \ln F_{ijk} - \mu,$$

$$\mu_{2(j)} = \frac{1}{r\ell} \sum_{i=1}^{r} \sum_{k=1}^{\ell} \ln F_{ijk} - \mu,$$

$$\mu_{3(k)} = \frac{1}{rc} \sum_{i=1}^{r} \sum_{j=1}^{c} \ln F_{ijk} - \mu,$$

$$\mu_{12(ij)} = \frac{1}{\ell} \sum_{k=1}^{\ell} \ln F_{ijk} - \mu_{1(i)} - \mu_{2(j)} - \mu,$$

$$\mu_{13(ik)} = \frac{1}{c} \sum_{j=1}^{c} \ln F_{ijk} - \mu_{1(i)} - \mu_{3(k)} - \mu,$$

$$\mu_{23(jk)} = \frac{1}{r} \sum_{i=1}^{r} \ln F_{ijk} - \mu_{2(j)} - \mu_{3(k)} - \mu,$$

$$\mu_{123(ijk)} = \ln F_{ijk} - \mu_{1(i)} - \mu_{2(j)} - \mu_{3(k)} - \mu_{12(ij)},$$
$$- \mu_{23(jk)} - \mu_{13(ik)} - \mu.$$

The following conditions follow from these definitions

$$\sum_{i=1}^{r} \mu_{1(i)} = \sum_{j=1}^{c} \mu_{2(j)} = \sum_{k=1}^{\ell} \mu_{3(k)} = 0,$$

$$\sum_{i=1}^{r} \sum_{j=1}^{c} \mu_{12(ij)} = \sum_{i=1}^{r} \sum_{k=1}^{\ell} \mu_{13(ik)} = \sum_{j=1}^{c} \sum_{k=1}^{\ell} \mu_{23(jk)} = 0,$$

$$\sum_{i=1}^{r} \sum_{j=1}^{c} \sum_{k=1}^{\ell} \mu_{123(ijk)} = 0.$$

In comparison to the saturated loglinear model for the two-way table the saturated model now contains a total of four interaction terms. Three of the interaction terms are two-way interactions, whereas the remaining term is a three-way interaction. For a two-way interaction, the interaction is independent of the third variable, but for a three-way interaction the two-way interaction varies over the categories of the third variable. When both two-way and three-way interactions are included, the two-way interaction is an average over the categories of the third variable, whereas the three-way interaction measures departures or residuals from this average.

Definitions of Parameters in Terms of Cell Frequencies

The μ parameters are functions of various marginal totals in the table of logarithms of the theoretical frequencies, $\ln F_{ijk}$. As in the case of the two-dimensional table discussed in Section 6.2, the μ parameters are also functions of the logarithms of various geometric means of the frequencies. The expressions for the μ parameters may also be written as $\mu = \ln \widetilde{F}_{...}$,

$$\mu_{1(i)} = \ln \widetilde{F}_{i..} - \ln \widetilde{F}_{...},$$
$$\mu_{2(j)} = \ln \widetilde{F}_{.j.} - \ln \widetilde{F}_{...},$$

$$\mu_{3(k)} \;=\; \ln \widetilde{F}_{..k} - \ln \widetilde{F}_{...},$$

$$\mu_{12(ij)} \;=\; \ln \widetilde{F}_{ij.} - \ln \widetilde{F}_{i..} - \ln \widetilde{F}_{.j.} + \ln \widetilde{F}_{...},$$

$$\mu_{13(ik)} \;=\; \ln \widetilde{F}_{i.k} - \ln \widetilde{F}_{i..} - \ln \widetilde{F}_{..k} + \ln \widetilde{F}_{...},$$

$$\mu_{23(jk)} \;=\; \ln \widetilde{F}_{.jk} - \ln \widetilde{F}_{.j.} - \ln \widetilde{F}_{..k} + \ln \widetilde{F}_{...},$$

$$\mu_{123(ijk)} \;=\; \ln F_{ijk} - \ln \widetilde{F}_{ij.} - \ln \widetilde{F}_{i.k} - \ln \widetilde{F}_{.jk},$$

$$+ \ln \widetilde{F}_{i..} + \ln \widetilde{F}_{.j.} + \ln \widetilde{F}_{..k} - \ln \widetilde{F}_{...},$$

where
$\widetilde{F}_{...}$ is the overall geometric mean of all the frequencies F_{ijk};
$\widetilde{F}_{i..}$ is the geometric mean of all the frequencies F_{ijk} holding i fixed;
$\widetilde{F}_{.j.}$ is the geometric mean of all the frequencies F_{ijk} holding j fixed;
$\widetilde{F}_{..k}$ is the geometric mean of all the frequencies F_{ijk} holding k fixed;
$\widetilde{F}_{ij.}$ is the geometric mean of all the frequencies F_{ijk} holding i, j fixed;
$\widetilde{F}_{.jk}$ is the geometric mean of all the frequencies F_{ijk} holding j, k fixed;
$\widetilde{F}_{i.k}$ is the geometric mean of all the frequencies F_{ijk} holding i, k fixed.

For each of the models introduced above for three-way tables the cell frequencies F_{ijk} have different properties. These properties imply that some of the μ parameters are zero. These models can be related to the loglinear model parameters as outlined below.

Independence Model

In the case of the independence model, $F_{ijk} = (F_{i..}F_{.j.}F_{..k})/n^2$ implies that $\ln F_{ijk} = \mu + \mu_{1(i)} + \mu_{2(j)} + \mu_{3(k)}$ with all remaining μ parameters zero.

Partial Independence Model

For the partial independence model the two-way interaction between i and j results in $\mu_{12(ij)}$ being non-zero. The other possible interactions are zero. The loglinear model for this particular partial independence model is therefore given by

$$\ln F_{ijk} = \mu + \mu_{1(i)} + \mu_{2(j)} + \mu_{3(k)} + \mu_{12(ij)}.$$

If the table is collapsed over k, the resulting two-dimensional table is fitted exactly. There are two other possible partial independence models that contain only one two-way interaction term.

Conditional Independence Model

In the conditional independence model, the relationship between i and k is captured by $\mu_{13(ik)}$, and the relationship between j and k is captured by $\mu_{23(jk)}$. Since i and j are independent at every level of k, $\mu_{12(ij)} = 0$. The loglinear model in this case is

$$\ln F_{ijk} = \mu + \mu_{1(i)} + \mu_{2(j)} + \mu_{3(k)} + \mu_{13(ik)} + \mu_{23(jk)}.$$

If the table is collapsed over i or over j, the resulting two-dimensional tables are fitted exactly. There are two other possible conditional independence models, one for each possible omitted two-way interaction.

No Three-way Interaction Model

In the no three-way interaction model all pairs are related, but these relationships are independent of the third variable. Only the term $\mu_{123(ijk)}$ is zero. The loglinear model is given by

$$\ln F_{ijk} = \mu + \mu_{1(i)} + \mu_{2(j)} + \mu_{3(k)} + \mu_{12(ij)} + \mu_{13(ik)} + \mu_{23(jk)}.$$

In this case the three two-dimensional tables obtained by collapsing the fitted table on the third variable have cell frequencies identical to the observed two-dimensional tables. This is precisely what is accomplished by the iterative proportional fitting algorithm.

Saturated Model

The saturated model given at the beginning of this section fits the three-dimensional table perfectly. Although this model is not needed to determine expected frequencies, it is often useful for characterizing the interactions in a three-way table. The estimated interaction parameters provide a systematic way of studying the relationships among the variables.

Multiplicative Form of the Loglinear Model

Taking the antilogarithm of both sides of the loglinear model yields a multiplicative model for the cell frequency F_{ijk}. The equation becomes

$$F_{ijk} = \beta_0 \beta_{1(i)} \beta_{2(j)} \beta_{3(k)} \beta_{12(ij)} \beta_{13(ik)} \beta_{23(jk)} \beta_{123(ijk)}.$$

The beta parameters are sometimes useful for characterizing the variation in the table. The beta parameters are defined by

$$\beta_0 = e^{\mu}, \quad \beta_{1(i)} = e^{\mu_{1(i)}}, \quad \beta_{2(j)} = e^{\mu_{2(j)}}, \quad \beta_{3(k)} = e^{\mu_{3(k)}},$$
$$\beta_{12(ij)} = e^{\mu_{12(ij)}}, \quad \beta_{13(ik)} = e^{\mu_{13(ik)}}, \quad \beta_{23(jk)} = e^{\mu_{23(jk)}},$$
$$\beta_{123(ijk)} = e^{\mu_{123(ijk)}}.$$

Hierarchical Models

The above collection of models does not include all possible variants using the parameters specified by the saturated model. Such models as

$$\ln F_{ijk} = \mu + \mu_{1(i)} + \mu_{2(j)} + \mu_{12(ij)} + \mu_{23(jk)} + \mu_{13(ik)}$$

and

$$\ln F_{ijk} = \mu + \mu_{1(i)} + \mu_{2(j)} + \mu_{3(k)} + \mu_{123(ijk)}$$

have not been considered. In order to maintain the practice of defining higher order terms using deviations of lower order terms, the *hierarchy principle* is followed. This principle requires that, if a given term is fitted, all lower order terms involving those variables must also be included. The above two models do not satisfy this hierarchy principle. The main difficulty with *non-hierarchical* models is the interpretation of the fitted parameters. An additional problem, however, is that the iterative proportional fitting procedure cannot be used to fit the model without some transformation of the model being carried out first. Only models that satisfy the hierarchy principle will be discussed in this text.

Notation for Loglinear Models

To simplify the notation for the remainder of this chapter, the various models in the hierarchical system will be denoted by symbols such as [1], [23] and [123]. Only the symbols for the highest order interaction for each variable will be used. All lower order terms containing that variable are automatically included in the hierarchical system. The model [12], [3], for instance, implies that the terms [1] and [2] are also present, whereas the parameters corresponding to [13] and [23] are not present. The saturated model is denoted by [123] and implies that all lower order terms are present.

Model Selection

Given a three-dimensional table of observed cell frequencies n_{ijk}, a variety of models in the hierarchical system can be fitted by replacing F_{ijk} by E_{ijk} in the above formulae for the loglinear model parameters. The expression for E_{ijk} depends on the model being fitted. The various formulae for E_{ijk} for the various models have been outlined above. The goodness of fit of a particular model can be judged using the χ^2 goodness of fit statistics G^2 and H^2. A probability level of 0.15 to 0.25 is usually required to confirm that the model adequately represents the interactions in the table. In practice we seek to fit the simplest model while maintaining a reasonable fit.

In addition to the overall measure of goodness of fit, the likelihood ratio statistic H^2 has the advantage that it can be used to compare nested models in the hierarchical system. Let H_1^2 and H_2^2 denote two likelihood chi-square statistics for two alternative models and assume that model 2 is the larger model which contains all the parameters of model 1. The *conditional likelihood chi-square statistic* $H_{2\cdot1}^2 = (H_1^2 - H_2^2)$ can be used to determine whether model 2 is superior to model 1. Under the null hypothesis that model 1 is equally as good as model 2, the statistic $H_{2\cdot1}^2$ is asymptotically a χ^2 distribution with degrees of freedom equal to the difference (d.f. model 1 – d.f. model 2). An example of such a test might involve a comparison of the model [13] [2] to the model [12] [13] [23]. The null hypothesis would be that the terms [12] and [23] are superfluous.

Standardized Estimates and Standardized Residuals

Expressions for the asymptotic variances of the loglinear model parameters can be obtained using expressions similar to those given in Section 6.2.2 for the two-dimensional table. The expressions can be generated using the theory presented in Bishop, Fienberg and Holland (1974).

The concept of standardized residuals introduced in Section 6.2.2 for the two-dimensional table can be extended to higher-dimensional tables and also to models other than the independence model. For each cell (i, j, k) in the table, the component of the Pearson χ^2 given by $(n_{ijk} - E_{ijk})^2/E_{ijk}$ provides the standardized residual $r_{1ijk} = (n_{ijk} - E_{ijk})/\sqrt{E_{ijk}}$.

The Freeman-Tukey standardized residuals introduced in Section 6.2.2 can also be extended to multidimensional tables using the expression

$$r_{2ijk} = \sqrt{n_{ijk}} + \sqrt{n_{ijk} + 1} - \sqrt{4E_{ijk} + 1}.$$

These residuals are useful for examining the quality of the fitted model on a cell by cell basis and also for detecting outliers.

Summary of Loglinear Model Fitting Procedure

The system of fitting loglinear models for the purpose of explaining interaction in a multidimensional contingency table is demonstrated by the diagram in Figure 6.4. It is useful to note that the estimates of F_{ijk} depend on the model being fitted.

For the three-dimensional table the simplest loglinear model is the constant cell density model that contains only one parameter μ. In this case the E_{ijk} are all equal to n/rc. The next three models in the hierarchy are the single effect models that fit the marginal frequencies for one variable and restrict the remaining two variables to constant marginal frequencies. The three models $(\mu + \mu_{1(i)})$, $(\mu + \mu_{2(j)})$ and $(\mu + \mu_{3(k)})$ fit the theoretical marginal frequencies so that $E_{i..} = n_{i..}$, $E_{.j.} = n_{.j.}$ and $E_{..k} = n_{..k}$ respectively. Similarly for the loglinear models which contain two of the three main effects two of the three sets of theoretical marginals, $E_{i..}$, $E_{.j.}$ and $E_{..k}$, are set equal to two of the three corresponding sample marginals, $n_{i..}$, $n_{.j.}$ and $n_{..k}$. Finally the independence model which contains all three main effects requires that $E_{i..} = n_{i..}$, $E_{.j.} = n_{.j.}$ and $E_{..k} = n_{..k}$.

For each interaction term that is fitted a set of two-dimensional marginals are fitted. If the parameter $\mu_{12(ij)}$ is included then $E_{ij.} = n_{ij.}$. Similarly for $\mu_{23(jk)}$ and $\mu_{13(ik)}$ the corresponding theoretical two-dimensional marginals are given by $E_{.jk} = n_{.jk}$ and $E_{i.k} = n_{i.k}$ respectively. For the no three-way interaction model all three sets of two-dimensional marginals are fitted as outlined in the iterative proportional fitting algorithm discussed above.

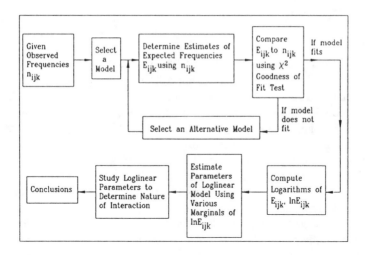

FIGURE 6.4. System for Fitting and Using Loglinear Models

Product Multinomial Sampling

In product multinomial sampling, certain marginals are held fixed. In the three dimensional table we consider the two cases corresponding to the fixing of the marginals for one or two of the three variables. If the marginals are fixed for the first variable, then the loglinear model must contain the term $\mu_{1(i)}$. This will ensure that the fitted marginals $E_{i..}$ are equal to the observed marginals $n_{i...}$ Similarly, if the marginals for both variables 1 and 2 are fixed, then the model must contain the parameters $\mu_{1(i)}$, $\mu_{2(j)}$ and $\mu_{12(ij)}$. In this case the fitted marginals $E_{i..}$, $E_{.j.}$ and $E_{ij.}$ are equivalent to the sample marginals $n_{i..}$, $n_{.j.}$ and $n_{ij..}$

In product multinomial sampling some of the variables can be viewed as response variables, whereas the remainder can be viewed as fixed or controlled. The control variables have the fixed marginals, and the marginals for the response variables are viewed as an outcome of the sampling process. The weighted least squares approach in Section 6.4 assumes product multinomial sampling.

Computer Software

The statistical software package BMDP will be used throughout Section 6.3 to perform the calculations.

6.3.2 SOME EXAMPLES

Accident Data

For the example presented in Table 6.23, the entire set of loglinear models was fitted using the maximum likelihood estimators E_{ijk} defined above and in Section 6.2. Table 6.24 summarizes the χ^2 goodness of fit statistics for the various models. The first three lines in the table present the goodness of fit statistics for the three single-effect models. In the model [2], for instance, the marginals for seatbelt usage are permitted to be different, but the marginals for driver condition and injury level are forced to have equal frequencies in all categories. Similarly in [1], only the marginal for driver condition is permitted to vary, whereas for [3] only the marginal for injury level is not constant.

The next three rows of Table 6.24 present the three possible models that fit two effects holding the third effect fixed. The seventh row is the independence model which permits all three marginals to vary but contains no interaction.

Rows 8, 9 and 10 of Table 6.24 summarize the results of fitting a saturated model to a marginal two-dimensional table while restricting the third variable to a constant marginal. Rows 11, 12 and 13 show the results for the fitting of the three possible partial independence models. In row 11 the model [2], [13] allows variables 1 and 3 to be related, but both are assumed to be independent of variable 2. Similarly, in row 12 variables 2 and 3 are independent of 1, and in row 13 variables 1 and 2 are independent of variable 3.

The three conditional independence models are shown in rows 14, 15 and 16. In row 14 the model [12], [23] requires that 1 and 3 be independent at each level of variable 2. Similarly, in row 15 variables 1 and 2 are independent at each level of 3, and in row 16 variables 2 and 3 are independent at each level of 1. The no three-way interaction model is fitted in the last row. In this model all two-way interactions among the three variables are assumed to explain all the interactions in the table.

An examination of the χ^2 goodness of fit statistics reveals that the no three-way interaction model can be used to explain the interactions among the three variables. Both the Pearson and likelihood χ^2 statistics show a *p*-value of 0.1705. The fitted parameters for this model are summarized in Table 6.25. The ratios of the loglinear model parameter estimates to their standard errors are also shown (with brackets) for selected parameters. Plots of the values of the parameter estimates are shown in Figure 6.5. The loglinear model being fitted is given by

$$\ln F_{ijk} = \mu + \mu_{1(i)} + \mu_{2(j)} + \mu_{3(k)} + \mu_{12(ij)} + \mu_{13(ik)} + \mu_{23(jk)}.$$

From the fitted parameters in Table 6.25 the logarithm of the geometric mean of the expected frequencies is 6.002. Very few of the parameter

TABLE 6.24. Summary of χ^2 Goodness of Fit Statistics for System of Hierarchical Models*

	Model	d.f.	Likelihood	Prob	Pearson	Prob
1.	[2]	14	25550.38	0.0000	428867.25	0.0000
2.	[1]	14	219138.75	0.0000	332231.00	0.0000
3.	[3]	12	125473.63	0.0000	136095.75	0.0000
4.	[1] [2]	13	175077.63	0.0000	201950.81	0.0000
5.	[1] [3]	11	45001.13	0.0000	41474.70	0.0000
6.	[2] [3]	11	81412.63	0.0000	67216.56	0.0000
7.	[1] [2] [3]	10	940.02	0.0000	1057.47	0.0000
8.	[12]	12	174680.25	0.0000	201019.50	0.0000
9.	[23]	8	81349.75	0.0000	67141.44	0.0000
10.	[13]	8	44505.84	0.0000	40381.77	0.0000
11.	[2] [13]	7	444.85	0.0000	372.21	0.0000
12.	[1] [23]	7	877.16	0.0000	967.92	0.0000
13.	[3] [12]	9	542.50	0.0000	682.37	0.0000
14.	[12] [23]	6	479.69	0.0000	610.75	0.0000
15.	[13] [23]	4	382.02	0.0000	317.32	0.0000
16.	[13] [12]	6	47.34	0.0000	44.51	0.0000
17.	[12] [13] [23]	3	5.02	0.1705	5.02	0.1705

*[2] = seatbelt usage, [1] = driver condition, [3] = injury level

estimates are not significantly different from zero. The driver condition effects indicate that the normal condition is much more frequent than the been drinking condition. The seatbelt usage effects indicate that many more drivers were not wearing seatbelts than were wearing them. The injury level parameters indicate that the large majority of drivers were not injured and that very few drivers sustained major or fatal injuries. The graph of the logarithms of the fitted frequencies shown in Figure 6.5 illustrates that at each injury level the frequencies are highest for the normal–no seatbelt category and lowest for the drinking–seatbelt category.

The interaction effects in Table 6.25 suggest that normal condition drivers were more likely to be wearing seatbelts than drivers who had been drinking. The driver condition-injury level interactions indicate that, in comparison to drivers who had been drinking, a larger proportion of drivers in the normal category had no injury and a smaller proportion of the normal category drivers were in the major or fatal injury category. For the minimal category, the interaction term was quite weak. The interaction between driver injury level and driver condition therefore seems to primarily affect only the two extremes of the injury level range. The seatbelt usage-injury level interaction appears to be relatively weak. There is some tendency, however, for seatbelt users to be over-represented in the no-injury category and under-represented in the minor injury category. The minimal injury

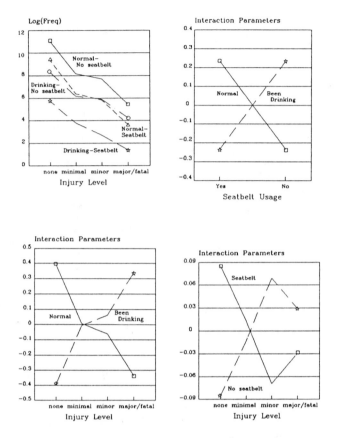

FIGURE 6.5. Parameter Estimates for Loglinear Model for Accident Data

category and the major/fatal category show only slight interactions with seatbelt usage. These interactions can also be seen in Figure 6.5.

In conclusion, we could say that a large majority of drivers appeared normal, had not been wearing seatbelts, and were not injured. For drivers wearing seatbelts, there were proportionately fewer who sustained an injury and proportionately more were in normal condition than for non-seatbelt users. Among those who had been drinking, proportionately more sustained a minor or major/fatal injury than among those who appeared normal.

A comparison of the observed frequencies to the expected frequencies under the no three-way interaction fitted model is shown in Table 6.26. The expected frequencies are shown in round brackets under the corresponding observed frequencies. The fit seems to be excellent with only minor differences in the minimal and minor categories for drivers who had been drinking and were wearing seatbelts. The values of the standardized residuals are shown in square brackets for each cell. The largest standardized

TABLE 6.25. Fitted Parameters for No Three-Way Interaction Loglinear Model

Overall Mean	$\hat{\mu} = 6.002$	

Driver Condition Effects $\hat{\mu}_{1(1)} = 1.212$ $\hat{\mu}_{1(2)} = \dfrac{-1.212}{(-53.906)}$

Seatbelt Usage Effects $\hat{\mu}_{2(1)} = -1.119$ $\hat{\mu}_{2(2)} = \dfrac{1.119}{(43.938)}$

Injury Level Effects

$\hat{\mu}_{3(1)} = \dfrac{2.626}{(97.270)}$ $\hat{\mu}_{3(2)} = \dfrac{0.072}{(2.152)}$ $\hat{\mu}_{3(3)} = \dfrac{-0.376}{(-10.315)}$ $\hat{\mu}_{3(4)} = \dfrac{-2.322}{(-32.387)}$

Driver Condition – Seatbelt Usage Interaction

$\hat{\mu}_{12(11)} = \dfrac{0.234}{(17.147)}$ $\hat{\mu}_{12(12)} = -0.234$ $\hat{\mu}_{12(21)} = -0.234$ $\hat{\mu}_{12(22)} = 0.234$

Driver Condition – Injury Level Interaction

$\hat{\mu}_{13(11)} = \dfrac{0.392}{(19.698)}$ $\hat{\mu}_{13(12)} = \dfrac{0.006}{(0.219)}$ $\hat{\mu}_{13(13)} = \dfrac{-0.061}{(-2.249)}$ $\hat{\mu}_{13(14)} = \dfrac{-0.337}{(-6.578)}$

$\hat{\mu}_{13(21)} = -0.392$ $\hat{\mu}_{13(22)} = -0.006$ $\hat{\mu}_{13(23)} = 0.061$ $\hat{\mu}_{13(24)} = 0.337$

Seatbelt Usage - Injury Level Interaction

$\hat{\mu}_{23(11)} = \dfrac{0.085}{(3.714)}$ $\hat{\mu}_{23(12)} = \dfrac{0.013}{(0.490)}$ $\hat{\mu}_{23(13)} = \dfrac{-0.069}{(-2.286)}$ $\hat{\mu}_{23(14)} = \dfrac{-0.029}{(-0.465)}$

$\hat{\mu}_{23(21)} = -0.085$ $\hat{\mu}_{23(22)} = -0.013$ $\hat{\mu}_{23(23)} = 0.069$ $\hat{\mu}_{23(24)} = 0.029$

residuals occurred in the minimal and minor categories for drivers who had been drinking. These residuals, however, were quite small indicating an excellent fit. In these two columns the frequencies are relatively small and hence the prediction errors are proportionately larger.

The fitted parameters in Table 6.25 were converted to multiplicative parameters and are summarized in Table 6.27. The multiplicative form of the fitted model is given by the equation

$$F_{ijk} = \beta_0 \beta_{1(i)} \beta_{2(j)} \beta_{3(k)} \beta_{12(ij)} \beta_{13(ik)} \beta_{23(jk)}.$$

The magnitudes of these multiplicative effects can be compared to 1.0 in order to interpret direction. The resulting interpretations will be the same as the interpretations derived from Table 6.25.

Three-Way Interaction

When a saturated model is required in order to obtain a good fit for a three-way table, the *three-way interaction* $\mu_{123(ijk)}$ is said to be signifi-

TABLE 6.26. Comparison of Observed and Expected Frequencies

Driver Condition	Seatbelt Usage	Driver Injury Level			
		None	Minimal	Minor	Major/Fatal
Normal	Yes	12500	604	344	38
		(12497.0)	(613.3)	(337.8)	(37.9)
		[0.0]	[-0.4]	[0.3]	[0.0]
	No	61971	3519	2272	237
		(61974.0)	(3509.7)	(2278.2)	(237.1)
		[0.0]	[0.2]	[-0.1]	[0.0]
Been Drinking	Yes	313	43	15	4
		(316.0)	(33.7)	(21.2)	(4.1)
		[-0.2]	[1.6]	[-1.3]	[-0.1]
	No	3992	481	370	66
		(3989.0)	(490.3)	(363.8)	(65.9)
		[0.0]	[-0.4]	[0.3]	[0.0]

cant. The presence of such an interaction indicates that each of the three two-way interactions cannot be assumed to be constant over the various levels of the third variable. As an example, consider the two-way interaction $\mu_{12(ij)}$. This parameter measures the interaction between variables 1 and 2 and is estimated using the marginal table obtained after summing over the subscript k. The two-way interaction $\mu_{12(ij)}$ therefore represents an average relationship between variables 1 and 2 summed over the categories of the third variable. The fact that $\mu_{123(ijk)}$ is nonzero indicates that the interaction between variables 1 and 2 varies over the levels of variable 3.

Example

To provide an example interpretation for three-way interaction parameters, the estimates $\hat{\mu}_{123(ijk)}$ for the data in Table 6.23 are shown in Table 6.28. Although these estimates are not significant they will be interpreted as if they were. The largest parameter estimate of 0.086 for the minor injury category allows us to conclude the following:

1. The two-way interaction between seatbelt usage and injury level indicates that the probability of a minor injury is greater for a seatbelt nonuser than for a seatbelt user. The three-way interaction with driver condition suggests that this two-way interaction in the case of drinking drivers is less pronounced, whereas for normal drivers it is stronger. In other words, the marginal effect of driver condition on

TABLE 6.27. Multiplicative Parameters for No Three-Way Interaction Model

Geometric Mean	$\hat{\beta} = 404.362$

Driver Condition Effects $\hat{\beta}_{1(1)} = 3.361$ $\hat{\beta}_{1(2)} = 0.298$

Seatbelt Usage Effects $\hat{\beta}_{2(1)} = 0.237$ $\hat{\beta}_{2(2)} = 3.061$

Injury Level Effects
$\hat{\beta}_{3(1)} = 13.824$ $\hat{\beta}_{3(2)} = 1.074$ $\hat{\beta}_{3(3)} = 0.687$ $\hat{\beta}_{3(4)} = 0.098$

Driver Condition – Seatbelt Usage Interaction
$\hat{\beta}_{12(11)} = 1.263$ $\hat{\beta}_{12(12)} = 0.792$ $\hat{\beta}_{12(21)} = 0.792$ $\hat{\beta}_{12(22)} = 1.263$

Driver Condition – Injury Level Interaction
$\hat{\beta}_{13(11)} = 1.481$ $\hat{\beta}_{13(12)} = 1.006$ $\hat{\beta}_{13(13)} = 0.941$ $\hat{\beta}_{13(14)} = 0.714$

$\hat{\beta}_{13(21)} = 0.675$ $\hat{\beta}_{13(22)} = 0.994$ $\hat{\beta}_{13(23)} = 1.063$ $\hat{\beta}_{13(24)} = 1.401$

Seatbelt Usage – Injury Level Interaction
$\hat{\beta}_{23(11)} = 1.088$ $\hat{\beta}_{23(12)} = 1.013$ $\hat{\beta}_{23(13)} = 0.934$ $\hat{\beta}_{23(14)} = 0.971$

$\hat{\beta}_{23(21)} = 0.919$ $\hat{\beta}_{23(22)} = 0.987$ $\hat{\beta}_{23(23)} = 1.071$ $\hat{\beta}_{23(24)} = 1.030$

the likelihood of minor injury is less for seatbelt nonusers than for seatbelt users.

2. The two-way interaction between driver condition and injury level indicates that the probability of a minor injury is greater for a driver who has been drinking than for a normal condition driver. The three-way interaction with seatbelt usage suggests that this two-way interaction in the case of seatbelt nonusers is less pronounced whereas for seatbelt users it is stronger. In other words, the marginal effect of seatbelt usage on the likelihood of minor injury is less for drinking drivers than for normal condition drivers.

3. The two-way interaction between seatbelt usage and driver condition indicates that the probability of seatbelt usage is greater for normal condition drivers than for drinking drivers. The three-way interaction with injury level suggests that for the minor injury category this interaction is less pronounced. In other words, among drivers who sustained a minor injury the relation between seatbelt usage and driver condition is different than when injury level is ignored.

Without the three-way interactions the two-way interactions are additive. With the three-way interactions included a correction is made for the fact

TABLE 6.28. Three-Way Interaction Terms from Saturated Model

Driver Condition	Seatbelt Usage	Driver Injury Level			
		None	Minimal	Minor	Major
Normal	Yes	−0.007	−0.080	+0.086	0.000
	No	+0.007	+0.080	−0.086	0.000
Been Drinking	Yes	+0.007	+0.080	−0.086	0.000
	No	−0.007	−0.080	+0.086	0.000

that the two-way interactions are not additive. In this example the seatbelt and driver condition interactions are not additive. The combined effect in the minor category is less than the sum. The reader is left to provide an interpretation for the minimal injury category.

Bus Driver Data

This example is based on a study of bus driver absenteeism in the transit system for the City of Edmonton, Alberta. The data is based on a survey of all shifts over a two-week period in each of four seasons of a calendar year. From this study, Table 6.29 relating Attendance (1), Day of Week (2) and Shift Type (3) was obtained. The frequencies in the table represent the total number of shifts in the cells. A swing shift is one that involves driving on the weekend, and a split shift implies that the seven hour driving day is split into two parts with several hours time off in between parts (i.e., morning rush period and evening rush period).

The χ^2 goodness of fit statistics are shown in Table 6.30 for all models of complexity greater than the independence model. The table shows that two models fit the data reasonably well: the conditional independence model [13] [23] and the no three-way interaction model [12] [13] [23]. The χ^2 p-values for the conditional independence model are 0.3882 for $H^2 = 43.97$ and 0.3103 for $G^2 = 45.91$. For the no three-way interaction model the χ^2 p-values are 0.5200 for $H^2 = 34.92$ and 0.3737 for $G^2 = 38.11$. Since the conditional independence model is less complex, it is usually the preferred model. Comparing the two models, $H^2_{2.1} = 43.97 - 34.92 = 9.05$ which has $42 - 36 = 6$ d.f. This χ^2 value is not significant at conventional probability values and hence we would conclude that the conditional independence model is as good a fit as the no three-way interaction model. We can conclude, therefore, that after controlling for type of shift there is no interaction between attendance and day of the week. Thus, for instance, after controlling for type of shift there is no tendency for drivers to be absent on Fridays or Mondays.

TABLE 6.29. Attendance versus Day of Week and Type of Shift

Attendance (1)	Day (2)	Type of Shift (3)						
		A.M.	Noon	P.M.	Swing	A Split	A Split /Swing	B Split /Swing
Absent	Sun	9	13	18	31	39	23	38
	Mon	99	31	67	34	252	27	43
	Tues	117	32	88	18	291	19	43
	Wed	13	39	96	13	290	24	50
	Thur	101	38	91	19	320	13	48
	Fri	91	34	82	19	292	16	47
	Sat	24	21	49	26	115	30	61
Present	Sun	135	59	230	409	393	273	458
	Mon	1029	361	605	2156	389	389	445
	Tues	1099	416	712	150	2293	221	445
	Wed	1103	409	704	155	2302	208	438
	Thur	1083	370	677	253	2264	227	440
	Fri	1093	374	662	253	2292	224	465
	Sat	264	123	399	414	821	386	659

TABLE 6.30. Goodness of Fit Statistics for Absenteeism by Day by Shift

Model	d.f.	H^2	Prob	G^2	Prob	R^2	A
[1] [2] [3]	84	3360.92	0.0000	3756.49	0.0000	0	3431
[1] [23]	48	132.42	0.0000	129.34	0.0000	0.961	130
[2] [13]	78	3272.65	0.0000	3643.53	0.0000	0.027	3331
[3] [12]	78	3343.08	0.0000	3720.47	0.0000	0.005	3401
[13] [23]	42	43.97	0.3882	45.91	0.3103	0.987	30
[13] [12]	72	3254.61	0.0000	3625.02	0.0000	0.032	3300
[12] [23]	42	114.41	0.0000	112.98	0.0000	0.966	100
[12] [13] [23]	36	34.92	0.5200	38.11	0.3737	0.990	9

The fitted parameters for the conditional independence model are summarized in Table 6.31. The numbers in brackets indicate the standardized values of the estimated parameters and hence can be compared to the standard normal to determine significance. The attendance parameters indicate that a large proportion of the drivers are not normally absent. The day effects indicate that the number of drivers required is lower on Saturday and much lower on Sunday than the other five days of the week. The shift effects indicate how the number of employees required per shift varies. The order from largest to smallest is A Split, A.M., P.M., B Split/Swing, Noon, Swing and A Split/Swing. The attendance by shift effects are relatively weak; however, there is some evidence that the P.M. and A Split shifts see somewhat smaller rates of absenteeism while the Swing and A Split/Swing shifts see somewhat larger rates of absenteeism. For the parameters that measure the interaction between day of the week and type of shift, we can conclude that in comparison to weekdays on the weekends relatively few drivers are on the A.M., Noon, P.M. and A Split shifts whereas a relatively large number of drivers are on the Swing, A Split/Swing and B Split/Swing shifts. During weekdays the opposite effects seem to occur. The large negative interaction parameters for the swing shift on Tuesdays and Wednesdays suggest that these days are the most common days off for the swing shift drivers who work on weekends.

Goodness of Fit and Model Selection

A measure of goodness of fit comparable to R^2 in multiple linear regression is given by

$$R^2 = 1 - H^2/H_0^2,$$

where H_0^2 is the value of the likelihood ratio χ^2 statistic for the independence model, and H^2 is the likelihood ratio χ^2 statistic for the model of interest.

An *adjusted* R^2 measure comparable to the adjusted R^2 in multiple linear regression is given by

$$R'^2 = 1 - \frac{H^2/(q-r)}{H_0^2/(q-r_0)},$$

where q = number of cells in the contingency table, and r and r_0 are the degrees of freedom associated with the models yielding H^2 and H_0^2 respectively.

An alternative approach to model comparison is provided by *Akaike's information criterion*. This criterion recommends choosing the model that minimizes the value of

$$A = H^2 - (q - 2r),$$

where H^2 is the likelihood ratio χ^2 statistic for the model, r is the number of degrees of freedom, and q is the number of cells in the table. The subtraction of the term $(q - 2r)$ is a method of compensating for the overfitting that can occur when the number of cells is relatively large.

Example

The values of the criteria R^2 and A for the bus data contingency table are shown in Table 6.30. The R^2 values for the models [1][23], [13][23], [12][23] and [12][13][23] are 0.961, 0.987, 0.966 and 0.990. For the Akaike criterion the values of A for these four models are 130, 30, 100 and 9 respectively. Although the maximum R^2 and minimum A correspond to the model [12][13][23] we have already seen from the likelihood ratio test

TABLE 6.31. Estimated Parameters for Conditional Independence Model for Absenteeism by Day by Shift

Mean	4.981						

Attendance Effects [1]

	Absent	Present					
	-1.159	1.159					
	(-97.978)						

Day Effects [2]

	Sun	Mon	Tues	Wed	Thurs	Fri	Sat
	-0.690	0.280	0.129	0.124	0.175	0.177	-0.195
	(-29.103)	(18.471)	(7.441)	(7.162)	(10.611)	(10.768)	(-10.547)

Shift Effects [3]

	A.M.	Noon	P.M.	Swing	A Split	A Split /Swing	B Split /Swing
	0.328	-0.598	0.249	-0.655	1.317	-0.664	0.022
	(13.327)	(-16.812)	(10.275)	(-17.429)	(75.445)	(-17.491)	(0.793)

Attendance by Shift Effects [13]

Present	-0.016	0.000	0.111	-0.114	0.130	-0.111	0.000
Absent	0.016	0.000	-0.111	0.114	-0.130	0.111	0.000
	(0.716)	(0.000)	(-4.758)	(3.108)	(-7.962)	(2.968)	(0.000)

Day by Shift Effects [23]

Sun	-0.915	-0.669	-0.190	1.103	-0.688	0.718	0.642
	(-13.917)	(-7.421)	(-3.584)	(24.049)	(-16.131)	(13.937)	(15.171)
Mon	0.174	0.055	-0.163	0.132	0.059	0.088	-0.345
	(5.798)	(1.234)	(-4.760)	(3.144)	(2.607)	(2.051)	(-8.988)
Tues	0.400	0.340	0.162	-0.679	0.281	-0.311	-0.194
	(13.096)	(7.771)	(4.838)	(-11.140)	(11.728)	(-5.891)	(-4.930)
Wed	0.404	0.345	0.176	-0.675	0.289	-0.341	-0.189
	(13.225)	(7.868)	(4.965)	(-11.065)	(12.025)	(-6.364)	(-4.814)
Thurs	0.327	0.201	0.076	-0.243	0.235	-0.357	-0.239
	(10.824)	(4.492)	(2.258)	(-4.838)	(10.074)	(-6.794)	(-6.154)
Fri	0.325	0.199	0.042	-0.246	0.233	-0.359	-0.194
	(10752)	(4.441)	(1.230)	(-4.885)	(9.981)	(-6.839)	(-5.064)
Sat	-0.716	-0.471	-0.093	0.608	-0.410	0.563	0.519
	(-14.846)	(-7.125)	(-2.285)	(14.008)	(-13.195)	(12.804)	(14.698)

that this model is not a significant improvement over the model [13][23]. This latter model has the second best values of the two goodness of fit criteria.

6.3.3   FOUR-DIMENSIONAL CONTINGENCY TABLES AND STEPWISE FITTING PROCEDURES

For a *four-dimensional contingency table* the independence model has the form

$$\ln F_{ijkh} = \mu + \mu_{1(i)} + \mu_{2(j)} + \mu_{3(k)} + \mu_{4(h)},$$

and the saturated model has the form

$$
\begin{aligned}
\ln F_{ijkh} = {} & \mu + \mu_{1(i)} + \mu_{2(j)} + \mu_{3(k)} + \mu_{4(h)} \\
& + \mu_{12(ij)} + \mu_{13(ik)} + \mu_{14(ih)} + \mu_{23(jk)} + \mu_{24(jh)} + \mu_{34(kh)} \\
& + \mu_{124(ijh)} + \mu_{123(ijk)} + \mu_{134(ikh)} + \mu_{234(jkh)} + \mu_{1234(ijkh)}.
\end{aligned}
$$

Between these two models there are over 100 possible models within the hierarchical system. Fitting all possible models to determine the simplest model that fits the data can therefore be an expensive and time-consuming process. For dimensions higher than four, the number of possible models is mind expanding. For four and higher-dimensional tables, therefore, stepwise search procedures are often used for selecting a suitable model.

Stepwise Model Selection

The stepwise approach to choosing a model begins with a particular model and either adds terms *(forward)* or deletes terms *(backward)* one at a time until the simplest model that fits the data is obtained. Since the stepwise procedure begins with a starting model, this model must be selected in a suitable manner.

A common approach to selecting a starting point is to fit all models of a *uniform order*. The uniform order models for a four-dimensional table are given below.

Order 1	[1] [2] [3] [4]
Order 2	[12] [13] [14] [23] [24] [34]
Order 3	[123] [234] [134] [124]
Order 4	[1234]

The simplest uniform order model that fits the table well makes an excellent upper bound. Usually, if a particular uniform order model fits the data well, all higher order models will also fit the data. A lower bound for a stepwise process would be the highest uniform order model that does not fit the table. Usually the upper and lower bounds differ by only one or two orders.

To describe the stepwise procedures, we denote the order of the upper bound by q and the lower bound by r. The forward procedure begins with the r uniform order model and adds terms one at a time in such a way that the change in the likelihood ratio statistic is maximized. Terms are added as long as the increase in the likelihood chi-square statistic is significant. For the backward procedure, terms are removed one at a time beginning with the q uniform order model. At each step, the term removed is the

TABLE 6.32. Observed Frequencies Injury Level, Condition of
Driver, Seatbelt Usage and Sex of Driver

Var (1) Injury Level	Var (2) Seatbelt Usage	Var (3) Sex	Var (4) Driver Condition	
			Normal	Been Drinking
None	Yes	Male	8312	263
		Female	4188	50
	No	Male	42476	3440
		Female	19495	552
Minimal	Yes	Male	313	37
		Female	291	6
	No	Male	1841	383
		Female	1678	98
Minor	Yes	Male	189	12
		Female	155	3
	No	Male	1214	290
		Female	1058	80
Major/Fatal	Yes	Male	24	3
		Female	14	1
	No	Male	146	51
		Female	91	15

one that yields the smallest change in the likelihood chi-square statistic.
Terms are removed as long as they are not considered to be significant.
These stepwise procedures are illustrated below using the auto accident
data introduced earlier in this chapter.

Example

The four-dimensional table showing the relationship between injury level,
condition of driver, seatbelt usage and sex of driver for a large number of
auto accidents is shown in Table 6.32. The goodness of fit statistics for the
uniform order models for $q = 1, 2$ and 3 are shown in Table 6.33. From the
χ^2 goodness of fit statistics it would appear that we should seek a model
between orders 1 and 3. The order 2 model seems to fit the data $(p = 0.30)$;
however, we would like to determine other models in the neighborhood of
this one for comparison purposes.

The results of a forward stepwise procedure are shown in Table 6.34.
Beginning with the uniform first-order model, the procedure adds first-
order interaction terms one at a time. A good fit of the table is not achieved
until the uniform second-order model is reached in Step 6. The goodness
of fit χ^2 has a p-value of 0.3025 for this model. The last term added before

TABLE 6.33. Goodness of Fit Statistics for Uniform Order Models

Model	d.f.	Likelihood Ratio χ^2	Prob	Pearson χ^2	Prob
[1] [2] [3] [4]	25	2448.86	0.0000	2498.04	0.0000
[12] [13] [14] [23] [24] [34]	13	15.24	0.3024	15.70	0.2656
[123] [234] [124] [134]	3	1.39	0.7081	1.31	0.7258

TABLE 6.34. Results of Forward Stepwise Procedure

Step	Model Fitted	Added Effect	d.f.	Likelihood Ratio χ^2	Prob
0	[1] [2] [3] [4]		25	2448.86	0.0000
1	[1] [2] [34]		24	1628.81	0.0000
		[34]	1	820.05	0.0000
2	[1] [24] [34]		23	1236.61	0.0000
		[24]	1	392.20	0.0000
3	[13] [24] [34]		20	698.46	0.0000
		[13]	3	538.15	0.0000
4	[13] [14] [24] [34]		17	73.78	0.0000
		[14]	3	624.68	0.0000
5	[12] [13] [14] [24] [34]		14	33.23	0.0027
		[12]	3	40.55	0.0000
6	[12] [13] [14] [23] [24] [34]		13	15.08	0.3025
		[23]	1	18.15	0.0000
7	[134] [12] [23] [24]		10	8.98	0.5340
		[134]	3	6.10	0.1069
8	[134] [124] [23]		7	4.12	0.7662
		[124]	3	4.86	0.1821
9	[134] [124] [123]		4	1.32	0.8587
		[123]	3	2.80	0.4232
10	[134] [124] [123] [234]		3	1.32	0.7246
		[234]	1	0.00	1.0

this model was reached, [23], was significant at the 0.0000 level. The first term added beyond the uniform second-order model was [134] which was significant at the 0.1069 level. A reasonable conclusion, therefore, is that the uniform second-order model provides a good fit to the table. The model cannot be simplified without deleting a significant second-order interaction, and adding a third-order interaction does not significantly improve the fit.

TABLE 6.35. Results of Backward Stepwise Procedure

Step	Model Fitted	Deleted Effect	d.f.	Likelihood Ratio χ^2	Prob
0	[123] [124] [134] [234]		3	1.32	0.7246
1	[123] [124] [134]		4	1.32	0.8587
		[234]	1	0.00	1.0
2	[124] [134] [23]		7	4.12	0.7662
		[123]	3	2.80	0.4232
3	[134] [24] [23]		10	8.98	0.5340
		[124]	3	4.86	0.1821
4	[12] [13] [14] [23] [24] [34]	[134]	3	6.10	0.1069

For comparison, the results obtained from using the backward procedure between the second and third orders is shown in Table 6.35. The results are simply the reverse of the forward procedure results in Table 6.34.

Tests of Partial and Marginal Association

Two additional procedures that can provide insight in model selection are tests of *partial and marginal association*. In partial association the partial significance of an effect is determined by comparing the uniform model of the same order to the model from which the effect in question has been removed. In the second order uniform model [12] [13] [14] [23] [24] [34], the test of partial association for [23] involves comparing the uniform second-order model to the model with [23] removed, given by [12] [13] [14] [24] [34]. The difference of the two likelihood ratio χ^2 statistics provides a test statistic for this partial association.

Example

From the results of the forward stepwise procedure in Table 6.34, we can conclude that the term [23] is significant at the 0.0000 level (see step 6). Thus we would conclude that, after fitting all other first-order interactions, the interaction [23] is significant. A second example is provided by step 10 in the same table. Removing [234] from the uniform third-order model does not result in any loss in the quality of the fit.

Marginal Association

A test of marginal association seeks to determine the importance of an effect by collapsing the table over all other effects and then determining whether the effect of interest is required to model the collapsed table. In the four-dimensional table a test for the marginal association of [123] is obtained by collapsing the table on variable 4 and then fitting the model

[12] [13] [23] to the collapsed table. If it fits the data well, then the effect [123] is not required. A difficulty with the test of marginal association is the potential problem of collapsibility. This topic is discussed in the next section.

Example

For the auto accident data, the three-dimensional table for variables 1, 2 and 4 was studied in Section 6.3.2. The model [12] [14] [24] was found to provide an adequate fit for this table. A test for the marginal association for the effect [124] would conclude that this effect is not important.

Estimated Parameters for the Four-Dimensional Auto Accident Table

The estimated parameters for the log linear model [12] [13] [14] [23] [24] [34] for the auto accident data are summarized in Table 6.36. The ratio of the parameter estimates to the standard errors is also shown for some of the parameters. The remaining ratios can be determined by symmetry. The main effect parameters indicate that the majority of the drivers were in normal condition and that more were male drivers than female drivers. A majority of the drivers were not wearing seatbelts. A large majority of the drivers were not injured and very few sustained major or fatal injuries.

The sex by driver condition interaction indicates that, in comparison to females, a larger proportion of male drivers had been drinking. The sex by seatbelt usage interaction indicates that males are less likely to be wearing seatbelts than females. The sex by injury level interaction suggests that, with the exception of the major/fatal category, females were more likely to sustain an injury (minimal or minor) than were males under the same conditions.

The seatbelt usage by driver condition interaction shows that drivers who had been drinking were less likely to be wearing seatbelts than normal condition drivers. The injury level by driver condition interaction indicates that drivers who had been drinking were more likely to sustain all levels of injury than normal condition drivers. For the injury level by seatbelt usage interaction the estimated parameters suggest that seatbelt wearers were less likely to have sustained minor or major/fatal injuries than seatbelt nonusers.

The expected frequencies for the fitted log linear model are displayed in Table 6.37. A comparison of the expected frequencies to the observed frequencies in Table 6.32 shows that the model fits the table very well. As in the case of the three-way table fitted in Section 6.3.2, the minimal and minor categories seem to be the most difficult to fit.

TABLE 6.36. Estimated Parameters for Loglinear Model

Overall Mean	5.1806			
	Normal	Been Drinking		
Driver Condition	1.320	-1.320 (-56.323)		
	Male	Female		
Sex of Driver	0.436	-0.436 (-24.229)		
	Yes	No		
Seatbelt Usage	-1.110	1.110 (43.499)		
	None	Minimal	Minor	Major/Fatal
Injury Level	2.540	0.119	-0.334	-2.325
	(91.425)	(3.522)	(-8.979)	(-31.479)

Sex by Driver Condition

	Male	Female
Normal	-0.270	0.270 (27.426)
Been Drinking	0.270	-0.270

Seatbelt Usage by Driver Condition

	Yes	No
Normal	0.229	-0.229 (-16.801)
Been Drinking	-0.229	0.229

Seatbelt Usage by Sex

	Male	Female
Yes	-0.021	0.021 (4.258)
No	0.021	-0.021

Injury Level by Driver Condition

	Normal	Been Drinking
None	0.435	-0.435 (-21.431)
Minimal	-0.022	0.022 (0.861)
Minor	-0.084	0.084 (3.034)
Major/Fatal	-0.329	0.329 (6.31)

Injury Level by Sex

	Male	Female
None	0.201	-0.201 (-12.685)
Minimal	-0.127	0.127 (6.767)
Minor	-0.106	0.106 (5.203)
Major/Fatal	0.032	-0.032 (-0.748)

Injury Level by Seatbelt

	Yes	No
None	0.088	-0.088 (-3.842)
Minimal	0.010	-0.010 (-0.368)
Minor	-0.071	0.071 (2.374)
Major/Fatal	-0.026	0.026 (0.418)

TABLE 6.37. Expected Frequencies Based on Uniform Second-Order Model

| Injury Level | Seatbelt Usage | Sex | Driver Condition | |
			Normal	Been Drinking
None	Yes	Male	8328.1	270.3
		Female	4168.8	45.9
	No	Male	42442.6	3450.1
		Female	19531.5	538.8
Minimal	Yes	Male	312.2	25.3
		Female	301.3	8.3
	No	Male	1859.7	376.9
		Female	1650.0	113.5
Minor	Yes	Male	175.5	16.1
		Female	162.4	5.0
	No	Male	1231.1	282.3
		Female	1047.1	81.5
Major/Fatal	Yes	Male	22.3	3.3
		Female	15.6	0.8
	No	Male	144.3	54.1
		Female	92.8	11.8

An Example with a Fitted Three-Way Interaction

To provide an example with a fitted three-way interaction, the previous example table will be fitted with the model [134] [12] [23] [24]. This model was the next model to be fitted in Table 6.34 (see step 7), immediately after the uniform second-order model [12] [13] [14] [23] [24] [34] described in Table 6.36. For the model containing the parameters $\mu_{134(ik\ell)}$, the estimates of these parameters are summarized in Table 6.38. The remaining parameter estimates for this model are very similar to those given in Table 6.36. The ratios of the parameter estimates to the standard errors are shown in brackets. These ratios generally indicate that the estimates are not significant. In order to illustrate how to interpret three-way interactions we shall proceed as if the parameter estimates are significant.

The two-way interaction between driver condition and injury level indicates that for drivers who have been drinking the probability of a minimal injury is greater than for drivers who are in normal condition. The three-way interaction with sex indicates that for females the driver condition effect on minimal injury is less pronounced while for males the driver condition effect on minimal injury is stronger. Alternatively the two-way interaction between sex and injury level indicates that in comparison to males, female drivers are more likely to be in the minimal and minor in-

TABLE 6.38. Three-Way Interaction Parameter
Estimates

| Injury Level | Sex | Driver Condition | |
		Normal	Been Drinking
None	Male	-0.002	0.002 (0.066)
	Female	0.002	-0.002
Minimal	Male	-0.063	0.063 (2.043)
	Female	0.063	-0.063
Minor	Male	-0.023	0.023 (0.677)
	Female	0.023	-0.023
Major/Fatal	Male	0.088	-0.088
	Female	-0.088	0.088 (1.378)

jury categories and less likely to be in the no injury category. The three-
way interaction parameter suggests that this difference between males and
famales is less pronounced for drinking drivers and more pronounced for
normal condition drivers. Finally although the two-way interaction between
sex and driver condition indicates that male drivers are less likely to be in
the normal category than female drivers, the three-way interaction with
injury level reduces these sex differences in the minimal and minor injury
categories.

6.3.4   THE EFFECTS OF COLLAPSING A CONTINGENCY TABLE AND STRUCTURAL ZEROES

Collapsing Contingency Tables

It is often the case in practice that the number of variables being studied is
less than the number of explanatory variables that actually have an impact
on the dependent variables of interest. In some cases variables cannot be
measured and in other cases variables are omitted to avoid complexity or
due to small sample size. The contingency table being analyzed should in
general be viewed as a collapsed table in that the cell frequencies in the table
represent sums of frequencies over the categories of the omitted variables.
In our study of two-dimensional tables in Section 6.2 it was demonstrated
that a Simpson's Paradox phenomenon can result from collapsing a table,
altering interactions among variables considerably. The important question,
therefore, is when may a table be collapsed over a particular variable?

In a three-dimensional table with variables A, B and C, the table may
be collapsed over the variable C if C is independent of at least one of the
variables A and B. If the model [AC] [AB] fits the table, then, since variable

TABLE 6.39. Observed Frequency Attendance by Sex by Type of Shift

[1] Attendance	[2] Sex	A.M.	Noon	P.M.	[3] Shift Swing	A Split	A Split /Swing	B Split /Swing
Absent	Male	545	183	347	133	1244	129	240
	Female	8	25	128	26	282	5	67
Present	Male	5735	2137	3653	1747	1116	1617	2720
	Female	32	55	272	294	1158	195	413

C is independent of B, the table may be collapsed over C and may also be collapsed over B. In this case, the interaction between A and B can be studied independently of C and similarly, the interaction between A and C can be studied independently of B. If, however, the three-dimensional table requires the more complex model [AB] [AC] [BC], then the table cannot be collapsed over any of the variables without changing the measurement of the interaction between any pair.

For a four-dimensional table with variables A, B, C and D, the table may be collapsed over D if the true model is [ABC] [AD]. Recall that this model contains the interactions [AB], [AC], [BC] as well as [AD]. Since D is independent of both B and C, none of the interactions among A, B and C will be affected by collapsing on D. Collapsing the table with respect to the variable C, however, will affect the interactions [ABC], [AC] [AB] and [BC].

If the true model is [AB] [AC] [BC] [AD] [BD], then the table cannot be collapsed on any of the variables. While collapsing on D will not affect [AC], [AB] will be affected. If the true model is [AB] [AC] [BC] [AD] [BD] [CD], then collapsing on any variable will affect all other interactions. The collapsibility of a contingency table therefore involves two considerations, the variable to be collapsed and the interaction to be studied.

Example

To illustrate the problem of collapsibility, we analyze a three-dimensional contingency table based on the bus data introduced in Section 6.3.2. A three-dimensional table relating attendance, sex and type of shift is shown in Table 6.39. A saturated loglinear model is required to explain the interactions as demonstrated by the goodness of fit statistics shown in Table 6.40. The estimated loglinear parameters for the saturated model are displayed in Table 6.41. From these parameter estimates we can obtain information about the interaction among the three variables.

TABLE 6.40. Goodness of Fit Statistics for Attendance by Sex by Type of Shift

Model	d.f.	Likelihood Ratio χ^2	Prob	Pearson χ^2	Prob
[1] [2] [3]	19	1602.39	0.0000	1430.39	0.0000
[12] [3]	18	1377.91	0.0000	1073.14	0.0000
[13] [2]	13	1517.17	0.0000	1303.71	0.0000
[23] [1]	13	400.90	0.0000	515.67	0.0000
[13] [23]	7	315.50	0.0000	408.07	0.0000
[12] [13]	12	1292.51	0.0000	989.57	0.0000
[12] [23]	12	176.21	0.0000	163.36	0.0000
[12] [13] [23]	6	92.63	0.0000	88.82	0.0000

Since this model contains a fitted second-order or three-way interaction term, we can interpret the first-order interactions as partial interactions between two variables while controlling for a third variable. The attendance by sex interaction parameters indicate that, after controlling for shift, the rate of absenteeism for females is greater than for males. From the attendance by shift interactions we can conclude that, after controlling for sex, the rate of absenteeism seems to be lowest for the Swing and A Split/Swing shifts and greatest for the Noon and P.M. shifts. From the sex by shift interaction we can conclude that there are proportionately more males on the A.M. shift and proportionately more females on the P.M., Swing, A Split/Swing and B Split/Swing shifts than one would expect under independence.

The second-order interaction parameters can be used to indicate how the first-order interactions are affected by the third variable. From the first-order interactions relating attendance to shift we found that certain shifts have higher rates of absenteeism. From the second-order interactions we have an indication of how this first order interaction differs by sex. For males, the rate of absenteeism of the A.M., P.M. and Noon shifts should be adjusted downward, whereas for the A Split/Swing shift the absenteeism must be adjusted upward. The opposite is true for females.

To examine the effects of collapsing a table when a second order interaction is present, we shall analyze the two-dimensional table Attendance by Type of Shift. The observed frequencies presented in Table 6.42 are obtained by collapsing on the sex variable. The independence model produces likelihood ratio and Pearson χ^2 values of 85.385 and 82.545 respectively, and so a saturated model is required to describe the interaction in the table. The estimated parameters are shown in Table 6.43. Based on a comparison of Tables 6.41 and 6.43 we can conclude that the main effects for attendance are similar, however for the shift effects there are a number of large differences. The A.M. effect changes from positive to negative whereas the B Split/Swing changes from negative to positive. In addition the Noon,

TABLE 6.41. Estimates of Loglinear Parameters for Saturated Model for Attendance by Sex by Type of Shift

Mean	5.655						

	Absent	Present					
Attendance Effects	-1.042	1.042					

	Male	Female					
Sex Effects	1.228	-1.228					

	A.M.	Noon	P.M.	Swing	A Split	A Split /Swing	B Split /Swing
Shift Effects	-0.530	-0.629	0.473	-0.330	1.630	-0.864	0.250

Attendance by Sex	Male	Female					
Absent	-0.167	0.167					
Present	0.167	-0.167					

	A.M.	Noon	P.M.	Swing	A /Split	A Split /Swing	B Split Swing
Attendance by Shift							
Absent	0.107	0.230	0.265	-0.208	0.141	-0.514	-0.020
Present	-0.107	-0.230	-0.265	0.208	-0.141	0.514	0.020

	A.M.	Noon	P.M.	Swing	A Split	A Split /Swing	B Split /Swing
Sex by Shift							
Male	1.125	0.185	-0.329	-0.374	-0.291	0.122	-0.437
Female	-1.125	-0.185	0.329	0.374	0.291	-0.122	0.437

Attendance by Sex by Shift							
		A.M.	Noon	P.M.	Swing	A Split	A Split /Swing	B Split /Swing
Absent	Male	-0.075	-0.250	-0.233	0.129	-0.028	0.442	0.015
	Female	0.075	0.250	0.233	-0.129	0.028	-0.442	-0.015
Present	Male	0.075	0.250	0.233	-0.129	0.028	-0.442	-0.015
	Female	-0.075	-0.250	-0.233	0.129	-0.028	0.442	0.015

P.M. and Swing effects also change in magnitude. Thus collapsing the table on the sex variable causes a number of changes in the shift effects.

From the estimated loglinear parameters for the interaction between attendance and shift in Table 6.43 we conclude that absenteeism is relatively high for the P.M. and A Split shifts and relatively low for the Swing and A Split/Swing shifts. This measure of interaction does not take into account the differences in the rates of absenteeism between the sexes, nor does it take into account the differences in proportions of females assigned to the different shifts. Comparing the attendance by shift parameter estimates in Table 6.43 to the same parameters in Table 6.41 shows that by

TABLE 6.42. Observed Frequencies Attendance by Type of Shift

Attendance	Type of Shift						
	A.M.	Noon	P.M.	Swing	A Split	A Split /Swing	B Split /Swing
Absent	553	208	475	159	1526	139	307
Present	5767	2192	3925	2041	2274	1812	3133

TABLE 6.43. Estimated Log-Linear Parameters for Saturated Model Attendance by Shift

Attendance Effects		Absent	Present				
		-1.172	1.172				
Shift Effects	A.M.	Noon	P.M.	Swing	A Split	A Split /Swing	B Split /Swing
	0.482	-0.491	0.213	-0.661	1.367	-0.791	-0.118
Attendance by Shift	A.M.	Noon	P.M.	Swing	A Split	A Split /Swing	B Split /Swing
Absent	0.0000	-0.006	0.116	-0.104	0.129	-0.145	0.010
Present	0.0000	0.006	-0.116	0.104	-0.129	0.145	-0.010

not controlling for sex the estimates change. Although the directions of the attendance by shift parameter estimates did not change, the magnitudes changed by more than a factor of 2 in most of the categories. By omitting the sex effect the attendance by shift interactions became weaker. Thus the interaction between attendance and shift depends on whether the sex factor has been fitted. The estimated parameters of this two-dimensional table relating attendance to shift cannot be used to predict rate of attendance if the male-female ratio or the distribution of males and females over the various shifts were changed. The two-dimensional model has not controlled for potential changes in the sex variable. The fitted model assumes that the interaction between absenteeism and shift type will not be affected by the sex of the driver.

Random Zeroes, Structural Zeroes and Incomplete Tables

An observation of zero in the cell of a contingency table may be due to chance *(random zero)* or it may be because it is impossible for that cell to occur *(a structural zero)*. For example, suppose in a study of automobile accidents drivers were classified according to whether they appeared to have been drinking, whether they were later convicted, and also by level of injury.

For those drivers who were killed in the accident the cell corresponding to later conviction and fatal injury must be a structural zero. Our study of contingency tables thus far has ignored the possibility of such structural zeroes. Any cell zeroes were treated as random zeroes and were assumed to have positive theoretical frequencies.

In large contingency tables with many cells it is often the case that there are many sampling or random zeroes. Depending on the distribution of the zeroes throughout the table, it is possible to have zero marginals resulting in undefined or negative expected cell frequencies. One approach to this problem is to set the expected cell frequencies to zero and fit the remainder of the table. The degrees of freedom must also be adjusted for the cells fitted by zeroes. A contingency table with structural zeroes is called an *incomplete table* if the cells with structural zeroes are removed from the table. Under certain conditions a loglinear model may be fitted to the incomplete table using the methods described above. A discussion of structural and random zeroes is given in Fienberg (1980).

Quasi-loglinear Models for Incomplete Tables

A quasi-loglinear model for an incomplete table is a loglinear model which is only defined over the cells of the table that do not contain structural zeros. For a three-dimensional table, an indicator variable δ_{ijk} is defined for each cell (i, j, k), and hence $\delta_{ijk} = 1$ if cell (i, j, k) is not a structural zero and $\delta_{ijk} = 0$ if cell (i, j, k) is a structural zero. For each cell, the observed frequencies n_{ijk} and expected frequencies F_{ijk} are replaced by the product quantities $\delta_{ijk} n_{ijk}$ and $\delta_{ijk} F_{ijk}$ respectively. The parameters and estimators can then be defined using these product quantities in place of the former quantities. If iterative proportional fitting is used, the starting estimate for each cell is δ_{ijk} which is either 0 or 1. Since the iterative proportional fitting algorithm uses ratios to revise the cell estimates, a starting value of zero for cells with structural zeros guarantees that these cells remain zero.

6.3.5   LOGIT MODELS FOR RESPONSE VARIABLES

Up to this point our discussion of modeling techniques for categorical data has not included the consideration of a dependent variable or *response variable*. If one of the variables in a contingency table can be regarded as a dependent variable, it is possible to construct a model for this dependent variable in terms of the remaining variables using the estimated loglinear model.

The Logit Function

Assume that a dependent variable is dichotomous and that, for given levels of the other variables, the probabilities for the two categories are p and $(1 - p)$. The function $\ln[p/(1 - p)]$ of the probabilities p and $(1 - p)$ is called a *logit function* and will serve as a convenient dependent variable. In a loglinear model for categorical data, it is easy to rewrite the model so that the left hand side is the logit function of a dichotomous response variable. To demonstrate, consider the saturated model for a three-dimensional contingency table.

$$\ln F_{ijk} = \mu + \mu_{1(i)} + \mu_{2(j)} + \mu_{3(k)}$$
$$+ \mu_{12(ij)} + \mu_{13(ik)} + \mu_{23(jk)} + \mu_{123(ijk)},$$
$$i = 1, 2, \ldots, r; \quad j = 1, 2, \ldots, c; \quad k = 1, 2, \ldots, \ell.$$

Assuming that the first variable is dichotomous $i = 1, 2$, the difference between the equations for the two values of i may be written as

$$\ln F_{2jk} - \ln F_{1jk} = \ln \frac{F_{2jk}}{F_{1jk}} = \ln \left(\frac{F_{2jk}}{F_{\cdot jk} - F_{2jk}} \right) = \ln \left(\frac{p_{jk}}{1 - p_{jk}} \right),$$

where $F_{\cdot jk} = (F_{1jk} + F_{2jk})$ is the total frequency in category (j, k) and p_{jk} is the probability of an observation in category (j, k) when $i = 2$. The right hand side of this model is given by

$$\left[\mu_{1(2)} - \mu_{1(1)} \right] + \left[\mu_{12(2j)} - \mu_{12(1j)} \right] + \left[\mu_{13(2k)} - \mu_{13(1k)} \right]$$
$$+ \left[\mu_{123(2jk)} - \mu_{123(1jk)} \right].$$

Since the pairs of parameters in brackets in the above expression must sum to zero, the two parameters in each bracket are equal in magnitude and opposite in sign. The right hand side therefore becomes

$$= 2\mu_{1(2)} + 2\mu_{12(2j)} + 2\mu_{13(2k)} + 2\mu_{123(2jk)}$$
$$= L + L_{2(j)} + L_{3(k)} + L_{23(jk)},$$

where $L = 2\mu_{1(2)}$, $L_{2(j)} = 2\mu_{12(2j)}$, $L_{3(k)} = 2\mu_{13(2k)}$ and $L_{23(jk)} = 2\mu_{123(2jk)}$.

Fitting a Logit Model

A fitted loglinear model can be used to fit a *logit model* provided certain terms are included in the loglinear model. We can see from the above example that in the logit model all the terms involving variable 1 are necessary but that all variables excluding 1 disappeared. The use of the logit model to explain the variation in the response variable therefore omits the interaction effects among the explanatory variables themselves. These interaction terms disappear by subtraction when the logit model is constructed from the loglinear model. In the logit model the primary purpose is to study the impact of the explanatory variables on the response variable.

Although the parameters in the model that do not contain the response variable do not appear in the final logit model, they may have to be included in the loglinear model being fitted. By including these parameters,

TABLE 6.44. Attendance by Garage by Type of Shift*

		Shift							
Attend.	Garage	A.M.	Noon	P.M.	Swing	A Split	A Split /Swing	B Split /Swing	Total
Present	A	2471	984	1702	920	4428	784	1614	12903
	B	1795	533	1146	552	4220	580	791	9617
	C	1500	675	1061	568	3593	484	745	8626
	Total	5766	2192	3909	2040	12241	1848	3150	31246
Absent	A	289	96	258	80	612	56	106	1497
	B	125	67	134	48	500	60	129	1063
	C	140	45	99	32	487	36	95	934
	Total	554	208	491	160	1599	152	330	3494
	TOTALS	6320	2400	4400	2200	18340	2000	3480	34520

*[1] = Attendance, [2] = Garage, [3] = Type of Shift

the associated marginals are fitted to the sample marginals. Recall that if the sampling scheme is product multinomial and the explanatory variables are control variables, the fitted marginals must be equal to the sample marginals. For explanatory variables that are not control variables and do not contain response variables if they are not significant, they can be omitted from the fitted loglinear model. In Section 6.4, logit models will be fitted to multidimensional contingency tables using weighted least squares assuming product multinomial sampling.

Example

The bus data introduced earlier in this chapter provides an example. From the bus data, the observed relationship between attendance, garage and type of shift is summarized in Table 6.44.

A saturated model was fitted to this three-dimensional table. The estimates of the parameters for the logit model for attendance are obtained by doubling the parameter estimates for $\mu_{1(2)}$, $\mu_{12(2j)}$, $\mu_{13(2k)}$ and $\mu_{23(2jk)}$ given in Table 6.45. The estimates for the model

$$\ln \frac{p_{jk}}{(1 - p_{jk})} = L + L_{2(j)} + L_{3(k)} + L_{23(jk)}$$

are summarized in Table 6.46.

The ratio $p/(1 - p)$ measures the ratio of the probability of Present to the probability of Absent. The estimate for the constant term, L, indicates that the category present is much more likely to occur than the category absent. From the parameter estimates in Table 6.46 we can conclude that the probability of Present is higher in Garage C than in Garages A and B.

TABLE 6.45. Fitted Parameters for Loglinear Model

Mean		5.891						

		Absent	Present					
Attendance Effects [1]		-1.159	1.159					

		A	B	C				
Garage [2]		0.301	-0.071	-0.230				

						A	A Split	B Split
Shift		A.M.	P.M.	Noon	Swing	Split	/Swing	/Swing
[3]		0.451	-0.514	0.192	-0.691	1.401	-0.739	-0.099

						A	A Split	B Split
		A.M.	P.M.	Noon	Swing	Split	/Swing	/Swing
Garage by	A	0.096	0.050	0.113	0.103	-0.186	-0.108	-0.068
Shift	B	-0.111	-0.064	-0.041	-0.037	0.060	0.147	0.046
[23]	C	0.015	0.014	-0.072	-0.066	0.126	-0.039	0.022

						A	A Split	B Split
		A.M.	P.M.	Noon	Swing	Split	/Swing	/Swing
Attendance	Absent	-0.038	-0.026	0.091	-0.135	0.149	-0.092	0.059
by Shift	Present	0.038	0.026	-0.091	0.135	-0.149	0.092	-0.059
[13]								

		A	B	C				
Attendance	Absent	0.006	0.049	-0.055				
by Garage	Present	-0.006	-0.049	0.055				
[12]								

Attendance by Garage by Shift						A	A Split	B Split
[123]		A.M.	P.M.	Noon	Swing	Split	/Swing	/Swing
	A	0.118	0.016	0.118	0.067	0.023	-0.074	-0.268
Absent	B	-0.184	0.099	-0.054	0.024	-0.097	0.068	0.144
	C	0.066	-0.115	-0.064	-0.090	0.074	0.006	0.124
	A	-0.118	-0.016	-0.118	-0.067	-0.023	0.074	0.268
Present	B	0.184	-0.099	0.054	-0.024	0.097	-0.068	-0.144
	C	-0.066	0.115	0.064	0.090	-0.074	-0.006	-0.124

From the shift parameter estimates we can conclude that the probability of Present is relatively high for Swing and A Split/Swing and relatively low for B Split/Swing, A Split and Noon. For the interaction between Garage and Shift we can conclude that, in Garage A, shifts B Split/Swing and A Split/Swing have better attendance records than in Garages B and C.

TABLE 6.46. Logit Model Parameter Estimates

$L = 2\mu_{1(2)}$	2.318						

	A	B	C				
$L_{2(j)} = 2\mu_{12(2j)}$	-0.012	-0.098	0.110				

					A Split	A Split	B Split
	A.M.	P.M.	Noon	Swing	Split	/Swing	/Swing
$L_{3(k)} = 2\mu_{13(2k)}$	0.076	0.052	-0.182	0.270	-0.280	0.184	-0.118

						A Split	A Split	B Split
		A.M.	P.M.	Noon	Swing	Split	/Swing	/Swing
$L_{23} = 2\mu_{123(2jk)}$	A	-0.236	-0.032	-0.236	-0.134	-0.046	0.148	0.536
	B	0.368	-0.198	0.108	-0.048	0.194	-0.136	-0.288
	C	-0.132	0.230	0.128	0.180	-0.148	-0.012	-0.248

For the P.M. shift, the opposite is true. For the A Split and A.M. shifts, Garage B has the superior attendance to A and C. For the Noon and Swing shifts, Garage C is superior in attendance to A or B.

Relationship to Logistic Regression

In Chapter 8 the logit function will be introduced in connection with logistic regression. In the logistic regression model the logit function is expressed as a function of a set of explanatory variables that may be interval scaled variables or dummy variables. In this section, the logit function has been expressed as a function of the cell frequencies of a contingency table. In the case of the contingency table model, a sample measure of goodness of fit based on a χ^2 statistic is available that is often not available for the logistic regression model because, with interval data, the cell frequencies are usually very small and often equal to one.

Polychotomous Response Variables

If the response variable has r, $(r > 2)$ categories, there are a variety of ways for constructing a set of logit models. If one category is a logical base case, a total of $(r - 1)$ models can be constructed by comparing the other categories to the base case. The logit functions would be given by $\ln(p_i/p_r)$, $i = 1, 2, \ldots, r - 1$ where the base case is denoted by $i = r$. In this case it should be noted that subtraction of the loglinear models does not result in the factor of 2 as was the case for the dichotomous model above.

If there is a natural order for the categories, a second alternative would be to use the logit function

$$
\ln \left[\frac{p_i}{\sum_{j=i+1}^{r} p_j} \right], \quad i = 1, 2, \ldots, r.
$$

The ratio $p_i / \sum_{j=i+1}^{r} p_j$ is called a *continuation ratio*. The advantage of the continuation ratio approach is that the likelihood χ^2 statistics for each fitted model can be added together to get an overall goodness of fit statistic for the complete set of $(r-1)$ models. Continuation ratios will be employed in Section 6.4 for the weighted least squares approach. This topic will also be discussed in connection with the logistic regression model in Chapter 8.

6.3.6   OTHER SOURCES OF INFORMATION

Extensive discussion of loglinear model techniques for three-dimensional tables are available in Andersen (1980, 1990), Fienberg (1980), Bishop, Fienberg and Holland (1975), Christensen (1991), Santner and Duffy (1989) and Reynolds (1977). Discussion of stepwise procedures, incomplete tables and structural zeroes is available in Fienberg (1980), Bishop, Fienberg and Holland (1975) and Christensen (1991). Logit response models are outlined in Fienberg (1980). Techniques available for ordinal variables are outlined in Agresti (1984).

6.4   The Weighted Least Squares Approach

So far our approach to modeling the variation among cell frequencies in multidimensional contingency tables has been restricted to logarithms of frequencies. In addition, the method used for estimation so far has been the maximum likelihood approach. It is possible to define ANOVA type models that do not involve the logarithms of the frequencies, and it is also possible to obtain estimators of model parameters by an alternative method called *weighted least squares*. This approach permits much more flexibility in both defining models and in the types of hypotheses that can be tested. This section provides an overview of this weighted least squares methodology.

6.4.1   THE WEIGHTED LEAST SQUARES THEORY

In the weighted least squares approach, the variables that combine to form the multidimensional contingency table are first classified into two categories, *response variables* and *factor* or *explanatory variables*. The cross-

TABLE 6.47. Contingency Table Showing Cross-Classification of Sample Frequency by Response Category and Subpopulation

Subpopulations	Response Levels				Totals
	1	2	...	c	
1	n_{11}	n_{12}	...	n_{1c}	$n_1.$
2	n_{21}	n_{22}	...	n_{2c}	$n_2.$
⋮	⋮	⋮		⋮	⋮
r	n_{r1}	n_{r2}	...	n_{rc}	$n_r.$
					n

classification of the entire set of factor variables yields a set of categories called *subpopulations*. Similarly, the cross-classification of the complete set of response variables produces a set of categories called *response levels*. Thus, regardless of the number of underlying variables, the multidimensional contingency table can be represented as a two-dimensional array representing the cross-classification of the response levels with the subpopulations.

Table 6.47 illustrates the allocation of a sample of size n to the rc cells. The number of response levels is c and the number of subpopulations is r. The cell frequency is denoted by n_{ij} for response level j and subpopulation i; $i = 1, 2, \ldots, r$; $j = 1, 2, \ldots, c$. The row totals $n_i.$, $i = 1, 2, \ldots, r$, represent the sample sizes for the r subpopulations.

The Product Multinomial Distribution Assumption

The underlying sampling distribution is assumed to be product multinomial. For each of the r subpopulations, the sampling process is multinomial and the r samples are assumed to be mutually independent. The theoretical cell densities for each subpopulation are shown in Table 6.48 and are denoted by p_{ij}, $i = 1, 2, \ldots, r$; $j = 1, 2, \ldots, c$. These distribution parameters satisfy the condition $\sum_{j=1}^{c} p_{ij} = 1$ and hence the row totals are unity. For each subpopulation the multinomial density is given by

$$f(n_{i1}, n_{i2}, \ldots, n_{ic}) = \frac{n_i.!}{\prod_{j=1}^{c} n_{ij}!} \prod_{j=1}^{c} p_{ij}^{n_{ij}} \quad i = 1, 2, \ldots, r.$$

TABLE 6.48. Population Densities for
the Independent Multinomial Components of the Product Multinomial

Subpopulations	Response Levels			
	1	2	...	c
1	p_{11}	p_{12}	$\cdots$	p_{1c}
2	p_{21}	p_{22}	$\cdots$	p_{2c}
$\vdots$	$\vdots$	$\vdots$		$\vdots$
r	p_{r1}	p_{r2}	$\cdots$	p_{rc}

The corresponding product multinomial density is therefore given by

$$f(n_{11}, n_{12}, \ldots, n_{rc}) = \prod_{i=1}^{r} \left[\frac{n_i.!}{\prod\limits_{j=1}^{c} n_{ij}!} \prod_{j=1}^{c} p_{ij}^{n_{ij}} \right].$$

The $(rc \times 1)$ vector of densities p_{ij}, $i = 1, 2, \ldots, r$; $j = 1, 2, \ldots, c$, will be denoted by $\mathbf{p} = [p_{11}, p_{12}, \ldots, p_{1c}, p_{21}, p_{22}, \ldots, p_{2c}, \ldots, p_{r1}, p_{r2}, \ldots, p_{rc}]$. In comparison to the two-dimensional contingency table discussed in Section 6.2 the densities p_{ij} are given by $p_{ij} = f_{ij}/f_i.$.

Example

The three-dimensional contingency table relating driver injury level to both seatbelt usage and driver condition first presented in Table 6.23 will be used in this section to provide an example for the weighted least squares approach. Table 6.49 contains both the cell frequencies and the corresponding row proportions. The response variable is driver injury level, which has four levels, and the cross-classification of the variables driver condition and seatbelt usage provide the four subpopulations. A comparison of the row proportions over the four subpopulations seems to suggest that the proportions are not homogeneous. We shall use the weighted least squares methodology to model the variation in row proportions.

Sampling Properties of the Row Proportions

The sample proportions $n_{ij}/n_i.$ which can be obtained from Table 6.47, provide estimators of the parameters p_{ij}. Under the product multinomial

TABLE 6.49. Driver Injury Level Response to Seatbelt Usage and Driver Condition – Cell Frequencies and Row Proportions

Subpopulations		Response Levels				
Driver	Seatbelt	Driver Injury Level				
Condition	Usage	None	Minimal	Minor	Major/Fatal	Totals
Normal	Yes	12500	604	344	38	13486
		0.9269	0.0448	0.0255	0.0028	
	No	61971	3519	2272	237	67999
		0.9114	0.0518	0.0334	0.0035	
Been	Yes	313	43	15	4	375
Drinking		0.8347	0.1146	0.0400	0.0107	
	No	3992	481	370	66	4909
		0.8132	0.0980	0.0754	0.0134	

assumption we have that

$$E[n_{ij}/n_{i\cdot}] = p_{ij}, \qquad\qquad i = 1, 2, \ldots, r; \ j = 1, 2, \ldots, c,$$

$$V[n_{ij}/n_{i\cdot}] = p_{ij}(1 - p_{ij})/n_{i\cdot}, \qquad i = 1, 2, \ldots, r; \ j = 1, 2, \ldots, c,$$

$$\text{Cov}\left[\frac{n_{ij}}{n_{i\cdot}}, \frac{n_{ik}}{n_{i\cdot}}\right] = -p_{ij}p_{ik}/n_{i\cdot}, \qquad i = 1, 2, \ldots, r; \ j = 1, 2, \ldots, c,$$

$$\text{Cov}\left[\frac{n_{ij}}{n_{i\cdot}}, \frac{n_{\ell k}}{n_{\ell\cdot}}\right] = 0, \qquad\qquad i \neq \ell \ \ i = 1, 2, \ldots, r; \ j = 1, 2, \ldots, c.$$

The $(rc \times 1)$ vector of estimators $n_{ij}/n_{i\cdot}$ will be denoted by

$$\hat{\mathbf{p}}' = \left[\frac{n_{11}}{n_{1\cdot}}, \frac{n_{12}}{n_{1\cdot}}, \frac{n_{13}}{n_{1\cdot}}, \ldots, \frac{n_{1c}}{n_{1\cdot}}, \frac{n_{21}}{n_{2\cdot}}, \ldots, \frac{n_{2c}}{n_{2\cdot}}, \ldots, \frac{n_{r1}}{n_{r\cdot}}, \ldots, \frac{n_{rc}}{n_{r\cdot}}\right]$$

and the individual elements by $\hat{p}_{ij} = n_{ij}/n_{i\cdot}$.

Example

The vector of observed proportions based on Table 6.49 is given by

$$\hat{\mathbf{p}}' = [0.9269, 0.0448, 0.0255, 0.0028, 0.9114, 0.0518, 0.0334, 0.0035,$$
$$0.8347, 0.1146, 0.0400, 0.0107, 0.8132, 0.0980, 0.0754, 0.0134].$$

Determining Linear Functions Among the Row Proportions

The weighted least squares approach is used to estimate relationships among linear functions of the elements of $\mathbf{p}$. A set of m linear functions of the elements of $\mathbf{p}$ are given by a system of equations

$$
\begin{array}{rcl}
g_1 &=& a_{111}p_{11} + a_{112}p_{12} + \ldots a_{1rc}p_{rc} \\
g_2 &=& a_{211}p_{11} + a_{212}p_{12} + \ldots a_{2rc}p_{rc} \\
&\vdots& \\
g_m &=& a_{m11}p_{11} + a_{m12}p_{12} + \ldots a_{mrc}p_{rc}.
\end{array}
$$

In matrix notation the system is given by $\mathbf{g} = \mathbf{Ap}$ where $\mathbf{g}$ is the $(m \times 1)$ vector of elements g_k, $k = 1, 2, \ldots, m$; $\mathbf{A}$ is the $(m \times rc)$ matrix of elements a_{kij}, $k = 1, 2, \ldots, m$; $i = 1, 2, \ldots, r$; $j = 1, 2, \ldots, c$; and $\mathbf{p}$ is the $(rc \times 1)$ vector of cell densities.

Since the proportions in each row of Table 6.48 add to unity, we need only use $(c - 1)$ relations among the proportions in a given row. One common approach is simply to omit one of the c columns so that the matrix $\mathbf{A}$ simply removes these elements from $\mathbf{p}$ to get $\mathbf{g}$. An alternative approach is to compare $(c - 1)$ of the column proportions to a particular column proportion using the differences $p_{ij} - p_{ik}$, $j \neq k$, $j = 1, 2, \ldots, c$. A third alternative would be to compute a score for each row based on a weighted sum of the p_{ij} values in each row such as $s_i = \sum_{j=1}^{c} w_j p_{ij}$. A simple case would be the mean scores with $w_j = 1/c$, $j = 1, 2, \ldots, c$.

Example

For the accident data in Table 6.49, a useful way of comparing the four subpopulations would be to use contrasts between the no injury level and the three injury levels. The $\mathbf{g}$ vector therefore contains 12 differences. The $\mathbf{A}$ matrix and $\mathbf{g}$ vector for the equation $\mathbf{g} = \mathbf{Ap}$ in this case are given by

$$
\mathbf{g} =
\begin{bmatrix}
p_{11} - p_{12} \\
p_{21} - p_{22} \\
p_{31} - p_{32} \\
p_{41} - p_{42} \\
p_{11} - p_{13} \\
p_{21} - p_{23} \\
p_{31} - p_{33} \\
p_{41} - p_{43} \\
p_{11} - p_{14} \\
p_{21} - p_{24} \\
p_{31} - p_{34} \\
p_{41} - p_{44}
\end{bmatrix},
$$

$$\mathbf{Ap} =
\begin{bmatrix}
1 & -1 & 0 & 0 & 0 & 0 & 0 & 0 & 0 & 0 & 0 & 0 & 0 & 0 & 0 & 0 \\
0 & 0 & 0 & 0 & 1 & -1 & 0 & 0 & 0 & 0 & 0 & 0 & 0 & 0 & 0 & 0 \\
0 & 0 & 0 & 0 & 0 & 0 & 0 & 0 & 1 & -1 & 0 & 0 & 0 & 0 & 0 & 0 \\
0 & 0 & 0 & 0 & 0 & 0 & 0 & 0 & 0 & 0 & 0 & 0 & 1 & -1 & 0 & 0 \\
1 & 0 & -1 & 0 & 0 & 0 & 0 & 0 & 0 & 0 & 0 & 0 & 0 & 0 & 0 & 0 \\
0 & 0 & 0 & 0 & 1 & 0 & -1 & 0 & 0 & 0 & 0 & 0 & 0 & 0 & 0 & 0 \\
0 & 0 & 0 & 0 & 0 & 0 & 0 & 0 & 1 & 0 & -1 & 0 & 0 & 0 & 0 & 0 \\
0 & 0 & 0 & 0 & 0 & 0 & 0 & 0 & 0 & 0 & 0 & 0 & 1 & 0 & -1 & 0 \\
1 & 0 & 0 & -1 & 0 & 0 & 0 & 0 & 0 & 0 & 0 & 0 & 0 & 0 & 0 & 0 \\
0 & 0 & 0 & 0 & 1 & 0 & 0 & -1 & 0 & 0 & 0 & 0 & 0 & 0 & 0 & 0 \\
0 & 0 & 0 & 0 & 0 & 0 & 0 & 0 & 1 & 0 & 0 & -1 & 0 & 0 & 0 & 0 \\
0 & 0 & 0 & 0 & 0 & 0 & 0 & 0 & 0 & 0 & 0 & 0 & 1 & 0 & 0 & -1
\end{bmatrix}
\begin{bmatrix}
p_{11} \\ p_{12} \\ p_{13} \\ p_{14} \\ p_{21} \\ p_{22} \\ p_{23} \\ p_{24} \\ p_{31} \\ p_{32} \\ p_{33} \\ p_{34} \\ p_{41} \\ p_{42} \\ p_{43} \\ p_{44}
\end{bmatrix}.$$

For each of the four no injury proportions each of the three injury proportions is subtracted to provide a contrast.

The Linear Model to Be Estimated

The vector **g** of linear functions is assumed to satisfy the linear model

$$\mathbf{g} = \mathbf{X}\boldsymbol{\beta}$$

where $\mathbf{X}$ $(m \times s)$ is a specified *design matrix* and $\boldsymbol{\beta}$ $(s \times 1)$ is an unknown parameter vector.

Computer Software

The statistical software package SAS, procedure CATMOD is used to perform the analyses in Section 6.4.

Example

For the example, a convenient model that can be used to describe the differences among the row proportions as a function of seatbelt usage and driver condition would be an ANOVA type model. Using *effect coding* the design matrix for a driver condition effect, a seatbelt effect and an interaction effect for each response is given by

$$\begin{bmatrix}
1 & 1 & 1 & 1 \\
1 & 1 & -1 & -1 \\
1 & -1 & 1 & -1 \\
1 & -1 & -1 & 1
\end{bmatrix}.$$

TABLE 6.50. Parameter Definitions for Model

Overall Mean		μ_j	
Driver Condition		Normal	Been Drinking
		α_j	$-\alpha_j$
Seatbelt Usage		Yes	No
		γ_j	$-\gamma_j$
Seatbelt Usage by			
Driver Condition		Normal	Been Drinking
	Yes	$(\alpha\gamma)_j$	$-(\alpha\gamma)_j$
	No	$-(\alpha\gamma)_j$	$(\alpha\gamma)_j$

The first column of the design matrix generates the mean, the second column yields the driver condition effect, the third column measures the seatbelt usage effect and the last column represents the interaction. This design matrix is repeated for each of the three response functions (corresponding to the three injury categories) in a block diagonal fashion to get the overall design matrix $\mathbf{X}$.

The parameter vector $\boldsymbol{\beta}$ contains three sets of four parameters. For the jth response function the elements of $\boldsymbol{\beta}$ are μ_j, α_j, γ_j and $(\alpha\gamma)_j$, $j = 2, 3$ and 4. For the jth response function $(j = 2, 3, 4)$ we have the four equations corresponding to the four subpopulations generated by the two seatbelt categories and the two driver conditions. The equations are given by

$$
\begin{bmatrix} p_{11} - p_{1j} \\ p_{21} - p_{2j} \\ p_{31} - p_{3j} \\ p_{41} - p_{4j} \end{bmatrix} = \begin{bmatrix} 1 & 1 & 1 & 1 \\ 1 & 1 & -1 & -1 \\ 1 & -1 & 1 & -1 \\ 1 & -1 & -1 & 1 \end{bmatrix} \begin{bmatrix} \mu_j \\ \alpha_j \\ \gamma_j \\ (\alpha\gamma)_j \end{bmatrix}
$$

$$
= \begin{matrix} \mu_j + \alpha_j + \gamma_j + (\alpha\gamma)_j \\ \mu_j + \alpha_j - \gamma_j - (\alpha\gamma)_j \\ \mu_j - \alpha_j + \gamma_j - (\alpha\gamma)_j \\ \mu_j - \alpha_j - \gamma_j - (\alpha\gamma)_j \end{matrix} \quad . \tag{6.1}
$$

These parameter estimates can also be summarized in tabular form as shown in Table 6.50. The parameters in each of the three categories sum to zero as is the case with ANOVA models.

Determining the Weighted Least Squares Estimator

Since the vector $\mathbf{g}$ is a function of the true proportions, it is not observable; we therefore replace $\mathbf{g}$ by $\hat{\mathbf{g}}$ where $\hat{\mathbf{g}} = \mathbf{A}\hat{\mathbf{p}}$, and $\hat{\mathbf{p}}$ is the corresponding vector of sample proportions. Defining the error vector $\mathbf{u}$ $(m \times 1)$ by $\mathbf{u} =$

$\hat{\mathbf{g}} - \mathbf{g}$, we have the linear model

$$\hat{\mathbf{g}} = \mathbf{X}\boldsymbol{\beta} + \mathbf{u}.$$

If the *covariance matrix* of $\mathbf{u}$ is denoted by $E[\mathbf{uu}'] = \mathbf{H}$, then from linear model theory the *weighted least squares estimator* of $\boldsymbol{\beta}$ is given by

$$\hat{\boldsymbol{\beta}} = (\mathbf{X}'\mathbf{H}^{-1}\mathbf{X})^{-1}\mathbf{X}'\mathbf{H}^{-1}\hat{\mathbf{g}}.$$

The covariance matrix $\mathbf{H}$ is defined by $\mathbf{H} = \mathbf{A}\boldsymbol{\Omega}\mathbf{A}'$ where $\boldsymbol{\Omega}$ is the covariance matrix of $\hat{\mathbf{p}}$.

The covariance matrix $\boldsymbol{\Omega}$ is *block diagonal* with block components $\boldsymbol{\Omega}_i$, $i = 1, 2, \ldots, r$,

$$\boldsymbol{\Omega} = \begin{bmatrix} \boldsymbol{\Omega}_1 & & & \\ & \boldsymbol{\Omega}_2 & & \\ & & \ddots & \\ & & & \boldsymbol{\Omega}_r \end{bmatrix}.$$

The block matrices $\boldsymbol{\Omega}_i$, $i = 1, 2, \ldots, r$ are functions of the p_{ij}, $j = 1, 2, \ldots, c$. The c diagonal elements of $\boldsymbol{\Omega}_i$ are given by $p_{ij}(1 - p_{ij})/n_{i\cdot}$, $j = 1, 2, \ldots, c$; and the off-diagonal elements by $-p_{ij}p_{ik}/n_{i\cdot}$, $j \neq k$, $j, k = 1, 2, \ldots, c$.

This covariance matrix is estimated by replacing the elements of $\mathbf{p}$ by the elements of $\hat{\mathbf{p}}$. The resulting estimator is denoted by $\widehat{\boldsymbol{\Omega}}$ and the estimator of $\mathbf{H}$ by $\widehat{\mathbf{H}} = \mathbf{A}\widehat{\boldsymbol{\Omega}}\mathbf{A}'$. The weighted least squares estimator of $\boldsymbol{\beta}$ therefore becomes

$$\hat{\boldsymbol{\beta}} = (\mathbf{X}'\widehat{\mathbf{H}}^{-1}\mathbf{X})^{-1}(\mathbf{X}'\widehat{\mathbf{H}}^{-1}\hat{\mathbf{g}}). \tag{6.2}$$

Example

Continuing the above example for the accident data, the value of $\hat{\mathbf{g}}$ is given by [0.8821, 0.8596, 0.7200, 0.7152, 0.9014, 0.8779, 0.7947, 0.7378, 0.9241, 0.9079, 0.8240, 0.7998]. These values are plotted in Figure 6.6 in such a way that the interaction between seatbelt usage and driver condition can be observed. As can be seen from the figure the greatest departure from two parallel lines occurs in the middle panel, which is concerned with the minor injury category. The bottom panel displays an almost parallel relationship. The resulting weighted least squares estimates for the two-way ANOVA model outlined above are given by

$$\hat{\mu} = \begin{bmatrix} 0.7942 \\ 0.8280 \\ 0.8639 \end{bmatrix}, \qquad \hat{\alpha} = \begin{bmatrix} 0.0766 \\ 0.0617 \\ 0.0520 \end{bmatrix},$$

$$\hat{\gamma} = \begin{bmatrix} 0.0068 \\ 0.0200 \\ 0.0101 \end{bmatrix}, \qquad (\hat{\alpha}\hat{\gamma}) = \begin{bmatrix} -0.0044 \\ 0.0083 \\ 0.0020 \end{bmatrix}.$$

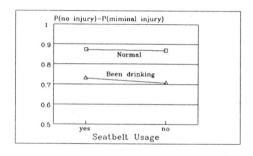

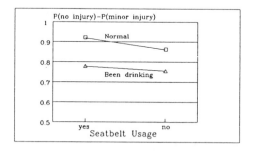

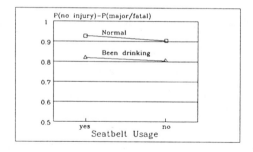

FIGURE 6.6. Relationship Between Injury Level Response and the Factors Seatbelt Usage

These three estimated models fit the contingency table perfectly since the model is a saturated model. In tabular form the parameter estimates are shown in Table 6.51. From the parameter estimates, we can conclude that the proportion of individuals in each injury category increases relative to the no injury category if the individual has been drinking and also if the individual was not wearing a seatbelt. The driver condition effects have larger magnitudes than the seatbelt usage effects. The interaction effects suggest that in the minor and major categories the effect is more pronounced if both are present, whereas in the minimal category the reverse is true. These effects are relatively weak.

TABLE 6.51. Estimates for Two-Way Model for Accident Data

			$(p_{i1} - p_{i2})$	$(p_{i1} - p_{i3})$	$(p_{i1} - p_{i4})$
Overall Mean			0.7942	0.8280	0.8639
Driver Condition	Normal		0.0766	0.0617	0.0520
	Been Drinking		-0.0766	-0.0617	-0.0520
Seatbelt Usage	Yes		0.0068	0.0200	0.0101
	No		-0.0068	-0.0200	-0.0101
Driver Condition	Normal	Yes	-0.0044	0.0083	0.0020
by	Normal	No	0.0044	-0.0083	-0.0020
Seatbelt Usage					
	Been Drinking	Yes	0.0044	-0.0083	-0.0020
	Been Drinking	No	-0.0044	0.0083	0.0020

6.4.2   STATISTICAL INFERENCE FOR THE WEIGHTED LEAST SQUARES PROCEDURE

Having obtained the weighted least squares estimator given by (6.2), a test of goodness of fit can be carried out using the chi-square test statistic given by

$$\hat{\mathbf{g}}'\widehat{\mathbf{H}}^{-1}\hat{\mathbf{g}} - \hat{\boldsymbol{\beta}}'(\mathbf{X}'\widehat{\mathbf{H}}^{-1}\mathbf{X})\hat{\boldsymbol{\beta}}.$$

In large samples this statistic has a χ^2 distribution with $(m - s)$ degrees of freedom if the model fits the data. [Recall that $\mathbf{X}$ is $(m \times s)$.] This statistic is sometimes referred to as a *Wald statistic*. This statistic is the minimum value of $(\hat{\mathbf{g}} - \mathbf{X}\boldsymbol{\beta})'\widehat{\mathbf{H}}^{-1}(\hat{\mathbf{g}} - \mathbf{X}\boldsymbol{\beta})$, which is minimum at $\boldsymbol{\beta} = \hat{\boldsymbol{\beta}}$.

If the model fits the data, hypotheses regarding linear functions of the parameter vector $\boldsymbol{\beta}$ can also be tested. Denoting the linear functions by $\mathbf{C}\boldsymbol{\beta}$ where $\mathbf{C}$ is $(q \times s)$, the test statistic

$$(\mathbf{C}\hat{\boldsymbol{\beta}})'[\mathbf{C}(\mathbf{X}'\widehat{\mathbf{H}}^{-1}\mathbf{X})^{-1}\mathbf{C}']^{-1}(\mathbf{C}\hat{\boldsymbol{\beta}})$$

has a χ^2 distribution with q degrees of freedom in large samples if $H_0: \mathbf{C}\boldsymbol{\beta} = 0$ is true.

Example

In the previous section a saturated model was fit to the accident data and hence the χ^2 goodness of fit test yields a χ^2 value of 0.0. The individual parameters in the model and the various effects can be tested using various forms of the matrix $\mathbf{C}$ in the test statistic for $H_0: \mathbf{C}\boldsymbol{\beta} = 0$. Using this

TABLE 6.52. Analysis of Variance Table for Effects

Source	d.f.	χ^2	p-Value
Intercepts	3	39371.44	0.0001
Seat Belts	3	16.31	0.0010
Driver Condition	3	91.42	0.0001
Interactions	3	7.93	0.0474
Residual	0	0	1.000

TABLE 6.53. Significance for Individual Parameter Estimates

Parameter	Estimate	χ^2	p-Value
μ_2	0.7942	8,085.8	0.0001
μ_3	0.8280	14,993.9	0.0001
μ_4	0.8639	24,437.2	0.0001
α_2	0.0068	0.60	0.4400
α_3	0.0201	8.81	0.0030
α_4	0.0101	3.35	0.0673
γ_2	0.0766	75.25	0.0001
γ_3	0.0617	83.28	0.0001
γ_4	0.0520	88.69	0.0001
$(\alpha\gamma)_2$	-0.0044	0.25	0.6161
$(\alpha\gamma)_3$	0.0083	1.52	0.2169
$(\alpha\gamma)_4$	0.0020	0.13	0.7160

procedure, the χ^2 statistics and p-values shown in Tables 6.52 and 6.53 can be produced. The ANOVA table suggests that all effects including the interaction are significant at the 0.05 level. The table, which summarizes the significance levels for the individual parameters, indicates that each of the interaction parameters is not significant at the margin. In other words, if any two of the interaction parameters are included, the third parameter does not contribute significantly to the overall goodness of fit. An examination of the significance of the individual parameter estimates for the main effects in Table 6.53 reveals that α_2 is not significant. It would appear that seatbelts on average do not affect the difference in proportions between no injury and minimal injury after controlling for driver condition effects. It would also appear that after controlling for driver condition, seatbelt usage had little impact on the difference for the major/fatal category (see α_4).

When the interaction terms are omitted from the model, the ANOVA table and parameter estimates for this reduced model are summarized in Tables 6.54 and 6.55. As can be seen from the table of parameter estimates, some changes occur in the estimates of the main effects. In general the

TABLE 6.54. Analysis of Variance Table for Effects

Source	d.f.	χ^2	p-Value
Intercepts	3	112727.68	0.0001
Seat Belts	3	46.42	0.0001
Driver Condition	3	320.59	0.0001
Residual	3	7.93	0.0474

TABLE 6.55. Significance for Individual Parameter Estimates

Parameter	Estimate	χ^2	p-Value
μ_2	0.7987	28,962.3	0.0001
μ_3	0.8225	38,125.2	0.0001
μ_4	0.8631	75,358.2	0.0001
α_2	0.0112	28.98	0.0001
α_3	0.0121	46.17	0.0001
α_4	0.0082	39.61	0.0001
γ_2	0.0721	264.10	0.0001
γ_3	0.0674	280.44	0.0001
γ_4	0.0530	312.51	0.0001

main effects appear to be stronger when the interaction is omitted. From the ANOVA table we can see that the residual χ^2 statistic is simply the χ^2 statistic for interaction observed in Table 6.52 since the former model was a saturated model.

6.4.3 SOME ALTERNATIVE ANALYSES

The previous discussion of the impact of seatbelt usage and driver condition on driver injury level focused on the comparison between the no injury level and each of the three levels of injury. Given that there are four response levels, it is possible to define a variety of other response functions using alternative linear transformation matrices **A**. Up to three linearly independent response functions can be defined. In this section several alternative analyses will be carried out using different transformations.

Marginal Analysis

Perhaps the simplest type of analysis that can be performed is to model all but one of the response proportions directly. In this case the response functions are called *marginal response* functions. The **A** matrix for this

TABLE 6.56. Significance for Individual Parameter Estimates for Marginal Analysis

Parameter	Estimate	χ^2	p-Value
μ_2	0.0723	1064.78	0.000
μ_3	0.0486	680.83	0.000
μ_4	0.0080	98.11	0.000
α_2	0.0033	11.57	0.000
α_3	0.0042	31.27	0.000
α_4	0.0000	1.82	0.1769
γ_2	0.0239	130.89	0.000
γ_3	0.0193	115.76	0.000
γ_4	0.0049	38.24	0.000

case contains rows with a single entry of 1 and the remaining entries 0. To illustrate using the accident data, the no injury level was omitted and the remaining three injury levels were related to seatbelt usage and driver condition using a two-way ANOVA model.

Even though the **A** matrix has changed, the overall significance levels for the effects are the same as in the analysis in Section 6.4.2. The individual effect parameters, however, now measure the impact of the factors on the injury level proportions rather than on the differences of the proportions from the no injury level. The equations are now

$$
\begin{aligned}
p_{1j} &= \mu_j + \alpha_j + \gamma_j + (\alpha\gamma)_j, \\
p_{2j} &= \mu_j + \alpha_j - \gamma_j - (\alpha\gamma)_j, \\
p_{3j} &= \mu_j - \alpha_j + \gamma_j - (\alpha\gamma)_j, \qquad j = 2,3,4; \\
p_{4j} &= \mu_j - \alpha_j - \gamma_j + (\alpha\gamma)_j,
\end{aligned}
$$

for the three injury levels. (Recall that α measures seatbelt effects and γ measures driver condition effects.) Notice that the design matrix is identical to (6.1). In this analysis we can determine how each proportion corresponding to a given injury level varies according to seatbelt usage and driver condition.

The results of this analysis are shown in Table 6.56. The model without the interaction term was fitted because of the marginal significance (0.0474) of this term. From this table it would appear that the two main effects are significant in the anticipated directions except for the seatbelt usage for the major/fatal level. In this case seatbelt usage does not have a significant effect on the proportion in this injury category.

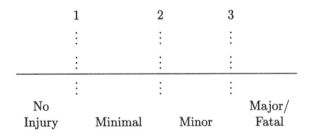

FIGURE 6.7. Three Partitions of Four Injury Level Groups

Continuation Differences

Since in the accident data example the four levels of injury are ordinal, it also makes sense to study the response probabilities using *continuation differences*. In this case the proportions are compared by dividing the four levels into two groups in three different ways by changing the line of division along the continuum from none to major/fatal. Figure 6.7 shows three partitions of the four groups.

In Figure 6.7 the three dotted vertical lines divide the four levels into two groups in three different ways. The division 1 compares the no injury category to the other three, and division 2 compares the two lower injury levels to the two higher injury levels. The final division 3 compares the major/fatal level to the other three.

As above, the overall significance of the effects is the same. The parameter estimates, however, provide an alternative way of measuring the effects. The three response functions in this case are given by $(3p_{i1} - p_{i2} - p_{i3} - p_{i4})$, $(p_{i1} + p_{i2} - p_{i3} - p_{i4})$ and $(p_{i1} + p_{i2} + p_{i3} - 3p_{i4})$. In each case, the right hand side has the same form as (6.1) for $i = 1, 2, 3, 4$.

The main effect parameter estimates obtained using a no interaction model are shown in Table 6.57. Once again all the main effects are significant except for the seatbelt usage effect in the major/fatal category. It would appear that the usage of a seatbelt in the major/fatal category does not yield a difference in proportion relative to the other three categories.

Averaging or Summing Response Functions

When there are two or more categories for the response variable, it is also possible to use a single response function based on a weighted average or weighted sum of the proportions. The resultant *average response* function or *sum of responses* is then related to the explanatory factors. In the case of the accident data, it would seem reasonable to use the sum of the three injury categories ignoring the no injury category. The response function would therefore reflect the injury proportion. The model would be given

TABLE 6.57. Significance for Individual Parameter Estimates for Continuation Differences

Parameter	Estimate	χ^2	p-Value
μ_2	2.4842	47482.9	0.000
μ_3	0.8867	48676.4	0.000
μ_4	0.9678	88771.1	0.000
α_2	0.0316	40.75	0.000
α_3	0.0091	33.02	0.000
α_4	0.0014	1.82	0.1769
γ_2	0.1925	316.00	0.000
γ_3	0.0484	155.92	0.000
γ_4	0.0196	38.24	0.000

by

$$
\begin{aligned}
(p_{21} + p_{31} + p_{41}) &= \mu + \alpha + \gamma + (\alpha\gamma) \\
(p_{22} + p_{32} + p_{42}) &= \mu + \alpha - \gamma - (\alpha\gamma) \\
(p_{23} + p_{33} + p_{43}) &= \mu - \alpha + \gamma - (\alpha\gamma) \\
(p_{24} + p_{34} + p_{44}) &= \mu - \alpha - \gamma + (\alpha\gamma).
\end{aligned}
$$

For the accident data the estimates of the parameters for the model above are $\hat{\mu} = 0.1297$, $\hat{\alpha} = 0.0078$ and $\hat{\gamma} = 0.0488$, all of which are significant at the 0.000 level. The interaction term was not significant and was omitted from the fit. We can conclude, therefore, that the proportion of drivers who sustain any level of injury is increased when a seatbelt is not used and also if the driver has been drinking. The impact of driver condition on injury level seems to be much greater than the impact of seatbelt usage.

Weighted Sums for Ordinal Responses

When the response levels are ordinal, *weighted averages* or *sums of the response functions* can be based on values that are attached to the levels. A common technique is simply to use integer values for the levels that reflect the rank order of the levels. In the case of the injury data, one may wish to attach weights that reflect the cost of the various injury levels. To provide an example we will use the weights 0, 5, 10 and 100 for the four injury levels. The model is given by $(5p_{i2} + 10p_{i3} + 100p_{i4})$. The right hand side of the model has the same form as (6.1).

For the accident data the estimated parameters for the model that excludes the interaction term are $\hat{\mu} = 1.6572$, $\hat{\alpha} = 0.0920$ and $\hat{\gamma} = 0.8070$. All three estimates were significant at the 0.000 level. A comparison of these results to the results obtained using the unweighted sum suggests that the magnitude of the driver condition parameter relative to the seatbelt usage parameter increases when higher values are placed on the more serious levels of injury.

6.4.4 WEIGHTED LEAST SQUARES ESTIMATION FOR LOGIT MODELS

The Logit Model as a Special Case of a Weighted Least Squares Model

The weighted least squares approach can also be used to estimate loglinear models based on the cell proportions. Since the transformation of the **p** vector is no longer linear in this case, the covariance matrix for the error term must be determined in an alternate manner.

We begin by denoting the vector containing the logarithms of the elements of **p** by $\ln \mathbf{p}$, and hence $\ln \mathbf{p}$ has the elements $\ln p_{ij}$, $i = 1, 2, \ldots, r$, $j = 1, 2, \ldots, c$. In a similar fashion the vector of elements $\ln \hat{p}_{ij}$ will be denoted by $\ln \hat{\mathbf{p}}$.

A covariance matrix for the vector $\ln \hat{\mathbf{p}}$ is required in order to perform the weighted least squares procedure. To determine the covariance matrix for $\ln \hat{\mathbf{p}}$, a Taylor Series expansion is used which yields the covariance matrix $\mathbf{D}^{-1} \Omega \mathbf{D}^{-1}$, where **D** is the diagonal matrix with the elements of **p** on the diagonal and Ω is the covariance matrix for $\hat{\mathbf{p}}$ defined in Section 6.4.1. Using the elements of $\hat{\mathbf{p}}$ as estimators of **p**, the estimated covariance matrix of $\ln \hat{\mathbf{p}}$ is given by $\widehat{\mathbf{D}}^{-1} \widehat{\Omega} \widehat{\mathbf{D}}^{-1}$.

A logit response model can be written as a linear transformation of the vector $\ln \mathbf{p}$ so that we have the same form as in the case of linear response models

$$\mathbf{g} = \mathbf{A} \ln \mathbf{p}.$$

When we relate the logit responses to the explanatory factors, the model becomes $\mathbf{g} = \mathbf{X}\beta$. Replacing **g** by $\hat{\mathbf{g}}$ when $\hat{\mathbf{p}}$ replaces **p**, we obtain the linear model

$$\hat{\mathbf{g}} = \mathbf{X}\beta + \mathbf{u}$$

and the weighted least squares estimator

$$\hat{\beta} = (\mathbf{X}' \, \widehat{\mathbf{H}}_L^{-1} \mathbf{X})^{-1} (\mathbf{X}' \widehat{\mathbf{H}}_L^{-1} \hat{\mathbf{g}}), \quad \text{where} \quad \widehat{\mathbf{H}}_L = \mathbf{A} \widehat{\mathbf{D}}^{-1} \widehat{\Omega} \widehat{\mathbf{D}}^{-1} \mathbf{A}'.$$

The inference procedures for this model are identical to those outlined in Section 6.4.2.

Example

The accident data table introduced in Section 6.4.1 is used here to illustrate the use of logit models. Two different types of response models are estimated. The logit model relating the no injury category to the remaining three injury categories has the three response functions $\ln p_{i1} - \ln p_{i2}$,

TABLE 6.58. Analysis of Variance for Logit Model

Source	d.f.	χ^2	p-Value
Intercepts	3	15427.26	0.000
Seat Belts	3	39.13	0.000
Driver Condition	3	561.73	0.000
Residual	3	4.96	0.1756

TABLE 6.59. Weighted Least Squares Parameter Estimates for Logit Model

Parameter	Estimate	χ^2	p-Value
μ_2	2.5534	7390.75	0.000
μ_3	3.0000	6572.24	0.000
μ_4	4.9515	2727.31	0.000
α_2	0.0707	10.51	0.001
α_3	0.1522	28.17	0.000
α_4	0.1168	1.99	0.159
γ_2	0.3887	251.30	0.000
γ_3	0.4559	255.33	0.000
γ_4	0.7311	116.54	0.000

$\ln p_{i1} - \ln p_{i3}$, and $\ln p_{i1} - \ln p_{i4}$. To relate these logits to the seatbelt usage and driver condition factors, a two-way ANOVA model is used. In this case the interaction term was not significant ($p = 0.1746$) and so the model was fitted without an interaction term. The analysis of variance results and parameter estimates are summarized in Tables 6.58 and 6.59 respectively. From the two tables we find that the main effects are significant in the expected direction and that only in the case of major/fatal was the seatbelt usage not significant.

The estimated logit model obtained above, using weighted least squares, can be compared to the estimated logit model, using maximum likelihood, by using the estimates given in Table 6.25 in Section 6.3.2. The estimates can be obtained as shown in Section 6.3.5. The resulting parameter estimates are summarized in Table 6.60. Comparison of Tables 6.60 and 6.59 shows that the parameter estimates obtained are virtually identical.

Continuation Ratios

In a manner similar to the continuation differences introduced above, the continuation differences based on $\ln p_{ij}$ are logarithms of ratios and are usually called *continuation ratios*. For the accident data, the parameter estimates for the continuation ratio response models are shown in Table 6.61.

TABLE 6.60. Parameter Estimates for Logit Model Based on Maximum Likelihood Estimates of the Loglinear Model

Parameter	
μ_2	$2.626 - 0.072 = 2.554$
μ_3	$2.626 + 0.376 = 3.002$
μ_4	$2.626 + 2.322 = 4.948$
α_2	$0.085 - 0.013 = 0.072$
α_3	$0.085 + 0.069 = 0.154$
α_4	$0.085 + 0.029 = 0.114$
γ_2	$0.392 - 0.006 = 0.386$
γ_3	$0.392 + 0.061 = 0.453$
γ_4	$0.392 + 0.337 = 0.729$

TABLE 6.61. Weighted Least Squares Parameter Estimates for Logit Model Using Continuation Ratios

Parameter	Estimate	χ^2	p-Value
μ_2	2.0020	7405.13	0.000
μ_3	2.9458	7281.99	0.000
μ_4	5.0838	2878.63	0.000
α_2	0.1030	35.31	0.000
α_3	0.1450	28.51	0.000
α_4	0.1077	1.69	0.1937
γ_2	0.4318	526.18	0.000
γ_3	0.4590	302.05	0.000
γ_4	0.6793	100.84	0.000

The first function compares the no injury case (numerator) to the three injury levels (denominator), the second function compares the no injury and minimal injury categories (numerator) to the minor and major/fatal categories (denominator). The third function compares the major/fatal category (denominator) to the other three categories (numerator). The fitted model in each case excludes the interaction term that is not significant.

6.4.5   Two or More Response Variables

Defining Response Functions

When there are two or more response variables, there are a variety of ways of defining response functions. We begin by examining the general case of two response variables. We then look at some specialized models for *repeated measures designs*.

If there are two response variables, with a and b levels respectively, the cross-classification of the two variables yields a total of $c = ab$ response categories. The cross-classification can now be viewed as a single response variable with c levels, and the weighted least squares methodology can be applied as outlined in Sections 6.4.1 and 6.4.2. Since the derived response variable represents a cross-classification of two variables, it is usually of interest to define the transformation matrix $\mathbf{A}$ in such a way that the response functions $\mathbf{g}$ represent separate effects for each of the two response variables, as well as interactions between the two sets of effects. The impact of the factor variables on the response variable main effects and interactions can then be measured by the model.

Example

For the accident data, both the injury level variable and seatbelt usage variable are treated as response variables, whereas the driver condition variable will be the only factor variable. The $\mathbf{p}$ vector now contains elements p_{ij}, $i = 1, 2$; $j = 1, 2, \ldots, 8$. We assume that $j = 1, 3, 5, 7$ corresponds to seatbelt usage = yes, and $j = 2, 4, 6, 8$ corresponds to seatbelt usage = no. The four levels of injury are none, minimal, minor and major/fatal, in that order.

The $\mathbf{A}$ matrix below uses effect coding to generate seven response functions for each of the two driver conditions. These response functions measure a seatbelt usage effect, three injury level effects, and the three interactions between these two sets of effects. The $\mathbf{p}$ vector is also shown below.

$$\mathbf{A} = \begin{bmatrix} 1 & -1 & 1 & -1 & 1 & -1 & 1 & -1 & 0 & 0 & 0 & 0 & 0 & 0 & 0 & 0 \\ -1 & -1 & 1 & 1 & 0 & 0 & 0 & 0 & 0 & 0 & 0 & 0 & 0 & 0 & 0 & 0 \\ -1 & -1 & 0 & 0 & 1 & 1 & 0 & 0 & 0 & 0 & 0 & 0 & 0 & 0 & 0 & 0 \\ -1 & -1 & 0 & 0 & 0 & 0 & 1 & 1 & 0 & 0 & 0 & 0 & 0 & 0 & 0 & 0 \\ -1 & 1 & 1 & -1 & 0 & 0 & 0 & 0 & 0 & 0 & 0 & 0 & 0 & 0 & 0 & 0 \\ -1 & 1 & 0 & 0 & 1 & -1 & 0 & 0 & 0 & 0 & 0 & 0 & 0 & 0 & 0 & 0 \\ -1 & 1 & 0 & 0 & 0 & 0 & 1 & -1 & 0 & 0 & 0 & 0 & 0 & 0 & 0 & 0 \\ 0 & 0 & 0 & 0 & 0 & 0 & 0 & 0 & 1 & -1 & 1 & -1 & 1 & -1 & 1 & -1 \\ 0 & 0 & 0 & 0 & 0 & 0 & 0 & 0 & -1 & -1 & 1 & 1 & 0 & 0 & 0 & 0 \\ 0 & 0 & 0 & 0 & 0 & 0 & 0 & 0 & -1 & -1 & 0 & 0 & 1 & 1 & 0 & 0 \\ 0 & 0 & 0 & 0 & 0 & 0 & 0 & 0 & -1 & -1 & 0 & 0 & 0 & 0 & 1 & 1 \\ 0 & 0 & 0 & 0 & 0 & 0 & 0 & 0 & -1 & 1 & 1 & -1 & 0 & 0 & 0 & 0 \\ 0 & 0 & 0 & 0 & 0 & 0 & 0 & 0 & -1 & 1 & 0 & 0 & 1 & -1 & 0 & 0 \\ 0 & 0 & 0 & 0 & 0 & 0 & 0 & 0 & -1 & 1 & 0 & 0 & 0 & 0 & 1 & -1 \end{bmatrix} \begin{bmatrix} p_{11} \\ p_{12} \\ p_{13} \\ p_{14} \\ p_{15} \\ p_{16} \\ p_{17} \\ p_{18} \\ p_{21} \\ p_{22} \\ p_{23} \\ p_{24} \\ p_{25} \\ p_{26} \\ p_{27} \\ p_{28} \end{bmatrix}.$$

The first and eighth rows of $\mathbf{A}$ derive the response function for seatbelt usage whereas rows 2, 3, 4 and 9, 10, 11 represent the three injury level effects. The remaining six rows represent the three interaction effects. These rows can be seen to be obtained by taking the products of elements in row 1 with each of rows 2, 3 and 4 and row 8 with each of rows 9, 10 and 11. On the right-hand side of the model, a design matrix is required to measure driver condition effects. For each response function the design matrix is $\begin{bmatrix} 1 & 1 \\ 1 & -1 \end{bmatrix}$. Denoting the parameter vector for response function i by $\begin{bmatrix} \mu_i \\ \alpha_i \end{bmatrix}$, the right-hand side has the form $(\mu_i + \alpha_i)$ for normal condition drivers and $(\mu_i - \alpha_i)$ for been drinking drivers. The μ_i parameter measures the category mean, and the α_i parameter measures the driver condition effect. The driver condition parameter is preceded by a positive sign for normal condition drivers and by a negative sign for been drinking drivers.

The equations for the normal condition drivers can be expressed by

$(p_{Yes} - p_{No})_{Normal}$ $= \mu_1 + \alpha_1$
All Injury Levels

$(p_{Minimal} - p_{None})_{Normal} = \mu_2 + \alpha_2$
Both Seatbelt Usages

$(p_{Minor} - p_{None})_{Normal}$ $= \mu_3 + \alpha_3$
Both Seatbelt Usages

$(p_{Major} - p_{None})_{Normal}$ $= \mu_4 + \alpha_4$
Both Seatbelt Usages

$(p_{Minimal} - p_{None})_{Normal} - (p_{Minimal} - p_{None})_{Normal}$
Seatbelt Yes   Seatbelt No

    or $= \mu_5 + \alpha_5$

$(p_{Yes} - p_{No})_{Normal}$ $- (p_{Yes} - p_{No})_{Normal}$
Minimal Injury   No Injury

$(p_{Minor} - p_{None})_{Normal} - (p_{Minor} - p_{None})_{Normal}$
Seatbelt Yes   Seatbelt No

    or $= \mu_6 + \alpha_6$

$(p_{Yes} - p_{No})_{Normal}$ $- (p_{Yes} - p_{No})_{Normal}$
Minor Injury   No Injury

$(p_{Major} - p_{None})_{Normal} - (p_{Major} - p_{None})_{Normal}$
Seatbelt Yes   Seatbelt No

    or $= \mu_7 + \alpha_7$

$(p_{Yes} - p_{No})_{Normal}$ $- (p_{Yes} - p_{No})_{Normal}$
Major/Fatal Injury   No Injury

For the been drinking drivers the equations are the same as above except the signs in front of the α parameters are all negative. Combining the equations for normal and been drinking drivers yields the parameter

relationships summarized below.

$$2\alpha_1 = (p_{Yes} - p_{No})_{Normal} - (p_{Yes} - p_{No})_{Drinking}$$
$$\text{All Injury Levels}$$

− a positive α_1 therefore implies that normal condition drivers have a greater tendency to wear seatbelts.

$$2\alpha_2 = (p_{Minimal} - p_{None})_{Normal} - (p_{Minimal} - p_{None})_{Drinking}$$
$$\text{Both Seatbelt Usages}$$

$$2\alpha_3 = (p_{Minor} - p_{None})_{Normal} - (p_{Minor} - p_{None})_{Drinking}$$
$$\text{Both Seatbelt Usages}$$

$$2\alpha_4 = (p_{Major} - p_{None})_{Normal} - (p_{Major} - p_{None})_{Drinking}$$
$$\text{Both Seatbelt Usages}$$

− negative values of α_2, α_3 and α_4 imply that normal condition drivers tend to have less injuries than drinking drivers.

$$2\alpha_5 = \{(p_{Minimal} - p_{None})_{Normal} - (p_{Minimal} - p_{None})_{Drinking}\}$$
$$\text{Seatbelt Yes}$$
$$- \{(p_{Minimal} - p_{None})_{Normal} - (p_{Minimal} - p_{None})_{Drinking}\}$$
$$\text{Seatbelts No}$$
$$\text{or}$$
$$= \{(p_{Yes} - p_{No})_{Normal} - (p_{Yes} - p_{No})_{Drinking}\}$$
$$\text{Minimal Injury}$$
$$- \{(p_{Yes} - p_{No})_{Normal} - (p_{Yes} - p_{No})_{Drinking}\}$$
$$\text{No Injury}$$

$$2\alpha_6 = \{(p_{Minor} - p_{None})_{Normal} - (p_{Minor} - p_{None})_{Drinking}\}$$
$$\text{Seatbelts Yes}$$
$$- \{(p_{Minor} - p_{None})_{Normal} - (p_{Minor} - p_{None})_{Drinking}\}$$
$$\text{Seatbelts No}$$
$$\text{or}$$
$$= \{(p_{Yes} - p_{No})_{Normal} - (p_{Yes} - p_{No})_{Drinking}\}$$
$$\text{Minor Injury}$$
$$- \{(p_{Yes} - p_{No})_{Normal} - (p_{Yes} - p_{No})_{Drinking}\}$$
$$\text{No Injury}$$

$$2\alpha_7 = \{(p_{Major} - p_{None})Normal - (p_{Major} - p_{None})Drinking\}$$
Seatbelts Yes

$$- \{(p_{Major} - p_{None})Normal - (p_{Major} - p_{None})Drinking\}$$
Seatbelts No

or

$$= \{(p_{Yes} - p_{No})Normal - (p_{Yes} - p_{No})Drinking\}$$
Major/Fatal Injury

$$- \{(p_{Yes} - p_{No})Normal - (p_{Yes} - p_{No})Drinking\}$$
No Injury

– negative values of α_5, α_6 and α_7 imply that the impact of driver condition on injury level is less pronounced among non-seatbelt users than among seatbelt users or, equivalently, the impact of driver condition on the level of seatbelt usage is less pronounced in the injury categories than in the no injury category.

The weighted least squares estimates for the resulting model are shown in Table 6.62. From the intercept parameter (μ parameter) estimates we can conclude that after averaging over the driver condition categories:

(a) a smaller proportion of the drivers wore seatbelts,

(b) for each of the three injury categories the proportions in the injury categories were much smaller than the proportion in the no injury category,

(c) the difference between the proportion of non-seatbelt wearers and seatbelt wearers is much larger in the no injury category than in the three injury categories. The magnitude of the parameter increases as the severity of injury increases reflecting the smaller proportion of injuries in the more serious injury categories.

The driver condition effects indicate how the intercept parameter responses are influenced by driver condition.

(a) The seatbelt usage response parameter indicates that for normal condition drivers the proportion wearing seatbelts tends to be higher than for drinking drivers.

(b) The injury level parameters indicate that for the three injury categories the proportions of normal drivers who incur injuries is less than the proportion of drinking drivers who incur injuries.

TABLE 6.62. Parameter Estimates for the Two-Response Model

Intercept Parameters	Estimate	χ^2	p-Value
(1) Seatbelt Usage	-0.7635	41133.8	0.000
(2) Minimal vs No Injury	-0.7894	31657.8	0.000
(3) Minor vs No Injury	-0.8118	40013.0	0.000
(4) Major/Fatal vs No Injury	-0.8560	80948.2	0.000
(5) Interaction (1) & (2)	0.5924	12807.5	0.000
(6) Interaction (1) & (3)	0.6063	14626.7	0.000
(7) Interaction (1) & (4)	0.6446	22223.2	0.000
Driver Condition Effects			
(1) Seatbelt Usage	0.0945	630.6	0.000
(2) Minimal vs No Injury	-0.0739	277.3	0.000
(3) Minor vs No Injury	-0.0700	297.3	0.000
(4) Major/Fatal vs No Injury	-0.0545	328.6	0.000
(5) Interaction (1) & (2)	-0.0210	16.10	0.000
(6) Interaction (1) & (3)	-0.0228	20.70	0.000
(7) Interaction (1) & (4)	-0.0399	85.2	0.000

(c) The interaction response parameters suggest that the difference in the proportion of seatbelt wearers between normal and drinking drivers is less at each of the injury levels than it is at the no injury level. The seatbelt response and injury level response are therefore not simply additive.

Example Using Logs

The above analysis can also be carried out using the log form $\ln p_{ij}$ in place of p_{ij}. When the analysis is carried out for the accident data, the interaction terms are not significant. This result is consistent with the loglinear model results summarized in Tables 6.24 and 6.25. The three-way interaction term had a p-value of 0.17 and was omitted from the model. It would seem that modeling the logarithm of the proportions yields different results from the model for the proportions.

Repeated Measurement Designs

In analysis of variance, the *repeated measures design* is commonly used. In the repeated measures design, more than one observation is obtained from each experimental unit. The multiple observations on each experimental unit may represent a variety of experimental conditions, such as treatments, or may represent responses to different items on a questionnaire. In such circumstances it is not only the interaction between the response variables that is of interest but also the symmetry among the response

distributions for the response variables. The repeated measures design will also be discussed in Chapters 7 and 8. In this section we examine some alternative models that can be used for tests of symmetry. Such tests are also referred to as tests of *marginal homogeneity*.

We assume that there are a total of r subpopulations of subjects and that the subjects in these subpopulations are observed on a total of d different measurement conditions. Each measurement or experimental condition yields a response on any one of b levels. The total number of response categories is therefore $c = db$. In place of the notation p_{ij}, $i = 1, 2, \ldots, r$; $j = 1, 2, \ldots, c$; used above for cell probabilities, it will be more convenient to use $q_{ik\ell}$ to denote these cell probabilities where k denotes the experimental condition and ℓ denotes the response level, $k = 1, 2, \ldots, d$; $\ell = 1, 2, \ldots, b$.

One hypothesis of interest is the *total symmetry hypothesis* given by

$$H_1: q_{1k\ell} = q_{2k\ell} = \ldots = q_{rk\ell}, \quad k = 1, 2, \ldots, d, \ \ell = 1, 2, \ldots, b,$$

which indicates that there are no differences between the r subpopulations with respect to the response probabilities in each of the $c = db$ cells. The distribution of the probability over the c cells is therefore assumed to be the same for all r subpopulations. A second hypothesis of interest is the *marginal symmetry* or marginal homogeneity hypothesis given by

$$H_2: q_{i1\ell} = q_{i2\ell} = \ldots = q_{id\ell}, \quad i = 1, 2, \ldots, r, \ \ell = 1, 2, \ldots, b,$$

which suggests that for a given subpopulation and a given response level, the cell probabilities are identical for all experimental conditions or response variables.

The two hypotheses are useful for comparing subpopulations and also for comparing experimental conditions. These two hypotheses will be tested by fitting a model to the cell proportions in such a way that the model parameters measure departures from the two hypotheses. For H_1 the parameters are similar to those used above to represent the variation in the factor variables that define the r subpopulations. For H_2 the parameters are designed to compare the degree of homogeneity among the distributions of the response variables. It is this latter hypothesis that requires a different type of model than those already discussed. An example will be used below to illustrate the procedure.

Example

To provide an example for the repeated measures case, the contingency table shown in Table 6.63 will be used. The contingency table is based on a sample of 1250 individuals who were asked to respond to a questionnaire dealing with the evaluation of police services. The two response variables CRIME 1 and CRIME 2 pertain to the individual perceptions of their safety while walking in their neighborhood at night (CRIME 1)

TABLE 6.63. Contingency Table Relating Crime
Perceptions to Education Level

| | | EDUC | | | |
CRIME 1	CRIME 2	1	2	3	TOTAL
	1	71	81	83	235
1	2	32	54	56	142
	3	5	8	6	19
	Total	108	143	145	396
	1	7	3	5	15
2	2	172	138	98	408
	3	66	48	33	147
	Total	245	189	136	570
	1	0	2	0	2
3	2	15	13	10	38
	3	119	79	46	244
	Total	134	94	56	284
					1250

and in the downtown region at night (CRIME 2). The codes 1, 2, and 3
refer to the three opinions very safe, somewhat safe and unsafe. The factor variable EDUC represents level of education and is coded 1, 2, or 3.
Level 1 corresponds to those with at most a high school diploma, level 2
corresponds to those who have some post secondary training, and level 3
corresponds to those who have a university degree. The **A** matrix used to
create 12 response functions for the cell probabilities is shown below. For
each of the two response variables there are two response probabilities to
model since there are three levels of response (three probabilities must sum
to one). Since there are three education subpopulations the total number
of response functions is $(3)(2)(2) = 12$.

The response functions are given by $\mathbf{g} = \mathbf{Ap}$. The **A** matrix is given by

$$\left[\begin{array}{ccccccccc|ccccccccc|ccccccccc}
1 & 1 & 1 & 0 & 0 & 0 & 0 & 0 & 0 & 0 & 0 & 0 & 0 & 0 & 0 & 0 & 0 & 0 & 0 & 0 & 0 & 0 & 0 & 0 & 0 & 0 & 0 \\
0 & 0 & 0 & 1 & 1 & 1 & 0 & 0 & 0 & 0 & 0 & 0 & 0 & 0 & 0 & 0 & 0 & 0 & 0 & 0 & 0 & 0 & 0 & 0 & 0 & 0 & 0 \\
1 & 0 & 0 & 1 & 0 & 0 & 1 & 0 & 0 & 0 & 0 & 0 & 0 & 0 & 0 & 0 & 0 & 0 & 0 & 0 & 0 & 0 & 0 & 0 & 0 & 0 & 0 \\
0 & 1 & 0 & 0 & 1 & 0 & 0 & 1 & 0 & 0 & 0 & 0 & 0 & 0 & 0 & 0 & 0 & 0 & 0 & 0 & 0 & 0 & 0 & 0 & 0 & 0 & 0 \\
\hline
0 & 0 & 0 & 0 & 0 & 0 & 0 & 0 & 0 & 1 & 1 & 1 & 0 & 0 & 0 & 0 & 0 & 0 & 0 & 0 & 0 & 0 & 0 & 0 & 0 & 0 & 0 \\
0 & 0 & 0 & 0 & 0 & 0 & 0 & 0 & 0 & 0 & 0 & 0 & 1 & 1 & 1 & 0 & 0 & 0 & 0 & 0 & 0 & 0 & 0 & 0 & 0 & 0 & 0 \\
0 & 0 & 0 & 0 & 0 & 0 & 0 & 0 & 0 & 1 & 0 & 0 & 1 & 0 & 0 & 1 & 0 & 0 & 0 & 0 & 0 & 0 & 0 & 0 & 0 & 0 & 0 \\
0 & 0 & 0 & 0 & 0 & 0 & 0 & 0 & 0 & 0 & 1 & 0 & 0 & 1 & 0 & 0 & 1 & 0 & 0 & 0 & 0 & 0 & 0 & 0 & 0 & 0 & 0 \\
\hline
0 & 0 & 0 & 0 & 0 & 0 & 0 & 0 & 0 & 0 & 0 & 0 & 0 & 0 & 0 & 0 & 0 & 0 & 1 & 1 & 1 & 0 & 0 & 0 & 0 & 0 & 0 \\
0 & 0 & 0 & 0 & 0 & 0 & 0 & 0 & 0 & 0 & 0 & 0 & 0 & 0 & 0 & 0 & 0 & 0 & 0 & 0 & 0 & 1 & 1 & 1 & 0 & 0 & 0 \\
0 & 0 & 0 & 0 & 0 & 0 & 0 & 0 & 0 & 0 & 0 & 0 & 0 & 0 & 0 & 0 & 0 & 0 & 1 & 0 & 0 & 1 & 0 & 0 & 1 & 0 & 0 \\
0 & 0 & 0 & 0 & 0 & 0 & 0 & 0 & 0 & 0 & 0 & 0 & 0 & 0 & 0 & 0 & 0 & 0 & 0 & 1 & 0 & 0 & 1 & 0 & 0 & 1 & 0
\end{array}\right]$$

This **A** matrix consists of three sections of four rows each corresponding
to the three education subpopulations. The four rows in each section gener-

ate the four response functions corresponding to the two response variables. The first row of $\mathbf{A}$ determines the probability for CRIME $1 = 1$, and the second row of $\mathbf{A}$ determines the probability for CRIME $1 = 2$. Similarly the third and fourth rows of $\mathbf{A}$ determine the probabilities for CRIME $2 = 1$ and CRIME $2 = 2$ respectively. The CRIME $1 = 3$ and CRIME $2 = 3$ categories are omitted since the proportions must sum to one for each variable. The same pattern is repeated in the four rows corresponding to each of the second and third subpopulations.

The design matrix, $\mathbf{X}$, for the model $\mathbf{g} = \mathbf{X}\boldsymbol{\beta}$ to be fitted is shown below. The parameter vector $\boldsymbol{\beta}$ is also shown. The $\mathbf{X}$ matrix contains three horizontal blocks corresponding to the three education subpopulations.

$$
\mathbf{X} =
\begin{bmatrix}
1 & 0 & 1 & 0 & 0 & 0 & 1 & 0 \\
0 & 1 & 0 & 1 & 0 & 0 & 0 & 1 \\
1 & 0 & 1 & 0 & 0 & 0 & -1 & 0 \\
0 & 1 & 0 & 1 & 0 & 0 & 0 & -1 \\
1 & 0 & 0 & 0 & 1 & 0 & 1 & 0 \\
0 & 1 & 0 & 0 & 0 & 1 & 0 & 1 \\
1 & 0 & 0 & 0 & 1 & 0 & -1 & 0 \\
0 & 1 & 0 & 0 & 0 & 1 & 0 & -1 \\
1 & 0 & -1 & 0 & -1 & 0 & 1 & 0 \\
0 & 1 & 0 & -1 & 0 & -1 & 0 & 1 \\
1 & 0 & -1 & 0 & -1 & 0 & -1 & 0 \\
0 & 1 & 0 & -1 & 0 & -1 & 0 & -1
\end{bmatrix},
\quad
\boldsymbol{\beta} =
\begin{bmatrix}
\mu_1 \\
\mu_2 \\
\alpha_{11} \\
\alpha_{12} \\
\alpha_{21} \\
\alpha_{22} \\
\gamma_1 \\
\gamma_2
\end{bmatrix}.
$$

In the $\boldsymbol{\beta}$ parameter vector there are two sets of parameters corresponding to the two levels of each response variable being modelled. Effect coding has been used to account for EDUC effects and the effects due to difference between CRIME 1 and CRIME 2. For the first level of the response variables (odd numbered rows of $\mathbf{A}$) the cell probabilities are described by $(\mu_1 + \alpha_{11} + \gamma_1)$, $(\mu_1 + \alpha_{11} - \gamma_1)$, $(\mu_1 + \alpha_{12} + \gamma_1)$, $(\mu_1 + \alpha_{12} - \gamma_1)$, $(\mu_1 - \alpha_{11} - \alpha_{12} + \gamma_1)$ and $(\mu_1 - \alpha_{11} - \alpha_{12} - \gamma_1)$, and similarly the second level of response by $(\mu_2 + \alpha_{21} + \gamma_2)$, $(\mu_2 + \alpha_{21} - \gamma_2)$, $(\mu_2 + \alpha_{22} + \gamma_2)$, $(\mu_2 + \alpha_{22} - \gamma_2)$, $(\mu_2 - \alpha_{21} - \alpha_{22} + \gamma_2)$ and $(\mu_2 - \alpha_{21} - \alpha_{22} - \gamma_2)$ (even numbered rows of $\mathbf{A}$). The parameters α_{11}, α_{12}, α_{21} and α_{22} represent the effects of EDUC while the parameters γ_1 and γ_2 account for differences between CRIME 1 and CRIME 2.

The analysis of variance table and parameter estimates are shown in Tables 6.64 and 6.65 respectively. From the analysis of variance table we can conclude that both the EDUC effects and the CRIME effects are significant. Since the EDUC effects are significant, the total symmetry hypothesis H_1 can be rejected, and hence the distribution of cell probabilities differs with respect to the three levels of EDUC. Similarly, since the CRIME effects are significant, the hypothesis of marginal symmetry H_2 must be rejected

TABLE 6.64. Analysis of Variance for Model Relating CRIME 1 and CRIME 2 to EDUC

Source	d.f.	χ^2	p-Value
Intercepts	2	4313.19	0.000
EDUC	4	27.35	0.000
CRIME	2	184.86	0.000
Residual	4	22.68	0.000

TABLE 6.65. Parameter Estimates and p-Values

Parameter	Estimate	χ^2	p-Value
μ_1	0.2653	546.53	0.000
μ_2	0.4607	1457.83	0.000
α_{11}	-0.0636	9.32	0.000
α_{12}	0.0142	0.78	0.392
α_{21}	-0.0043	0.08	0.781
α_{22}	-0.0021	0.01	0.903
γ_1	0.0511	111.56	0.000
γ_2	-0.0014	0.04	0.844

and therefore the distribution of probabilities over the response levels of CRIME 1 is different from the distribution of these probabilities over the levels of CRIME 2. The parameter estimates in Table 6.65 suggest that, at the margin, some of the parameter estimates are not significantly different from zero. It would appear that for the first response level of the two crime variables the proportion is lower in the first education category. Thus individuals from the lowest education category feel less safe in general. Outside of the first response category and the first education category there does not appear to be any other differences. Also, for the first levels of CRIME 1 and CRIME 2 the cell probability is higher for CRIME 1 than for CRIME 2. This indicates that at night a larger proportion of people tend to feel very safe in their neighborhood than in the downtown region. For the somewhat safe category the proportions are about the same for both crime variables. The parameters μ_1 and μ_2 indicate that on average 26.5% of the respondents choose the first response whereas 46% choose the second response.

Adding Interaction Effects

The analysis of variance shown in Table 6.64 suggests that the residual is significant and hence that the fitted model does not fit the data well. To obtain a better fit we consider the impact of adding interaction ef-

TABLE 6.66. Analysis of Variance

Source	d.f.	χ^2	p-Value
Intercepts	2	4333.32	0.000
EDUC	4	37.00	0.000
CRIME	2	203.29	0.000
CRIME-EDUC Interaction	4	22.68	0.000
Residual	0	0	1.000

TABLE 6.67. Parameter Estimates and p-Values

Parameter	Estimate	χ^2	p-Value
μ_1	0.2658	557.52	0.000
μ_2	0.4613	1461.30	0.000
α_{11}	-0.0775	27.36	0.000
α_{12}	0.0151	0.83	0.363
α_{21}	0.0000	0.00	0.985
α_{22}	0.0011	0.00	0.947
γ_1	0.0608	131.58	0.000
γ_2	-0.0112	2.21	0.137
$(\alpha_{11}\gamma_1)$	-0.0300	20.94	0.000
$(\alpha_{12}\gamma_1)$	0.0379	14.60	0.000
$(\alpha_{21}\gamma_2)$	0.0061	0.68	0.409
$(\alpha_{22}\gamma_2)$	-0.0076	0.52	0.472

fects between the EDUC levels and the CRIME effects. Columns can be added to the $\mathbf{X}$ matrix to account for the interaction between CRIME and EDUC. Since there are two EDUC parameters and one CRIME parameter in each equation, there are two interaction parameters for each equation. The columns corresponding to these interaction parameters can be obtained by taking the product of corresponding elements in the main effect columns. Tables 6.66 and 6.67 summarize the results obtained from fitting the model which includes the interaction terms.

From Table 6.67 we can conclude that fitting the interaction parameters leaves the main effect parameters virtually unchanged. Two significant interaction parameters are also obtained for CRIME response level 1. The interaction between the first level of EDUC and CRIME response level 1 is negative, and between EDUC level 2 and CRIME response level 1 the interaction is positive. We can conclude, therefore, that the difference in cell proportions between CRIME 1 and CRIME 2 for level 1 depends on the level of EDUC. At EDUC level 1 it would seem that there is less difference between the two crime variables than for EDUC level 2.

6.4.6   OTHER SOURCES OF INFORMATION

The most comprehensive outline of the weighted least squares methodology is contained in Forthofer and Lehnen (1981). This methodology is also discussed in Reynolds (1977) and Freeman (1987). A number of papers by Koch and others can also be used to gain further understanding. References to these papers can be found in the three texts listed above. Two useful papers are Grizzle, Starmer and Koch (1969) and Koch, Landis, Freeman, Freeman and Lehnen (1977).

Cited Literature and References

1. Agresti, Alan (1984). *Analysis of Ordinal Categorical Data*. New York: John Wiley and Sons.

2. Andersen, Erling B. (1980). *Discrete Statistical Models With Social Science Applications*. Amsterdam: North-Holland Publishing Company.

3. Andersen, Erling B. (1990). *The Statistical Analysis of Categorical Data*. Berlin: Springer–Verlag.

4. Bishop, Yvonne M.M., Fienberg, Stephen E., and Holland, Paul W. (1975). *Discrete Multivariate Analysis: Theory and Practice*. Cambridge, Ma.: MIT Press.

5. Christensen, Ronald (1991). *Log-Linear Models*. New York: Springer–Verlag.

6. Everitt, B.S. (1977). *The Analysis of Contingency Tables*. London: Chapman and Hall.

7. Fienberg, Stephen E. (1980). *The Analysis of Cross-Classified Categorical Data*, Second Edition. Cambridge, Ma.: MIT Press.

8. Forthofer, Ron N. and Lehnen, Robert G. (1981). *Public Program Analysis*. Belmont, Ca.: Lifetime Learning Publications.

9. Freeman, Daniel H. Jr. (1987). *Applied Categorical Data Analysis*. New York: Marcel Dekker.

10. Grizzle, James E., Starmer, C. and Koch, Gary G. (1969). "Analysis of Categorical Data by Linear Models," *Biometrics* 25, 489–504.

11. Koch, G.G., Landis, J.R., Freeman, J.L., Freeman, D.H., and Lehnen, R.G. (1977). "A General Methodology for the Analysis of Experiments with Repeated Measurement of Categorical Data," *Biometrics* 33, 133–158.

12. Reynolds, H.T. (1977). *The Analysis of Cross-Classifications*. New York: The Free Press.

13. Santner, Thomas J. and Duffy, Dianne E. (1989). *The Statistical Analysis of Discrete Data*. New York: Springer–Verlag.

14. Upton, Graham J.G. (1978). *The Analysis of Cross-Tabulated Data*. New York: John Wiley and Sons.

Exercises For Chapter 6

1. This exercise is based on the Bus Data in Table V1 in the Data
   Appendix.

   (a) Using the three three-dimensional tables determine the three
   two-dimensional tables relating the variable ATTEND to each
   of the variables SEX, DAY and GARAGE. For each of the three
   tables analyze the relationships and discuss the results.

   (b) For the three three-dimensional tables construct two-dimensional
   tables that relate SEX and DAY, SEX and GARAGE, and DAY
   and GARAGE. For each of the tables analyze the relationships
   and discuss the results.

   (c) For each of the three three-dimensional tables use the maximum likelihood approach to determine a loglinear model that
   adequately fits the table. Are the findings consistent with the
   results obtained in (a) and (b)? Discuss the fitted model in each
   case. Include graphs of the interaction effects as part of your
   discussion.

   (d) For each of the three three-dimensional tables in (c) use the
   fitted saturated model to determine a logit model relating the
   dependent variable ATTEND to the remaining two variables.
   Discuss and interpret the results.

   (e) A four-dimensional table relating ATTEND, SEX, DAY and
   GARAGE could not be obtained from company records. Assuming that the loglinear model for this table does not require
   interaction terms with the variable ATTEND of greater than
   second order (three-way) a logit model for ATTEND can be estimated using the three logit models estimated in (d). Write out
   the saturated versions of the three logit models corresponding
   to the three tables and add the right hand side terms together
   to produce a logit model for ATTEND for the four dimensional
   table. In cases where the same term appears in more than one
   model the terms can be replaced by a simple average. Estimate
   the three saturated models for the three tables and combine the
   estimates to obtain an estimate for the four dimensional table
   logit model for ATTEND. Use the fitted logit model to obtain
   a table giving P[PRESENT] for each of the $2 \times 7 \times 3 = 42$
   cells. Explain how this table could be used by the bus company
   management to plan staff requirements.

   (f) In parts (a) and (b) collapsed tables were used to relate the
   four variables two at a time. In part (c) the four variables were
   studied three at a time. Did collapsing the tables change the

conclusions about the relationships? Examine the collapsibility conditions given in the text and comment on the results.

(g) Use weighted least squares to fit logit models for the variable ATTEND to each of the three three-dimensional tables and compare the results to (d).

(h) Use weighted least squares to fit a model to each of the three tables using P[PRESENT] as the dependent variable. In this case model the proportions rather than the logarithms of the proportions as in (g). Interpret the results in each case and compare to the logit model results in (g).

2. This exercise is based on the Accident Data in Table V2 of the Data Appendix. Because of empty cells you should combine the DRIVER INJURY categories for MINOR and MAJFAT when using weighted least squares.

   (a) Table V2 contains a four-dimensional contingency table relating SEATBELT, DRIVER INJURY LEVEL, DRIVER CONDITION and POINT OF IMPACT. Determine the three three-dimensional tables relating DRIVER INJURY LEVEL with two other variables. The three-dimensional table relating DRIVER INJURY LEVEL to DRIVER CONDITION and SEATBELT is identical to Table 6.23 discussed in the text. Fit loglinear models to each of the three tables using maximum likelihood. In each case obtain the fitted parameters and discuss the results. Include graphs of the interaction effects as part of your discussion.

   (b) Fit a loglinear model to the four-dimensional table which includes all three-way interaction terms using maximum likelihood. Discuss the results. Compare the results to (a). Did collapsing the table have any effect?

   (c) Use maximum likelihood stepwise methods to fit a model to the four dimensional table. Estimate the fitted model and discuss the results. Compare the results to the results obtained in (a) for the three collapsed tables. Did the collapsing of the table change the results?

   (d) Using the fitted model in (b) determine logit models relating INJURY LEVEL to the other three variables. Compare each INJURY category to the NO INJURY category. Discuss the fitted models.

   (e) Use weighted least squares to fit logit models relating INJURY LEVEL to NO INJURY for the four-dimensional table. Compare the results to the results obtained in (d).

   (f) Repeat the analysis in (e) using continuation ratios as outlined in the text.

(g) Use weighted least squares to fit a model that relates the differences between injury level proportions and the no injury proportion to the other three variables. Discuss the results and compare to the logit model results in (e).

(h) Repeat the analysis in (g) using continuation differences as outlined in the text. Compare the results to the results in (f).

(i) Use weighted least squares to relate a weighted sum of two injury category proportions to the other three variables. Use the weights for the three injury categories 1 (minimal) and 10 (minor). Discuss the results.

(j) Use weighted least squares to fit a model that combines the INJURY LEVEL and SEATBELT variables into one dependent variable and relate to the other two variables. Use models based on proportions and on log proportions. Is there any interaction between INJURY LEVEL and SEATBELT usage? Discuss the results.

3. This exercise is based on the Accident Data in Table V3 of the Data Appendix.

(a) Table V3 contains a four-dimensional contingency table relating SEATBELT, DRIVER INJURY LEVEL, DRIVER CONDITION and SPEED LIMIT. Determine the three three-dimensional tables relating DRIVER INJURY LEVEL with two other variables. The three-dimensional table relating DRIVER INJURY LEVEL to DRIVER CONDITION and SEATBELT is identical to Table 6.23 discussed in the text. Fit loglinear models to each of the three tables using maximum likelihood. In each case obtain the fitted parameters and discuss the results. Include graphs of the interaction effects as part of your discussion.

(b) Fit a loglinear model to the four-dimensional table which includes all three-way interaction terms using maximum likelihood. Discuss the results. Compare the results to (a). Did collapsing the table have any effect?

(c) Use maximum likelihood stepwise methods to fit a model to the four-dimensional table. Estimate the fitted model and discuss the results. Compare the results to the results obtained in (a) for the three collapsed tables. Did the collapsing of the table change the results?

(d) Using the fitted model in (b) determine logit models relating INJURY LEVEL to the other three variables. Compare each INJURY category to the NO INJURY category. Discuss the fitted models.

(e) Use weighted least squares to fit logit models relating INJURY LEVEL to NO INJURY for the four-dimensional table. Compare the results to the results in (d).

(f) Repeat the analysis in (e) using continuation ratios as outlined in the text.

(g) Use weighted least squares to fit a model that relates the differences between injury level proportions and the no injury proportion. Discuss the results and compare to the logit model results in (e).

(h) Repeat the analysis in (g) using continuation differences as outlined in the text. Compare the results to the results in (f).

(i) Use weighted least squares to relate a weighted sum of three injury category proportions to the other three variables. Use the weights for the three injury categories 1 (minimal), 10 (minor) and 100 (major/fatal). Discuss the results.

(j) Use weighted least squares to fit a model that combines the INJURY LEVEL and SEATBELT variables into one dependent variable and relate to the other two variables. Use models based on proportions and on log proportions. Is there any interaction between INJURY LEVEL and SEATBELT usage? Discuss the results.

Questions for Chapter 6

1. (a) Given that n_1 and n_2 are independent Poisson random variables with densities

$$f(n_i) = F_i^{n_i} e^{-F_i}/n_i!, \qquad i = 1, 2,$$

show that the joint density of n_1, n_2 is given by

$$f(n_1, n_2) = F_1^{n_1} F_2^{n_2} e^{-(F_1+F_2)}/n_1! n_2!.$$

   (b) Show that the probability $P[n_1 = a, \; n_2 = s - a]$ is given by $F_1^a F_2^{(s-a)} e^{-(F_1+F_2)}/a!(s-a)!$.

   (c) Show that the density of $n = (n_1 + n_2)$ is given by a Poisson distribution with mean parameter $F = (F_1 + F_2)$ by determining $P[n = s]$. (HINT: use (b) and sum the expression from $a = 0$ to $a = s$; also recall from the binomial theorem that $(X + Y)^n = \sum_{r=0}^{n} \binom{n}{r} X^r Y^{n-r}$.)

   (d) The conditional density of $n_1 = a$ and $n_2 = (s - a)$ given $n = (n_1+n_2) = s$ can be obtained by combining the results in (b) and (c). Recall that a conditional density is obtained by dividing the joint density by a marginal density. Show that the conditional density for $n_1 = a$, $n_2 = (s - a)$ given $(n_1 + n_2) = s$ is given by

$$\binom{s}{a} \left(\frac{F_1}{F}\right)^a \left(\frac{F_2}{F}\right)^{(s-a)}.$$

   What is the density called?

   (e) Generalize the result obtained in (d) to an $(r \times c)$ contingency table by writing an expression for the conditional density of the n_{ij} given the total $n = \sum_{i=1}^{r} \sum_{j=1}^{c} n_{ij}$ is fixed. Show that the resulting density is the multinomial given in Chapter 6. Begin by writing the joint density for the n_{ij} as a product of Poisson densities, and repeat the steps in (a) through (d). (HINT: Use the result from (b) and (c) that the sum of independent Poisson random variables is also Poisson.)

2. (a) Assume that in a 2×2 contingency table the cell frequencies n_{ij}, $i = 1, 2$, $j = 1, 2$, satisfy the multinomial with fixed $n = (n_{11} + n_{12} + n_{21} + n_{22})$. Show that the density is given by

$$f(n_{11}, n_{12}, n_{21}, n_{22}) = \frac{n!}{n_{11}! n_{12}! n_{21}! n_{22}!} \; f_{11}^{n_{11}} f_{12}^{n_{12}} f_{21}^{n_{21}} f_{22}^{n_{22}}$$

   where f_{11}, f_{12}, f_{21} and f_{22} are the theoretical probabilities for the four cells respectively.

(b) Assume that the 2×2 contingency table is collapsed over the columns and let $n_1. = (n_{11} + n_{12})$ and $n_2. = (n_{21} + n_{22})$. Show that the multinomial density for $(n_1., n_2.)$ for fixed $n = (n_1. + n_2.)$ is given by

$$f(n_1., n_2.) = \frac{n!}{n_1.! n_2.!} f_1.^{n_1.} f_2.^{n_2.}$$

where $f_1. = (f_{11} + f_{12})$ and $f_2. = (f_{21} + f_{22})$.

(c) Use the marginal density in (b) and the joint density in (a) to show that the conditional density for $(n_{11}, n_{12}, n_{21}, n_{22})$ given $(n_1., n_2.)$ is given by the product multinomial

$$f(n_{11}, n_{12}, n_{21}, n_{22}/n_1., n_2.) =$$
$$\frac{n_1.! n_2.!}{n_{11}! n_{12}! n_{21}! n_{22}!} \left(\frac{f_{11}}{f_1.}\right)^{n_{11}} \left(\frac{f_{12}}{f_1.}\right)^{n_{12}} \left(\frac{f_{21}}{f_{.2}}\right)^{n_{21}} \left(\frac{f_{22}}{f_{.2}}\right)^{n_{22}}.$$

(HINT: Joint density = marginal density × conditional density.)

(d) Generalize the result in (c) to an $(r \times c)$ contingency table and give the multinomial density for the n_{ij}, the marginal density for the $n_i.$ and the conditional density for the n_{ij} given the $n_i..$ Check that the latter density is the product multinomial given in Chapter 6.

3. (a) Show that the multinomial density for the $(r \times c)$ contingency table given in Section 6.2.2 can be written as

$$f(n_{11}, n_{12}, \ldots, n_{ij}, \ldots, n_{rc}) = \frac{n!}{\prod\limits_{i=1}^{r} \prod\limits_{j=1}^{c} n_{ij}!} \prod_{i=1}^{r} \prod_{j=1}^{c} e^{n_{ij} \ln f_{ij}}.$$

(HINT: $a = e^{\ln a}$.)

(b) Show that the logarithm of the density in (a) is given by

$$\ln f = \ln \left[\frac{n!}{\prod\limits_{i=1}^{r} \prod\limits_{j=1}^{c} n_{ij}!} \right] + \sum_{i=1}^{r} \sum_{j=1}^{c} n_{ij} \ln f_{ij}.$$

(c) Given that the logarithm of the density is equivalent to the logarithm of the likelihood as given in (b) show that the maximum likelihood estimator for f_{ij} is n_{ij}/n. Use a Lagrange multiplier and the condition $\sum_{i=1}^{r} \sum_{j=1}^{c} f_{ij} = 1$.

(d) Show that the logarithm of the likelihood function in (c) evaluated at $f_{ij} = n_{ij}/n$ is given by

$$\ln L = \ln \left[\frac{n!}{\prod\limits_{i=1}^{r} \prod\limits_{j=1}^{c} n_{ij}!} \right] + \sum_{i=1}^{r} \sum_{j=1}^{c} n_{ij} \ln[n_{ij}/n].$$

(e) The maximum likelihood estimator of a function of parameters can be obtained by evaluating the function with the parameters replaced by their maximum likelihood estimators. Given that the maximum likelihood estimator of f_{ij} obtained in (c) is given by n_{ij}/n show that the maximum likelihood estimators of $f_{i\cdot}$ and $f_{\cdot j}$ are given by $n_{i\cdot}/n$ and $n_{\cdot j}/n$.

(f) Show that the maximum likelihood function evaluated under the independence model assumption $(f_{ij} = f_{i\cdot} f_{\cdot j})$ is given by

$$\ln L = \ln \left[\frac{n!}{\prod\limits_{i=1}^{r} \prod\limits_{j=1}^{c} n_{ij}!} \right] + \sum_{i=1}^{r} \sum_{j=1}^{c} n_{ij} \ln[n_{i\cdot} n_{\cdot j}/n^2].$$

(g) The likelihood ratio test for the independence model compares the likelihood in (f) with the likelihood in (d). In large samples the logarithm of the likelihood ratio multiplied by (-2) has a χ^2 distribution. Show that the likelihood ratio statistic in this case is given by $2\sum_{i=1}^{r}\sum_{j=1}^{c} n_{ij} \ln[(n n_{ij})/n_{i\cdot} n_{\cdot j}]$.

(h) The number of degrees of freedom for the χ^2 distribution is $(q - p - 1)$, where q is the number of cells in the table and p is the number of independent parameters estimated in the fitted model. Show that $(q - p - 1) = (r - 1)(c - 1)$ for the χ^2 test in (g).

4. (a) Show that the Poisson joint density for the $r \times c$ contingency table given in Section 6.2.2 can be written as

$$f(n_{11}, n_{12}, \ldots, n_{ij}, \ldots, n_{rc}) = \prod_{i=1}^{r} \prod_{j=1}^{c} [e^{-F_{ij}}/n_{ij}!][e^{n_{ij} \ln F_{ij}}].$$

(HINT: $a = e^{\ln a}$.)

(b) Given that the logarithm of the joint density is the logarithm of the likelihood show that the log likelihood for (a) is given by

$$
\begin{aligned}
\ln L &= \sum_{i=1}^{r}\sum_{j=1}^{c}[-\ln n_{ij}! - F_{ij}] + \sum_{i=1}^{r}\sum_{j=1}^{c} n_{ij}\ln F_{ij} \\
&= \sum_{i=1}^{r}\sum_{j=1}^{c} n_{ij}\ln F_{ij} - \sum_{i=1}^{r}\sum_{j=1}^{c}\ln n_{ij}! - \sum_{i=1}^{r}\sum_{j=1}^{c} F_{ij}.
\end{aligned}
$$

(c) Show using (b) that the maximum likelihood estimator of F_{ij} is n_{ij}.

(d) Show that the logarithm of the likelihood function in (b) evaluated at $F_{ij} = n_{ij}$ is given by

$$
\ln L = \sum_{i=1}^{r}\sum_{j=1}^{c} n_{ij}\ln n_{ij} - \sum_{i=1}^{r}\sum_{j=1}^{c}\ln n_{ij}! - n.
$$

(e) The maximum likelihood estimator of a function of parameters can be obtained by evaluating the function with the parameters replaced by their maximum likelihood estimators. Given that the maximum likelihood estimator of F_{ij} obtained in (c) is given by n_{ij} show that the maximum likelihood estimators for $F_{i.}$ and $F_{.j}$ are given by $n_{i.}$ and $n_{.j}$, and also that $F_{..}$ is estimated by n.

(f) Show that the maximum likelihood function evaluated under the independence model assumption is given by

$$
\ln L = \sum_{i=1}^{r}\sum_{j=1}^{c} n_{ij}\ln\left[\frac{n_{i.}n_{.j}}{n}\right] - \sum_{i=1}^{r}\sum_{j=1}^{c}\ln n_{ij}! - n.
$$

(g) The likelihood ratio test for the independence model compares the likelihood in (f) with the likelihood in (d). In large samples the logarithm of the likelihood ratio multiplied by (-2) has a χ^2 distribution. Show that this likelihood ratio statistic is given by

$$
2\sum_{i=1}^{r}\sum_{j=1}^{c} n_{ij}\ln\left[\frac{nn_{ij}}{n_{i.}n_{.j}}\right].
$$

(h) The number of degrees of freedom for the χ^2 distribution is $(q - p - 1)$, where p is the number of independent parameters estimated in the fitted model and q is the number of cells in the table. Show that $(q - p - 1) = (r - 1)(c - 1)$ for the χ^2 test in (g). Note that in this case the total sample size n is not fixed.

5. In a two-dimensional contingency table denote the true cell frequency by F_{ij}, $i = 1, 2, \ldots, r$, $j = 1, 2, \ldots, c$.

(a) Let $\widetilde{F}..$ denote the geometric mean of the rc cell frequencies and show that

$$\ln \widetilde{F}.. = \frac{1}{rc}\sum_{i=1}^{r}\sum_{j=1}^{c}\ln F_{ij}.$$

(b) Let $\widetilde{F}_{i.}$ and $\widetilde{F}._{j}$ denote the geometric means of the cell frequencies in row i and column j respectively and show that

$$\ln \widetilde{F}_{i.} = \frac{1}{c}\sum_{j=1}^{c}\ln F_{ij}$$

$$\ln \widetilde{F}._{j} = \frac{1}{r}\sum_{i=1}^{r}\ln F_{ij}.$$

(c) Let $F_{i.}$ and $F._{j}$ denote the row and column total frequencies and denote the geometric means of the row totals and column totals respectively by $\widetilde{F}_{0.}$ and $\widetilde{F}._{0}$. Show that

$$\ln \widetilde{F}_{0.} = \frac{1}{r}\sum_{i=1}^{r}\ln F_{i.}$$

$$\ln \widetilde{F}._{0} = \frac{1}{c}\sum_{j=1}^{c}\ln F._{j}.$$

(d) Given that the independence model can be written as $F_{ij} = (F_{i.})(F._{j})/n$ show that

$$\ln F_{ij} = \ln F_{i.} + \ln F._{j} - \ln n \qquad (1)$$

and that after summing over i and j and dividing by rc

$$\ln \widetilde{F}.. = \ln \widetilde{F}._{0} + \ln \widetilde{F}_{0.} - \ln n; \qquad (2)$$

hence show that

$$\widetilde{F}.. = \widetilde{F}_{0.}\widetilde{F}._{0}/n.$$

(e) Show that by summing over the subscript j and dividing by c in equation (1) yields

$$\ln \widetilde{F}_{i.} = \ln F_{i.} + \ln \widetilde{F}._{0} - \ln n \qquad (3)$$

and similarly by summing over the subscript i and dividing by r in equation (1) yields

$$\ln \widetilde{F}._{j} = \ln \widetilde{F}_{0.} + \ln F._{j} - \ln n. \qquad (4)$$

(f) Combine equations (2), (3) and (4) to show that

$$[\ln F_{i.} + \ln F_{.j}] = \ln \widetilde{F}_{i.} + \ln \widetilde{F}_{.j} - \ln \widetilde{F}_{..} + \ln n.$$

and then use the independence result (1) to obtain

$$\ln F_{ij} = \ln \widetilde{F}_{i.} + \ln \widetilde{F}_{.j} - \ln \widetilde{F}_{..}$$

and hence

$$F_{ij} = \widetilde{F}_{i.} \widetilde{F}_{.j} / \widetilde{F}_{..}$$

(g) Use the result in (f) to show that under independence

$$F_{ij} = \widetilde{F}_{..} \left[\frac{\widetilde{F}_{i.}}{\widetilde{F}_{..}}\right]\left[\frac{\widetilde{F}_{.j}}{\widetilde{F}_{..}}\right],$$

and give a verbal description of the three terms on the right hand side.

(h) If independence does not hold show that the following equation holds

$$\ln F_{ij} \quad = \quad [\ln \widetilde{F}_{..}] + [\ln \widetilde{F}_{i.} - \ln \widetilde{F}_{..}] + [\ln \widetilde{F}_{.j} - \ln \widetilde{F}_{..}]$$
$$+ [\ln F_{ij} - \ln \widetilde{F}_{i.} - \ln \widetilde{F}_{.j} + \ln \widetilde{F}_{..}]$$

and provide an interpretation for the last term.

6. (a) Construct an example of Simpson's Paradox with different numbers and different variables than the examples in Chapter 6.

   (b) For the contingency tables shown below derive a set of conditions that would represent Simpson's Paradox.

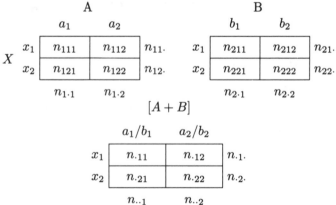

   (c) Use your result to explain why Simpson's Paradox occurs.

7. For the three-dimensional contingency table the observed cell frequencies are denoted by n_{ijk} and the theoretical cell frequencies are denoted by F_{ijk}, $i = 1, 2, \ldots, r$, $j = 1, 2, \ldots, c$, $k = 1, 2, \ldots, \ell$. If the maximum likelihood estimators of the F_{ijk} are given by the n_{ijk} show that the maximum likelihood estimators for the functions

(a) $F_{ijk} = F_{i..}F_{.j.}F_{..k}/n^2$,

(b) $F_{ijk} = F_{ij.}F_{..k}/n$,

(c) $F_{ijk} = F_{i.k}F_{.jk}/F_{..k}$,

are given respectively by

(d) $E_{ijk} = n_{i..}n_{.j.}n_{..k}/n^2$,

(e) $E_{ijk} = n_{ij.}n_{..k}/n$,

(f) $E_{ijk} = n_{i.k}n_{.jk}/n_{..k}$.

(HINT: Use the fact that maximum likelihood estimators of functions of parameters are the functions of the maximum likelihood estimators of the parameters.)

8. The saturated loglinear model for a three-dimensional contingency table is given by

$$
\begin{aligned}
\ln F_{ijk} \;=\; & \mu + \mu_{1(i)} + \mu_{2(j)} + \mu_{3(k)} + \mu_{12(ij)} + \mu_{13(ik)} \\
& + \mu_{23(jk)} + \mu_{123(ijk)}, \\
& i = 1, 2, \ldots, r, \quad j = 1, 2, \ldots, c, \quad k = 1, 2, \ldots, \ell,
\end{aligned}
$$

where F_{ijk} = true frequency in cell (i, j, k) and

$$
\mu = \frac{1}{rc\ell} \sum_{i=1}^{r} \sum_{j=1}^{c} \sum_{k=1}^{\ell} \ln F_{ijk},
$$

$$
\mu_{1(i)} = \frac{1}{c\ell} \sum_{j=1}^{c} \sum_{k=1}^{\ell} \ln F_{ijk} - \mu,
$$

$$
\mu_{2(j)} = \frac{1}{r\ell} \sum_{i=1}^{r} \sum_{k=1}^{\ell} \ln F_{ijk} - \mu,
$$

$$
\mu_{3(k)} = \frac{1}{rc} \sum_{i=1}^{r} \sum_{j=1}^{c} \ln F_{ijk} - \mu,
$$

$$
\mu_{12(ij)} = \frac{1}{\ell} \sum_{k=1}^{\ell} \ln F_{ijk} - \mu_{1(i)} - \mu_{2(j)} - \mu,
$$

$$
\mu_{13(ik)} = \frac{1}{c} \sum_{j=1}^{c} \ln F_{ijk} - \mu_{1(i)} - \mu_{3(k)} - \mu,
$$

$$\mu_{23(jk)} \;=\; \frac{1}{r}\sum_{i=1}^{r} \ln F_{ijk} - \mu_{2(j)} - \mu_{3(k)} - \mu,$$

$$\mu_{123(ijk)} \;=\; \ln F_{ijk} - \mu_{1(i)} - \mu_{2(j)} - \mu_{3(k)} - \mu_{12(ij)},$$

$$-\mu_{23(jk)} - \mu_{13(ik)} - \mu.$$

(a) Show that the properties below follow from these definitions

$$\sum_{i=1}^{r} \mu_{1(i)} = \sum_{j=1}^{c} \mu_{2(j)} = \sum_{k=1}^{\ell} \mu_{3(k)} = 0,$$

$$\sum_{i=1}^{r}\sum_{j=1}^{c} \mu_{12(ij)} = \sum_{i=1}^{r}\sum_{k=1}^{\ell} \mu_{13(ik)} = \sum_{j=1}^{c}\sum_{k=1}^{\ell} \mu_{23(jk)} = 0,$$

$$\sum_{i=1}^{r}\sum_{j=1}^{c}\sum_{k=1}^{\ell} \mu_{123(ijk)} = 0.$$

(b) Given the following notation for various geometric means based on the F_{ijk} determine expressions for the logarithms of these quantities in terms of summations of $\ln F_{ijk}$.

$\widetilde{F}_{...}$ is the overall geometric mean of all the frequencies F_{ijk};

$\widetilde{F}_{i..}$ is the geometric mean of all the frequencies F_{ijk} holding i fixed;

$\widetilde{F}_{.j.}$ is the geometric mean of all the frequencies F_{ijk} holding j fixed;

$\widetilde{F}_{..k}$ is the geometric mean of all the frequencies F_{ijk} holding k fixed;

$\widetilde{F}_{ij.}$ is the geometric mean of all the frequencies F_{ijk} holding i, j fixed;

$\widetilde{F}_{.jk}$ is the geometric mean of all the frequencies F_{ijk} holding j, k fixed;

$\widetilde{F}_{i.k}$ is the geometric mean of all the frequencies F_{ijk} holding i, k fixed;

(c) Use the expressions derived in (b) and the parameter definitions to show that the following expressions hold.

$$\mu \;=\; \ln \widetilde{F}_{...},$$

$$\mu_{1(i)} \;=\; \ln \widetilde{F}_{i..} - \ln \widetilde{F}_{...},$$

$$\mu_{2(j)} \;=\; \ln \widetilde{F}_{.j.} - \ln \widetilde{F}_{...}$$

$$\mu_{3(k)} \;=\; \ln \widetilde{F}_{..k} - \ln \widetilde{F}_{...}$$

$$\mu_{12(ij)} \;=\; \ln \widetilde{F}_{ij.} - \ln \widetilde{F}_{i..} - \ln \widetilde{F}_{.j.} + \ln \widetilde{F}_{...}$$

$$\mu_{13(ik)} \;=\; \ln \widetilde{F}_{i.k} - \ln \widetilde{F}_{i..} - \ln \widetilde{F}_{..k} + \ln \widetilde{F}_{...}$$

$$\mu_{23(jk)} \;=\; \ln \widetilde{F}_{\cdot jk} - \ln \widetilde{F}_{\cdot j\cdot} - \ln \widetilde{F}_{\cdot\cdot k} + \ln \widetilde{F}_{\cdots}$$

$$\mu_{123(ijk)} \;=\; \ln \widetilde{F}_{ijk} - \ln \widetilde{F}_{ij\cdot} - \ln \widetilde{F}_{\cdot jk} - \ln \widetilde{F}_{i\cdot k}$$
$$+ \ln \widetilde{F}_{i\cdot\cdot} + \ln \widetilde{F}_{\cdot j\cdot} + \ln \widetilde{F}_{\cdot\cdot k} - \ln \widetilde{F}_{\cdots}$$

9. (a) Using the saturated loglinear model for a three-dimensional table derive the expression for the logit model assuming one of the variables is a response variable (see Section 6.3.5). Assume that the response variable has only two categories.

   (b) Repeat the exercise in (a) assuming the response variable has three categories. Obtain expressions for $\ln(p_i/p_j)$ for all possible pairs.

   (c) Repeat the exercise in (b) using continuation ratios for the logit model. (HINT: Sum the frequencies over two of the three categories and then repeat the steps in (a)).

   (d) Assume that the logit model derived in (a) pertains to the categories Present and Absent. Use the model to derive an expression for the probability of Present.

10. (a) In the weighted least squares approach to fitting contingency tables the covariance matrix for the sample proportions is block diagonal as outlined in Section 6.4.1. Use the expressions for the covariances among the elements of $\hat{\mathbf{p}}$ given in Section 6.4.1 to show that the covariance matrix of $\hat{\mathbf{p}}$ given by $\boldsymbol{\Omega}$ in Section 6.4.1 is block diagonal.

   (b) Given that $\hat{\mathbf{g}} = \mathbf{A}\hat{\mathbf{p}}$ as in Section 6.4.1 show that the covariance matrix for the elements of $\hat{\mathbf{g}}$ is given by $\mathbf{A}\boldsymbol{\Omega}\mathbf{A}'$, where $\boldsymbol{\Omega}$ is given in (a).

   (c) Review your knowledge of the weighted least squares estimator in linear models and discuss the rationale for the weighted least squares estimator given by (6.2).

   (d) Assume $\widehat{\mathbf{H}} = \mathbf{H}$ and show that the covariance matrix for $\hat{\boldsymbol{\beta}}$ given by (6.2) is $(\mathbf{X}'\widehat{\mathbf{H}}^{-1}\mathbf{X})^{-1}$.

   (e) Use the result in (d) to show that the covariance matrix for $\mathbf{C}\hat{\boldsymbol{\beta}}$ is given by
$$[\mathbf{C}(\mathbf{X}'\widehat{\mathbf{H}}^{-1}\mathbf{X})^{-1}\mathbf{C}'].$$

   (f) Show that if $\hat{\boldsymbol{\beta}} = (\mathbf{X}'\widehat{\mathbf{H}}^{-1}\mathbf{X})^{-1}(\mathbf{X}'\widehat{\mathbf{H}}^{-1}\hat{\mathbf{g}})$ minimizes the quadratic form $(\hat{\mathbf{g}}-\mathbf{X}\boldsymbol{\beta})'\widehat{\mathbf{H}}^{-1}(\hat{\mathbf{g}}-\mathbf{X}\boldsymbol{\beta})$ with respect to $\boldsymbol{\beta}$ then the quadratic form has the value $\hat{\mathbf{g}}'\widehat{\mathbf{H}}^{-1}\hat{\mathbf{g}} - \hat{\boldsymbol{\beta}}'(\mathbf{X}'\widehat{\mathbf{H}}^{-1}\mathbf{X})\hat{\boldsymbol{\beta}}$.

7

Multivariate Distributions, Inference, Regression and Canonical Correlation

Before we introduce additional techniques for multivariate analysis, it is necessary to explain notation for multivariate random variables and samples. Since many multivariate inference procedures require a multivariate normal distribution assumption, an introduction to this distribution is also provided here. In addition, the chapter includes an outline of inference procedures for the mean vector and covariance matrix. In some applications multivariate random variables are partitioned into two or more subsets. The relationship between the variables in different sets is often of interest. In the last section of this chapter we outline the techniques of multivariate regression and canonical correlation in order to study the relationships between subsets of random variables.

7.1 Multivariate Random Variables and Samples

The $(n \times p)$ data matrix $\mathbf{X}$ is viewed as a sample of n observations on each of the p random variables $X_1, X_2, \ldots, X_p$. The $\mathbf{X}$ matrix therefore contains the p $(n \times 1)$ observation vectors $\mathbf{x}_1, \mathbf{x}_2, \ldots, \mathbf{x}_p$, where

$$\mathbf{x}_j = \begin{bmatrix} x_{1j} \\ x_{2j} \\ \vdots \\ x_{nj} \end{bmatrix}, \quad j = 1, 2, \ldots, p, \text{ and } \mathbf{X} = [\mathbf{x}_1, \ \mathbf{x}_2, \ \ldots, \ \mathbf{x}_p].$$

Thus each column of $\mathbf{X}$ is a $(n \times 1)$ vector of observations on one of the p variables.

Each row of $\mathbf{X}$ contains observations on the p variables, $X_1, X_2, \ldots, X_p$, corresponding to a particular individual or object. The p random variables together form a $(p \times 1)$ *vector random variable* $\mathbf{x}$, where

$$\mathbf{x} = \begin{bmatrix} X_1 \\ X_2 \\ \vdots \\ X_p \end{bmatrix}.$$

The $\mathbf{X}$ matrix therefore consists of n observations on the *multivariate random variable* $\mathbf{x}$, denoted by $\mathbf{x}_1, \mathbf{x}_2, \ldots, \mathbf{x}_n$ and hence

$$\mathbf{X} = \begin{bmatrix} \mathbf{x}_1' \\ \mathbf{x}_2' \\ \vdots \\ \mathbf{x}_n' \end{bmatrix}.$$

The vector $\mathbf{x}_i$ therefore will be used to denote an $(n \times 1)$ vector of observations on the variable X_i or a $(p \times 1)$ vector of observations on the variables $X_1, X_2, \ldots, X_p$ for individual i. The choice between these two possibilities will usually be clear from the context. An outline of properties for multivariate random variables and multivariate samples is provided in the next section.

Examples

Two examples of data matrices are shown in Tables 7.1 and 7.2. Table 7.1 contains 50 weekly return observations (in percents) on each of ten stock portfolios. The portfolios were constructed from stocks on the Toronto Stock Exchange from 1982. Each portfolio is an equal weight average of 50 stocks. For all practical purposes, weekly stock returns can be assumed to be independent over time. Table 7.2 contains the responses of 50 police officers to eight questions regarding the stress they felt in various work situations. The stress is measured on a five-point scale with the value 1 indicating low stress and the value 5 indicating high stress. These two data matrices will be used in this chapter to illustrate various types of analysis.

7.1.1   MULTIVARIATE DISTRIBUTIONS AND MULTIVARIATE RANDOM VARIABLES

Joint Distribution

The *joint distribution function* for the $(p \times 1)$ vector random variable $\mathbf{x}$ is denoted by $F_{\mathbf{x}}(\mathbf{x}^*)$, where

$$\begin{aligned} F_{\mathbf{x}}(\mathbf{x}^*) = F_{\mathbf{x}}(x_1^*, x_2^*, \ldots, x_p^*) &= P[\mathbf{x} \le \mathbf{x}^*] \\ &= P[X_1 \le x_1^*, \ X_2 \le x_2^*, \ldots, X_p \le x_p^*]. \end{aligned}$$

The *joint density function* for $\mathbf{x}$ is denoted by

TABLE 7.1. Portfolio Returns

X_1	X_2	X_3	X_4	X_5	X_6	X_7	X_8	X_9	X_{10}
1.99	3.00	3.55	6.94	6.51	9.91	12.71	8.57	9.27	7.67
4.46	6.57	9.28	13.68	15.00	18.35	19.23	18.93	17.13	16.88
1.78	2.76	3.63	9.30	15.19	11.67	20.20	20.59	22.49	30.52
−0.16	−1.86	−0.34	−1.66	1.89	3.72	3.27	3.38	8.60	3.27
−0.35	−0.44	1.33	1.34	3.89	2.91	7.37	8.19	11.57	41.35
1.52	1.12	−0.22	1.24	7.32	5.59	0.96	−0.12	0.17	6.98
2.13	2.75	1.19	3.00	0.28	−0.44	1.55	−1.57	0.65	5.62
1.29	1.77	2.75	1.98	−0.14	−2.10	−2.42	2.09	−1.01	−0.02
1.31	−0.64	−2.97	−4.84	−6.54	−5.16	−5.87	−4.49	−5.43	−5.04
1.83	1.21	−0.93	−0.42	1.93	1.71	1.80	1.15	4.11	10.89
1.15	0.67	−0.93	0.55	−1.54	−1.90	−2.09	−0.30	3.01	176.80
−0.48	−0.02	−3.10	−1.77	−0.52	−2.14	−4.05	−5.60	−3.50	−2.77
0.41	1.28	0.01	2.84	0.39	0.34	0.01	−0.72	1.93	−1.72
1.51	0.20	1.64	0.52	1.81	−0.31	1.63	2.63	4.78	−0.74
0.11	0.45	1.97	−0.38	0.54	0.15	2.77	−1.85	0.16	−2.01
3.24	3.32	4.16	2.10	1.74	5.20	1.13	5.48	2.38	2.31
2.25	2.22	0.83	5.38	0.29	−1.34	−0.78	4.74	−3.04	0.28
1.94	0.94	0.74	−0.83	−2.21	3.29	2.57	−0.21	−2.69	5.54
2.77	2.98	4.76	4.32	8.66	6.15	8.24	11.14	7.67	7.11
1.70	2.79	3.68	5.52	3.94	0.38	4.34	2.72	5.64	8.37
−0.14	−0.80	−0.56	1.55	0.27	0.56	1.42	−2.46	−0.67	0.89
1.21	2.99	1.73	2.54	4.12	4.17	−1.17	4.68	1.59	1.63
0.35	−1.59	−1.31	0.75	0.57	1.15	1.19	0.43	−0.62	0.07
−0.27	−0.90	1.15	0.89	−0.39	1.81	−1.19	−1.23	1.68	−2.46
0.84	3.25	1.41	0.76	0.64	−0.12	1.07	0.49	1.37	1.46
1.52	1.48	1.81	−1.20	−2.88	−1.79	−4.56	−3.02	−4.81	−5.11
−0.31	−0.17	−0.32	−0.54	1.18	−0.26	−3.02	−2.56	−0.67	−0.18
2.16	1.39	2.51	−0.08	1.52	−0.32	−0.41	−0.59	2.27	24.86
−0.67	0.54	−0.66	−0.07	−1.31	−1.10	−0.29	3.66	−2.75	−2.54
1.02	2.53	2.12	2.40	0.75	4.21	0.78	−0.65	−1.51	1.19
0.10	0.04	−0.30	−0.54	−1.11	−2.50	2.13	6.39	1.16	1.01
−0.16	−1.33	−2.58	−2.12	0.37	−2.45	−0.77	−3.79	−3.77	−2.85
−0.21	−0.33	1.24	−0.16	−2.23	−3.74	1.79	−1.03	2.09	−0.37
0.93	0.17	−0.18	0.60	4.35	0.96	−0.25	−1.32	1.36	−2.31
0.73	0.78	0.31	0.76	−1.15	3.61	0.78	−2.15	−0.16	−1.42
1.00	−0.24	−0.01	2.43	4.46	−0.82	4.71	2.41	4.67	3.54
2.13	1.69	1.89	−0.78	−3.34	1.30	−1.35	−1.20	−0.20	−0.74
−0.73	−1.75	−1.37	−0.74	−3.90	−5.26	−5.53	−2.55	−5.19	−0.59
2.12	0.83	−0.43	−0.52	0.34	−1.60	−3.25	−4.34	−5.17	−1.76
−0.54	−2.58	−2.68	−3.22	−3.32	−1.80	−2.95	−3.32	−4.10	−2.34
0.25	0.01	−1.02	−2.55	0.17	1.76	1.93	1.14	−0.11	−4.09
−0.17	−0.53	0.05	0.35	−3.29	−0.23	−2.07	−0.11	−5.92	1.62
−0.65	−2.12	−1.54	−3.30	−3.90	−4.67	−4.93	−3.26	−3.85	−4.06
−1.16	−1.76	−2.66	−1.70	−3.80	−6.79	−4.75	−1.45	−4.63	−3.40
0.99	0.50	−1.17	−2.30	1.01	−1.28	−2.14	−3.00	−2.50	16.03
1.36	2.37	0.50	0.59	4.43	−1.96	1.57	1.91	1.01	1.36
0.93	0.04	−1.09	−0.49	−2.20	2.31	8.88	−0.89	−0.82	0.85
1.30	−0.28	0.33	−2.84	−2.31	−3.43	−2.41	−3.96	−1.88	−2.63
1.82	2.00	5.00	4.42	3.30	2.45	4.50	10.86	4.57	−3.31
0.31	−0.28	−0.50	−3.15	1.72	−1.43	−1.86	−4.43	−3.05	−4.87

TABLE 7.2. Police Officer Stress Data

X_1	X_2	X_3	X_4	X_5	X_6	X_7	X_8	X_1	X_2	X_3	X_4	X_5	X_6	X_7	X_8
2	1	3	3	4	3	3	4	4	3	3	4	4	3	5	4
4	2	3	4	3	5	5	5	3	2	2	4	4	3	1	2
1	1	2	4	5	4	4	1	3	3	4	4	4	4	5	3
1	3	3	4	4	5	3	2	3	5	3	3	2	2	3	2
5	4	1	5	5	5	4	4	3	4	3	4	5	5	5	4
2	1	5	5	5	5	4	5	1	1	2	3	4	3	5	2
2	3	1	3	3	4	3	3	1	1	1	3	3	3	2	1
1	2	1	4	5	5	4	3	4	5	1	4	5	4	5	2
1	1	1	2	4	4	4	1	3	2	3	1	3	3	4	3
2	2	2	3	3	2	3	2	1	3	2	4	1	3	4	4
1	2	2	2	3	2	2	2	4	1	4	4	3	4	3	3
1	3	1	4	4	5	3	1	3	3	5	4	5	5	5	4
1	2	3	1	2	4	2	1	1	5	1	2	2	1	3	3
1	2	3	2	3	3	1	1	1	1	1	2	1	3	5	1
3	2	2	4	3	4	3	5	3	5	1	4	3	4	4	2
2	1	3	3	3	4	4	1	3	1	1	3	1	4	3	3
4	3	4	5	4	4	4	3	2	3	1	5	5	1	3	4
3	2	2	4	5	4	4	3	4	4	1	4	4	3	4	4
1	2	3	3	3	4	3	3	1	1	2	3	3	4	4	2
2	2	2	2	3	2	2	2	2	2	3	4	3	4	4	2
2	3	3	4	4	3	3	2	3	5	3	4	4	4	4	3
1	2	3	4	2	1	3	3	1	3	2	4	2	2	3	3
3	1	1	5	4	4	3	2	1	1	4	5	3	5	4	3
1	1	2	4	4	4	3	3	1	1	2	3	1	3	5	1
2	1	1	4	2	3	1	1	1	3	3	4	4	4	4	3

Variable Descriptions

X_1 Handling an investigation where there is serious injury or fatality.
X_2 Dealing with obnoxious or intoxicated people.
X_3 Tolerating verbal abuse in public.
X_4 Being unable to solve a continuing series of serious offences.
X_5 Resources such as doctors, ambulances etc. not being available when needed.
X_6 Poor presentation of a case by the prosecutor leading to dismissal of charge.
X_7 Unit members not getting along with unit commander.
X_8 Investigating domestic quarrels.

$$f_{\mathbf{X}}(\mathbf{x}^*) = f_{\mathbf{X}}(x_1^*, x_2^*, \ldots, x_p^*),$$

where

$$F_{\mathbf{X}}(\mathbf{x}^*) = \int_{-\infty}^{x_1^*} \int_{-\infty}^{x_2^*} \ldots \int_{-\infty}^{x_p^*} f_{\mathbf{X}}(\mathbf{x}^*) dx_1 dx_2 \ldots dx_p.$$

Partitioning the Random Variable

The random variable **x** can be partitioned into two mutually exclusive subsets, where

$$\mathbf{x} = \begin{bmatrix} \mathbf{x}_1 \\ \mathbf{x}_2 \end{bmatrix}, \quad \mathbf{x}_1 \text{ is } (q \times 1), \ \mathbf{x}_2 \text{ is } (s \times 1) \text{ and } p = (q + s).$$

Thus $\mathbf{x}_1$ and $\mathbf{x}_2$ are also vector random variables but of lower dimension than $\mathbf{x}$. The joint distribution function $F_{\mathbf{x}_1}(\mathbf{x}_1^*)$ for $\mathbf{x}_1$ can be obtained from $F_{\mathbf{x}}(\mathbf{x}^*)$ by integrating the joint density $f_{\mathbf{x}}(\mathbf{x}^*)$ over the entire range of the variables in $\mathbf{x}_2$. Denoting the joint density by $f_{\mathbf{x}}(\mathbf{x}_1, \mathbf{x}_2)$, the joint distribution function $F_{\mathbf{x}_1}(\mathbf{x}_1^*)$ is given by

$$
\begin{aligned}
F_{\mathbf{x}_1}(\mathbf{x}_1^*) &= \int_{-\infty}^{x_1^*} \cdots \int_{-\infty}^{x_q^*} \int_{-\infty}^{\infty} \cdots \int_{-\infty}^{\infty} f_{\mathbf{x}}(\mathbf{x}_1, \mathbf{x}_2) dx_1 dx_2 \ldots dx_p, \\
&= \int_{-\infty}^{x_1^*} \cdots \int_{-\infty}^{x_q^*} f_{\mathbf{x}_1}(\mathbf{x}_1) dx_1 dx_2 \ldots dx_q,
\end{aligned}
$$

where $f_{\mathbf{x}_1}(\mathbf{x}_1)$ is the joint density of $\mathbf{x}_1$. The joint density for $\mathbf{x}_1$ is obtained from the joint density for $\mathbf{x}$ by integrating $f_{\mathbf{x}}(\mathbf{x})$ over the range of the variables in $\mathbf{x}_2$.

$$f_{\mathbf{x}_1}(x_1, x_2, \ldots, x_q) = f_{\mathbf{x}_1}(\mathbf{x}_1) = \int_{-\infty}^{-\infty} \cdots \int_{-\infty}^{\infty} f_{\mathbf{x}}(\mathbf{x}_1, \mathbf{x}_2) dx_{q+1} \ldots dx_p.$$

A special case of the distribution for $\mathbf{x}_1$ occurs when $q = 1$. In this case $\mathbf{x}_1$ is equivalent to the scalar random variable X_1 and the distribution is called the marginal distribution of X_1.

Conditional Distributions and Independence

The *conditional distribution* for $\mathbf{x}_2$ given $\mathbf{x}_1$ is obtained from $f_{\mathbf{x}}(\mathbf{x})$ by determining

$$f_{\mathbf{x}_2|\mathbf{x}_1}(\mathbf{x}_2 \mid \mathbf{x}_1 = \mathbf{x}_1^*) = f_{\mathbf{x}}(\mathbf{x}_1^*, \mathbf{x}_2)/f_{\mathbf{x}_1}(\mathbf{x}_1^*)$$

where $f_{\mathbf{x}_1}(\mathbf{x}_1^*)$ is the joint density for $\mathbf{x}_1$ evaluated at $\mathbf{x}_1^*$

The two vector random variables $\mathbf{x}_1$ and $\mathbf{x}_2$ are *independent* if and only if

$$f_{\mathbf{x}_2|\mathbf{x}_1}(\mathbf{x}_2 \mid \mathbf{x}_1 = \mathbf{x}_1^*) = f_{\mathbf{x}_2}(\mathbf{x}_2) \text{ for all } \mathbf{x}_1^* \text{ and all } \mathbf{x}_2,$$

or equivalently $f_{\mathbf{x}}(\mathbf{x}) = f_{\mathbf{x}_1}(\mathbf{x}_1)f_{\mathbf{x}_2}(\mathbf{x}_2)$ for all $\mathbf{x}$. A special case of this result for bivariate independence is given by $f_{xy}(x, y) = f_x(x)f_y(y)$.

Mean Vector and Covariance Matrix

The *mean vector* $\boldsymbol{\mu}$ corresponding to the $(p \times 1)$ random variable $\mathbf{x}$ is the $(p \times 1)$ vector of elements $\mu_j = E[X_j]$, $j = 1, 2, \ldots, p$, and we write $\boldsymbol{\mu} = E[\mathbf{x}]$. The *covariance matrix* for $\mathbf{x}$ is the $(p \times p)$ matrix $\boldsymbol{\Sigma}$ with diagonal elements $\sigma_j^2 = V[X_j]$, $j = 1, 2, \ldots, p$, and off diagonal elements $\sigma_{jk} =$

$\text{Cov}(X_j, X_k)$, $j \neq k$, $j, k = 1, 2, \ldots, p$. The mean vector $\boldsymbol{\mu}$ and covariance matrix $\boldsymbol{\Sigma}$ are given by

$$
\boldsymbol{\mu} = \begin{bmatrix} \mu_1 \\ \mu_2 \\ \vdots \\ \mu_p \end{bmatrix}, \quad
\boldsymbol{\Sigma} = \begin{bmatrix}
\sigma_1^2 & \sigma_{12} & \sigma_{13} & \cdots & \sigma_{1p} \\
\sigma_{12} & \sigma_2^2 & \sigma_{23} & \cdots & \sigma_{2p} \\
\vdots & & \sigma_3^2 & \ddots & \vdots \\
\sigma_{1p} & \sigma_{2p} & \cdots & \cdots & \sigma_p^2
\end{bmatrix}.
$$

The covariance matrix $\boldsymbol{\Sigma}$ can also be expressed as

$$
E[(\mathbf{x} - \boldsymbol{\mu})(\mathbf{x} - \boldsymbol{\mu})'] = \boldsymbol{\Sigma}.
$$

Correlation Matrix

The *correlation matrix* $\boldsymbol{\rho}$ is obtained from the elements of the covariance matrix $\boldsymbol{\Sigma}$ by determining the off diagonal elements from

$$
\rho_{jk} = \sigma_{jk} / \sqrt{\sigma_j^2 \sigma_k^2}, \quad j \neq k, \ j, k = 1, 2, \ldots, p.
$$

The matrix $\boldsymbol{\rho}$ is given by

$$
\boldsymbol{\rho} = \begin{bmatrix}
1 & \rho_{12} & \rho_{13} & \cdots & \rho_{1p} \\
\rho_{12} & 1 & \rho_{23} & \cdots & \rho_{2p} \\
\vdots & & 1 & \ddots & \vdots \\
\rho_{1p} & \rho_{2p} & \cdots & \cdots & 1
\end{bmatrix}.
$$

The covariance matrix $\boldsymbol{\Sigma}$ can also be expressed as

$$
\boldsymbol{\Sigma} = \begin{bmatrix}
\sigma_1^2 & \rho_{12}\sigma_1\sigma_2 & \rho_{13}\sigma_1\sigma_3 & \cdots & \rho_{1p}\sigma_1\sigma_p \\
\rho_{12}\sigma_1\sigma_2 & \sigma_2^2 & \rho_{23}\sigma_2\sigma_3 & \cdots & \rho_{2p}\sigma_2\sigma_p \\
\vdots & & \sigma_3^2 & \ddots & \vdots \\
\rho_{1p}\sigma_1\sigma_p & \rho_{2p}\sigma_2\sigma_p & \cdots & \cdots & \sigma_p^2
\end{bmatrix}.
$$

7.1.2 MULTIVARIATE SAMPLES

The data matrix $\mathbf{X}$ represents a sample of n observations on $\mathbf{x}$ from the multivariate population and is called a *multivariate sample*.

Sample Mean Vector and Covariance Matrix

Each row of the $(n \times p)$ data matrix $\mathbf{X}$ represents an observation on the $(p \times 1)$ random vector $\mathbf{x}$. For row i the $(1 \times p)$ observation vector is denoted by $\mathbf{x}'_i$. The $(p \times 1)$ *sample mean vector* is denoted by $\bar{\mathbf{x}}$ and is defined by $\bar{\mathbf{x}} = \sum_{i=1}^{n} \mathbf{x}_i/n$. The elements of the $(p \times 1)$ vector $\bar{\mathbf{x}}$ are the individual sample means for each variable

$$\bar{\mathbf{x}} = \begin{bmatrix} \bar{x}_{\cdot 1} \\ \bar{x}_{\cdot 2} \\ \vdots \\ \bar{x}_{\cdot p} \end{bmatrix}, \quad \text{where } \bar{x}_{\cdot j} = \frac{1}{n}\sum_{i=1}^{n} x_{ij}, \ j = 1, 2, \ldots, p.$$

The *sample covariance matrix* is denoted by $\mathbf{S}$ where $\mathbf{S} = \sum_{i=1}^{n}(\mathbf{x}_i - \bar{\mathbf{x}})(\mathbf{x}_i - \bar{\mathbf{x}})'/(n-1)$. The diagonal elements are given by $s_j^2 = \sum_{i=1}^{n}(x_{ij} - \bar{x}_{\cdot j})^2/(n-1)$, $j = 1, 2, \ldots, p$, and the off-diagonal elements have the form $s_{jk} = \sum_{i=1}^{n}(x_{ij} - \bar{x}_{\cdot j})(x_{ik} - \bar{x}_{\cdot k})/(n-1)$, $j \neq k$, $j, k = 1, 2, \ldots, p$. The matrix $(n-1)\mathbf{S}$ therefore has the form

$$\begin{bmatrix} \sum_{i=1}^{n}(x_{i1} - \bar{x}_{\cdot 1})^2 & \sum_{i=1}^{n}(x_{i1} - \bar{x}_{\cdot 1})(x_{i2} - \bar{x}_{\cdot 2}) \ldots \sum_{i=1}^{n}(x_{i1} - \bar{x}_{\cdot 1})(x_{ip} - \bar{x}_{\cdot p}) \\ \sum_{i=1}^{n}(x_{i1} - \bar{x}_{\cdot 1})(x_{i2} - \bar{x}_{\cdot 2}) & \sum_{i=1}^{n}(x_{i2} - \bar{x}_{\cdot 2})^2 \quad \ldots \sum_{i=1}^{n}(x_{i2} - \bar{x}_{\cdot 2})(x_{ip} - \bar{x}_{\cdot p}) \\ \vdots & \vdots & \vdots \\ \sum_{i=1}^{n}(x_{i1} - \bar{x}_{\cdot 1})(x_{ip} - \bar{x}_{\cdot p}) \ \sum_{i=1}^{n}(x_{i2} - \bar{x}_{\cdot 2})(x_{ip} - \bar{x}_{\cdot p}) \ldots & \sum_{i=1}^{n}(x_{ip} - \bar{x}_{\cdot p})^2 \end{bmatrix}$$

which can be written as $\mathbf{X}^{*\prime}\mathbf{X}^*$ where $\mathbf{X}^*$ is the *mean-corrected* or *mean-centered* $\mathbf{X}$ matrix given by

$$\mathbf{X}^* = \begin{bmatrix} x_{11} - \bar{x}_{\cdot 1} & x_{12} - \bar{x}_{\cdot 2} & \ldots & x_{1p} - \bar{x}_{\cdot p} \\ x_{21} - \bar{x}_{\cdot 1} & x_{22} - \bar{x}_{\cdot 2} & \ldots & x_{2p} - \bar{x}_{\cdot p} \\ \vdots & & & \\ x_{n1} - \bar{x}_{\cdot 1} & x_{n2} - \bar{x}_{\cdot 2} & \ldots & x_{np} - \bar{x}_{\cdot p} \end{bmatrix}.$$

Sample Correlation Matrix

The *sample correlations* among the p variables are given by

$$r_{jk} = s_{jk}/\sqrt{s_j^2 s_k^2} = s_{jk}/s_j s_k, \quad j, k = 1, 2, \ldots, p.$$

The correlation matrix that summarizes the correlations is given by

TABLE 7.3. Portfolio Data — Mean Vector, Covariance and Correlation Matrices

| Mean Vector | \multicolumn{10}{c}{Covariance and Correlation Matrices} |
	X_1	X_2	X_3	X_4	X_5	X_6	X_7	X_8	X_9	X_{10}
0.93	1.32	1.68	1.92	2.27	2.54	2.95	2.99	3.13	2.72	3.76
0.74	0.82	3.13	3.34	4.43	4.65	5.07	5.19	5.81	4.95	4.80
0.65	0.71	0.81	5.42	6.18	6.46	7.30	8.06	9.39	8.60	2.52
0.81	0.59	0.76	0.80	10.86	11.00	11.14	13.90	14.78	13.64	12.53
1.05	0.52	0.62	0.67	0.79	17.71	14.91	18.31	18.18	19.61	12.84
0.77	0.58	0.65	0.71	0.77	0.81	19.12	19.66	18.04	18.61	10.02
1.09	0.48	0.54	0.64	0.78	0.81	0.83	28.74	25.01	26.34	17.97
1.18	0.49	0.60	0.73	0.82	0.79	0.75	0.85	29.90	26.03	22.48
1.06	0.42	0.50	0.66	0.74	0.83	0.76	0.88	0.85	30.99	41.52
6.41	0.12	0.10	0.04	0.14	0.11	0.08	0.12	0.15	0.28	686.59

$$
\mathbf{R} =
\begin{bmatrix}
1 & r_{12} & r_{13} & \cdots & r_{1p} \\
r_{12} & 1 & r_{23} & \cdots & r_{2p} \\
r_{13} & r_{23} & 1 & \cdots & r_{3p} \\
 & & \vdots & & \vdots \\
r_{1p} & r_{2p} & r_{3p} & \cdots & 1
\end{bmatrix} .
$$

Example

The sample mean vectors, covariance matrices and correlation matrices for the data matrices in Table 7.1 and 7.2 are shown in Tables 7.3 and 7.4. The correlation coefficients are shown in the lower left triangle and the covariance matrix in the upper right including the diagonal.

Sums of Squares and Cross Product Matrices

The matrices $\mathbf{S}$ and $\mathbf{R}$ are both examples of *sums of squares and cross product matrices*. As indicated above, $\mathbf{S} = \mathbf{X}^{*\prime}\mathbf{X}^{*}/(n-1)$ where $\mathbf{X}^{*}$ is the matrix of mean corrected X values given above. The correlation matrix $\mathbf{R}$ can be written as $\widetilde{\mathbf{X}}'\widetilde{\mathbf{X}}/(n-1)$ where $\widetilde{\mathbf{X}}$ is the $n \times p$ matrix of *standardized observations*

TABLE 7.4. Stress Data — Mean Vector, Covariance and Correlation Matrices

	Mean Vector	Covariance and Correlation Matrices							
		X_1	X_2	X_3	X_4	X_5	X_6	X_7	X_8
X_1	2.12	1.33	0.59	0.14	0.42	0.42	0.32	0.30	0.61
X_2	2.32	0.40	1.61	−0.09	0.19	0.28	−0.06	0.23	0.34
X_3	2.30	0.11	−0.06	1.23	0.16	0.19	0.31	0.24	0.40
X_4	3.54	0.36	0.15	0.14	0.98	0.54	0.41	0.24	0.51
X_5	3.38	0.31	0.19	0.14	0.47	1.34	0.58	0.28	0.37
X_6	3.50	0.25	−0.04	0.25	0.37	0.45	1.23	0.40	0.21
X_7	3.52	0.24	0.17	0.20	0.22	0.23	0.34	1.15	0.40
X_8	2.62	0.46	0.23	0.31	0.44	0.27	0.16	0.32	1.34

$$
\widetilde{\mathbf{X}} = \begin{bmatrix}
\dfrac{x_{11} - \bar{x}_{\cdot 1}}{s_1} & \dfrac{x_{12} - \bar{x}_{\cdot 2}}{s_2} & \cdots & \dfrac{x_{1p} - \bar{x}_{\cdot p}}{s_p} \\[2mm]
\dfrac{x_{21} - \bar{x}_{\cdot 1}}{s_1} & \dfrac{x_{22} - \bar{x}_{\cdot 2}}{s_2} & \cdots & \dfrac{x_{2p} - \bar{x}_{\cdot p}}{s_p} \\[2mm]
\vdots & & & \\[2mm]
\dfrac{x_{n1} - \bar{x}_{\cdot 1}}{s_1} & \dfrac{x_{n2} - \bar{x}_{\cdot 2}}{s_2} & \cdots & \dfrac{x_{np} - \bar{x}_{\cdot p}}{s_p}
\end{bmatrix}.
$$

The correlation matrix can be written as

$$
\mathbf{R} = \mathbf{D}^{-1/2} \mathbf{S} \mathbf{D}^{-1/2}
$$

where $\mathbf{D}$ is the diagonal matrix containing the diagonal elements of $\mathbf{S}$.

A third type of sums of squares and cross products matrix is given by $\mathbf{X}'\mathbf{X}$. This matrix contains the *raw sums of squares and cross products* given by

$$
\mathbf{X}'\mathbf{X} = \begin{bmatrix}
\sum_{i=1}^{n} x_{i1}^2 & \sum_{i=1}^{n} x_{i1}x_{i2} & \cdots & \sum_{i=1}^{n} x_{i1}x_{ip} \\[2mm]
\sum_{i=1}^{n} x_{i1}x_{i2} & \sum_{i=1}^{n} x_{i2}^2 & \cdots & \sum_{i=1}^{n} x_{i2}x_{ip} \\[2mm]
\vdots & & & \vdots \\[2mm]
\sum_{i=1}^{n} x_{i1}x_{ip} & \sum_{i=1}^{n} x_{i2}x_{ip} & \cdots & \sum_{i=1}^{n} x_{ip}^2
\end{bmatrix}.
$$

This matrix will be referred to as the sums of squares and cross products matrix and will sometimes be abbreviated by SSCP.

Multivariate Central Limit Theorem

If the rows of the data matrix $\mathbf{X}$ denoted by $\mathbf{x}_1', \mathbf{x}_2', \ldots, x_n'$ represent a multivariate random sample from a multivariate distribution with $E[\mathbf{x}] = \boldsymbol{\mu}$ and $\mathrm{Cov}(\mathbf{x}) = \boldsymbol{\Sigma}$, then the asymptotic distribution of $\bar{\mathbf{x}} = \sum_{i=1}^n \mathbf{x}_i/n$ is multivariate normal with mean vector $\boldsymbol{\mu}$ and covariance matrix $\boldsymbol{\Sigma}/n$. In other words, in large samples $\sqrt{n}(\bar{\mathbf{x}} - \boldsymbol{\mu})$ is multivariate normal with mean $\mathbf{0}$ and covariance matrix $\boldsymbol{\Sigma}$.

The *multivariate central limit theorem* also applies to the elements s_{ij} of $\mathbf{S}$. The elements of $\sqrt{n}[\mathbf{S} - \boldsymbol{\Sigma}]$ converge in distribution to a multivariate normal with mean $\mathbf{0}$ and covariance matrix $\mathbf{H}$ where a typical element of $\mathbf{H}$ is given by $\mathrm{Cov}[\sqrt{n}(s_{ij} - \sigma_{ij}), \ \sqrt{n}(s_{k\ell} - \sigma_{k\ell})] = \sigma_{ik}\sigma_{j\ell} + \sigma_{jk}\sigma_{i\ell}$, $i = 1, 2, \ldots, p$.

7.1.3 GEOMETRIC INTERPRETATIONS FOR DATA MATRICES

Geometric interpretations can be applied to multivariate samples in two different ways. The columns of $\mathbf{X}$ generate a p-dimensional space and the rows of $\mathbf{X}$ generate a n-dimensional space.

p-Dimensional Space

The most common geometric interpretation for the data matrix $\mathbf{X}$ is to regard each row as a point in a p-dimensional space. Thus the $(n \times p)$ matrix $\mathbf{X}$ summarizes the coordinates of n points in a p-dimensional space. The amount of scatter among the n points depends on the interrelationship among the p variables and on the mean and variance of the variables. Depending on the scatter it may be possible to find a smaller number of axes or dimensions $(< p)$ that could be used to locate the n points with fewer than p coordinates; for example, for three variables X_1, X_2 and X_3, it may be possible to represent the points adequately in a two-dimensional plane given by $X_3 = aX_1 + bX_2$. Most readers will be familiar with the two-dimensional scatterplot $(p = 2)$ used in simple linear regression and correlation.

n-Dimensional Space

An alternative geometrical interpretation can be obtained by viewing the p columns of $\mathbf{X}$ as coordinates of p points in an n-dimensional space. Each of the p variables can be represented by a vector drawn from the origin to the point denoted by the values of the n coordinates. If the p variables have mean zero, the angle between any two variables is related to the correlation between the variables. The strength of linear association between any two mean zero variables is measured by the cosine of the angle between the two vectors.

Mahalanobis Distance and Generalized Variance

In the discussion of bivariate samples the quantity

$$m^2 = \begin{bmatrix} x_1 - \bar{x}_{.1} \\ x_2 - \bar{x}_{.2} \end{bmatrix}' \begin{bmatrix} s_1^2 & s_{12} \\ s_{12} & s_2^2 \end{bmatrix}^{-1} \begin{bmatrix} x_1 - \bar{x}_{.1} \\ x_2 - \bar{x}_{.2} \end{bmatrix}$$

is often used to describe the locus of an ellipse in two-dimensional space with centre $(\bar{x}_{.1}, \bar{x}_{.2})$. This quantity also measures the square of the *Mahalanobis distance* between the point (x_1, x_2) and the centre $(\bar{x}_{.1}, \bar{x}_{.2})$. All points on this ellipse have the same distance m^2 from $(\bar{x}_{.1}, \bar{x}_{.2})$. This squared distance m^2 is the square of the radius of the circle that would be obtained after transforming X_1 and X_2 into new variables Z_1 and Z_2 with zero means, unit variances and zero correlation. The Mahalanobis distance therefore takes into account the variances and covariances. In comparison, the Euclidean distance is given by d, where

$$d^2 = \begin{bmatrix} x_1 - \bar{x}_{.1} \\ x_2 - \bar{x}_{.2} \end{bmatrix}' \begin{bmatrix} x_1 - \bar{x}_{.1} \\ x_2 - \bar{x}_{.2} \end{bmatrix}.$$

An alternative way to view the Mahalanobis distance is to begin with the circle located at the origin and given by $Z_1^2 + Z_2^2 = m^2$. If the variables Z_1 and Z_2 are transformed using linear combinations $X_1 = (a_1 Z_1 + b_1 Z_2 + c_1)$ and $X_2 = (a_2 Z_1 + b_2 Z_2 + c_2)$, the value of

$$\begin{bmatrix} x_1 - \bar{x}_{.1} \\ x_2 - \bar{x}_{.2} \end{bmatrix}' \begin{bmatrix} s_1^2 & s_{12} \\ s_{12} & s_2^2 \end{bmatrix}^{-1} \begin{bmatrix} x_1 - \bar{x}_{.1} \\ x_2 - \bar{x}_{.2} \end{bmatrix}$$

will still be m^2. The mean vector and covariance matrix corresponding to $\begin{bmatrix} X_1 \\ X_2 \end{bmatrix}$ is denoted by $\begin{bmatrix} \bar{x}_{.1} \\ \bar{x}_{.2} \end{bmatrix}$ and $\begin{bmatrix} s_1^2 & s_{12} \\ s_{12} & s_2^2 \end{bmatrix}$.

p-Dimensional Ellipsoid

For multivariate samples the Mahalanobis distance of $\mathbf{x}' = (x_1, x_2, \ldots, x_p)$ from the mean $\bar{\mathbf{x}}' = (\bar{x}_{.1}, \bar{x}_{.2}, \ldots, \bar{x}_{.p})$ is given by m, where $m^2 = (\mathbf{x} - \bar{\mathbf{x}})' \mathbf{S}^{-1} (\mathbf{x} - \bar{\mathbf{x}})$, which describes an *ellipsoid* in p-dimensional space. The sample squared Mahalanobis distance from the mean for each of the observations for the data matrix in Table 7.1 is shown in Table 7.5. These values can be used to indicate the distance of each sample observation from the centre of the data. Thus observation 11 in the table appears to be furthest from the centre of the data while observation 43 is closest to the centre.

Generalized Variance

The volume of the sample ellipsoid defined above is given by $m^p |\mathbf{S}|^{1/2} C(p)$, where $C(p)$ is a constant that depends on the number of variables p and

TABLE 7.5. Squared Mahalanobis Distances for Portfolio Data

Obs. No.	Squared Mahalanobis Distance	Obs. No.	Squared Mahalanobis Distance	Obs. No.	Squared Mahalanobis Distance	Obs. No.	Squared Mahalanobis Distance
1	7.72	14	5.41	26	4.73	38	6.20
2	19.16	15	6.67	27	4.13	39	7.09
3	20.33	16	9.63	28	6.69	40	5.21
4	14.67	17	18.28	29	11.20	41	6.41
5	10.67	18	8.81	30	7.12	42	8.79
6	13.40	19	9.48	31	9.30	43	3.62
7	8.62	20	10.08	32	5.57	44	7.06
8	5.08	21	5.86	33	10.71	45	4.26
9	13.25	22	11.28	34	5.06	46	10.44
10	9.17	23	7.00	35	5.80	47	21.44
11	44.86	24	8.60	36	8.26	48	5.65
12	8.09	25	12.19	37	8.59	49	10.69
13	8.58					50	9.13

$|\mathbf{S}|$ denotes the determinant of $\mathbf{S}$. From this expression we can see that for given values of p and m^2 the volume of the ellipsoid depends on $|\mathbf{S}|$. The quantity $|\mathbf{S}|$ is usually called the *generalized variance* since it is related to the overall variation among the p variables. For the portfolio data the generalized variance is 18992767.

If the n-dimensional geometrical representation for the sample is used, the columns of $\mathbf{X}^*/\sqrt{n-1}$, where $\mathbf{S} = \mathbf{X}^{*\prime}\mathbf{X}^*/(n-1)$, are represented by mean corrected vectors $(\mathbf{x}_j - \bar{x}_{.j}\mathbf{e})/\sqrt{n-1}$, $j = 1, 2, \dots, p$, eminating from the origin. The p vectors can be used to generate a p-dimensional *trapezoid* or a *parallelepiped*. The volume of the p-dimensional figure is given by $(n-1)^{p/2}|\mathbf{S}|^{1/2}$. Thus the volume is influenced by the lengths of the vectors in $\mathbf{X}^*$ and the angles among them. This provides an alternative characterization for the generalized variance $|\mathbf{S}|$. The generalized variance increases if the magnitudes of the elements of $\mathbf{X}^*$ increase and/or if the columns of $\mathbf{X}^*$ become less collinear. Obviously, if the columns of $\mathbf{X}^*$ are linearly dependent, $|\mathbf{S}| = 0$. If the columns are almost collinear, $|\mathbf{S}|$ will be very small.

Trace Measure of Overall Variance

An alternative measure of overall variance is the sum of the diagonal elements of $\mathbf{S}$, $tr\,\mathbf{S}$. This measure is simply the total of the variances for the p random variables. Unlike $|\mathbf{S}|$, this measure is not sensitive to the degree of collinearity among the columns of $\mathbf{X}^*$. The trace of the portfolio covariance matrix is 833.824.

Generalized Variance for Correlation Matrices

The generalized variance determined from the sample correlation matrix is given by $|\mathbf{R}|$. Because the variables are standardized, this quantity is not influenced by the magnitudes of the sample variances $s_1^2, s_2^2, \ldots, s_p^2$. The diagonal elements of $\mathbf{R}$ are necessarily 1, and the off-diagonal elements of $\mathbf{R}$ must lie in the interval $(-1, 1)$. If all of the off-diagonal elements are zero, the variables are mutually uncorrelated and $|\mathbf{R}| = 1$. As the off-diagonal elements increase in absolute value away from zero, the magnitude of $|\mathbf{R}|$ decreases. If any one of the off-diagonal elements is close to 1 or -1, then $|\mathbf{R}|$ will be negligible. The generalized variance based on $|\mathbf{R}|$ is therefore a measure of the lack of correlation among the variables. The generalized variance for the portfolio correlation matrix is 0.0000125.

The quantity $|\mathbf{R}|$ can be related to the volume generated in n space by the standardized variables vectors. The volume is given by $(n-1)^{p/2}|\mathbf{R}|^{1/2}$ as in the case of $|\mathbf{S}|$ above. This volume is a function of the angles among the p vectors. The quantity $|\mathbf{R}|$ can be related to $|\mathbf{S}|$ using $|\mathbf{R}| = |\mathbf{S}|/s_1 s_2 \ldots s_p$ and hence $|\mathbf{S}|$ also includes the impact of scale given by $s_1, \ldots, s_p$.

Eigenvalues and Eigenvectors for Sums of Squares and Cross Product Matrices

In Chapter 9 the eigenvectors and eigenvalues of matrices of the form $\mathbf{X}'\mathbf{X}$ will be used to achieve dimension reduction by defining new variables called *principal components*. The principal components are linear transformations of the form $\mathbf{Y} = \mathbf{A}\mathbf{X}$, with the transformation matrix $\mathbf{A}$ provided by the matrix of eigenvectors of $\mathbf{X}'\mathbf{X}$. The principal components are designed to retain most of the variation described by $\mathbf{X}'\mathbf{X}$, while reducing the number of dimensions or variables. The eigenvalues and eigenvectors of $\mathbf{X}'\mathbf{X}$ therefore provide important information about the structure of $\mathbf{X}'\mathbf{X}$.

The matrix $\mathbf{S}$ is a special case of a matrix of the form $\mathbf{X}'\mathbf{X}$ and hence the eigenvalues and eigenvectors provide important information about the structure of $\mathbf{S}$. As outlined in the Appendix, the eigenvectors and eigenvalues of $\mathbf{X}'\mathbf{X}$ satisfy the equations

$$\mathbf{X}'\mathbf{X}\mathbf{v}_j = \lambda_j \mathbf{v}_j, \quad j = 1, 2, \ldots, p.$$

In this case since $\mathbf{X}'\mathbf{X}$ is positive definite and symmetric, the eigenvectors $\mathbf{v}_j$ are *mutually orthogonal* and are usually scaled so that $\mathbf{v}_j'\mathbf{v}_j = 1$. The eigenvalues λ_j, $j = 1, 2, \ldots, p$ satisfy the properties $\Pi_{j=1}^p \lambda_j = |\mathbf{X}'\mathbf{X}|$ and $tr(\mathbf{X}'\mathbf{X}) = \sum_{j=1}^p \lambda_j$. Since $|\mathbf{X}'\mathbf{X}|$ and $tr(\mathbf{X}'\mathbf{X})$ are measures of overall variation, the eigenvalues can be seen to represent such information. The arithmetic mean of the eigenvalues represents the average of the diagonal elements of $\mathbf{X}'\mathbf{X}$ (or variances if $\mathbf{S} = \mathbf{X}'\mathbf{X}$). The geometric mean of the eigenvalues of $\mathbf{X}'\mathbf{X}$ reflects the pth root of the generalized variance of $\mathbf{X}'\mathbf{X}$.

TABLE 7.6. Eigenvalues for Portfolio Covariance Matrix

Eigenvalue No.	1	2	3	4	5	6	7	8	9	10
Eigenvalue	691.75	117.57	7.07	5.99	4.17	2.59	2.49	1.32	0.55	0.27
Arithmetic Mean	83.38			Geometric Mean	5.34					

The eigenvalues for the portfolio return covariance matrix are summarized in Table 7.6. The arithmetic and geometric means of the eigenvalues are also shown in Table 7.6.

7.1.4   OTHER SOURCES OF INFORMATION

More extensive coverage of the topics in Section 7.1 is available in Mardia, Kent and Bibby (1979), Johnson and Wichern (1988) and Kryzanowski (1988).

7.2   The Multivariate Normal Distribution

The univariate and bivariate normal distributions play an important role in statistical inference. For multivariate random variables, the *multivariate normal distribution* is a convenient and easy generalization of these two distributions. As in the univariate and bivariate normal distributions, the multivariate normal is completely defined by its first and second moments. The marginal distribution of any one variable from the multivariate normal random variable is univariate normal, and the joint distribution of any pair of variables from the multivariate normal is bivariate normal. Therefore the equivalence between independence and zero correlation for bivariate normal random variables holds for all pairs of multivariate normal random variables. More generally, any subset of q variables in a p-dimensional multivariate normal, $q < p$, has a q-dimensional multivariate normal distribution. Also, a linear combination of the variables from a multivariate normal is univariate normal. Finally, many procedures based on the assumption of multivariate normality are robust to departures from normality, and many multivariate statistics used in practice converge in distribution to a multivariate normal (multivariate central limit theorem).

In this section we introduce the multivariate normal distribution. Inference techniques for the mean vector μ and the covariance matrix Σ will be introduced in Section 7.4.

7.2.1 THE MULTIVARIATE NORMAL

Multivariate Normal Density

The random vector $\mathbf{x}$ $(p \times 1)$ has a p-dimensional multivariate normal distribution if its density is given by

$$f(\mathbf{x}) = (2\pi)^{-p/2} |\mathbf{\Sigma}|^{-1/2} \exp[-\left(\frac{1}{2}\right)(\mathbf{x} - \boldsymbol{\mu})' \mathbf{\Sigma}^{-1}(\mathbf{x} - \boldsymbol{\mu})], \qquad (7.1)$$

where the elements of $\mathbf{x}$ are in the range $(-\infty, \infty)$ and $\mathbf{\Sigma}$ is of rank p. The mean of $\mathbf{x}$ is given by $E[\mathbf{x}] = \boldsymbol{\mu}$, and the covariance matrix for $\mathbf{x}$ is given by $E[(\mathbf{x} - \boldsymbol{\mu})(\mathbf{x} - \boldsymbol{\mu})'] = \mathbf{\Sigma}$. The correlation matrix $\boldsymbol{\rho}$ relating the variables in $\mathbf{x}$ is given by

$$\boldsymbol{\rho} = \mathbf{D}_\sigma^{-1/2} \mathbf{\Sigma} \mathbf{D}_\sigma^{-1/2},$$

where $\mathbf{D}_\sigma$ is the diagonal matrix of elements $\sigma_1^2, \sigma_2^2, \ldots, \sigma_p^2$. The density is usually denoted by $N_p(\boldsymbol{\mu}, \mathbf{\Sigma})$.

The elements of $\boldsymbol{\mu}$ and $\mathbf{\Sigma}$ are denoted by

$$\boldsymbol{\mu} = \begin{bmatrix} \mu_1 \\ \mu_2 \\ \vdots \\ \mu_p \end{bmatrix}, \qquad \text{and}$$

$$\mathbf{\Sigma} = \begin{bmatrix} \sigma_1^2 & \sigma_{12} & \cdots & \sigma_{1p} \\ \sigma_{12} & \sigma_2^2 & \cdots & \sigma_{2p} \\ \vdots & \vdots & & \vdots \\ \sigma_{1p} & \sigma_{2p} & & \sigma_p^2 \end{bmatrix} = \begin{bmatrix} \sigma_1^2 & \rho_{12}\sigma_1\sigma_2 & \cdots & \rho_{1p}\sigma_1\sigma_p \\ \rho_{12}\sigma_1\sigma_2 & \sigma_2^2 & \cdots & \rho_{2p}\sigma_2\sigma_p \\ \vdots & & & \vdots \\ \rho_{1p}\sigma_1\sigma_p & \cdots & \cdots & \sigma_p^2 \end{bmatrix}.$$

Constant Probability Density Contour

The quantity $(\mathbf{x} - \boldsymbol{\mu})' \mathbf{\Sigma}^{-1}(\mathbf{x} - \boldsymbol{\mu}) = c^2$, which is the squared Mahalanobis distance between $\mathbf{x}$ and $\boldsymbol{\mu}$, describes the surface of an ellipsoid centered at $\boldsymbol{\mu}$. The density of $\mathbf{x}$ is therefore a constant over the ellipsoidal surface $(\mathbf{x} - \boldsymbol{\mu})' \mathbf{\Sigma}^{-1}(\mathbf{x} - \boldsymbol{\mu}) = c^2$. This surface is called a *constant probability density contour*. As in the case of the univariate normal, the density is maximum at $\mathbf{x} = \boldsymbol{\mu}$.

Linear Transformations

A linear combination of the p variables given by $y = \mathbf{c}'\mathbf{x}$, $\mathbf{c}$ $(p \times 1)$, has a univariate normal distribution with mean $\mu_y = \mathbf{c}'\boldsymbol{\mu}$ and variance $\sigma_y^2 = \mathbf{c}'\mathbf{\Sigma}\mathbf{c}$, hence $y \sim N(\mathbf{c}'\boldsymbol{\mu}, \mathbf{c}'\mathbf{\Sigma}\mathbf{c})$. Similarly, if $\mathbf{C}$ is a $(q \times p)$ *linear transformation matrix*, the random variable $\mathbf{y} = \mathbf{C}\mathbf{x}$ has a q-dimensional multivariate normal distribution with mean vector $\boldsymbol{\mu}_y = \mathbf{C}\boldsymbol{\mu}$ and covariance matrix $\boldsymbol{\Sigma}_\mathbf{y} = \mathbf{C}\boldsymbol{\Sigma}\mathbf{C}'$, hence $\mathbf{y} \sim N_q(\mathbf{C}\boldsymbol{\mu}, \mathbf{C}\boldsymbol{\Sigma}\mathbf{C}')$.

Distribution of Probability Density Contour

A useful property of the multivariate normal is that, for a random observation $\mathbf{x}$ from $N_p(\boldsymbol{\mu}, \boldsymbol{\Sigma})$, the quantity $(\mathbf{x}-\boldsymbol{\mu})'\boldsymbol{\Sigma}^{-1}(\mathbf{x}-\boldsymbol{\mu})$ has a χ^2 distribution with p degrees of freedom. Since $(\mathbf{x}-\boldsymbol{\mu})'\boldsymbol{\Sigma}^{-1}(\mathbf{x}-\boldsymbol{\mu})$ describes an ellipsoid with center $\boldsymbol{\mu}$, the probability is α that a random $\mathbf{x}$ will be outside the ellipsoid $(\mathbf{x}-\boldsymbol{\mu})'\boldsymbol{\Sigma}^{-1}(\mathbf{x}-\boldsymbol{\mu}) = \chi^2_{\alpha;p}$. In Section 7.3 this property will be used to check for normal goodness of fit and outliers for multivariate samples.

7.2.2   PARTITIONING THE NORMAL

The multivariate random variable $\mathbf{x}$ $(p \times 1)$ can be partitioned into two subvectors $\mathbf{x} = \begin{bmatrix} \mathbf{x}_1 \\ \mathbf{x}_2 \end{bmatrix}$, where $\mathbf{x}_1$ denotes the first q elements of $\mathbf{x}$, and $\mathbf{x}_2$ denotes the last $s = (p - q)$ elements of $\mathbf{x}$. The corresponding partitions of $\boldsymbol{\mu}$ and $\boldsymbol{\Sigma}$ are given by

$$\boldsymbol{\mu} = \begin{bmatrix} \boldsymbol{\mu}_1 \\ \boldsymbol{\mu}_2 \end{bmatrix} \quad \text{and} \quad \boldsymbol{\Sigma} = \begin{bmatrix} \boldsymbol{\Sigma}_{11} & \boldsymbol{\Sigma}_{12} \\ \boldsymbol{\Sigma}_{21} & \boldsymbol{\Sigma}_{22} \end{bmatrix},$$

where $\boldsymbol{\mu}_1$ is $(q \times 1)$, $\boldsymbol{\mu}_2$ is $(s \times 1)$, $\boldsymbol{\Sigma}_{11}$ is $(q \times q)$, $\boldsymbol{\Sigma}_{22}$ is $(s \times s)$, $\boldsymbol{\Sigma}_{12}$ is $(q \times s)$, and $\boldsymbol{\Sigma}_{21} = \boldsymbol{\Sigma}'_{12}$.

Marginal Distributions

The *marginal distribution* for $\mathbf{x}_1$ is $N_q(\boldsymbol{\mu}_1, \boldsymbol{\Sigma}_{11})$, and the *marginal distribution* for $\mathbf{x}_2$ is $N_s(\boldsymbol{\mu}_2, \boldsymbol{\Sigma}_{22})$. If $\boldsymbol{\Sigma}_{12} = \boldsymbol{\Sigma}_{21} = 0$, then the elements of the vector $\mathbf{x}_1$ are uncorrelated with the elements of $\mathbf{x}_2$, and hence under multivariate normality $\mathbf{x}_1$ and $\mathbf{x}_2$ are independent.

Conditional Distributions

The conditional density of the random variable $\mathbf{x}_2$ given $\mathbf{x}_1 = \mathbf{x}_1^*$ is normal with mean vector

$$\boldsymbol{\mu}_{2 \cdot 1}(\mathbf{x}_1^*) = (\boldsymbol{\mu}_2 - \boldsymbol{\Sigma}_{21}\boldsymbol{\Sigma}_{11}^{-1}\boldsymbol{\mu}_1) + \boldsymbol{\Sigma}_{21}\boldsymbol{\Sigma}_{11}^{-1}\mathbf{x}_1^*$$

and covariance matrix

$$\boldsymbol{\Sigma}_{22 \cdot 1} = (\boldsymbol{\Sigma}_{22} - \boldsymbol{\Sigma}_{21}\boldsymbol{\Sigma}_{11}^{-1}\boldsymbol{\Sigma}_{12}).$$

Therefore the *conditional mean vector* is a function of $\mathbf{x}_1^*$, but the covariance matrix of the conditional distribution is independent of $\mathbf{x}_1^*$. The conditional density of $\mathbf{x}_2$ given $\mathbf{x}_1 = \mathbf{x}_1^*$ is given by

$$f(\mathbf{x}_2 \mid \mathbf{x}_1 = \mathbf{x}_1^*) = (2\pi)^{-s/2} |\boldsymbol{\Sigma}_{22\cdot1}|^{-1/2} \exp[-\left(\frac{1}{2}\right)(\mathbf{x}_2-\boldsymbol{\mu}_{2\cdot1})' \boldsymbol{\Sigma}_{22\cdot1}^{-1}(\mathbf{x}_2-\boldsymbol{\mu}_{2\cdot1})].$$

Multivariate Regression Function

The conditional mean vector $\boldsymbol{\mu}_{2\cdot1}$ is called the *multivariate regression function* for $\mathbf{x}_2$ on $\mathbf{x}_1$. If $q = (p-1)$, then $s = (p-q) = 1$, and $\mathbf{x}_2 = X_2$ is a univariate random variable. In this case $\boldsymbol{\Sigma}_{21}$ is a $[(p-1)\times1]$ vector, say $\boldsymbol{\sigma}_2$. The conditional mean of X_2 given $\mathbf{x}_1$ is given by

$$\boldsymbol{\mu}_{2\cdot1}(\mathbf{x}_1) = (\boldsymbol{\mu}_2 - \boldsymbol{\Sigma}_{21}\boldsymbol{\Sigma}_{11}^{-1}\boldsymbol{\mu}_1) + \boldsymbol{\Sigma}_{21}\boldsymbol{\Sigma}_{11}^{-1}\mathbf{x}_1 = (\boldsymbol{\mu}_2 - \boldsymbol{\sigma}_2'\boldsymbol{\Sigma}_{11}^{-1}\boldsymbol{\mu}_1) + \boldsymbol{\sigma}_2'\boldsymbol{\Sigma}_{11}^{-1}\mathbf{x}_1$$

and is called the regression function of X_2 on $\mathbf{x}_1$. For the regression of X_2 on $\mathbf{x}_1$, the true intercept and slope parameters β_0 and $\boldsymbol{\beta}^*$ are given by $(\boldsymbol{\mu}_2 - \boldsymbol{\sigma}_2'\boldsymbol{\Sigma}_{11}^{-1}\boldsymbol{\mu}_1)$ and $\boldsymbol{\sigma}_2'\boldsymbol{\Sigma}_{11}^{-1}$ respectively. The variance of X_2 can be partitioned into two terms, $\sigma_{x_2}^2 = (\sigma_{x_2}^2 - \boldsymbol{\sigma}_2'\boldsymbol{\Sigma}_{11}^{-1}\boldsymbol{\sigma}_2) + \boldsymbol{\sigma}_2'\boldsymbol{\Sigma}_{11}^{-1}\boldsymbol{\sigma}_2$. The first term is the residual variance, and the second term is the variance of X_2 explained by the regression relationship with $\mathbf{x}_1$. The *coefficient of determination* is given by $R^2 = \boldsymbol{\sigma}_2'\boldsymbol{\Sigma}_{11}^{-1}\boldsymbol{\sigma}_2/\sigma_{x_2}^2$, which is the square of the *multiple correlation* between X_2 and $\mathbf{x}_1$.

If $s \geq 2$ and each of the elements of $\mathbf{x}_2$ is regressed on $\mathbf{x}_1$, the set of regression coefficients of $\mathbf{x}_1$ is given by the $(s \times q)$ matrix $\boldsymbol{\Sigma}_{21}\boldsymbol{\Sigma}_{11}^{-1}$, with the $(s \times 1)$ intercept vector $(\boldsymbol{\mu}_2 - \boldsymbol{\Sigma}_{21}\boldsymbol{\Sigma}_{11}^{-1}\boldsymbol{\mu}_1)$. The *intercept vector* and matrix of regression coefficients will be referred to later in this chapter as the *multivariate regression coefficients* for the *multivariate regression* of $\mathbf{x}_2$ on $\mathbf{x}_1$. The intercept vector will be denoted by $\boldsymbol{\beta}_0$ and the matrix of regression coefficients by $\mathbf{B}'^*$. The combined *parameter matrix* $[s \times (q+1)]$ is given by $\mathbf{B}' = [\boldsymbol{\beta}_0\ \mathbf{B}'^*]$.

Partial Correlation

The *conditional covariance matrix* $\boldsymbol{\Sigma}_{22\cdot1} = (\boldsymbol{\Sigma}_{22} - \boldsymbol{\Sigma}_{21}\boldsymbol{\Sigma}_{11}^{-1}\boldsymbol{\Sigma}_{12})$ contains the elements necessary to construct the matrix of *partial correlations* between the variables in $\mathbf{x}_2$ controlling for the variables in $\mathbf{x}_1$. The partial correlation matrix is given by

$$\mathbf{D}_{\sigma_{2\cdot1}}^{-1/2}(\boldsymbol{\Sigma}_{22} - \boldsymbol{\Sigma}_{21}\boldsymbol{\Sigma}_{11}^{-1}\boldsymbol{\Sigma}_{12})\mathbf{D}_{\sigma_{2\cdot1}}^{-1/2},$$

where $\mathbf{D}_{\sigma_{2\cdot1}}$ denotes the diagonal matrix containing the diagonal elements of $(\boldsymbol{\Sigma}_{22} - \boldsymbol{\Sigma}_{21}\boldsymbol{\Sigma}_{11}^{-1}\boldsymbol{\Sigma}_{12})$. The elements of the conditional covariance matrix $\boldsymbol{\Sigma}_{22\cdot1}$ are usually denoted by

$$\sigma_{ij\cdot1,2,\dots,q}, \quad \text{where } i,j = 1,2,\dots,s.$$

The elements of the partial correlation matrix are given by

$$\rho_{ij\cdot1,2,\ldots,q} = \sigma_{ij\cdot1,2,\ldots,q}/\sqrt{\sigma_{ii\cdot1,2,\ldots,q}}\ \sqrt{\sigma_{jj\cdot1,2,\ldots,q}},$$
$$i,j = 1,2,\ldots,s,$$

which are usually referred to as *q*th *order partial correlations*. For the *first-order partial q* = 1, recall that

$$\rho_{ij\cdot1} = \frac{\rho_{ij} - \rho_{i1}\rho_{j1}}{\sqrt{1 - \rho_{i1}^2}\ \sqrt{1 - \rho_{j1}^2}}.$$

The *higher order partial correlations* can be obtained in a recursive manner using the relationship

$$\rho_{ij\cdot1,2,\ldots,q} = \frac{\rho_{ij\cdot1,2,\ldots,q-1} - \rho_{iq\cdot1,2,\ldots,q-1}\ \rho_{jq\cdot1,2,\ldots,q-1}}{\sqrt{1 - \rho_{iq\cdot1,2,\ldots,q-1}^2}\ \sqrt{1 - p_{jq\cdot1,2,\ldots,q-1}^2}}.$$

7.3   Testing for Normality, Outliers and Robust Estimation

The multivariate normal distribution has the property that, for all subsets of variables, multivariate normality holds; however, the converse is not necessarily true. Ensuring univariate normality for all individual variables and/or ensuring bivariate normality for all possible pairs does not therefore guarantee multivariate normality. Tests for multivariate normality are discussed in this section.

The detection of univariate outliers is relatively straightforward in the sense that outliers are generally observations that are somewhat distant from the remainder of the data. To guard against the effects of outliers, robust estimators can be obtained by trimming extreme observations. For the bivariate distribution case, scatterplots, regression residuals and measures of influence can be used for both detecting and measuring the impact of outliers. Robust estimators for covariances, correlations and regression parameters are also available. In this section techniques are discussed for detecting outliers in multivariate distributions. In addition, the robust estimation of the mean vector and covariance matrix is also studied.

7.3.1   TESTING FOR NORMALITY

Mahalanobis Distances from the Sample Mean

For a bivariate distribution with random variables (X, Y), in large samples the ordered distances $m_{(i)}^2$, $i = 1, 2, \ldots, n$, can be compared to the χ^2 distribution, $\chi_{(1-\alpha_i);2}^2$, where $(1 - \alpha_i) = (i - .5)/n$. The distance m_i^2 is the squared Mahalanobis distance between (x_i, y_i) and $(\bar{x}, \bar{y})$ given by

$$m_i^2 = \frac{1}{(1-r^2)}\left[\frac{(x_i - \bar{x})^2}{s_x^2} + \frac{(y_i - \bar{y})^2}{s_y^2} - \frac{2r(x_i - \bar{x})(y_i - \bar{y})}{s_x s_y}\right].$$

A plot of the points $\left[m_{(i)}^2, \chi_{(1-\alpha_i);2}^2\right]$ should yield a straight line.

This plotting technique can be extended to the multivariate normal by computing the squared Mahalanobis distances

$$m_i^2 = (\mathbf{x}_i - \bar{\mathbf{x}})'\mathbf{S}^{-1}(\mathbf{x}_i - \bar{\mathbf{x}})$$

for the n multivariate observations $i = 1, 2, \ldots, n$. The ordered distances $m_{(i)}^2$ are plotted against the χ^2 distribution percentiles, $\chi_{(1-\alpha_i);p}^2$ where $(1 - \alpha_i) = (i - .5)/n$, $i = 1, 2, \ldots, n$.

Multivariate Skewness and Kurtosis

Tests for multivariate normality can also be based on measures of multivariate skewness and kurtosis. The Mardia (1970) sample measures of *multivariate skewness and kurtosis* are given by

$$\hat{\gamma}_{1p} = \frac{1}{n^2}\sum_{i=1}^n\sum_{j=1}^n m_{ij}^3 \quad \text{and} \quad \hat{\gamma}_{2p} = \frac{1}{n}\sum_{i=1}^n m_i^4,$$

where

$$\begin{aligned} m_i^2 &= (\mathbf{x}_i - \bar{\mathbf{x}})'\mathbf{S}^{-1}(\mathbf{x}_i - \bar{\mathbf{x}}) \quad \text{and} \\ m_{ij} &= (\mathbf{x}_i - \bar{\mathbf{x}})'\mathbf{S}^{-1}(\mathbf{x}_j - \bar{\mathbf{x}}). \end{aligned}$$

In large samples from a multivariate normal, $n\hat{\gamma}_{1p}/6$ has a χ^2 distribution with $p(p + 1)(p + 2)/6$ degrees of freedom, and $\hat{\gamma}_{2p}$ is normally distributed with mean $p(p + 2)$ and variance $8p(p + 2)/n$.

For the multivariate normal population, the Mardia (1970) measures of skewness and kurtosis are

$$\begin{aligned} \gamma_{1,p} &= E[(\mathbf{x} - \boldsymbol{\mu})'\boldsymbol{\Sigma}^{-1}(\mathbf{y} - \boldsymbol{\mu})]^3 \quad \text{and} \\ \gamma_{2,p} &= E[(\mathbf{x} - \boldsymbol{\mu})'\boldsymbol{\Sigma}^{-1}(\mathbf{x} - \boldsymbol{\mu})]^2, \end{aligned}$$

where $\mathbf{x}$ and $\mathbf{y}$ are independently distributed with mean vector $\boldsymbol{\mu}$ and covariance matrix $\boldsymbol{\Sigma}$.

Example

Table 7.7 contains the squared Mahalanobis distances for the fifty observations on the ten portfolios. Table 7.7 also contains the χ^2 distribution $(1 - p)$-values based on 10 degrees of freedom, and also the values of $(i - 0.5)/n$ for $i = 1, 2, \ldots, n$. Since the values of χ^2 corresponding to

$(i - 0.5)/n$ are difficult to obtain, they do not appear in the table. A comparison of the $(1 - p)$-values to $(i - 0.5)/n$ would suggest that there are generally fewer observations in both tails of the distribution of m^2 than could be expected for a theoretical χ^2. There also appear to be a few outliers with relatively large values of m^2, which is confirmed later on in this section.

The values of the measures of multivariate skewness and multivariate kurtosis were determined to be $\hat{\gamma}_1 = 58.49$ and $\hat{\gamma}_2 = 137.34$ respectively. The value of $n\hat{\gamma}_1/6 = 487.49$ when compared to a 220 degree of freedom χ^2 yielded a p-value less than 0.0000. For $\hat{\gamma}_2$ the Z value was determined to be 3.96, once again suggesting an extremely small p-value. The two measures therefore suggest that the multivariate normality assumption is questionable.

If each of the ten individual portfolio distributions is tested for normality, all but the first two result in rejection of the univariate normality hypothesis at p-values less than 0.01.

If the observation corresponding to $m^2 = 44.86$ (observation 11) in Table 7.7 is removed from the data set, the values of the skewness and kurtosis statistics are reduced to $\hat{\gamma}_1 = 33.45$ and $\hat{\gamma}_2 = 117.69$. The test statistics in this case yield p-values of 0.0084 for $\hat{\gamma}_1$ and greater than 0.5000 for $\hat{\gamma}_2$. Removal of this potential outlier results in a considerable reduction in both skewness and kurtosis. The detection of outliers is discussed below.

Transforming to Normality

A variety of procedures are available for transforming univariate random variables to normality. For multivariate random variables the simplest procedure is to transform each random variable using the appropriate univariate technique. Although this approach does not guarantee multivariate normality, it is usually good enough in practice.

7.3.2   Multivariate Outliers

The procedures commonly available for detecting ouliers in univariate and bivariate distributions should be used as a preliminary step to identifying potential outliers for multivariate data. Since it is possible for a *multivariate outlier* not to be an outlier with respect to any one of the underlying univariate distributions, the detection of extreme observations in multivariate distributions is more difficult. A more general approach for detecting multivariate outliers is discussed here.

Multivariate Outliers and Mahalanobis Distance

A useful way of detecting multivariate outliers is to measure the distance of each observation from the centre of the data using the Mahalanobis distance. Each observation $\mathbf{x}_i$ can be ordered or ranked in terms of its

TABLE 7.7. Mahalanobis Distances and Chi Square p-Values

m^2	$(1-p)$-Value	$(i-0.5)/n$
3.624	0.037	0.01
4.135	0.058	0.03
4.264	0.065	0.05
4.736	0.091	0.07
5.068	0.113	0.09
5.080	0.114	0.11
5.210	0.123	0.13
5.414	0.138	0.15
5.570	0.150	0.17
5.656	0.156	0.19
5.806	0.168	0.21
5.869	0.173	0.23
6.206	0.202	0.25
6.411	0.220	0.27
6.669	0.243	0.29
6.696	0.246	0.31
7.001	0.274	0.33
7.066	0.280	0.35
7.091	0.283	0.37
7.126	0.286	0.39
7.720	0.343	0.41
8.094	0.380	0.43
8.264	0.397	0.45
8.583	0.427	0.47
8.589	0.428	0.49
8.604	0.429	0.51
8.623	0.431	0.53
8.793	0.448	0.55
8.818	0.450	0.57
9.133	0.480	0.59
9.169	0.483	0.61
9.296	0.495	0.63
9.481	0.512	0.65
9.634	0.526	0.67
10.083	0.566	0.69
10.446	0.597	0.71
10.674	0.616	0.73
10.690	0.617	0.75
10.709	0.619	0.77
11.202	0.658	0.79
11.284	0.664	0.81
12.191	0.727	0.83
13.257	0.790	0.85
13.402	0.797	0.87
14.470	0.847	0.89
18.288	0.949	0.91
19.168	0.961	0.93
20.339	0.973	0.95
21.411	0.981	0.97
44.867	1.000	0.99

value of $m_i^2 = (\mathbf{x}_i - \bar{\mathbf{x}})'\mathbf{S}^{-1}(\mathbf{x}_i - \bar{\mathbf{x}})$ which is the equation of a p-dimensional ellipsoid. An equivalent procedure is to compute the ratio of the generalized variance $r_i^2 = |\mathbf{S}_{(i)}|/|\mathbf{S}|$, where $\mathbf{S}_{(i)}$ denotes the sample covariance matrix with $\mathbf{x}_i$ omitted. A relatively small value of r_i^2 would indicate that $\mathbf{x}_i$ is a potential outlier. Since $r_i^2 = 1 - nm_i^2/(n-1)$, the two methods of ordering extreme observations are equivalent. Recall from Section 7.1 that both of these measures can be related to the volume of a p-dimensional ellipsoid.

Although the measure m_i^2 is relatively easy to use, in practice it is worth pointing out that this measure is related to the measure

$$b_i^2 = (\mathbf{x}_i - \bar{\mathbf{x}}_{(i)})'\mathbf{S}_{(i)}^{-1}(\mathbf{x}_i - \bar{\mathbf{x}}_{(i)})$$

where $\bar{\mathbf{x}}_{(i)}$ denotes the sample mean vector with $\mathbf{x}_i$ omitted. The relationship between m_i^2 and b_i^2 is given by $m_i^2 = (n-1)^3 b_i^2 / [n^2(n-2) + (n-1)b_i^2]$. Ordering based on m_i^2 is therefore equivalent to ordering based on b_i^2. The presence of a single oultier in the calculation of $\bar{\mathbf{x}}$ and $\mathbf{S}$ does not therefore affect the ordering.

Testing for Multivariate Outliers

Under the assumption of multivariate normality and the null hypothesis that $\mathbf{x}_k \sim N_p(\boldsymbol{\mu}, \boldsymbol{\Sigma})$, $k = 1, 2, \ldots, n$, the statistic given by b_i^2 above, is related to the two-sample Hotelling T^2 statistic for testing the null hypothesis against the alternative $H_A : \mathbf{x}_i \sim N_p(\boldsymbol{\mu}_i, \boldsymbol{\Sigma})$, $\mathbf{x}_k \sim N_p(\boldsymbol{\mu}, \boldsymbol{\Sigma})$, $k \neq i, k = 1, 2, \ldots, n$. Hotelling's T^2 will be discussed in Section 7.4. The alternative hypothesis permits a single mean shift. (This test is a special case of the test for equality of means across g groups to be discussed in Chapter 8 in MANOVA.) The Hotelling T^2 statistic is given by $(n-1)b_i^2/n = T_i^2$.

The largest value of T_i^2 over the sample is given by $T_{\max}^2$ which can be used to test for the presence of a single outlier. This test statistic is also useful in testing the null hypothesis against the alternative $H_A : \mathbf{x}_i \sim N_p(\boldsymbol{\mu}, c_i \boldsymbol{\Sigma})$, $\mathbf{x}_k \sim N_p(\boldsymbol{\mu}, \boldsymbol{\Sigma})$, $k \neq i, k = 1, 2, \ldots, n$. In other words the outlier is obtained from a multivariate normal distribution with a different mean $\boldsymbol{\mu}$ (mean shift) or a different covariance matrix $\boldsymbol{\Sigma}$ (scale shift). The statistic $(n - p - 1)T_i^2/(n-2)p$ is distributed as an F distribution with p and $(n-p-1)$ degrees of freedom if the value of $\mathbf{x}_i$ corresponding to $\max T_i^2$ has the same multivariate normal distribution as the remaining $(n-1)$ observations. The computation of T_i^2 can be carried out more simply using the relationship of T_i^2 to the squared Mahalanobis distance m_i^2 discussed above. The F-test statistic can be written as $m_i^2(n - p - 1)/p[(n-1) - m_i^2]$.

Multiple Outliers

The identification of subsets of outliers is a more difficult problem. If the subset cannot be prespecified independently of the data, there is no inference procedure for detection. For any prespecified subset believed to contain outliers, however, the two-sample T^2 test can be used. Similarly, for groups of observations believed to have different means, the techniques of multivariate ANOVA to be discussed in Chapter 8 can be used.

The measure r_i^2 introduced above for single outliers can be extended for multiple outliers. We denote by $\mathbf{S}_{(\mathbf{i})}$ the covariance matrix without the t observations $x_{i1}, x_{i2}, \ldots, x_{it}$, where $\mathbf{i}$ denotes the vector of subscripts $(i1, i2, \ldots, it)$. The critical ratio r_i^2 is given by $|\mathbf{S}_{(\mathbf{i})}|/|\mathbf{S}|$. A subset of observations for which r_i^2 is relatively small is an indication that outliers may be present.

The F-test statistic based on the Mahalanobis distance given above can be used to detect multiple outliers. Although the F distribution only applies to the maximum value of m_i^2 if very small p-values are used and if F is computed recursively, other ouliers can be detected. Beginning with the entire sample, the observation yielding the largest value of m_i^2 is removed from the sample if the corresponding F-statistic is considered significant. The values of m_i^2 are then recomputed and a new maximum value of m_i^2 is compared to F. By restricting the significance level to very low p-values such as 0.000, this process can provide some evidence of multivariate outliers over and above the outliers detected by univariate and bivariate methods.

Example

Table 7.8 contains the squared Mahalanobis distances, F-values and p-values for the successive application of the T^2 max procedure applied to the portfolio data of Table 7.1. The table presents the results for five interations of this procedure. For step 1, observation 11 yields the largest m^2, a value of 44.8673, and a corresponding p-value of 0.0000. After removing observation 11, the m^2-values are computed again in step 2. Observation 5 has the largest value of m^2 (26.0082) in step 2 even though its value of m^2 in step 1 was relatively small. This suggests that removal of observation 11 in step 1 moved the centre of the data away from observation 5. In step 3, observation 3 yields the largest value of m^2 (24.9623). This observation had a relatively large value of m^2 in all three steps. The p-value for this m^2 is 0.0000. In step 4, after the removal of observations 3, 5 and 11, the largest value of m^2 occurs for observation 28 ($m^2 = 24.1450$, $p = 0.0000$). It is interesting to note that once again an observation with the largest m^2 did not have a relatively large m^2 in the first step. This phenomenon occurs again in step 5 where observation 45 yields the largest m^2-value (23.0393). Up until step 5, observation 45 had had relatively small m^2 values. The p-value for observation 45 in step 5 was 0.0000. In step 6 (not shown) observation 45 was deleted. The smallest p-values in step 6 were in the neighborhood of 0.0002.

TABLE 7.8. Mahalanobis Distances and F-Values for Two Sample max T^2 Test for Outliers

Obs. No.	Step 1			Step 2			Step 3			Step 4			Step 5		
	m^2	F	p-value	m^2	F	p-value	m^2	F	p-value	m^2	F	p-value	m^2	F	p-value
1	7.720	0.729	0.928	7.819	0.739	0.917	7.656	0.719	0.931	8.767	0.847	0.775	8.740	0.843	0.779
2	19.168	2.505	0.000	18.844	2.456	0.000	18.466	2.394	0.000	20.763	2.961	0.000	20.408	2.904	0.000
3	20.339	2.767	0.000	20.075	2.731	0.000	24.962	4.191	0.000						
4	14.470	1.634	0.030	15.467	1.806	0.013	15.750	1.864	0.010	18.289	2.37	0.001	18.047	2.343	0.001
5	10.675	1.086	0.389	26.008	4.493	0.000									
6	13.402	1.468	0.069	13.375	1.467	0.072	13.128	1.434	0.089	12.877	1.399	0.108	12.736	1.381	0.120
7	8.623	0.832	0.805	8.799	0.853	0.772	8.621	0.831	0.803	8.425	0.807	0.833	8.611	0.828	0.801
8	5.080	0.451	0.999	5.980	0.439	0.999	4.863	0.427	0.999	4.939	0.433	0.999	5.152	0.452	0.999
9	13.257	1.446	0.077	12.987	1.409	0.096	12.971	1.410	0.099	12.987	1.416	0.100	12.824	1.394	0.114
10	9.169	0.897	0.702	9.289	0.911	0.676	9.457	0.932	0.641	11.367	1.181	0.276	12.798	1.391	0.116
11	44.867	42.340	0.000												
12	8.094	0.771	0.886	7.929	0.751	0.904	7.808	0.737	0.916	7.622	0.715	0.933	7.986	0.755	0.893
13	8.583	0.828	0.812	8.620	0.831	0.804	8.589	0.827	0.808	8.426	0.807	0.833	9.007	0.875	0.730
14	5.414	0.484	0.999	6.539	0.599	0.991	6.563	0.600	0.990	6.467	0.588	0.991	6.308	0.570	0.994
15	6.669	0.614	0.988	6.695	0.615	0.987	6.599	0.604	0.989	6.806	0.625	0.982	6.809	0.624	0.982
16	9.634	0.954	0.604	9.428	0.928	0.647	9.346	0.918	0.664	9.397	0.924	0.653	9.252	0.905	0.681
17	18.288	2.322	0.000	18.332	2.348	0.000	18.622	2.428	0.000	18.222	2.361	0.001	18.222	2.381	0.001
18	8.818	0.855	0.770	9.481	0.935	0.636	9.476	0.934	0.637	9.407	0.925	0.651	10.133	1.017	0.500
19	9.481	0.935	0.637	9.712	0.963	0.587	9.529	0.941	0.626	10.074	1.009	0.512	10.081	1.010	0.511
20	10.083	1.010	0.508	9.929	0.991	0.541	9.819	0.977	0.565	9.907	0.988	0.546	11.274	1.170	0.292
21	5.869	0.530	0.998	5.744	0.516	0.999	5.619	0.502	0.999	5.479	0.486	0.999	5.339	0.471	0.999
22	11.284	1.166	0.283	11.041	1.135	0.325	10.800	1.103	0.369	10.675	1.087	0.393	10.569	1.074	0.415
23	7.001	0.650	0.977	6.843	0.631	0.982	6.703	0.615	0.986	6.623	0.605	0.988	6.559	0.597	0.989
24	8.604	0.830	0.808	8.658	0.836	0.798	8.516	0.818	0.820	8.381	0.802	0.840	8.179	0.777	0.868
25	12.191	1.291	0.163	12.011	1.268	0.185	11.764	1.235	0.218	11.540	1.205	0.250	11.268	1.169	0.293

TABLE 7.8. Mahalanobis Distances and F-Values for Two Sample max T^2 Test for Outliers (continued)

Obs. No.	Step 1 m^2	Step 1 F	Step 1 p-value	Step 2 m^2	Step 2 F	Step 2 p-value	Step 3 m^2	Step 3 F	Step 3 p-value	Step 4 m^2	Step 4 F	Step 4 p-value	Step 5 m^2	Step 5 F	Step 5 p-value
26	4.736	0.417	0.999	4.631	0.405	0.999	4.746	0.415	0.999	7.351	0.684	0.955	7.170	0.663	0.965
27	4.135	0.359	1.000	4.040	0.349	1.000	4.019	0.345	1.000	3.913	0.334	1.000	3.860	0.334	1.000
28	6.696	0.617	0.988	15.565	1.823	0.012	23.129	3.585	0.000	24.145	3.977	0.000			
29	11.202	1.155	0.296	10.980	1.127	0.335	10.918	1.119	0.348	10.995	1.130	0.336	11.184	1.157	0.306
30	7.126	0.663	0.971	7.101	0.659	0.972	6.934	0.640	0.978	7.300	0.679	0.958	7.125	0.658	0.968
31	9.296	0.913	0.676	9.243	0.906	0.686	9.273	0.909	0.679	10.678	1.088	0.393	10.431	1.056	0.441
32	5.570	0.500	0.999	5.570	0.498	0.999	5.606	0.501	0.999	5.490	0.487	0.999	5.619	0.499	0.999
33	10.709	1.090	0.383	10.823	1.106	0.363	10.916	1.119	0.348	10.702	1.091	0.388	10.898	1.118	0.354
34	5.068	0.449	0.999	5.780	0.520	0.998	6.530	0.597	0.990	6.408	0.582	0.992	6.764	0.619	0.983
35	5.806	0.524	0.998	5.709	0.512	0.999	5.613	0.501	0.999	5.702	0.509	0.999	5.894	0.527	0.998
36	8.264	0.791	0.862	8.483	0.815	0.827	8.410	0.806	0.837	9.219	0.902	0.689	9.006	0.875	0.730
37	8.589	0.829	0.811	8.405	0.806	0.840	8.232	0.785	0.864	8.770	0.848	0.775	8.641	0.831	0.796
38	6.206	0.565	0.996	6.668	0.613	0.988	7.103	0.658	0.971	6.933	0.638	0.978	7.341	0.682	0.954
39	7.091	0.660	0.973	6.974	0.646	0.977	7.951	0.753	0.900	9.377	0.921	0.657	9.176	0.896	0.697
40	5.210	0.464	0.999	5.089	0.450	0.999	5.075	0.447	0.999	4.947	0.433	0.999	4.820	0.419	0.999
41	6.411	0.587	0.993	7.508	0.704	0.945	7.440	0.695	0.950	7.295	0.678	0.959	8.268	0.787	0.855
42	8.793	0.852	0.775	10.019	1.002	0.522	10.662	1.085	0.394	10.510	1.066	0.425	10.800	1.105	0.372
43	3.624	0.311	1.000	3.559	0.304	1.000	3.626	0.309	1.000	3.580	0.303	1.000	3.564	0.301	1.000
44	7.066	0.657	0.974	6.911	0.639	0.980	6.966	0.643	0.977	6.995	0.645	0.975	6.927	0.625	0.981
45	4.264	0.371	1.000	9.138	0.893	0.707	12.197	1.296	0.167	12.838	1.393	0.111	23.039	3.671	0.000
46	10.446	1.056	0.434	10.574	1.073	0.410	10.626	1.080	0.401	10.540	1.070	0.419	10.315	1.040	0.464
47	21.411	3.026	0.000	21.062	2.971	0.000	20.655	2.901	0.000	20.196	2.817	0.000	20.134	2.833	0.000
48	5.656	0.508	0.999	5.726	0.514	0.999	5.687	0.509	0.999	6.305	0.571	0.994	6.369	0.577	0.993
49	10.690	1.088	0.386	12.118	1.283	0.173	12.173	1.283	0.168	11.894	1.255	0.203	12.087	1.285	0.183
50	9.133	0.893	0.709	9.696	0.961	0.591	9.863	0.982	0.556	10.457	1.059	0.435	10.347	1.045	0.458

7.3.3 ROBUST ESTIMATION

To obtain a *robust estimator* of a mean vector, it is a simple matter to employ a vector of univariate robust estimators for the vector components. For the covariance matrix and correlation matrix $(p > 2)$, however, using individual robust estimators for the parameters of the matrix will not guarantee that the matrix of estimators is positive definite.

Obtaining Robust Estimators of Covariance and Correlation Matrices

If the correlation matrix is not positive definite, adjustments can be made to make the matrix positive definite. The approach is to *shrink* the absolute magnitude of the off-diagonal elements relative to the diagonal elements. A method proposed by Devlin et al. (1975) is to compute revised estimates of each correlation coefficient using the *Fisher Z-transformation* as outlined below. For a given positive constant Δ compute

$$
\begin{aligned}
Z_{ij} &= \tfrac{1}{2}\ln\left(\frac{1+r_{ij}}{1-r_{ij}}\right) + \Delta; \quad \text{if } \tfrac{1}{2}\ln\left(\frac{1+r_{ij}}{1-r_{ij}}\right) < -\Delta, \quad i,j = 1,2,\ldots,p, \\
&= \tfrac{1}{2}\ln\left(\frac{1+r_{ij}}{1-r_{ij}}\right) - \Delta; \quad \text{if } \tfrac{1}{2}\ln\left(\frac{1+r_{ij}}{1-r_{ij}}\right) > \Delta, \\
&= 0 \qquad\qquad\qquad\qquad \text{otherwise.}
\end{aligned}
$$

Then determine $\tilde{r}_{ij}$ from the inverse Fisher transformation of Z_{ij}. A robust positive definite estimate of the covariance matrix can be obtained using $\tilde{r}_{ij}\tilde{s}_i\tilde{s}_j$, where $\tilde{s}_i^2, \tilde{s}_j^2$ are robust estimators of σ_i^2, σ_j^2 respectively, $i,j = 1,2,\ldots,p$.

Multivariate Trimming

An alternative multivariate approach to obtaining a robust estimator of the covariance matrix is called *multivariate trimming*. The Mahalanobis metric is used to identify the $100\alpha\%$ extreme observations that are to be trimmed. The calculation is done iteratively each time determining the proportion α of extreme observations. Beginning with the conventional estimators $\bar{\mathbf{x}}$ and $\mathbf{S}$ for each observation $\mathbf{x}_i$, $i = 1,2,\ldots,n$, the distance $m_i^2 = (\mathbf{x}_i - \bar{\mathbf{x}})'\mathbf{S}^{-1}(\mathbf{x}_i - \bar{\mathbf{x}})$ is computed. For a given α the observations corresponding to the largest proportion α of the values of m_i^2 are determined. New trimmed estimators $\bar{\mathbf{x}}^*$ and $\mathbf{S}^*$ of μ and Σ are then determined using the remaining observations.

Once again for all n observations $\mathbf{x}_i$ the distances $m_i^2 = (\mathbf{x}_i - \bar{\mathbf{x}}^*)'\mathbf{S}^{*-1}(\mathbf{x}_i - \bar{\mathbf{x}}^*)$, $i = 1,2,\ldots,n$, are determined and a proportion α of the largest m_i^2 are trimmed. The remaining observations are used to recompute the estimators $\bar{\mathbf{x}}^*$ and $\mathbf{S}^*$. As long as the number of observations remaining after trimming exceeds p, the dimension of the vector $\bar{\mathbf{x}}$, the estimator $\mathbf{S}^*$ determined by multivariate trimming will be positive definite. A robust estimator of the correlation matrix is obtained using the elements of $\mathbf{S}^*$.

Example

The largest outlier for the portfolio data in Table 7.8 was observation 11. Removal of this observation results in a reduction in the mean and variance for portfolio 10 from $\overline{X}_{10} = 6.41$, $s_{10}^2 = 686.59$ to $\overline{X}_{10} = 2.93$ and $s_{10}^2 = 83.74$. With the exception of some changes in the covariances between portfolio 10 and the remaining portfolios, only very minor changes in the remaining elements of the mean vector and covariance matrix occurred after trimming observation 11. A perusal of the original data shows that for portfolio 10 observation 11 was extremely large in comparison to the remaining observations.

7.3.4   OTHER SOURCES OF INFORMATION

More extensive coverage of the topics in this section can be found in Hawkins (1980), Gnanadesikan (1977) and Mardia, Kent and Bibby (1979).

7.4   Inference for the Multivariate Normal

The purpose of this section is to outline a variety of inference procedures for the mean vector $\boldsymbol{\mu}$ and covariance matrix $\boldsymbol{\Sigma}$ for a multivariate normal distribution. We assume that a random sample of size n from a multivariate normal has been obtained and is given by the $n \times p$ data matrix $\mathbf{X}$. The sample mean vector and covariance matrix were defined in Section 7.1 and denoted by $\bar{\mathbf{x}}$ and $\mathbf{S}$ respectively.

7.4.1   INFERENCE PROCEDURES FOR THE MEAN VECTOR

Sample Likelihood Function

The *likelihood function* for a random sample of size n from a multivariate normal is given by the product of the densities evaluated at each of the n observations $(\mathbf{x}_1, \mathbf{x}_2, \ldots, \mathbf{x}_n)$. Using the expression for the multivariate normal density given by (7.1) the likelihood function is therefore given by

$$L = \prod_{i=1}^{n} f(\mathbf{x}_i) = (2\pi)^{-np/2} |\boldsymbol{\Sigma}|^{-n/2} \exp\left[-\frac{1}{2} \sum_{i=1}^{n} (\mathbf{x}_i - \boldsymbol{\mu})' \boldsymbol{\Sigma}^{-1} (\mathbf{x}_i - \boldsymbol{\mu}) \right].$$

Maximizing this likelihood function with respect to $\boldsymbol{\mu}$ and $\boldsymbol{\Sigma}$ yields the maximum likelihood estimators of $\boldsymbol{\mu}$ and $\boldsymbol{\Sigma}$ given respectively by $\bar{\mathbf{x}}$ and $(n-1)\mathbf{S}/n$. This likelihood function will be the basis for many test procedures in this chapter.

If $\bar{\mathbf{x}}$ is obtained from a multivariate normal random sample, then $\bar{\mathbf{x}} \sim N_p(\boldsymbol{\mu}, \boldsymbol{\Sigma}/n)$. The statistic $n(\bar{\mathbf{x}} - \boldsymbol{\mu})' \boldsymbol{\Sigma}^{-1} (\bar{\mathbf{x}} - \boldsymbol{\mu})$ therefore has a χ^2 distribution with p degrees of freedom and can be used to make inferences about $\boldsymbol{\mu}$.

Given a sample mean vector $\bar{\mathbf{x}}$, the equation $n(\bar{\mathbf{x}} - \boldsymbol{\mu})' \boldsymbol{\Sigma}^{-1}(\bar{\mathbf{x}} - \boldsymbol{\mu}) = \chi^2_{\alpha;p}$ describes an ellipsoid with center at $\bar{\mathbf{x}}$. This equation provides a $100(1-\alpha)\%$ *confidence ellipsoid for* $\boldsymbol{\mu}$.

Hotelling's T^2

If $\boldsymbol{\Sigma}$ is unknown, we can replace $\boldsymbol{\Sigma}$ by $\mathbf{S}$ and use the fact that

$$T^2 = n(\bar{\mathbf{x}} - \boldsymbol{\mu})' \mathbf{S}^{-1}(\bar{\mathbf{x}} - \boldsymbol{\mu})$$

is distributed as $(n-1)pF_{p,(n-p)}/(n-p)$, where $F_{p,n-p}$ denotes an F distribution with p and $(n-p)$ degrees of freedom. The quantity T^2 is usually referred to as *Hotelling's* T^2. Therefore, the statistic $(n-p)T^2/(n-1)p$ has an F distribution with p and $(n-p)$ d.f. The elliposid $n(\bar{\mathbf{x}}-\boldsymbol{\mu})' \mathbf{S}^{-1}(\bar{\mathbf{x}}-\boldsymbol{\mu}) = (n-1)pF_{\alpha;p,(n-p)}/(n-p)$ provides a $100(1-\alpha)\%$ *confidence ellipsoid for* $\boldsymbol{\mu}$.

Inference

The confidence ellipsoids for $\boldsymbol{\mu}$ given above can also be used to test hypotheses regarding $\boldsymbol{\mu}$. To test $H_0: \boldsymbol{\mu} = \boldsymbol{\mu}_0$, we can use one of the test statistics

$$n(\bar{\mathbf{x}} - \boldsymbol{\mu}_0)' \boldsymbol{\Sigma}^{-1}(\bar{\mathbf{x}} - \boldsymbol{\mu}_0) \quad \text{or}$$
$$n(\bar{\mathbf{x}} - \boldsymbol{\mu}_0)' \mathbf{S}^{-1}(\bar{\mathbf{x}} - \boldsymbol{\mu}_0)$$

depending on whether $\boldsymbol{\Sigma}$ is known or unknown. For an α level test of H_0, the critical values of the test statistics are $\chi^2_{\alpha;p}$ and $(n-1)pF_{\alpha;p,(n-p)}/(n-p)$ respectively. These two tests are equivalent to the tests that would be obtained using a likelihood ratio approach.

Example

For the portfolio returns given in Table 7.1, a test of the null hypothesis that the mean return vector is the zero vector requires the test statistic $(n-p)n\bar{\mathbf{x}}' \mathbf{S}^{-1}\bar{\mathbf{x}}/(n-1)p$, which has an F distribution with p and $(n-p)$ degrees of freedom if the null hypothesis is true. For the portfolio data, $F = 0.9850$ which has a p-value of 0.472 for F with 10 and 40 degrees of freedom. We therefore cannot reject the hypothesis that the mean vector is zero.

For the police officer stress data of Table 7.2, a test of the null hypothesis that the elements of the mean return vector are all three requires the test statistic $(n-p)n(\bar{\mathbf{x}} - 3\mathbf{i})' \mathbf{S}^{-1}(\bar{\mathbf{x}} - 3\mathbf{i})/(n-1)p$ where $\mathbf{i}$ is a vector of unities. This statistic has the value 2.603 for the stress data and has an F distribution with 8 and 42 degrees of freedom if the null hypothesis is true. The p-value in this case is 0.021. It is therefore difficult to accept the null hypothesis that the elements of the mean vector are all 3.

Simultaneous Confidence Regions

Given that $\bar{\mathbf{x}} \sim N_p(\boldsymbol{\mu}, \boldsymbol{\Sigma}/n)$, the distribution of $\boldsymbol{\ell}'\bar{\mathbf{x}}$ where $\boldsymbol{\ell}$ $(p \times 1)$ is a vector of constants is $N(\boldsymbol{\ell}'\boldsymbol{\mu}, \boldsymbol{\ell}'\boldsymbol{\Sigma}\boldsymbol{\ell}/n)$. An important property of the Hotelling T^2 statistic is that, simultaneously for **All** linear combinations of the vector $\bar{\mathbf{x}}$, say $\boldsymbol{\ell}'\bar{\mathbf{x}}$, where $\boldsymbol{\ell}$ $(p \times 1)$, the probability is $(1 - \alpha)$ that the interval

$$\boldsymbol{\ell}'\bar{\mathbf{x}} - \sqrt{\frac{p(n-1)}{n(n-p)}}\, F_{\alpha;p,(n-p)}\boldsymbol{\ell}'\mathbf{S}\boldsymbol{\ell}, \qquad \boldsymbol{\ell}'\bar{\mathbf{x}} + \sqrt{\frac{p(n-1)}{n(n-p)}}\, F_{\alpha;p,(n-p)}\boldsymbol{\ell}'\mathbf{S}\boldsymbol{\ell}$$

contains $\boldsymbol{\ell}'\boldsymbol{\mu}$. In addition to making comparisons among elements of $\boldsymbol{\mu}$, this result can be used to give individual confidence intervals for the elements of $\boldsymbol{\mu}$. By choosing $\boldsymbol{\ell}$ to contain only one nonzero element, a unity corresponding to j, a $(1 - \alpha)$ probability interval for μ_j is given by

$$\bar{\mathbf{x}}_j \pm \sqrt{\frac{p(n-1)}{(n-p)}\, F_{\alpha;p,(n-p)}}\, \sqrt{s_j^2/n}.$$

As in the case of *Scheffé's multiple comparison procedure* in ANOVA, this procedure is a conservative approach to comparing means. The *experimentwise error rate* α is preserved over all possible linear combinations $\boldsymbol{\ell}'\boldsymbol{\mu}$.

Examples

For the ten portfolios in Table 7.1, a single portfolio was constructed using the weights $\boldsymbol{\ell}' = [5, 4, 3, 2, 1, 0, -2, -3, -4, -5]$. (Note that the weights must add to one for a portfolio.) The estimated variance is given by $\boldsymbol{\ell}'\mathbf{S}\boldsymbol{\ell} = 20022.8$. A 95% confidence interval for the mean of this portfolio is given by -29.83 ± 108.05. The mean of this portfolio is therefore not significantly different from zero.

For the eight stress indicators in Table 7.2 a weight vector indicating the frequency of occurrence is given by $\boldsymbol{\ell}' = [0.20, 0.20, 0.15, 0.05, 0.10, 0.05, 0.10, 0.15]$. The estimated weighted average stress is given by $\boldsymbol{\ell}'\bar{\mathbf{x}} = 2.668$ and the estimated variance is given by $\boldsymbol{\ell}'\mathbf{S}\boldsymbol{\ell} = 0.4805$. A 95% confidence interval for the average stress measure is given by 2.668 ± 0.487. The interval $(2.181, 3.155)$ therefore contains the true average stress measure with a probability of 0.95.

Inferences For Linear Functions

It is sometimes of interest to make inferences about a linear transformation $\mathbf{y} = \mathbf{C}\mathbf{x}$, which we have already indicated is multivariate normal with mean $\mathbf{C}\boldsymbol{\mu}$ and covariance matrix $\mathbf{C}\boldsymbol{\Sigma}\mathbf{C}'$, where $\mathbf{C}$ is $(q \times p)$. *Confidence ellipsoids* and tests of hypotheses regarding $\mathbf{C}\boldsymbol{\mu}$ can therefore be obtained using the fact that

$$n(C\bar{x} - C\mu)'(C\Sigma C')^{-1}(C\bar{x} - C\mu) \quad \text{is distributed as } \chi_q^2 \qquad (7.2)$$

and

$$n(C\bar{x} - C\mu)'(CSC')^{-1}(C\bar{x} - C\mu) \quad \text{is distributed as } \frac{(n-1)q}{(n-q)} F_{q,(n-q)}. \qquad (7.3)$$

Examples

The four stress indicators X_1, X_2, X_3 and X_8 are related to direct dealings with the public, whereas the four stress indicators X_4, X_5, X_6 and X_7 are related to the structure of the organization. The two sets of stress indicators can be used to form two indices of stress, one for public stress and one for organizational stress. The two different indices of stress can be constructed using the transformation matrix

$$C = \begin{bmatrix} 0.25 & 0.25 & 0.25 & 0.0 & 0.0 & 0.0 & 0.0 & 0.25 \\ 0.0 & 0.0 & 0.0 & 0.25 & 0.25 & 0.25 & 0.25 & 0.0 \end{bmatrix}.$$

To test the hypothesis that $C\mu = \begin{bmatrix} 2.5 \\ 3.5 \end{bmatrix}$ the sample value is $C\bar{x} = \begin{bmatrix} 2.340 \\ 3.485 \end{bmatrix}$. The test statistic value is given by 2.547 which is distributed as $(2)(49)F_{2,48}/(48)$. The value of $F_{2,48}$ is therefore 1.248 which has a p-value of 0.296. The null hypothesis therefore cannot be rejected.

7.4.2 REPEATED MEASURES COMPARISONS

Repeated Measurements on a Single Variable

An important application of the linear function property discussed in the previous section is the *repeated measurements comparison*. In this application we assume that the same n individuals or objects are measured on a particular response variable (i.e., some measure of performance) under a variety of p conditions or treatments (i.e., p different time periods). We denote the response vector for each individual over the p conditions by x $(p \times 1)$ and assume that $x \sim N_p(\mu, \Sigma)$. It is often of interest in such situations to determine if $\mu_1 = \mu_2 = \ldots = \mu_p$ and hence that the mean level of performance over the n individuals is the same for the p conditions. Testing the hypothesis $H_0: \mu_1 = \mu_2 = \ldots = \mu_p$ is equivalent to testing $H_0: (\mu_2 - \mu_1) = (\mu_3 - \mu_2) = \ldots = (\mu_p - \mu_{p-1}) = 0$. Using the transformed

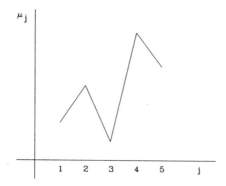

FIGURE 7.1. Profile Plot for Five Variables

vector random variable $\mathbf{Cx}$, where $\mathbf{C}$ is the $(p-1) \times p$ matrix given by

$$
\mathbf{C} =
\begin{bmatrix}
-1 & 1 & 0 & \cdots & 0 & 0 \\
0 & -1 & 1 & & 0 & 0 \\
0 & 0 & -1 & & 0 & 0 \\
\vdots & \vdots & \vdots & & \vdots & \vdots \\
0 & 0 & 0 & & 1 & 0 \\
0 & 0 & 0 & & -1 & 1
\end{bmatrix},
\tag{7.4}
$$

the test statistic given by (7.2) or (7.3) can be used to test H_0.

An alternative approach to computing the test statistic (7.3) is available using the expressions $\bar{\mathbf{x}}'\mathbf{S}^{-1}\bar{\mathbf{x}}$, $\bar{\mathbf{x}}'\mathbf{S}^{-1}\mathbf{i}$ and $\mathbf{i}'\mathbf{S}^{-1}\mathbf{i}$, where $\mathbf{i}$ is a $(p \times 1)$ vector of unities. The statistic (7.3) can be written as

$$
n\bar{\mathbf{x}}'\mathbf{S}^{-1}\bar{\mathbf{x}} - n(\bar{\mathbf{x}}'\mathbf{S}^{-1}\mathbf{i})^2 / \mathbf{i}'\mathbf{S}^{-1}\mathbf{i}.
\tag{7.5}
$$

Profile Characterization

A plot of the components of $\boldsymbol{\mu}$ in order from 1 to p is called a *profile*, as illustrated in Figure 7.1, for a mean vector with five components. Sometimes adjacent points are joined to show the changes in level among the elements of $\boldsymbol{\mu}$.

Examples

For the officer stress data of Table 7.2, the test statistic (7.5) has the value 131.35, which has a p-value of 0.0000 when compared to an F distribution with 7 and 43 degrees of freedom. The hypothesis of equal stress means is therefore rejected at conventional levels.

An alternative application of this test is given by the comparison of the mean returns on p securities over n time periods. Weekly security returns can usually be assumed to be independent over time, but the covariance between the returns on any two securities in a given time period is usually nonzero. Denoting the vector of returns on the p securities in a given time period by $\mathbf{x}$, we assume that $\mathbf{x} \sim N_p(\mu, \Sigma)$. It is sometimes of interest to test the hypothesis that the mean returns on the p securities are equal. The test statistic is again given by (7.3) with $\mathbf{C}$ defined above and with $q = (p-1)$.

For the ten portfolios in Table 7.1 the test statistic for the hypothesis of equal mean returns is given by 8.8429. Under the null hypothesis this statistic has an F distribution with 9 and 41 degrees of freedom and hence has a p-value of 0.0000. We cannot, therefore, accept the hypothesis of equal portfolio mean returns. It is interesting to note that earlier we could not reject the hypothesis that the mean vector was the zero vector.

Repeated Measures in a Randomized Block Design

In analysis of variance the repeated measures experiment is often applied as a special case of a randomized block design. Since in the repeated measures experiment the same experimental unit (block) is used for each treatment, the responses in each block may no longer be independent. As outlined above, the covariance matrix for treatment responses could be assumed to be some arbitrary Σ. If Σ does not have the form $\sigma^2 \mathbf{I}$, as assumed in the randomized block design, then the conventional F-test for equality of means may no longer be valid.

The conventional ANOVA F-statistic can be used for the repeated measures, randomized block experiment for some Σ structures other than $\sigma^2 \mathbf{I}$. The most commonly used assumption is that Σ has an *equal variance-equal covariance structure* given by

$$
\Sigma = \sigma^2 \begin{bmatrix}
1 & \rho & \rho & \cdots & \rho \\
\rho & 1 & \rho & \cdots & \rho \\
\vdots & & & & \vdots \\
\rho & \cdots & \rho & \cdots & 1
\end{bmatrix}.
$$

Necessary and Sufficient Conditions for Validity of Univariate F Test

Although the equal variance-equal covariance structure is sufficient for the F-statistic to be valid, it is not necessary. Necessary and sufficient conditions are that the elements of Σ satisfy

$$\sigma_j^2 \quad = 2\lambda_j + \alpha, \qquad j = 1, 2, \ldots, p,$$

$$\sigma_{jk} \quad = \lambda_j + \lambda_k, \qquad j \neq k, \; j, k = 1, 2, \ldots, p,$$

or

$$\Sigma = \lambda \mathbf{i}' + \mathbf{i}\lambda' + \alpha \mathbf{I}, \tag{7.6}$$

where $\lambda = \{\lambda_j\}$, $j = 1, 2, \ldots, p$, is a vector of constants, and $\mathbf{i}$ $(p \times 1)$ is a vector of unities. This variance-covariance structure is commonly called the *Huynh-Feldt pattern* (1970) and is equivalent to the condition that the differences between all pairs of responses $(x_j - x_k)$ have the same variance and the same covariances.

The variance-covariance structure given by (7.6) is equivalent to the condition $\varepsilon = 1$ in the expression

$$\varepsilon = \frac{p^2(\bar{\sigma}_d - \bar{\sigma}_{..})^2}{(p-1)\left(S - 2p \sum_{k=1}^{p} \bar{\sigma}_{.k}^2 + p^2\bar{\sigma}_{..}^2\right)},$$

where S denotes the sum of squares of the elements of Σ, $\bar{\sigma}_d$ is the mean of the diagonal elements of $\Sigma = \{\sigma_{jk}\}$, $\bar{\sigma}_{.k}$ is the mean of the elements in row k, and $\bar{\sigma}_{..}$ is the mean of all p^2 elements of Σ. This constant lies in the range $(p-1)^{-1} < \varepsilon \leq 1$ and can be estimated using the elements of the sample covariance matrix $\mathbf{S}$. The *Geisser–Greenhouse procedure* (1958) for the repeated measurements design consists of adjusting the degrees of freedom for the conventional ANOVA F test to take into account the departure of ε from 1. The adjusted degrees of freedom are $\hat{\varepsilon}(p-1)$ and $\hat{\varepsilon}(p-1)(n-1)$ instead of the usual $(p-1)$ and $(p-1)(n-1)$. $\hat{\varepsilon}$ is the estimate of ε obtained by replacing the elements of Σ by the elements of $\mathbf{S}$ in the above expression for ε. This procedure is meant as a substitute for the more general multivariate approach outlined above. The multivariate procedure is more reliable but has less power than the Geisser–Greenhouse procedure.

7.4.3 MAHALANOBIS DISTANCE OF THE MEAN VECTOR FROM THE ORIGIN

Mahalanobis Distance of Mean Vector from the Origin

A measure of the distance from the origin in a p-dimensional space for the mean vector μ $(p \times 1)$ is provided by the Mahalanobis distance given by $\mu'\Sigma^{-1}\mu$. The equation $\mu'\Sigma^{-1}\mu = k$ describes an ellipsoid centered at the origin. For a subset of the elements of μ denoted by μ_1 $(q \times 1)$, $q < p$, with corresponding covariance matrix Σ_{11} $(q \times q)$, we can compute $\mu_1'\Sigma_{11}^{-1}\mu_1$ which is the Mahalanobis distance from the origin in a reduced q-dimensional space. If $\mu'\Sigma^{-1}\mu = \mu_1'\Sigma_{11}^{-1}\mu_1$, then $\mu_2 = \Sigma_{21}\Sigma_{11}^{-1}\mu_1$ and hence the distance of μ from the origin can be captured by the distance of

μ_1 from the origin. The remaining $(p - q)$ dimensions are not required to describe the distance.

Application to Financial Portfolios

An example of an application of this measure comes from the evaluation of financial portfolios. Let $\mathbf{x}_t$ denote the returns on p financial assets earned in time period t, where $\mathbf{x}_t \sim N_p(\boldsymbol{\mu}, \boldsymbol{\Sigma})$, and where the $\mathbf{x}_t$ are independent over time. The return on a portfolio of these assets at time is given by $y_t = \mathbf{c}'\mathbf{x}_t$ (where $\sum_{i=1}^p c_i = 1$). The mean and variance of the portfolio return y_t are $\mu_y = \mathbf{c}'\boldsymbol{\mu}$ and $\sigma_y^2 = \mathbf{c}'\boldsymbol{\Sigma}\mathbf{c}$. If the performance of the portfolio is measured using μ_y^2/σ_y^2, then the maximum value of this measure is given by $\boldsymbol{\mu}'\boldsymbol{\Sigma}^{-1}\boldsymbol{\mu}$, which is the Mahalanobis distance from the origin. For a subset of q assets with return vector $z_t = \mathbf{c}_1'\mathbf{x}_{1t}$, $\mu_z = \mathbf{c}_1'\boldsymbol{\mu}_1$ and $\sigma_z^2 = \mathbf{c}_1'\boldsymbol{\Sigma}\mathbf{c}_1$, the maximum value of μ_z^2/σ_z^2 is given by $\boldsymbol{\mu}_1'\boldsymbol{\Sigma}_{11}^{-1}\boldsymbol{\mu}_1$. If $\boldsymbol{\mu}'\boldsymbol{\Sigma}^{-1}\boldsymbol{\mu} = \boldsymbol{\mu}_1'\boldsymbol{\Sigma}_{11}^{-1}\boldsymbol{\mu}_1$, then there is no improvement in performance to be achieved by adding the $s = (p - q)$ additional assets to the portfolio.

Given a random sample of observations on $\mathbf{x}$ with sample mean return vector $\bar{\mathbf{x}}$ and covariance matrix $\mathbf{S}$, a test statistic for the hypothesis $H_0: \boldsymbol{\mu}'\boldsymbol{\Sigma}^{-1}\boldsymbol{\mu} = \boldsymbol{\mu}_1'\boldsymbol{\Sigma}_{11}^{-1}\boldsymbol{\mu}_1$ is provided by

$$\frac{(n - p)}{s}[\bar{\mathbf{x}}'\mathbf{S}^{-1}\bar{\mathbf{x}} - \bar{\mathbf{x}}_{.1}'\mathbf{S}_{11}^{-1}\bar{\mathbf{x}}_{.1}]/[1 + \bar{\mathbf{x}}_{.1}'\mathbf{S}_{11}^{-1}\bar{\mathbf{x}}_{.1}]. \tag{7.7}$$

The quantities $\bar{\mathbf{x}}_{.1}$ and $\mathbf{S}_{11}$ denote the portions of the sample mean vector $\bar{\mathbf{x}}$ and the sample covariance matrix $\mathbf{S}$ corresponding to $\boldsymbol{\mu}_1$ and $\boldsymbol{\Sigma}_{11}$. If H_0 is true, this statistic has an F distribution with s and $(n - p)$ degrees of freedom.

It can be shown that if $(\boldsymbol{\mu}'\boldsymbol{\Sigma}^{-1}\boldsymbol{\mu} - \boldsymbol{\mu}_1'\boldsymbol{\Sigma}_{11}^{-1}\boldsymbol{\mu}_1) = 0$ then $(\boldsymbol{\mu}_2 - \boldsymbol{\Sigma}_{21}\boldsymbol{\Sigma}_{11}^{-1}\boldsymbol{\mu}_1) = 0$ and hence the test $H_0: \boldsymbol{\mu}'\boldsymbol{\Sigma}^{-1}\boldsymbol{\mu} = \boldsymbol{\mu}_1'\boldsymbol{\Sigma}_{11}^{-1}\boldsymbol{\mu}_1$ is equivalent to $H_0: (\boldsymbol{\mu}_2 - \boldsymbol{\Sigma}_{21}\boldsymbol{\Sigma}_{11}^{-1}\boldsymbol{\mu}_1) = 0$. As we shall see later in this chapter, this test is equivalent to a test for zero intercept vector in the multivariate regression of $\mathbf{x}_2$ on $\mathbf{x}_1$. A more detailed discussion of this applicatoin is available in Jobson and Korkie (1982, 1989).

Example

For the portfolio data, the value of $\bar{\mathbf{x}}'\mathbf{S}^{-1}\bar{\mathbf{x}}$ for all ten portfolios is 1.0416. For the subset of five portfolios 1, 2, 3, 6 and 9, the value of $\bar{\mathbf{x}}_1'\mathbf{S}_{11}^{-1}\bar{\mathbf{x}}_1$ is 0.9880. The value of the test statistic (7.7) is therefore 0.2157. In comparison to an F random variable with 5 and 40 degrees of freedom, the p-value is 0.9538. There is very little improvement in performance to be achieved by adding the remaining five portfolios to the first subset of five.

7.4.4   INFERENCE FOR THE COVARIANCE AND CORRELATION MATRICES

Wishart Distribution

If $\mathbf{X}$ $(n \times p)$ is a data matrix obtained as a random sample from $N_p(\mathbf{0}, \boldsymbol{\Sigma})$, then $\mathbf{X}'\mathbf{X}$ has a p-dimensional Wishart distribution $W_p(\boldsymbol{\Sigma}, n)$ with n degrees of freedom. The sample covariance matrix $\mathbf{S}$ multiplied by $(n-1)$ has the Wishart distribution $W_p[\boldsymbol{\Sigma}, (n-1)]$. The Wishart distribution can be used to make inferences about $\boldsymbol{\Sigma}$.

Sphericity Test and Test for Independence

If the covariance matrix $\boldsymbol{\Sigma}$ for the multivariate normal has the form $\boldsymbol{\Sigma} = \sigma^2 \mathbf{I}$, then the p random variables in $\mathbf{x}$ are mutually independent with common variance σ^2. In this case, the multivariate distribution is said to be *spherical* since the ellipsoid of constant density $(\mathbf{x} - \boldsymbol{\mu})' \boldsymbol{\Sigma}^{-1} (\mathbf{x} - \boldsymbol{\mu}) = c^2$ is equivalent to $(\mathbf{x} - \boldsymbol{\mu})'(\mathbf{x} - \boldsymbol{\mu}) = \sigma^2 c^2$, which is the equation of a p-dimensional sphere centered at $\boldsymbol{\mu}$.

A test statistic for testing the null hypothesis $H_0 \colon \boldsymbol{\Sigma} = \sigma^2 \mathbf{I}$ is given by $np \ln[A/B]$, where $A = \sum_{j=1}^p \lambda_j / p$ and $B = \left[\Pi_{j=1}^p \lambda_j \right]^{1/p}$ are the arithmetic and geometric means of the eigenvalues $\lambda_1, \lambda_2, \ldots, \lambda_p$ of $\mathbf{S}$ respectively. If H_0 is true, the statistic has a χ^2 distribution with $\frac{1}{2}(p-1)(p+2)$ d.f. in large samples.

Example

For the portfolio data of Table 7.1, the eigenvalues are shown in Table 7.6. The arithmetic mean and geometric mean for the ten eigenvalues are also shown in Table 7.6. Using these two means the test statistic for testing sphericity is given by $(50)(10) \ln[83.382/5.344] = 164.85$. Under the null hypothesis of sphericity, the test statistic has a χ^2 distribution with $\frac{1}{2}(p-1)(p+2) = 54$ degrees of freedom. The p-value for 164.85 is therefore 0.0000. In the case of the portfolio data, the sphericity hypothesis for the covariance matrix cannot be justified.

A Test for Zero Correlation

For the correlation matrix $\mathbf{R}$, if $\mathbf{R} = \mathbf{I}$ the covariance matrix $\boldsymbol{\Sigma}$ is diagonal. To test $H_0 \colon \mathbf{R} = \mathbf{I}$, in large samples, the test statistic $-[n - (1/6)(2p + 11)]\sum_{j=1}^p \ln \lambda_j$ has a χ^2 distribution with $(1/2)p(p-1)$ d.f. if H_0 is true. In this case the λ_j, $j = 1, 2, \ldots, p$ are the eigenvalues of the sample correlation matrix $\mathbf{R}$.

Example

For the stress data the eight eigenvalues of the correlation matrix are given by 2.8959, 1.2741, 0.9853, 0.8156, 0.6516, 0.5694, 0.4427 and 0.3655. The test statistic for testing $H_0\colon \mathbf{R} = \mathbf{I}$ has the value 77.3823 which when compared to a χ^2 distribution with 45 degrees of freedom has a p-value of 0.0019. It is therefore very unlikely that the correlation matrix for the eight stress variables is an identity matrix.

Test Statistics for Repeated Measures Designs

In the discussion of the repeated measures experiment in Section 7.4.2, two special variance-covariance structures were found to be useful in permitting the more powerful F-test used for the univariate ANOVA randomized block design. Test statistics for the equal variance-equal covariance structure and the Huynh-Feldt pattern are outlined below.

Test for Equal Variance-Equal Covariance Structure

For the equal variance-equal covariance structure, shown in Section 7.4.2, the test statistic

$$-\left[(n-1) - \frac{p(p+1)^2(2p-3)}{6(p-1)(p^2+p-4)}\right] \ln L$$

in large samples is approximately distributed as χ^2 with $\frac{1}{2}p(p+1) - 2$ degrees of freedom, if the equal variance-equal covariance hypothesis holds. The likelihood ratio statistic L is given by

$$L = |\mathbf{S}|/(\bar{s}^2)^p(1-\bar{r})^{p-1}[1 + (p-1)\bar{r}],$$

where $\bar{s}^2 = \sum_{j=1}^p s_j^2/p$ and $\bar{r} = \dfrac{1}{p(p-1)} \sum_{\substack{j=1 \\ j\neq k}}^p \sum_{k=1}^p s_{jk}/\bar{s}^2$.

Test for the Hyunh-Feldt Pattern

For the Hyunh-Feldt pattern the test statistic is given by

$$-\left[(n-1) - \frac{2p^2 - 3p + 3}{6(p-1)}\right] \ln W,$$

where $W = (p-1)^{p-1}|\mathbf{CSC}'|/(tr\mathbf{CXC}')^{p-1}$, and where $\mathbf{C}$ $(p-1) \times p$ is a matrix of rows that are mutually orthogonal and are also orthogonal to the row vector of unities $\mathbf{i}'$, and whose elements sum to one. The matrix $\mathbf{C}$ is a submatrix of the *Helmert matrix* which is given by $\mathbf{H}' = [\mathbf{i}'/\sqrt{p}\ \mathbf{C}]$. The rows of $\mathbf{C}$ have the same structure as the contrast coefficients in analysis of variance. In large samples the test statistic has a χ^2 distribution with $[\frac{1}{2}p(p-1) - 1]$ degrees of freedom if the hypothesized Huynh-Feldt pattern is true.

Example

For the stress data in Table 7.2 the equal variance-equal covariance statistic has a value of 35.6982. In comparison to a χ^2 distribution with 34 degrees of freedom, the p-value is 0.3885. The stress data covariance matrix therefore seems to be consistent with an equal variance-equal covariance structure. The test statistic for the Huynh-Feldt structure has a value of 31.4994. A comparison of this value to a χ^2 distribution with 27 degrees of freedom yields the p-value of 0.2954. Since the equal variance-equal covariance structure implies the Huynh-Feldt structure, the test for Huynh-Feldt structure was not necessary, given that the first test did not reject the equal variance-covariance structure.

The above two tests for structure when carried out for the portfolio data yield extremely large values of the χ^2 test statistics and hence are not reported here. The lack of homogeneity among the sample variances in this case is a major contributor to the magnitude of the test statistics.

Equal Correlation Structure

A very useful additional test for covariance matrix structure is provided by the test for *equal correlation structure*. In this case the correlation matrix is given by

$$\rho = (1-\rho)\mathbf{I} + \rho\mathbf{ii}' = \begin{bmatrix} 1 & \rho & \rho & \cdots & \rho \\ \rho & 1 & \rho & \cdots & \rho \\ \vdots & & & & \vdots \\ & & & & \rho \\ \rho & \cdots & \cdots & \rho & 1 \end{bmatrix}.$$

The covariance matrix is given by

$$\Sigma = \mathbf{D}_\sigma^{1/2}[(1-\rho)\mathbf{I} + \rho\mathbf{ii}']\mathbf{D}_\sigma^{1/2} = \begin{bmatrix} \sigma_1^2 & \sigma_1\sigma_2\rho & \cdots & \sigma_1\sigma_p\rho \\ \sigma_1\sigma_2\rho & \sigma_2^2 & & \sigma_2\sigma_p\rho \\ \vdots & & & \vdots \\ \sigma_1\sigma_p\rho & \cdots & \cdots & \sigma_p^2 \end{bmatrix},$$

where $\mathbf{D}_\sigma$ is the diagonal matrix of variances σ_j^2, $j = 1, 2, \ldots, p$.

Using the off-diagonal elements of the sample correlation matrix $\mathbf{R} = \{r_{jk}\}$ $j, k = 1, 2, \ldots, p$, the test statistic is given by

$$\frac{(n-1)}{(1-\bar{r})^2}\left[\frac{1}{2}\sum_{\substack{j=1\\j\neq k}}^{p}\sum_{\substack{k=1}}^{p}(r_{jk}-\bar{r})^2 - q\sum_{k=1}^{p}(\bar{r}_k-\bar{r})^2\right],$$

where

$$\bar{r}_k = \frac{1}{p-1}\sum_{\substack{j=1\\j\neq k}}^{p}r_{jk}, \qquad \bar{r} = \frac{1}{p(p-1)}\sum_{\substack{j=1\\j\neq k}}^{p}\sum_{\substack{k=1}}^{p}r_{jk}$$

and

$$q = \frac{(p-1)^2[1-(1-\bar{r})^2]}{p-(p-2)(1-\bar{r})^2}.$$

Under the null hypothesis of equal correlation structure, this statistic has a χ^2 distribution with $(p+1)(p-2)/2$ degrees of freedom.

Example

For the police officer stress data the value of the above test statistic is 21.4303. For the χ^2 distribution with 27 degrees of freedom the p-value is 0.7657. The null hypothesis of equal correlation therefore cannot be rejected.

Independent Blocks

For the partitioned model of Section 7.2 with

$$\begin{bmatrix} \mathbf{x}_1 \\ \mathbf{x}_2 \end{bmatrix} \sim N_p\left(\begin{bmatrix} \boldsymbol{\mu}_1 \\ \boldsymbol{\mu}_2 \end{bmatrix}, \begin{bmatrix} \boldsymbol{\Sigma}_{11} & \boldsymbol{\Sigma}_{12} \\ \boldsymbol{\Sigma}_{21} & \boldsymbol{\Sigma}_{22} \end{bmatrix}\right),$$

it is sometimes of interest to test for independence between the two subsets, $H_0\colon \boldsymbol{\Sigma}_{12} = \mathbf{0}$. Using the notation of Section 7.2, we assume $\mathbf{x}_1$ ($q \times 1$) and $\mathbf{x}_2$ ($s \times 1$) where $(q+s) = p$. The mean vector $\boldsymbol{\mu}$ and covariance matrix $\boldsymbol{\Sigma}$ are partitioned accordingly.

The likelihood ratio test of H_0 is based on the statistic

$$-\left[n - \frac{1}{2}(q+s+3)\right]\ln|\mathbf{I} - \mathbf{S}_{22}^{-1}\mathbf{S}_{21}\mathbf{S}_{11}^{-1}\mathbf{S}_{12}|$$

which has a χ^2 distribution with sq degrees of freedom in large samples. The sample covariance matrix $\mathbf{S}$ has been partitioned to conform to the partitioning of $\mathbf{x}$. This test will also be used in canonical correlation in Section 7.5 and will be exemplified there.

Partial and Multiple Correlation

The maximum likelihood estimators of the partial correlation coefficients are obtained by replacing the elements of $\boldsymbol{\Sigma}$ by the elements of $\mathbf{S}$ in the definitions of Section 7.2. The matrix of sample partial correlation coefficients is given by $\mathbf{R}_{22\cdot 1} = \mathbf{D}_{22\cdot 1}^{-1/2}\mathbf{S}_{22\cdot 1}\mathbf{D}_{22\cdot 1}^{-1/2}$ where $\mathbf{D}_{22\cdot 1}$ is the diagonal matrix containing the diagonal elements of $\mathbf{S}_{22\cdot 1}$. Inferences for the elements of $\rho_{22\cdot 1}$ can be made using the conventional Fisher transformation for zero-order correlations. The transformation is given by

$$Z = \left(\frac{1}{2}\right)\ln\left(\frac{1 + r_{ij\cdot 1,2,\ldots,q}}{1 - r_{ij\cdot 1,2,\ldots,q}}\right),$$

where $r_{ij\cdot 1,2,\ldots,q}$ is the partial correlation between variables i and j controlling for the variables $1, 2, \ldots, q$.

The sample multiple correlation coefficient $\sqrt{R^2}$ relating a single variable X_2 and a vector of variables $\mathbf{x}_1$ is commonly used in a multiple regression model. If the true multiple correlation coefficient is zero, then the statistic $(n - p - 1)R^2/(1 - R^2)(p - 1)$ has an F distribution with $(p - 1)$ and $(n - p - 1)$ degrees of freedom.

7.4.5   OTHER SOURCES OF INFORMATION

More extensive coverage of the multivariate normal distribution and inference for the multivariate normal can be found in Anderson (1984), Seber (1984), Morrison (1976), Mardia, Kent and Bibby (1979), Johnson and Wichern (1988) and Stevens (1986).

7.5   Multivariate Regression and Canonical Correlation

In Section 7.2, the multivariate normal random vector $\mathbf{x}$ was partitioned into two subvectors $\mathbf{x}_1$ ($q \times 1$) and $\mathbf{x}_2$ ($s \times 1$), where $p = (q + s)$. The conditional distribution of $\mathbf{x}_2$ given $\mathbf{x}_1$ was introduced to measure the relationship between the two random vectors. The mean of this conditional distribution was called the multivariate regression function of $\mathbf{x}_2$ on $\mathbf{x}_1$. This function gives the mean value of $\mathbf{x}_2$ at specific values of $\mathbf{x}_1$. In this section we are concerned with the use of random samples to make inferences about the multivariate regression relationship between $\mathbf{x}_2$ and $\mathbf{x}_1$. As outlined below, the estimators involved are simple extensions of the ordinary least squares estimators used for multiple linear regression.

The relationship between the two vector random variables $\mathbf{x}_1$ and $\mathbf{x}_2$ depends on the elements of the covariance matrix $\mathbf{\Sigma}_{12}$ and their magnitudes relative to the elements of $\mathbf{\Sigma}_{11}$ and $\mathbf{\Sigma}_{22}$. If $\mathbf{\Sigma}_{12} = \mathbf{0}$, then the two random vectors are uncorrelated and hence under the normality assumption would be independent. If, however, the elements of $\mathbf{\Sigma}_{12}$ are relatively large, it may be of interest to determine relationships between the two random vectors $\mathbf{x}_1$ and $\mathbf{x}_2$. The *multivariate regression model* determines linear functions of the variables in $\mathbf{x}_1$ that are related to each of the variables in $\mathbf{x}_2$ (one linear function for each variable in $\mathbf{x}_2$). An alternative approach to relating the two vectors, is to determine a linear function of the variables of $\mathbf{x}_2$ and a linear function of the variables of $\mathbf{x}_1$ in such a way that the two linear functions are strongly related. This technique is commonly called *canonical correlation*. Canonical correlation is discussed in the latter part of this section.

7.5.1   MULTIVARIATE REGRESSION

The Multivariate Regression Function

In Section 7.2 the multivariate regression function relating $\mathbf{x}_2$ $(s \times 1)$ to $\mathbf{x}_1$ $(q \times 1)$, $p = s + q$, was given by the $(s \times 1)$ vector $E[\mathbf{x}_2 \mid \mathbf{x}_1] = \boldsymbol{\mu}_{2\cdot 1}$, where

$$\boldsymbol{\mu}_{2\cdot 1}' = (\boldsymbol{\mu}_2 - \boldsymbol{\Sigma}_{21}\boldsymbol{\Sigma}_{11}^{-1}\boldsymbol{\mu}_1)' + \mathbf{x}_1'\boldsymbol{\Sigma}_{11}^{-1}\boldsymbol{\Sigma}_{12}.$$

The intercept vector $\boldsymbol{\beta}_0'$ $(1 \times s)$ and the matrix of slope coefficients $\mathbf{B}^*(q \times s)$, were given by $(\boldsymbol{\mu}_2 - \boldsymbol{\Sigma}_{21}\boldsymbol{\Sigma}_{11}^{-1}\boldsymbol{\mu}_1)'$ and $\boldsymbol{\Sigma}_{11}^{-1}\boldsymbol{\Sigma}_{12}$ respectively with

$$\mathbf{B} = \begin{bmatrix} \boldsymbol{\beta}_0' \\ \mathbf{B}^* \end{bmatrix}.$$

To relate this to the conventional regression notation, we denote the $(s \times 1)$ vector $\mathbf{x}_2$ by $\mathbf{y}$ and the $[(q + 1) \times 1]$ vector $\begin{bmatrix} 1 \\ \mathbf{x}_1 \end{bmatrix}$ by $\mathbf{x}$. The multivariate regression relationship relating $\mathbf{y}$ to $\mathbf{x}$ is therefore given by

$$\mathbf{y}' = \mathbf{x}'\mathbf{B} + \mathbf{u}', \tag{7.8}$$

where $\mathbf{u}' = \mathbf{y}' - \boldsymbol{\mu}_{2\cdot 1}'$, and $\boldsymbol{\mu}_{2\cdot 1}' = \begin{bmatrix} 1 \\ \mathbf{x}_1 \end{bmatrix}' \mathbf{B}$. In Section 7.2 the conditional covariance matrix for $\mathbf{u}$ was given by $\boldsymbol{\Sigma}_{22\cdot 1} = (\boldsymbol{\Sigma}_{22} - \boldsymbol{\Sigma}_{21}\boldsymbol{\Sigma}_{11}^{-1}\boldsymbol{\Sigma}_{12})$, or equivalently in our alternate notation $\boldsymbol{\Sigma}_{\mathbf{yy}\cdot\mathbf{x}_1} = (\boldsymbol{\Sigma}_{\mathbf{yy}} - \boldsymbol{\Sigma}_{\mathbf{yx}_1}\boldsymbol{\Sigma}_{\mathbf{x}_1\mathbf{x}_1}^{-1}\boldsymbol{\Sigma}_{\mathbf{x}_1\mathbf{y}})$. This covariance matrix is also denoted by $\boldsymbol{\Gamma}$ later in this section.

Estimation of the Multivariate Regression Model

Given a random sample of n observations on $\mathbf{y}$ and $\mathbf{x}$, the maximum likelihood estimators of $\boldsymbol{\beta}_0'$, $\mathbf{B}^*$ and $\boldsymbol{\Sigma}_{\mathbf{yy}\cdot\mathbf{x}_1}$ are given by $\hat{\boldsymbol{\beta}}_0' = \bar{\mathbf{y}}' - \bar{\mathbf{x}}_{\cdot 1}'\mathbf{S}_{\mathbf{x}_1\mathbf{x}_1}^{-1}\mathbf{S}_{\mathbf{x}_1\mathbf{y}}$, $\hat{\mathbf{B}}^* = \mathbf{S}_{\mathbf{x}_1\mathbf{x}_1}^{-1}\mathbf{S}_{\mathbf{x}_1\mathbf{y}}$ and $\mathbf{S}_{\mathbf{yy}\cdot\mathbf{x}_1} = [\mathbf{S}_{\mathbf{yy}} - \mathbf{S}_{\mathbf{yx}_1}\mathbf{S}_{\mathbf{x}_1\mathbf{x}_1}^{-1}\mathbf{S}_{\mathbf{x}_1\mathbf{y}}]$. The elements of the estimator $\hat{\boldsymbol{\beta}}_0$ and the rows of the estimator $\hat{\mathbf{B}}^*$ are equivalent to the maximum likelihood estimators of the intercept and slope coefficients in the multiple regression model relating each element of $\mathbf{y}$ separately to the vector $\mathbf{x}$. The elements of $\hat{\boldsymbol{\beta}}_0$ are given by

$$\hat{\beta}_{0j} = \bar{y}_j - \bar{\mathbf{x}}_1' \hat{\boldsymbol{\beta}}_j^*, \qquad j = 1, 2, \ldots, s,$$

and the columns of $\widehat{\mathbf{B}}^*$ by

$$\hat{\boldsymbol{\beta}}_j^* = \mathbf{S}_{\mathbf{x}_1\mathbf{x}_1}^{-1} \mathbf{S}_{\mathbf{x}_1\mathbf{y}_j}, \qquad j = 1, 2, \ldots, s.$$

These estimators are equivalent to the *ordinary least squares estimators* for the *univariate multiple regression model* $\mathbf{y}_j = \mathbf{X}\boldsymbol{\beta}_j + \mathbf{u}_j$, $j = 1, 2, \ldots, s$, given by $\hat{\boldsymbol{\beta}}_j = (\mathbf{X}'\mathbf{X})^{-1}\mathbf{X}'\mathbf{y}_j$, where $\hat{\boldsymbol{\beta}}_j = \begin{bmatrix} \hat{\beta}_{0j} \\ \hat{\boldsymbol{\beta}}_j^* \end{bmatrix}$. The $[n \times (q+1)]$ matrix $\mathbf{X}$ denotes the n observations on the $(q+1) \times 1$ vector $\mathbf{x}$.

Relationship to Ordinary Least Squares

For the multivariate regression model involving s univariate multiple regressions, we can write the model for all n observations as

$$\mathbf{Y} = \mathbf{X}\mathbf{B} + \mathbf{U}, \tag{7.9}$$

where $\mathbf{Y}$ is $(n \times s)$, $\mathbf{X}$ is $[n \times (q+1)]$, $\mathbf{B} = \begin{bmatrix} \boldsymbol{\beta}_0' \\ \mathbf{B}^* \end{bmatrix}$ is $[(q+1) \times s]$, and $\mathbf{U}$ is $(n \times s)$. The ordinary least squares estimator given by

$$\widehat{\mathbf{B}} = (\mathbf{X}'\mathbf{X})^{-1}\mathbf{X}'\mathbf{Y} \tag{7.10}$$

is an unbiased estimator of $\mathbf{B}$. The columns of $\widehat{\mathbf{B}}$ are equivalent to the s univariate ordinary least squares estimators $\hat{\boldsymbol{\beta}}_j$, $j = 1, 2, \ldots, s$, corresponding to the s univariate multiple regressions. This ordinary least squares estimator can also be written as

$$\widehat{\mathbf{B}} = \begin{bmatrix} \bar{\mathbf{y}}' - \bar{\mathbf{x}}_{.1}' \mathbf{S}_{\mathbf{x}_1\mathbf{x}_1}^{-1} \mathbf{S}_{\mathbf{x}_1\mathbf{y}} \\ \mathbf{S}_{\mathbf{x}_1\mathbf{x}_1}^{-1} \mathbf{S}_{\mathbf{x}_1\mathbf{y}} \end{bmatrix},$$

and is therefore equivalent to the maximum likelihood estimator.

Residuals, Influence, Outliers and Cross Validation

As outlined above the multivariate regression model can be estimated using the ordinary least squares estimator for univariate multiple regression models. Therefore, the procedures available for the study of residuals, detection of outliers and the measurement of influence available in multiple linear regression can be employed in multivariate linear regression. Similarly the techniques available for cross validation in multiple linear regression can also be extended.

Estimation of the Error Covariance Matrix

The maximum likelihood estimator $(n-1)(\mathbf{S_{yy}} - \mathbf{S_{yx_1}}\mathbf{S_{x_1x_1}^{-1}}\mathbf{S_{x_1y_1}})/n$ of $\mathbf{\Sigma_{yy \cdot x_1}}$ is related to the *multivariate residual sum-of-squares matrix* $(\mathbf{Y} - \mathbf{X\widehat{B}})$ and is given by

$$(\mathbf{Y} - \mathbf{X\widehat{B}})'(\mathbf{Y} - \mathbf{X\widehat{B}})/n.$$

An unbiased estimator of $\mathbf{\Gamma} = \mathbf{\Sigma_{yy \cdot x_1}}$, however, is given by

$$\widehat{\mathbf{\Gamma}} = (\mathbf{Y} - \mathbf{X\widehat{B}})'(\mathbf{Y} - \mathbf{X\widehat{B}})/(n - q - 1).$$

Relationship to Multiple Linear Regression

For each of the dependent or response variables $Y_1, Y_2, \ldots, Y_s$ a multiple linear regression model is given by

$$\mathbf{y}_j = \mathbf{X}\boldsymbol{\beta}_j + \mathbf{u}_j, \quad j = 1, 2, \ldots, s,$$

where $\mathbf{y}_j$ is $(n \times 1)$, $\mathbf{X}$ is $[n \times (q+1)]$, $\boldsymbol{\beta}_j$ is $[(q+1) \times 1]$, $\mathbf{u}_j$ is $(n \times 1)$ and $E[\mathbf{u}_j\mathbf{u}_j'] = \gamma_j^2\mathbf{I}$. The ordinary least squares estimator is given by $\widehat{\boldsymbol{\beta}}_j = (\mathbf{X'X})^{-1}\mathbf{X'y}_j$ and the residuals are given by $\mathbf{e}_j = (\mathbf{y}_j - \mathbf{X}\widehat{\boldsymbol{\beta}}_j)$. The variance γ_j^2 is estimated using $\widehat{\gamma}_j^2 = (\mathbf{y}_j - \mathbf{X}\widehat{\boldsymbol{\beta}}_j)'(\mathbf{y}_j - \mathbf{X}\widehat{\boldsymbol{\beta}}_j)/(n - q - 1)$.

The multivariate regression model combines the s multiple regression models into a single model

$$\mathbf{Y} = \mathbf{XB} + \mathbf{U},$$

where $\mathbf{Y} = [\mathbf{y}_1\mathbf{y}_2 \ldots \mathbf{y}_s]$, $\mathbf{B} = [\boldsymbol{\beta}_1\boldsymbol{\beta}_2 \ldots \boldsymbol{\beta}_s]$ and $\mathbf{U} = [\mathbf{u}_1\mathbf{u}_2 \ldots \mathbf{u}_s]$. In addition to $E[\mathbf{u}_j\mathbf{u}_j'] = \gamma_j^2\mathbf{I}$ we also have a relationship between error terms from different multiple regression, $E[\mathbf{u}_j\mathbf{u}_k'] = \gamma_{jk}\mathbf{I}$, $j \neq k$, $j, k = 1, 2, \ldots, s$. The $(s \times s)$ matrix of elements γ_j^2, γ_{jk} is denoted by $\mathbf{\Gamma}$ and represents the covariance matrix among the error terms for individuals.

The residuals are denoted by $(\mathbf{Y} - \mathbf{X\widehat{B}})$ and the covariance matrix $\mathbf{\Gamma}$ is estimated by $\widehat{\mathbf{\Gamma}} = (\mathbf{Y} - \mathbf{X\widehat{B}})'(\mathbf{Y} - \mathbf{X\widehat{B}})/(n - q - 1)$. The diagonal elements of $\widehat{\mathbf{\Gamma}}$ are equivalent to the $\widehat{\gamma}_j^2$ obtained from the multiple regression for $\mathbf{y}_j$. The off-diagonal elements are equivalent to $\mathbf{e}_j'\mathbf{e}_k/(n - q - 1)$, where $\mathbf{e}_j$ and $\mathbf{e}_k$ are obtained from the multiple regressions for $\mathbf{y}_j$ and $\mathbf{y}_k$.

The important distinction between the set of single multiple regressions and multivariate regression is that in the multivariate regression model there are nonzero correlations among error terms from the different multiple regression models. If joint inferences are required involving two or more of the multiple regression models these correlations must be taken into account. These joint inference procedures are discussed next.

Testing the Hypothesis that Some Coefficients are Zero

As in the case of univariate regression, it is sometimes useful to be able to test the null hypothesis that a subset of the columns of $\mathbf{X}$ is superfluous. A *reduced model* is given by

$$\mathbf{Y} = \mathbf{X}_v \mathbf{B}_v + \mathbf{U}$$

where $\mathbf{X}_v$ is $[n \times (v+1)]$, $\mathbf{B}_v$ is $[(v+1) \times s]$ and $v = (q+1-r)$. The matrices $\mathbf{B}$ and $\mathbf{X}$ are partitioned as $\begin{bmatrix} \mathbf{B}_r \\ \mathbf{B}_v \end{bmatrix}$ and $[\mathbf{X}_r \; \mathbf{X}_v]$ respectively where $\mathbf{X}_r$ is $(n \times r)$ and $\mathbf{B}_r$ is $(r \times s)$. We therefore require a test of $H_0 \colon \mathbf{B}_r = \mathbf{0}$ and hence that the first r X variables are superfluous.

For the full or complete model, we denote the *residual sum of squares matrix* by $\mathbf{E} = (\mathbf{Y} - \mathbf{X}\widehat{\mathbf{B}})'(\mathbf{Y} - \mathbf{X}\widehat{\mathbf{B}})$. For the reduced model corresponding to $\mathbf{B}_r = \mathbf{0}$, the least squares estimator $\widetilde{\mathbf{B}}_v = (\mathbf{X}_v'\mathbf{X}_v)^{-1}\mathbf{X}_v'\mathbf{Y}$ is determined, and the residual sum of squares matrix is given by $\mathbf{E}_0 = (\mathbf{Y} - \mathbf{X}_v\widehat{\mathbf{B}}_v)'(\mathbf{Y} - \mathbf{X}_v\widehat{\mathbf{B}}_v)$. The *likelihood ratio criterion* is a function of the ratio given by $\Lambda = |\mathbf{E}|/|\mathbf{E}_0|$ which is called *Wilk's Lambda* and has parameters $[s, r, (n-q-1)]$. The dimension s refers to $\mathbf{Y}$, the dimension r reflects the number of X variables deleted if H_0 is true and $(n - q - 1)$ refers to the degrees of freedom for $\mathbf{E}$.

If H_0 is true, then in large samples the distribution of Λ can be approximated by the statistic $m_2(1 - \Lambda^{1/\nu})/m_1\Lambda^{1/\nu}$ which has an F distribution with m_1 and m_2 degrees of freedom, where

$$\nu = \sqrt{\frac{s^2r^2 - 4}{s^2 + r^2 - 5}}, \quad m_1 = rs$$

and

$$m_2 = \nu[(n - q - 1) - \tfrac{1}{2}(s - r + 1)] - \frac{sr}{2} + 1.$$

This F approximation is often referred to as *Rao's F*. If $s = 1$ or 2 or if $r = 1$ or 2 this F distribution is exact.

Other Tests

An alternative large sample statistic for H_0 is based on the χ^2 distribution. In large samples the statistic $-[n - q - 1 - \tfrac{1}{2}(s - r + 1)]\ln\Lambda$ has a χ^2 distribution with sr degrees of freedom.

There are several asymptotic tests available that are not based on the likelihood ratio criterion. These statistics are based on such measures as $tr[\mathbf{E}(\mathbf{E}_0-\mathbf{E})^{-1}]$ (Lawley–Hotelling), $tr[\mathbf{E}\mathbf{E}_0^{-1}]$ (Pillai) and the largest eigenvalue of $\mathbf{E}(\mathbf{E}_0 - \mathbf{E})^{-1}$ (Roy) all of which require special tables. Monte Carlo studies have shown that none of these alternative criteria is uniformly superior to the likelihood ratio criterion. Throughout the text only the likelihood ratio criterion will be used for the multivariate linear model.

Inferences for Linear Functions

The condition $\mathbf{B}_r = \mathbf{0}$ discussed above is a special case of a more general set of constraints on the coefficient matrix $\mathbf{B}$ given by $\mathbf{AB} = \mathbf{K}$, where $\mathbf{A}$ is a $[a \times (q+1)]$ matrix of known constants of rank a and $\mathbf{K}$ is an $(a \times s)$ matrix of known constants. The *restricted least squares estimator* of $\mathbf{B}$ subject to $\mathbf{AB} = \mathbf{K}$ is given by

$$\widehat{\mathbf{B}}_A = \widehat{\mathbf{B}} - (\mathbf{X}'\mathbf{X})^{-1}\mathbf{A}'[\mathbf{A}(\mathbf{X}'\mathbf{X})^{-1}\mathbf{A}']^{-1}(\mathbf{A}\widehat{\mathbf{B}} - \mathbf{K}).$$

The likelihood ratio test of the hypothesis $H_0 \colon \mathbf{AB} = \mathbf{K}$ is carried out by using the Wilk's Lambda statistic $\Lambda = |\mathbf{E}|/|\mathbf{E}_0|$ with parameters $[s, a, (n - q - 1)]$ where $\mathbf{E}_0 = (\mathbf{Y} - \mathbf{X}\widehat{\mathbf{B}}_A)'(\mathbf{Y} - \mathbf{X}\widehat{\mathbf{B}}_A)$ and $\mathbf{E} = (\mathbf{Y} - \mathbf{X}\widehat{\mathbf{B}})'(\mathbf{Y} - \mathbf{X}\widehat{\mathbf{B}})$. Using Rao's F approximation if H_0 is true then, in large samples, the statistic

$$\frac{1 - \Lambda^{1/\nu}}{\Lambda^{1/\nu}} \cdot \frac{m_2}{m_1} \tag{7.11}$$

has an F distribution with m_1 and m_2 degrees of freedom where

$$\nu = \sqrt{\frac{s^2 a^2 - 4}{s^2 + a^2 - 5}}, \quad m_1 = sa,$$

$$m_2 = \nu[n - q - 1 - \tfrac{1}{2}(s - a + 1)] - \tfrac{sa}{2} + 1.$$

For the multivariate linear model

$$\mathbf{Y} = \mathbf{XB} + \mathbf{U}$$

a more general hypothesis is given by $H_0 \colon \mathbf{ABM} = \mathbf{K}$ where $\mathbf{A}\ [a \times (q+1)]$, $\mathbf{M}\ (s \times b)$, and $\mathbf{K}\ (a \times b)$ are matrices of specified constants. The relationship to be tested can be converted to the form of the previous test by writing

$$\mathbf{YM} = \mathbf{XBM} + \mathbf{UM}$$

or

$$\mathbf{Y}^* = \mathbf{XB}^* + \mathbf{U}^*,$$

and hence we wish to test $H_0 \colon \mathbf{AB}^* = \mathbf{K}$. This hypothesis can be tested using the statistic given by (7.11) by revising m_1, m_2 and r accordingly.

Computer Software

The calculations for the examples in this section were performed using SAS PROC REG.

Example

The data in Table 7.9, which provide an example for Section 7.5, represent observations on 100 bank employees on each of six variables. Two variables are the logarithm of two salary variables LCURRENT = ln(CURRENT SALARY) and LSTART = ln(STARTING SALARY). The remaining four variables are background variables consisting of EDUC = level of education in years, AGE, EXPER = years of relevant work experience at time of hiring, and SENIOR = level of seniority with the bank. The two salary variables are the dependent variables, and the four background variables are explanatory variables.

The multivariate linear regression model is denoted by

$$
\begin{aligned}
\text{LCURRENT} \;=\;& \beta_0 + \beta_1 \text{ EDUC} + \beta_2 \text{ AGE} \\
& +\beta_3 \text{ EXPER} + \beta_4 \text{ SENIOR} + u_1 \\
\text{LSTART} \;=\;& \alpha_0 + \alpha_1 \text{ EDUC} + \alpha_2 \text{ AGE} \\
& +\alpha_3 \text{ EXPER} + \alpha_4 \text{ SENIOR} + u_2.
\end{aligned}
$$

In matrix notation the model is given by

$$
\mathbf{y}' = \mathbf{x}'\mathbf{B} + \mathbf{u},
$$

where

$$
\mathbf{y}' = [\text{LCURRENT LSTART}], \quad \mathbf{u} = [u_1, u_2], \quad
\mathbf{B} = \begin{bmatrix} \beta_0 & \alpha_0 \\ \beta_1 & \alpha_1 \\ \beta_2 & \alpha_2 \\ \beta_3 & \alpha_3 \\ \beta_4 & \alpha_4 \end{bmatrix},
$$

$$
\mathbf{x} = [1 \text{ EDUC AGE EXPER SENIOR}].
$$

With no restrictions the estimated regression relationships are given by

$$
\begin{aligned}
\text{LCURRENT} = \underset{(0.000)}{8.699} \;+\;& \underset{(0.000)}{0.083} \text{ EDUC} - \underset{(0.000)}{0.015} \text{ AGE} + \underset{(0.001)}{0.016} \text{ EXPER} \\
-\;& \underset{(0.487)}{0.002} \text{ SENIOR} \qquad R^2 = 0.528 \quad \text{and} \\
\text{LSTART} = \underset{(0.000)}{8.285} \;+\;& \underset{(0.000)}{0.081} \text{ EDUC} - \underset{(0.003)}{0.010} \text{ AGE} + \underset{(0.000)}{0.016} \text{ EXPER} \\
-\;& \underset{(0.202)}{0.003} \text{ SENIOR} \qquad R^2 = 0.543.
\end{aligned}
$$

The *p*-values for the coefficients appear in brackets below the coefficient estimates. These *p*-values are based on the usual multiple linear regression statistics.

TABLE 7.9. Bank Employee Salary Data

LCURRENT	LSTART	EDUC	EXPER	AGE	SENIOR	LCURRENT	LSTART	EDUC	EXPER	AGE	SENIOR
9.685	9.035	16	0.25	28.50	81	9.105	8.384	8	0.42	24.33	81
10.252	9.764	19	13.00	41.92	83	9.007	8.411	12	0.00	23.42	69
10.034	9.287	15	12.00	41.17	98	9.259	8.438	12	4.00	53.92	97
9.417	8.637	12	20.00	46.25	94	9.362	8.537	17	31.25	55.58	73
9.928	9.304	16	5.75	35.17	83	9.752	8.853	16	1.67	28.42	79
10.212	9.457	19	2.92	30.08	78	9.623	8.748	16	0.58	29.92	90
9.105	8.794	12	18.00	44.50	70	9.078	8.268	12	13.00	52.00	86
9.303	8.679	12	3.42	27.83	70	8.984	8.313	12	4.67	52.17	72
9.681	9.230	15	11.08	35.42	78	8.890	8.268	12	6.00	62.00	86
9.996	9.304	16	5.67	34.33	90	9.529	8.961	16	6.58	30.75	65
10.085	9.158	16	4.92	34.00	94	9.838	8.839	16	1.58	32.00	86
9.647	9.158	15	14.67	38.92	78	9.021	8.648	15	20.08	59.50	65
9.535	9.314	16	12.42	44.42	82	10.012	8.986	16	3.00	29.00	64
10.373	8.748	15	2.83	29.50	82	9.497	8.699	12	0.25	44.50	90
9.392	8.748	12	29.00	63.58	69	8.977	8.313	8	6.25	59.08	83
9.504	8.748	12	3.92	27.42	66	9.028	8.594	12	15.08	51.50	67
9.455	8.699	8	36.50	58.08	85	9.764	8.961	16	0.92	27.58	68
10.096	9.104	16	3.67	33.75	91	8.821	8.313	8	26.58	56.92	72
9.137	8.648	12	3.83	29.33	95	9.335	8.476	8	2.17	27.83	98
10.292	9.259	18	4.67	39.67	94	9.230	8.658	12	3.08	28.67	86
10.165	9.615	21	22.00	56.67	88	9.532	8.748	8	20.50	48.50	79
9.341	8.748	8	14.50	59.42	69	9.417	8.699	8	36.00	60.67	92
9.903	9.472	16	22.00	48.33	77	9.402	8.748	8	12.92	32.92	68
9.455	8.748	8	34.00	63.50	74	9.377	8.748	8	12.00	35.25	76
9.392	8.748	12	25.67	54.17	79	9.407	8.699	12	4.98	30.33	91

TABLE 7.9. Bank Employee Salary Data (continued)

LCURRENT	LSTART	EDUC	EXPER	AGE	SENIOR	LCURRENT	LSTART	EDUC	EXPER	AGE	SENIOR
9.553	8.748	12	32.25	59.83	78	10.075	9.510	12	22.67	48.25	74
9.811	9.137	16	5.00	32.25	86	9.431	8.748	12	12.92	38.67	73
10.116	9.595	19	16.58	42.58	80	9.230	8.425	12	0.75	24.75	81
10.126	9.546	18	6.17	35.42	75	10.354	9.349	17	5.58	32.08	70
10.513	9.680	19	14.58	44.92	96	9.422	8.699	12	8.50	34.58	94
9.417	8.594	12	25.58	56.00	95	9.417	8.188	12	26.17	53.50	91
10.633	8.306	16	5.83	37.08	97	10.596	9.777	16	10.67	35.33	66
9.417	8.699	8	37.58	64.25	87	9.392	8.748	8	25.42	47.25	67
10.240	9.547	19	6.25	36.92	67	9.156	8.594	15	29.92	57.50	88
10.026	9.210	16	2.83	33.42	93	9.320	8.699	14	3.92	30.33	79
9.575	8.922	15	9.50	37.17	98	9.181	8.699	15	15.92	42.17	90
10.116	9.349	19	2.17	28.42	65	9.392	8.699	8	12.00	37.83	92
9.057	8.476	12	31.75	64.50	98	9.417	8.699	8	25.17	48.83	83
9.071	8.411	12	0.00	23.00	65	9.105	8.268	8	3.00	29.17	94
9.084	8.476	12	0.50	40.17	76	9.241	8.594	15	4.25	48.67	78
9.392	8.794	12	19.00	55.25	95	9.188	8.648	12	9.75	45.17	74
9.144	8.594	12	16.50	45.50	64	8.821	8.188	12	10.33	60.67	97
9.441	8.922	12	17.08	61.67	90	8.830	8.313	8	0.00	51.50	81
9.367	8.438	15	0.92	26.83	93	9.131	8.476	16	17.83	47.58	84
9.175	8.537	8	8.00	46.17	66	9.387	8.699	15	6.58	40.50	86
9.314	8.961	15	22.08	58.75	82	9.084	8.384	12	0.42	25.50	92
8.757	8.313	8	3.58	55.25	74	9.014	8.411	12	12.83	47.92	73
9.218	8.658	12	6.92	33.50	91	8.946	8.188	15	1.92	60.50	96
9.064	8.594	8	19.00	64.25	66	9.303	8.313	12	8.42	54.17	85
10.030	9.392	16	1.25	31.92	73	9.188	8.384	12	2.67	27.58	75

To structure several hypotheses to be tested using this data, the notation $H_0: \mathbf{ABM} = \mathbf{K}$ is used. To test the null hypothesis that the coefficient of SENIOR is zero in both equations the relationship for H_0 is given by

$$[0\ 0\ 0\ 0\ 1] \begin{bmatrix} \beta_0 & \alpha_0 \\ \beta_1 & \alpha_1 \\ \beta_2 & \alpha_2 \\ \beta_3 & \alpha_3 \\ \beta_4 & \alpha_4 \end{bmatrix} = [0\ 0].$$

Therefore, $\mathbf{A} = [0\ 0\ 0\ 0\ 1]$, $\mathbf{K} = [0\ 0]$ and $\mathbf{M} = \mathbf{I}$. The F-statistic with 2 and 94 degrees of freedom derived from Wilks' Lambda has a value of 4.795 and a p-value of 0.0104. Thus, at best, H_0 can only be weakly rejected. The variable SENIOR seems to be of only minor importance after the other three explanatory variables.

To test the null hypothesis that the coefficients of the EDUC and EXPER variables are the same in both equations, the relationship $\mathbf{ABM} = \mathbf{K}$ is given by

$$\begin{bmatrix} 0 & 1 & 0 & 0 & 0 \\ 0 & 0 & 0 & 1 & 0 \end{bmatrix} \begin{bmatrix} \beta_0 & \alpha_0 \\ \beta_1 & \alpha_1 \\ \beta_2 & \alpha_2 \\ \beta_3 & \alpha_3 \\ \beta_4 & \alpha_4 \end{bmatrix} \begin{bmatrix} 1 \\ -1 \end{bmatrix} = \begin{bmatrix} 0 \\ 0 \end{bmatrix}$$

or equivalently $\beta_1 - \alpha_1 = 0$ and $\beta_3 - \alpha_3 = 0$. The F-statistic with 2 and 95 degrees of freedom derived from Wilks' Lambda has a value of 0.048 and a p-value of 0.9531. The hypothesis that the coefficients of EDUC and EXPER are equal in both equations is therefore consistent with the data.

If the variable SENIOR is omitted from both equations, the unrestricted ordinary least squares estimation yields

$$\begin{array}{llll} \text{LCURRENT} = & 8.854 & +0.084 \text{ EDUC} & - 0.015 \text{ AGE} \\ & (0.000) & (0.000) & (0.000) \\ & & +0.016 \text{ EXPER} & R^2 = 0.526 \quad \text{and} \\ & & (0.001) & \end{array}$$

$$\begin{array}{llll} \text{LSTART} = & 8.031 & +0.081 \text{ EDUC} & - 0.011 \text{ AGE} \\ & (0.000) & (0.000) & (0.002) \\ & & +0.016 \text{ EXPER} & R^2 = 0.535. \\ & & (0.000) & \end{array}$$

A test of the hypothesis that the coefficients of the three explanatory variables are equal yields an F-statistic of 2.551. Comparing this statistic to an

F distribution with 3 and 96 degrees of freedom yields a p-value of 0.0601. Using the same estimated coefficients for the three variables in both regressions is therefore a reasonable procedure for the sampled population.

Relationship to Generalized Least Squares

An alternative way of writing equation (7.9) is given by

$$\mathbf{y}^* = \mathbf{X}^*\boldsymbol{\beta}^* + \mathbf{u}^*, \tag{7.12}$$

where $\mathbf{y}^*$, $\boldsymbol{\beta}^*$ and $\mathbf{u}^*$ are the vectors formed by stacking the columns of $\mathbf{Y}$, $\mathbf{B}$ and $\mathbf{U}$, and where $\mathbf{X}^*$ is the block diagonal matrix formed by repeating the matrix $\mathbf{X}$. The quantities are defined by

$$\mathbf{y}^* \ (ns \times 1) = \begin{bmatrix} \mathbf{y}_1 \\ \mathbf{y}_2 \\ \vdots \\ \mathbf{y}_s \end{bmatrix}, \quad \text{where } \mathbf{y}_j \ (n \times 1), \ j = 1, 2, \ldots, s, \text{ denotes the } j\text{th column of } \mathbf{Y},$$

$$\boldsymbol{\beta}^* \ [s(q+1) \times 1] = \begin{bmatrix} \boldsymbol{\beta}_1 \\ \boldsymbol{\beta}_2 \\ \vdots \\ \boldsymbol{\beta}_s \end{bmatrix}, \quad \text{where } \boldsymbol{\beta}_j \ [(q+1) \times 1], \ j = 1, 2, \ldots, s, \text{ denotes the } j\text{th column of } \mathbf{B},$$

$$\mathbf{u}^* \ (ns \times 1) = \begin{bmatrix} \mathbf{u}_1 \\ \mathbf{u}_2 \\ \vdots \\ \mathbf{u}_s \end{bmatrix}, \quad \text{where } \mathbf{u}_j, \ j = 1, 2, \ldots, s, \text{ denotes the } j\text{th column of } \mathbf{U},$$

and

$$\mathbf{X}^* = \begin{bmatrix} \mathbf{X} & 0 & \cdots & 0 \\ 0 & \mathbf{X} & & 0 \\ \vdots & & \ddots & \\ 0 & \cdots & 0 & \mathbf{X} \end{bmatrix}, \quad \text{where } \mathbf{X} \ [n \times (q+1)]$$

$$= [\mathbf{I}_x \otimes \mathbf{X}] \quad \text{the } \textit{direct product} \text{ of the identity matrix } \mathbf{I}_s \ (s \times s) \text{ with } \mathbf{X} \text{ (see Appendix).}$$

The covariance matrix for the error term $\mathbf{u}^*$ is given by the direct product $\boldsymbol{\Omega} = [\boldsymbol{\Gamma} \otimes \mathbf{I}_n]$, where $\boldsymbol{\Gamma} = \boldsymbol{\Sigma}_{22 \cdot 1} = \{\gamma_{jk}\}$, $j, k = 1, 2, \ldots, s$, is the covariance matrix for $\mathbf{u}_j, \quad j = 1, 2, \ldots, s,$

$$\boldsymbol{\Omega} = \begin{bmatrix} \gamma_1^2 \mathbf{I}_n & \gamma_{12} \mathbf{I}_n & \cdots & \gamma_{1s} \mathbf{I}_n \\ \gamma_{12} \mathbf{I}_n & \gamma_2^2 \mathbf{I}_n & & \gamma_{2s} \mathbf{I}_n \\ \vdots & \vdots & & \vdots \\ \gamma_{1s} \mathbf{I}_n & \gamma_{2s} \mathbf{I}_n & & \gamma_s^2 \mathbf{I}_n \end{bmatrix}$$

and $\mathbf{I}_n$ is the $(n \times n)$ identity matrix.

From the theory of the general linear model, the generalized least squares estimator for $\boldsymbol{\beta}^*$ is given by

$$\hat{\boldsymbol{\beta}}^* = (\mathbf{X}^* \boldsymbol{\Omega}^{-1} \mathbf{X}^*)^{-1} (\mathbf{X}^* \boldsymbol{\Omega}^{-1} \mathbf{y}^*), \tag{7.13}$$

which can be simplified in this case to

$$\hat{\boldsymbol{\beta}}^* = [\mathbf{I}_{(q+1)} \otimes (\mathbf{X}'\mathbf{X})^{-1}\mathbf{X}']\mathbf{y}^*. \tag{7.14}$$

From (7.14) we can see the matrix pre-multiplying $\mathbf{y}^*$ is block diagonal and hence that $\boldsymbol{\beta}^*$ is simply the stacked vector of columns of $\mathbf{B}$, which are the ordinary least squares estimators of the $\boldsymbol{\beta}_j$ in the s models $\mathbf{y}_j = \mathbf{X}_j \boldsymbol{\beta}_j + \mathbf{u}_j$, $j = 1, 2, \ldots, s$. The generalized least squares estimator and the ordinary least squares estimator are therefore equivalent for the multivariate regression model.

Zellner's Seemingly Unrelated Regression Model

The multivariate regression model (7.9) written in the form of (7.12) represents s separate univariate regression models all having the same $\mathbf{X}$ matrix. The models are related in that the error vectors $\mathbf{u}_j, \quad j = 1, 2, \ldots, s$, are mutually correlated. Even though the $\mathbf{u}_j$ vectors are correlated, the individual equation ordinary least squares estimators are equivalent to the system ordinary least squares estimator given by (7.10) and also to the generalized least squares estimator given by (7.13).

A more generalized multivariate regression model allows the individual equation $\mathbf{X}$ matrices to be different, say $\mathbf{X}_j, \quad j = 1, 2, \ldots, s$. In this case $\mathbf{X}^*$ has the block diagonal form with distinct block diagonal elements $\mathbf{X}_j$, $j = 1, 2, \ldots, s$. Thus $\mathbf{X}^*$ can no longer be written in the simplified form of $[\mathbf{I}_s \otimes \mathbf{X}]$. As a result, the generalized least squares estimator (7.13) is no longer equivalent to the ordinary least squares estimator $\hat{\mathbf{B}}$ in (7.10). This type of model has appeared in the economics literature and is usually referred to as *Zellner's Seemingly Unrelated Regression Model* (1962). The name "seemingly unrelated" is derived from the property that the equations in the model are only related through the error terms $\mathbf{u}_j, \quad j = 1, 2, \ldots, s$.

To obtain an estimator for this model, a feasible generalized least squares procedure is used. If $\boldsymbol{\Gamma}$, the covariance matrix for $\mathbf{u}_j$, is unknown, it can be estimated using the residuals from the multivariate ordinary least squares estimated model $\widehat{\mathbf{U}} = \mathbf{Y} - \mathbf{X}\widehat{\mathbf{B}}$. In this case $\widehat{\boldsymbol{\Gamma}} = (\mathbf{Y} - \mathbf{X}\widehat{\mathbf{B}})'(\mathbf{Y} - \mathbf{X}\widehat{\mathbf{B}})/n = \mathbf{E}/n$, where $\mathbf{E}$ is defined in the expression for Wilk's Lambda in multivariate regression. This estimator of $\boldsymbol{\Gamma}$ is consistent and hence can be used to yield the feasible generalized least squares estimator

$$\tilde{\boldsymbol{\beta}}^* = (\mathbf{X}^* \widehat{\boldsymbol{\Omega}}^{-1} \mathbf{X}^*)^{-1}(\mathbf{X}^* \widehat{\boldsymbol{\Omega}}^{-1} \mathbf{y}^*),$$

where $\widehat{\boldsymbol{\Omega}} = [\widehat{\boldsymbol{\Gamma}} \otimes \mathbf{I}_n]$.

The feasible generalized least squares estimator $\tilde{\boldsymbol{\beta}}^*$ is consistent, and, under the assumption of multivariate normality, inferences for $\boldsymbol{\beta}^*$ can be made using the fact that the expression

$$(\tilde{\boldsymbol{\beta}}^* - \boldsymbol{\beta}^*)'(\mathbf{X}^* \widehat{\boldsymbol{\Omega}}^{-1} \mathbf{X}^*)^{-1}(\tilde{\boldsymbol{\beta}}^* - \boldsymbol{\beta}^*)$$

has a χ^2 distribution with s degrees of freedom in large samples.

7.5.2   CANONICAL CORRELATION

Given two random variable vectors $\mathbf{y}$ $(s \times 1)$ and $\mathbf{x}$ $(q \times 1)$, we have already studied two ways of relating the variable elements of $\mathbf{y}$ to the variable elements of $\mathbf{x}$. One way is to examine the degree of linear association between all possible pairs consisting of one element of $\mathbf{y}$ and one element of $\mathbf{x}$ using the covariance matrix $\boldsymbol{\Sigma}_{\mathbf{xy}}$ or the corresponding correlation matrix $\boldsymbol{\rho}_{\mathbf{xy}}$. Alternatively, multivariate regression can be used to relate each element of $\mathbf{y}$ to all the elements of $\mathbf{x}$ and vice versa. The multivariate linear regression model determines linear combinations of the $\mathbf{x}$ variables that are *maximally correlated* with a particular $\mathbf{y}$ variable. In this section, we introduce *canonical correlation*, which is used to find linear combinations of both sets of variables $\mathbf{y}$ and $\mathbf{x}$ that are maximally correlated. Often in practice one vector of variables is a criterion set and the other vector of variables is a predictor set. The objective in canonical correlation analysis is to determine simultaneous relationships between the two sets of variables.

Derivation of Canonical Relationships

As in multivariate regression, we begin with the two random variable vectors $\mathbf{y}$ $(s \times 1)$ and $\mathbf{x}$ $(q \times 1)$ which have zero-valued mean vectors $\boldsymbol{\mu}_{\mathbf{y}} = \boldsymbol{\mu}_{\mathbf{x}} = \mathbf{0}$ and covariance matrix $\boldsymbol{\Sigma} = \begin{bmatrix} \boldsymbol{\Sigma}_{\mathbf{yy}} & \boldsymbol{\Sigma}_{\mathbf{yx}} \\ \boldsymbol{\Sigma}_{\mathbf{xy}} & \boldsymbol{\Sigma}_{\mathbf{xx}} \end{bmatrix}$. In this case there is no intercept term because the variables are assumed to have zero means.

Let $W = \boldsymbol{\beta}'\mathbf{x}$ and $Z = \boldsymbol{\alpha}'\mathbf{y}$ denote linear combinations of the $\mathbf{x}$ and $\mathbf{y}$ variables respectively. For each single variable in $\mathbf{y}$, say Y_j, we can use multiple regression to determine the vector $\boldsymbol{\beta}$ that maximizes the correlation between Y_j and W. Similarly, for any single variable in $\mathbf{x}$, say X_k, we can use multiple regression to determine the vector α that maximizes the correlation between X_k and Z. In canonical correlation we simultaneously determine the vectors α and β in such a way that the correlation between the two linear combinations Z and W is maximized.

The covariance between Z and W is given by $\alpha' \Sigma_{yx} \beta$, and the variances of Z and W are given by $\alpha' \Sigma_{yy} \alpha$ and $\beta' \Sigma_{xx} \beta$ respectively. The correlation between Z and W is therefore given by

$$r_{ZW} = \alpha' \Sigma_{yx} \beta / (\alpha' \Sigma_{yy} \alpha)^{1/2} (\beta' \Sigma_{xx} \beta)^{1/2}.$$

To determine unique values of α and β in order to maximize r_{ZW}, side conditions on the scales of Z and W must also be included. It is convenient to use the conditions $\alpha' \Sigma_{yy} \alpha = \beta' \Sigma_{xx} \beta = 1$.

An Eigenvalue Problem

To maximize r_{ZW} subject to $\alpha' \Sigma_{yy} \alpha = \beta' \Sigma_{xx} \beta = 1$ we require solutions to the two *systems of homogeneous equations*

$$(\Sigma_{xx}^{-1} \Sigma_{xy} \Sigma_{yy}^{-1} \Sigma_{yx} - \lambda_b I_b) \beta = 0 \quad \text{and}$$

$$(\Sigma_{yy}^{-1} \Sigma_{yx} \Sigma_{xx}^{-1} \Sigma_{xy} - \lambda_a I_a) \alpha = 0,$$

where I_b $(q \times q)$ and I_a $(s \times s)$ are identity matrices. The solution is obtained by determining the eigenvalues and eigenvectors of the matrices

$$\Sigma_{xx}^{-1} \Sigma_{xy} \Sigma_{yy}^{-1} \Sigma_{yx} \quad \text{and} \quad \Sigma_{yy}^{-1} \Sigma_{yx} \Sigma_{xx}^{-1} \Sigma_{xy}. \quad (7.15)$$

The eigenvalues of the two matrices are identical, $\lambda_a = \lambda_b = \lambda$, and the number of positive eigenvalues is t, where $t = \min(s, q)$ is the rank of the two matrices in (7.15). Corresponding to each eigenvalue, λ, is a unique pair of eigenvectors α and β. Denoting by $\lambda_1, \lambda_2, \ldots, \lambda_t$ the eigenvalues in order of magnitude from largest to smallest, the corresponding eigenvectors are denoted by $\alpha_1, \alpha_2, \ldots, \alpha_t$ and $\beta_1, \beta_2, \ldots, \beta_t$. The correlation between the two corresponding linear functions $\alpha'_j y$ and $\beta'_j x$ is given by $\sqrt{\lambda_j}$, $j = 1, 2, \ldots, t$.

The maximum correlation solution corresponds to λ_1, the largest eigenvalue, and hence the correlation is maximized by using $Z_1 = \alpha'_1 y$ and $W_1 = \beta'_1 x$. The remaining linear combinations for x given by $W_2, W_3, \ldots, W_t$ are mutually uncorrelated and uncorrelated with W_1. Similarly, the remaining linear combinations for y given by $Z_2, Z_3, \ldots, Z_t$ are also mutually uncorrelated and uncorrelated with Z_1. In addition, non-corresponding members of the two sets are uncorrelated; that is, Z_j is uncorrelated with W_k, $k \neq j, k$, $j = 1, 2, \ldots, t$.

The Canonical Variables

As a result of determining the eigenvalues and eigenvectors of

$$\Sigma_{xx}^{-1}\Sigma_{xy}\Sigma_{yy}^{-1}\Sigma_{yx} \quad \text{and}$$

$$\Sigma_{yy}^{-1}\Sigma_{yx}\Sigma_{xx}^{-1}\Sigma_{xy},$$

we have t pairs of *canonical variables* (Z_j, W_j) with correlations $\sqrt{\lambda_j}$, $j = 1, 2, \ldots, t$. Each successive pair of canonical variables maximizes the correlation subject to being uncorrelated with the previously determined pairs. In practice all but a small number of pairs usually have negligible correlations. Typically the eigenvalues λ_j, $j = 1, 2, \ldots, t$ decline in a rapid geometric fashion.

The canonical variables Z and W have been derived using the covariance matrices and the expressions for Z and W are in terms of the variables $\mathbf{y}$ and $\mathbf{x}$ respectively. If the correlation matrices ρ_{yy}, ρ_{xx} and ρ_{yx} are used, the same eigenvalues would be obtained. If, however, the correlation matrices are used, the canonical variables are expressed as functions of the standardized variables. The eigenvectors are not the same, therefore, when standardized data are used.

Sample Canonical Correlation Analysis

The canonical variates can be estimated using the sample covariance or correlation matrices $\mathbf{S_{xx}}$, $\mathbf{S_{yy}}$, $\mathbf{S_{xy}}$ and $\mathbf{S_{yx}}$, or $\mathbf{R_{xx}}$, $\mathbf{R_{yy}}$, $\mathbf{R_{xy}}$ and $\mathbf{R_{yx}}$ respectively. We assume in this discussion that the correlation matrices are used. The sample eigenvalues and eigenvectors are therefore determined from the matrices $\mathbf{R_{xx}^{-1}R_{xy}R_{yy}^{-1}R_{yx}}$ and $\mathbf{R_{yy}^{-1}R_{yx}R_{xx}^{-1}R_{xy}}$ and are denoted by $\lambda_1, \lambda_2, \ldots, \lambda_t$, $\mathbf{b_1, b_2, \ldots, b_t}$, and $\mathbf{a_1, a_2, \ldots, a_t}$, respectively.

Canonical Weights and Canonical Variables

The eigenvectors $\mathbf{a}_j$ and $\mathbf{b}_j$ are usually referred to as the *canonical weights*. These weights can be used to determine the values of the canonical variates Z_j and W_j, where $Z_j = \mathbf{a}_j'\mathbf{y}$, and $W_j = \mathbf{b}_j'\mathbf{x}$. The n values of the two new variables (Z_j, W_j) corresponding to the n observations are called the *canonical variate scores*. The canonical weights can also be used to interpret the canonical variables and the relationship between the canonical variables. The canonical variables are interpreted like regression functions. Each canonical weight gives the marginal impact of that variable on the canonical variable holding the other variables in the equation fixed. After each canonical variable of the pair is interpreted, the relationship between the pair is interpreted.

Inference For Canonical Correlation

Under the assumption that the Xs and Ys are multivarite normal, we can test the hypothesis that the correlations between the canonical variates are not significantly different from zero. To test the hypothesis that none of the λ_j are significantly different from zero, we use the test statistic $\chi^2 = -[n - \left(\frac{1}{2}\right)(s+q+3)]\log \Lambda$, which has approximately a χ^2 distribution with sq d.f. if the null hypothesis is true. The statistic Λ which is given by $\Lambda = \Pi_{j=1}^{t}(1 - \lambda_j)$ is called Wilk's Lambda. This statistic is equivalent to the statistic used to test the independence between two sets of variables introduced in Section 7.4. If the first hypothesis is rejected, we remove λ_1, the largest eigenvalue from Λ and compute $\Lambda_1 = \Pi_{j=2}^{t}(1 - \lambda_j)$. We then test the hypothesis that all remaining λ_j are not significantly different from zero, using the test statistic $\chi^2 = -[n - \left(\frac{1}{2}\right)(s + q + 3)]\log \Lambda_1$ which has a χ^2 distribution with $(s - 1)(q - 1)$ d.f. if the null hypothesis is true. To test the hypothesis that all remaining λ_j after the first k are not significantly different from zero, we compute $\Lambda_k = \Pi_{j=(k+1)}^{t}(1 - \lambda_j)$ where χ^2 now has $(s - k)(q - k)$ d.f. This process continues until the null hypothesis is accepted.

An Alternative Test Statistic

An alternative large sample approximation for the distribution of Wilk's Lamda under the hypothesis of independence is based on Rao's F used in multivariate regression above. The statistic is given by $F = m_{2k}(1 - \Lambda_k)^{1/\nu_k}/m_{1k}\Lambda_k^{1/\nu_k}$ where

$$\nu_k = \sqrt{\frac{(s - k)^2(q - k)^2 - 4}{(s - k)^2 + (q - k)^2 - 5}}$$

$$m_{1k} = (s - k)(q - k)$$

$$m_{2k} = \nu_k[n - \frac{1}{2}(s + q + 3)] - \frac{(s - k)(q - k)}{2} + 1,$$

which has m_{1k} and m_{2k} degrees of freedom if all but the first k eigenvectors are zero. Some computer software for canonical correlation analysis uses this F-approximation claiming that it is superior to the χ^2 approximation in small samples.

Computer Software

The calculations for the example in this section were performed using SAS PROC CANCORR.

Example

Using the bank salary data from Table 7.9, a canonical correlation analysis was carried out to relate the two salary variables LCURRENT and

TABLE 7.10. Correlation Matrix for Bank Data

	LCURRENT	LSTART	EDUC	AGE	EXPER	SENIOR
LCURRENT	1.000	0.889	0.666	-0.333	-0.099	0.050
LSTART	0.889	1.000	0.673	-0.003	-0.080	-0.234
EDUC	0.666	0.673	1.000	-0.294	-0.254	0.054
AGE	-0.333	-0.003	-0.294	1.000	0.730	0.070
EXPER	-0.099	-0.080	-0.254	0.730	1.000	-0.013
SENIOR	0.050	-0.234	0.054	0.070	-0.013	1.000

LSTART to the four background variables EDUC, AGE, EXPER and SE-
NIOR. The correlation matrix for the six variables is shown in Table 7.10.

The two eigenvalues obtained from the canonical correlation analysis are
$\lambda_1 = 0.559$ and $\lambda_2 = 0.142$. The correlations between the two corresponding canonical functions, which are the square roots of the eigenvalues, are
therefore 0.748 and 0.377 respectively. The values of Rao's F likelihood
ratio statistic are 14.71 with 8 and 188 degrees of freedom and 5.26 with
3 and 95 degrees of freedom. The resulting p-values are 0.000 and 0.002
respectively.

The two pairs of canonical functions using standardized coefficients are
given by

$$
\begin{aligned}
Z_1 &= 0.43 \text{ LCURRENT} + 0.60 \text{ LSTART} \\
W_1 &= 0.91 \text{ EDUC} + 0.54 \text{ EXPER} - 0.50 \text{ AGE} - 0.04 \text{ SENIOR} \\
Z_2 &= -2.14 \text{ LCURRENT} + 2.10 \text{ LSTART} \\
W_2 &= 0.22 \text{ EDUC} + 0.16 \text{ EXPER} + 0.59 \text{ AGE} - 0.78 \text{ SENIOR}.
\end{aligned}
$$

The first canonical function Z_1 for the salary variables almost represents a
simple average of the two salary variables and hence is a measure of salary
level. The first canonical function W_1 for the background variables contains
relatively large positive coefficients for EDUC and EXPER and a relatively
large negative coefficient for AGE. The function W_1 therefore measures a
contrast between the variables EDUC and EXPER and the variable AGE.
Therefore, from the canonical correlation relationship the higher the values
of EDUC and EXPER relative to AGE, the greater the value of W_1. The
positive correlation between Z_1 and W_1 therefore suggests that salary level
is higher when EDUC and EXPER are large relative to AGE.

The second canonical function Z_2 for the salary data measures a contrast
between LSTART and LCURRENT. The value of Z_2 increases as LSTART
increases relative to LCURRENT. The second canonical function W_2 for
the background variables is primarily a function of the variables AGE and
SENIOR. As AGE increases relative to SENIOR, the function W_2 increases.
The positive correlation between Z_2 and W_2 suggests therefore that, as

TABLE 7.11. Correlations Between Canonical Functions and Original Variables

	Z_1	Z_2	W_1	W_2
LCURRENT	0.96	-0.27	0.72	-0.10
LSTART	0.98	0.20	0.73	0.07
EDUC	0.69	-0.01	0.92	-0.03
AGE	-0.28	0.22	-0.38	0.59
EXPER	-0.04	0.21	-0.06	0.55
SENIOR	-0.03	-0.28	-0.04	-0.73

seniority with the bank increases relative to AGE, current salary increases relative to beginning salary. From the two estimated canonical relationships we have determined that salary level is high when education and experience are large relative to age and that salary growth is large when seniority with the bank is large relative to age.

Structure Correlations or Canonical Loadings

It is also useful to determine the correlation coefficients between the canonical variables and each of the constituent variables used to define the canonical variable. These correlations are called *structure correlations* or *canonical loadings*. By examining these canonical loadings the canonical variate can also be interpreted. The matrix of structure correlations between the **x** variables and the canonical variates $W_1, W_2, \ldots, W_t$ is given by $\mathbf{R_{xw}} = \mathbf{R_{xx}B}$, and similarly for the **y** variables and the canonical variates $Z_1, Z_2, \ldots, Z_t$ the matrix of structure correlations is given by $\mathbf{R_{yz}} = \mathbf{R_{yy}A}$. The matrices **B** and **A** contain the columns of eigenvectors $\mathbf{b}_j$, $j = 1, 2, \ldots, t$ and $\mathbf{a}_j$, $j = 1, 2, \ldots, t$ respectively. Figure 7.2 illustrates the relationships among the various correlation matrices.

Example

The correlations between the canonical functions and the original variables are shown in Table 7.11. For the first salary canonical function Z_1, a very strong positive correlation exists with both salary variables. The correlation between Z_1 and the background variables shows a relatively strong positive correlation with EDUC and a weak negative correlation with AGE. Thus, as EDUC increases, salary level tends to increase as well. For the second salary canonical function Z_2, the correlation with the two salary variables are quite weak. This results because the function Z_2 measures the difference between the two salary variables. The correlation between Z_2 and the background variables indicates weak positive correlations with AGE and EXPER and a weak negative correlation with SENIOR.

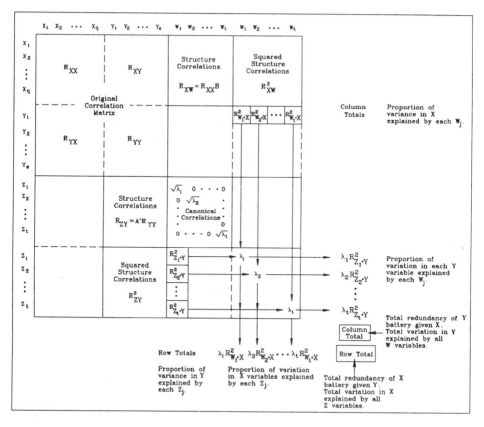

FIGURE 7.2. Summary of Canonical Correlation Terminology

For the background variables, the first canonical function W_1 is strongly positively correlated with the two salary variables and the variable EDUC and weakly negatively related to the variable AGE. For the second canonical function W_2, the correlations with the two salary variables are negligible. For the background variables, W_2 is positively related to AGE and EXPER and negatively related to SENIOR.

Redundancy Analysis and Proportion of Variance Explained

The square of any element of a *structure correlation matrix* gives the proportion of the variance of an original variable explained by a canonical variate. The sum of the squares in any column of a structure correlation matrix gives the total variation in the original variables explained by the canonical variate. For the jth column of $\mathbf{R_{XW}}$ the sum of squares of the elements gives the total of the proportions of variance explained by W_j. For a correlation matrix this total divided by q is the proportion of the total variation in the $\mathbf{x}$ variables explained by W_j (and is denoted by $R^2_{W_j\mathbf{x}}$). See

Figure 7.2. Similarly, the total of the elements squared in the jth column of $\mathbf{R_{yz}}$ divided by s denoted by $R^2_{Z_j \cdot \mathbf{y}}$ gives the proportion of the total variation in the $\mathbf{y}$ variables explained by Z_j.

Redundancy Measure for a Given Canonical Variate

It is possible for the canonical correlation $\sqrt{\lambda_j}$ to be relatively large, even though the proportion of variance of the underlying variables explained by the canonical variates Z_j and W_j is relatively small. Thus λ_j does not measure the strength of the correlation between the canonical variables Z_j and W_j and the underlying variables. The eigenvalue λ_j gives the proportion of variation in W_j explained by Z_j, and $R^2_{W_j \cdot \mathbf{x}}$ gives the proportion of variance in the $\mathbf{x}$ variables explained by W_j. The product $\lambda_j R^2_{W_j \cdot \mathbf{y}}$ therefore gives the proportion of the variation in $\mathbf{x}$ variables explained by the canonical variate Z_j. This product is called the *redundancy measure* and measures the quality of Z_j as a predictor of the $\mathbf{x}$ variables. Similarly, the redundancy measure $\lambda_j R^2_{Z_j \cdot \mathbf{y}}$ gives the proportion of the variation in the $\mathbf{y}$ variables explained by the canonical variate W_j. The relationships are shown in Figure 7.2.

Total Redundancy

A measure of *total redundancy* of the $\mathbf{x}$ battery given the $\mathbf{y}$ battery is given by the total $\sum_{j=1}^{t} \lambda_j R^2_{W_j \cdot \mathbf{x}}$, and similarly a measure of the total redundancy for the $\mathbf{y}$ battery given the $\mathbf{x}$ battery is $\sum_{j=1}^{t} \lambda_j R^2_{Z_j \cdot \mathbf{y}}$ (see Figure 7.2). It is possible for one of these totals to be high and the other low. A set of macroeconomic variables might be excellent predictors of certain microeconomic variables in a particular sector of the economy, but the reverse may not be so. Another example might be provided by student grades in a set of courses and a set of scores on an aptitude test. The grades may be more predictable from the aptitude scores than the reverse.

Relation to Multiple Regression

If each element of the $\mathbf{y}$ vector variable, say Y_j, is regressed separately on the $\mathbf{x}$ vector variable, $R^2_{Y_j \cdot \mathbf{x}}$ gives the proportion of the variation in Y_j, explained by the $\mathbf{x}$ set. If these R^2 values are averaged over all $\mathbf{y}$ variables, the result is equivalent to the total redundancy measure of the $\mathbf{y}$ battery given the $\mathbf{x}$ battery; thus

$$\sum_{j=1}^{t} \lambda_j R^2_{Z_j \cdot \mathbf{y}} = \sum_{j=1}^{r} R^2_{Y_j \cdot \mathbf{x}} / r.$$

Similarly, the total redundancy measure of the $\mathbf{x}$ battery given the $\mathbf{y}$ battery is given by

$$\sum_{j=1}^{t} \lambda_j R^2_{W_j \cdot \mathbf{x}} = \sum_{j=1}^{q} R^2_{X_j \cdot \mathbf{y}}/q,$$

which is the average of the squared multiple correlations relating each of the X variables to the Y set.

Example

A redundancy analysis can be carried out using the eigenvalues determined above and the correlations in Table 7.11. For the canonical function Z_1, the cumulative proportion of variance explained by the salary variables is $\left(\frac{1}{2}\right)[(0.96)^2 + (0.98)^2] = 0.94$ and, after multiplying by $\lambda_1 = 0.56$, the redundancy measure is $(0.94)(0.56) = 0.53$. For the second function Z_2, the variance explained is $\left(\frac{1}{2}\right)[(-0.27)^2 + (0.20)^2] = 0.06$ and, on multiplication by $\lambda_2 = 0.14$, the redundancy measure is $(0.06)(0.14) = 0.01$. The proportion of variation in the salary variables explained by the canonical functions of the four background variables W_1 and W_2 is $0.53 + 0.01 = 0.54$.

For the two canonical functions W_1 and W_2, the redundancy measures are

$$\left(\frac{1}{4}\right)[(0.92)^2 + (-0.38)^2 + (-0.06)^2 + (0.04)^2](0.56) \quad = \quad 0.14 \quad \text{and}$$

$$\left(\frac{1}{4}\right)[(-0.03)^2 + (0.59)^2 + (0.55)^2 + (-0.73)^2](0.14) \quad = \quad 0.04$$

and hence a total of 0.18. The two canonical functions Z_1 and Z_2 based on the salary variables explain a proportion 0.18 of the variation in the background variables. Thus the redundancy analysis indicates that the background variables explain a large portion of the variation in the salary variables but that the reverse is not true.

Residuals, Influence, Outliers and Cross Validation

The techniques available for studying residuals, detecting outliers and measuring influence in multiple linear regression can be used in canonical correlation analyses. By relating each variable in one group to all of the variables in the other group using multiple linear regression, conventional regression software can be used.

A cross validation can be carried out by splitting the sample randomly into g mutually exclusive groups. Leaving out one group at a time a canonical correlation is carried out using the combined data for the $(g-1)$ groups. The canonical weights obtained are then applied to the omitted group and the correlations determined. The correlations should be similar to the correlations determined in the canonical correlation analysis based on the $(g-1)$ groups. This procedure is repeated for each group as an omitted group.

7.5.3 OTHER SOURCES OF INFORMATION

More extensive discussion of the topics of Section 7.5 can be found in Anderson (1984), Seber (1984), Mardia, Kent and Bibby (1979) and Press (1972).

Cited Literature and References

1. Anderson, T.W. (1984). *An Introduction to Multivariate Statistical Analysis*, Second Edition. New York: John Wiley and Sons.

2. Devlin, S.J., Gnanadesikan, R. and Kettering, J.R. (1975), "Robust Estimation and Outlier Detection with Correlation Coefficients," *Biometrika*, 62, 531–545.

3. Geisser, S. and Greenhouse, S.W. (1958). "An Extension of Box's Results on the Use of the F-Distribution in Multivariate Analysis," *Annals of Mathematical Statistics*, 29, 885–891.

4. Gnanadesikan, R. (1977). *Methods for Statistical Data Analysis of Multivariate Observations*. New York: John Wiley and Sons.

5. Hawkins, D.M. (1980). *Identification of Outliers*. London: Chapman and Hall.

6. Huynh, H. and Feldt, L.S. (1970), "Conditions under which Mean Square Ratios in Repeated Measurement Designs have Exact F-Distributions," *Journal of the American Statistical Association*, 65, 1582–1589.

7. Jobson, J.D. and Korkie, R. (1982). "Potential Performance and Tests of Portfolio Efficiency," *Journal of Financial Economics* 10, 433–456.

8. Jobson, J.D. and Korkie, R. (1989). "A Performance Interpretation of Multivariate Tests of Asset Set Intersection, Spanning and Mean-Variance Efficiency," *Journal of Financial and Quantitative Analysis* 24, 185–204.

9. Johnson, Richard A. and Wichern, Dean W. (1988). *Applied Multivariate Statistical Analysis*, Second Edition. Englewood Cliffs, NJ: Prentice–Hall.

10. Kryzanowski, W.J. (1988). *Principles of Multivariate Analysis: A User's Perspective*. Oxford: Oxford University Press.

11. Mardia, K.V. (1970). "Measures of Multivariate Skewness and Kurtosis with Applications," *Biometrika*, 57, 519–530.

12. Mardia, K.V., Kent, J.T. and Bibby, J.M. (1979). *Multivariate Analysis*. London: Academic Press.

13. Morrison, Donald F. (1976). *Multivariate Statistical Methods*, Second Edition. New York: McGraw–Hill.

14. Press, S. James (1972). *Applied Multivariate Analysis*. New York: Holt, Rinehart, Winston.

15. Seber, G.A.F. (1984). *Multivariate Observations*. New York: John Wiley and Sons.

16. Stevens, James (1986). *Applied Multivariate Statistics for the Social Sciences*. Hillsdale, NJ: Lawrence Erlbaum Associates.

17. Zellner, A. (1962). "An Efficient Method of Estimating Seemingly Unrelated Regressions and Tests of Aggregation Bias," *Journal of the American Statistical Association*, 57, 348–368.

Exercises for Chapter 7

1. This exercise is based on the Real Estate Data in Table V4 of the Data Appendix.

   (a) Estimate the multivariate regression model with the two equations given below. Discuss the results of the analysis.

   $$\begin{aligned} \text{LISTP} \;=\; & \beta_0 + \beta_1 \text{SQF} + \beta_2 \text{AGE} + \beta_3 \text{ROOMS} + \beta_4 \text{BATH} \\ & + \beta_5 \text{EXTRAS} + \beta_6 \text{GARAGE} + \beta_7 \text{CHATTELS} \\ & + \beta_8 \text{BEDR} + \beta_9 \text{SELLDAYS} \\ \text{SELLP} \;=\; & \alpha_0 + \alpha_1 \text{SQF} + \alpha_2 \text{AGE} + \alpha_3 \text{ROOMS} + \alpha_4 \text{BATH} \\ & + \alpha_5 \text{EXTRAS} + \alpha_6 \text{GARAGE} + \alpha_7 \text{CHATTELS} \\ & + \alpha_8 \text{BEDR} + \alpha_9 \text{SELLDAYS}. \end{aligned}$$

   (b) Carry out a test of the hypothesis

   $$\alpha_j = \beta_j, \quad j = 1, 2, \ldots, 9, \text{ for each variable separately.}$$

   Discuss the outcome of the test.

   (c) Carry out a test of the hypothesis

   $$\alpha_j = \beta_j = 0, \quad j = 1, 2, \ldots, 9, \text{ for each variable separately.}$$

   Discuss the outcome of the test.

   (d) If you were to estimate a reduced model what variables would you include? Estimate the reduced model and compare it to the complete model using a test statistic.

   (e) Carry out a canonical correlation analysis relating the two price variables to the nine explanatory variables and discuss the results. Also provide an interpretation for each of the canonical functions using both the canonical function coefficients and the correlations between the canonical functions and the original variables. Also discuss the results of a redundancy analysis.

2. This exercise is based on the Automobile Data in Table V5 of the Data Appendix.

   (a) Estimate the multivariate regression model with the two equations given below. Discuss the results of the analysis.

   $$\begin{aligned} \text{URBRATE} \;=\; & \beta_0 + \beta_1 \text{ENGSIZE} + \beta_2 \text{WEIGHT} + \beta_3 \text{FOR} \\ & + \beta_4 \text{AUTOMAT} + \beta_5 \text{FWEIGHT} + \beta_6 \text{AWEIGHT} \\ & + \beta_7 \text{FENGSIZE} + \beta_8 \text{AENGSIZE} \end{aligned}$$

$$\begin{aligned}
\text{HWRATE} \;=\;\; & \alpha_0 + \alpha_1\text{ENGSIZE} + \alpha_2\text{WEIGHT} + \alpha_3\text{FOR} \\
& + \alpha_4\text{AUTOMAT} + \alpha_5\text{FWEIGHT} + \alpha_6\text{AWEIGHT} \\
& + \alpha_7\text{FENGSIZE} + \alpha_8\text{AENGSIZE}
\end{aligned}$$

(b) Carry out a test of the hypothesis

$$\alpha_j = \beta_j, \quad j = 1, 2, \ldots, 8, \text{ for each variable separately.}$$

Discuss the outcome of the test.

(c) Carry out a test of the hypothesis

$$\alpha_j = \beta_j = 0, \quad j = 1, 2, \ldots, 8, \text{ for each variable separately.}$$

(d) If you were to estimate a reduced model what variables would you include? Estimate the reduced model and compare it to the full model using a test statistic.

(e) Carry out a canonical correlation analysis relating the two rate variables to the nine explanatory variables and discuss the results. Also provide an interpretation of the canonical functions using both the canonical function coefficients and the correlations between the canonical functions and the original variables. Also discuss the results of a redundancy analysis.

Questions for Chapter 7

1. Let the joint density of the random variables X_1, X_2 and X_3 be given by

$$f(x_1, x_2, x_3) = 2/3(x_1 + x_2 + x_3), \quad 0 \le x_1, x_2, x_3 \le 1$$
$$= 0 \quad \text{otherwise.}$$

(a) Show that $f(x_1, x_2, x_3)$ is a density by showing that

$$\int_0^1 \int_0^1 \int_0^1 2/3(x_1 + x_2 + x_3) dx_1 dx_2 dx_3 = 1.$$

(b) Show that the distribution function is given by

$$F(x_1, x_2, x_3) = 1/3[x_1^2 x_2 x_3 + x_1 x_2^2 x_3 + x_1 x_2 x_3^2].$$

Use this function to determine $P[x_1 \le \frac{1}{2}, \ x_2 \le \frac{1}{2}, \ x_3 \le \frac{1}{2}]$.

(c) Show that the marginal density of X_1 is given by

$$f(x_1) = \int_0^1 \int_0^1 2/3(x_1 + x_2 + x_3) dx_2 dx_3 = \left(\frac{2}{3}\right)(x_1 + 1).$$

Plot the density for X_1.

(d) Show that the conditional density for X_2, X_3 given $X_1 = x_1$ is given by

$$f_{X_2, X_3 | X_1}(x_2, x_3 \mid X_1 = x_1) = (x_1 + x_2 + x_3)/(x_1 + 1)$$

and show that

$$\int_0^1 \int_0^1 (x_1 + x_2 + x_3)/(x_1 + 1) dx_2 dx_3 = 1.$$

Is X_1 independent of X_2 and X_3?

(e) Show that $E[x_1] = \mu_1$ is given by

$$E[X_1] = \int_0^1 \left(\frac{2}{3}\right)(x_1 + 1)x_1 dx_1 = \left(\frac{5}{9}\right).$$

What are the values of $E[X_2]$ and $E[X_3]$?

(f) Show that

$$E[X_2/X_1 = x_1] = \int_0^1 \int_0^1 \frac{(x_1 + x_2 + x_3)}{(x_1 + 1)} \cdot x_2 dx_2 dx_3$$
$$= \left(\frac{1}{2}\right)(x_1 + 7/6)/(x_1 + 1).$$

Use this function to evaluate $E[X_2 \mid X_1 = \frac{1}{2}]$. Is the regression function for X_2 on X_1 linear?

2. Given

$$
\mathbf{x} = \begin{bmatrix} x_1 \\ x_2 \\ x_3 \end{bmatrix}, \quad E[\mathbf{x}] = \begin{bmatrix} \mu_1 \\ \mu_2 \\ \mu_3 \end{bmatrix} = \boldsymbol{\mu}, \text{ and}
$$

$$
\mathrm{Cov}(\mathbf{x}) = \begin{bmatrix} \sigma_1^2 & \sigma_{12} & \sigma_{13} \\ \sigma_{12} & \sigma_2^2 & \sigma_{23} \\ \sigma_{13} & \sigma_{23} & \sigma_3^2 \end{bmatrix} = \boldsymbol{\Sigma}.
$$

(a) Let

$$
\begin{aligned}
z_1 &= a_{11}x_1 + a_{12}x_2 + a_{13}x_3 \\
z_2 &= a_{21}x_1 + a_{22}x_2 + a_{23}x_3,
\end{aligned}
$$

and show that $\mathbf{z} = \mathbf{A}\mathbf{x}$, where $\mathbf{z} = \begin{bmatrix} z_1 \\ z_2 \end{bmatrix}$,

$$
\mathbf{A} = \begin{bmatrix} a_{11} & a_{12} & a_{13} \\ a_{21} & a_{22} & a_{23} \end{bmatrix}.
$$

(b) Show that $E[\mathbf{z}] = \mathbf{A}\boldsymbol{\mu}$ and $\mathrm{Cov}(\mathbf{z}) = \mathbf{A}\boldsymbol{\Sigma}\mathbf{A}'$.

(c) Given that $\boldsymbol{\Sigma}$ $(p \times p)$ is the covariance matrix for $\mathbf{x}$ $(p \times 1)$ and that $\boldsymbol{\rho}$ $(p \times p)$ is the corresponding correlation matrix show that

$$
\boldsymbol{\rho} = \boldsymbol{\delta}^{-1/2}\boldsymbol{\Sigma}\boldsymbol{\delta}^{-1/2}
$$

where $\boldsymbol{\delta}$ is a diagonal matrix whose diagonal elements are equal to the diagonal elements of $\boldsymbol{\Sigma}$.

3. Given that the sample covariance matrix is given by $\mathbf{S} = \sum_{i=1}^{n}(\mathbf{x}_i - \bar{\mathbf{x}})(\mathbf{x}_i - \bar{\mathbf{x}})'/(n-1)$, show that $\mathbf{S} = [\mathbf{X}'\mathbf{X} - n\bar{\mathbf{x}}\bar{\mathbf{x}}']/(n-1)$ where $\mathbf{X}$ is the $(n \times p)$ data matrix and $\bar{\mathbf{x}}$ $(p \times 1)$ is the vector of sample means.

4. (a) Show that

$$
(n-1)^{-1}\mathbf{S}^{-1} = (\mathbf{X}'\mathbf{X})^{-1} + \frac{n(\mathbf{X}'\mathbf{X})^{-1}\bar{\mathbf{x}}\bar{\mathbf{x}}'(\mathbf{X}'\mathbf{X})^{-1}}{1 - n\bar{\mathbf{x}}'(\mathbf{X}'\mathbf{X})^{-1}\bar{\mathbf{x}}}
$$

using the identity

$$
[\mathbf{A} + \mathbf{a}\mathbf{a}']^{-1} = \mathbf{A}^{-1} - \frac{\mathbf{A}^{-1}\mathbf{a}\mathbf{a}'\mathbf{A}^{-1}}{1 + \mathbf{a}'\mathbf{A}^{-1}\mathbf{a}}.
$$

(b) Use the result in (a) to show that

$$
(n-1)^{-1}\bar{\mathbf{x}}'\mathbf{S}^{-1}\bar{\mathbf{x}} = \bar{\mathbf{x}}'(\mathbf{X}'\mathbf{X})^{-1}\bar{\mathbf{x}}/[1 - n\bar{\mathbf{x}}'(\mathbf{X}'\mathbf{X})^{-1}\bar{\mathbf{x}}].
$$

5. Let $(\mathbf{x}_i - \bar{\mathbf{x}})'\mathbf{S}^{-1}(\mathbf{x}_i - \bar{\mathbf{x}})$ denote the square of the sample Mahalanobis distance from $\mathbf{x}_i$ to $\bar{\mathbf{x}}$ where $\bar{\mathbf{x}}$ and $\mathbf{S}$ are the sample mean vector and covariance matrix. Use the fact that for $\mathbf{a}$ ($p \times 1$) and $\mathbf{A}$ ($p \times p$) the scalar $\mathbf{a}'\mathbf{A}\mathbf{a}$ can be expressed as $\mathbf{a}'\mathbf{A}\mathbf{a} = tr\mathbf{a}'\mathbf{A}\mathbf{a} = tr\mathbf{A}\mathbf{a}\mathbf{a}'$ to show that $\sum_{i=1}^{n}(\mathbf{x}_i - \bar{\mathbf{x}})'\mathbf{S}^{-1}(\mathbf{x}_i - \bar{\mathbf{x}}) = (n-1)p$. NOTE: $\mathbf{S} = \sum_{i=1}^{n}(\mathbf{x}_i - \bar{\mathbf{x}})(\mathbf{x}_i - \bar{\mathbf{x}})'/(n-1)$.

6. (a) Let $\mathbf{z}_i = \mathbf{D}^{-1/2}(\mathbf{x}_i - \bar{\mathbf{x}})$, $i = 1, 2, \ldots, n$, denote a transformation of $\mathbf{x}$ where $\mathbf{D}$ is the diagonal matrix whose diagonal elements are the diagonal elements of $\mathbf{S}$. If $\mathbf{Z} = \begin{bmatrix} \mathbf{z}'_1 \\ \mathbf{z}'_2 \\ \vdots \\ \mathbf{z}'_n \end{bmatrix}$ denotes an $(n \times p)$ matrix of transformed observations show that $\mathbf{Z}'\mathbf{Z}/(n-1)$ is the sample correlation matrix $\mathbf{R}$. The transformed variables are the standardized variables. In what way are the new variables standardized?

   (b) Let $\mathbf{w}_i = \mathbf{S}^{-1/2}(\mathbf{x}_i - \bar{\mathbf{x}})$ denote a transformation of $\mathbf{x}_i$ and let $\mathbf{W} = \begin{bmatrix} \mathbf{w}'_1 \\ \mathbf{w}'_2 \\ \vdots \\ \mathbf{w}'_n \end{bmatrix}$ denote the $(n \times p)$ matrix of transformed observations. Show that $\mathbf{W}'\mathbf{W}/(n-1) = \mathbf{I}$ (the identity matrix). What is the covariance matrix for the variables in $\mathbf{W}$?

7. Let $\mathbf{X}(n)$ denote the data matrix $\mathbf{X}$ with the nth row $\mathbf{x}'_n$ deleted. Show that

   (a) $\mathbf{X}'\mathbf{X} = \mathbf{X}'(n)\mathbf{X}(n) + \mathbf{x}_n\mathbf{x}'_n$;

   (b) $\bar{\mathbf{x}}(n) = \frac{1}{(n-1)}[n\bar{\mathbf{x}} - \mathbf{x}_n]$, where $\bar{\mathbf{x}}(n)$ denotes the sample mean vector based on the first $(n-1)$ rows of $\mathbf{X}$.

   (c) Use the relationships $(n-1)\mathbf{S} = \mathbf{X}'\mathbf{X} - n\bar{\mathbf{x}}\bar{\mathbf{x}}'$, $(n-2)\mathbf{S}(n) = \mathbf{X}'(n)\mathbf{X}(n) - (n-1)\bar{\mathbf{x}}(n)\bar{\mathbf{x}}'(n)$ and results (a) and (b) to show that $\mathbf{S}(n) = \frac{(n-1)}{(n-2)}\mathbf{S} - \frac{n}{(n-1)(n-2)}(\mathbf{x}_n - \bar{\mathbf{x}})(\mathbf{x}_n - \bar{\mathbf{x}})'$ where $\mathbf{S}(n)$ denotes the sample covariance matrix based on the first $(n-1)$ rows of $\mathbf{X}$. What can you conclude about the difference between $\mathbf{S}(n)$ and $\mathbf{S}$ if $\mathbf{x}_n$ is an outlier which is large relative to $\bar{\mathbf{x}}$.

   (d) Show that

   $$\mathbf{S}^{-1}(n) = \frac{(n-2)}{(n-1)}\mathbf{S}^{-1} + \frac{\frac{n(n-2)}{(n-1)^3}\mathbf{S}^{-1}(\mathbf{x}_n - \bar{\mathbf{x}})(\mathbf{x}_n - \bar{\mathbf{x}})'\mathbf{S}^{-1}}{\left[1 - \frac{n}{(n-1)^2}(\mathbf{x}_n - \bar{\mathbf{x}})'\mathbf{S}^{-1}(\mathbf{x}_n - \bar{\mathbf{x}})\right]}$$

   by using the relation

$$[\mathbf{A} - \mathbf{d}\mathbf{d}']^{-1} = \mathbf{A}^{-1} + \mathbf{A}^{-1}\mathbf{d}\mathbf{d}'\mathbf{A}^{-1}/(1 - \mathbf{d}'\mathbf{A}^{-1}\mathbf{d}).$$

What can you conclude about the difference between $\mathbf{S}^{-1}(n)$ and $\mathbf{S}^{-1}$ if $\mathbf{x}_n$ is an outlier which is large relative to $\bar{\mathbf{x}}$?

8. Partition the random variable $\mathbf{x}$ into $\mathbf{x} = \begin{bmatrix} \mathbf{x}_1 \\ \mathbf{x}_2 \end{bmatrix}$ and let the corresponding partitions for the covariance matrix $\boldsymbol{\Sigma}$ and the mean vector $\boldsymbol{\mu}$ be denoted by

$$\boldsymbol{\Sigma} = \begin{bmatrix} \boldsymbol{\Sigma}_{11} & \boldsymbol{\Sigma}_{12} \\ \boldsymbol{\Sigma}_{21} & \boldsymbol{\Sigma}_{22} \end{bmatrix} \quad \text{and} \quad \boldsymbol{\mu} = \begin{bmatrix} \mu_1 \\ \mu_2 \end{bmatrix}.$$

(a) Use the formula for the inverse of a partitioned matrix to show that

$$\boldsymbol{\Sigma}^{-1} = \begin{bmatrix} \boldsymbol{\Theta}_{11} & \boldsymbol{\Theta}_{12} \\ \boldsymbol{\Theta}_{21} & \boldsymbol{\Theta}_{22} \end{bmatrix},$$

where

$$\begin{aligned}
\boldsymbol{\Theta}_{11} &= \boldsymbol{\Sigma}_{11}^{-1} + \boldsymbol{\Sigma}_{11}^{-1}\boldsymbol{\Sigma}_{12}[\boldsymbol{\Sigma}_{22} - \boldsymbol{\Sigma}_{21}\boldsymbol{\Sigma}_{11}^{-1}\boldsymbol{\Sigma}_{12}]^{-1}\boldsymbol{\Sigma}_{21}\boldsymbol{\Sigma}_{11}^{-1} \\
\boldsymbol{\Theta}_{12} &= -\boldsymbol{\Sigma}_{11}^{-1}\boldsymbol{\Sigma}_{12}[\boldsymbol{\Sigma}_{22} - \boldsymbol{\Sigma}_{21}\boldsymbol{\Sigma}_{11}^{-1}\boldsymbol{\Sigma}_{12}]^{-1} \\
\boldsymbol{\Theta}_{22} &= [\boldsymbol{\Sigma}_{22} - \boldsymbol{\Sigma}_{21}\boldsymbol{\Sigma}_{11}^{-1}\boldsymbol{\Sigma}_{12}]^{-1}.
\end{aligned}$$

Recall (see Appendix) the formula for the inverse of a partitioned symmetric matrix is given by

$$\begin{bmatrix} \mathbf{A} & \mathbf{B} \\ \mathbf{B}' & \mathbf{D} \end{bmatrix}^{-1} = \begin{bmatrix} \alpha & \beta \\ \beta' & \gamma \end{bmatrix},$$

where

$$\begin{aligned}
\alpha &= \mathbf{A}^{-1}\mathbf{B}[\mathbf{D} - \mathbf{B}'\mathbf{A}^{-1}\mathbf{B}]^{-1}\mathbf{B}'\mathbf{A}^{-1} + \mathbf{A}^{-1} \\
\beta &= -\mathbf{A}^{-1}\mathbf{B}[\mathbf{D} - \mathbf{B}'\mathbf{A}^{-1}\mathbf{B}]^{-1} \\
\gamma &= [\mathbf{D} - \mathbf{B}'\mathbf{A}^{-1}\mathbf{B}]^{-1}.
\end{aligned}$$

(b) Use the expression for the inverse of the partitioned matrix $\boldsymbol{\Sigma}$ to show that

$$\begin{aligned}
(\mathbf{x} - \boldsymbol{\mu})'\boldsymbol{\Sigma}^{-1}(\mathbf{x} - \boldsymbol{\mu}) &= (\mathbf{x}_1 - \mu_1)'\boldsymbol{\Sigma}_{11}^{-1}(\mathbf{x}_1 - \mu_1) \\
&+ [(\mathbf{x}_2 - \mu_2) - \boldsymbol{\Sigma}_{21}\boldsymbol{\Sigma}_{11}^{-1}(\mathbf{x}_1 - \mu_1)]' \\
&\times [\boldsymbol{\Sigma}_{22} - \boldsymbol{\Sigma}_{21}\boldsymbol{\Sigma}_{11}^{-1}\boldsymbol{\Sigma}_{12}]^{-1}[(\mathbf{x}_2 - \mu_2) - \boldsymbol{\Sigma}_{21}\boldsymbol{\Sigma}_{11}^{-1}(\mathbf{x}_1 - \mu_1)].
\end{aligned}$$

(c) Use the result in (b) and the fact that $|\boldsymbol{\Sigma}| = |\boldsymbol{\Sigma}_{11}| \, |\boldsymbol{\Sigma}_{22} - \boldsymbol{\Sigma}_{21}\boldsymbol{\Sigma}_{11}^{-1}\boldsymbol{\Sigma}_{12}|$ to show that the multivariate normal density for x can be expressed as

$$(2\pi)^{-q/2}|\Sigma_{11}|^{-1/2}\exp\left[-\left(\frac{1}{2}\right)(x_1-\mu_1)'\Sigma_{11}^{-1}(x_1-\mu_1)\right]$$
$$\times(2\pi)^{-(p-q)/2}|\Sigma_{22}-\Sigma_{21}\Sigma_{11}^{-1}\Sigma_{12}|^{-1/2}$$
$$\times\exp\left\{-\left(\frac{1}{2}\right)[(x_2-\mu_2)-\Sigma_{21}\Sigma_{11}^{-1}(x_1-\mu_1)]'\right.$$
$$\left.\times[\Sigma_{22}-\Sigma_{21}\Sigma_{11}^{-1}\Sigma_{12}]^{-1}[(x_2-\mu_2)-\Sigma_{21}\Sigma_{11}^{-1}(x_1-\mu_1)]\right\}$$

and hence that $f(x) = f_1(x_1)f_2(x_2 \mid x_1)$.

9. If x $(p \times 1)$ has a multivariate normal distribution with mean vector μ and covariance matrix Σ then the new random variable $y(q \times 1) = Ax$ has a multivariate normal with mean vector $A\mu$ and covariance matrix $A\Sigma A'$, where A $(q \times p)$ has rank q.

(a) Partition x into $\begin{bmatrix} x_1 \\ x_2 \end{bmatrix}$ where x_2 is $(q \times 1)$ and define $y = \Sigma_{12}\Sigma_{22}^{-1}x_2$ where Σ_{12}, Σ_{22} are the partitions of Σ corresponding to $\begin{bmatrix} x_1 \\ x_2 \end{bmatrix}$. Show that y is multivariate normal with mean vector $\Sigma_{12}\Sigma_{22}^{-1}\mu_2$ and covariance matrix $\Sigma_{12}\Sigma_{22}^{-1}\Sigma_{21}$.

(b) Given that $y = Ax$ the covariance matrix $\text{Cov}(y, x)$ is given by $A\Sigma$ whereas the covariance matrix $\text{Cov}(x, y)$ is given by $\Sigma A'$. Show that $\text{Cov}(y, x_2)$ and $\text{Cov}(x_2, y)$ for y given in (a) are Σ_{12} and Σ_{21} respectively.

(c) Given that $y = Ax_2$, the covariance $\text{Cov}(y, x_1)$ is given by $A\Sigma_{21}$ whereas the covariance matrix $\text{Cov}(x_1, y)$ is given by $\Sigma_{12}A'$. Show that $\text{Cov}(y, x_1)$ and $\text{Cov}(x_1, y)$ for y given in (a) is $\Sigma_{12}\Sigma_{22}^{-1}\Sigma_{21}$ in each case.

(d) Define $z = x_1 - \Sigma_{12}\Sigma_{22}^{-1}x_2$ and use (b) and (c) to show that $\text{Cov}(x_2, z) = 0$ and $\text{Cov}(x_1, z) = \Sigma_{11} - \Sigma_{12}\Sigma_{22}^{-1}\Sigma_{21}$.

(e) Recognizing that $x_1 = y + z$, where y is defined in (a) and z is defined in (d) use the results in (b), (c) and (d) to show that y and z divide x_1 into two components such that z is uncorrelated with x_2.

(f) Relate the variables $\Sigma_{12}\Sigma_{22}^{-1}x_2$ and $(x_1 - \Sigma_{12}\Sigma_{22}^{-1}x_2)$ to the multivariate regression of x_1 on x_2.

10. In the Appendix, it is shown that if Σ $(p \times p)$ is a full rank symmetric matrix then Σ can be written as $\Sigma = V\Lambda V'$ where Λ is a diagonal matrix of eigenvalues of Σ and V is the corresponding orthogonal matrix of eigenvectors $(V' = V^{-1})$.

(a) Use this result to show that $Q = (\mathbf{x} - \boldsymbol{\mu_x})' \boldsymbol{\Sigma}^{-1}(\mathbf{x} - \boldsymbol{\mu_x})$ can be written as $Q = (\mathbf{y} - \boldsymbol{\mu_y})' \boldsymbol{\Lambda}^{-1}(\mathbf{y} - \boldsymbol{\mu_y})$ where $\mathbf{y} = \mathbf{V}'\mathbf{x}$ and $\boldsymbol{\mu_y} = \mathbf{V}'\boldsymbol{\mu_x}$.

(b) Show that Q in (a) can be written as $\sum_{i=1}^{p}(y_i - \mu_{y_i})^2/\lambda_i$ where y_i, $\mu_{\mathbf{y}_i}$ and λ_i are elements of $\mathbf{y}$, $\boldsymbol{\mu_y}$ and $\boldsymbol{\Lambda}$ respectively.

(c) If $\mathbf{x} \sim N(\boldsymbol{\mu}, \boldsymbol{\Sigma})$ and $\mathbf{y} = \mathbf{A}'\mathbf{x}$ with $\mathbf{A}$ $(p \times p)$ of rank p then $\mathbf{y} \sim N(\mathbf{A}\boldsymbol{\mu}, \mathbf{A}'\boldsymbol{\Sigma}\mathbf{A})$. Use this result to show that $\mathbf{y} = \mathbf{V}'\mathbf{x}$ defined in (a) is a normal distribution with mean vector $\mathbf{V}'\boldsymbol{\mu}$ and diagonal covariance matrix $\boldsymbol{\Lambda}$. Why are the elements of $\mathbf{y}$ statistically independent?

(d) Given that the sum of squares of p mutually independent standard normal random variables has a χ^2 distribution with p degrees of freedom, show that Q in (a) and (b) has a χ^2 distribution with p degrees of freedom.

11. The density of the multivariate normal random variable is given by

$$f(\mathbf{x}_i) = |2\pi\boldsymbol{\Sigma}|^{-1/2}\exp[-\frac{1}{2}(\mathbf{x}_i - \boldsymbol{\mu})'\boldsymbol{\Sigma}^{-1}(\mathbf{x}_i - \boldsymbol{\mu})].$$

(a) Show that the joint density and hence the likelihood function for the random sample $\mathbf{x}_1, \mathbf{x}_2, \ldots, \mathbf{x}_n$ is given by

$$L = \prod_{i=1}^{n} f(\mathbf{x}_i) = |2\pi\boldsymbol{\Sigma}|^{-n/2}\exp\left[-\frac{1}{2}\sum_{i=1}^{n}(\mathbf{x}_i - \boldsymbol{\mu})\boldsymbol{\Sigma}^{-1}(\mathbf{x}_i - \boldsymbol{\mu})\right].$$

(b) Show that the logarithm of the likelihood function in (a) is given by

$$\ln L = -\frac{n}{2}\ln[|2\pi\boldsymbol{\Sigma}|] - \frac{1}{2}\sum_{i=1}^{n}(\mathbf{x}_i - \boldsymbol{\mu})'\boldsymbol{\Sigma}^{-1}(\mathbf{x}_i - \boldsymbol{\mu}).$$

(c) Show that $\sum_{i=1}^{n}(\mathbf{x}_i - \boldsymbol{\mu})'\boldsymbol{\Sigma}^{-1}(\mathbf{x}_i - \boldsymbol{\mu})$ can be written as

$$\sum_{i=1}^{n}(\mathbf{x}_i - \bar{\mathbf{x}})'\boldsymbol{\Sigma}^{-1}(\mathbf{x}_i - \bar{\mathbf{x}}) + n(\bar{\mathbf{x}} - \boldsymbol{\mu})'\boldsymbol{\Sigma}^{-1}(\bar{\mathbf{x}} - \boldsymbol{\mu}).$$

(HINT: Use the fact that $\sum_{i=1}^{n}(\mathbf{x}_i - \bar{\mathbf{x}}) = \mathbf{0}$.)

(d) Use the fact that $tr(\mathbf{x}_i - \bar{\mathbf{x}})'\boldsymbol{\Sigma}^{-1}(\mathbf{x}_i - \bar{\mathbf{x}}) = tr\boldsymbol{\Sigma}^{-1}(\mathbf{x}_i - \bar{\mathbf{x}})(\mathbf{x}_i - \bar{\mathbf{x}})'$ to show that $\sum_{i=1}^{n}(\mathbf{x}_i - \bar{\mathbf{x}})'\boldsymbol{\Sigma}^{-1}(\mathbf{x}_i - \bar{\mathbf{x}}) = tr\boldsymbol{\Sigma}^{-1}\mathbf{S}^*n$, where $\mathbf{S}^* = (n-1)\mathbf{S}/n$.

(e) Use the results of (c) and (d) to show that $\ln L$ in (b) can be written as

$$\ln L = -\frac{n}{2}\ln[|2\pi\boldsymbol{\Sigma}|] - \frac{n}{2}tr\boldsymbol{\Sigma}^{-1}\mathbf{S}^* - \frac{n}{2}(\bar{\mathbf{x}} - \boldsymbol{\mu})'\boldsymbol{\Sigma}^{-1}(\bar{\mathbf{x}} - \boldsymbol{\mu}).$$

(f) Since the only term of $\ln L$ in (e) that depends on μ is the last term, show that $\ln L$ is maximized with respect to μ if $\mu = \bar{x}$ and hence that the maximum likelihood estimator of μ is $\bar{x}$.

(g) For any fixed matrix S^* the function $-\frac{n}{2}\ln[\|\Sigma\|] - \frac{1}{2}tr\Sigma^{-1}S^*$ is maximized with respect to Σ if $\Sigma = S^*$. Use this result to obtain that the maximum likelihood estimator of Σ in (e) is S^*.

(h) Show that the value of the likelihood function evaluated at $\mu = \bar{x}$ and $\Sigma = S^*$ is given by $\ln L = -\frac{n}{2}\ln[\|2\pi S^*\|] - np/2$.

12. The maximum likelihood estimator of Σ given $\mu = \mu_0$ is given by

$$\tilde{S} = \frac{1}{n}\sum_{i=1}^{n}(x_i - \mu_0)(x_i - \mu_0)'.$$

(a) Show that this expression for $\tilde{S}$ can be written as $\tilde{S} = S^* + (\bar{x} - \mu_0)(\bar{x} - \mu_0)'$ where $S^* = (n-1)S/n$.

(b) Use the fact that $|S^* + (\bar{x} - \mu_0)(\bar{x} - \mu_0)'| = |S^*\|1 + (\bar{x} - \mu_0)'S^{*-1}(\bar{x} - \mu_0)|$ to show that the value of the logarithm of the likelihood function of 11(b) in this case is given by

$$\ln L = -\frac{np}{2}\ln 2\pi - \frac{n}{2}\ln|S^*| - \frac{np}{2} - \frac{n}{2}\ln[1 + (\bar{x}-\mu_0)'S^{*-1}(\bar{x}-\mu_0)].$$

(c) The logarithm of the likelihood ratio test for testing $H_0: \mu = \mu_0$ is obtained from the difference of the logarithms of the likelihoods in 12(b) and 11(h). Show that the difference is given by

$$-\frac{n}{2}\ln[1 + (\bar{x} - \mu_0)'S^{*-1}(\bar{x} - \mu_0)]$$

and hence that the test of $H_0: \mu = \mu_0$ depends on

$$(\bar{x} - \mu_0)'S^{*-1}(\bar{x} - \mu_0)$$

which is proportional to Hotelling's T^2.

13. In the multivariate normal distribution $x \sim N(\mu, \Sigma)$, the distribution is partitioned so that

$$x = \begin{bmatrix} x_1 \\ x_2 \end{bmatrix}, \quad \mu = \begin{bmatrix} \mu_1 \\ \mu_2 \end{bmatrix} \quad \text{and} \quad \Sigma = \begin{bmatrix} \Sigma_{11} & \Sigma_{12} \\ \Sigma_{21} & \Sigma_{22} \end{bmatrix}.$$

(a) Use the formula for the inverse of a partitioned matrix (see Question 8) to show that

$$[\mu'\Sigma^{-1}\mu - \mu_1'\Sigma_{11}^{-1}\mu_1]$$
$$= [\mu_2 - \Sigma_{21}\Sigma_{11}^{-1}\mu_1][\Sigma_{22} - \Sigma_{21}\Sigma_{11}^{-1}\Sigma_{12}]^{-1}$$
$$\times [\mu_2 - \Sigma_{21}\Sigma_{11}^{-1}\mu_1].$$

(b) Use the result in (a) to show that if the intercept is 0 in the regression of x_2 and x_1 then $\mu'\Sigma^{-1}\mu = \mu_1'\Sigma_{11}^{-1}\mu_1$.

14. The equal variance-equal covariance structure covariance matrix is given by

$$\Sigma(n \times n) = \sigma^2 \begin{bmatrix} 1 & \rho & \rho & \cdots & \rho \\ \rho & 1 & \cdots & \cdots & \rho \\ \vdots & & \ddots & & \rho \\ \rho & \cdots & \cdots & \rho & 1 \end{bmatrix}.$$

(a) Show that this can be written as $\sigma^2\rho\mathbf{ii}' + \sigma^2(1-\rho)\mathbf{I}$, where $\mathbf{i}(n \times 1)$ is a vector of unities.

(b) Use the formula for the inverse of the matrix $[\mathbf{A} + \mathbf{aa}']$ given by $\mathbf{A}^{-1} - \mathbf{A}^{-1}\mathbf{aa}'\mathbf{A}^{-1}/(1 + \mathbf{a}'\mathbf{A}^{-1}\mathbf{a})$ to show that

$$\Sigma^{-1} = \frac{1}{\sigma^2(1-\rho)}\mathbf{I} - \frac{\rho}{\sigma^2(1-\rho)}\frac{\mathbf{ii}'}{[1+\rho(n-1)]}.$$

(c) Show that for $\mathbf{X}(n \times p)$ and $\mathbf{y}(n \times 1)$

$$\mathbf{X}'\Sigma^{-1}\mathbf{X} = \frac{1}{\sigma^2(1-\rho)}\mathbf{X}'\mathbf{X} - \frac{\rho}{\sigma^2(1-\rho)}\frac{(\mathbf{X}'\mathbf{i})(\mathbf{i}'\mathbf{X})}{[1+\rho(n-1)]}$$

$$(\mathbf{X}'\Sigma^{-1}\mathbf{X})^{-1} = \sigma^2(1-\rho)(\mathbf{X}'\mathbf{X})^{-1}$$

$$+ \frac{\rho(1-\rho)\sigma^2(\mathbf{X}'\mathbf{X})^{-1}\mathbf{X}'\mathbf{ii}'\mathbf{X}(\mathbf{X}'\mathbf{X})^{-1}}{[1+\rho(n-1)-\rho\mathbf{i}'\mathbf{X}(\mathbf{X}'\mathbf{X})^{-1}\mathbf{X}'\mathbf{i}]}$$

$$\text{and } (\mathbf{X}'\Sigma^{-1}\mathbf{y}) = \frac{1}{\sigma^2(1-\rho)}\mathbf{X}'\mathbf{y} - \frac{\rho}{\sigma^2(1-\rho)}\frac{\mathbf{X}'\mathbf{ii}'\mathbf{y}}{[1+\rho(n-1)]}.$$

(d) Assume that the first column of $\mathbf{X}$ contains the vector of unities, $\mathbf{i}$, and recall that, for any $\mathbf{x}_j$ in $\mathbf{X}$, $\mathbf{X}(\mathbf{X}'\mathbf{X})^{-1}\mathbf{X}'\mathbf{x}_j = \mathbf{x}_j$ and hence $\mathbf{X}(\mathbf{X}'\mathbf{X})^{-1}\mathbf{X}'\mathbf{i} = \mathbf{i}$. Use this property to show that

$$(\mathbf{X}'\Sigma^{-1}\mathbf{X})^{-1}(\mathbf{X}'\Sigma^{-1}\mathbf{y}) = (\mathbf{X}'\mathbf{X})^{-1}\mathbf{X}'\mathbf{y}.$$

(e) For the regression of $\mathbf{y}$ on $\mathbf{X}$ if the conditional covariance of $\mathbf{y}$ given $\mathbf{X}$ is given by Σ, the generalized least squares estimator of the regression parameters is given by the left hand side of the equation in (d). What does the equation in (d) say about the property of this estimator if Σ has the equal variance–equal covariance structure?

15. In the multivariate regression model $\mathbf{Y} = \mathbf{XB} + \mathbf{U}$, $\mathbf{Y}(n \times s)$ denotes

n observations on the $(s \times 1)$ random vector $\mathbf{y}$, $\mathbf{Y} = \begin{bmatrix} \mathbf{y}_1' \\ \mathbf{y}_2' \\ \vdots \\ \mathbf{y}_n' \end{bmatrix}$ and

$\mathbf{X}[n \times (q+1)]$ denotes the corrsponding n observations on the q X

variables plus a column of unities, $\mathbf{X} = \begin{bmatrix} 1 & \mathbf{x}_1' \\ 1 & \mathbf{x}_2' \\ \vdots & \vdots \\ 1 & \mathbf{x}_n' \end{bmatrix}$. The $(n \times s)$

matrix $\mathbf{U}$ denotes the matrix of unobserved error terms and the $(q+1) \times s$ matrix $\mathbf{B}$ denotes the unknown parameters. The i-th row of $\mathbf{U}$, $\mathbf{u}_i'$ is assumed to be multivariate normal with mean $\mathbf{0}$ and covariance matrix $\boldsymbol{\Gamma}$ $(s \times s)$ where $\mathbf{u}_i' = \mathbf{y}_i' - [1 \; \mathbf{x}_i']\mathbf{B}$.

(a) Show that the log likelihood for the n independent $\mathbf{u}_i'$, $i = 1, 2, \ldots, n$, is given by

$$-\frac{n}{2}\ln(|2\pi\boldsymbol{\Gamma}|) - \left(\frac{1}{2}\right)tr(\mathbf{Y} - \mathbf{XB})\boldsymbol{\Gamma}^{-1}(\mathbf{Y} - \mathbf{XB})'.$$

HINT: See the multivariate normal likelihood function in Question 11.

(b) Let $\widehat{\mathbf{B}} = (\mathbf{X}'\mathbf{X})^{-1}\mathbf{X}'\mathbf{Y}$ denote the ordinary least squares estimator of $\mathbf{B}$ and denote by $\widehat{\mathbf{U}} = \mathbf{Y} - \mathbf{X}\widehat{\mathbf{B}}$ the matrix of residuals. Show that if $\widehat{\mathbf{B}}$ is substituted for $\mathbf{B}$ and $\widehat{\boldsymbol{\Gamma}} = \widehat{\mathbf{U}}'\widehat{\mathbf{U}}/n$ is substituted for $\boldsymbol{\Gamma}$, then the likelihood function in (a) becomes $-\frac{n}{2}\ln(|2\pi\widehat{\boldsymbol{\Gamma}}|) - \frac{1}{2}ns$, which has the same form as the sample value of the likelihood function in 11(h). HINT: Use the relationship

$$tr(\mathbf{Y} - \mathbf{X}\widehat{\mathbf{B}})\widehat{\boldsymbol{\Gamma}}^{-1}(\mathbf{Y} - \mathbf{X}\widehat{\mathbf{B}})' = tr\widehat{\boldsymbol{\Gamma}}^{-1}(\mathbf{Y} - \mathbf{X}\widehat{\mathbf{B}})(\mathbf{Y} - \mathbf{X}\widehat{\mathbf{B}})' = ns.$$

16. Given $\begin{bmatrix} \mathbf{y} \\ \mathbf{x} \end{bmatrix}$, $\mathbf{y}(s \times 1)$, $\mathbf{x}(q \times 1)$ and $E\begin{bmatrix} \mathbf{y} \\ \mathbf{x} \end{bmatrix} = \begin{bmatrix} \mathbf{0} \\ \mathbf{0} \end{bmatrix}$, $\mathrm{Cov}\begin{pmatrix} \mathbf{y} \\ \mathbf{x} \end{pmatrix} = \begin{bmatrix} \boldsymbol{\Sigma}_{yy} & \boldsymbol{\Sigma}_{yx} \\ \boldsymbol{\Sigma}_{xy} & \boldsymbol{\Sigma}_{xx} \end{bmatrix}$ let $z = \mathbf{a}'\mathbf{y}$ and $w = \mathbf{b}'\mathbf{x}$ denote linear transformations of $\mathbf{y}$ and $\mathbf{x}$ respectively. The steps outlined below are designed to derive the canonical correlation results of Section 7.5.2.

(a) Show that the correlation between z and w is given by $r_{zw} = \mathbf{a}'\boldsymbol{\Sigma}_{yx}\mathbf{b}/(\mathbf{b}'\boldsymbol{\Sigma}_{xx}\mathbf{b})^{1/2}(\mathbf{a}'\boldsymbol{\Sigma}_{yy}\mathbf{a})^{1/2}$.

(b) To determine the values of $\mathbf{a}$ and $\mathbf{b}$ such that r_{zw} is maximized subject to the conditions $\mathbf{a}'\boldsymbol{\Sigma}_{yy}\mathbf{a} = 1$ and $\mathbf{b}'\boldsymbol{\Sigma}_{xx}\mathbf{b} = 1$, the

Lagrangian expression is given by

$$S = \mathbf{a}' \Sigma_{\mathbf{yx}} \mathbf{b} - u_1(\mathbf{a}' \Sigma_{\mathbf{yy}} \mathbf{a} - 1) - u_2(\mathbf{b}' \Sigma_{\mathbf{xx}} \mathbf{b} - 1).$$

Use differentiation of S with respect to $\mathbf{a}$ and $\mathbf{b}$ to obtain the equations

$$\Sigma_{\mathbf{yx}} \mathbf{b} - 2u_1 \Sigma_{\mathbf{yy}} \mathbf{a} \;=\; 0 \tag{1}$$

$$\Sigma_{\mathbf{xy}} \mathbf{a} - 2u_2 \Sigma_{\mathbf{xx}} \mathbf{b} \;=\; 0. \tag{2}$$

NOTE: Formulae for differentiation of matrix expressions are given in the Appendix.

(c) Multiply through (1) by $\mathbf{a}'$ and (2) by $\mathbf{b}'$ and use the conditions $\mathbf{a}' \Sigma_{\mathbf{yy}} \mathbf{a} = 1$, $\mathbf{b}' \Sigma_{\mathbf{xx}} \mathbf{b} = 1$ to show that $u_1 = u_2 = \mathbf{a}' \Sigma_{\mathbf{yx}} \mathbf{b}/2$ is the correlation between z and w.

(d) Letting $2u_1 = 2u_2 = \lambda^{1/2}$ and solving (1) for $\mathbf{a}$ and (2) for $\mathbf{b}$, show by substitution that (1) and (2) can be expressed as

$$(\Sigma_{\mathbf{yx}} \Sigma_{\mathbf{xx}}^{-1} \Sigma_{\mathbf{xy}} - \lambda \Sigma_{\mathbf{yy}})\mathbf{a} \;=\; 0 \tag{3}$$

$$(\Sigma_{\mathbf{xy}} \Sigma_{\mathbf{yy}}^{-1} \Sigma_{\mathbf{yx}} - \lambda \Sigma_{\mathbf{xx}})\mathbf{b} \;=\; 0. \tag{4}$$

(e) Rewriting (3) and (4) as

$$(\Sigma_{\mathbf{yy}}^{-1} \Sigma_{\mathbf{yx}} \Sigma_{\mathbf{xx}}^{-1} \Sigma_{\mathbf{xy}} - \lambda \mathbf{I})\mathbf{a} \;=\; 0 \tag{5}$$

$$(\Sigma_{\mathbf{xx}}^{-1} \Sigma_{\mathbf{xy}} \Sigma_{\mathbf{yy}}^{-1} \Sigma_{\mathbf{yx}} - \lambda \mathbf{I})\mathbf{b} \;=\; 0, \tag{6}$$

use the theory of eigenvectors and eigenvalues given in the Appendix to obtain that $\mathbf{a}$ and $\mathbf{b}$ are eigenvectors of

$$\Sigma_{\mathbf{yy}}^{-1} \Sigma_{\mathbf{yx}} \Sigma_{\mathbf{xx}}^{-1} \Sigma_{\mathbf{xy}} \quad \text{and}$$

$$\Sigma_{\mathbf{xx}}^{-1} \Sigma_{\mathbf{xy}} \Sigma_{\mathbf{yy}}^{-1} \Sigma_{\mathbf{yx}}$$

respectively and that λ the corresponding eigenvalue is common to both equations.

(f) Given that the matrices

$$\Sigma_{\mathbf{yy}}^{-1} \Sigma_{\mathbf{yx}} \Sigma_{\mathbf{xx}}^{-1} \Sigma_{\mathbf{xy}} \quad \text{and}$$

$$\Sigma_{\mathbf{xx}}^{-1} \Sigma_{\mathbf{xy}} \Sigma_{\mathbf{yy}}^{-1} \Sigma_{\mathbf{yx}}$$

are both positive definite of rank $t = \min(s, q)$ use the theory of eigenvalues and eigenvectors to establish that there are two sets of canonical functions $z_i = \mathbf{a}_i' \mathbf{y}$, $i = 1, 2, \ldots, t$, and $w_i = \mathbf{b}_i' \mathbf{x}$, $i = 1, 2, \ldots, t$, where $\mathbf{a}_i$ and $\mathbf{b}_i$ satisfy (3) and (4) respectively.

(g) Use equations (5) and (6) to obtain the four characteristic equations for a_i, a_j and b_i, b_j and show that by premultiplication and subtraction

$$(\lambda_i - \lambda_j)a_i'\Sigma_{yy}a_j = 0$$
$$\text{and} \quad (\lambda_i - \lambda_j)b_i'\Sigma_{xx}b_j = 0,$$

and hence that the z_i are mutually uncorrelated and the w_i are mutually uncorrelated.

(h) Using two of the four characteristic equations determined in 16(g) premultiply by $b_j'\Sigma_{xy}$ and $a_i'\Sigma_{yx}$ and subtract to get that $(\lambda_i - \lambda_j)b_j'\Sigma_{xy}a_i = 0$. What property does this result establish?

(i) Use equation (1) or (2) to show that if $2u_1 = 2u_2 = \lambda^{1/2}$ then $\lambda^{1/2} = r_{zw}$ and hence r_{zw} is maximized if λ is the largest eigenvalue.

17. Let $x^*(q \times 1) = Ax + g$ and $y^*(s \times 1) = By + h$, where $A(q \times q)$ and $B(s \times s)$ are nonsingular matrices and $g(q \times 1)$ and $h(s \times 1)$ are constant vectors. Denote the covariance matrix for

$$\begin{bmatrix} x^* \\ y^* \end{bmatrix} \quad \text{by} \quad \Sigma^* = \begin{bmatrix} \Sigma_{xx}^* & \Sigma_{xy}^* \\ \Sigma_{yx}^* & \Sigma_{yy}^* \end{bmatrix}$$

and for

$$\begin{bmatrix} x \\ y \end{bmatrix} \quad \text{by} \quad \Sigma = \begin{bmatrix} \Sigma_{xx} & \Sigma_{xy} \\ \Sigma_{yx} & \Sigma_{yy} \end{bmatrix}.$$

Use the fact that $A\Sigma_{xx}A' = \Sigma_{xx}^*$, $B\Sigma_{yy}B' = \Sigma_{yy}^*$, $A\Sigma_{xy}B' = \Sigma_{xy}^*$ and $B\Sigma_{yx}A' = \Sigma_{yx}^*$ to show that

$$(\Sigma_{xy}^* \Sigma_{yy}^{*-1} \Sigma_{yx}^* b - \lambda\Sigma_{xx}^* b) = 0$$

yields the same solution as

$$(\Sigma_{xy} \Sigma_{yy}^{-1} \Sigma_{yx} b - \lambda\Sigma_{xx} b) = 0$$

and

$$(\Sigma_{yx}^* \Sigma_{xx}^{*-1} \Sigma_{xy}^* a - \lambda\Sigma_{yy}^* a) = 0$$

yields the same solution as

$$(\Sigma_{yx} \Sigma_{xx}^{-1} \Sigma_{xy} a - \lambda\Sigma_{yy} a) = 0.$$

What does this result imply about the relationship between the canonical correlation analysis for x^* and y^* and the canonical correlation analysis for x and y.

18. For the multivariate regression model $\mathbf{Y} = \mathbf{XB} + \mathbf{U}$ where $\mathbf{Y}(n \times s)$, $\mathbf{X}[n \times (q+1)]$, $\mathbf{B}[(q+1) \times s]$ and $\mathbf{U}(n \times s)$ the least squares estimator of $\mathbf{B}$ is given by $\hat{\mathbf{B}} = (\mathbf{X'X})^{-1}\mathbf{X'Y}$.

   (a) Show that $\hat{\mathbf{Y}} = \mathbf{X}\hat{\mathbf{B}}$ and $(\mathbf{Y} - \hat{\mathbf{Y}})$ can be written as $\mathbf{HY}$ and $(\mathbf{I} - \mathbf{H})\mathbf{Y}$ respectively, where $\mathbf{H} = \mathbf{X}(\mathbf{X'X})^{-1}\mathbf{X'}$ and $\mathbf{I}$ is the identity matrix.

   (b) Show that $\hat{\mathbf{B}}'\mathbf{X'X}\hat{\mathbf{B}}$ and $(\mathbf{Y} - \hat{\mathbf{Y}})'(\mathbf{Y} - \hat{\mathbf{Y}})$ are given by $\mathbf{Y'HY}$ and $\mathbf{Y'}(\mathbf{I}-\mathbf{H})\mathbf{Y}$ respectively using the fact that $\mathbf{H'} = \mathbf{H}$, $\mathbf{HH} = \mathbf{H}$, $(\mathbf{I} - \mathbf{H})' = (\mathbf{I} - \mathbf{H})$ and $(\mathbf{I} - \mathbf{H})(\mathbf{I} - \mathbf{H}) = (\mathbf{I} - \mathbf{H})$, which are properties of idempotent matrices.

   (c) Show that $\mathbf{Y'Y} = \mathbf{Y'HY} + \mathbf{Y'}(\mathbf{I} - \mathbf{H})\mathbf{Y}$ and hence that $\mathbf{Y'Y} = \hat{\mathbf{B}}'\mathbf{X'X}\hat{\mathbf{B}} + (\mathbf{Y} - \mathbf{X}\hat{\mathbf{B}})'(\mathbf{Y} - \mathbf{X}\hat{\mathbf{B}})$ using (b).

   (d) Partition the multivariate regression model as

   $$\mathbf{Y} = \mathbf{X}_1\mathbf{B}_1 + \mathbf{X}_2\mathbf{B}_2 + \mathbf{U} = [\mathbf{X}_1\mathbf{X}_2] \begin{bmatrix} \mathbf{B}_1 \\ \mathbf{B}_2 \end{bmatrix} + \mathbf{U},$$

   where $\mathbf{X}_1 [n \times (v + 1)]$, $\mathbf{X}_2 (n \times r)$ and $q = (v + r)$. Let $\tilde{\mathbf{B}}_1$ denote the least squares estimator for the reduced model $\mathbf{Y} = \mathbf{X}_1\mathbf{B}_1 + \mathbf{U}$, $\tilde{\mathbf{B}}_1 = (\mathbf{X}_1'\mathbf{X}_1)^{-1}\mathbf{X}_1'\mathbf{Y}$ and show that

   $$\begin{aligned} \mathbf{Y'Y} &= \mathbf{Y'H}_1\mathbf{Y} + \mathbf{Y'}(\mathbf{I} - \mathbf{H}_1)\mathbf{Y} \\ &= \tilde{\mathbf{B}}_1'\mathbf{X}_1'\mathbf{X}_1\tilde{\mathbf{B}}_1 + (\mathbf{Y} - \mathbf{X}_1\tilde{\mathbf{B}}_1)'(\mathbf{Y} - \mathbf{X}_1\tilde{\mathbf{B}}_1), \end{aligned}$$

   where $\mathbf{H}_1 = \mathbf{X}_1(\mathbf{X}_1'\mathbf{X}_1)^{-1}\mathbf{X}_1'$ and $\mathbf{I}$ $(n \times n)$ is an identity matrix.

   (e) Show that $\mathbf{Y'}(\mathbf{I} - \mathbf{H}_1)\mathbf{Y} = \mathbf{Y'}(\mathbf{I} - \mathbf{H})\mathbf{Y} + \mathbf{Y'}(\mathbf{H} - \mathbf{H}_1)\mathbf{Y}$ and hence show that

   $$\begin{aligned} &(\mathbf{Y} - \mathbf{X}_1\tilde{\mathbf{B}}_1)'(\mathbf{Y} - \mathbf{X}_1\tilde{\mathbf{B}}_1) \\ &= (\mathbf{Y} - \mathbf{XB})'(\mathbf{Y} - \mathbf{XB}) + \hat{\mathbf{B}}'\mathbf{X'X}\hat{\mathbf{B}} - \tilde{\mathbf{B}}_1'\mathbf{X}_1'\mathbf{X}_1\tilde{\mathbf{B}}_1. \end{aligned}$$

   (f) The Wilk's Lambda test statistic for testing $H_0: \mathbf{B}_2 = \mathbf{0}$ in the model

   $$\mathbf{Y} = \mathbf{X}_1\mathbf{B}_1 + \mathbf{X}_2\mathbf{B}_2 + \mathbf{U}$$

   is given by $\Lambda = |\mathbf{E}|/|\mathbf{E}_0|$, where $\mathbf{E} = (\mathbf{Y} - \mathbf{X}\hat{\mathbf{B}})'(\mathbf{Y} - \mathbf{X}\hat{\mathbf{B}})$ and $\mathbf{E}_0 = (\mathbf{Y} - \mathbf{X}_1\tilde{\mathbf{B}}_1)'(\mathbf{Y} - \mathbf{X}_1\tilde{\mathbf{B}}_1)$. Use the result of (e) to show that $\Lambda = |\mathbf{E}|/|\mathbf{E} + \mathbf{G}|$, where $\mathbf{G} = (\hat{\mathbf{B}}'\mathbf{X'X}\hat{\mathbf{B}} - \tilde{\mathbf{B}}_1'\mathbf{X}_1'\mathbf{X}_1\tilde{\mathbf{B}}_1)$.

19. For the multivariate regression model $\mathbf{Y} = \mathbf{XB} + \mathbf{U}$ assume

$$\mathbf{B} = \begin{bmatrix} \beta_{01} & \beta_{02} & \beta_{03} \\ \beta_{11} & \beta_{12} & \beta_{13} \\ \beta_{21} & \beta_{22} & \beta_{23} \\ \beta_{31} & \beta_{32} & \beta_{33} \\ \beta_{41} & \beta_{42} & \beta_{43} \end{bmatrix}.$$

For each of the hypotheses of the form $H_0: \mathbf{ABM} = \mathbf{0}$ given below explain what is being tested and the practical significance in each case.

(a) $\begin{bmatrix} 0 & 1 & 0 & 0 & 0 \\ 0 & 0 & 1 & 0 & 0 \\ 0 & 0 & 0 & 1 & 0 \\ 0 & 0 & 0 & 0 & 1 \end{bmatrix} \mathbf{B} = \mathbf{0}.$

(b) $\mathbf{B} \begin{bmatrix} 1 & 1 \\ -1 & 0 \\ 0 & -1 \end{bmatrix} = \mathbf{0}.$

(c) $\begin{bmatrix} 1 & -1 & 0 & 0 & 0 \\ 1 & 0 & -1 & 0 & 0 \\ 1 & 0 & 0 & -1 & 0 \\ 1 & 0 & 0 & 0 & -1 \end{bmatrix} \mathbf{B} = \mathbf{0}.$

(d) $\begin{bmatrix} 1 & -1 & 0 & 0 & 0 \\ 1 & 0 & -1 & 0 & 0 \\ 1 & 0 & 0 & -1 & 0 \\ 1 & 0 & 0 & 0 & -1 \end{bmatrix} \mathbf{B} \begin{bmatrix} 1 & 1 \\ -1 & 0 \\ 0 & -1 \end{bmatrix} = \mathbf{0}.$

8

MANOVA, Discriminant Analysis and Qualitative Response Models

The first part of this chapter extends Chapter 7 by specializing the multivariate linear regression model to the case where the explanatory variables represent design variables. In the same manner that ANOVA is a special case of multiple regression, we see here that multivariate analysis of variance (MANOVA) can be viewed as a special case of multivariate linear regression.

A special case of canonical correlation discussed in Chapter 7 occurs if one of the two sets of variables are dummy variables. This specialized technique is called discriminant analysis and is useful for characterizing group differences obtained from a significant MANOVA. Discriminant analysis is also useful as a technique for classifying unknowns.

If in a multivariate regression model the dependent variables are categorical, the model is said to be a qualitative response model. The qualitative response model, like discriminant analysis, can also be used to characterize group differences and classify unknowns. Special cases of this type of model presented in this chapter are called logistic regression, probit analysis and multinomial logit.

This chapter presents a summary of MANOVA, discriminant analysis, and qualitative response models.

8.1 Multivariate Analysis of Variance

8.1.1 ONE-WAY MULTIVARIATE ANALYSIS OF VARIANCE

Comparison to Univariate Analysis of Variance

Multivariate analysis of variance (MANOVA) is an extension of the concept of analysis of variance to the case of more than one dependent variable. Given g groups of individuals, a set of p variables $X_1, X_2, \ldots, X_p$, denoted by the $(p \times 1)$ vector $\mathbf{x}$, is observed in each group rather than a single variable X as in ANOVA. In ANOVA, the mean of X in group k was denoted by μ_k, $k = 1, 2, \ldots, g$, whereas in MANOVA the mean vector for

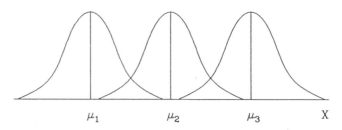

FIGURE 8.1. Comparison of Three Univariate Normal Distributions with Same Variance

$\mathbf{x}$ in each group is denoted by the vector $\boldsymbol{\mu}_k$, $k = 1, 2, \ldots, g$. In ANOVA we are interested in testing the null hypothesis of equality of means on the single variable X, $H_{01}: \mu_1 = \mu_2 = \ldots = \mu_g$, whereas in MANOVA we wish to test the null hypothesis that for the random vector $\mathbf{x}$ the mean vectors are equal, $H_{0p}: \boldsymbol{\mu}_1 = \boldsymbol{\mu}_2 = \ldots = \boldsymbol{\mu}_g$. Thus acceptance of H_{0p} in MANOVA implies acceptance of H_{01} in ANOVA for each of the p components of $\mathbf{x}$ separately.

Graphically, a one-way ANOVA on a variable X over three groups is a comparison of the location of several normal distributions with common variance as in Figure 8.1. In comparison, a one-way MANOVA over three groups for $p = 2$ can be viewed in two-dimensional space as a comparison of the centroids of three ellipsoids with equal dimensions but possibly different centres (see Figure 8.2.). In Figure 8.2, the major axes of the three ellipsoids are parallel, which follows if the covariance structure for the three groups is the same. A fundamental assumption of MANOVA is that the covariance matrices are equal.

In univariate analysis of variance, a random sample of n_k individuals is selected from group k, $k = 1, 2, \ldots, g$ and observations are obtained on X for each individual in the sample. The variation of individuals around the grand mean is then divided into two portions: SSA, the variation among groups and SSW, the variation within groups. The mean squares MSA and MSW derived from these sums of squares are then compared using an F test. Large values of MSA relative to MSW indicate large differences among the group sample means and hence a large value of F and rejection of H_{01}. In multivariate analysis of variance, sums of squares are computed as matrices. Multivariate analysis of variance is based on a comparison of these sums of squares matrices. Before outlining the MANOVA methodology a data example is introduced.

Example

The data matrix in Table 8.1 is used to provide an example for multivariate analysis of variance. The 100 observations represent a sample of 25

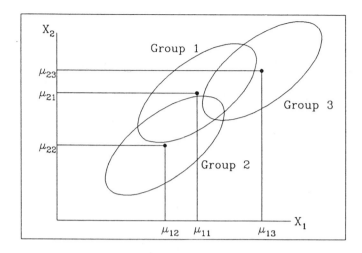

FIGURE 8.2. Comparison of Three Bivariate Normal Ellipsoids

observations from each of four different communities served by different administrative units of the Royal Canadian Mounted Police (R.C.M.P). The 100 individuals were asked to respond to six different questions regarding how safe they felt in their community. The individuals were asked to respond using one of six responses from $1 =$ very safe to $6 =$ very unsafe.

The six items are

X1: How safe do you feel in your town as a whole?

X2: How safe do you feel in your home?

X3: How safe do you feel walking alone in your neighborhood during the day?

X4: How safe do you feel walking in your neighborhood at night?

X5: How safe do you feel in downtown during the day?

X6: How safe do you feel in downtown at night?

A MANOVA will be used to compare the means on the six variables over the four groups (administrative units). The observations on the variables INC and EDCAT will be used later in this section. Before discussing the MANOVA technique additional notation is required.

TABLE 8.1. Observations From Public Safety Questionnaire

Community	X1	X2	X3	X4	X5	X6	INC	EDCAT	Community	X1	X2	X3	X4	X5	X6	INC	EDCAT
1	3	1	1	2	1	6	7	5	3	4	2	1	2	3	5	4	3
1	2	2	1	2	1	3	9	5	3	1	1	1	1	1	1	9	5
1	2	1	1	1	1	2	9	5	3	2	1	1	2	2	4	9	4
1	1	1	1	1	1	1	9	5	3	3	2	3	4	5	5	8	1
1	2	2	1	4	2	5	8	4	3	4	2	1	2	1	4	9	4
1	2	2	2	2	2	4	9	4	3	1	1	1	1	1	2	4	5
1	5	3	2	6	2	6	8	1	3	2	1	2	2	1	2	9	5
1	2	2	2	3	2	4	2	4	3	3	2	1	4	1	6	6	3
1	3	1	2	2	2	3	5	5	3	3	3	2	4	2	4	4	2
1	5	2	2	3	2	4	9	3	3	2	2	2	4	2	6	1	2
1	3	2	2	5	2	5	9	2	3	2	2	2	4	3	6	7	2
1	3	4	2	5	2	5	6	2	3	2	1	1	2	1	3	2	4
1	2	2	2	5	3	5	5	2	3	2	3	4	6	5	6	3	1
1	4	4	4	5	2	6	6	1	3	1	1	1	2	1	2	7	5
1	2	2	2	4	2	6	8	3	3	1	1	1	1	1	2	7	5
1	5	2	2	6	2	6	9	1	3	2	1	1	2	2	4	6	4
1	4	2	2	4	2	5	7	2	3	2	1	1	2	2	4	9	4
1	2	2	2	3	3	4	6	4	3	4	2	1	3	2	5	5	3
1	4	1	3	4	3	4	9	2	3	2	2	2	3	2	4	5	3
1	4	4	3	6	3	6	3	1	3	5	2	1	3	1	6	4	2
1	5	6	2	4	4	5	9	1	3	4	2	3	2	2	4	6	3
1	2	3	2	4	2	5	4	3	3	5	2	3	5	2	5	8	1
1	3	1	2	5	2	5	9	3	3	3	3	3	4	3	4	9	2
1	4	2	2	4	2	4	7	3	3	6	5	3	6	4	6	2	1
1	2	2	2	4	3	4	9	4	3	4	3	3	4	3	4	4	1
2	2	2	2	2	1	1	6	2	4	1	1	1	1	1	1	9	5
2	2	1	1	1	1	1	6	4	4	1	1	1	1	1	1	6	5
2	2	2	2	2	2	2	8	1	4	2	2	1	2	1	2	1	2
2	2	1	2	3	2	3	6	1	4	2	2	1	2	1	2	6	2
2	1	1	1	1	1	1	7	5	4	1	1	1	1	1	1	9	5
2	1	1	1	2	1	2	7	4	4	2	2	1	2	1	3	5	1
2	2	1	1	4	1	4	7	1	4	1	1	1	1	1	1	6	5
2	2	2	2	2	2	3	9	1	4	1	1	1	1	1	1	3	5
2	1	1	1	1	1	2	6	4	4	1	1	1	1	1	1	9	4
2	1	1	1	1	1	1	6	5	4	1	1	1	1	1	1	8	4
2	1	1	1	2	1	2	8	3	4	1	1	1	1	1	1	8	4
2	1	1	1	1	1	1	9	5	4	1	1	1	1	1	1	4	4
2	2	1	1	2	1	2	9	3	4	2	1	1	1	1	2	6	2
2	2	2	1	2	1	2	6	2	4	1	1	1	1	1	1	8	4
2	1	2	1	2	1	3	7	2	4	2	1	1	1	1	2	6	2
2	1	1	1	1	1	1	6	5	4	3	2	2	4	1	3	6	1
2	1	1	1	1	1	1	5	5	4	1	1	1	1	1	1	9	3
2	2	2	1	2	1	2	6	2	4	3	1	3	3	1	4	7	1
2	1	2	1	2	1	1	9	3	4	2	1	1	2	2	3	7	1
2	2	2	1	2	1	2	6	2	4	1	1	1	1	1	1	6	3
2	1	1	1	1	1	1	6	4	4	1	1	1	1	1	1	7	3
2	2	2	1	1	1	2	7	3	4	1	1	1	1	1	1	8	3
2	2	1	1	1	1	2	7	3	4	1	1	1	1	1	2	6	3
2	1	1	1	1	1	1	7	4	4	2	1	1	2	1	2	5	2
2	2	2	1	2	1	2	6	1	4	2	2	1	4	1	2	4	1

Notation for Several Multivariate Populations

Let the $(p \times 1)$ random vector $\mathbf{x}_k$ denote a vector of observations on the p variables for group k which is one of g groups,

$$\mathbf{x}_k = \begin{bmatrix} x_{1k} \\ x_{2k} \\ \vdots \\ x_{pk} \end{bmatrix} \quad k = 1, 2, \ldots, g.$$

Mean Vector for Group k and Common Covariance Matrix

Denote the *mean vector for group k* by $\boldsymbol{\mu}_k = \begin{bmatrix} \mu_{1k} \\ \vdots \\ \mu_{pk} \end{bmatrix}$, $k = 1, 2, \ldots, g$, and

assume the covariance matrix is the same for all groups $k = 1, 2, \ldots, g$, and is given by

$$\boldsymbol{\Sigma} = \begin{bmatrix} \sigma_1^2 & \sigma_{12} & & \sigma_{1p} \\ \sigma_{12} & \sigma_2^2 & & \\ \vdots & \vdots & & \\ \sigma_{1p} & \sigma_{2p} & \cdots & \sigma_p^2 \end{bmatrix}.$$

Grand Mean Vector

The *grand mean vector* over all groups is given by $\boldsymbol{\mu} = \begin{bmatrix} \mu_1 \\ \mu_2 \\ \vdots \\ \mu_p \end{bmatrix}$. If the

groups are ignored, the expected value of a vector of observations $\mathbf{x}$ obtained at random from the entire population of g groups is given by $E[\mathbf{x}] = \boldsymbol{\mu}$.

Notation for Samples

Given a random sample of n_k observations from group k, we denote the vector of observations on individual i for group k by

$$\mathbf{x}_{ik} = \begin{bmatrix} x_{i1k} \\ x_{i2k} \\ \vdots \\ x_{ipk} \end{bmatrix}, \quad i = 1, 2, \ldots, n_k.$$

Sample Mean Vector and Sample Covariance Matrix for Group k

The *sample mean vector for group k* is denoted by

$$\bar{\mathbf{x}}_{.k} = \begin{bmatrix} \bar{x}_{.1k} \\ \bar{x}_{.2k} \\ \vdots \\ \bar{x}_{.pk} \end{bmatrix},$$

where

$$\bar{x}_{.jk} = \sum_{i=1}^{n_k} x_{ijk}/n_k, \qquad j = 1, 2, \ldots, p$$

$$k = 1, 2, \ldots, g,$$

and the sample covariance matrix by $\mathbf{S}_k = \sum_{i=1}^{n_k} (\mathbf{x}_{ik} - \bar{\mathbf{x}}_{.k})(\mathbf{x}_{ik} - \bar{\mathbf{x}}_{.k})'/(n_k - 1)$.

Sample Grand Mean Vector

The *sample grand mean vector* over all g groups is given by

$$\bar{\mathbf{x}} = \begin{bmatrix} \bar{x}_{.1.} \\ \bar{x}_{.2.} \\ \vdots \\ \bar{x}_{.p.} \end{bmatrix}$$

where

$$\bar{x}_{.j.} = \sum_{k=1}^{g} \sum_{i=1}^{n_k} x_{ijk} / \sum_{k=1}^{g} n_k, \quad j = 1, 2, \ldots, p,$$

$$= \sum_{k=1}^{g} n_k \bar{x}_{.jk}/n \quad \text{where } n = \sum_{k=1}^{g} n_k.$$

Figure 8.3 illustrates the notation.

The Multivariate Analysis of Variance Model

The MANOVA model is a generalization of the univariate analysis of variance model and is given by

$$\begin{bmatrix} x_{i1k} \\ x_{i2k} \\ \vdots \\ x_{ipk} \end{bmatrix} = \begin{bmatrix} \mu_1 \\ \mu_2 \\ \vdots \\ \mu_p \end{bmatrix} + \begin{bmatrix} \mu_{1k} - \mu_1 \\ \mu_{2k} - \mu_2 \\ \vdots \\ \mu_{pk} - \mu_p \end{bmatrix} + \begin{bmatrix} x_{i1k} - \mu_{1k} \\ x_{i2k} - \mu_{2k} \\ \vdots \\ x_{ipk} - \mu_{pk} \end{bmatrix}, \quad \begin{array}{l} i = 1, 2, \ldots, n_k, \\ k = 1, 2, \ldots, g; \end{array}$$

	Group 1	Group 2	...	Group g
Parameters	$\boldsymbol{\mu}_1, \boldsymbol{\Sigma}$;	$\boldsymbol{\mu}_2, \boldsymbol{\Sigma}$;	...	$\boldsymbol{\mu}_g, \boldsymbol{\Sigma}$;
Sample	$\mathbf{x}_{11}, \mathbf{x}_{21}, \ldots, \mathbf{x}_{n_11}$;	$\mathbf{x}_{12}, \mathbf{x}_{22}, \ldots, \mathbf{x}_{n_22}$;	...	$\mathbf{x}_{1g}, \mathbf{x}_{2g}, \ldots, \mathbf{x}_{n_gg}$;
Mean Vector	$\bar{\mathbf{x}}_{.1}$;	$\bar{\mathbf{x}}_{.2}$	...	$\bar{\mathbf{x}}_{.g}$;
Covariance Matrix	$\mathbf{S}_1$;	$\mathbf{S}_2$;	...	$\mathbf{S}_g$.

FIGURE 8.3. Notation for Group Parameters and Sample Statistics

or $\mathbf{x}_{ik} = \boldsymbol{\mu} + \boldsymbol{\alpha}_k + \boldsymbol{\varepsilon}_{ik}$, where $\mathbf{x}_{ik}$, $\boldsymbol{\mu}$, $\boldsymbol{\alpha}_k$ and $\boldsymbol{\varepsilon}_{ik}$ are the $(p \times 1)$ vectors illustrated above. The hypothesis of equality of mean vectors, $H_{op}\colon \boldsymbol{\mu}_1 = \boldsymbol{\mu}_2 = \ldots = \boldsymbol{\mu}_g$, is equivalent to the hypothesis $H_{op}\colon \boldsymbol{\alpha}_1 = \boldsymbol{\alpha}_2 = \ldots = \boldsymbol{\alpha}_g = \mathbf{0}$.

As in the univariate case we may write $\mathbf{x}_{ik} = \bar{\mathbf{x}} + (\bar{\mathbf{x}}_{.k} - \bar{\mathbf{x}}) + (\mathbf{x}_{ik} - \bar{\mathbf{x}}_{.k})$ as an estimate of the model.

Within Group Sum of Squares Matrix

The *within group sum of squares matrix* is defined as

$$\mathbf{W} = \sum_{k=1}^{g} \sum_{i=1}^{n_k} (\mathbf{x}_{ik} - \bar{\mathbf{x}}_{.k})(\mathbf{x}_{ik} - \bar{\mathbf{x}}_{.k})'.$$

This matrix can also be expressed as a weighted sum of the individual group covariance matrices

$$\mathbf{W} = (n_1 - 1)\mathbf{S}_1 + (n_2 - 1)\mathbf{S}_2 + \ldots + (n_g - 1)\mathbf{S}_g.$$

The matrix $\bar{\mathbf{S}} = \mathbf{W}/(n-g)$ provides an estimator of the common covariance matrix $\boldsymbol{\Sigma}$ under the homogeneity of covariance matrix assumption.

Among Group Sum of Squares Matrix

The *among group sum of squares matrix* is defined as

$$\mathbf{G} = \sum_{k=1}^{g} \sum_{i=1}^{n_k} (\bar{\mathbf{x}}_{.k} - \bar{\mathbf{x}})(\bar{\mathbf{x}}_{.k} - \bar{\mathbf{x}})' = \sum_{k=1}^{g} n_k (\bar{\mathbf{x}}_{.k} - \bar{\mathbf{x}})(\bar{\mathbf{x}}_{.k} - \bar{\mathbf{x}})'.$$

Total Sum of Squares Matrix

The *total sum of squares matrix* is given by $\mathbf{T} = \sum_{k=1}^{g} \sum_{i=1}^{n_k} (\mathbf{x}_{ik} - \bar{\mathbf{x}})(\mathbf{x}_{ik} - \bar{\mathbf{x}})'$ and $\mathbf{T} = \mathbf{G} + \mathbf{W}$.

Computer Software

The calculations for the examples in this section were performed using SAS PROC GLM and SAS PROC REG.

Example

For the public safety data in Table 8.1, the mean vector and sum of squares and cross products matrices are shown in Table 8.2. The p-values for the univariate ANOVA F statistics are also shown in this table. In all cases the means on the six items are significantly different over the four different communities. Table 8.3 shows the correlations among the six variables for the combined sample of 100. The p-values for the null hypothesis of zero correlation are also shown in Table 8.3. From the correlation coefficients and p-values we see that the correlations are strongly positive and significantly different from zero. Because of the strong correlations among the six variables any joint inferences regarding the six means should be made by using multivariate methods. The simplest joint inference procedure regarding the mean vector is the MANOVA procedure discussed below.

Statistical Inference for MANOVA

For statistical inference purposes we assume that for each group k, the random vector $\mathbf{x}_k$ has a multivariate normal distribution with mean vector $\boldsymbol{\mu}_k$ and covariance matrix $\boldsymbol{\Sigma}$, $\mathbf{x}_k \sim N_p(\boldsymbol{\mu}_k, \boldsymbol{\Sigma})$, $k = 1, 2, \ldots, g$. Note that we are assuming a homogeneous covariance matrix $\boldsymbol{\Sigma}$ over the g groups.

Wilk's Lambda Likelihood Ratio F-Statistic

A likelihood ratio test of the hypothesis $H_{op}\!: \boldsymbol{\mu}_1 = \boldsymbol{\mu}_2 = \ldots = \boldsymbol{\mu}_g$, or equivalently $H_{op}\!: \boldsymbol{\alpha}_1 = \boldsymbol{\alpha}_2 = \ldots = \boldsymbol{\alpha}_g = \mathbf{0}$ is a function of Wilk's Lambda which was introduced in Chapter 7. The statistic is given by

$$\Lambda = |\mathbf{W}|/|\mathbf{W} + \mathbf{G}|. \tag{8.1}$$

In this case Λ has parameters p, $(g-1)$ and $(n-g)$ referring to the dimension of $\mathbf{x}$ and the degrees of freedom for $\mathbf{G}$ and $\mathbf{W}$ respectively.

In large samples if H_{op} is true, the test statistic

$$F = \frac{(1-y)}{y} \frac{m_2}{m_1}$$

has an F distribution with m_1 and m_2 degrees of freedom, where

$$y = \Lambda^{1/s}, \qquad s = \sqrt{\frac{p^2(g-1)^2 - 4}{p^2 + (g-1)^2 - 5}},$$

$$m_1 = p(g-1), \quad m_2 = s[n - (p+g+2)/2] - \frac{p(g-1)}{2} + 1,$$

TABLE 8.2. Mean Vector and Sum of Squares Matrices

Group	Sample Size	Means					
		X1	X2	X3	X4	X5	X6
1	25	3.04	2.24	1.96	3.76	2.12	4.52
2	25	1.52	1.40	1.16	1.72	1.12	1.80
3	25	2.80	1.92	1.80	3.00	2.12	4.16
4	25	1.48	1.20	1.12	1.52	1.04	1.64
Univariate	F Stat.						
ANOVA	p Value	0.0001	0.0001	0.0001	0.0001	0.0001	0.0001

Among Sum of Squares Matrix

	X1	X2	X3	X4	X5	X6
X1	51.15	28.75	26.77	64.86	36.96	94.17
X2	28.75	16.99	15.17	37.86	20.48	52.93
X3	26.77	15.17	14.03	34.18	19.28	49.27
X4	64.86	37.86	34.18	85.16	45.96	119.10
X5	36.96	20.48	19.28	45.96	27.12	68.28
X6	94.17	52.93	49.27	119.10	68.28	173.55

Within Sum of Squares Matrix

	X1	X2	X3	X4	X5	X6
X1	97.44	42.76	26.52	63.64	20.44	65.20
X2	42.76	66.40	24.64	55.64	29.12	43.00
X3	26.52	24.64	40.96	46.32	29.12	30.20
X4	63.64	55.64	46.32	135.84	44.04	100.40
X5	20.44	29.12	29.12	44.04	50.88	34.92
X6	65.20	43.00	30.20	100.40	34.92	127.36

and $n = \sum_{k=1}^{g} n_k$. This approximation is usually referred to as Rao's F. If $p = 1$, this statistic is equivalent to the F-statistic used in ANOVA which

TABLE 8.3. Correlation Matrix and Associated p-Values*

	X1	X2	X3	X4	X5	X6
X1	1.000	0.642	0.590	0.709	0.533	0.754
X2	0.000	1.000	0.588	0.689	0.615	0.606
X3	0.000	0.000	1.000	0.730	0.739	0.618
X4	0.000	0.000	0.000	1.000	0.685	0.851
X5	0.000	0.000	0.000	0.000	1.000	0.674
X6	0.000	0.000	0.000	0.000	0.000	1.000

*Correlations upper right, p-values lower left

is an exact F distribution. It is also true that if $p = 2$ the F-statistic has an exact F distribution under H_{op} for all $g \geq 2$. In the special cases that $(g - 1) = 1$ for all p and $(g - 1) = 2$ for all p, the F-statistic is also exact under H_{op}. In all other cases for p and $(g - 1)$, the F distribution is only approximate.

An Alternative Test Statistic

An alternative approximation to the distribution of Λ is Bartlett's χ^2 statistic. If H_{op} is true, then in large samples the statistic $-[(n - 1) - (p + g)/2] \ln \Lambda$ has a χ^2 distribution with $p(g - 1)$ degrees of freedom. The F approximation is preferred to the χ^2 approximation when g is small.

Example

For the public safety data, the likelihood ratio test of equality of mean vectors yields a Wilks' Lambda of 0.3597. Rao's F approximation is given by 6.24 and has 18 and 257.9 degrees of freedom. The p-value for this statistic is 0.0001, and hence the four (6×1) mean vectors would be declared significantly different at conventional levels. The four groups therefore seem to differ with respect to the average values of the six public safety measures. A quick perusal of Table 8.2 suggests that groups 1 and 3 tend to have higher means than groups 2 and 4. Additional comparisons among the four groups will be made later in this section.

Correlation Ratio

In multiple regression and in univariate analysis of variance, a useful measure of strength of relationship is R^2 where $R^2 = 1 - \text{SSW}/\text{SST}$ or SSR/SST. An extension of this measure to MANOVA is provided by $(1 - \Lambda)$ where $\Lambda = |\mathbf{W}|/|\mathbf{W} + \mathbf{G}|$. The ratio Λ may be interpreted as the ratio of the generalized within sum of squares to the generalized total sum of squares. The larger the ratio $(1 - \Lambda)$, the greater the proportion of generalized variance that can be attributed to the variation among groups. This measure of strength of association is strongly positively biased. A much less biased alternative is given by $1 - n\Lambda/[(n - g) + \Lambda] = w^2$. An approximately unbiased measure [see Tatsuoka (1988)] of the criterion is given by

$$w_c^2 = w^2 - \frac{p^2 + (g - 1)^2}{3n}(1 - w^2).$$

Using the Wilks' Lambda value obtained above for the public safety data, $(1 - \Lambda)$ is given by 0.6403. The less biased versions of the R^2 type measure are $w^2 = 0.6267$ and $w_c^2 = 0.5707$. We can conclude, therefore, that approximately 60% of the total variation among the observations is due to the variation among the four mean vectors.

The Special Case of Two Groups

If there are only two groups, $g = 2$, the MANOVA can be simplified considerably. Since $\mathbf{x} \sim N_p(\boldsymbol{\mu}_1, \Sigma)$ for group 1 and $\mathbf{x} \sim N_p(\boldsymbol{\mu}_2, \Sigma)$ for group 2, we may write $(\bar{\mathbf{x}}_{.1} - \bar{\mathbf{x}}_{.2}) \sim N_p[\boldsymbol{\mu}_1 - \boldsymbol{\mu}_2, (\frac{1}{n_1} + \frac{1}{n_2})\Sigma]$. A test of the hypothesis $H_{0p}: (\boldsymbol{\mu}_1 - \boldsymbol{\mu}_2) = \mathbf{0}$ can therefore be carried out using the Hotelling's T^2 procedure for testing $H_0: \boldsymbol{\mu} = \mathbf{0}$ introduced in Section 7.3. In this case the test statistic is given by

$$T^2 = (\bar{\mathbf{x}}_{.1} - \bar{\mathbf{x}}_{.2})'\overline{\mathbf{S}}^{-1}(\bar{\mathbf{x}}_{.1} - \bar{\mathbf{x}}_{.2})\left(\frac{n_1 n_2}{n_1 + n_2}\right),$$

where $\overline{\mathbf{S}}$ is the estimator of the common Σ given by

$$\overline{\mathbf{S}} = [(n_1 - 1)\mathbf{S}_1 + (n_2 - 1)\mathbf{S}_2]/(n_1 + n_2 - 2).$$

If H_{0p} is true, the statistic $(n_1 + n_2 - p - 1)T^2/(n_1 + n_2 - 2)p$ has an F distribution with p and $(n_1 + n_2 - p - 1)$ degrees of freedom. The F-test statistic given above may also be used to test the hypothesis that the squared Mahalanobis distance between the two groups is zero (under the assumption of common covariance matrix). The true squared Mahalanobis distance $(\boldsymbol{\mu}_1 - \boldsymbol{\mu}_2)'\Sigma^{-1}(\boldsymbol{\mu}_1 - \boldsymbol{\mu}_2)$ is estimated by $(\bar{\mathbf{x}}_{.1} - \bar{\mathbf{x}}_{.2})'\mathbf{S}^{-1}(\bar{\mathbf{x}}_{.1} - \bar{\mathbf{x}}_{.2})$, which is proportional to T^2. This test is the basis for the test for outliers discussed in Section 7.3.2.

The Hotelling's T^2 statistic can also be used to provide a simultaneous confidence interval for the elements of $(\boldsymbol{\mu}_1 - \boldsymbol{\mu}_2)$. An important property of Hotelling's T^2 statistic is that a $(1-\alpha)$ probability interval can be expressed for All linear combinations $\boldsymbol{\ell}'(\bar{\mathbf{x}}_{.1} - \bar{\mathbf{x}}_{.2})$ of $(\bar{\mathbf{x}}_{.1} - \bar{\mathbf{x}}_{.2})$ where $\boldsymbol{\ell}$ is a $(p \times 1)$ vector of constants. The interval is given by

$$\boldsymbol{\ell}'(\bar{\mathbf{x}}_{.1} - \bar{\mathbf{x}}_{.2}) \pm c\sqrt{\boldsymbol{\ell}'\overline{\mathbf{S}}\boldsymbol{\ell}(1/n_1 + 1/n_2)}$$

where

$$c^2 = \frac{(n_1 + n_2 - 2)p}{(n_1 + n_2 - p - 1)}F_{\alpha;p,(n_1+n_2-p-1)}.$$

The probability is $(1 - \alpha)$ that this interval covers $\boldsymbol{\ell}'(\boldsymbol{\mu}_1 - \boldsymbol{\mu}_2)$ simultaneously for all possible $\boldsymbol{\ell}$.

This property can be used to get confidence interval estimates for the individual elements of $(\boldsymbol{\mu}_1 - \boldsymbol{\mu}_2)$. For the jth component of $(\boldsymbol{\mu}_1 - \boldsymbol{\mu}_2)$ we have the interval $(\bar{x}_{.j1} - \bar{x}_{.j2}) \pm c(\bar{s}_{jj}^2(1/n_1 + 1/n_2))^{1/2}$ where $\bar{s}_{jj}^2$ is the jth diagonal element of $\overline{\mathbf{S}}$. As in the case of the Scheffe multiple comparison procedure used in ANOVA, this confidence interval procedure is very conservative.

A Bonferroni Approximation

An alternative approach involves the Bonferroni approximation

$$(\bar{x}_{\cdot j1} - \bar{x}_{\cdot j2}) \pm t_{\alpha^*,(n_1+n_2-p-1)}\sqrt{\bar{s}_{jj}^2(1/n_1 + 1/n_2)}\,,$$

where $\alpha^* = \alpha/2p$ and $t_{\alpha^*,(n_1+n_2-p-1)}$ denotes the α^* critical value of the t distribution for $(n_1 + n_2 - p - 1)$ degrees of freedom. The Bonferroni intervals are shorter than those based on T^2; particularly for larger values.

Multiple Comparison Procedures Based on Two Group Comparisons

If the null hypothesis $H_{0p}: \boldsymbol{\mu}_1 = \boldsymbol{\mu}_2 = \ldots = \boldsymbol{\mu}_g$ is rejected, there is usually a need to obtain additional information about the nature of the departure from equality. A useful first step is to use a two independent group comparison procedure to compare all possible group pairs. For each pair the Hotelling's T^2 test can be used to compare the vector means. To compare the means for groups r and s the statistic

$$(\bar{\mathbf{x}}_{\cdot r} - \bar{\mathbf{x}}_{\cdot s})'\overline{\mathbf{S}}^{-1}(\bar{\mathbf{x}}_{\cdot r} - \bar{\mathbf{x}}_{\cdot s})\left(\frac{n_r n_s}{n_r + n_s}\right)\frac{(n_r + n_s - p - 1)}{(n_r + n_s - 2)p}$$

is compared to $F_{\alpha^*;p,(n_s+n_r-p-1)}$, where

$$\overline{\mathbf{S}} = \frac{(n_r - 1)\mathbf{S}_r + (n_s - 1)\mathbf{S}_s}{(n_r + n_s - 2)}$$

and $\alpha^* = 2\alpha/g(g - 1)$ is the Bonferroni approximate critical value for an α level test over the $g(g - 1)/2$ comparisons.

For pairs in which the mean vectors are declared to be significantly different, a comparison of the component means can be carried out using the univariate multiple comparison procedures from ANOVA. A simple t-statistic procedure with a Bonferroni approximation for multiple comparisons would involve the intervals

$$(\bar{x}_{\cdot jr} - \bar{x}_{\cdot js}) \pm t_{\alpha^{**};(n-g)}\sqrt{\bar{s}_{jj}^2\left(\frac{1}{n_r} + \frac{1}{n_s}\right)}\,,$$

where $\alpha^{**} = \alpha^*/p$, $\bar{s}_{jj}^2$ is the jth diagonal element of

$$\overline{\mathbf{S}} = \frac{(n_1 - 1)\mathbf{S}_1 + (n_2 - 1)\mathbf{S}_2 + \ldots + (n_g - 1)\mathbf{S}_g}{(n - g)}\,,$$

and $t_{\alpha^{**};(n-g)}$ is the α^{**} critical value of the t-distribution for $(n - g)$ degrees of freedom. Other procedures such as Tukey's method for multiple comparisons could also be used.

In Section 8.2 we introduce a multivariate approach to group comparisons called discriminant analysis. This technique seeks to find linear functions of the p-dimensional vector random variable $\mathbf{x}$ that highlight the differences among the groups.

Testing for the Equality of Covariance Matrices

The assumption of equality for the covariance matrices for the g groups can be critical, particularly if the group sample sizes n_k, $k = 1, 2, \ldots, g$, are radically different. It is therefore of value to test the null hypothesis

$$H_0: \Sigma_1 = \Sigma_2 = \ldots = \Sigma_g,$$

given that x_k is normally distributed, $N(\mu_k, \Sigma_k)$, $k = 1, 2, \ldots, g$.
We define the quantity M as follows

$$M = \left[\prod_{k=1}^{g} |S_k|^{(n_k-1)/2} \right] \Big/ |\bar{S}|^{(n-g)/2},$$

where S_k is the sample covariance matrix for group k, and $\bar{S} = \sum_{k=1}^{g}(n_k - 1)S_k/(n-g)$ and $n = \sum_{k=1}^{g} n_k$.
There are two asymptotic approximations to the distribution of M.

1. $-2(1-c_1)\ln M$ is approximately χ^2 with $(\frac{1}{2})p(p+1)(g-1)$ d.f. where

$$c_1 = \frac{(2p^2 + 3p - 1)}{6(p+1)(g-1)} \left(\sum_{k=1}^{g} \frac{1}{(n_k-1)} - \frac{1}{(n-g)} \right).$$

2. $-2b\log M$ is approximately distributed as an F distribution with v_1 and v_2 degrees of freedom where

$$
\begin{aligned}
v_1 &= (1/2)p(p+1)(g-1), \quad c_1 \text{ is given in (1)}, \\
v_2 &= (v_1+2)/|c_2 - c_1^2|, \quad b = [1 - c_1 - v_1/v_2]/v_1, \\
c_2 &= \frac{(p-1)(p+2)}{6(g-1)} \left(\sum_{k=1}^{g} \frac{1}{(n_k-1)^2} - \frac{1}{(n-g)^2} \right).
\end{aligned}
$$

If $c_2 - c_1^2 < 0$, then $\frac{-2b_1 v_2 \ln M}{v_1 + 2b_1 v_1 \ln M}$ is approximately distributed as an F distribution with v_1 and v_2 d.f. where $b_1 = \frac{1-c_1-(2/v_2)}{v_2}$. The F approximation is usually better than the χ^2 approximation.

Example

For the public safety data, a test of the equality of the four group covariance matrices was carried out using the χ^2-test. The χ^2-value obtained was 273.23 with 63 degrees of freedom. The p-value of this statistic is 0.0001. An examination of the elements of the four covariance matrices (not shown here) reveals that the elements for the covariance matrices for groups 1 and 3 are much larger than the elements in the covariance matrices for groups 2 and 4.

8.1.2 INDICATOR VARIABLES, MULTIVARIATE REGRESSION AND ANALYSIS OF COVARIANCE

Analysis of variance can be viewed as a special case of multiple linear regression. The explanatory variables are indicator variables which are usually derived from dummy coding or effect coding for the various categories. The resulting matrix of explanatory variables is called the design matrix. In a similar fashion, multivariate analysis of variance is a special case of multivariate regression with the explanatory variables provided by the identical design matrix patterns used in ANOVA. The design matrices and corresponding parameter matrices for both dummy coding and effect coding are illustrated in Figures 8.4 and 8.5 respectively.

For the multivariate regression model discussed in Section 7.5, the dependent variables are denoted by $\mathbf{X}$ and the explanatory variables by the design matrix $\mathbf{D}$. The design matrix $\mathbf{D}$ contains observations on indicator variables $D_1, D_2, \ldots, D_{g-1}$. The column vectors in $\mathbf{D}$ are denoted by $\mathbf{d}_1, \mathbf{d}_2, \ldots, \mathbf{d}_{g-1}$, and the observations on $D_1, D_2, \ldots, D_{g-1}$ are denoted by $d_1, d_2, \ldots, d_{g-1}$. The multivariate regression likelihood ratio test of the hypothesis that the $(g-1)$ indicator variables are superfluous is equivalent to the MANOVA procedure outlined above. More complex MANOVA models can also be analyzed using multivariate regression with indicator variables.

Example

For the public safety data, a multivariate regression of the six safety variables on three group dummy variables plus an intercept would be equivalent to the MANOVA analysis discussed above. Using dummy variables for the first three groups, the model is given by

$$\mathbf{X} = \mathbf{DB} + \mathbf{U} \qquad (8.2)$$

where $\mathbf{X}$ is the (100×6) data matrix, $\mathbf{D}$ (100×4) is the matrix $\mathbf{D} = [\mathbf{i}\ \mathbf{d}_1 \mathbf{d}_2 \mathbf{d}_3]$ containing $\mathbf{i}$, a vector of unities, plus the three dummy variable vectors $\mathbf{d}_1, \mathbf{d}_2$ and $\mathbf{d}_3$, $\mathbf{B}$ (4×6) is the matrix of regression coefficient vectors $\boldsymbol{\beta}_1, \boldsymbol{\beta}_2, \ldots, \boldsymbol{\beta}_6$, and $\mathbf{U}$ (100×6) is the error matrix. Each regression coefficient vector $\boldsymbol{\beta}_j$ contains an intercept element β_{j0} representing the mean μ_{j4} for group 4 and three dummy coefficients β_{j1}, β_{j2} and β_{j3} denoting the mean differences $(\mu_{j1} - \mu_{j4})$, $(\mu_{j2} - \mu_{j4})$ and $(\mu_{j3} - \mu_{j4})$.

The multivariate regression yields the six equations

$$\begin{aligned}
\hat{x}_1 &= 1.48 + 1.56 d_1 + 0.04 d_2 + 1.32 d_3 \\
\hat{x}_2 &= 1.20 + 1.04 d_1 + 0.20 d_2 + 0.72 d_3 \\
\hat{x}_3 &= 1.12 + 0.84 d_1 + 0.04 d_2 + 0.68 d_3 \\
\hat{x}_4 &= 1.52 + 2.24 d_1 + 0.20 d_2 + 1.48 d_3 \\
\hat{x}_5 &= 1.04 + 1.08 d_1 + 0.08 d_2 + 1.08 d_3 \\
\hat{x}_6 &= 1.64 + 2.88 d_1 + 0.16 d_2 + 2.52 d_3 .
\end{aligned}$$

$$
\mathbf{D} \\
\begin{bmatrix}
1\ 1\ 0 \ldots 0\ 0 \\
1\ 1\ 0 \quad\ 0\ 0 \\
\vdots\ \vdots\ \vdots \quad\ \vdots\ \vdots \\
1\ 1\ 0 \quad\ 0\ 0 \\
1\ 0\ 1 \quad\ 0\ 0 \\
1\ 0\ 1 \ldots 0\ 0 \\
\vdots\ \vdots\ \vdots \quad\ \vdots\ \vdots \\
1\ 0\ 1 \quad\ 0\ 0 \\
\vdots\ \vdots\ \vdots \quad\ \vdots\ \vdots \\
1\ 0\ 0 \ldots 0\ 1 \\
1\ 0\ 0 \quad\ 0\ 1 \\
\vdots\ \vdots\ \vdots \quad\ \vdots\ \vdots \\
1\ 0\ 0 \quad\ 0\ 1 \\
1\ 0\ 0 \quad\ 0\ 0 \\
1\ 0\ 0 \quad\ 0\ 0 \\
\vdots\ \vdots\ \vdots \quad\ \vdots\ \vdots \\
1\ 0\ 0 \quad\ 0\ 0
\end{bmatrix}
$$

$$
\mathbf{B} \\
\begin{bmatrix}
\mu_{1g} & \mu_{2g} & \cdots & \mu_{pg} \\
(\mu_{11}-\mu_{1g}) & (\mu_{21}-\mu_{2g}) & \cdots & (\mu_{p1}-\mu_{pg}) \\
(\mu_{12}-\mu_{1g}) & (\mu_{22}-\mu_{2g}) & \cdots & (\mu_{p2}-\mu_{pg}) \\
\vdots & \vdots & & \vdots \\
(\mu_{1(g-1)}-\mu_{1g}) & (\mu_{2(g-1)}-\mu_{2g}) & \cdots & (\mu_{p(g-1)}-\mu_{pg})
\end{bmatrix}
$$

FIGURE 8.4. Dummy Coding

The p-values for the coefficients of d_2 vary from 0.9 to 0.4, and hence we can conclude that groups 2 and 4 have mean vectors that are not significantly different. The p-values for all remaining coefficients are less than 0.00, and hence the mean vectors for groups 1 and 3 are significantly different from group 4, and also the elements of the mean vector for group 4 are significantly different from zero. Changing the base case to group 1 shows that groups 1 and 3 are not significantly different but that groups 2 and 4 are significantly different from 1 (not shown here).

Some Relationships to the Multivariate Regression Test for $H_0\colon \mathbf{ABM} = \mathbf{0}$

It is also possible to use multivariate regression to test hypotheses regarding relationships among the means. Using the multivariate test procedures given in Chapter 7, a test that is equivalent to the MANOVA test $H_{0p}\colon \boldsymbol{\mu}_1 = \boldsymbol{\mu}_2 = \boldsymbol{\mu}_3 = \boldsymbol{\mu}_4$, is the test $H_0\colon \mathbf{AB} = \mathbf{0}$ where $\mathbf{A} = \begin{bmatrix} 0 & 1 & 0 & 0 \\ 0 & 0 & 1 & 0 \\ 0 & 0 & 0 & 1 \end{bmatrix}$ and $\mathbf{B}$ is the coefficient matrix given in the multivariate regression with indicator variables. If this hypothesis is true, the coefficients for all three indicator

$$
\mathbf{D}
$$

$$
\begin{bmatrix}
1 & 1 & 0 & \cdots & 0 & 0 \\
1 & 1 & 0 & & 0 & 0 \\
\vdots & \vdots & \vdots & & \vdots & \vdots \\
1 & 1 & 0 & & 0 & 0 \\
1 & 0 & 1 & & 0 & 0 \\
1 & 0 & 1 & \cdots & 0 & 0 \\
\vdots & \vdots & \vdots & & \vdots & \vdots \\
1 & 0 & 1 & & 0 & 0 \\
\vdots & \vdots & \vdots & & \vdots & \vdots \\
1 & 0 & 0 & \cdots & 0 & 1 \\
1 & 0 & 0 & & 0 & 1 \\
\vdots & \vdots & \vdots & & \vdots & \vdots \\
1 & 0 & 0 & & 0 & 1 \\
1 & -1 & -1 & \cdots & -1 & -1 \\
1 & -1 & -1 & & -1 & -1 \\
\vdots & \vdots & \vdots & & \vdots & \vdots \\
1 & -1 & -1 & & -1 & -1
\end{bmatrix}
$$

$$
\mathbf{B}
$$

$$
\begin{bmatrix}
\bar{\mu}_{1\cdot} & \bar{\mu}_{2\cdot} & \cdots & \bar{\mu}_{p\cdot} \\
(\mu_{11} - \bar{\mu}_{1\cdot}) & (\mu_{21} - \bar{\mu}_{2\cdot}) & \cdots & (\mu_{p1} - \bar{\mu}_{p\cdot}) \\
(\mu_{12} - \bar{\mu}_{1\cdot}) & (\mu_{22} - \bar{\mu}_{2\cdot}) & \cdots & (\mu_{p2} - \bar{\mu}_{p\cdot}) \\
\vdots & \vdots & & \vdots \\
(\mu_{1(g-1)} - \bar{\mu}_{1\cdot}) & (\mu_{2(g-1)} - \bar{\mu}_{2\cdot}) & \cdots & (\mu_{p(g-1)} - \bar{\mu}_{p\cdot})
\end{bmatrix}
$$

FIGURE 8.5. Effect Coding

variables in all equations are zero. This approach is illustrated next using the previous example.

Example

The multivariate regression results above suggest that the mean vectors for groups 1 and 3 are similar and the mean vectors for groups 2 and 4 are similar. A test of the null hypothesis $H_0: \mu_1 = \mu_3; \ \mu_2 = \mu_4$ can be carried out using $H_0: \mathbf{AB} = \mathbf{0}$ where $\mathbf{A} = \begin{bmatrix} 0 & 0 & 1 & 0 \\ 0 & 1 & 0 & -1 \end{bmatrix}$ and $\mathbf{B}$ is given in (8.2). For each of the six equations, the null hypothesis is that the coefficient of d_2 is zero and that the coefficients of d_1 and d_3 are equal. The F-value for this test is 0.723, which has a p-value of 0.728 when compared to an F-distribution with 12 and 182 degrees of freedom. The null hypothesis H_0 therefore cannot be rejected.

An interesting alternative test is to determine whether the four coefficients in the six equations in (8.2) are equal. The null hypothesis is given by $H_0: \mathbf{BM} = \mathbf{0}$ where

$$
\mathbf{M} = \begin{bmatrix}
1 & 1 & 1 & 1 & 1 \\
-1 & 0 & 0 & 0 & 0 \\
0 & -1 & 0 & 0 & 0 \\
0 & 0 & -1 & 0 & 0 \\
0 & 0 & 0 & -1 & 0 \\
0 & 0 & 0 & 0 & -1
\end{bmatrix}.
$$

In this case, the null hypothesis is that the four coefficients in the first equation are equal to the four coefficients in each of the five remaining equations. For the public safety data this test yields an F-value of 9.303 with 20 and 306.1 degrees of freedom. The p-value of this statistic is 0.0001; hence we cannot assume the same equation for all six variables. In this case the hypothesis being tested is that for each group the components of the mean vector are equal but that the magnitudes of the means can vary across the groups. We see later in Section 8.1.3 that this test is useful in profile analysis for repeated measurement designs and is equivalent to a test for horizontal profiles. In the context of the example the hypothesis implies that in each community the average response on each of the six items is the same. Comparing across communities however, differences in average response are permitted.

From this example we can see that the elements in a given column of $\mathbf{B}$ correspond to a given variable and that each row of $\mathbf{B}$ corresponds to a group. The matrix $\mathbf{A}$ is used to obtain comparisons among groups (within an equation), and the matrix $\mathbf{M}$ is used to obtain comparisons among variables (across equations).

Cell Parameter Coding

With dummy coding and effect coding the design matrix $\mathbf{D}$ contains a column of unities $\mathbf{i}$ $(n \times 1)$ and indicator variables for all but one of the g groups. The group without the indicator variable is usually referred to as the base case. Since many statistical software packages now permit inference procedures for functions of model parameters it is also possible to use cell parameter coding. With this approach the column of unities $\mathbf{i}$ is eliminated from the design matrix, and a dummy variable is used for all g groups. In this case the design matrix and parameter vector are illustrated in Figure 8.6.

In the previous example involving six variables and four groups a test of $H_0: \boldsymbol{\mu}_1 = \boldsymbol{\mu}_2 = \boldsymbol{\mu}_3 = \boldsymbol{\mu}_4$ can be carried out using a test of $H_0: \mathbf{AB} = \mathbf{0}$, where

$$
\mathbf{A} = \begin{bmatrix}
1 & -1 & 0 & 0 \\
1 & 0 & -1 & 0 \\
1 & 0 & 0 & -1
\end{bmatrix}.
$$

$$
\mathbf{D} = \begin{bmatrix}
1 & 0 & 0 & \cdots & 0 \\
1 & 0 & 0 & & 0 \\
\vdots & \vdots & \vdots & & \vdots \\
1 & 0 & 0 & & 0 \\
0 & 1 & 0 & \cdots & 0 \\
0 & 1 & 0 & & 0 \\
\vdots & \vdots & \vdots & & \vdots \\
0 & 1 & 0 & & 0 \\
0 & 0 & 1 & \cdots & 0 \\
0 & 0 & 1 & & 0 \\
\vdots & \vdots & \vdots & & \vdots \\
0 & 0 & 1 & & 0 \\
\vdots & \vdots & \vdots & & \vdots \\
0 & 0 & 0 & & 1 \\
0 & 0 & 0 & \cdots & 1 \\
\vdots & \vdots & \vdots & & \vdots \\
0 & 0 & 0 & & 1
\end{bmatrix}
\qquad
\mathbf{B} = \begin{bmatrix}
\mu_{11} & \mu_{12} & \mu_{13} & \cdots & \mu_{1p} \\
\mu_{21} & \mu_{22} & \mu_{23} & \cdots & \mu_{2p} \\
\mu_{31} & \mu_{32} & \mu_{33} & \cdots & \mu_{3p} \\
\vdots & \vdots & \vdots & & \vdots \\
\mu_{g1} & \mu_{g2} & \mu_{g3} & \cdots & \mu_{gp}
\end{bmatrix}
$$

FIGURE 8.6. Cell Parameter Coding

A test of equality for the four coefficients across the six equations could be carried out by testing $H_0: \mathbf{BM} = \mathbf{0}$ where

$$
\mathbf{M} = \begin{bmatrix}
1 & 1 & 1 & 1 & 1 \\
-1 & 0 & 0 & 0 & 0 \\
0 & -1 & 0 & 0 & 0 \\
0 & 0 & -1 & 0 & 0 \\
0 & 0 & 0 & -1 & 0 \\
0 & 0 & 0 & 0 & -1
\end{bmatrix}.
$$

This test is equivalent to a test of equal means for the six variables within each group. The four groups are permitted to differ with respect to the overall level of the means.

The Non-Full Rank Design Matrix

The group effects form of the MANOVA model given in Section 8.1.1 can also be written as a multivariate regression model with a design matrix $\mathbf{D}$ of indicator variables that is no longer of full rank. The model is given by

$$
\mathbf{X} = \mathbf{DB} + \mathbf{V}
$$

$$D = \begin{bmatrix} 1 & 1 & 0 & \cdots & 0 \\ 1 & 1 & 0 & & \\ \vdots & \vdots & \vdots & \cdots & \\ 1 & 1 & 0 & & 0 \\ 1 & 0 & 1 & & 0 \\ & 0 & 1 & & \\ \vdots & \vdots & & \cdots & \\ 1 & 0 & 1 & & 0 \\ \vdots & \vdots & & \cdots & \vdots \\ 1 & 0 & 0 & & 1 \\ & & & & 1 \\ \vdots & \vdots & \vdots & \cdots & \vdots \\ 1 & 0 & 0 & & 1 \end{bmatrix} \qquad B = \begin{bmatrix} \mu_1 & \mu_2 & \cdots & \mu_p \\ \alpha_{11} & \alpha_{12} & \cdots & \alpha_{1p} \\ \alpha_{21} & \alpha_{22} & \cdots & \alpha_{2p} \\ \vdots & \vdots & \cdots & \vdots \\ \alpha_{g1} & \alpha_{g2} & \cdots & \alpha_{gp} \end{bmatrix}.$$

FIGURE 8.7. Group Effects Parameter Coding

where the *group effects* design matrix D and the corresponding parameter matrix B are shown in Figure 8.7. In this model the design matrix D contains an additional column ($g + 1$ columns) compared to the dummy coding and effect coding versions of D shown in Figures 8.4 and 8.5. To test the hypothesis $H_{0p}: \alpha_1 = \alpha_2 = \ldots = \alpha_g = 0$ that is equivalent to $H_{0p}: \mu_1 = \mu_2 = \ldots = \mu_g$ in the multivariate linear model, a test of $H_0: \begin{bmatrix} b_0' \\ B_1 \end{bmatrix} = \begin{bmatrix} b_0' \\ 0 \end{bmatrix}$ is required where b_0' denotes the first row of B, and B_1 denotes the remainder of B. The test of H_{0p} is equivalent to the test discussed in Section 8.1.1.

As described above for dummy and effect coding other tests can be carried out for the general form $H_0: ABM = 0$. In this case, however, the matrices A and M must be adjusted to reflect the group effects model. To test the hypothesis that $\mu_1 = \mu_3$ and $\mu_2 = \mu_4$ when $g = 4$ the matrix A has the form $A = \begin{bmatrix} 0 & 1 & 0 & 1 & 0 \\ 0 & 0 & 1 & 0 & 1 \end{bmatrix}$. To test the hypothesis that the group effects are identical for all $p = 6$ variables the M matrix is defined by

$$M = \begin{bmatrix} 0 & 0 & \cdots & 0 \\ 1 & 1 & & 1 \\ -1 & 0 & & 0 \\ 0 & -1 & \cdots & \\ & 0 & & \vdots \\ \vdots & \vdots & & 0 \\ 0 & 0 & \cdots & -1 \end{bmatrix}.$$

The matrices $\mathbf{A}$ and $\mathbf{M}$ illustrated above can sometimes be employed with MANOVA software. This approach is an alternative to the multivariate regression approach discussed above for use with dummy and effect coding.

A summary of the various types of coding and corresponding design matrices is provided in Volume I, Chapters 4 and 5.

Multivariate Analysis of Covariance

For the multivariate analysis of covariance the inference theory for multivariate regression can be used. The $\mathbf{D}$ matrix now consists of the column of unities, $(g-1)$ indicator variables $D_1, D_2, \ldots, D_{g-1}$ and, in addition, the columns corresponding to a set of b concomittant variables $Z_1, Z_2, \ldots, Z_b$. Thus the $\mathbf{D}$ matrix is given by $\mathbf{D} = [\mathbf{i} \ \mathbf{d}_1\mathbf{d}_2 \ldots \mathbf{d}_{g-1}, \ \mathbf{z}_1\mathbf{z}_2 \ldots \mathbf{z}_b]$, where $\mathbf{i} \ (n \times 1)$ is a column of unities, $\mathbf{d}_1, \mathbf{d}_2, \ldots, \mathbf{d}_{g-1}$ are columns of indicator variables corresponding to $(g-1)$ of the g groups, and $\mathbf{z}_1, \mathbf{z}_2, \ldots, \mathbf{z}_b$ are the columns corresponding to the covariates. The test procedure outlined in 7.5 for $H_0\colon \mathbf{ABM} = \mathbf{K}$ can be used to carry out various test procedures for the analysis of covariance.

Example

To provide an example for MANOVA with a covariate, the variable INC was added to the right hand side of the multivariate regression model discussed above. The income variable is a measure of family income and uses nine categories coded from 1 to 9. The observations are shown in Table 8.1. The income variable will be treated as an interval scaled variable. A test of the hypothesis that the coefficient of INC is zero in all six equations yields an F-statistic of 2.1956 with 6 and 90 degrees of freedom. The p-value for this test result was 0.051. For the six individual regression equations the coefficient of INC was negative in each case and for X2, X4 and X6 the regression coefficient was significant at the 0.02 level. It would appear that individuals with higher income feel more safe in their homes and at night than those with less income.

A comparison of the regression coefficients with and without the income variable is shown in Table 8.4. In general, the addition of the income variable tended to increase the other coefficients. In particular, group 4 (see intercept) had large changes in the means after the adjustment for INC. Since the intercepts also reflect the overall levels, this result suggests that the increases in the elements of the mean vector are fairly uniform over the four groups. A variety of comparisons can be carried out using the multivariate regression test $H_0\colon \mathbf{ABM} = \mathbf{K}$.

TABLE 8.4. Comparison of Multivarite Regression Coefficients for Models With and Without Income

Dependent Variable	Intercept		D1		D2		D3	
	Without Income	With Income	Without Income	With Income	Without Income	With Income	Without Income	With Income
X1	1.48	1.82	1.56	1.61	0.04	0.07	1.32	1.29
X2	1.20	1.96	1.04	1.15	0.20	0.26	0.72	0.66
X3	1.12	1.40	0.84	0.88	0.04	0.06	0.68	0.66
X4	1.52	2.44	2.24	2.37	0.20	0.28	1.48	1.41
X5	1.04	1.31	1.08	1.12	0.08	0.10	1.08	1.06
X6	1.64	2.58	2.88	3.01	0.16	0.24	2.52	2.45

8.1.3 PROFILE ANALYSIS WITH REPEATED MEASUREMENTS

In the MANOVA discussed in Section 8.1.1, the observations on the p-dimensional random vector $\mathbf{x}$ may actually represent repeated observations on the same individual or object under p different conditions. In each group j, $j = 1, 2, \ldots, g$, a total of n_j individuals are observed on each of the p different conditions. In this repeated measurements environment, it is usually of interest to compare the means on the p conditions over the g groups as well as the means over the g groups on the p conditions. In each group the variation in the p means is usually characterized by the profile. The comparison of group means and condition means is equivalent to a comparison of profile shapes.

In Section 7.4.2 the repeated measurements comparison was introduced for a single group of n individuals observed under each of p conditions. The test of $H_0: \mu_1 = \mu_2 = \ldots = \mu_p$ was of interest to determine whether the p condition means are equal and hence a horizontal profile. Figure 7.1 in Chapter 7 shows an example profile based on the means of n objects over five conditions. Thus in Chapter 7 we were concerned with the shape of a single profile (a within subjects comparison). In the case of g groups here, we are also concerned with a comparison of the profiles across groups (a between groups comparison).

Comparing Profiles

If the repeated measures experiment is carried out for a set of g different groups, there are a total of g profiles to be analyzed. If the g profiles are horizontal, there are no condition effects, whereas if the g profiles are equal, there are no group effects. If the g profiles are neither horizontal nor equal, they may still be parallel, which is an indication that there is no interaction between the group effects and the condition effects.

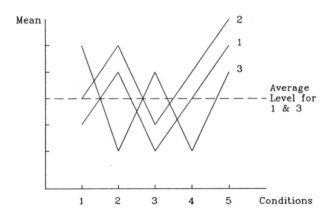

FIGURE 8.8. Comparison of Three Group Profiles Over Five Conditions

The hypothesis of equality of mean vectors in MANOVA given by H_{0p}: $\boldsymbol{\mu}_1 = \boldsymbol{\mu}_2 = \ldots = \boldsymbol{\mu}_g$ is also a test for equality of the g profiles. This test, however, is not concerned with the profile shape. If this hypothesis is true it may also be of interest to know if the profiles are horizontal. If the hypothesis H_{0p} is rejected indicating that there are group effects, it may still be true that the profiles are parallel and also perhaps horizontal. Figure 8.8 shows profiles for three groups over five conditions. Profiles 1 and 2 are parallel whereas profiles 1 and 3 have the same average level. A variety of test procedures for the study of profiles is outlined next.

Parallel Profiles

If the g profiles being compared in H_{0p} are not equal, they may still be parallel. This weaker hypothesis is given by

$$H_{0I} \colon [\mu_{(j+1)1} - \mu_{j1}] = [\mu_{(j+1)2} - \mu_{j2}]$$
$$= \ldots [\mu_{(j+1)g} - \mu_{jg}],$$
$$j = 1, 2, \ldots, (p-1).$$

This hypothesis can be tested by defining the $(p-1) \times p$ transformation matrix $\mathbf{C}$ where

$$\mathbf{C} = \begin{bmatrix} -1 & 1 & 0 & \ldots & 0 & 0 \\ 0 & -1 & 1 & & 0 & 0 \\ 0 & 0 & -1 & \ldots & 0 & 0 \\ \vdots & & \vdots & & & \vdots \\ 0 & 0 & 0 & \ldots & 1 & 0 \\ 0 & 0 & 0 & \ldots & -1 & 1 \end{bmatrix}$$

which is equal to the matrix $\mathbf{C}$ used in Section 7.4.2. This test for parallel profiles is equivalent to a test for no interaction between the groups and the conditions. In this case the test is given by

$$H_{0I}: \mathbf{C}\boldsymbol{\mu}_1 = \mathbf{C}\boldsymbol{\mu}_2 = \ldots = \mathbf{C}\boldsymbol{\mu}_g.$$

This test can be carried out by applying the Wilk's Lambda procedure of Section 8.1.1 to the transformed variables $\mathbf{y} = \mathbf{Cx}$ to test $H_{0I}: \boldsymbol{\mu}_{y_1} = \boldsymbol{\mu}_{y_2} = \ldots = \boldsymbol{\mu}_{y_g}$, where $\boldsymbol{\mu}_y$ denotes the mean vector for the transformed variable $\mathbf{y}$.

Alternatively, the multivariate regression approach outlined in Section 8.1.2 can be applied using matrices $\mathbf{A}$ and $\mathbf{M}$ and testing $H_{0I}: \mathbf{ABM} = \mathbf{0}$. The matrices $\mathbf{A}$ and $\mathbf{M}$ are given by

$$\mathbf{A}_{(g-1)\times g} = \begin{bmatrix} 0 & 1 & 0 & 0 & \cdots & 0 \\ 0 & 0 & 1 & 0 & \cdots & 0 \\ 0 & 0 & 0 & 1 & & \vdots \\ \vdots & & & & & 0 \\ 0 & 0 & \cdots & \cdots & 0 & 1 \end{bmatrix},$$

$$\mathbf{M}_{p\times(p-1)} = \begin{bmatrix} 1 & 1 & 1 & \cdots & 1 \\ -1 & 0 & 0 & \cdots & 0 \\ 0 & -1 & 0 & \cdots & 0 \\ \vdots & & & & \vdots \\ 0 & 0 & \cdots & 0 & -1 \end{bmatrix},$$

whereas the matrix $\mathbf{B}$ is either the dummy coding matrix in Figure 8.4 or the effect coding matrix in Figure 8.5.

Example

We use the public safety data and the dummy variable multivariate regression to demonstrate the tests outlined above. Figure 8.9 shows a comparison of the four profiles. It would appear that groups 1 and 3 and groups 2 and 4 have similar profiles. To test the hypothesis H_{0I} that the g profiles are parallel we use the dummy variables defined in Section 8.1.2. The matrices $\mathbf{A}$ and $\mathbf{M}$ are given by

$$\mathbf{A} = \begin{bmatrix} 0 & 1 & 0 & 0 \\ 0 & 0 & 1 & 0 \\ 0 & 0 & 0 & 1 \end{bmatrix} \quad \text{and} \quad \mathbf{M} = \begin{bmatrix} 1 & 1 & 1 & 1 & 1 \\ -1 & 0 & 0 & 0 & 0 \\ 0 & -1 & 0 & 0 & 0 \\ 0 & 0 & -1 & 0 & 0 \\ 0 & 0 & 0 & -1 & 0 \\ 0 & 0 & 0 & 0 & -1 \end{bmatrix}.$$

Thus $H_{0I}: \mathbf{ABM} = \mathbf{0}$ implies that the coefficients for the dummy variables $D1$, $D2$ and $D3$ in the equations for $X2$ through $X6$ are identical

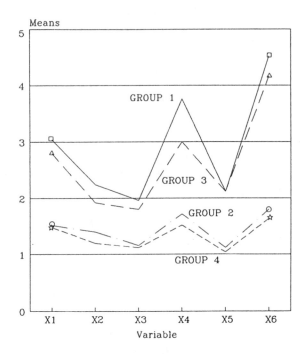

FIGURE 8.9. Profile Analysis for Public Safety Data

to the coefficients for these dummy variables in the equation for $X1$. We are allowing the overall mean level to be different for each variable, but the differences between the profile points in each group must be the same. For the public safety data, the F-statistic has a value of 5.334 with 15 and 254.4 degrees of freedom. The p-value of this statistic is less than 0.0001 and hence we cannot assume that the profiles are parallel.

Equal Profiles Given Parallel Profiles

If the hypothesis H_{0I} of parallel profiles is true, then the hypothesis H_{0p} of equal profiles is equivalent to the hypothesis $H_{0\bar{p}}: \mathbf{i}'\boldsymbol{\mu}_1 = \mathbf{i}'\boldsymbol{\mu}_2 = \ldots = \mathbf{i}'\boldsymbol{\mu}_g$, where $\mathbf{i}(p \times 1)$ is a vector of unities. This hypothesis indicates that the average profiles for each of the g groups are equal. If the hypothesis H_{0I} is not true, then the hypothesis $H_{0\bar{p}}$ is usually of little value. Using the multivariate regression model, a test of $H_{0\bar{p}}$ is given by $H_{0\bar{p}}: \mathbf{ABM} = \mathbf{0}$ where $\mathbf{M}(p \times 1)$ is given by $\mathbf{M}' = [1\ 1\ 1\ \ldots\ 1]$, $\mathbf{A}$ is identical to the $\mathbf{A}$ matrix used in the multivariate regression test for parallel profiles, and $\mathbf{B}$ is one of the two design matrices shown in Figures 8.4 and 8.5.

Horizontal Profiles Given Parallel Profiles

If the hypothesis H_{0I} of parallel profiles is true, then the hypothesis of horizontal profiles is equivalent to the hypothesis of equal means for the p conditions given by $H_{0e}: \bar{\mu}_{1\cdot} = \bar{\mu}_{2\cdot} = \ldots = \bar{\mu}_{p\cdot}$. This test can be carried out using the multivariate regression model by defining **A** and **M** as follows:

$$
\mathbf{A} = [g\ 1\ 1 \ldots 1]; \quad \mathbf{M} =
\begin{bmatrix}
1 & 1 & 1 & \cdots & 1 \\
-1 & 0 & 0 & \cdots & 0 \\
0 & -1 & 0 & \cdots & \vdots \\
\vdots & \vdots & \vdots & & 0 \\
0 & 0 & 0 & \cdots & -1
\end{bmatrix}.
$$

The test of $H_{0e}: \mathbf{ABM} = \mathbf{0}$ is carried out by employing one of the **B** matrices given in Figures 8.4 and 8.5.

Horizontal Profiles

The hypothesis of horizontal profiles can also be tested without regard to H_{0I}. This hypothesis is given by

$$
\begin{aligned}
H_{0H}: [\mu_{(j+1)1} - \mu_{j1}] &= [\mu_{(j+1)2} - \mu_{j2}] = \ldots \\
&= [\mu_{(j+1)g} - \mu_{jg}] = 0, \\
&\qquad j = 1, 2, \ldots, (p-1)
\end{aligned}
$$

and can also be tested using the multivariate regression test $H_{0H}: \mathbf{BM} = \mathbf{0}$. The **M** matrix is given by

$$
\mathbf{M} =
\begin{bmatrix}
-1 & 0 & 0 & \cdots & 0 & 0 \\
1 & -1 & 0 & \cdots & 0 & 0 \\
0 & 1 & -1 & & \vdots & \vdots \\
\vdots & \vdots & & & -1 & 0 \\
0 & 0 & 0 & \cdots & 1 & -1 \\
0 & 0 & 0 & \cdots & 0 & 1
\end{bmatrix}.
$$

Example

For the public safety data a test of the hypothesis that the average of the six variables is the same for each group is a test for equal profiles if the profiles are known to be parallel. Since the parallel profile hypothesis has already been rejected, this test cannot be used in this case to test for equal profiles. The test is carried out here for illustrative purposes even though it has no practical value. The test statistic is determined from the multivariate regression test $H_{0\bar{p}}: \mathbf{ABM} = \mathbf{0}$, where **A** is equivalent to the **A** matrix used for the above test for parallel profiles and **M** is given by

$\mathbf{M}' = [1\ 1\ 1\ 1\ 1\ 1]$. The regression therefore becomes a univariate multiple regression, and the hypothesis $H_{0\bar{p}}$ is tested by using the conventional F-test for the significance of all explanatory variables. For the public safety data, the F-statistic is given by 31.392, which has 3 and 96 d.f. and a p-value of 0.0001.

The test that the average of the elements of the mean vector over the g groups is the same for all six variables is a test for horizontal profiles if the profiles are parallel. If however the profiles are not parallel this hypothesis simply indicates that the average profile is horizontal. This test allows variation among the four groups over the six variables provided that the overall average is the same for each of the variables. Using the multivariate regression approach, the $\mathbf{A}$ and $\mathbf{M}$ matrices for this test are given by

$$\mathbf{A} = [4\ 1\ 1\ 1], \quad \mathbf{M} = \begin{bmatrix} 1 & 1 & 1 & 1 & 1 \\ -1 & 0 & 0 & 0 & 0 \\ 0 & -1 & 0 & 0 & 0 \\ 0 & 0 & -1 & 0 & 0 \\ 0 & 0 & 0 & -1 & 0 \\ 0 & 0 & 0 & 0 & -1 \end{bmatrix}.$$

The F-value for this test is 42.522 which has a p-value less than 0.0001 at 5 and 92 degrees of freedom. We cannot conclude therefore that the average profile is horizontal. As can be seen from Figure 8.9, variables 1, 4 and 6 tend to have larger values than variables 2, 3 and 5.

8.1.4   BALANCED TWO-WAY MANOVA

The Model

The procedures for one-way MANOVA can be easily extended to the case of two-way MANOVA. We assume that individuals are classified according to two different classification variables. As in the case of ANOVA, we assume that the classifications are based on g groups and b blocks. The cross-classification by group and block produces a total of bg cells. A total of c observations of the $(p \times 1)$ multivariate vector $\mathbf{x}$ are obtained from each cell (k, ℓ). The multivariate linear model relating $\mathbf{x}$ to both the group and block effects is given by

$$\mathbf{x}_{k\ell} = \boldsymbol{\mu} + \boldsymbol{\alpha}_k + \boldsymbol{\beta}_\ell + (\boldsymbol{\alpha\beta})_{k\ell} + \boldsymbol{\varepsilon}_{k\ell}, \qquad \begin{aligned} k &= 1, 2, \ldots, g, \\ \ell &= 1, 2, \ldots, b. \end{aligned}$$

The $(p \times 1)$ vectors $\boldsymbol{\mu}$, $\boldsymbol{\alpha}_k$, $\boldsymbol{\beta}_\ell$, $(\boldsymbol{\alpha\beta})_{k\ell}$ denote the grand mean vector, the group effects vector, the block effects vector and the interaction effects vector. The effects vectors satisfy $\sum_{k=1}^{g} \boldsymbol{\alpha}_k = \sum_{\ell=1}^{b} \boldsymbol{\beta}_\ell = \sum_{k=1}^{g} (\boldsymbol{\alpha\beta})_{k\ell} = \sum_{\ell=1}^{b} (\boldsymbol{\alpha\beta})_{k\ell} = 0$. The $(p \times 1)$ vector $\boldsymbol{\varepsilon}_{k\ell}$ is the residual vector.

The complete data matrix $\mathbf{X}$ $(n \times p)$ consists of n observations on p variables where $n = cgb$. Each element of the data matrix is denoted by

$x_{ijk\ell}$, $i = 1, 2, \ldots, c$; $j = 1, 2, \ldots, p$; $k = 1, 2, \ldots, g$; $\ell = 1, 2, \ldots, b$. The subscript j denotes the column of $\mathbf{X}$, whereas the subscripts i, k, ℓ combine to denote the row of $\mathbf{X}$. For each combination of k and ℓ, there are a total of c rows of $\mathbf{X}$ observations for cell (k, ℓ).

Sum of Squares Matrices

Denoting the $(p \times 1)$ sample grand mean vector by $\bar{\mathbf{x}}$ the total sum of squares matrix is given by

$$\mathbf{T} = \sum_{\ell=1}^{b} \sum_{k=1}^{g} \sum_{i=1}^{c} (\mathbf{x}_{ik\ell} - \bar{\mathbf{x}})(\mathbf{x}_{ik\ell} - \bar{\mathbf{x}})'.$$

As in the case of univariate two-way ANOVA, this sum of squares can be decomposed into parts representing groups, blocks, interaction and error. The sum of squares matrices are given by

$$\mathbf{G} = \sum_{k=1}^{g} bc(\bar{\mathbf{x}}_{\cdot k \cdot} - \bar{\mathbf{x}})(\bar{\mathbf{x}}_{\cdot k \cdot} - \bar{\mathbf{x}})' \quad \text{for groups,}$$

$$\mathbf{B} = \sum_{\ell=1}^{b} gc(\bar{\mathbf{x}}_{\cdot \cdot \ell} - \bar{\mathbf{x}})(\bar{\mathbf{x}}_{\cdot \cdot \ell} - \bar{\mathbf{x}})' \quad \text{for blocks,}$$

$$\mathbf{W} = \sum_{k=1}^{g} \sum_{\ell=1}^{b} c(\bar{\mathbf{x}}_{\cdot k\ell} - \bar{\mathbf{x}}_{\cdot k \cdot} - \bar{\mathbf{x}}_{\cdot \cdot \ell} + \bar{\mathbf{x}})(\bar{\mathbf{x}}_{\cdot k\ell} - \bar{\mathbf{x}}_{\cdot k \cdot} - \bar{\mathbf{x}}_{\cdot \cdot \ell} + \bar{\mathbf{x}})'$$

$$\text{for interaction, and}$$

$$\mathbf{E} = \sum_{\ell=1}^{b} \sum_{k=1}^{g} \sum_{i=1}^{c} (\mathbf{x}_{ik\ell} - \bar{\mathbf{x}}_{\cdot k\ell})(\mathbf{x}_{ik\ell} - \bar{\mathbf{x}}_{\cdot k\ell})' \quad \text{for error.}$$

The cell mean vectors are denoted by $\bar{\mathbf{x}}_{\cdot k\ell}$, whereas the group and block mean vectors are denoted by $\bar{\mathbf{x}}_{\cdot k \cdot}$ and $\bar{\mathbf{x}}_{\cdot \cdot \ell}$ respectively. The sum of squares matrices have the property that

$$\mathbf{T} = \mathbf{G} + \mathbf{B} + \mathbf{W} + \mathbf{E}.$$

Inference

To test the hypothesis of zero interaction

$$H_0 : (\alpha\beta)_{k\ell} = 0, \quad k = 1, 2, \ldots, g, \quad \ell = 1, 2, \ldots, b,$$

the Wilk's Lambda statistic is given by

$$\Lambda_{\alpha\beta} = |\mathbf{E}|/|\mathbf{E} + \mathbf{W}|,$$

with $bg(c - 1)$ degrees of freedom for $\mathbf{E}$ and $(b - 1)(g - 1)$ degrees of freedom for $\mathbf{W}$. Tests for main effects can also be carried out for groups

$H_0: \boldsymbol{\alpha}_k = \mathbf{0}$, $k = 1, 2, \ldots, g$, and for blocks $H_0: \boldsymbol{\beta}_\ell = \mathbf{0}$, $\ell = 1, 2, \ldots, b$. The corresponding Wilk's Lambda statistics for groups and blocks are given by

$$\Lambda_\alpha = |\mathbf{E}|/|\mathbf{G} + \mathbf{E}| \quad \text{with } bg(c - 1) \text{ degrees of freedom for } \mathbf{E}$$
$$\text{and } (g - 1) \text{ degrees of freedom for } \mathbf{G} \text{ and}$$

$$\Lambda_\beta = |\mathbf{E}|/|\mathbf{B} + \mathbf{E}| \quad bg(c - 1) \text{ degrees of freedom for } \mathbf{E}$$
$$\text{and } (b - 1) \text{ degrees of freedom for } \mathbf{B}.$$

Either the F or χ^2 approximations to the distribution of Wilk's Lambda can be used to test these hypotheses.

Example

For the public safety data introduced in Table 8.1, an additional categorical variable was generated by dividing the individuals into five education categories. This variable is denoted by EDCAT. Using the four COMMUNITY classifications and the five EDCAT classifications, a total of 20 cells were generated. The total of 100 observations were distributed equally over the 20 cells yielding a balanced two-way design with five observations per cell. The cell means are shown in Table 8.5. The results for the six univariate ANOVAs and the MANOVA are shown in Tables 8.6 and 8.7.

In Table 8.6 it can be seen that for all six variables the COMMUNITY effects and EDCAT effects are significant at the 0.000 level whereas the interaction is significant at the 0.04 level or lower. From the MANOVA results in Table 8.7 we can see that the two main effects and the interaction are significant at 0.000 level.

Graphs showing the relationships between the cell means and the levels of EDCAT and COMMUNITY for each of the six variables are shown in Figure 8.10. From the graphs, it is easily seen that the levels of unsafe are higher in communities 1 and 3, and that the relationship between unsafe and education level is downward sloping. The nonzero interaction in each case is due primarily to the larger negative slopes for communities 1 and 3. It would appear that the variation in the unsafe variables could be captured by a dummy variable that distinguishes communities 2 and 4 from communities 1 and 3. Treating EDCAT as an interval variable and using a slope shifter also appears to be reasonable. Using this latter approach, the error sums of squares for the six variables are (50.7, 36.5, 22.8, 44.9, 30.1 and 56.3). By comparison to the error column in Table 8.6, we can conclude that in each case there is some loss of information; however, the proportion lost is quite small. The EDCAT variable could also have been introduced as a covariate in the MANOVA as in the case of INC in Section 8.1.2.

TABLE 8.5. Cell Means for Public Safety Data

COMMUNITY	EDCAT	X1	X2	X3	X4	X5	X6
	1	4.6	3.8	2.6	5.4	2.6	5.8
	2	3.2	2.2	2.2	4.6	2.4	4.8
1	3	3.2	2.0	2.0	4.0	2.0	4.8
	4	2.0	2.0	1.8	3.2	2.4	4.2
	5	2.2	1.2	1.2	1.6	1.2	3.0
	1	2.0	1.6	1.6	3.6	1.6	2.8
	2	1.8	2.0	1.2	2.0	1.0	2.0
2	3	1.6	1.4	1.0	1.6	1.0	1.8
	4	1.2	1.0	1.0	1.4	1.0	1.4
	5	1.0	1.0	1.0	1.0	1.0	1.0
	1	4.0	3.0	3.2	5.0	3.8	5.2
	2	3.0	2.4	2.0	3.8	2.2	5.2
3	3	3.4	2.0	1.6	2.8	2.0	4.8
	4	2.4	1.2	1.0	2.0	1.6	3.8
	5	1.2	1.0	1.2	1.4	1.0	1.8
	1	2.4	1.6	1.6	3.0	1.2	3.0
	2	2.0	1.4	1.0	1.6	1.0	2.0
4	3	1.0	1.0	1.0	1.0	1.0	1.2
	4	1.0	1.0	1.0	1.0	1.0	1.0
	5	1.0	1.0	1.0	1.0	1.0	1.0

The Multivariate Paired Comparison Test

A random sample of n individuals is observed on the random vector $\mathbf{x}$ $(p \times 1)$ under two different conditions or regimes. The vector $\mathbf{x}$ is assumed to be distributed as a multivariate normal with means $\boldsymbol{\mu}_1$ and $\boldsymbol{\mu}_2$ respectively for the two regimes. A test of equality of means in this case is a special case of the balanced two-way MANOVA with two blocks.

Denoting the two observation vectors by $\mathbf{x}_1$ and $\mathbf{x}_2$, the difference between the two vectors is given by $\mathbf{d} = (\mathbf{x}_1 - \mathbf{x}_2)$ and between the corresponding means by $\boldsymbol{\delta} = (\boldsymbol{\mu}_1 - \boldsymbol{\mu}_2)$. Thus if $\text{Cov}(\mathbf{d}) = \boldsymbol{\Sigma}_d$ then $\mathbf{d} \sim N(\boldsymbol{\delta}, \boldsymbol{\Sigma}_d)$.

For a sample of n observations on $\mathbf{x}_1$ and $\mathbf{x}_2$ given by $\mathbf{X}_1$ $(n \times p)$ and $\mathbf{X}_2$ $(n \times p)$, the sample statistics are given by

$$\bar{\mathbf{d}} = (\bar{\mathbf{x}}_{.1} - \bar{\mathbf{x}}_{.2}) \quad \text{and} \quad \mathbf{S}_d = \sum_{i=1}^{n}(\mathbf{d}_i - \bar{\mathbf{d}})(\mathbf{d}_i - \bar{\mathbf{d}})'/(n-1).$$

Under $H_0: \boldsymbol{\delta} = \mathbf{0}$, the statistic $T^2 = n\bar{\mathbf{d}}'\mathbf{S}_d^{-1}\bar{\mathbf{d}}$ is distributed as $(n-1)pF_{p;(n-p)}/(n-p)$. A $100(1-\alpha)\%$ confidence interval for $\boldsymbol{\delta}$ is given

TABLE 8.6. ANOVA Results for Public Safety Data

Variable		Community	Education	Interaction	Error	Total
X_1	Sum of Squares	51.15	44.54	12.10	40.8	148.59
	F-Ratio	33.43	21.83	1.98		
	p-Value	(0.000)	(0.000)	(0.037)		
X_2	Sum of Squares	16.99	26.44	10.76	29.20	83.39
	F-Ratio	15.52	18.11	2.46		
	p-Value	(0.000)	(0.000)	(0.009)		
X_3	Sum of Squares	14.03	16.64	6.72	17.60	54.99
	F-Ratio	21.26	18.91	2.55		
	p-Value	(0.000)	(0.000)	(0.007)		
X_4	Sum of Squares	85.16	88.90	16.94	30.00	221.00
	F-Ratio	75.70	59.27	3.76		
	p-Value	(0.000)	(0.000)	(0.000)		
X_5	Sum of Squares	27.12	16.30	13.38	21.20	78.00
	F-Ratio	34.11	15.38	4.21		
	p-Value	(0.000)	(0.000)	(0.000)		
X_6	Sum of Squares	173.55	71.16	15.40	40.80	300.91
	F-Ratio	113.43	34.88	2.52		
	p-Value	(0.000)	(0.000)	(0.007)		

TABLE 8.7. MANOVA Results for Public Safety Data

Effect	Wilk's Lambda	F	p Value
Community	0.0728	18.02	0.000
Education	0.0889	10.96	0.000
Interaction	0.1620	2.28	0.000

by $\bar{\mathbf{d}} \pm \mathbf{q}$ where the elements of $\mathbf{q}$ are given by

$$q_i = \sqrt{\frac{(n-1)p}{(n-p)} F_{p,(n-p)} \left(\frac{s_{di}^2}{n} \right)}$$

and where s_{di}^2 is the ith diagonal element of the covariance matrix $\mathbf{S}_d$.

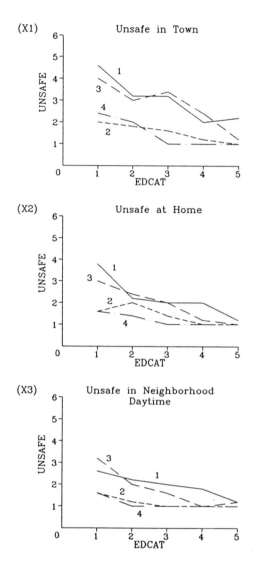

FIGURE 8.10. Variation Among Means for Public Safety Data

8.1.5   AN UNBALANCED MANOVA WITH COVARIATE

In this section, an example is used to illustrate how the multivariate linear regression model can be used to analyze data derived from unbalanced designs. Indicator variables are defined to represent the levels of the factors, and interaction variables are then derived from the cross product of the nonrelated indicator variables. Since the design is unbalanced, the various effects are no longer orthogonal, and hence significance tests must be car-

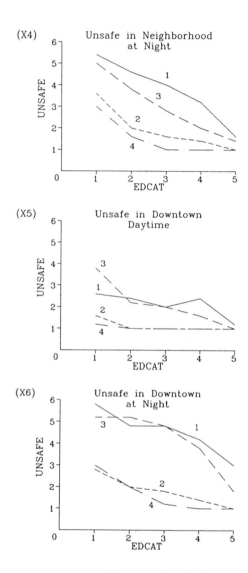

FIGURE 8.10. Variation Among Means for Public Safety Data (continued)

ried out in a conventional manner as in regression analysis. A two-factor MANOVA model with a covariate is used in this section to illustrate the analysis.

For the bank employee salary data introduced in Table 7.9 of Chapter 7, information on two additional variables SEX (0 = male, 1 = female) and RACE (0 = white, 1 = nonwhite) was added to the variables LCURRENT, LSTART and EDUC to provide an example. The data in Table 7.9 in Chapter 7 is arranged so that observations 1–37 are RACE = 0, SEX = 0; observations 38–68 are RACE = 0, SEX = 1; observations 69–89 are RACE = 1, SEX = 0; and observations 90–100 are RACE = 1 and SEX = 1. Since the data is unbalanced (unequal observations in the four sex-race categories), a multivariate regression model is used to analyze the data.

Using effect coding the variables SX and RC are defined by

$$
\begin{array}{llll}
SX = & 1 & \text{females} & RC = & 1 & \text{nonwhite} \\
SX = & -1 & \text{males} & RC = & -1 & \text{white.}
\end{array}
$$

Interaction variables SXRC, SXED, RCED and SXRCED are defined as

$$
\begin{array}{ll}
SXRC = SX*RC, & SXED = SX*EDUC, \quad RCED = RC*EDUC \\
\text{and} & SXRCED = SX*RC*EDUC.
\end{array}
$$

Table 8.8 illustrates the results of multivariate tests for various effects. Row 1 of the table indicates that at least some of the effects are required; whereas row 2 shows that at least one of the SEX and RACE effects is required. Rows 3 and 4 suggest that the various interactions are not significant. Rows 5 and 7 show that the SEX effect is important and rows 6 and 8 show that the RACE effect is marginal. Row 9 illustrates that the EDUC effect is significant. It would appear that the variables SX and EDUC are sufficient to explain the variation in LCURRENT and LSTART. The variable RC does add marginally to the variation explained as shown in the results of row 6.

To demonstrate the marginal impact of the various effects on the dependent variables LCURRENT and LSTART, Tables 8.9 and 8.10 show the results of a variety of regression models. The p-values for the coefficients in the rows of the two tables illustrate that, when the interaction terms are present, the SX and RC variables are not significant. The R^2 values in rows 7 and 9 in both tables illustrate that omission of the sex effect reduces R^2 considerably. Row 8 of both tables shows that the variation in both LCURRENT and LSTART can be explained adequately by the variables RC, SX and EDUC. In both cases the interactions are not significant over and above these main effects. From the coefficients shown in row 8, it would appear that average salaries are lower for females and for nonwhites even after controlling for the level of education.

8.1.6 OTHER SOURCES OF INFORMATION

Additional information on MANOVA can be found in Anderson (1984), Seber (1984), Press (1972), Morrison (1976), Johnson and Wichern (1988) and Stevens (1986).

TABLE 8.8. Results of Various Multivariate Regression Tests for Effects*

Effects	SX	RC	SXRC	SXED	RCED	SXRCED	EDUC	Wilk's Lambda F	p-Value	d.f.
1. All	X	X	X	X	X	X	X	12.088	0.000	14 and 182
2. Sex and Race	X	X	X	X	X	X		6.049	0.000	12 and 182
3. All Interactions			X	X	X	X		1.283	0.255	8 and 182
4. Interaction with EDUC				X	X	X		1.394	0.219	6 and 182
5. Interaction plus Sex	X		X	X	X	X		6.973	0.000	10 and 182
6. Interaction plus Race		X	X	X	X	X		1.730	0.077	10 and 182
7. Sex Effect	X		X	X		X		6.880	0.000	8 and 182
8. Race Effect		X	X		X	X		1.520	0.153	8 and 182
9. EDUC Effect				X	X	X	X	11.331	0.000	8 and 182

*Crosses show which terms are omitted from the full model for the test.

8.2 Discriminant Analysis

If the hypothesis of equality of mean vectors $H_{0p}: \boldsymbol{\mu}_1 = \boldsymbol{\mu}_2 = \ldots = \boldsymbol{\mu}_g$ is rejected in MANOVA, it is usually of interest to characterize the differences among the mean vectors over the g groups. Discriminant analysis is useful for highlighting such group differences. It can also be used to assist in classifying observations whose group membership is unknown. Discriminant analysis can therefore be used both as a descriptive tool and as a classification tool. We first discuss discriminant analysis as a descriptive device.

TABLE 8.9. Relationships Between LSTART and Various Explanatory Variables

	Intercept	SX	RC	SXRC	SXED	RCED	SXRCED	EDUC	R^2
1	7.977	0.001	0.148	−0.038	−0.014	−0.017	0.000	0.057	0.693
	(0.000)	(0.995)	(0.257)	(0.773)	(0.160)	(0.090)	(0.971)	(0.000)	
2	8.016	−0.002	0.167	—	−0.013	−0.018	—	0.054	0.688
	(0.000)	(0.986)	(0.134)	—	(0.106)	(0.035)	—	(0.000)	
3	7.824	−0.198	−0.076	−0.032	—	—	—	0.070	0.672
	(0.000)	(0.000)	(0.006)	(0.243)	—	—	—	(0.000)	
4	7.913	−0.051	−0.063	—	−0.010	—	—	0.064	0.673
	(0.000)	(0.634)	(0.023)	—	(0.200)	—	—	(0.000)	
5	7.940	−0.173	0.130	—	—	−0.016	—	0.060	0.680
	(0.000)	(0.000)	(0.236)	—	—	(0.061)	—	(0.000)	
6	7.890	0.011	—	—	−0.015	—	—	0.067	0.655
	(0.000)	(0.915)	—	—	(0.072)	—	—	(0.000)	
7	7.956	—	0.360	—	—	−0.033	—	0.061	0.514
	(0.000)	—	(0.006)	—	—	(0.002)	—	(0.000)	
8	7.861	−0.186	−0.071	—	—	—	—	0.068	0.668
	(0.000)	(0.000)	(0.009)	—	—	—	—	(0.000)	
9	7.786	—	−0.041	—	—	—	—	0.077	0.461
	(0.000)	—	(0.225)	—	—	—	—	(0.000)	
10	7.809	−0.177	—	—	—	—	—	0.074	0.643
	(0.000)	(0.000)	—	—	—	—	—	(0.000)	

Example

The data in Table 8.11 will be analyzed to provide an example of discriminant analysis as a descriptive tool. The data was obtained from a sample of 100 avid readers of mystery novels who were asked to respond to a number of questions regarding their preferences for style of novel. The responses to ten items are shown in Table 8.11. The ten questions are also summarized in the table. The respondents were asked to rate the importance of each of ten characteristics on a scale from 0 to 20. The sex of the respondent and the responses to the variable, level of education (high school diploma or less, some post high school training, undergraduate university degree and advanced university degree) were used to define a classification variable SEXED with eight categories. For the four levels of education for males, SEXED was coded 1 to 4, and for the four levels of education for females, SEXED was coded 5 to 8. The codes for the variable SEXED are also shown in Table 8.11.

For the mystery data in Table 8.11, the MANOVA comparison of the ten mean vectors across the eight SEXED groups yielded a Wilk's Lambda of 0.2418 which yields an F-statistic of 1.932. Comparison of the F-statistic to an F distribution with 70 and 490.8 degrees of freedom yields a p-value of 0.0001. We can therefore conclude that the eight mean vectors are

TABLE 8.10. Relationships Between LCURRENT and Various Explanatory Variables

	Intercept	SX	RC	SXRC	SXED	RCED	SXRCED	EDUC	R^2
1	8.534	−0.132	0.204	0.001	−0.006	−0.021	−0.003	0.068	0.668
	(0.000)	(0.381)	(0.177)	(0.994)	(0.637)	(0.077)	(0.802)	(0.000)	
2	8.558	−0.159	0.202	—	−0.002	−0.020	—	0.066	0.663
	(0.000)	(0.204)	(0.115)	—	(0.807)	(0.043)	—		
3	8.399	−0.219	−0.057	−0.041	—	—	—	0.080	0.654
	(0.000)	(0.000)	(0.065)	(0.187)	—	—	—	(0.000)	
4	8.444	−0.213	−0.051	—	0.0007	—	—	0.077	0.648
	(0.000)	(0.087)	(0.105)	—	(0.941)	—	—		
5	8.545	−0.188	0.196	—	—	−0.020	—	0.067	0.663
	(0.000)	(0.000)	(0.112)	—	—	(0.043)	—	(0.000)	
6	8.426	−0.162	—	—	−0.003	—	—	0.079	0.638
	(0.000)	(0.181)	—	—	(0.765)	—	—	(0.000)	
7	8.562	—	0.446	—	—	−0.038	—	0.068	0.503
	(0.000)	—	(0.002)	—	—	(0.001)	—	(0.000)	
8	8.448	−0.204	−0.050	—	—	—	—	0.076	0.648
	(0.000)	(0.000)	(0.098)	—	—	—	—	(0.000)	
9	8.366	—	−0.017	—	—	—	—	0.086	0.445
	(0.000)	—	(0.641)	--	—	—	—	(0.000)	
10	8.411	−0.197	—	—	—	—	—	0.081	0.638
	(0.000)	(0.000)	—	—	—	—	—	(0.000)	

significantly different. The individual mean values for each of the ten items are summarized for the eight groups in Table 8.12. This table also provides the univariate ANOVA results for each of the items. From this table, we can see that item C10 yields the largest difference in means among the eight groups whereas item C7 yields the least differences. Five of the ten items have univariate F-statistics with p-values less than 0.05. Discriminant analysis will now be used to characterize the differences among the eight groups. The theory of discriminant analysis is presented first.

8.2.1 FISHER'S DISCRIMINANT CRITERION AND CANONICAL DISCRIMINANT ANALYSIS

Fisher's Discriminant Criterion

If the outcome of a MANOVA suggests that the means on the vector $\mathbf{x}$ differ over the g groups, it is valuable to examine the variation over the g groups for linear combinations of $\mathbf{x}$ given by $y = \mathbf{b}'\mathbf{x}$. In particular we might ask if there is a linear combination y that emphasizes the differences among the groups with respect to elements of $\boldsymbol{\mu}$. Since the new variable y is a univariate random variable, we may compare the means on y over the g groups $\mu_{y_1}, \mu_{y_2}, \ldots, \mu_{y_g}$ using univariate ANOVA.

TABLE 8.11. Mystery Novel Preference Data

C1	C2	C3	C4	C5	C6	C7	C8	C9	C10	SEXED
12	6	11	12	12	6	13	16	20	17	7
19	3	10	3	3	3	11	11	4	11	5
15	12	15	15	12	14	11	11	12	13	3
14	4	11	17	10	9	9	20	20	9	8
12	7	12	12	11	11	11	11	15	13	7
12	12	18	18	18	5	16	18	11	19	6
12	12	18	18	18	5	16	18	11	19	6
12	12	19	19	19	12	12	14	19	12	8
16	11	18	3	11	10	15	15	20	2	8
13	7	19	16	13	11	11	18	17	14	1
11	12	3	6	11	16	14	16	19	15	8
11	4	10	18	10	10	10	10	18	10	8
12	8	9	10	9	10	10	17	16	16	7
12	12	11	11	11	11	11	11	11	11	5
11	4	17	17	3	4	5	17	18	17	6
11	4	11	11	11	11	11	11	11	11	7
12	5	13	13	13	12	20	20	12	13	8
12	4	18	18	12	11	11	18	18	17	6
11	10	20	18	8	20	16	20	17	8	7
11	11	14	13	13	12	11	13	14	11	8
16	4	15	15	4	7	11	17	17	14	6
11	10	18	15	10	13	19	20	18	13	6
11	11	17	13	11	12	14	14	18	11	4
11	11	17	15	16	11	11	17	14	11	8
11	3	18	11	11	11	11	18	11	3	5
12	12	12	20	12	20	12	12	12	20	6
4	1	20	1	1	10	1	11	1	20	6
12	5	13	12	12	12	13	12	13	13	5
11	4	10	10	3	9	9	17	17	9	8
12	11	19	11	12	19	10	19	11	19	7
8	2	12	9	11	3	14	18	18	15	8
14	3	3	9	3	10	3	18	13	10	7
4	1	8	1	1	1	11	15	20	11	6
12	9	12	12	13	12	12	13	13	13	8
11	11	11	11	11	11	11	11	11	14	7
11	3	11	16	10	11	11	8	18	8	3
7	3	20	8	10	10	10	12	20	12	7
12	10	19	19	11	13	10	17	14	10	4
12	9	12	10	10	11	10	14	15	11	8
12	1	13	11	2	15	11	17	20	11	7
18	11	6	11	6	6	11	19	19	19	7
9	9	4	12	12	11	14	8	14	11	8
11	6	16	15	10	10	15	18	14	10	8
16	15	15	15	11	11	11	16	17	16	3
17	13	16	15	7	13	10	11	15	12	3
11	12	12	13	13	14	14	14	15	15	6
11	16	17	14	10	10	10	10	14	10	1
10	6	9	7	10	8	10	13	11	11	5
13	16	17	11	17	10	9	15	15	8	4
19	19	19	11	11	4	11	4	4	19	5

How important are each of the following

C1: More than one murder or crime

C2: Mouthy, obstinate detective

C3: Powerful Opponents

C4: Detective gets into impossible jam

C5: Large amounts of money involved in the crime

C6: Detective is a loner

C7: Murder by non physical means such as poisoning

C8: Many possible suspects

C9: Puzzle is "fair play" in that clues are given

C10: Suspects appear to be average people

TABLE 8.11. Mystery Novel Preference Data (continued)

C1	C2	C3	C4	C5	C6	C7	C8	C9	C10	SEXED
19	6	12	6	13	13	13	20	20	20	7
12	12	15	13	13	14	14	14	14	14	7
19	12	12	12	12	11	11	11	11	11	7
12	8	15	15	10	11	6	14	15	9	7
10	11	11	11	8	13	10	13	17	10	8
11	14	20	1	20	15	12	20	17	1	2
12	12	19	19	11	11	11	19	11	11	7
11	12	7	10	10	9	14	13	7	11	1
14	2	18	13	7	11	14	19	14	6	4
11	7	15	14	9	11	16	15	17	16	8
16	10	15	12	14	9	14	7	9	5	5
19	19	11	5	12	5	12	18	5	12	2
11	11	18	15	16	11	13	17	11	10	2
12	1	10	6	12	11	10	15	7	10	5
12	4	14	7	3	19	16	18	16	14	7
11	16	7	6	18	17	3	3	1	3	8
12	17	17	15	12	14	10	15	16	18	7
16	5	11	14	11	9	11	13	19	16	7
11	8	17	9	10	12	12	16	11	18	7
11	11	12	12	12	12	12	12	12	12	6
11	3	19	19	3	11	11	19	11	11	6
11	20	20	19	4	11	11	18	20	20	7
11	11	19	3	9	9	18	18	18	19	7
13	8	8	5	7	15	11	14	16	14	5
11	6	18	10	10	17	3	8	17	8	4
11	5	11	8	2	14	10	10	10	10	8
12	15	15	12	12	12	12	15	15	12	7
18	16	16	16	11	15	15	15	15	10	3
12	4	11	20	11	19	12	20	19	11	8
17	17	16	14	4	10	10	18	12	13	4
12	19	19	12	12	19	12	12	12	13	3
5	18	0	18	4	11	11	11	19	18	8
12	2	11	11	10	19	10	19	19	10	8
15	4	16	15	7	11	11	17	20	16	3
17	4	16	11	9	11	11	12	11	15	7
12	19	19	12	12	13	13	13	13	13	7
11	3	3	3	5	10	15	10	10	15	8
11	1	17	11	6	11	11	20	11	17	6
12	3	17	17	11	10	16	18	19	10	8
12	14	19	12	12	12	12	20	20	8	3
11	16	16	15	8	6	17	15	3	7	8
11	11	20	20	11	11	10	20	20	10	7
11	1	16	10	10	10	10	14	18	10	6
11	4	16	11	11	17	10	18	18	20	2
12	7	9	10	7	9	9	13	14	15	3
11	1	15	14	10	17	10	19	15	18	8
11	5	13	7	12	11	10	14	15	13	8
11	1	6	14	9	10	10	15	10	11	5
11	3	15	11	7	13	10	10	17	12	3
9	17	17	15	4	11	4	7	17	15	3

How important are each of the following
C1: More than one murder or crime
C2: Mouthy, obstinate detective
C3: Powerful Opponents
C4: Detective gets into impossible jam
C5: Large amounts of money involved in the crime
C6: Detective is a loner
C7: Murder by non physical means such as poisoning
C8: Many possible suspects
C9: Puzzle is "fair play" in that clues are given
C10: Suspects appear to be average people

TABLE 8.12. Summary of Item Means by Group with Univariate ANOVA Statistics

| Variable | Group Means | | | | | | | | F | p-Value |
	1	2	3	4	5	6	7	8		
C1	11.66	13.00	13.45	12.85	13.50	10.53	12.73	11.11	2.24	0.037
C2	8.33	12.00	11.27	10.57	6.80	5.84	9.23	7.69	1.73	0.111
C3	14.33	14.75	15.27	16.71	11.90	15.53	14.46	11.57	2.32	0.031
C4	13.33	8.00	13.81	13.14	9.20	13.07	11.96	12.11	1.54	0.162
C5	11.00	14.75	9.09	10.28	10.00	8.07	9.50	10.38	1.57	0.152
C6	10.00	12.00	12.63	13.14	9.40	9.92	11.92	11.76	1.29	0.261
C7	11.66	11.75	10.54	10.28	11.20	11.07	11.38	12.15	0.48	0.844
C8	13.66	18.25	12.72	14.71	12.00	15.92	15.65	14.61	2.26	0.036
C9	12.66	12.75	16.09	14.57	9.60	14.00	15.50	15.03	2.66	0.014
C10	11.66	10.75	12.63	9.71	10.80	15.07	14.23	11.26	2.96	0.007

To compare the means on y over the g groups we require the F ratio

$$F_y = (n - g)\text{SSA}_y/\text{SSW}_y(g - 1),$$

where SSA_y and SSW_y denote the sums of squares among the groups and within the groups for the variable y. If F_y is sufficiently large, we reject the hypothesis of equality of means $\mu_{y_1}, \mu_{y_2}, \ldots, \mu_{y_g}$. The sums of squares are given by

$$\text{SSA}_y = \mathbf{b}'\mathbf{Gb} \quad \text{and} \quad \text{SSW}_y = \mathbf{b}'\mathbf{Wb},$$

and hence

$$F_y = (n - g)\mathbf{b}'\mathbf{Gb}/\mathbf{b}'\mathbf{Wb}(g - 1).$$

To determine a variable y that should characterize differences among groups, we determine the values of the elements of the vector $\mathbf{b}$ that maximize F_y. Ignoring the constant $(n - g)/(g - 1)$, we need to determine $\mathbf{b}$ to maximize the ratio criterion $\mathbf{b}'\mathbf{Gb}/\mathbf{b}'\mathbf{Wb}$. This criterion is usually called *Fisher's discriminant criterion*. Since the solution involves the eigenvalue problem the reader may wish to review the theory in Section 3 of the Appendix.

An Eigenvalue Problem

Determination of the vector $\mathbf{b}$ that maximizes Fisher's criterion involves solving the system of equations given by

$$(\mathbf{G} - \lambda\mathbf{W})\mathbf{b} = \mathbf{0} \tag{8.3}$$

or equivalently $(\mathbf{W}^{-1}\mathbf{G} - \lambda\mathbf{I})\mathbf{b} = \mathbf{0}$.

The solution requires the eigenvalues and eigenvectors of the matrix $\mathbf{W}^{-1}\mathbf{G}$. Denoting the rank of $\mathbf{W}^{-1}\mathbf{G}$ by $r = \min[(g-1), p]$, the r eigenvalues are denoted by $\lambda_1, \lambda_2, \ldots, \lambda_r$ in decreasing order of magnitude. The corresponding eigenvectors are denoted by $\mathbf{b}_1, \mathbf{b}_2, \ldots, \mathbf{b}_r$ respectively. These eigenvectors are mutually orthogonal.

From (8.3) we can see that the largest eigenvalue λ_1 maximizes the ratio $\lambda = \mathbf{b}'\mathbf{G}\mathbf{b}/\mathbf{b}'\mathbf{W}\mathbf{b}$. Since the eigenvalue/eigenvector solution is only unique up to a constant, it is customary to choose $\mathbf{b}$ so that $\mathbf{b}'\mathbf{W}\mathbf{b} = 1$. This condition is convenient in that $\lambda = \mathbf{b}'\mathbf{G}\mathbf{b}$, and hence the eigenvalue measures the among group sum of squares for the variable $y = \mathbf{b}'\mathbf{x}$.

Canonical Discriminant Functions

Corresponding to the r eigenvectors $\mathbf{b}_1, \mathbf{b}_2, \ldots, \mathbf{b}_r$ are the r variables, $y_1, y_2, \ldots, y_r$, which have unit variances since $\mathbf{b}_j'\mathbf{W}\mathbf{b}_j = 1$, $j = 1, 2, \ldots, r$, and which are mutually orthogonal. These variables are usually called the *canonical discriminant functions*.

The sum of squares and cross product matrices for the r canonical discriminant functions are given by

$$\mathbf{G_y} = \mathbf{B}'\mathbf{G}\mathbf{B}, \quad \mathbf{W_y} = \mathbf{B}'\mathbf{W}\mathbf{B} \quad \text{and} \quad \mathbf{T_y} = \mathbf{G_y} + \mathbf{W_y}$$

where $\mathbf{B}$ is the $(p \times r)$ matrix of eigenvectors $\mathbf{b}_1, \mathbf{b}_2, \ldots, \mathbf{b}_r$. Since the eigenvectors are orthogonal, the matrices $\mathbf{W_y}$ and $\mathbf{G_y}$ are diagonal with diagonal elements 1 and λ_j, $j = 1, 2, \ldots, r$ respectively. The total sum of squares and cross products matrix $\mathbf{T_y}$ is therefore diagonal with elements $(1 + \lambda_j)$, $j = 1, 2, \ldots, r$. The sum of the diagonal elements of $\mathbf{G_y}$ is $\sum_{j=1}^{r} \lambda_j$ and hence the ratio $\lambda_j / \sum_{j=1}^{r} \lambda_j$ denotes the proportion of among group sum of squares accounted for by the discriminant function y_j.

Inferences for Canonical Discriminant Functions

Bartlett's Test
The eigenvalues for the canonical discriminant functions can be related to the Wilk's Lambda ratio for MANOVA. For $\Lambda = |\mathbf{W}|/|\mathbf{T}|$ it can be shown that $\Lambda^{-1} = \Pi_{j=1}^{r}(1 + \lambda_j)$, and hence Bartlett's χ^2 statistic given in Section 8.1.1 can be written as

$$[n - 1 - (p + g)/2] \sum_{j=1}^{r} \ln(1 + \lambda_j),$$

which has a χ^2 distribution with $p(g-1)$ degrees of freedom if H_{0p} is true. The test of $H_{0p}: \boldsymbol{\mu}_1 = \boldsymbol{\mu}_2 = \ldots = \boldsymbol{\mu}_g$ is equivalent to the test $H_0: \lambda_1 = \lambda_2 = \ldots = \lambda_r = 0$. Rejection of H_{0p} suggests, therefore, that at least one of the r eigenvalues, namely λ_1, is positive and hence that at least one of the discriminant functions will be significant.

If H_0 is rejected, then, ignoring λ_1, an approximate test of $H_{01}: \lambda_2 = \ldots = \lambda_r = 0$ can be carried out using the revised test statistic

$$[n - 1 - (p + g)/2] \sum_{j=2}^{r} \ln(1 + \lambda_j),$$

which has a χ^2 distribution with $(p - 1)(g - 2)$ degrees of freedom. If H_{01} is rejected, λ_2 can be dropped from H_{01}, and a revised test statistic can be computed to test $H_{02}: \lambda_3 = \ldots = \lambda_r = 0$.

After a rejection of the hypothesis that the first k eigenvalues are zero, the null hypothesis becomes $H_{0k}: \lambda_{k+1} = \ldots = \lambda_r = 0$, which is tested using the test statistic

$$[n - 1 - (p + g)/2] \sum_{j=(k+1)}^{r} \ln(1 + \lambda_j).$$

This statistic has a χ^2 distribution with $(p - k)(g - k - 1)$ d.f. if H_{0k} is true.

An Alternative Test Statistic - F

As in Section 7.5 the F approximation to the distribution of Wilk's Lambda can also be used. A test of H_{0k} is carried out using

$$F = \frac{1 - \Lambda_k^{1/\nu_k}}{\Lambda_k^{1/\nu_k}} \cdot \frac{m_{2k}}{m_{1k}} \quad \text{where} \quad \nu_k = \sqrt{\frac{(p - k)^2(g - k - 1)^2 - 4}{(p - k)^2 + (q - k)^2 - 5}},$$

$$m_{1k} = (p-k)(g-k-1), \quad m_{2k} = \nu_k\left[n-1-\frac{1}{2}(p+g)\right] - \frac{(p - k)(g - k - 1)}{2} + 1,$$

$$\Lambda_k = \sum_{j=k+1}^{r} \ln(1 + \lambda_j).$$

In large samples the statistic has an F distribution with m_{1k} and m_{2k} d.f. if H_{0k} is true.

Interpretation of the Discriminant Analysis Solution

Given a set of r canonical discriminant functions $y_1, y_2, \ldots, y_r$, the functions can be interpreted in two ways. The function coefficients given by the eigenvectors $\mathbf{b}_1, \mathbf{b}_2, \ldots, \mathbf{b}_r$ are the *"raw" discriminant function coefficients*. If the variables were not standardized before the analysis, then the "raw" coefficients can be standardized by multiplying by the variable standard deviations. The diagonal elements of the matrix $\mathbf{W}$ divided by $(n - g)$, $\bar{\mathbf{S}} = \mathbf{W}/(n - g)$, represent unbiased estimators of the common covariance matrix $\boldsymbol{\Sigma}$. Let $\mathbf{F}$ denote the diagonal matrix of elements that are the square

TABLE 8.13. Standardized Coefficients for First Three Discriminant Functions

	$C1$	$C2$	$C3$	$C4$	$C5$	$C6$	$C7$	$C8$	$C9$	$C10$
Y_1	0.388	0.436	0.556	0.070	-0.530	0.351	-0.270	-0.323	0.652	0.202
Y_2	0.415	0.487	-0.305	-0.378	0.284	0.299	-0.142	0.080	0.203	-0.592
Y_3	-0.199	0.312	0.122	-0.486	0.406	0.006	-0.343	0.980	-0.059	0.359

roots of the diagonal elements of $\overline{\mathbf{S}}$. The *standardized coefficients* are given by $\mathbf{b}^* = \mathbf{F}\mathbf{b}$.

The discriminant function usually contains both positive and negative coefficients and hence for interpretation purposes it is best to group the coefficients into the positive group and the negative group. As in multiple regression analysis each coefficient b_j should be interpreted as the marginal impact of x_j on the function y holding the other variables fixed.

Interpretation Using Correlations

An alternative approach to interpretation uses the correlation coefficients between the canonical discriminant functions and the original variables. In this case the correlations measure how the variables relate jointly rather than marginally. Thus, given a strong positive correlation between the discriminant function y and the variable x_j, we can conclude that the function y and x_j tend to move together.

Regardless of the approach used, it should be noted that like regression analysis the coefficients have been determined to maximize the sample relationship between the functions and the groups. Large samples are required therefore to insure the stability of the coefficients over different samples.

Computer Software

The calculations for discriminant analysis required for the examples in this section were performed using SAS PROC CANDISC, SAS PROC DISCRIM, SAS PROC STEPDISC and SAS PROC NEIGHBOR.

Example

Using Fisher's criterion, a discriminant analysis was carried out for the data in Table 8.11. The standardized coefficients for the first three discriminant functions are summarized in Table 8.13. The results for the successive likelihood ratio tests for significance of the eigenvalues is shown in Table 8.14. From Table 8.14 the first three discriminant functions have p-values of 0.000, 0.004 and 0.056 respectively. The remaining four discriminant functions have p-values that exceed 0.4.

TABLE 8.14. Sequential Likelihood Ratio Test for Significance of Discriminant Functions

Likelihood Ratio	0.242	0.385	0.542	0.723	0.904	0.984	0.998
F	1.932	1.649	1.410	1.045	0.500	0.145	0.041
Numerator d.f.	70	54	40	28	18	10	4
Denominator d.f.	490.8	432.9	373.5	311.5	246.6	176	89
p-Value	0.0001	0.0039	0.0560	0.4066	0.9568	0.9990	0.9968

The first three discriminant functions contrast various groups of variables. The first discriminant function is given by

$$Y_1 = 0.388C1 + 0.436C2 + 0.556C3 + 0.351C6 + 0.652C9 + 0.202C10$$
$$-0.530C5 - 0.270C7 - 0.323C8$$

with $C4$ omitted because of the small standardized coefficient. The function Y_1 contrasts the two sets of variables A and B below:

A	B
more than one murder	large amount of money involved
mouthy, obstinate detective	murder by nonphysical means
powerful opponents	many possible suspects
detective is a loner	
clues are given	
suspects are average people	

The items in A reflect a desire for complexity with fair clues and a detective who has certain characteristics. The items in B reflect items that make the story more complex or interesting. The discriminant function $Y1$ therefore tends to have a high value when the items in A are more important than the elements in B.

The second and third discriminant functions are given by

$$Y_2 = 0.415C1 + 0.487C2 + 0.284C5 + 0.299C6 + 0.203C9$$
$$-0.305C3 - 0.378C4 - 0.592C10 \quad \text{and}$$
$$Y_3 = 0.312C2 + 0.406C5 + 0.980C8 + 0.359C10 - 0.199C1$$
$$-0.486C4 - 0.343C7.$$

In Y_2 items $C7$ and $C8$ with small coefficients were omitted while in Y_3 items $C6$ and $C9$ were omitted due to small coefficients. The function Y_2 tends to increase if there is more than one murder, money is involved and if

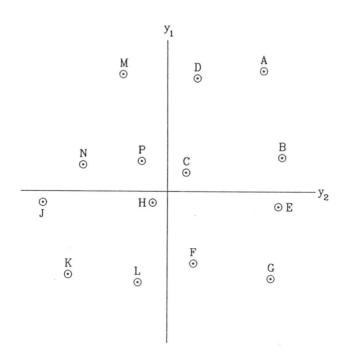

FIGURE 8.11. Variation of Group Means for Two Canonical Discriminant Functions

many clues are given. Y_2 also tends to increase if the detective is an unusual character. Y_2 decreases if the crime is difficult to solve and if the suspects are average people.

The value of Y_3 is relatively large particularly if there are many suspects. Y_3 will also be relatively large if the detective is mouthy and if large amounts of money are involved. On the negative side Y_3 tends to decrease if there is more than one murder, if the detective gets into an impossible jam, or if the murder is by nonphysical means.

Graphical Approach to Group Characterization

Two-dimensional graphs are useful for relating the canonical discriminant functions to the original groups. For each group, the group mean is determined on each of the discriminant functions. For various pairs of discriminant functions, the group means can be plotted to determine the pattern of variation of the group means over the two functions. Figure 8.11 illustrates how such a graph may appear for the means of y_1 and y_2 over 14 groups labeled A through P.

Having obtained interpretations for y_1 and y_2, we can characterize group differences with respect to these two dimensions. From the Figure 8.11, it

TABLE 8.15. Discriminant Function Means by SEXED Category Group Means

Disc. Function	1	2	3	4	5	6	7	8
$Y1$	−0.555	−0.860	1.362	0.728	−1.057	−0.139	0.487	−0.587
$Y2$	−0.266	1.403	0.299	0.660	0.245	−1.353	−0.068	0.161
$Y3$	−0.352	1.909	−0.617	−0.066	−0.734	0.297	0.386	−0.226

would appear that groups A, M and D tend to have relatively high values of y_1 whereas groups F, K, L and G have relatively low values of y_1. Similarly, y_2 separates J, K and N from E, A, B and G along the dimension y_2. Groups P, C and H have middle-of-the-road values along both dimensions.

Example

The means on the three discriminant functions over the eight SEXED groups can be studied to determine the relationship between the ten variables and the SEXED categories. The means are summarized in Table 8.15. A useful way to examine the relationship is to plot the category means on a two-dimensional graph as illustrated in Figure 8.12. The two axes correspond to discriminant functions Y_1 and Y_2. From the graph, we can see that the eight groups are well dispersed with respect to the two characteristics Y_1 and Y_2. Some interesting comparisons can be made using the two dimensions provided by Y_1 and Y_2.

The two groups defined by SEXED = 2 and SEXED = 6 correspond to males and females with some post high school education. The males on average tend to have very high values of Y_2 whereas the females have very low values of Y_2. It would appear that for this education group females had a relatively high preference for mysteries characterized by suspects who are average people, and detectives who encounter much difficulty. Males on the other hand seem to be more concerned with mysteries involving many murders, money, a mouthy obstinate detective and lots of clues.

For males and females with university degrees, the two sexes also seem to differ in the same way with respect to Y_2, but the difference is less pronounced than above. For SEXED groups 3 and 4 the values of Y_2 are larger than for SEXED groups 7 and 8. The two different education groups can also be compared in this case. It would seem that the two groups having the most education have lower values of Y_1 than their less educated counterparts. Thus group 4 has a lower value of Y_1 than group 3, and group 8 has a lower value of Y_1 than group 7. In addition with the exception of groups 2 and 6 the female groups tend to be lower on Y_1 than the corresponding male groups.

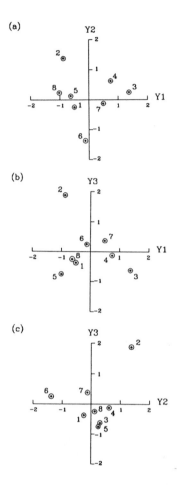

FIGURE 8.12. Location of Group Means with Respect to Discriminant Functions

The interpretation of the remaining graphs in Figure 8.12 is left as an exercise for the reader.

Comparison of Correlation Coefficients and Discriminant Function Coefficients

It is useful to compare the matrix of correlations between the original variables and the discriminant functions to the matrix of discriminant function coefficients. If there is a strong correlation between two of the original variables, we can expect that for at least some of the discriminant functions the coefficients corresponding to the two correlated variables will be similar in sign and magnitude. In some situations it is possible for the coefficients to be opposite in sign even though the correlation is strongly positive. This is particularly true if there are several significant discriminant functions. This result can occur if there are observations that are not consistent with the underlying correlation structure and if these observations tend to belong to a particular group.

Effect of Correlation Structure on Discriminant Analysis

In Figure 8.13 scatterplots are shown indicating various types of relationships between two hypothetical variables X_1 and X_2. In panel (a) the centres of the four ellipses are positioned in such a way that two orthogonal dimensions are required to identify the four points. In this case we might expect one discriminant function to have two coefficients of the same sign and the second discriminant function to have two coefficients of opposite sign. In panel (b) the centres of the four ellipses lie along a line and hence a single discriminant function with the same sign coefficients would be sufficient.

Panels (c) and (d) are designed to shown the impact of outliers on the discriminant function coefficients. In panel (c) there are a small number of outliers that are not consistent with the overall correlation structure. If the points to the left of the ellipse all belong to a particular group and the points to the right all belong to another group, a discriminant function can be based on the contrast between X_2 and X_1. In panel (d) the outliers are responsible for a strong correlation between X_1 and X_2. If these outliers are ignored, the correlation between X_1 and X_2 disappears. If the outliers belong to two groups such that the high values are in one group and the low values in another group, we would expect one discriminant function to show coefficents of the same size and sign for X_1 and X_2, whereas the remaining discriminant function coefficients for X_1 and X_2 would tend to reflect a zero correlation between X_1 and X_2.

Discriminant Analysis and Canonical Correlation

We have already indicated that MANOVA can be viewed as a special case of multivariate regression with dummy type explanatory variables. In a similar fashion, discriminant analysis can be viewed as a special case of canonical correlation. If dummy variables are defined for the groups say $D_1, D_2, \ldots, D_{g-1}$, the canonical correlation analysis between the X variables and the dummy variables is equivalent to the discriminant analysis obtained by the Fisher criterion. The canonical correlations are given by $\lambda_j/(1 + \lambda_j)$ for the eigenvalues λ_j, $j = 1, 2, \ldots, r$, from discriminant analysis. This relationship between the Fisher criterion discriminant analysis and canonical correlation led to the term canonical discriminant function introduced earlier.

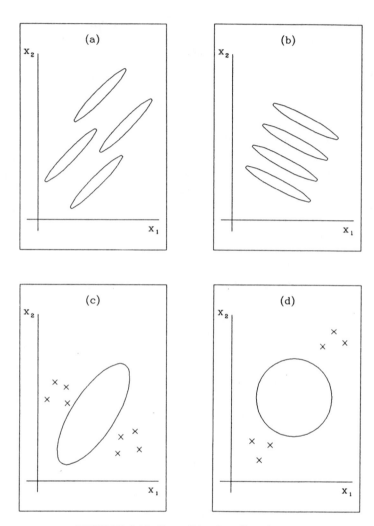

FIGURE 8.13. Some Bivariate Populations

Discriminant Analysis and Dimension Reduction

We have already introduced an inference procedure for selecting a subset of
"significant" discriminant functions from the total of r available. Retaining
only those, say f, significant functions, it is of interest to determine if they
can be used to correctly classify the n individuals into the g groups. To
classify the n observations using the f discriminant functions, a useful rule
is to place an individual in the group that is closest as measured by the
Mahalanobis distance from the individual to the group mean. Since the
functions have variance 1 and are mutually uncorrelated, these distances
are also Euclidean distances. To evaluate the quality of such a discriminant

procedure the proportion of observations misclassified by this approach can be determined. The proportion misclassified can be compared to what might be misclassified just by chance using random allocation.

To place observation i, let $y_{i1}, y_{i2}, \ldots, y_{if}$ denote the values of the canonical discriminant functions. Since the discriminant functions are mutually uncorrelated, the Euclidean distance from individual i to group k is given by

$$\delta_{ik}^2 = \sum_{j=1}^{f}(y_{ij} - \bar{y}_{\cdot j}^{(k)})^2 = \sum_{j=1}^{f}[\mathbf{b}_j'(\mathbf{x}_i - \bar{\mathbf{x}}_k)]^2$$

where $\bar{y}_{\cdot j}^{(k)}$ is the mean of discriminant function y_j for group k. Thus if $\delta_{im}^2 = \min_k \delta_{ik}^2$, then observation i is closest to group m. If all r discriminant functions are used, the distance δ_{ik}^2 can also be written as the Mahalanobis distance between $\mathbf{x}_i$ and the mean for group k, $\bar{\mathbf{x}}_k$, given by

$$\delta_{ik}^2 = (\mathbf{x}_i - \bar{\mathbf{x}}_k)'\mathbf{S}^{-1}(\mathbf{x}_i - \bar{\mathbf{x}}_k).$$

Note that this allocation method does not employ any assumptions about prior probabilities on group membership. The use of discriminant analysis to classify objects is discussed in the next section.

Example

Table 8.16 shows the Mahalanobis distance among the eight groups based on all seven discriminant functions. The distances are given to the right of the main diagonal. The p-values associated with Hotelling's T^2 test of equality of means are shown to the left of the main diagonal. Group 1 seems to represent a middle-of-the-road group in that its Mahalanobis distance from all other groups is not significant. Groups 2, 5 and 6 are significantly different from all other groups except group 1. An examination of the data matrix reveals that there are only two observations for group 1. This results in some difficulty in discriminating between group 1 and the remaining groups. Comparing the male groups to the corresponding female groups at the same education level, it would appear that groups 2 and 6 are significantly different ($p = 0.003$) but that groups 3 and 7 ($p = 0.193$) and groups 4 and 8 ($p = 0.102$) are less so.

For groups with university degrees a comparison of groups 3 and 4 shows a p-value of 0.783 whereas a comparison of groups 7 and 8 yields a p-value of 0.010. Comparing group 7 to groups 3 and 4 yields p-values of 0.193 and 0.459 respectively. For the four groups with university degrees it would appear that the females with advanced degrees are different from the remaining three groups.

The use of discriminant analysis to classify unknowns and the use of prior probabilities is discussed in the next section.

TABLE 8.16. Mahalanobis Distances Between Groups (upper right) and Associated F-Test p-Values (lower left)

	1	2	3	4	5	6	7	8
1	—	2.870	2.099	1.786	1.429	1.434	1.560	1.125
2	0.258	—	3.562	2.747	2.987	3.287	2.584	2.711
3	0.504	0.001	—	1.280	2.615	2.433	1.419	2.124
4	0.806	0.087	0.783	—	2.333	2.298	1.410	1.832
5	0.931	0.020	0.002	0.041	—	2.318	2.159	1.820
6	0.916	0.003	0.002	0.028	0.006	—	1.534	1.868
7	0.817	0.034	0.193	0.459	0.003	0.066	—	1.478
8	0.978	0.020	0.002	0.102	0.029	0.006	0.010	—

8.2.2 DISCRIMINANT FUNCTIONS AND CLASSIFICATION

Our discussion of discriminant analysis has thus far been concerned with its use as a descriptive device in characterizing differences among groups with respect to a vector of p random variables $\mathbf{x}$. An alternative use for discriminant analysis is in the classification of an observation of unknown origin into one of several possible groups.

For example, a banker measures a number of variables in order to decide whether a client should be given a loan. On the basis of the observations the banker classifies the client as either a safe loan prospect or not. A doctor observes the results of several laboratory tests on a patient and then decides whether or not the patient requires surgery for the detection and removal of cancer.

Discriminant analysis can be used to develop decision criteria for the assignment of unknowns to one of several possible groups. The determination of the discriminant criterion requires that the function either be known or estimated on the basis of a prior sample of "knowns". Initially, we assume that the population group mean vectors and covariance matrices are known. Our concern here is in using this information to classify unknowns. We begin with a discussion of the two group problem with known parameters.

Discrimination Between Two Groups With Parameters Known

Assume that the population is divided into two groups and that the distributions of the random variable $\mathbf{x}$ in the two groups are $N(\boldsymbol{\mu}_1, \boldsymbol{\Sigma})$ and $N(\boldsymbol{\mu}_2, \boldsymbol{\Sigma})$ respectively. For two groups the Fisher discriminant criterion is equivalent to

$$\max_{\mathbf{b}}[\mathbf{b}'\boldsymbol{\mu}_1 - \mathbf{b}'\boldsymbol{\mu}_2]^2/\mathbf{b}'\boldsymbol{\Sigma}\mathbf{b}.$$

In this case the solution vector $\mathbf{b}$ is given by

$$\mathbf{b} = k\boldsymbol{\Sigma}^{-1}(\boldsymbol{\mu}_1 - \boldsymbol{\mu}_2),$$

where k is an arbitrary constant. Using the condition $\mathbf{b}'\boldsymbol{\Sigma}\mathbf{b} = 1$ employed earlier in this section, the constant k is given by

$$k = [(\boldsymbol{\mu}_1 - \boldsymbol{\mu}_2)'\boldsymbol{\Sigma}^{-1}(\boldsymbol{\mu}_1 - \boldsymbol{\mu}_2)]^{-1/2}.$$

The discriminant function is therefore given by

$$\mathbf{x}'\mathbf{b} = \mathbf{x}'\boldsymbol{\Sigma}^{-1}(\boldsymbol{\mu}_1 - \boldsymbol{\mu}_2)/[(\boldsymbol{\mu}_1 - \boldsymbol{\mu}_2)'\boldsymbol{\Sigma}^{-1}(\boldsymbol{\mu}_1 - \boldsymbol{\mu}_2)]^{1/2}.$$

The means of the discriminant functions in the two groups are given by

$$\boldsymbol{\mu}_1'\mathbf{b} = \boldsymbol{\mu}_1'\boldsymbol{\Sigma}^{-1}(\boldsymbol{\mu}_1 - \boldsymbol{\mu}_2)/[(\boldsymbol{\mu}_1 - \boldsymbol{\mu}_2)'\boldsymbol{\Sigma}^{-1}(\boldsymbol{\mu}_1 - \boldsymbol{\mu}_2)]^{1/2}$$

and

$$\boldsymbol{\mu}_2'\mathbf{b} = \boldsymbol{\mu}_2'\boldsymbol{\Sigma}^{-1}(\boldsymbol{\mu}_1 - \boldsymbol{\mu}_2)/[(\boldsymbol{\mu}_1 - \boldsymbol{\mu}_2)'\boldsymbol{\Sigma}^{-1}(\boldsymbol{\mu}_1 - \boldsymbol{\mu}_2)]^{1/2}$$

respectively.

Classification of an Unknown

Given an observation $\mathbf{x}$ whose group membership is unknown, an approach to classification would be to place $\mathbf{x}$ in the closest group. The midpoint between $\mathbf{b}'\boldsymbol{\mu}_1$ and $\mathbf{b}'\boldsymbol{\mu}_2$ is

$$
\begin{aligned}
c &= (\mathbf{b}'\boldsymbol{\mu}_1 + \mathbf{b}'\boldsymbol{\mu}_2)/2 \\
&= \frac{1}{2}(\boldsymbol{\mu}_1 + \boldsymbol{\mu}_2)\boldsymbol{\Sigma}^{-1}(\boldsymbol{\mu}_1 - \boldsymbol{\mu}_2)/[(\boldsymbol{\mu}_1 - \boldsymbol{\mu}_2)'\boldsymbol{\Sigma}^{-1}(\boldsymbol{\mu}_1 - \boldsymbol{\mu}_2)]^{1/2}.
\end{aligned}
$$

Since from the definition of $\mathbf{b}$, $(\mathbf{b}'\boldsymbol{\mu}_1 - \mathbf{b}'\boldsymbol{\mu}_2) = [(\boldsymbol{\mu}_1 - \boldsymbol{\mu}_2)'\boldsymbol{\Sigma}^{-1}(\boldsymbol{\mu}_1 - \boldsymbol{\mu}_2)]^{1/2} > 0$, $\mathbf{b}'\boldsymbol{\mu}_1$ must be greater than $\mathbf{b}'\boldsymbol{\mu}_2$. Therefore we would place $\mathbf{x}$ in group 1 if $\mathbf{b}'\mathbf{x} > c$. Thus the discriminant criterion becomes

if $\mathbf{x}'\boldsymbol{\Sigma}^{-1}(\boldsymbol{\mu}_1 - \boldsymbol{\mu}_2) > 1/2(\boldsymbol{\mu}_1 + \boldsymbol{\mu}_2)'\boldsymbol{\Sigma}^{-1}(\boldsymbol{\mu}_1 - \boldsymbol{\mu}_2)$,   then $\mathbf{x}$ in group 1
and
if $\mathbf{x}'\boldsymbol{\Sigma}^{-1}(\boldsymbol{\mu}_1 - \boldsymbol{\mu}_2) < 1/2(\boldsymbol{\mu}_1 + \boldsymbol{\mu}_2)'\boldsymbol{\Sigma}^{-1}(\boldsymbol{\mu}_1 - \boldsymbol{\mu}_2)$,   then $\mathbf{x}$ in group 2.

Fisher Criterion and Mahalanobis Distance

The difference between the means on the discriminant function for the two groups is

$$(\mathbf{b}'\boldsymbol{\mu}_1 - \mathbf{b}'\boldsymbol{\mu}_2) = [(\boldsymbol{\mu}_1 - \boldsymbol{\mu}_2)'\boldsymbol{\Sigma}^{-1}(\boldsymbol{\mu}_1 - \boldsymbol{\mu}_2)]^{1/2},$$

which is the Mahalanobis distance between the two groups. Thus, using the Fisher criterion, an observation whose group membership is unknown is placed in the group whose Mahalanobis distance from the observation is smallest. It can be shown that under the assumption of multivariate normality with common covariance matrix, this criterion is equivalent to the criterion that assigns $\mathbf{x}$ to group 1 if $f_1(\mathbf{x})/f_2(\mathbf{x}) > 1$ and otherwise to group 2 where $f_1(\mathbf{x})$ and $f_2(\mathbf{x})$ are the multivariate normal densities for groups 1 and 2 respectively.

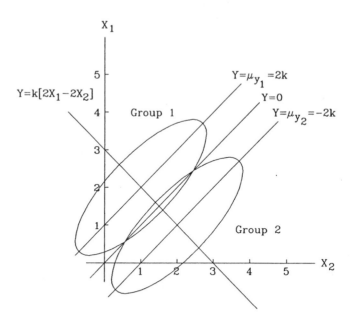

FIGURE 8.14. Discriminant Function for Bivariate Normal – Two Groups

Example

A numerical example with two variables is useful for showing the discriminant function solution graphically. For the two populations assume

$$\mu_1 = \begin{bmatrix} 2 \\ 1 \end{bmatrix}, \quad \mu_2 = \begin{bmatrix} 1 \\ 2 \end{bmatrix}, \quad \Sigma = \begin{bmatrix} 1 & 0.5 \\ 0.5 & 1 \end{bmatrix}.$$

The discriminant function $Y = k\mathbf{x}' \Sigma^{-1}(\mu_1 - \mu_2)$ is obtained from

$$\Sigma^{-1} = 4/3 \begin{bmatrix} 1 & -.5 \\ -.5 & 1 \end{bmatrix}, \quad \mu_1 - \mu_2 = \begin{bmatrix} 1 \\ -1 \end{bmatrix}$$

yielding $Y = k[2X_1 - 2X_2]$. Figure 8.14 shows confidence ellipsoids for the two populations as well as the discriminant function Y. The value of Y at the group 1 centroid is $\mu_{y_1} = 2k$ and at the group 2 centroid $\mu_{y_2} = -2k$. The midpoint between the two centroids is therefore $Y = 0$. We therefore choose group 1 if $Y > 0$ and group 2 if $Y < 0$. The discriminant function contrasts the observations on X_1 and X_2. If X_1 is large relative to X_2 then the observation is more likely to be from group 1, whereas if X_1 is small relative to X_2 the observation is more likely to be from group 2.

Maximum Likelihood Criterion

An alternative criterion for allocating $\mathbf{x}$ to one of two groups is the maximum likelihood criterion. The likelihood function for $\mathbf{x}$ is determined for each group assuming μ_1, μ_2 and Σ are known. The unclassified observation is then assigned to the group with the larger of the two likelihoods. Under the assumption of multivariate normality with common covariance matrix, this criterion is equivalent to the Fisher criterion outlined above.

Minimum Total Probability of Misclassification Criterion

Both the Fisher and maximum likelihood criteria assume no prior information about the distribution of the population between the two groups. A useful criterion for choosing between two groups is to minimize the overall probability of misclassification. Suppose that p_1 and $p_2 = (1 - p_1)$ are known to be the proportions of the population that are in groups 1 and 2 respectively and let $f_1(\mathbf{x})$ and $f_2(\mathbf{x})$ denote the respective densities. The minimum total probability of misclassification criterion assigns $\mathbf{x}$ to group 1 if $f_1(\mathbf{x})/f_2(\mathbf{x}) > p_2/p_1$. Under multivariate normality with common covariance matrix this criterion becomes assign $\mathbf{x}$ to group 1 if

$$\mathbf{x}' \Sigma^{-1}(\mu_1 - \mu_2) > 1/2(\mu_1 + \mu_2)' \Sigma^{-1}(\mu_1 - \mu_2) + \ln(p_2/p_1).$$

If the priors are equal, $p_2 = p_1$, this classification criterion is equivalent to both the Fisher and maximum likelihood criteria.

Bayes Theorem Criterion

Bayes theorem can also be used to obtain a classification criterion. Let p_1 and $(1 - p_1) = p_2$ denote the prior probabilities for group membership. Let $f_1(\mathbf{x})$ and $f_2(\mathbf{x})$ denote the densities for the two groups. The probabilities (posterior) for membership in the two groups are given by

$$\begin{aligned} P[\text{Group } 1 \mid \mathbf{x}] &= p_1 f_1(\mathbf{x})/[p_1 f_1(\mathbf{x}) + p_2 f_2(\mathbf{x})] \quad \text{and} \\ P[\text{Group } 2 \mid \mathbf{x}] &= p_2 f_2(\mathbf{x})/[p_1 f_1(\mathbf{x}) + p_2 f_2(\mathbf{x})], \end{aligned}$$

and $\mathbf{x}$ is assigned to the group with the maximum posterior probability. Therefore, $\mathbf{x}$ is assigned to group 1 if $f_1(\mathbf{x})/f_2(\mathbf{x}) > p_2/p_1$. Assigning an observation on the basis of the larger of the two posterior probabilities is therefore equivalent to the minimum total probability of misclassification rule given above.

Minimax Criterion

In classification we may be concerned with the probability of misclassification for a group which has low probability (e.g., a rare disease). Minimizing the total probability of misclassification approach may not be ideal in the case of such a rare group. An alternative rule is the minimax criterion which minimizes the maximum probability of misclassification. In the case of normal distributions with common covariance matrix, this rule is equivalent to the maximum likelihood criterion and hence also to the Fisher discriminant criterion. Given $\mathbf{x}$, assign it to group 1 if $f_1(\mathbf{x})/f_2(\mathbf{x}) > 1$; otherwise, assign it to group 2.

Minimum Cost Criterion

In some instances the costs of misclassification are not equal for both errors. In this case a minimum total cost of misclassification criterion is in order. This criterion results in the rule "assign $\mathbf{x}$ to group 1" if

$$\mathbf{x}'\boldsymbol{\Sigma}^{-1}(\boldsymbol{\mu}_1 - \boldsymbol{\mu}_2) > 1/2(\boldsymbol{\mu}_1 + \boldsymbol{\mu}_2)'\boldsymbol{\Sigma}^{-1}(\boldsymbol{\mu}_1 - \boldsymbol{\mu}_2) + \ln(w_2/w_1),$$

where $w_2 = c_2 p_2/[c_1 p_1 + c_2 p_2]$, $w_1 = c_1 p_1/[c_1 p_1 + c_2 p_2]$, $c_1 = $ cost of misclassifying a unit of group 1 and $c_2 = $ cost of misclassifying a unit of group 2. This criterion can be stated as assign $\mathbf{x}$ to group 1 if $f_1(\mathbf{x})/f_2(\mathbf{x}) > c_2 p_2/c_1 p_1$ otherwise to group 2. Notice that the cost factors result in an adjustment to the effective priors for the two groups.

Summary

Ignoring costs, we can conclude from the above outline of discriminant criteria that there are two different approaches to classification. One approach does not involve prior probabilities of group membership and the other does. The approach that is based on knowledge of priors is equivalent to the approach that does not use priors when the two prior probabilities are equal.

In summary, we may say that the discriminant criteria are based on a comparison of the two quantities

$$\mathbf{x}'\boldsymbol{\Sigma}^{-1}\boldsymbol{\mu}_1 - \tfrac{1}{2}\boldsymbol{\mu}_1'\boldsymbol{\Sigma}^{-1}\boldsymbol{\mu}_1 + \ln p_1 \qquad \text{for group 1 and}$$

$$\mathbf{x}'\boldsymbol{\Sigma}^{-1}\boldsymbol{\mu}_2 - \tfrac{1}{2}\boldsymbol{\mu}_2'\boldsymbol{\Sigma}^{-1}\boldsymbol{\mu}_2 + \ln p_2 \qquad \text{for group 2.}$$

(8.4)

The observation $\mathbf{x}$ is classified into the group corresponding to the largest of the two quantities. The Fisher and maximum likelihood criteria employ the assumption that the priors are equal. If the costs c_1 and c_2 are unequal, the priors p_1 and p_2 in the above expressions are replaced by $c_1 p_1$ and $c_2 p_2$. Equivalently, the classification of the unknown $\mathbf{x}$ is based on the ratio $f_1(\mathbf{x})/f_2(\mathbf{x})$. If $f_1(\mathbf{x})/f_2(\mathbf{x}) > c_2 p_2/c_1 p_1$, then assign $\mathbf{x}$ to group 1 otherwise to group 2.

Quadratic Discriminant Function and Unequal Covariance Matrices

Under the normality assumption, if the covariance matrices are not equal, $\boldsymbol{\Sigma}_1 \neq \boldsymbol{\Sigma}_2$, the optimal classification rule is the quadratic discriminant function

$$Q(\mathbf{x}) > \ln p_2/p_1,$$

where

$$Q(\mathbf{x}) = \frac{1}{2}\ln(|\boldsymbol{\Sigma}_2|/|\boldsymbol{\Sigma}_1|) - \frac{1}{2}(\mathbf{x} - \boldsymbol{\mu}_1)'\boldsymbol{\Sigma}_1^{-1}(\mathbf{x} - \boldsymbol{\mu}_1)$$
$$+ \frac{1}{2}(\mathbf{x} - \boldsymbol{\mu}_2)'\boldsymbol{\Sigma}_2^{-1}(\mathbf{x} - \boldsymbol{\mu}_2).$$

An unknown is classified into group 1 if $\frac{1}{2}\mathbf{x}'\boldsymbol{\Sigma}_1^{-1}\mathbf{x} + \mathbf{x}'\boldsymbol{\Sigma}_1^{-1}\boldsymbol{\mu}_1 + \ln|\boldsymbol{\Sigma}_1| - \ln p_1$ exceeds $\frac{1}{2}\mathbf{x}'\boldsymbol{\Sigma}_2^{-1}\mathbf{x} + \mathbf{x}'\boldsymbol{\Sigma}_2^{-1}\boldsymbol{\mu}_2 + \ln|\boldsymbol{\Sigma}_2| - \ln p_2$. This function is called quadratic because of the terms in $\mathbf{x}'\boldsymbol{\Sigma}_1^{-1}\mathbf{x}$ and $\mathbf{x}'\boldsymbol{\Sigma}_2^{-1}\mathbf{x}$ which are quadratic forms in $\mathbf{x}$.

Classification in Practice

The above outline of classification procedures presumes that the distributions are multivariate normal with known parameters $\boldsymbol{\mu}_1, \boldsymbol{\mu}_2$ and known common covariance matrix $\boldsymbol{\Sigma}$. The prior probabilities of group membership are also assumed known in some cases. In practice the true parameters are rarely known and hence are usually replaced by the conventional unbiased estimators, $\bar{\mathbf{x}}_{.1}$, $\bar{\mathbf{x}}_{.2}$ and S. The sample versions of (8.4) are then used for classification.

In some applications the assumption of multivariate normality is not tenable. The discriminant criteria outlined above do not generally perform well in the absence of normality. For discrete data in small samples, however, the discriminant criteria perform as well or better than methods based on various discrete distributions. For continuous data, transformations to normality do not in general result in improved classification error rates and in some cases the error rates are larger than for the nontransformed data. One important reason for the poor performance of the transformation approach is that the covariance matrices rarely remain equal.

In the case of unequal covariance matrices the use of the quadratic discriminant criterion is not robust to departures from normality. The quadratic function is particularly usesful when the covariance matrices are radically different and when sample sizes are large. In the case of small differences among covariance matrices the usual linear function is satisfactory.

Evaluation of a Discriminant Function as a Classification Mechanism

The usefulness of a discriminant function as a classification tool can be assessed by estimating the error rate or misclassification probability. The sample discriminant function can be used to determine the apparent error rate, which is the fraction of the sample observations that would be misclassified by the sample discriminant function. The number of observations correctly classified for each group is usually reported in a *confusion*

Predicted
Group

		1	2
Actual	1	n_{11}	n_{12}
Group	2	n_{21}	n_{22}

$$n = (n_{11} + n_{12} + n_{21} + n_{22})$$

FIGURE 8.15. Confusion Matrix for Two Groups

matrix. Figure 8.15 gives the confusion matrix for a two group case that illustrates the calculation. The probability of misclassification for the sample is $(n_{12} + n_{21})/n$. For confusion matrices determined from the same sample data used to estimate the discriminant criterion, the apparent error rate tends to underestimate the true error rate. The data used to evaluate the discriminant criterion should not be the same data that was used to estimate the criterion.

Split Sample

An alternative estimate of the error rate can be determined by splitting the sample into two parts. The first part of the sample is used to estimate the discriminant function which is then applied to classify the second part of the data. The proportion of misclassified observations for the second part provides an estimate of the error rate. The two roles can then be reversed by using the second part to predict the first part. This technique requires a large sample in order that the two subsamples be representative of the entire sample.

Jackknife Procedure

A more labor-intensive technique is to use a jackknife procedure that removes only one observation and uses the remaining $(n-1)$ observations to determine the discriminant function. The estimated discriminant function is then used to classify the omitted observation. This process is repeated $(n-1)$ times so that each observation is left out once. The misclassification error rate obtained from this process is a nearly unbiased estimate of the expected error rate from samples with n_1 and n_2 observations from groups 1 and 2 respectively.

Example

In this example the shape of the yield curve for interest rates is related to a set of macroeconomic variables. At any given point in time, the term structure of interest rates is described by a yield curve that shows the relationship between yield to maturity on the vertical axis and term to maturity on the horizontal axis. In most periods, the yield curve is upward sloping indicating that yield to maturity increases with increases in the term to maturity of the debt instrument. Occasionally the yield curve is humped or downward sloping. A humped or downward sloping yield curve is an indication that future interest rates are expected to fall, which is believed to be an indication of a recession or a decline in economic growth. The shape of the yield curve therefore should be related to other economic variables. In this example the slope of the yield curve will be related to several economic variables using a discriminant analysis.

For the eighty quarters from $Q3$ 1968 to $Q2$ 1988, the yield curve was determined for Canadian government bonds using the yields on T-bills, 1–3 year bonds, 3–5 year bonds, 5–10 year bonds, and bonds whose maturity is greater than 10 years. For the 80 yield curves, 44 were classified as upward sloping, 34 were classified as humped, and 2 were classified as downward sloping. For this analysis two groups will be defined: upward sloping and non-upward sloping.

The quantities unemployment rate (UNEMPLR), unemployment rate for males in the age group 25–54 (UNEMPLRM), gross domestic product in 1981 prices (GDP), Bank of Canada rate (BANKRT), industrial capacity utilization (CAPUTIL), and gross business investment in 1981 prices (BUS-INV) were used as explanatory variables. The data was obtained from the CANSIM data tape produced by Statistics Canada. Monthly figures were averaged over three months to get quarterly data. The data is displayed in Table 8.17.

A random selection of 65 of the 80 observations was used to obtain a discriminant function under the assumption of equal priors. Table 8.17 indicates which observations were used for estimation. The 65 observations used for estimation are classified as prediction class 1, and the 15 observations used for the holdout sample are classified as prediction class 2.

The discriminant function has an eigenvalue of 0.71294, a canonical correlation of 0.645, and a Wilk's Lambda value of 0.584 with 6 d.f. The discriminant function has a p-value of less than 0.000. The discriminant function with standardized coefficients is given by

$$Y = 5.81 \text{ UNEMPLR} + 3.41 \text{ BUSINV} + 1.54 \text{ CAPUTIL}$$
$$-0.83 \text{ BANKRT} - 3.72 \text{ UNEMPLRM} - 3.76 \text{ GDP}.$$

The discriminant function if evaluated at the group means has the value 0.75 for upward-sloping yield curves and −0.93 for non-upward-sloping yield curves. From the coefficients in Y we can conclude that the yield curve tends to be upward sloping when the unemployment rate for males is low relative to the overall unemployment rate and when the level of business investment and capacity utilization are relatively high. In addition, an upward sloping curve is more likely when the bank rate is relatively low and when GDP is relatively low.

TABLE 8.17. Yield Curve Data

Prediction Class	Yield Curve Class	BANKRT	UNEMPLR	UNEMPLRM	CAP-UTIL	GDP	BUS-INV
1	2	6.33	4.53	3.10	88.90	205088	8592
1	1	6.17	4.47	3.20	89.60	209076	8768
1	2	6.67	4.27	2.93	90.50	210604	9228
1	2	7.17	4.37	3.03	89.60	211748	9288
1	2	8.00	4.40	3.07	88.30	214584	9520
1	2	8.00	4.60	3.13	88.20	218848	9740
2	2	8.00	4.87	3.37	88.80	218504	9696
1	2	7.50	5.77	4.07	85.20	217400	9588
1	1	6.83	6.17	4.23	84.70	221176	9880
1	1	6.17	6.10	4.17	85.40	220912	9464
1	1	5.50	6.23	4.20	83.40	222884	9380
2	1	5.25	6.27	4.23	83.30	229796	9784
1	1	5.25	6.10	4.07	85.10	236684	9996
2	1	4.75	6.13	4.03	85.30	239184	10536
2	1	4.75	5.97	3.77	86.30	238420	10600
2	1	4.75	6.07	3.93	87.80	244928	10764
1	1	4.75	6.37	4.20	88.70	245864	10700
1	1	4.75	6.47	4.13	92.00	252552	10776
1	1	4.75	5.87	3.57	94.40	260360	12052
1	1	5.75	5.40	3.17	95.60	262716	12604
1	1	6.75	5.40	3.23	94.50	263604	13328
1	1	7.25	5.53	3.27	94.50	270796	14352
1	1	7.25	5.23	3.30	94.90	273488	14700
1	1	8.58	5.20	3.17	93.20	275072	14436
2	2	9.25	5.27	2.90	90.90	276736	14452
1	2	8.92	5.67	3.47	87.30	278728	14572
1	2	8.25	6.70	4.30	83.10	278432	15172
1	1	8.25	6.83	4.13	80.00	281048	15428
1	1	8.50	7.00	4.23	81.50	285228	15772
1	1	9.00	7.13	4.40	80.90	288040	14984
2	2	9.17	6.87	4.17	81.90	294624	15344
1	2	9.50	6.97	4.20	84.60	301344	16120
1	2	9.50	7.20	4.30	84.10	303416	15736
1	2	9.00	7.43	4.77	84.40	303168	16732
2	2	8.17	7.90	4.93	85.60	308204	16508
2	2	7.67	7.90	5.00	85.40	309360	15716
1	1	7.50	8.23	5.00	84.80	311344	16252
1	1	7.50	8.40	4.97	84.70	317108	15696
1	1	7.67	8.40	5.23	83.40	320172	15968
1	1	8.50	8.43	5.40	83.30	324836	17092

To evaluate the discriminant function as a predictor of yield curve type, the function Y was determined for all 65 observations used to estimate the function. Using the estimated function, 30 of the 36 upward-sloping curves were predicted, while for the non-upward-sloping curves 23 of 29

TABLE 8.17. Yield Curve Data (continued)

Prediction Class	Yield Curve Class	BANKRT	UNEMPLR	UNEMPLRM	CAP-UTIL	GDP GDP	BUS-INV
1	2	9.17	8.43	5.37	83.10	327040	17928
1	2	10.58	8.10	5.03	86.30	330956	18396
1	1	11.25	7.87	4.93	88.10	334800	18768
2	1	11.25	7.57	4.63	89.30	336708	19044
1	2	11.92	7.10	4.30	89.90	340096	21036
1	1	14.00	7.20	4.43	87.30	341844	21768
2	2	14.26	7.53	4.93	85.10	342776	22552
1	2	12.72	7.70	4.97	80.70	342264	23232
1	2	10.55	7.40	4.93	79.50	340716	25484
1	2	14.03	7.17	4.80	80.80	347780	26544
1	2	16.91	7.23	4.57	79.90	354836	28088
1	2	18.51	7.13	4.70	80.90	359352	29452
1	2	20.18	7.47	5.13	78.40	356152	29264
2	2	16.12	8.33	5.73	75.60	353636	28676
1	2	14.86	8.83	6.33	73.00	349568	26952
1	2	15.74	10.43	7.83	69.60	345284	24964
1	2	14.35	12.00	9.33	68.20	343028	23896
1	2	10.89	12.63	10.53	67.60	340292	23528
1	2	9.55	12.57	10.33	69.50	346072	23176
2	1	9.42	12.20	9.87	72.10	353860	23116
2	1	9.53	11.50	9.23	75.30	359544	24028
1	1	9.71	11.07	8.87	77.50	362304	24928
1	1	10.26	11.27	9.47	79.20	368280	25072
1	1	11.47	11.40	9.47	82.40	376768	24868
1	2	12.64	11.17	9.17	83.50	381016	25156
1	1	10.88	11.07	9.07	83.90	385396	25404
1	1	10.60	11.00	9.17	85.20	390240	26860
1	1	9.64	10.57	8.50	85.70	391580	27704
1	1	9.27	10.17	8.10	86.40	396384	28968
1	1	9.08	10.10	8.00	87.10	405308	29296
1	2	10.87	9.63	7.57	85.90	405680	30524
1	2	8.85	9.53	7.77	84.50	408116	31544
1	2	8.61	9.53	7.80	82.80	409160	31700
1	1	8.53	9.40	7.73	82.90	409616	32720
1	1	7.49	9.50	7.70	84.10	416484	33676
1	1	8.46	9.03	7.27	84.70	422916	35524
2	1	9.19	8.67	7.00	86.40	429980	37536
1	1	8.47	8.13	6.47	87.50	436264	40764
1	1	8.66	7.83	6.10	87.60	440592	42412
1	1	9.21	7.67	6.07	88.00	446680	45668

were correctly predicted. The overall success rate was therefore 81.5%. For the holdout sample of 15 observations the function correctly predicted all eight upward sloping curves and in addition correctly predicted five of the seven non-upward sloping curves. The success rate for the holdout sample was 86.7%.

Multiple Group Classification

The general criterion (8.4) used to compare two groups can be easily extended to the case of g groups. For a given observation $\mathbf{x}$ the criterion

$$E_k = \mathbf{x}' \boldsymbol{\Sigma}^{-1} \boldsymbol{\mu}_k - \frac{1}{2} \boldsymbol{\mu}'_k \boldsymbol{\Sigma}^{-1} \boldsymbol{\mu}_k + \ln p_k \qquad (8.5)$$

is computed for all groups $k = 1, 2, \ldots, g$. The observation $\mathbf{x}$ is then classified into the group m corresponding to the maximum value of E_k, $E_m = \max\limits_k E_k$. Since in practice the values of $\boldsymbol{\mu}_k$ and $\boldsymbol{\Sigma}$ are usually unknown, these parameters are replaced by the sample mean vector $\bar{\mathbf{x}}_k$ and the within group covariance matrix $\overline{\mathbf{S}} = \mathbf{W}/(n-g)$. The prior probabilities $p_1, p_2, \ldots, p_g$ can be replaced by the observed proportions n_k/n, $k = 1, 2, \ldots, g$.

As in the case of two groups, the Fisher criterion assumes that the priors are equal and hence the classification is based on a comparison of

$$E_k^* = \mathbf{x}' \boldsymbol{\Sigma}^{-1} \boldsymbol{\mu}_k - \frac{1}{2} \boldsymbol{\mu}'_k \boldsymbol{\Sigma}^{-1} \boldsymbol{\mu}_k, \quad k = 1, 2, \ldots, g.$$

This criterion is equivalent to a comparison of the squared Mahalanobis distances from $\mathbf{x}$ to $\boldsymbol{\mu}_k$

$$d^2 = (\mathbf{x} - \boldsymbol{\mu}_k)' \boldsymbol{\Sigma}^{-1} (\mathbf{x} - \boldsymbol{\mu}_k), \quad k = 1, 2, \ldots, g.$$

The Fisher criterion is also equivalent to the criterion that uses the r canonical discriminant functions and places $\mathbf{x}$ in the group whose centroid $(\mathbf{b}'_1 \boldsymbol{\mu}_1, \mathbf{b}'_2 \boldsymbol{\mu}_2, \ldots, \mathbf{b}'_r \boldsymbol{\mu}_r)$ is closest to the discriminant function value of the observation $(\mathbf{b}'_1 \mathbf{x}, \mathbf{b}'_2 \mathbf{x}, \ldots, \mathbf{b}'_r \mathbf{x})$.

Bias When Parameters Are Unknown

The optimal classification rule for unknown $\mathbf{x}$ assumes that $\boldsymbol{\Sigma}$ and $\boldsymbol{\mu}_j$, $j = 1, 2, \ldots, g$, are known. If $\boldsymbol{\mu}_j$ and $\boldsymbol{\Sigma}$ are replaced by the estimators $\overline{\mathbf{S}}$ and $\bar{\mathbf{x}}_j$, $j = 1, 2, \ldots, g$, then the misclassification probabilities are no longer minimized. One approach to obtaining improved classification results is to replace $\overline{\mathbf{S}}^{-1}$ by $[\overline{\mathbf{S}} + k\mathbf{I}]^{-1}$ in the classification rule. The value of the constant k can be determined from the calibration sample using a trial and error process. The value of k that minimizes the number of misclassified observations is chosen. Alternatively, the optimum k can be determined using a jackknife procedure by omitting one observation at a time and using the remaining observations to determine $\overline{\mathbf{S}}$ and $\bar{\mathbf{x}}_j$. The range of optimum values of k can then be used to determine an average k value. An alternative rule to choosing k to minimize the misclassification error is to choose k in a range where the discriminant coefficients are stable.

Example

An application of discriminant analysis that has been used frequently is the classification of corporate bonds into various bond rating categories. Bond ratings have an impact on the cost of borrowing and on the company's ability to borrow. Bond ratings are used by investors as measures of the financial health of the corporation. Various financial ratios determined from corporate annual reports are often used to help determine a company's bond rating. A sample of 95 companies was selected from the COMPUSTAT financial data tapes. For each company selected, the 1979 value for the financial ratios listed below were obtained. To obtain variables whose distributions were more normal-like, the logarithmic transformation was used in all but one case. Only the variable LTDCAP did not require the transformation. The bond ratings were obtained from Moody's Bond Ratings (June 1980). The data is summarized in Table 8.18. The ten variables are labeled as follows:

OPMAR =	operating margin
FIXCHAR =	pretax fixed charge coverage
LTDCAP =	long-term debt to capitalization
GEARRAT =	total long-term debt/total equity
LEVER =	leverage
CASHLTD =	cashflow to long-term debt
ACIDRAT =	acid test ratio
CURRAT =	current assets to current liabilities
RECTURN =	receivable turnover
ASSLTD =	net tangible assets to long-term debt.

The data shown in Table 8.18 is the transformed data and hence the variable names begin with 'L'.

The bond rating data in Table 8.18 will be used to provide an example of discriminant analysis as a classification technique. To provide a hold out sample of unknowns, the last two observations in each rating category (14 in total) were omitted from the determination of the discriminant functions. For the remaining 81 observations, the sequential F-test applied to the six discriminant functions showed the first three to be significant at the 0.10 level. The Mahalanobis distances between classes and the associated F statistic p-values are summarized in Table 8.19. Table 8.20 contains the standardized discriminant function coefficients. These functions are left to the reader for interpretation.

An examination of the Mahalanobis distances (upper right) and their p-values (lower left) in Table 8.19 gives some indication of how different the seven ratings are with respect to the ten financial ratios. In general the distances are consistent with the order of the ratings. It would appear that ratings B and C are definitely below the other five ratings. For the first five ratings, adjacent ratings are not generally significantly different. It may therefore be very difficult to classify companies precisely.

TABLE 8.18. Bond Rating Data

OBS	RATING	LOP-MAR	LFIX-CHAR	LGEARRAT	LTD-CAP	LLEVER	LCASH-LTD	LACID-RAT	LCUR-RAT	LREC-TURN	LASS-LTD
1	AAA	-1.663	0.749	-0.491	0.378	0.160	-1.225	0.433	1.120	1.629	1.277
2	AAA	-2.382	0.814	0.147	0.534	1.188	-1.552	-1.008	0.553	2.415	1.357
3	AAA	-1.401	2.561	-1.797	0.142	-0.531	0.496	0.314	1.014	1.728	2.273
4	AAA	-2.040	2.514	-1.528	0.178	-0.325	0.019	0.149	0.773	2.612	2.070
5	AAA	-1.360	2.432	-1.118	0.246	-0.085	-0.083	0.033	0.344	1.854	1.772
6	AAA	-1.687	2.891	-1.637	0.162	0.025	0.183	-0.051	0.328	2.197	2.361
7	AAA	-1.694	0.499	0.054	0.513	0.474	-1.539	0.745	0.897	1.949	0.907
8	AAA	-1.323	0.998	-0.936	0.281	-0.042	-0.187	0.001	0.863	1.349	1.704
9	AAA	-2.100	1.516	-1.654	0.159	0.251	0.342	-0.077	0.347	1.762	2.515
10	AAA	-1.888	2.484	-1.015	0.265	-0.099	-0.393	0.274	0.926	1.727	1.662
11	AAA	-1.633	1.589	-0.966	0.275	-0.105	-0.724	-0.287	0.204	2.128	1.610
12	AA	-2.041	2.636	-1.714	0.152	-0.240	0.537	0.393	0.634	1.911	2.296
13	AA	-2.434	2.193	-0.779	0.314	0.113	-0.798	-0.215	0.686	2.324	1.592
14	AA	-2.473	1.155	-0.925	0.283	-0.177	-0.732	0.074	0.737	2.257	1.541
15	AA	-2.632	2.342	-1.387	0.199	-0.182	-0.233	-0.189	0.824	2.728	2.022
16	AA	-2.285	1.767	-1.493	0.183	-0.049	-0.154	-0.321	0.642	2.335	2.222
17	AA	-2.363	1.807	-0.770	0.316	0.007	-0.707	-0.072	0.644	2.778	1.498
18	AA	-2.108	1.783	-0.829	0.303	-0.004	-0.414	0.030	0.494	2.227	1.540
19	AA	-2.064	1,742	-0.598	0.354	0.089	-0.837	0.011	0.628	3.252	1.337
20	AA	-1.884	2.291	-1.339	0.206	-0.235	-0.102	-0.164	0.525	2.182	1.941
21	AA	-1.740	1.606	-0.832	0.303	-0.147	-0.636	0.502	0.953	1.637	1.466
22	AA	-1.684	1.354	-1.368	0.202	-0.364	0.137	0.205	0.883	1.886	1.912
23	AA	-1.743	1.626	-1.207	0.230	-0.066	-0.266	-0.229	0.543	1.718	1.917
24	AA	-1.776	1.153	-0.450	0.389	0.171	-0.898	-0.073	0.440	2.227	1.251
25	AA	-2.371	2.451	-1.491	0.183	-0.024	-0.010	0.040	0.268	1.758	2.173
26	AA	-1.545	2.560	-0.820	0.302	0.306	-0.334	0.135	0.395	2.016	1.693
27	A	-1.720	1.239	-0.586	0.357	0.151	-0.935	0.110	0.876	1.758	1.377
28	A	-2.429	2.254	-1.221	0.226	-0.240	-0.297	0.061	0.751	2.325	1.812
29	A	-2.841	1.600	-1.235	0.225	-0.116	-0.413	0.070	0.932	2.724	1.915
30	A	-2.114	1.587	-0.833	0.302	-0.138	-0.776	-0.248	0.552	2.365	1.470
31	A	-1.416	1.353	-0.287	0.428	0.379	-0.750	-0.098	0.402	1.504	1.235
32	A	-1.466	1.679	-0.589	0.356	0.372	-0.487	-0.054	0.291	1.643	1.490

TABLE 8.18. Bond Rating Data (continued)

OBS	RATING	LOP-MAR	LFIX-CHAR	LGEARRAT	LTD-CAP	LLEVER	LCASH-LTD	LACID-RAT	LCUR-RAT	LREC-TURN	LASS-LTD
33	A	-1.850	1.490	-0.540	0.364	0.287	-0.546	0.156	0.619	1.899	1.400
34	A	-2.499	2.004	-0.492	0.378	0.525	-0.450	0.035	0.412	2.069	1.487
35	A	-2.388	1.248	-0.390	0.403	0.635	-0.664	-0.389	0.185	2.712	1.460
36	A	-2.014	1.736	-1.429	0.193	-0.417	-0.130	0.375	0.866	1.968	1.950
37	A	-1.704	3.691	-3.155	0.040	-0.936	1.573	0.122	0.998	2.033	3.493
38	A	-1.774	0.887	-0.532	0.369	0.013	-0.929	0.070	0.781	1.891	1.232
39	A	-2.219	1.776	-1.760	0.146	0.099	0.231	-0.003	0.818	1.801	2.539
40	A	-1.999	1.580	-1.059	0.257	0.122	-1.487	0.328	0.691	0.897	1.868
41	BAA	-3.323	1.021	-0.912	0.286	-0.049	-0.863	0.110	0.934	2.827	1.675
42	BAA	-2.147	1.373	-0.861	0.297	-0.233	-0.803	0.508	1.087	1.706	1.453
43	BAA	-1.844	2.238	-1.391	0.199	-0.450	-0.171	0.239	1.006	1.662	1.892
44	BAA	-2.145	1.834	-1.857	0.134	-0.300	0.219	0.182	0.808	1.675	2.428
45	BAA	-2.443	0.505	-0.622	0.348	0.136	-1.243	0.154	0.828	1.860	1.400
46	BAA	-2.195	1.546	-1.122	0.244	-0.057	-0.492	0.038	0.583	1.833	1.790
47	BAA	-2.353	0.816	-0.884	0.292	-0.047	-0.791	-0.034	0.716	2.086	1.569
48	BAA	-2.296	1.283	-0.695	0.332	-0.020	-0.984	-0.083	0.588	2.281	1.381
49	BAA	-2.403	1.597	-1.130	0.243	0.441	-0.148	-0.730	0.499	2.545	2.097
50	BAA	-2.194	1.601	-0.790	0.311	0.272	-0.675	-0.224	0.612	1.923	1.649
51	BAA	-1.288	1.727	-0.734	0.324	0.177	-0.502	-0.048	0.957	1.575	1.544
52	BAA	-2.163	1.097	-1.099	0.249	-0.101	-0.453	-0.709	0.787	3.011	1.763
53	BAA	-1.987	0.528	-1.059	0.257	-0.049	-0.531	-0.225	0.476	1.433	1.742
54	BAA	-1.494	2.986	-1.429	0.193	-0.316	-0.007	0.539	1.074	1.351	1.976
55	BAA	-1.308	0.526	0.427	0.605	0.898	-1.990	-0.244	0.230	2.385	0.846
56	BA	-1.595	0.936	-0.620	0.349	0.130	-1.301	-0.315	0.784	1.625	1.411
57	BA	-2.024	2.363	-1.897	0.130	-0.318	0.506	0.317	0.662	1.992	2.458
58	BA	-1.282	1.293	-0.169	0.457	0.398	-0.980	0.563	0.770	1.557	1.082
59	BA	-1.629	2.101	-0.418	0.395	0.703	-0.743	-0.084	0.233	2.047	1.534
60	BA	-2.297	2.629	-1.387	0.199	-0.452	-0.330	0.678	1.243	1.738	1.880
61	BA	-2.486	1.719	-0.739	0.323	-0.154	-1.383	0.590	1.224	1.556	1.358
62	BA	-1.674	2.229	-0.605	0.353	0.355	-0.539	-0.713	0.299	2.438	1.495
63	BA	-2.205	1.385	-0.744	0.322	-0.077	-1.057	0.306	0.884	1.883	1.419
64	BA	-2.451	1.290	-0.330	0.418	0.345	-1.252	0.204	0.730	1.898	1.211

TABLE 8.18. Bond Rating Data (continued)

OBS	RATING	LOP-MAR	LFIX-CHAR	LGEARRAT	LTD-CAP	LLEVER	LCASH-LTD	LACID-RAT	LCUR-RAT	LREC-TURN	LASS-LTD
65	BA	-1.552	2.089	-0.405	0.399	0.176	-0.760	0.216	0.759	1.935	1.202
66	BA	-2.281	1.055	-0.452	0.388	0.128	-1.138	0.004	0.830	2.295	1.233
67	BA	-1.901	2.523	-1.187	0.233	-0.244	-0.166	0.022	0.726	2.083	1.767
68	BA	-2.387	0.583	-0.901	0.288	-0.024	-0.952	-0.006	0.569	2.303	1.581
69	BA	-2.201	0.659	-0.136	0.465	0.396	-1.583	-0.005	0.899	2.009	1.052
70	BA	-1.936	1.642	-0.310	0.422	0.451	-1.068	0.156	0.809	1.177	1.290
71	B	-0.661	0.177	0.101	0.525	0.638	-1.473	-0.205	0.189	0.586	0.960
72	B	-0.486	1.345	0.480	0.617	0.733	-1.242	0.683	1.186	0.900	0.727
73	B	-3.399	0.518	-0.149	0.462	0.096	-1.709	-0.009	0.756	3.025	0.925
74	B	-2.475	0.847	-0.480	0.382	0.189	-1.892	0.265	0.960	1.092	1.409
75	B	-2.116	1.390	-0.218	0.445	0.499	-0.983	0.233	0.649	1.735	1.253
76	B	-2.102	0.705	0.328	0.581	0.394	-1.739	-0.039	0.789	2.294	0.673
77	B	-2.757	0.494	0.757	0.680	1.386	-1.568	-0.487	0.152	2.569	0.922
78	B	-2.287	1.233	-1.057	0.257	0.192	-0.719	-0.492	0.425	2.022	1.952
79	B	-1.841	1.939	-0.553	0.365	0.030	-0.992	0.174	1.078	1.806	1.286
80	B	-2.305	0.655	-0.100	0.474	0.164	-1.715	-0.299	0.705	1.960	0.930
81	B	-2.435	1.179	0.034	0.506	0.626	-1.482	-0.189	0.846	2.011	1.098
82	B	-3.467	0.811	-1.300	0.214	-0.372	-1.398	0.234	1.075	1.554	1.826
83	B	-2.013	1.535	-0.148	0.462	0.591	-0.985	0.193	0.833	1.519	1.200
84	C	-2.274	0.631	-0.414	0.396	0.324	-1.343	-0.261	0.763	2.047	1.336
85	C	-0.384	0.373	0.248	0.561	0.750	-1.385	0.247	0.481	-0.134	0.898
86	C	-1.201	0.412	0.949	0.721	1.274	-1.909	-0.020	0.301	1.446	0.582
87	C	-1.394	1.392	-0.107	0.473	0.587	-0.748	-0.133	0.654	1.629	1.160
88	C	-2.412	0.277	0.227	0.556	0.815	-2.327	0.041	0.726	0.984	0.958
89	C	-2.395	0.498	0.298	0.574	0.938	-1.560	-0.368	0.484	1.627	1.089
90	C	-2.053	1.723	-0.174	0.456	0.540	-0.790	-0.091	0.192	2.418	1.173
91	C	-2.098	0.004	0.134	0.533	0.538	-1.961	0.571	0.661	2.221	0.863
92	C	-1.147	0.000	0.024	0.503	0.381	-1.484	-0.348	0.088	1.903	0.886
93	C	-2.478	0.563	-0.099	0.475	0.290	-1.213	0.046	0.292	2.359	0.992
94	C	-1.698	1.535	-0.394	0.402	0.463	-0.244	0.545	0.848	1.219	1.373
95	C	-2.397	1.021	0.010	0.502	0.662	-0.933	-0.145	0.328	2.361	1.127

TABLE 8.19. Mahalanobis Distance Between Bond Rating Classes (upper right) and Associated F Statistic p-Values (lower left)

	AAA	AA	A	BAA	BA	B	C
AAA	—	1.6570	1.9551	2.0331	1.9017	3.2990	3.4616
AA	0.2589	—	1.2275	1.6030	1.4460	3.0064	3.4564
A	0.0934	0.6052	—	1.5188	1.3717	2.6035	2.6932
BAA	0.0565	0.1722	0.2691	—	1.8749	3.1254	3.4425
BA	0.1024	0.3117	0.4278	0.0467	—	2.5443	2.8917
B	0.0001	0.0001	0.0012	0.0001	0.0013	—	1.6181
C	0.0001	0.0001	0.0005	0.0001	0.0002	0.3047	—

TABLE 8.20. Standardized Discriminant Function Coefficients for Bond Data

Variable	Standardized Coefficients					
	D1	D2	D3	D4	D5	D6
LOPMAR	0.41	1.60	0.56	0.60	-0.01	-0.56
LFIXCHAR	-0.25	0.73	-1.51	-0.09	-0.40	0.32
LGEARRAT	-1.40	8.98	-2.26	20.48	12.87	-5.69
LTDCAP	-3.76	-2.14	-0.26	-4.10	0.10	3.07
LLEVER	2.15	-1.87	1.94	-5.76	-5.12	1.81
LCASHLTD	0.55	-2.46	0.44	-0.99	1.58	1.41
LACIDRAT	0.10	1.11	0.21	-0.05	0.16	-0.31
LCURRAT	0.53	-0.53	0.31	0.40	-0.71	0.87
LRECTURN	0.60	1.20	0.33	0.42	0.35	-0.47
LASSLTD	-2.58	7.15	-0.28	12.77	7.97	-2.68

For the 81 observations used in the discriminant analysis, the observations were classified using the criterion given by (8.5) using equal priors. This classification is equivalent to using the Mahalanobis distances. The confusion matrix for this classification is shown in Table 8.21. The proportion of misclassified observations is $30/81 = 0.37$. Using priors proportional to the group sample sizes yields an almost identical result since in this case the group sample sizes were almost identical. An examination of Table 8.21 reveals that the proportion $15/81 = 0.185$ of the observations were misclassified outside a category adjacent to the correct category.

The discriminant functions determined above were used to classify the holdout sample of 14 companies. The results of this classification for the equal prior assumption are shown in Table 8.22. Of the 14 observations, 9 were classified incorrectly. Using random allocation under equal priors, we would expect 11 of the 14 observations to be classified incorrectly. It is interesting to determine from Table 8.22 the frequency of misclassification beyond one rating class from the correct class. In this case only 3 of the 14

TABLE 8.21. Confusion Matrix for Training Sample

Actual Rating	Predicted Rating							Total
	AAA	AA	A	BAA	BA	B	C	
AAA	5	2	0	1	0	1	0	9
AA	1	7	1	2	2	0	0	13
A	0	3	6	2	1	0	0	12
BAA	0	1	0	11	1	0	0	13
BA	2	1	0	2	8	0	0	13
B	1	0	0	0	1	8	1	11
C	0	0	2	1	0	1	6	10
Total	9	14	9	19	13	10	7	81

TABLE 8.22. Confusion Matrix for Holdout Sample

Actual Rating	Predicted Rating							Total
	AAA	AA	A	BAA	BA	B	C	
AAA	0	1	0	0	1	0	0	2
AA	1	0	1	0	0	0	0	2
A	0	0	0	2	0	0	0	2
BAA	0	0	0	0	1	0	1	2
BA	0	0	0	0	2	0	0	2
B	0	0	1	1	0	0	0	2
C	0	0	1	0	0	0	1	2
Total	1	1	3	3	4	0	2	14

companies were incorrectly classified into a rating class beyond a class that is adjacent to the correct class. Using a jackknife procedure to estimate the true classification probabilities 27 of the 81 observations were correctly classified. A total of 57 of the 81 observations were correctly placed within one class of the correct class.

8.2.3 TESTS OF SUFFICIENCY AND VARIABLE SELECTION

Given the sample of n observations on $\mathbf{x}$, an important question is whether all variables are necessary to provide good discrimination. If the function is going to be used as a classification device in the future, the cost will be affected by the number of variables employed. In addition, the greater the number of variables the greater must be the sample size in order to achieve the same level of precision. In the two group case, a statistical test can be employed to test the hypothesis that the Mahalanobis distance between the two groups is the same for a subset of $q < p$ of the variables as it is for the full set of p variables.

Two Groups

Partitioning the variables into two sets we have

$$\mu = \begin{bmatrix} \mu_1 \\ \mu_2 \end{bmatrix} \quad \text{and} \quad \Sigma = \begin{bmatrix} \Sigma_{11} & \Sigma_{12} \\ \Sigma_{21} & \Sigma_{22} \end{bmatrix},$$

where μ_1 is $(q \times 1)$, μ_2 is $[(p - q) \times 1]$ and Σ is partitioned to conform. For the two groups we have mean vectors μ_1, μ_2 and assume a common covariance matrix Σ.

The squared Mahalanobis distance between the two groups is

$$\delta_p^2 = (\mu_1 - \mu_2)' \Sigma^{-1} (\mu_1 - \mu_2)$$

based on all p variables, and

$$\delta_q^2 = (\mu_{11} - \mu_{21})' \Sigma_{11}^{-1} (\mu_{11} - \mu_{21})$$

if based on the first q variables. To test $H_0: \delta_p^2 = \delta_q^2$ we have a test statistic

$$\frac{(n_1 + n_2 - p - 1)}{(p - q)} \frac{(d_p^2 - d_q^2)}{v + d_q^2}$$

where $v = (n_1 + n_2)(n_1 + n_2 - 2)/(n_1 n_2)$ and d_p^2 and d_q^2 are obtained by replacing the true parameters by their maximum likelihood estimators in the expressions for δ_p^2 and δ_q^2. If H_0 is true, this statistic has an F distribution with $(p - q)$ and $(n_1 + n_2 - p - 1)$ degrees of freedom.

This test is equivalent to a test that the coefficients are zero for a subset of the X variables in a discriminant function. This statistic can be used to provide a criterion for entry and exit in a stepwise discriminant procedure. As in the case of multiple regression, a forward selection or backward elimination method can be developed using this F-statistic. If p is not too large, an all possible subsets variable selection approach is preferable.

More than Two Groups

If there are more than two groups, say $g > 2$, with $n_1, n_2, \ldots, n_g$ observations respectively and total observations $n = n_1 + n_2 + \ldots + n_g$, the stepwise process can be extended by using Wilk's Lambda. Denoting by Λ_j the value of Wilk's Lambda for MANOVA based on the first j variables, a test statistic for the value of the $(j + 1)$th variable is given by

$$\frac{(n - g - j)}{(g - 1)} \left(\frac{\Lambda_j}{\Lambda_{j+1}} - 1 \right),$$

which in large samples has an F distribution with $(g - 1)$ and $(n - g - j)$ degrees of freedom if the $(j + 1)$th variable does not bring about a significant improvement in discrimination among the groups. Unlike the

TABLE 8.23. Results of Forward Selection Procedure

	p-Values for Entry				
Variable	Step #1	Step #2	Step #3	Step #4	Step #5
LOPMAR	0.3574	0.4559	0.4989	0.5428	0.5763
LFIXCHAR	0.0001	0.0628	0.0461	–	–
LGEARRAT	0.0001	0.0565	–	–	–
LTDCAP	0.0001	–	–	–	–
LLEVER	0.0001	0.4781	0.4915	0.3777	0.6821
LCASHLTD	0.0001	0.2094	0.1854	0.0830	–
LACIDRAT	0.7976	0.8144	0.8242	0.8796	0.9280
LCURRAT	0 1636	0.4123	0.3778	0.3767	0.5796
LRECTURN	0.1534	0.5389	0.6606	0.7301	0.6611
LASSLTD	0.0001	0.1501	0.5939	0.4710	0.6790

two group case, a disadvantage of this F-statistic is its tendency to favor the increased separation of well-separated groups rather than the improvement in separation for poorly separated groups. A comparison of all possible subsets of variables approach is better if p is not too large.

Example

The results of forward selection and backward elimination procedures are shown in Tables 8.23 and 8.24 respectively. In both cases the p-values for entry or exit were 0.15. For the forward selection procedure only four of the variables, LTDCAP, LGEARRAT, LFIXCHAR and LCASHLTD were entered before all of the p-values were above 0.15. Thus, after these four variables are included in the discriminant analysis, the addition of any one of the remaining six variables does not provide a significant improvement in the discrimination. For the backward elimination procedure, after the variables LACIDRAT and LCASHLTD are removed, the remaining eight variables have p-values less than 0.15. In this case, removal of any one of the eight variables will result in a significant loss in discrimination.

The two stepwise discriminant functions were evaluated by classifying both the 81 observations used to develop the function and also the 14 test observations. The forward stepwise solution misclassified 41 of the 81 observations, whereas the backward solution misclassified 40 of these observations. For the test sample the forward solution correctly classified 4 of the 14 observations whereas the backward solution classified only two observations correctly. It is interesting to note that the two stepwise discriminant functions only agreed on 6 of the 14 test observations, and only one of these was the correct classification.

TABLE 8.24. Results of Backward Elimination Procedure

	p-Values in the Equation		
Variable	Step #1	Step #2	Step #3
LOPMAR	0.0260	0.1418	0.0741
LFIXCHAR	0.0368	0.0273	0.0525
LGEARRAT	0.1253	0.1277	0.1301
LTDCAP	0.0117	0.0117	0.0091
LLEVER	0.0700	0.0556	0.0587
LCASHLTD	0.1170	0.2466	–
LACIDRAT	0.2119	–	–
LCURRAT	0.1346	0.0929	0.0777
LRECTURN	0.0450	0.2047	0.1179
LASSLTD	0.0698	0.0729	0.0823

8.2.4 DISCRIMINATION WITHOUT NORMALITY

Throughout our discussion of discriminant analysis it has been assumed that the distribution of **x** is multivariate normal. In addition we have usually assumed that the covariance matrices in each group are homogeneous. If these assumptions do not hold, other discrimination techniques may be preferable. The logistic regression model and the probit model described in Section 8.3 below are generally superior to discriminant analysis when the multivariate normality assumption does not hold.

In the absence of any assumptions about the underlying density nonparametric methods of discrimination can be used. In this section the nearest neighbor approach and a method based on ranks are presented.

Discrimination Using Ranks

In the absence of multivarite normality it is possible to use the discriminant analysis methods discussed above after replacing the original observations by their ranks. For each univariate random variable X_j, $j = 1, 2, \ldots, p$ [or component of **x** $(p \times 1)$], the observed data is ranked from smallest to largest over all groups simultaneously. The discriminant criterion is then computed using the sample mean vector and covariance matrix derived from the ranked data. To classify an unknown observation vector **x**, the rank position of each component of **x** is determined by interpolation with the rank transformation already determined for the original data.

Example

For the bond rating data the 95 observations on the ten variables were rank ordered and the ranks were used in place of the original data. Once again the last two observations in each rating class were omitted from the discriminant analysis. The classification with equal priors for the 81 training sample points yielded 43 misclassifications; for the 14 holdout companies 11 were misclassified. Using the jackknife procedure an almost unbiased estimate of the misclassification rate is 61 out of 81. In comparison to the nonranked data, the classification performance using ranks for this example is much inferior.

Nearest Neighbor Method

We have already seen that for the multivariate normal distribution classification using the Fisher discriminant criterion is equivalent to classification using the Mahalanobis distance. Each observation is classified into the closest group where the Mahalanobis distance is used to measure "closeness".

The nearest neighbor classification rule classifies unknowns into the closest group where closeness is measured using Euclidean distance. The square of the Euclidean distance between $\mathbf{x}_j$ and group k is given by $(\mathbf{x}_j - \bar{\mathbf{x}}_k)'(\mathbf{x}_j - \bar{\mathbf{x}}_k)$. Thus, in comparison to the Mahalanobis distance, the Euclidean distance makes no allowance for the covariance structure. In this case, therefore, random variables with large variances dominate the distance measure. The nearest neighbor concept will be discussed more fully in Chapter 10 under cluster analysis.

Example

The nearest neighbor method with Euclidean distance was used to classify the companies into bond rating classes. For the first 81 observations all but 17 observations were misclassified, and for the holdout sample of 14, 11 observations were misclassified. For this example, the nearest neighbor method is much inferior to the Fisher discriminant criterion. It would seem that the two methods that do not take into account the variance of the ten variables are much less reliable than the Mahalanobis distance type criterion which standardizes the variables.

8.2.5 OTHER SOURCES OF INFORMATION

Extensive discussions of discriminant analysis are available in Lachenbruch (1975) and Seber (1984).

8.3 Qualitative Response Regression Models and Logistic Regression

As outlined above, discriminant analysis can be used to determine the relationship between a categorical variable and a set of interval scaled variables. In this section we consider the use of a regression model to relate a categorical response variable to the explanatory variables.

In the multiple linear regression model the dependent variable Y is always assumed to have an interval scale. The explanatory variables in $\mathbf{x}$, however, can be either interval scaled or categorical. If the explanatory variables are categorical they are usually constructed using dummy coding or effect coding. If the dependent variable is categorical, then the multiple regression-type linear model is called a qualitative response regression model. This section begins with the simple case of a binary response model and then extends the techniques to the polychotomous case. The section is devoted mainly to variations of the logit model, although probit analysis is also introduced. An example is introduced next before outlining the theory.

Example

To provide examples for the discussion of qualitative response regression models the data summarized in Table 8.25 is used. The data represents a sample of 100 observations on married women selected from the Michigan Panel Study of Income Dynamics. The variables THISYR and LASTYR are indicator variables for whether the wife worked (=1) or did not work (=0) in the current year and the previous year respectively. The variables CHILD1, CHILD2, and BLACK are dummy variables indicating whether the wife has children under 2 (CHILD1), children between age 2 and age 6 (CHILD2) or is BLACK respectively. Finally the three variables AGE, EDUC and HUBINC are measures of the years of age and years of education of the wife and the income of the husband, respectively. The variables THISYR and LASTYR will be used as response variables and the remaining variables will be used as explanatory variables.

8.3.1   THE DICHOTOMOUS RESPONSE MODEL

The Point Binomial

We assume that individuals or objects can be classified into one of two mutually exclusive categories A or B, and that the probabilities associated with these two categories are p and $(1-p)$ respectively. As an example, the categories A and B might represent the events that a business firm will or will not go bankrupt in the next year.

We define the dummy random variable Y to indicate the two categories by letting $Y = 1$ for category A and $Y = 0$ for category B. The probability density for Y given the parameter p is therefore given by

$$f(Y \mid p) = p^Y (1 - p)^{(1-Y)}$$

which is the density of a *point binominal*.

TABLE 8.25. Full-Time Work Outside the Home for Married Women

OBS	LASTYR	THISYR	CHILD1	CHILD2	BLACK	HUBINC	EDUC	AGE
1	0	0	0	0	0	4.340	12	42
2	0	0	0	1	0	13.648	12	31
3	1	1	0	1	1	4.973	10	38
4	0	1	0	0	0	8.427	12	46
5	0	1	0	0	0	18.320	18	46
6	0	1	0	1	1	7.680	10	29
7	1	1	0	1	0	5.612	12	25
8	0	0	0	1	0	13.554	12	32
9	1	0	0	0	0	5.329	12	26
10	1	1	0	0	0	10.511	12	29
11	1	1	0	0	0	10.486	12	34
12	0	1	0	0	0	14.071	16	38
13	1	1	0	0	0	9.024	12	32
14	1	1	0	1	0	14.329	12	36
15	1	1	1	0	0	5.118	18	28
16	1	1	0	0	1	3.044	12	37
17	1	1	0	0	1	2.640	7	38
18	1	1	0	0	1	2.050	7	43
19	0	0	0	1	1	6.750	12	23
20	1	0	0	0	0	3.383	12	24
21	1	1	0	0	0	6.630	12	40
22	1	1	0	0	0	7.000	12	46
23	0	1	0	0	0	8.815	12	42
24	1	1	0	0	0	3.450	12	46
25	0	0	0	0	0	12.031	12	42
26	1	1	0	0	1	6.144	12	31
27	0	0	0	1	0	11.513	12	39
28	0	1	0	1	0	12.167	12	46
29	0	0	1	0	0	9.968	16	28
30	0	0	1	0	0	5.888	12	23
31	1	1	0	0	0	10.232	12	32
32	1	1	0	0	0	8.017	12	40
33	1	1	0	0	0	11.686	12	45
34	1	0	0	1	0	28.363	12	31
35	1	1	0	0	1	4.343	7	46
36	1	1	0	0	0	10.554	12	38
37	1	1	0	1	0	2.484	10	29
38	0	0	0	0	0	5.672	12	44
39	1	1	0	0	1	13.319	18	31
40	1	1	0	0	1	7.678	18	35
41	1	1	0	0	0	7.162	12	24
42	0	0	0	0	0	7.804	12	34
43	0	1	0	1	0	13.648	16	28
44	0	0	0	1	0	9.311	12	27
45	1	1	0	0	0	27.938	12	46
46	1	1	0	1	0	6.704	12	27
47	1	1	0	0	0	7.711	12	32
48	1	1	0	0	0	8.576	16	38
49	0	1	0	1	0	7.223	16	26
50	0	0	1	0	0	11.259	16	31

TABLE 8.25. Full-Time Work Outside the Home for Married Women (continued)

OBS	LASTYR	THISYR	CHILD1	CHILD2	BLACK	HUBINC	EDUC	AGE
51	0	0	0	1	0	26.063	12	30
52	1	1	0	0	0	11.776	12	42
53	1	1	0	0	0	12.793	18	46
54	1	1	0	0	0	11.080	12	44
55	1	1	0	0	0	7.074	12	31
56	1	1	0	1	0	6.679	12	36
57	0	1	0	0	0	15.868	12	45
58	1	1	0	0	0	7.972	16	42
59	0	0	1	0	1	0.000	12	29
60	1	1	0	0	0	3.030	10	43
61	1	1	0	0	0	2.970	16	27
62	1	1	0	0	0	9.305	12	40
63	1	1	0	0	0	8.125	12	30
64	1	0	0	1	1	13.033	10	29
65	1	1	0	0	1	0.000	12	39
66	1	1	0	1	1	2.781	12	30
67	1	1	0	0	1	3.010	12	35
68	0	0	0	0	0	26.056	12	40
69	0	0	0	0	0	5.795	12	46
70	1	1	0	1	0	0.000	12	36
71	1	1	0	1	0	2.639	12	28
72	1	1	0	0	0	9.087	12	24
73	1	0	0	0	0	12.312	12	34
74	1	0	0	0	0	7.325	12	33
75	1	1	0	0	0	3.517	10	26
76	1	1	0	0	0	17.140	12	35
77	1	1	0	0	0	24.054	12	40
78	1	1	0	0	1	6.144	12	42
79	0	1	0	0	1	13.211	12	34
80	0	1	0	0	0	9.309	12	45
81	1	1	0	0	0	3.135	10	40
82	1	1	0	0	0	2.935	10	45
83	1	1	0	0	0	9.607	12	41
84	0	1	0	0	0	10.629	12	44
85	1	1	0	0	0	8.207	12	24
86	1	1	0	0	0	9.772	12	42
87	1	1	0	0	0	8.955	12	46
88	1	1	0	0	0	6.204	10	46
89	0	1	0	0	1	9.378	12	32
90	0	0	0	0	0	54.281	12	45
91	1	1	0	1	0	7.525	12	31
92	0	0	1	0	0	11.504	12	32
93	0	1	0	0	0	5.763	12	42
94	0	0	0	1	0	5.683	12	32
95	0	1	0	0	0	10.937	12	40
96	1	1	0	0	0	9.361	12	45
97	0	0	0	1	0	6.342	12	35
98	1	0	0	0	0	7.160	10	31
99	1	0	0	1	0	7.788	12	31
100	1	1	0	0	1	2.402	10	25

Probability as a Function of Other Variables

To continue the example of business firms and bankruptcy, we assume that the probability of bankruptcy depends on a measure of financial health D, where D is a linear function given by $D = \beta_0 + \beta_1 X$ and where X is a measure of a company's ability to repay its debts, such as debt-equity ratio. In other words the probability of bankruptcy is a function of D and will be denoted by $p(D)$. For an individual firm i with debt-equity ratio x_i, the firms value of D is given by $d_i = \beta_0 + \beta_1 x_i$ and the conditional probability density for y_i given $p(d_i)$ has the form

$$f\left(y_i \mid p(d_i)\right) = [p(d_i)]^{y_i}[1 - p(d_i)]^{(1-y_i)}.$$

For a random sample of n firms we have $(d_1, d_2, \ldots, d_n)$, and the joint conditional density for $(y_1, y_2, \ldots, y_n)$ given $[p(d_1), p(d_2), \ldots, p(d_n)]$ has the form

$$
\begin{aligned}
f&\left(y_1, y_2, \ldots, y_n \mid p(d_1), p(d_2), \ldots, p(d_n)\right) \\
&= [p(d_1)]^{y_1}[1 - p(d_1)]^{(1-y_1)}[p(d_2)]^{y_2}[1 - p(d_2)]^{(1-y_2)} \\
&\quad \ldots [p(d_n)]^{y_n}[1 - p(d_n)]^{(1-y_n)} \\
&= \prod_{i=1}^{n}[p(d_i)]^{y_i}[1 - p(d_i)]^{(1-y_i)}.
\end{aligned}
$$

Note here that the parameters β_0 and β_1 are assumed to be constant across the complete sample.

Alternative Response Functions

To be able to relate the value y of the response variable Y to the value d of the variable D, a more specific assumption about the form of the function $p(d)$ is required. Three alternatives for $p(d)$ are discussed below. These are

1. the *linear probability model:*

$$
\begin{aligned}
p(d) &= \alpha_0 + \alpha_1 d & d_0 \le d \le d_1 \\
&= 0 & d < d_0 \\
&= 1 & d > d_1.
\end{aligned}
$$

2. the *probit model:*

$$p(d) = F(d) \qquad -\infty \le d \le \infty \text{ where } F(d) \text{ is}$$

the distribution function

for the normal density.

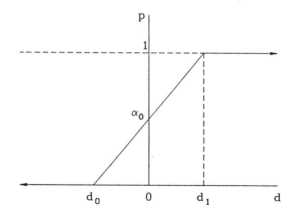

FIGURE 8.16. Linear Probability Model

3. the *logistic or logit model:*

$$p(d) = G(d) \qquad -\infty \le d \le \infty \text{ where } G(d) \text{ is}$$
the distribution function
for the logistic density.

For the linear probability model the relationship between $p(d)$ and d is given in Figure 8.16. The linear probability model is usually estimated by defining a dummy variable for the dependent variable and using ordinary least squares. In this case heteroscedasticity exists since the variance at each point d is a function of $p(d)$. If $p(d)$ can be estimated reliably a weighted least squares estimator can then be constructed. The ordinary least squares estimator can be used as a preliminary estimator for the required weights. To obtain a reliable estimator of $p(d)$ before using weighted least squares several observations are required at each value of d. Unfortunately in many applications only one observation is available at each value of d. A weighted least squares procedure is discussed in Section 8.3.4. The ordinary least squares estimator of $p(d)$ can also be used as a preliminary estimator, which is required for the logistic and probit models discussed below. The ordinary least squares estimator of $p(d)$ can result in estimates outside the interval $(0, 1)$. Although the linear probability model is simple to handle mathematically, it is difficult to justify the "kinks" or discontinuities at $p(d_0) = 0$ and $p(d_1) = 1$ in Figure 8.16.

A more realistic shape for $p(d)$ is the normal distribution function shown in Figure 8.17. In this case the function $p(d)$ is not truncated but instead can converge to its upper and lower limits as d becomes very small or very large. The function $p(d)$ in this case is called the *probit transformation* and

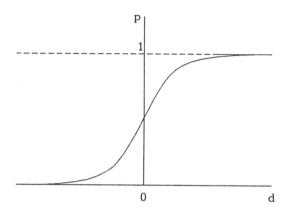

FIGURE 8.17. Shape of Normal and Logistic Distribution Functions

is given by

$$p(d) = \int_{-\infty}^{d} \frac{1}{\sqrt{2\pi}} \, e^{-w^2/2} dw.$$

The model in this case is commonly called a probit model. Unfortunately, unlike the linear probability model, in the normal distribution case $p(d)$ cannot be written explicitly as a function of d. The probit model will be discussed more extensively in Section 8.3.2.

The third alternative for $p(d)$ given by (3) above is called the *logistic transformation*. In this case

$$p(d) = e^d/(1 + e^d),$$

which is the *logistic distribution function*. The shape of $p(d)$ for the logistic is quite similar to the shape for the normal given in Figure 8.17. As we outline next, the logistic transformation lends itself to a useful explicit functional relationship between $p(d)$ and d.

The standardized logistic density is given by

$$f(w) = e^{-w}/(1 + e^{-w})^2$$

and the mean and variance of this density are 0 and $\pi^2/3$ respectively. The standard normal and standardized logistic distribution yield very similar shaped densities and distribution functions. Like the standard normal density, the standardized logistic density has a median and mode of zero and a skewness of zero. The kurtosis of the logistic density is 4.2, which indicates fatter tails than the normal, which has a kurtosis of 3. The standardized logistic distribution with $w^* = w/\sqrt{\pi^2/3}$ has slightly heavier tails than the standard normal distribution.

An important advantage of the logistic distribution in this context is that the logit transformation $\ln[p/(1-p)]$ has the form

$$\ln\left[\frac{p(d)}{1-p(d)}\right] = \ln\left[\frac{e^d/(1+e^d)}{1/(1+e^d)}\right] = d.$$

Therefore, if d is assumed to be a linear function of x, $d = \alpha + \beta x$, the logit has the familiar linear model form. This logit model is usually referred to as the *logistic regression model*. Like the multiple linear regression model the x observation is fixed. The logistic regression model is based on the conditional distribution of Y given x.

Logistic Regression With c Explanatory Variables

The logistic regression model can be extended to include c explanatory variables. In this case it is assumed that $p = p(d)$, where $d = \beta_0 + \sum_{j=1}^{c}\beta_j x_j$, is a linear function of observations $(x_1, x_2, \ldots, x_c)$ on c explanatory variables $(X_1, X_2, \ldots, X_c)$. Thus for the bankruptcy example we assume that there are a total of c variables that are related to the probability of bankruptcy.

The logit of p is given by

$$\ln[p/(1-p)] = d = \beta_0 + \sum_{j=1}^{c}\beta_j x_j$$

which has the form of a multiple linear regression model. The estimation of the parameters $\beta_0, \beta_1, \ldots, \beta_c$ is discussed next.

Maximum Likelihood Estimation for Dichotomous Logistic Regression

Given a random sample of n observations from the population, a dummy variable Y, $(Y = 0$ or $1)$, is used to indicate to which group each observation belongs. Thus we observe $y_i = 0$ or 1 for $i = 1, 2, \ldots, n$. The observations on the corresponding c explanatory variables are denoted by $(x_{i1}, x_{i2}, .., x_{ic})$. Given these x observations, the likelihood for the ith observation can be expressed by

$$p_i^{y_i}[1-p_i]^{(1-y_i)} = [p(d_i)]^{y_i}[1-p(d_i)]^{1-y_i},$$

where $d_i = \beta_0 + \sum_{j=1}^{c}\beta_j x_{ij} = \mathbf{x}_i'\boldsymbol{\beta}$ and where $\boldsymbol{\beta}' = (\beta_0, \beta_1, \ldots, \beta_c)$ and $\mathbf{x}_i' = (1, x_{i1}, x_{i2}, \ldots, x_{ic})$.

For the entire sample, the likelihood function conditional on $\mathbf{x}_i$, $i = 1, 2, \ldots, n$, is given by

$$L = \prod_{i=1}^{n}[p(\mathbf{x}_i'\boldsymbol{\beta})]^{y_i}[1-p(\mathbf{x}_i'\boldsymbol{\beta})]^{1-y_i},$$

and the logarithm of the likelihood by

$$\ln L = \sum_{i=1}^{n}\{y_i \ln[p(\mathbf{x}_i'\boldsymbol{\beta})] + (1-y_i)\ln[1-p(\mathbf{x}_i'\boldsymbol{\beta})]\}.$$

Newton–Raphson Procedure

The maximum likelihood estimator $\widetilde{\beta}$ of β is determined by maximizing $\ln L$ with respect to β. Unfortunately there is no analytical solution for β. The Newton–Raphson iterative procedure can be used to obtain $\widetilde{\beta}$ based on a preliminary estimator, say $\hat{\beta}$. This procedure will converge to the maximum likelihood estimator regardless of the choice of the preliminary estimator. A common choice for the preliminary estimator is the ordinary least squares estimator $\hat{\beta} = (\mathbf{X'X})^{-1}\mathbf{X'y}$ where $\mathbf{y}$ is the vector of y_i values, $i = 1, 2, \ldots, n$, and $\mathbf{X}(n \times c)$ is the matrix of X observations. The Newton–Raphson procedure is outlined in Judge *et al.* (1985, pp. 955–958).

The maximum likelihood estimator for β in the logistic regression model is obtained by solving the system of $(c + 1)$ equations given by

$$\sum_{i=1}^{n} p_i \mathbf{x}_i = \sum_{i=1}^{n} y_i \mathbf{x}_i$$

where $p_i = e^{\mathbf{x}_i'\beta}/(1 + e^{\mathbf{x}_i'\beta})$. The solutions to these equations given by $\widetilde{\beta}$ can be used to obtain the estimator $\tilde{p}_i = e^{\mathbf{x}_i'\widetilde{\beta}}/(1 + e^{\mathbf{x}_i'\widetilde{\beta}})$ for each of the n observations and hence the fitted sum $\sum_{i=1}^{n} \tilde{p}_i \mathbf{x}_i$ is equal to the observed sum $\sum_{i=1}^{n} y_i \mathbf{x}_i$.

Inference for the Dichotomous Logistic Regression Model

The dichotomous logistic regression model assumes that the logit function $\ln[p/(1 - p)]$ can be modeled as a linear function of a set of explanatory variables $\beta_0 + \sum_{j=1}^{c} X_j \beta_j$. Given a random sample of observations $(y_i, x_{i1}, x_{i2}, \ldots, x_{ic})$, $i = 1, 2, \ldots, n$, the maximum likelihood estimator $\widetilde{\beta}$ can be obtained as outlined above. In comparison to the multiple linear regression model, the coefficient vector $\widetilde{\beta}$ in this case must be interpreted differently. A marginal one unit increase in X_j brings about an increase in $\ln[p/(1-p)]$ of the amount $\widetilde{\beta}_j$. The magnitude of the increase in p, however, depends on the initial value of p.

Comparing Nested Models and Inference for Coefficients

Inferences regarding the coefficients in the logistic regression model can be made by comparing models and submodels using a likelihood ratio test. To compare a full model with c explanatory variables plus an intercept to a reduced model with $(c - q)$ explanatory variables plus an intercept, the logarithm of the likelihood ratio yields the statistic $-2[\ln L_1 - \ln L_2]$, which has a χ^2 distribution with q degrees of freedom if the q deleted variables are superfluous. L_2 is the likelihood function for the full model, and L_1 is the likelihood function for the reduced model.

The reader may recall that this approach was also used for loglinear models in Chapter 6. In Chapter 6 the likelihood ratios were determined with reference to a multinomial distribution with no explanatory variables. The likelihood ratio χ^2 statistic H^2 of Chapter 6 can be written as $H^2 = -2[\ln L_1 - \ln L]$, where L is the likelihood of the saturated model and L_1 is the likelihood for the simpler model.

Computer Software

The calculations for the logistic regression examples in this section were carried out using the program BMDP LR and the program SAS LOGISTIC.

Example

To examine the bivariate relationships in Table 8.25 between THISYR and LASTYR and each of the six explanatory variables, single variable logistic regression models were estimated. The results are summarized in Table 8.26. The p-values for the coefficients are shown in brackets underneath the coefficient estimates.

To illustrate the information contained in Table 8.26, we examine in detail the relationship between THISYR and each of the interval scaled variable HUBINC and the categorical variable CHILD1. For the variable HUBINC the fitted logistic regression model has the equation

$$\ln[p/(1-p)] = 1.6001 - 0.0675 \text{ HUBINC}$$

where p is the probability that the wife will choose to work THISYR. The log of the likelihood ratio for the model is given by $\ln L_2 = -56.982$ whereas the log of the likelihood ratio with HUBINC omitted is $\ln L_1 = -59.295$. The likelihood ratio χ^2 statistic is therefore $-2[\ln L_1 - \ln L_2] = 4.62$ which has a p-value of 0.0315 for a 1 d.f. χ^2. The fitted equation indicates that as HUBINC decreases the value of p increases. The wife is therefore less likely to work if her husband's income is relatively high. From Table 8.25 the range of HUBINC is 0 to 54.3 and hence the value of the logit varies from $+1.6001$ to -2.0652. The range of values for p is therefore given by

$$p = \exp[+1.6001]/(1 + \exp[+1.6001]) = 0.83 \quad \text{and}$$
$$p = \exp[-2.0652]/(1 + \exp[-2.0652]) = 0.11.$$

Thus for wives whose husband's income is zero the probability that the wife chooses to work is 0.83, whereas at the other extreme for wives whose husband's income was in the highest category the probability that the wife chooses to work is only 0.11.

For the variable CHILD1 the fitted model is given by

$$\ln[p/(1-p)] = 1.1272 - 2.7366 \text{ CHILD1}.$$

The probability that the woman chooses to work therefore is either $p = \exp[1.1272]/(1+\exp[1.1272]) = 0.76$ for CHILD1 $= 0$ or $p = \exp[-1.6094]/$

TABLE 8.26. Bivariate Relationships Between THISYR, LASTYR and Each of the Six Explanatory Variables

	THISYR					LASTYR				
Variable	Intercept	Coefficient	Model Log Likelihood	Coefficient χ^2	χ^2 Hosmer-Lemeshow p-Value	Intercept	Coefficient	Model Log Likelihood	Coefficient χ^2	χ^2 Hosmer-Lemeshow p-Value
AGE	-1.8260 (0.1151)	+0.0794 (0.0147)	-56.318	5.95	0.49	+1.1585 (0.2836)	-0.0150 (.6123)	-64.616	0.26	0.21
EDUC	+0.5050 (0.7127)	+0.0360 (0.7432)	-59.242	0.11	0.13	+2.5742 (0.0434)	-0.1585 (0.1209)	-63.542	2.41	0.19
HUBINC	+1.6001 (0.0000)	-0.0675 (0.0315)	-56.982	4.62	0.30	+1.4264 (0.0001)	-0.8545 (0.0096)	-61.394	6.70	0.71
BLACK	+0.8065 (0.0005)	+0.8675 (0.1682)	-58.346	1.90	—	0.5306 (0.0189)	0.4990 (0.3693)	-64.342	0.81	—
CHILD1	1.1272 (0.0000)	-2.7366 (0.0034)	-55.006	8.58	—	0.7577 (0.0004)	-2.3671 (0.0117)	-61.569	6.35	—
CHILD2	1.2272 (0.0000)	-0.9861 (0.0452)	-57.290	4.01	—	0.8158 (0.0007)	-0.7357 (0.1205)	-63.539	2.41	—
Log Likelihood for Constant Only	-59.295					-64.745				

$(1+\exp[-1.6094]) = 0.17$ for CHILD1 $= 1$. This logistic regression model is equivalent to the logit model that would be derived from fitting a saturated loglinear model to the two-way contingency table as discussed in Chapter 6. The significance of the coefficient of CHILD1 is obtained from $-2[\ln L_1 - \ln L_2] = -2[-59.295-(-55.006)] = 8.58$ which has a p-value of 0.0034 for a 1 d.f. χ^2. The fact that CHILD1 is significant in the prediction of THISYR is equivalent to the statement that the hypothesis of independence between the two variables is rejected. The χ^2 value for CHILD1 here is equal to the likelihood ratio χ^2 in the test of independence discussed in Chapter 6.

An examination of the coefficient p-values in Table 8.26 reveals that the most important variables in predicting whether a woman will choose to work THISYR are AGE, HUBINC, CHILD2 and CHILD1. The coefficients of these variables indicate that p tends to be larger if AGE is large, HUBINC is small, CHILD1 $= 0$ and CHILD2 $= 0$. For the variable LASTYR the most significant explanatory variables are HUBINC, CHILD1 and CHILD2. The coefficients in these three models indicate that p increases with decreasing HUBINC, and that p is larger if CHILD1 $= 0$ and if CHILD2 $= 0$.

Goodness of Fit

The goodness of fit χ^2 statistics introduced in Chapter 6 for loglinear models cannot usually be used in logistic regression models containing interval scale explanatory variables. Because the observations are almost all unique the asymptotic distribution properties of the two statistics cannot be expected to hold with only one observation per cell. In Section 8.3.4 the explanatory variables are assumed to be design variables and multiple observations are assumed for each design combination. In that case the χ^2 goodness of fit tests can be used. For the more general case other measures have been developed to measure goodness of fit.

A pseudo-measure of goodness of fit is given by

$$R^2 = 1 - (\ln L / \ln L_0^*)$$

where L_0^* denotes the likelihood function value when all variables are excluded except the constant term β_0. Thus the sample value of L_0^* is the value of L evaluated using the sample proportion for the maximum likelihood estimator of p. This R^2 measures the proportion of uncertainty in the data that is explained by the model. If the full model is a perfect indicator, then $L = 1$, $\ln L = 0$, and $R^2 = 1$. If the reduced model yields the same likelihood as the full model, then $\ln L = \ln L_0^*$ and $R^2 = 0$. In this case the explanatory variables contribute nothing to the likelihood.

Hosmer-Lemeshow Goodness of Fit Test

Since in many applications a large majority of the observations $(y_i, x_{i1}, x_{i2}, \ldots, x_{ic})$, $i = 1, 2, \ldots, n$, are unique in the sense that in general no two observations yield identical values on all variables, the fitted model cannot be evaluated using the χ^2 goodness of fit tests introduced in Chapter 6. To obtain a goodness of fit test, Hosmer and Lemeshow (1980) suggest that the range of p $[0, 1]$ be divided into s mutually exclusive categories and then a comparison of the observed and predicted frequencies be carried out using a χ^2 statistic. The categories can be determined by ranking the n $\hat{p}$-values and then dividing them into s equal groups or by dividing the range of p into s equal intervals.

We denote the actual frequency in group j by o_j, the predicted frequency by n_j, and the average value of $\hat{p}$ in group j by $\bar{p}_j$. The statistic $\sum_{j=1}^{s} (o_j - n_j \bar{p}_j)^2 / n_j \bar{p}_j (1 - \bar{p}_j)$ is approximately χ^2 with $(s - 2)$ degrees of freedom if the fitted logistic regression model is correct.

Example

For the fitted bivariate relationships in Table 8.26 the R^2 measure can be determined from the loglikelihood information given in the table. For the relationship between THISYR and HUBINC, R^2 is given by $1 - (-56.982)/(-59.295) = 0.40$. For the CHILD1 variable R^2 is given by $1 - (-55.006)/(-59.295) = 0.072$. For a logistic regression model with only a single explanatory variable, if this variable is a dummy variable the fitted model is equivalent to the saturated loglinear model for a (2×2) contingency table. The fit is therefore perfect (see Chapter 6). In the case of the non-categorical variables HUBINC, AGE and EDUC the Hosmer–Lemeshow goodness of fit p-values are shown in Table 8.26.

Covariance Matrix for Estimated Coefficients

The variance-covariance matrix for the estimated regression coefficients $\widetilde{\beta}$ is obtained from the expression $[\mathbf{X}' \widetilde{\mathbf{W}} \mathbf{X}]^{-1}$, where $\widetilde{\mathbf{W}}$ is the $(n \times n)$ diagonal matrix with diagonal elements $\tilde{p}_i(1 - \tilde{p}_i)$ determined from the estimates of p_i, $i = 1, 2, \ldots, n$, and $\mathbf{X}$ is the $(n \times c)$ matrix of observations with n rows given by $\mathbf{x}'_i$, $i = 1, 2, \ldots, n$. The (j, k)th element of $[\mathbf{X}' \widetilde{\mathbf{W}} \mathbf{X}]$ is therefore given by

$$\sum_{i=1}^{n} \tilde{p}_i (1 - \tilde{p}_i) x_{ij} x_{ik} = \sum_{i=1}^{n} [x_{ij} x_{ik} e^{\mathbf{x}'_i \widetilde{\beta}} / (1 + e^{\mathbf{x}'_i \widetilde{\beta}})^2].$$

The square roots of the diagonal elements of this covariance matrix can be used to provide approximate standard errors for the regression coefficients.

Example – Logistic Regression With Multiple Explanatory Variables

To determine how the explanatory variables together predict p, a logistic regression model was fitted using all six explanatory variables. The fitted models for THISYR and LASTYR are given by

THISYR

$$\ln[p/(1-p)] = \underset{(0.0472)}{-\ 6.0624} - \underset{(0.0040)}{0.1079}\ \text{HUBINC} + \underset{(0.0148)}{0.4777}\ \text{EDUC}$$
$$+\ \underset{(0.0765)}{0.0773}\ \text{AGE} + \underset{(0.0708)}{1.5451}\ \text{BLACK}$$
$$-\ \underset{(0.0003)}{4.5179}\ \text{CHILD1} - \underset{(0.0621)}{1.1238}\ \text{CHILD2},$$

LASTYR

$$\ln[p/(1-p)] = \underset{(0.0076)}{+\ 6.3641} - \underset{(0.0257)}{0.0799}\ \text{HUBINC} - \underset{(0.4638)}{0.0911}\ \text{EDUC}$$
$$-\ \underset{(0.0423)}{0.0870}\ \text{AGE} - \underset{(0.8937)}{0.0879}\ \text{BLACK}$$
$$-\ \underset{(0.0008)}{3.6948}\ \text{CHILD1} - \underset{(0.0057)}{1.6928}\ \text{CHILD2}.$$

The p-values for the coefficients are shown in brackets. The p-values were obtained using the differences of loglikelihood ratios.

The fitted logistic regression model for THISYR indicates that at the margin the probability that a woman will choose to work increases with decreases in HUBINC, but decreases with decreases in AGE and EDUC. For the dummy variables, a woman of the black race is more likely to work, whereas if children are present the woman is less likely to work. For the variable LASTYR the variables HUBINC, CHILD1 and CHILD2 have the same impact as in the case of THISYR; the remaining variables are insignificant.

The log likelihoods for the two models are -44.044 and -53.314 for THISYR and LASTYR respectively. Excluding all six variables yields log likelihoods of -59.295 and -64.745. The likelihood ratio χ^2 statistics for the significance of all six variables are given by $-2[-59.295 - (-44.044)] = 30.502$ and $-2[-64.745 - (-53.314)] = 22.862$ which have p-values of 0.000 and 0.001 when compared to a χ^2 distribution with 6 d.f. The pseudo R^2 values are given by $[1 - 44.044/59.295] = 0.26$ and $[1 - 53.314/64.745] = 0.18$, respectively. The Hosmer–Lemeshow χ^2 p-values are 0.55 and 0.26 for the variables THISYR and LASTYR respectively.

The Role of the Intercept and Categorical Variables

Before the addition of any explanatory variables, the logistic regression model has the form $\ln[p/(1-p)] = \beta_0$. The maximum likelihood estimator of the probability p is the sample proportion $\hat{p} = m/n$, where m is the number of observations corresponding to $Y = 1$ and $(n - m)$ is the number of observations in the category corresponding to $Y = 0$. The likelihood function in this case is given by $L = m^m(n-m)^{n-m}/n^n$ and the logarithm by $\ln L = m\ln(m) + (n - m)\ln(n - m) - n\ln(n)$.

If $p = 0.5$, then $\beta_0 = 0$ and the likelihood function is given by $(0.5)^m \times (0.5)^{n-m}$. The logarithm of the likelihood is given by $\ln L_0 = n \ln(0.5)$. The likelihood ratio χ^2 statistic for testing $H_0: \beta_0 = 0$ is therefore given by $2[\ln L - \ln L_0]$ which is given by $+2m \ln[m/n] + 2(n-m) \ln[(n-m)/n] - 2n \ln(0.5)$. If $H_0: \beta_0 = 0$ is true, this statistic has a χ^2 distribution with 1 degree of freedom in large samples. This test is equivalent to a test of equal proportions.

If a categorical variable Z is included as the only explanatory variable, then the logistic regression model is equivalent to the loglinear model for a 2×2 contingency table discussed in Chapter 6. If the variable Z is a dummy variable and hence takes on the values 0 or 1, the sample can be summarized in a two-dimensional table as follows.

Z

		0	1	Totals
Y	0	n_{11}	n_{12}	$n_{1.}$
	1	n_{21}	n_{22}	$n_{2.}$
Totals		$n_{.1}$	$n_{.2}$	n

The logistic regression model in this case has the form

$$\ln[p/(1-p)] = \beta_0 + \beta_1 Z \quad = \beta_0 + \beta_1 \quad \text{if } Z = 1$$
$$= \beta_0 \qquad \text{if } Z = 0.$$

Recalling that the maximum likelihood estimators are given by $n_{21}/n_{.1}$ for p when $Z = 0$ and $n_{22}/n_{.2}$ for p when $Z = 1$, we have that

$$\ln[n_{21}/n_{11}] = \hat{\beta}_0 + \hat{\beta}_1 \quad \text{and} \quad \ln[n_{22}/n_{12}] = \hat{\beta}_0.$$

The likelihood function for this fitted model has the value

$$\left(\frac{n_{21}}{n_{.1}}\right)^{n_{21}} \left(\frac{n_{11}}{n_{.1}}\right)^{n_{11}} \left(\frac{n_{22}}{n_{.2}}\right)^{n_{22}} \left(\frac{n_{12}}{n_{.2}}\right)^{n_{12}},$$

with the logarithm given by

$$L_1 = n_{21} \ln(n_{21}) + n_{11} \ln(n_{11}) + n_{22} \ln(n_{22})$$
$$+ n_{12} \ln(n_{12}) - n_{.1} \ln(n_{.1}) - n_{.2} \ln(n_{.2}).$$

The likelihood ratio test of $H_0: \beta_1 = 0$ involves the difference between $\ln L_0$ and $\ln L_1$, where $\ln L_0$ refers to the likelihood function when only the intercept is fitted. The likelihood ratio statistic is given by

$$2[\ln L_1 - \ln L_0] = 2\left[n_{11} \ln\left(\frac{n_{11}n}{n_{.1}n_{1.}}\right) + n_{12} \ln\left(\frac{n_{12}n}{n_{.2}n_{.1}}\right)\right.$$
$$\left. + n_{21} \ln\left(\frac{n_{21}n}{n_{.1}n_{2.}}\right) + n_{22} \ln\left(\frac{n_{22}n}{n_{.2}n_{2.}}\right)\right]$$

and has a χ^2 distribution with one degree of freedom if H_0 is true. This statistic is equivalent to the likelihood ratio χ^2 used for testing independence in a two-by-two contingency table discussed in Chapter 6.

Testing for Zero Intercept

A test for zero intercept term $\beta_0 = 0$ in the logistic regression model is not usually of practical value. If the explanatory variable Z is a classification variable, the interpretation of this test depends on the type of coding used for Z. If $\beta_0 = 0$ and Z is coded 0 or 1, then the logistic regression model is given by

$$\ln[p/(1-p)] = \beta_1 Z \quad = \beta_1 \quad \text{if } Z = 1$$
$$= 0 \quad \text{if } Z = 0.$$

The condition $\beta_0 = 0$ therefore implies that $p = 0.5$ when $Z = 0$, whereas p is free to vary when $Z = 1$. If, however, Z is coded using effect coding, -1 and $+1$, then the model is given by

$$\ln[p/(1-p)] = \beta_1 Z \quad = \beta_1 \quad \text{if } Z = 1$$
$$= -\beta_1 \quad \text{if } Z = -1.$$

In this case the restriction $\beta_0 = 0$ implies that the value of p when $Z = 1$ is equal to the value of $(1-p)$ when $Z = -1$. Although these two forms of the hypothesis $H_0: \beta_0 = 0$ may have particular specialized uses this test is not in general useful.

If an explanatory variable, say X, is interval scaled, the restriction $\beta_0 = 0$ implies that $\ln[p/(1-p)] = \beta_1 X$. In this case the value of p is very sensitive to X and at $X = 0$, $p = 0.5$. As in the case of the simple linear regression model, the intercept term simply locates the value of the function when $X = 0$. The intercept is only zero if $p = 0.5$ at $X = 0$, which is similar to the zero intercept linear regression model that has the value 0 when $X = 0$. Omitting the intercept in a logistic regression model is therefore only done when the particular situation suggests that $p = 0.5$ when $X = 0$ (see Example below).

If the logistic regression model includes a classification variable Z, as well as an interval scaled variable X, then the condition $\beta_0 = 0$ can have a variety of different implications depending on the coding used for the categorical variable. When the interval scaled variable X has the value 0, the logistic regression model has the form $\ln[p/(1-p)] = \beta_0 + \beta_1 Z + \beta_2 X = \beta_0 + \beta_1 Z$. The condition $\beta_0 = 0$ has an impact therefore on the point where the function crosses the p axis. For instance, suppose Z is coded 0 or 1; then $p = 0.5$ for $Z = 0$ and $X = 0$, whereas for $Z = 1$, p may have any value at $X = 0$. Or suppose Z is coded -1 or $+1$; then at $X = 0$ the value of p at $Z = +1$ is equal to the value of $(1-p)$ at $Z = -1$. Once again such restrictions are only of value in specialized situations and hence in general the intercept should be retained.

Example

A comparison of the intercepts in the estimated models for THISYR and LASTYR reveals a startling difference in that they are almost equal in magnitude but opposite in sign. An explanation for this difference can be explained by substituting the mean values for HUBINC, AGE and EDUC into the two models and evaluating the terms. The sums of the three terms are 7.3815 and −4.7301 for THISYR and LASTYR respectively. Adding these values to the intercepts yields new intercepts of 1.3191 and 1.6341 respectively. Therefore we can conclude that for points near the average values of the three interval variables the two functions are similar in value. This example helps to illustrate that the intercept itself is simply an adjustment to the scale. In this case the intercept represents a point which is a large distance away from the observed data. By comparing the two functions in the centre of the data we can obtain a more meaningful interpretation. As we shall see below in the discussion of DFBETA the coefficient of EDUC in the equation for THISYR is unusually large due to one particular observation. The intercept provides the necessary scale adjustment.

Dummy Variables as Explanatory Variables – A Caution

In logistic regression analysis one or more of the explanatory variables can be dummy variables. Usually such dummy variables have several "1" observations and the remaining observations are zero. In order for the coefficient of the dummy variable to be defined in logistic regression, at least two of the "1" observations must correspond to opposites. In other words, when the dummy variable has the value "1", at least one of the Y values must be 0 and at least one of the Y values must be 1. A special case of this is the observation specific dummy, where all but one of the observations yield a dummy variable value of 0 and only one observation yields a dummy variable value of 1. In the case of several observations with the dummy variable 1, it may happen that all such individuals have the same value of Y. For example, if 1 indicates a medical doctor and $Y = 1$ for employed, it may be that all individuals in the sample who are medical doctors have the value of $Y = 1$. In such circumstances the parameter estimate obtained from a computer program for the coefficient of the dummy will tend to be large. Removing this dummy from the model will not affect the remainder of the parameter estimates.

The Fitted Model and Classification

The fitted logistic regression model can be used to obtain the value of $\hat{p}_i$ for each observation by determining the value of $\ln[\hat{p}_i/(1 - \hat{p}_i)] = \tilde{\beta}_0 + \sum_{j=1}^{r} \tilde{\beta}_j x_{ij}$ and then solving for $\hat{p}_i$. The value of $\hat{p}_i$ is given by $\hat{p}_i = e^{\mathbf{x}_i'\tilde{\boldsymbol{\beta}}}/(1 + e^{\mathbf{x}_i'\tilde{\boldsymbol{\beta}}})$. Assume that the observation is placed in the category $Y = 0$ if $\hat{p}_i < 0.50$, and otherwise the observation is placed in the category $Y = 1$. A prediction success matrix or confusion matrix in this case can be constructed as shown below.

<table>
<tr><td></td><td colspan="2">True Category</td></tr>
<tr><td></td><td>$Y = 0$</td><td>$Y = 1$</td></tr>
<tr><td>$\hat{p}_i < 0.50$</td><td>n_{00}</td><td>n_{01}</td></tr>
<tr><td>$\hat{p} \geq 0.50$</td><td>n_{10}</td><td>n_{11}</td></tr>
</table>

$$n = (n_{00} + n_{01} + n_{10} + n_{11})$$

This table shows the distribution of the predictions for each of the two categories. The proportion of correctly classified observations is given by $(n_{00} + n_{11})/n$. The logistic regression model therefore provides a discriminant function which can be used to classify unknowns.

A Jackknife Approach

In Section 8.2.2 a jackknife procedure was outlined for obtaining a nearly unbiased estimator of the true prediction probabilities. This procedure omits one observation at a time and uses the remaining $(n-1)$ observations to fit the model and classify the one omitted observation. The process is repeated for all n observations.

Example

It is of interest to examine the abilities of the two fitted models to predict the values of THISYR and LASTYR. If no explanatory variables are included in the model, the probabilities based on the observations are $p[\text{THISYR} = 1] = 0.72$ and $p[\text{LASTYR} = 1] = 0.65$. Prediction success tables based on these probabilities are therefore given by

<table>
<tr><td></td><td colspan="2">Predicted
THISYR</td><td></td><td></td><td colspan="2">Predicted
LASTYR</td><td></td></tr>
<tr><td></td><td>0</td><td>1</td><td></td><td></td><td>0</td><td>1</td><td></td></tr>
<tr><td>THISYR</td><td></td><td></td><td></td><td>LASTYR</td><td></td><td></td><td></td></tr>
<tr><td>0</td><td>8</td><td>20</td><td>28</td><td>0</td><td>12</td><td>23</td><td>35</td></tr>
<tr><td>1</td><td>20</td><td>52</td><td>72</td><td>1</td><td>23</td><td>42</td><td>65</td></tr>
<tr><td></td><td>28</td><td>72</td><td></td><td></td><td>35</td><td>65</td><td></td></tr>
</table>

We would therefore expect to predict correctly 60% of the values of THISYR and 54% of the values of LASTYR. An equal priors model would only be expected to predict 50% correctly.

To determine the predictions based on the fitted logistic models, values of $\hat{p}$ were determined for both models for all 100 observations. If $\hat{p} < 0.50$ for a particular individual, then that individual was placed in the "not work" category; otherwise the individual was placed in the "work" category. The prediction success tables are shown below.

	Predicted THISYR		
	0	1	
THISYR			
0	13	15	28
1	5	67	72
	18	82	

	Predicted LASTYR		
	0	1	
LASTYR			
0	14	21	35
1	8	57	65
	22	78	

For the variable THISYR, the use of the fitted logistic regression model results in a correct classification for 80% of the cases, whereas for the variable LASTYR, use of the fitted model results in a correct classification for 71%. The increases in percentage correctly classified as a result of the fitted logistic regression model are 20% and 17% respectively. In other words, in the case of THISYR an additional 20 of the 100 cases were correctly classified, whereas for LASTYR an additional 17 of the 100 cases were correctly classified.

Using the jackknife approach for the variable THISYR 11 of the 28 were correctly coded "0" and 65 of the 72 were correctly coded "1". For the variable LASTYR 14 of the 35 were correctly coded "0" and 56 of the 65 were correctly coded "1". These estimates of classification success are only slightly smaller than those obtained above using all n observations. In a later section of this chapter a multivariate qualitative response model will be used to predict both the variables LASTYR and THISYR simultaneously.

Stepwise Logistic Regression

As in the case of multiple linear regression and discriminant analysis, it is possible to use variable selection methods to choose a subset of explanatory variables in logistic regression. A forward, backward or stepwise approach can be used by calculating the chi-square statistic $-2[\ln L_0 - \ln L_1]$ at each step. For the variable LASTYR, a forward selection procedure is illustrated below.

Table 8.27 shows the results of successive steps of the forward procedure applied to the model for LASTYR. In comparison to the full model fitted for LASTYR above, at step 4 of the forward procedure, the variables that are significant at the 0.05 level in the full model are the ones included in this model. Some of the coefficients however are quite different. Excluding two variables seems to have an impact on the estimated coefficients.

Influence Diagnostics

The Chi Statistic
Given that the observed values of the dependent variable are denoted by y_i (0 or 1), and the fitted values are denoted by $\hat{p}_i = e^{x_i'\tilde{\beta}}/(1 + e^{x_i'\tilde{\beta}})$, the

TABLE 8.27. Forward Selection Process for LASTYR Model

Variable	Step 0	Step 1	Step 2	Step 3	Step 4
INTER-CEPT	+0.6190 IN (0.0025)	+1.4264 IN (0.0001)	+0.3672 IN (0.5861)	+0.0159 IN (0.9789)	+2.3766 IN (0.0913)
AGE	– OUT (0.6123)	– OUT (0.9718)	– OUT (0.4156)	– OUT (0.0479)	-0.0823 IN (0.0479)
EDUC	– OUT (0.1209)	– OUT (0.2591)	– OUT (0.7011)	– OUT (0.5953)	– OUT (0.4713)
HUBINC	– OUT (0.0096)	-0.0855 IN (0.0096)	-0.0965 IN (0.0046)	-0.0974 IN (0.0043)	-0.0861 IN (0.0135)
BLACK	– OUT (0.3693)	– OUT (0.7960)	– OUT (0.8548)	– OUT (0.7960)	– OUT (1.0000)
CHILD1	– OUT (0.0117)	– OUT (0.0056)	-1.3254 IN (0.0056)	-1.4785 IN (0.0022)	-1.9093 IN (0.0003)
CHILD2	– OUT (0.1205)	– OUT (0.1256)	– OUT (0.0449)	-0.5098 IN (0.0449)	-0.8163 IN (0.0070)

logistic regression model residuals are given by $e_i = (y_i - \hat{p}_i)$ $i = 1, 2, \ldots, n$. Recalling that the variance of $\hat{p}_i$ is given by $p_i(1 - p_i)$, a goodness of fit statistic is given by $\sum_{i=1}^{n} e_i^2 / [\hat{p}_i(1 - \hat{p}_i)]$. If the model is adequate, then this statistic has a χ^2 distribution with $(n - c)$ degrees of freedom. Each component of this statistic $e_i / \sqrt{\hat{p}_i(1 - \hat{p}_i)}$ can be used as an indicator of lack of fit for each observation. This statistic is called the *chi statistic*.

The Deviance Statistic

An alternative residual measure is based on the likelihood ratio goodness of fit statistic defined in Chapter 6. Comparing the likelihood of the fitted model to the likelihood based on fitting each point exactly, the value of $2[\ln L - \ln L_0]$ is given by

$$2 \sum_{\substack{i=1 \\ (y_i = 1)}}^{m} y_i \ln[y_i/\hat{p}_i]] + 2 \sum_{\substack{i=1 \\ (y_i = 0)}}^{n-m} [(1 - y_i) \ln[(1 - y_i)/(1 - \hat{p}_i)]]$$

where m is the frequency for the event $(y_i = 1)$. Each component of this goodness of fit statistic can be used to indicate a lack of fit for each observation. The resulting *deviance residual* is therefore given by $\{2[y_i \ln(y_i/\hat{p}_i)]\}^{1/2}$ or $\{2[(1 - y_i) \ln[(1 - y_i)/(1 - \hat{p}_i)]]\}^{1/2}$ depending on whether $y_i = 1$ or 0

respectively. The *deviance statistic* is usually preferred as a measure of goodness of fit because it tends to have greater power against alternative models.

Leverage

Since the covariance matrix for $\widetilde{\beta}$ in the logistic regression model is given by $(\mathbf{X}'\mathbf{W}\mathbf{X})^{-1}$, a measure analogous to the 'hat' matrix used to measure *leverage* in multiple regression is given by

$$\mathbf{H}^* = \widetilde{\mathbf{W}}^{1/2}\mathbf{X}(\mathbf{X}'\widetilde{\mathbf{W}}\mathbf{X})^{-1}\mathbf{X}'\widetilde{\mathbf{W}}^{1/2}.$$

The diagonal elements of $\mathbf{H}^*$, h_{ii}^*, $i = 1, 2, \ldots, n$, lie in the interval $[0, 1]$, and the average of the diagonal elements is c/n. These diagonal elements are often useful for detecting influential observations. Unlike the linear model, however, the "hat" values depend on the $\hat{p}_i$ as well as on $\mathbf{x}$. In the case of one observation at each $\mathbf{x}$, h_{ii}^* is less than or equal to 1, otherwise $h_{ii}^* \leq 1/m_i$, where m_i is the frequency of $\mathbf{x}_i$. The matrix $\widetilde{\mathbf{W}}$ denotes the estimator of $\mathbf{W}$ based on $\widetilde{\beta}$, and $\mathbf{W}$ is the diagonal matrix with diagonal elements $p_i(1 - p_i)$, $i = 1, 2, \ldots, n$.

Influence

In the linear regression model, the *Cook's D* measure of influence is related to a confidence ellipsoid for β based on $\hat{\beta}_{(i)}$ and the F-statistic, where $\hat{\beta}_{(i)}$ denotes the estimator of β obtained after deleting observation i. In the case of logistic regression the statistic is based on χ^2 and a comparison of likelihood ratios. The *influence statistic* is given by

$$\left(\frac{1}{c}\right)(\widetilde{\beta} - \widetilde{\beta}_{(i)})'\mathbf{X}'\widetilde{\mathbf{W}}\mathbf{X}(\widetilde{\beta} - \widetilde{\beta}_{(i)}) = \frac{e_i^2}{c(1 - h_{ii}^2)}\mathbf{x}_i'(\mathbf{X}'\widetilde{\mathbf{W}}\mathbf{X})^{-1}\mathbf{x}_i$$

$$= \frac{h_{ii}e_i^2}{c(1 - h_{ii})^2\hat{p}_i(1 - \hat{p}_i)}.$$

As in the case of the linear regression model this influence statistic is a function of the residual measure and the leverage measure.

The DFBETA Measure

The DFBETA measure is designed to reflect the influence that individual observations have on particular regression coefficients. This measure was introduced in Volume I in the context of linear regression models. For the coefficient β_j the DFBETA value for observation i is given by $(\widetilde{\beta}_j - \widetilde{\beta}_{j(i)})/s(\widetilde{\beta}_j)$, where $\widetilde{\beta}_j$ and $\widetilde{\beta}_{j(i)}$ denote elements of $\widetilde{\beta}$ and $\widetilde{\beta}_{(i)}$ and $s(\widetilde{\beta}_j)$ denotes the standard error of $\widetilde{\beta}_j$. A value of DFBETA which is relatively large in absolute value would indicate that observation i influences the estimate of β_j. Observations corresponding to DFBETA values which exceed 0.5 in absolute value should be examined to determine the reason for the unusual level of influence.

Example

The residual statistics and influence diagnostic statistics for the logistic regression model for THISYR are summarized in Table 8.28. The residual statistics CHI and DEVIANCE show three relatively large values (3.40, 2.25), (3.42, 2.26) and (3.67, 2.31) corresponding to observations 1, 38 and 69 respectively. From Table 8.25, we can see that observation 1 corresponds to a married woman who did not choose to work, who had no children, and whose husband's income was relatively low. Observations 38 and 69 also have these same characteristics. According to the logistic regression model such married women tend to choose to work. A perusal of the INFLUENCE measure in Table 8.28 reveals two values that exceed 0.40. Corresponding to observation 15, this measure is 1.63, while for observation 19 this measure is 0.71. For observation 19 the residual statistics are moderately large (1.77, 1.68) and the LEVERAGE value is also moderately large (0.16). Observation 15 has the largest LEVERAGE value (0.40) while the residual is also moderately large (−1.20, −1.34). In addition to observation 15, there are two observations having LEVERAGE values that exceed 0.19. These are 0.36 and 0.22 corresponding to observations 59 and 90 respectively. Neither of these observations, however, has a large enough residual to yield a very large measure of INFLUENCE.

The DFBETA values for observations 15 and 19 (not shown) in Table 8.28 reveal values of 0.59 for CHILD1 on observation 15 and −0.67 and 0.48 for BLACK and EDUC respectively on observation 19. An examination of the data in Table 8.25 indicates that observation 15 is the only instance in which both CHILD1 = 1 and THISYR = 1. The positive value of DFBETA suggests that omitting observation 15 results in a larger negative coefficient for CHILD1. For observation 19 THISYR = 0, BLACK = 1 and EDUC = 12. The DFBETA values suggest that omitting observation 19 would result in a larger positive coefficient for BLACK and smaller positive or negative coefficient for EDUC. Thus omitting observation 19 would result in a model in which a black woman is more likely to work outside the home. In the case of the EDUC coefficient its significance may be reduced as a result of omitting observation 19.

8.3.2   THE PROBIT MODEL

Given a random sample of n observations on the point binomial random variable y_i ($y_i = 0$ or 1), the joint density is given by $\Pi_{i=1}^n p_i^{y_i} (1 - p_i)^{(1-y_i)}$, where $p_i = P[y_i = 1]$ and $(1-p_i) = P[y_i = 0]$. In 8.6.1 the logistic regression model was derived from the assumption that $p_i = e^{\mathbf{x}_i'\boldsymbol{\beta}}/[1 + e^{\mathbf{x}_i'\boldsymbol{\beta}}]$, where

TABLE 8.28. Residuals and Influence Diagnostics

OBS	THISYR	CHI	DEVIANCE	LEVERAGE	INFLUENCE
1	0	<u>3.40</u>	<u>2.25</u>	0.02	0.28
2	0	0.77	0.96	0.06	0.04
3	1	-0.46	-0.62	0.11	0.03
4	1	-0.31	-0.43	0.03	0.00
5	1	-0.13	-0.18	0.04	0.00
6	1	-0.76	-0.95	0.15	0.12
7	1	-1.07	-1.23	0.08	0.10
8	0	0.80	1.00	0.06	0.04
9	0	1.74	1.67	0.06	0.21
10	1	-0.68	-0.87	0.05	0.03
11	1	-0.56	-0.74	0.02	0.01
12	1	-0.22	-0.31	0.05	0.00
13	1	-0.56	-0.74	0.03	0.01
14	1	-1.11	-1.27	0.07	0.10
15	1	-1.20	-1.34	<u>0.40</u>	<u>1.63</u>
16	1	-0.15	-0.22	0.02	0.00
17	1	-0.48	-0.64	0.15	0.05
18	1	-0.38	-0.52	0.12	0.02
19	0	1.77	1.68	0.16	<u>0.71</u>
20	0	1.79	1.69	0.08	0.29
21	1	-0.36	-0.49	0.02	0.00
22	1	-0.29	-0.40	0.03	0.00
23	1	-0.37	-0.51	0.03	0.00
24	1	-0.24	-0.33	0.03	0.00
25	0	2.25	1.90	0.03	0.14
26	1	-0.23	-0.32	0.05	0.00
27	0	1.17	1.32	0.08	0.12
28	1	-0.67	-0.87	0.12	0.07
29	0	0.40	0.54	0.17	0.04
30	0	0.16	0.22	0.06	0.00
31	1	-0.59	-0.78	0.03	0.01
32	1	-0.39	-0.53	0.02	0.00
33	1	-0.39	-0.53	0.03	0.01
34	0	0.35	0.48	0.09	0.01
35	1	-0.38	-0.52	0.12	0.02
36	1	-0.48	-0.64	0.02	0.01
37	1	-1.24	-1.37	0.13	0.25
38	0	<u>3.42</u>	<u>2.26</u>	0.02	0.30
39	1	-0.08	-0.11	0.03	0.00
40	1	-0.05	-0.07	0.01	0.00
41	1	-0.69	-0.88	0.09	0.05
42	0	2.07	1.83	0.02	0.10
43	1	-0.56	-0.74	0.16	0.07
44	0	0.83	1.02	0.06	0.05
45	1	-0.90	-1.09	0.28	0.21
46	1	-1.05	-1.22	0.07	0.08
47	1	-0.52	-0.69	0.03	0.01
48	1	-0.17	-0.23	0.03	0.00
49	1	-0.43	-0.58	0.13	0.03
50	0	0.42	0.56	0.18	0.05

TABLE 8.28. Residuals and Influence Diagnostics (continued)

OBS	THISYR	CHI	DEVIANCE	LEVERAGE	INFLUENCE
51	0	0.38	0.52	0.09	0.01
52	1	-0.44	-0.59	0.03	0.01
53	1	-0.09	-0.13	0.02	0.00
54	1	-0.39	-0.53	0.03	0.00
55	1	-0.52	-0.69	0.03	0.01
56	1	-0.74	-0.93	0.06	0.04
57	1	-0.49	-0.65	0.04	0.01
58	1	-0.14	-0.19	0.02	0.00
59	0	0.59	0.77	0.36	0.30
50	1	-0.42	-0.58	0.06	0.01
61	1	-0.19	-0.26	0.04	0.00
62	1	-0.42	-0.56	0.02	0.00
63	1	-0.57	-0.75	0.04	0.01
64	0	0.99	1.17	0.19	0.28
65	1	-0.12	-0.17	0.02	0.00
66	1	-0.35	-0.48	0.08	0.01
67	1	-0.17	-0.23	0.03	0.00
68	0	0.98	1.16	0.14	0.18
69	0	3.67	2.31	0.03	0.38
70	1	-0.51	-0.69	0.08	0.02
71	1	-0.81	-1.00	0.07	0.06
72	1	-0.76	-0.96	0.09	0.07
73	0	1.62	1.61	0.03	0.08
74	0	2.05	1.81	0.02	0.11
75	1	-0.84	-1.03	0.13	0.12
76	1	-0.77	-0.96	0.05	0.04
77	1	-0.92	-1.11	0.11	0.12
78	1	-0.15	-0.21	0.02	0.00
79	1	-0.30	-0.41	0.07	0.01
80	1	-0.34	-0.47	0.03	0.00
81	1	-0.48	-0.64	0.06	0.01
82	1	-0.39	-0.53	0.06	0.01
83	1	-0.39	-0.54	0.02	0.00
84	1	-0.38	-0.52	0.03	0.00
85	1	-0.73	-0.92	0.09	0.06
86	1	-0.39	-0.54	0.02	0.00
87	1	-0.32	-0.45	0.03	0.00
88	1	-0.45	-0.61	0.06	0.01
89	1	-0.26	-0.36	0.06	0.00
90	0	0.26	0.36	0.22	0.02
91	1	-0.94	-1.12	0.06	0.05
92	0	0.16	0.23	0.06	0.00
93	1	-0.32	-0.44	0.02	0.00
94	0	1.23	1.35	0.06	0.10
95	1	-0.45	-0.61	0.02	0.01
96	1	-0.34	-0.47	0.03	0.00
97	0	1.33	1.43	0.06	0.12
98	0	1.18	1.32	0.09	0.15
99	0	1.05	1.22	0.06	0.07
100	1	-0.38	-0.52	0.12	0.02

$\mathbf{x}_i\ [(c+1) \times 1]$ denotes a vector of observations on a set of c explanatory variables. Rather than the logistic density assumption, a similar shaped function for p_i is provided by the *probit model* given by

$$p_i = F(\mathbf{x}'_i\beta) = \int_{-\infty}^{\mathbf{x}'_i\beta} f(z)dz,$$

where $f(z)$ denotes the standard normal density. Denoting the cumulative standard normal density by $F(\mathbf{x}'_i\beta)$ and the standard normal density by $f(\mathbf{x}'_i\beta)$, the derivative of the log of the likelihood function

$$\ln L = \sum_{i=1}^{n}\{y_i \ln[F(\mathbf{x}'_i\beta)] + (1-y_i)\ln[1 - F(\mathbf{x}'_i\beta)]\},$$

is given by

$$\frac{\partial \ln L}{\partial \beta} = \sum_{i=1}^{n}\left[y_i \frac{f(\mathbf{x}'_i\beta)}{F(\mathbf{x}'_i\beta)} - (1-y_i)\frac{f(\mathbf{x}'_i\beta)}{1 - F(\mathbf{x}'_i\beta)}\right]\mathbf{x}_i.$$

Setting this derivative equal to zero yields the equations for the maximum likelihood estimator. As in the case of the logistic regression model the estimator is usually obtained by the Newton-Raphson interative procedure. This procedure necessarily converges to the maximum regardless of the initial starting estimate of β.

Given the function $\mathbf{x}'\beta$, the probability that the standard normal random variable is less than or equal to $\mathbf{x}'\beta$ is given by $F(\mathbf{x}'\beta)$. Since $p = F(\mathbf{x}'\beta)$, then for a given $\mathbf{x}$ the probability that the individual belongs to the $Y = 1$ category is given by $F(\mathbf{x}'\beta)$.

Computer Software

The calculations for the probit analysis example were carried out using the SPSSX program PROBIT and the SAS program PROBIT.

Example

For the labor force participation example discussed in Section 8.3.1 the probit models for THISYR and LASTYR were estimated. The estimated linear model obtained for THISYR is given by

$$\hat{p}\left[\begin{array}{c}\text{THISYR}\\=1\end{array}\right] = \underset{(.077)}{2.616} - \underset{(.026)}{0.062}\text{ HUBINC} + \underset{(.284)}{0.025}\text{ AGE} + \underset{(.020)}{0.242}\text{ EDUC}$$
$$-\underset{(.841)}{0.085}\text{ BLACK} - \underset{(.037)}{1.370}\text{ CHILD1} - \underset{(.001)}{1.033}\text{ CHILD2}$$

where the p-values appear in brackets under the coefficients. The p-values were obtained by dividing the coefficients by their standard errors, and by using the two tails of the standard normal distribution. We can conclude from the model that the probability that the wife will choose to work THISYR increases with AGE and EDUC and decreases with HUBINC. Also, if there are pre-school children, then the probability that she chooses to work THISYR also decreases.

For the dependent variable LASTYR, the estimated probit model is given by

$$\hat{p}\left[\begin{array}{c} \text{LASTYR} \\ = 1 \end{array}\right] = \underset{(.019)}{3.092} - \underset{(.031)}{0.062} \text{ HUBINC} + \underset{(.53)}{0.014} \text{ AGE} + \underset{(.029)}{0.209} \text{ EDUC}$$
$$+ \underset{(.262)}{0.289} \text{ BLACK} - \underset{(.073)}{0.678} \text{ CHILD1} - \underset{(.029)}{0.556} \text{ CHILD2}$$

where the p-values appear brackets under the coefficients.

The prediction success tables for the two models are shown below.

	Predicted THISYR 0	1	
THISYR			
0	20	8	28
1	12	60	72
	32	68	

	Predicted LASTYR 0	1	
LASTYR			
0	11	17	28
1	15	57	72
	26	74	

For the variable THISYR, 80% were correctly classified, whereas for the variable LASTYR, 68% were correctly classified. For the logistic regression model fitted in Section 8.3.1 these percentages were 80% and 71% respectively.

8.3.3 LOGISTIC REGRESSION AND PROBIT ANALYSIS: A SECOND EXAMPLE

A second example for logistic regression and probit analysis is provided by the yield curve data analyzed using discriminant analysis in Section 8.2.2. Using all 80 cases, the logistic regression and probit analysis models are given below. The p-values for the coefficients are shown in brackets.

Logistic Regression

$$\ln[\hat{p}/(1-\hat{p})] = \underset{(0.368)}{56.040} - \underset{(0.004)}{9.052} \text{ BANKRT} + \underset{(0.038)}{29.822} \text{ UNEMPLR}$$
$$- \underset{(0.110)}{11.062} \text{ UNEMPLRM} + \underset{(0.015)}{39.319} \text{ CAPUTIL}$$
$$- \underset{(0.133)}{37.522} \text{ GDP} + \underset{(0.114)}{16.385} \text{ BUSINV}$$

Probit Analysis

$$\hat{p} = \underset{(0.038)}{-14.947} - \underset{(0.001)}{0.404} \text{ BANKRT} + \underset{(0.016)}{2.566} \text{ UNEMPLR}$$
$$\underset{(0.040)}{-1.565} \text{ UNEMPLRM} + \underset{(0.006)}{0.271} \text{ CAPUTIL}$$
$$\underset{(0.085)}{-0.00006} \text{ GDP} + \underset{(0.077)}{0.00039} \text{ BUSINV}$$

In comparison to the discriminant function obtained in Section 8.2.2, the signs of the coefficients in all three models are the same. All three models therefore provide the same characterization for the two different types of yield curve.

A comparison of the prediction successes of the two models is provided below. For the probit analysis the prediction success is 86.3%, whereas for the logistic regression it is 80.0%. In comparison, the discriminant analysis model showed success rates of 81.5% and 86.7% for the estimation and holdout sample respectively.

Probit Analysis			
Actual	Predicted		
	Upward	Not Upward	Totals
Upward	39	6	44
Not Upward	6	30	36
Totals	45	36	80

Logistic Regression			
Actual	Predicted		
	Upward	Not Upward	Totals
Upward	37	7	44
Not Upward	9	27	36
Totals	46	34	80

8.3.4 MULTIPLE OBSERVATIONS AND DESIGN VARIABLES

The Model and Maximum Likelihood Estimation

In some applications all of the explanatory variables are categorical (as in Chapter 6), and/or the X variables are preselected design variables that are used for multiple observations on Y. In this case the number of cells is fixed, say $i = 1, 2, \ldots, s$, m_i is the number of observations in the ith cell for which $Y = 1$, n_i is the total number of observations from the ith cell. The binomial density for the ith cell is given by

$$f(m_i \mid p_i, n_i) = \left(\begin{array}{c} n_i \\ m_i \end{array} \right) (p_i)^{m_i} (1 - p_i)^{(n_i - m_i)},$$

where p_i is the probability that $Y = 1$ in cell i. The likelihood function for the sample of $n = \sum_{i=1}^{s} n_i$ is therefore given by

$$L = \prod_{i=1}^{s} \left(\begin{array}{c} n_i \\ m_i \end{array} \right) (p_i)^{m_i} (1 - p_i)^{(n_i - m_i)}.$$

In this case we assume that s is fixed, and the asymptotic properties are derived by assuming that as the n_i become large, m_i/n_i converges to p_i.

For the logistic regression model, the probabilities p_i are assumed to satisfy the logistic model

$$p_i(\mathbf{x}_i, \boldsymbol{\beta}) = e^{\mathbf{x}_i' \boldsymbol{\beta}} / (1 + e^{\mathbf{x}_i' \boldsymbol{\beta}})$$

where $\boldsymbol{\beta}$ $(c \times 1)$ is a vector of unknown parameters and $\mathbf{x}_i$ is the $(c \times 1)$ vector of explanatory variables.

The likelihood function is given by

$$L = \prod_{i=1}^{s} \left(\begin{array}{c} n_i \\ m_i \end{array} \right) e^{m_i \mathbf{x}_i' \boldsymbol{\beta}} / (1 + e^{\mathbf{x}_i' \boldsymbol{\beta}})^{n_i}$$

and the log of the likelihood by

$$\ln L = \sum_{i=1}^{s} \left[\ln \left(\begin{array}{c} n_i \\ m_i \end{array} \right) + m_i \mathbf{x}_i' \boldsymbol{\beta} - n_i \ln(1 + e^{\mathbf{x}_i' \boldsymbol{\beta}}) \right].$$

Once again the Newton-Raphson method can be used to compute the maximum likelihood estimator $\widetilde{\boldsymbol{\beta}}$.

Given $\widetilde{\boldsymbol{\beta}}$, the maximum likelihood estimator of $\boldsymbol{\beta}$, the maximum likelihood estimator of p_i is given by

$$\hat{p}_i = e^{\mathbf{x}_i' \widetilde{\boldsymbol{\beta}}} / (1 + e^{\mathbf{x}_i' \widetilde{\boldsymbol{\beta}}}),$$

and this maximum likelihood estimator $\hat{p}_i$ of p_i satisfies

$$\sum_{i=1}^{s} m_i x_i = \sum_{i=1}^{s} n_i \hat{p}_i x_i.$$

Under the assumption of fixed s, the χ^2 goodness of fit test statistics can be used to evaluate the model. The Pearson and likelihood ratio statistics respectively are given by

$$\sum_{i=1}^{s} [m_i - n_i \hat{p}_i]^2 / n_i \hat{p}_i (1 - \hat{p}_i),$$

and

$$\sum_{i=1}^{s} \left[m_i \ln \left(\frac{m_i}{n_i \hat{p}_i} \right) + (n_i - m_i) \ln \left(\frac{n_i - m_i}{n_i(1 - \hat{p}_i)} \right) \right],$$

both of which are distributed as χ^2 with $(s - c)$ degrees of freedom if the model is correct.

The Chi and Deviance Statistics

In the case of multiple observations, the chi and deviance statistics introduced in 8.3.1 are given by $(m_i - n_i \hat{p}_i)/[n_i \hat{p}_i(1 - \hat{p}_i)]^{1/2}$ and $\left[2\{m_i \ln(m_i/n_i \hat{p}_i)\} + 2\{(n_i - m_i) \ln[(n_i - m_i)/n_i(1 - \hat{p}_i)]\}\right]^{1/2}$ respectively.

Weighted Least Squares or Minimum Logit Chi-Square Estimation

When the number of cells is fixed, the weighted least squares type estimator discussed in Chapter 6 can also be employed.

In practice the empirical logits are usually defined by $\ln[(m_i + 1/2)/(n_i - m_i + 1/2)] = \ell_i$, $i = 1, 2, \ldots, s$. The weighted least squares estimator minimizes

$$\sum_{i=1}^{s} n_i \hat{p}_i (1 - \hat{p}_i)[\ell_i - \mathbf{x}_i' \boldsymbol{\beta}]^2$$

and hence is often called the minimum logit chi-square estimator.

8.3.5   OTHER SOURCES OF INFORMATION

Summaries of logistic regression can be found in Andersen (1990), McCullagh and Nelder (1989), Fomby, Hill and Johnson (1984), Judge et al. (1985), Santner and Duffy (1989), Fienberg (1980), Fox (1984) and Hosmer and Lemeshow (1989). Probit analysis is discussed in Fomby et al. (1984) and Judge et al. (1985).

8.3.6   THE MULTINOMIAL LOGIT MODEL

In the previous section the logistic regression model and the probit model were introduced as techniques for modeling a dichotomous response variable. In this section we extend the logit model to the case of dependent variables with more than two categories. This model is commonly referred to as a *multinomial logit model*. The corporate bond classification example discussed in Section 8.2.2 will be used as an example for multinomial logit models.

Parameterization of the Model

If the categorical dependent variable has more than two possible categories, then the logistic regression model introduced in Section 8.3.1 could be employed by comparing each category to all of the remaining categories. If there are g categories, we could estimate the g models

$$\ln[p_j/(1 - p_j)] = \mathbf{x}'\boldsymbol{\beta}_j, \qquad j = 1, 2, \ldots, g,$$

separately using logistic regression. As in Section 8.3.1 we assume that $\mathbf{x}$ and $\boldsymbol{\beta}_j$ are $(c+1) \times 1$ vectors.

Since the probabilities p_j must satisfy $\sum_{j=1}^{g} p_j = 1$, however, it is necessary to estimate these models subject to this condition. A more appropriate method would be to incorporate this condition by re-parameterization.

For the binary dependent variable, the probabilities for the two categories $Y = 0$ and $Y = 1$ were given by

$$\begin{aligned}
p[Y = 1] &= p = e^{\mathbf{x}'\boldsymbol{\beta}}/(1 + e^{\mathbf{x}'\boldsymbol{\beta}}) \\
p[Y = 0] &= (1 - p) = 1/(1 + e^{\mathbf{x}'\boldsymbol{\beta}}) \text{ and} \\
p/(1 - p) &= e^{\mathbf{x}'\boldsymbol{\beta}}.
\end{aligned}$$

Similarly, for the case of g categories, we define dummy variables Y_j, $j = 1, 2, \ldots, g$, where $Y_j = 1$ if the observation is in category j, and $Y_j = 0$ otherwise. We define conditional probabilities $p_1, p_2, p_3, \ldots, p_g$, where

$$p[Y_j = 1] = p_j = e^{\mathbf{x}'\boldsymbol{\beta}_j}/\left(1 + \sum_{j=1}^{(g-1)} e^{\mathbf{x}'\boldsymbol{\beta}_j}\right), \qquad j = 1, 2, \ldots, (g-1)$$

$$p[Y_g = 1] = p_g = 1/\left(1 + \sum_{j=1}^{(g-1)} e^{\mathbf{x}'\boldsymbol{\beta}_j}\right), \qquad j = g.$$

Thus $\sum_{j=1}^{g} p_j = 1$. In this case the p_j, $j = 1, 2, \ldots, g$ satisfy a multivariate logistic distribution function.

Comparisons of various categories are conveniently carried out as follows:

$$p_r/p_s = e^{\mathbf{x}'\boldsymbol{\beta}_r}/e^{\mathbf{x}'\boldsymbol{\beta}_s} = e^{\mathbf{x}'(\boldsymbol{\beta}_r - \boldsymbol{\beta}_s)}$$

and hence

$$\ln[p_r/p_s] = \sum_{k=1}^{(c+1)} x_k(\beta_{rk} - \beta_{sk}), \qquad r, s \neq g.$$

For r or $s = g$ only one $\boldsymbol{\beta}$ vector appears in the above expressions. We can also write

$$\begin{aligned}
p_r/(p_r + p_s) &= e^{\mathbf{x}'\boldsymbol{\beta}_r}/(e^{\mathbf{x}'\boldsymbol{\beta}_r} + e^{\mathbf{x}'\boldsymbol{\beta}_s}) = e^{\mathbf{x}'(\boldsymbol{\beta}_r - \boldsymbol{\beta}_s)}/(1 + e^{\mathbf{x}'(\boldsymbol{\beta}_r - \boldsymbol{\beta}_s)}) \\
p_s/(p_r + p_s) &= e^{\mathbf{x}'\boldsymbol{\beta}_s}/(e^{\mathbf{x}'\boldsymbol{\beta}_r} + e^{\mathbf{x}'\boldsymbol{\beta}_s}) = 1/(1 + e^{\mathbf{x}'(\boldsymbol{\beta}_r - \boldsymbol{\beta}_s)}).
\end{aligned}$$

Thus any two categories can be compared as in the binary case.

These equations indicate that the ratio of any pair of category probabilities is independent of the parameters pertaining to the other categories. In some applications this property is unacceptable. In the theory of consumer choice this property has been called the *independence of irrelevant alternatives*. It therefore does not matter what other categories are included in the model since the parameters corresponding to these categories do not affect the probability ratio of the pair being studied. An extension of the probit model called the multinomial probit model can be used to avoid this property (see Maddala 1983).

Inference for the Multinomial Logit

The likelihood function for a random sample of n observations is given by

$$L = \prod_{i=1}^{n} p_{i1}^{y_{i1}} p_{i2}^{y_{i2}} \cdots p_{ig}^{y_{ig}}$$

where y_{ij} denotes the value of Y_j for observation i, and p_{ij} denotes the value of p_j for observations i, $i = 1, 2, \ldots, n$. The logarithm of this likelihood function is given by

$$
\begin{aligned}
\ln L &= \sum_{i=1}^{n} \sum_{j=1}^{g} y_{ij} \ln p_{ij} \\
&= \sum_{i=1}^{n} \left[\sum_{j=1}^{(g-1)} y_{ij} \ln[e^{\mathbf{x}_i'\boldsymbol{\beta}_j} / \left(1 + \sum_{j=1}^{(g-1)} e^{\mathbf{x}_i'\boldsymbol{\beta}_j}\right)] \right. \\
&\qquad \left. + y_{ig} \ln\left[1 / \left(1 + \sum_{j=1}^{(g-1)} e^{\mathbf{x}_i'\boldsymbol{\beta}_j}\right)\right] \right., \\
&= \sum_{i=1}^{n} \left[\sum_{j=1}^{g-1} y_{ij} \ln e^{\mathbf{x}_i'\boldsymbol{\beta}_j} - \sum_{j=1}^{g} y_{ij} \ln\left(1 + \sum_{j=1}^{(g-1)} e^{\mathbf{x}_i'\boldsymbol{\beta}_j}\right) \right].
\end{aligned}
$$

The vectors $\mathbf{x}_i$ and $\boldsymbol{\beta}_j$ are $(c+1) \times 1$.

The first derivative of the log likelihood can be expressed as

$$\frac{\partial \ln L}{\partial \boldsymbol{\beta}_k} = \sum_{i=1}^{n} (y_{ik} - p_{ik}) \mathbf{x}_i, \qquad k = 1, 2, \ldots, g,$$

which is a simple extension of the equation obtained for the maximum likelihood estimator in the dichotomous logit or logistic regression model in Section 8.3.1.

As in the case of the logistic regression model, Newton–Raphson procedures must be used to approximate the maximum likelihood solution.

The properties of the estimators obtained in this fashion are similar to the properties outlined in Section 8.3.1.

A common multiple equation approach is to estimate models for the logits

$$\ln[p_{ij}/p_{ig}] = \mathbf{x}'_i\boldsymbol{\beta}_j, \quad j = 1, 2, \ldots, (g-1), \; i = 1, 2, \ldots, n,$$

and hence each category j is compared to the last category $j = g$. Any other pair of categories can then be compared using

$$\ln[p_{ir}/p_{is}] = \mathbf{x}'_i(\boldsymbol{\beta}_r - \boldsymbol{\beta}_s), \quad i = 1, 2, \ldots, n.$$

Computer Software

The calculations for the examples in this section were performed using SAS PROC CATMOD.

Example

The bond classification data of Table 8.18 in Section 8.2 is used here to provide an example. To demonstrate the simultaneous equation estimation for a multinomial logit model, we use only two explanatory variables LFIX-CHAR and LASSLTD. Table 8.29 shows the estimated coefficients for the six logit models based on the base case category 7. The dependent variables are therefore $\ln[p_j/p_7]$, $j = 1, 2, \ldots, 6$. In all six equations, the coefficients of LFIXCHAR and LASSLTD are positive, indicating that larger values of both of these explanatory variables increase the likelihood of categories other than 7 and decrease the likelihood of category 7. In the case of category 6, however, the two coefficients were not significant at conventional levels.

Table 8.30 shows the overall significance of the variables to the simultaneous system. Each chi-square statistic is based on the value of the statistic $-2[\ln L_0 - \ln L_1]$ outlined in Section 8.3.1. From this table we can determine the overall importance of the variables to the system. The overall likelihood ratio in Table 8.30 has a value of $L_0 = 251.83$ with $[(81)(6) - (6)(3)] = 468$ degrees of freedom.

The estimated coefficients shown in Table 8.29 can be used to determine estimated coefficients for other logits. As an example, the model for $\ln[p_1/p_2]$ is given by

$$\begin{aligned}
\ln[p_1/p_2] &= \ln[p_1/p_7] - \ln[p_2/p_7] \\
&= -0.97 - 0.92 \text{ LFIXCHAR} + 1.25 \text{ LASSLTD}.
\end{aligned}$$

To determine the significance of the derived coefficients, the covariance matrix for the estimated coefficients shown in Table 8.31 can be used. Using the fact that $V(\hat{\beta}_1 - \hat{\beta}_2) = V(\hat{\beta}_1) + V(\hat{\beta}_2) - 2\text{Cov}(\hat{\beta}_1, \hat{\beta}_2)$, the variances of

TABLE 8.29. Simultaneous Logit Model Estimates

Equation Number	Logit	INTERCEPT	LFIXCHAR	LASSLTD
1	p_1/p_7	-10.84 (0.00)	1.37 (0.31)	7.08 (0.00)
2	p_2/p_7	-9.87 (0.00)	2.29 (0.08)	5.83 (0.02)
3	p_3/p_7	-9.60 (0.00)	2.14 (0.10)	5.79 (0.02)
4	p_4/p_7	-10.08 (0.00)	0.24 (0.85)	7.81 (0.00)
5	p_5/p_7	-7.48 (0.00)	2.96 (0.02)	3.62 (0.14)
6	p_6/p_7	-1.21 (0.54)	1.46 (0.20)	0.19 (0.93)

TABLE 8.30. Significance of Marginal Likelihood Ratios

Variable	d.f.	χ^2	p-Value
INTERCEPT	6	18.63	0.0048
LFIXCHAR	6	11.61	0.0713
LASSLTD	6	16.58	0.0110
Likelihood Ratio		468	251.83

the three derived coefficients are given by

$$8.02 + 7.47 - 2(6.35) = 2.79$$
$$1.82 + 1.70 - 2(1.34) = 0.84$$
$$6.35 + 6.05 - 2(5.34) = 1.72.$$

The asymptotic "t" values for the three coefficients are given by

$$\frac{-0.97}{\sqrt{2.79}} = -0.58$$
$$\frac{-0.92}{\sqrt{0.84}} = -1.00$$
$$\frac{1.25}{\sqrt{1.72}} = 0.95$$

TABLE 8.31. Covariance Matrices for Logit
Model Coefficients

			INTERCEPTS			
	1	2	3	4	5	6
1	8.02	6.35	6.26	6.22	5.67	2.57
2	6.35	7.47	6.22	6.11	5.70	2.58
3	6.26	6.22	7.51	6.06	5.67	2.57
4	6.22	6.11	6.06	7.57	5.53	2.56
5	5.67	5.70	5.67	5.53	7.16	2.59
6	2.57	2.58	2.57	2.56	2.59	3.97

		Coefficient of LFIXCHAR				
	1	2	3	4	5	6
1	1.82	1.34	1.33	1.34	1.27	0.92
2	1.34	1.70	1.33	1.30	1.30	0.93
3	1.33	1.33	1.70	1.30	1.29	0.93
4	1.34	1.30	1.30	1.72	1.24	0.91
5	1.27	1.29	1.29	1.24	1.71	0.94
6	0.92	0.93	0.93	0.91	0.94	1.31

		Coefficient of LASSLTD				
	1	2	3	4	5	6
1	6.35	5.34	5.31	5.38	4.96	2.97
2	5.34	6.05	5.29	5.28	4.99	2.98
3	5.31	5.29	6.09	5.25	4.97	2.98
4	5.38	5.28	5.25	6.18	4.90	2.96
5	4.96	4.99	4.97	4.90	6.07	3.01
6	2.97	2.98	2.98	2.96	3.01	4.66

and hence none of the three derived coefficients are considered significant
at conventional levels. It would appear therefore that the two variables
LFIXCHAR and LASSLTD cannot be used to distinguish between the first
two bond categories.

Using Multinomial Logit Models

As in the case of the dichotomous logit model discussed above, the multinomial logit model may also be used for prediction. After obtaining the
estimated coefficients for the logistic regression models for $\ln[p_r/p_g]$, $r =
1, 2, \ldots, g-1$, we can solve the estimated equations for p_r, $r = 1, 2, \ldots, (g-
1)$ and $p_g = \left(1 - \sum_{r=1}^{g-1} p_r\right)$. The resulting g equations, one for each p_r,
$r = 1, 2, \ldots, g$, can then be used to determine the values of p_r for each
unknown. The unknown is then placed in the category corresponding to
the largest value of p_r obtained.

Example

For the purpose of comparison to the discriminant analysis example in Section 8.2, the system of six logit model equations was estimated for $\ln[p_r/p_s]$, $r \neq s$, $r = 1, 2, \ldots, 7$, for all different bases, $s = 1, 2, \ldots, 7$. In this case all ten variables are used. The 21 different equations are shown in Table 8.32. Coefficients with a p-value < 0.10 appear with an $*$ in the table. The significance level is shown under each coefficient. The interpretation of each equation will not be discussed here.

The significance of each of the variables is summarized in Table 8.33. The overall likelihood ratio of 133.66 has $[(6)(81) - (6)(11)] = 420$ degrees of freedom. The incremental likelihood ratios yield the χ^2-values and p-values shown in the table. All ten explanatory variables are significant at the 0.10 level.

The estimated values of the probabilities p_r, $r = 1, 2, \ldots, 7$, were determined for the 81 observations used to estimate the models. The values of p_r are summarized in Table 8.34. This table also contains the correct group classification. The largest value of p_r for each observation is underlined in the table, and an indication of whether the classification is correct is provided in the last column of the table. Table 8.35 uses Table 8.34 to obtain a confusion or prediction success matrix. From this matrix we can conclude that 51 of the 81 observations were correctly classified using the logit models. In addition we can conclude that an additional 13 of the 81 observations were misclassified into an adjacent class. In discriminant analysis in Section 8.2 we also found that 51 of the 81 observations were correctly classified and that an additional 15 were misclassified into an adjacent class. A comparison of the two confusion matrices reveals an almost identical pattern.

Since the bond data example employs explanatory variables which are approximately normally distributed, it is perhaps not surprising that the two techniques yield similar results. Logit models tend to be superior when the set of explanatory variables does not satisfy the normality assumption, particularly if the sample size used for estimation is relatively large.

Estimation Using Single Equation Methods

The maximum likelihood estimators defined above require the solution of a more complex equation than the equation solved in the binary case. If the software is not available for the multinomial logit model but is available for the logistic regression model, it is possible to estimate the multinomial logit model using one equation at a time. The binary logits may be defined in such a way that the likelihood for the multinomial logit model can be written as the product of the $(g-1)$ likelihoods corresponding to the $(g-1)$ binary logit models. The binary logits are obtained from a *nested sequence of partitions* of the g categories.

TABLE 8.32. Simultaneous Logit Model Estimates

Logit	INTERCEPT	LOPMAR	LFIXCHAR	LGEARRAT	LTDCAP	LLEVER
p_1/p_7	-148.52	24.74*	-1.41	17.41	101.80	-29.95
	(0.44)	(0.00)	(0.73)	(0.84)	(0.78)	(0.49)
p_2/p_7	-103.33	6.96*	1.55	-23.97	136.45	-11.90
	(0.60)	(0.10)	(0.67)	(0.80)	(0.72)	(0.69)
p_3/p_7	-124.17	1.91	1.12	-54.90	229.30	-5.26
	(0.52)	(0.62)	(0.75)	(0.55)	(0.52)	(0.85)
p_4/p_7	3.61	-11.77	-1.15	-118.40	92.34	42.75
	(0.99)	(0.11)	(0.76)	(0.22)	(0.81)	(0.19)
p_5/p_7	-119.29	5.95*	6.74*	-134.27	236.90	40.88
	(0.54)	(0.10)	(0.07)	(0.16)	(0.52)	(0.19)
p_6/p_7	-194.31	4.26	5.36	14.92	256.15	-55.47*
	(0.34)	(0.20)	(0.17)	(0.85)	(0.48)	(0.06)
p_1/p_6	45.79	20.48*	-6.77*	2.49	-154.35*	35.52
	(0.12)	(0.01)	(0.05)	(0.96)	(0.01)	(0.17)
p_2/p_6	90.97*	2.70	-3.80	-38.88	-119.70	43.57
	(0.05)	(0.53)	(0.15)	(0.43)	(0.14)	(0.13)
p_3/p_6	70.14*	-2.35	-4.23*	-69.81	-26.84	50.21*
	(0.02)	(0.56)	(0.10)	(0.14)	(0.57)	(0.06)
p_4/p_6	197.92*	-16.03*	-6.50*	-133.31*	-163.80	
	(0.00)	(0.03)	(0.02)	(0.01)	(0.12)	(0.00)
p_5/p_6	75.02*	1.69	1.39	-149.19*	-19.24	96.35*
	(0.01)	(0.64)	(0.59)	(0.01)	(0.70)	(0.00)
p_1/p_5	-29.23	18.79*	-8.15*	151.68*	-135.11*	-60.83*
	(0.22)	(0.01)	(0.01)	(0.00)	(0.01)	(0.02)
p_2/p_5	15.95	1.01	-5.19*	110.31*	-100.45	-52.78*
	(0.69)	(0.72)	(0.00)	(0.02)	(0.17)	(0.04)
p_3/p_5	-4.88	-4.04	-5.62*	79.37*	-7.60	-46.14*
	(0.82)	(0.19)	(0.00)	(0.09)	(0.83)	(0.06)
p_4/p_5	122.90*	-17.73*	-7.89*	15.87	-144.56	1.87
	(0.03)	(0.01)	(0.00)	(0.76)	(0.15)	(0.94)
p_1/p_4	-152.13	36.52*	-0.26	135.81*	9.45	-62.70*
	(0.01)	(0.00)	(0.92)	(0.00)	(0.93)	(0.01)
p_2/p_4	-106.95*	18.73*	2.70	94.43*	44.11	-54.64*
	(0.09)	(0.01)	(0.13)	(0.02)	(0.69)	(0.02)
p_3/p_4	-127.78*	13.69*	2.27	63.50*	136.96	-48.00*
	(0.03)	(0.05)	(0.23)	(0.10)	(0.15)	(0.03)
p_1/p_3	-24.35	22.83*	- 2.53	72.31*	-127.51*	-14.69
	(0.26)	(0.00)	(0.33)	(0.07)	(0.00)	(0.43)
p_2/p_3	20.83	5.05	0.43	30.93	-92.85	-6.64
	(0.58)	(0.14)	(0.75)	(0.33)	(0.18)	(0.70)
p_1/p_2	-45.18	17.78*	-2.96	41.37	-34.65	-8.05
	(0.25)	(0.01)	(0.25)	(0.31)	(0.67)	(0.68)

TABLE 8.32. Simultaneous Logit Model Estimates (continued)

Logit	LCASHLTD	LACIDRAT	LCURRAT	LRECTURN	LASSLTD
p_1/p_7	-11.51	10.78*	6.80	19.20*	86.88*
	(0.17)	(0.03)	(0.31)	(0.00)	(0.08)
p_2/p_7	1.13	1.27	4.18	9.82*	25.32
	(0.87)	(0.79)	(0.50)	(0.03)	(0.60)
p_3/p_7	4.49	-2.45	4.87	4.03	6.51
	(0.49)	(0.59)	(0.35)	(0.35)	(0.88)
p_4/p_7	12.69	-22.69*	26.89*	-10.47	-87.37
	(0.17)	(0.02)	(0.01)	(0.16)	(0.12)
p_5/p_7	-14.25*	3.33	2.54	5.88	-51.15
	(0.05)	(0.39)	(0.67)	(0.14)	(0.31)
p_6/p_7	-10.06	1.74	-4.11	1.73	81.74*
	(0.18)	(0.64)	(0.47)	(0.49)	(0.07)
p_1/p_6	-1.46	9.03*	10.91	17.47*	5.14
	(0.84)	(0.07)	(0.11)	(0.00)	(0.90)
p_2/p_6	11.19	-0.47	8.29	8.09*	-56.42
	(0.11)	(0.92)	(0.22)	(0.08)	(0.20)
p_3/p_6	14.55*	-4.19	8.98	2.29	-75.23*
	(0.03)	(0.39)	(0.15)	(0.61)	(0.07)
p_4/p_6	22.75*	-24.43*	31.00*	-12.20	-169.10*
	(0.01)	(0.01)	(0.00)	(0.11)	(0.00)
p_5/p_6	-4.19	1.58	6.65	4.15	-132.88*
	(0.50)	(0.68)	(0.31)	(0.32)	(0.00)
p_1/p_5	2.74	7.45*	4.26	13.32*	138.03*
	(0.60)	(0.07)	(0.40)	(0.00)	(0.00)
p_2/p_5	15.38*	-2.05	1.64	3.94	76.47*
	(0.00)	(0.56)	(0.72)	(0.14)	(0.07)
p_3/p_5	18.74*	-5.78	2.33	-1.85	57.65
	(0.00)	(0.15)	(0.61)	(0.59)	(0.15)
p_4/p_5	26.94*	-26.02*	24.35*	-16.35*	-36.22
	(0.00)	(0.01)	(0.01)	(0.02)	(0.46)
p_1/p_4	-24.20*	33.46*	-20.09*	29.67*	174.24*
	(0.01)	(0.00)	(0.02)	(0.00)	(0.00)
p_2/p_4	-11.56	23.96*	-22.71*	20.29*	112.68*
	(0.12)	(0.01)	(0.00)	(0.00)	(0.01)
p_3/p_4	-8.20	20.24*	-22.02*	14.50*	93.87*
	(0.26)	(0.03)	(0.01)	(0.03)	(0.03)
p_1/p_3	-16.00*	13.22*	-0.69	15.17*	80.37*
	(0.01)	(0.02)	(0.87)	(0.00)	(0.03)
p_2/p_3	-3.36	3.72	22.02*	5.80*	18.81
	(0.43)	(0.36)	(0.01)	(0.09)	(0.51)
p_1/p_2	-12.64	9.51*	2.62	9.38*	61.56*
	(0.04)	(0.07)	(0.62)	(0.03)	(0.10)

TABLE 8.33. Significance of Marginal Likelihood Ratios

Variable	df	χ^2	p-Value
INTERCEPT	6	11.89	0.0646
LOPMAR	6	14.63	0.0233
LFIXCHAR	6	15.37	0.0176
LGEARRAT	6	15.07	0.0198
LTDCAP	6	12.33	0.0549
LLEVER	6	15.43	0.0171
LCASHLTD	6	16.70	0.0104
LACIDRAT	6	12.53	0.0512
LCURRAT	6	10.77	0.0956
LRECTURN	6	16.13	0.0131
LASSLTD	6	17.60	0.0073
Overall Likelihood Ratio	420	133.66	

Note: 486 =	$420 + 66 \leftarrow 11(6)$
486 =	6(81)
	($\uparrow$ 6 equations, 81 cells)

Continuation Ratios

A special case of such a nested sequence is provided by the set of *continuation ratios*. The first logit compares category $j = 1$ to the remaining categories using $\ln\left[p_1 / \sum_{j=2}^{g} p_j\right]$. Thus the first partition consists of the first category in the first set and the remaining categories in the second set. The second logit then compares the category $j = 2$ to the remaining categories from the second set of the first partition. Thus the second logit is given by $\ln\left[p_2 / \sum_{j=3}^{g} p_j\right]$. The second logit uses a partition of the second set obtained from the first partition. The second partition therefore consists of the second category in the first set and the remaining categories from the second set of the first partition in the second set. This process is repeated each time comparing one category to a steadily decreasing set of remaining categories. The diagram below illustrates the pattern for a total

TABLE 8.34. Estimated Values of p_r for Observed Data

	Group	p_1	p_2	p_3	p_4	p_5	p_6	p_7	Correct Class
1	1	<u>0.67</u>	0.00	0.00	0.11	0.31	0.00	0.00	Y
2	1	0.28	0.00	0.02	0.00	0.00	0.33	<u>0.36</u>	N
3	1	<u>0.93</u>	0.05	0.02	0.00	0.00	0.00	0.00	Y
4	1	0.32	<u>0.54</u>	0.01	0.00	0.13	0.00	0.00	N
5	1	0.07	<u>0.52</u>	0.05	0.00	0.35	0.00	0.00	N
6	1	<u>0.62</u>	0.00	0.00	0.00	0.38	0.00	0.00	Y
7	1	<u>0.56</u>	0.00	0.00	0.00	0.14	0.03	0.26	Y
8	1	<u>0.73</u>	0.06	0.09	0.12	0.00	0.00	0.00	Y
9	1	<u>0.91</u>	0.01	0.04	0.01	0.03	0.00	0.00	Y
10	2	0.00	0.29	0.45	0.00	0.26	0.00	0.00	N
11	2	0.00	0.14	0.11	0.01	<u>0.49</u>	0.25	0.00	N
12	2	0.00	0.08	0.15	<u>0.77</u>	0.00	0.00	0.00	N
13	2	0.00	<u>0.48</u>	0.09	0.28	0.14	0.00	0.00	Y
14	2	<u>0.54</u>	0.32	0.08	0.04	0.03	0.00	0.00	N
15	2	0.00	<u>0.87</u>	0.08	0.00	0.05	0.00	0.00	Y
16	2	0.00	<u>0.54</u>	0.45	0.00	0.01	0.00	0.00	Y
17	2	<u>0.54</u>	0.38	0.00	0.00	0.07	0.00	0.00	N
18	2	0.01	<u>0.81</u>	0.13	0.00	0.05	0.00	0.00	Y
19	2	0.01	<u>0.36</u>	0.28	0.01	0.18	0.16	0.00	Y
20	2	0.08	<u>0.57</u>	0.26	0.09	0.00	0.00	0.00	Y
21	2	0.08	<u>0.39</u>	0.33	0.18	0.01	0.01	0.00	Y
22	2	0.01	<u>0.40</u>	0.38	0.00	0.06	0.12	0.04	Y
23	3	0.04	0.10	0.18	0.05	<u>0.59</u>	0.92	0.01	N
24	3	0.00	<u>0.43</u>	0.24	0.21	0.12	0.00	0.00	N
25	3	0.00	<u>0.49</u>	0.09	0.40	0.01	0.00	0.00	N
26	3	0.00	<u>0.50</u>	0.30	0.01	0.09	0.11	0.00	N
27	3	0.00	0.01	0.18	0.00	0.00	<u>0.51</u>	0.30	N
28	3	0.01	0.15	0.34	0.00	<u>0.48</u>	0.01	0.02	N
29	3	0.00	0.24	<u>0.67</u>	0.00	0.04	0.00	0.04	Y
30	3	0.00	0.02	<u>0.76</u>	0.04	0.11	0.00	0.07	Y
31	3	0.00	0.33	<u>0.61</u>	0.00	0.01	0.00	0.04	Y
32	3	0.06	<u>0.74</u>	0.19	0.00	0.00	0.01	0.00	N
33	3	0.09	0.00	<u>0.82</u>	0.00	0.07	0.02	0.00	Y
34	3	0.00	0.11	<u>0.59</u>	0.20	0.03	0.02	0.05	Y
35	4	0.00	<u>0.44</u>	0.24	0.30	0.00	0.01	0.00	N
36	4	0.00	0.02	0.05	<u>0.87</u>	0.01	0.04	0.00	Y
37	4	0.00	0.14	0.14	<u>0.67</u>	0.01	0.05	0.00	Y
38	4	0.00	0.01	0.14	<u>0.77</u>	0.07	0.00	0.00	Y
39	4	0.00	0.00	0.01	<u>0.96</u>	0.02	0.00	0.01	Y
40	4	0.00	0.02	0.05	<u>0.77</u>	0.17	0.00	0.00	Y
41	4	0.00	0.00	0.01	<u>0.98</u>	0.00	0.00	0.00	Y
42	4	0.00	0.12	0.24	0.03	<u>0.54</u>	0.06	0.00	N
43	4	0.00	0.00	0.00	<u>1.00</u>	0.00	0.00	0.00	Y
44	4	0.00	0.00	0.01	<u>0.95</u>	0.03	0.00	0.00	Y
45	4	0.38	0.05	0.05	<u>0.50</u>	0.02	0.00	0.00	Y
46	4	0.01	0.16	0.01	<u>0.82</u>	0.00	0.00	0.00	Y
47	4	0.00	0.00	0.00	<u>1.00</u>	0.00	0.00	0.00	Y
48	5	0.07	0.01	0.01	0.11	<u>0.64</u>	0.16	0.00	Y
49	5	0.20	0.25	<u>0.32</u>	0.00	0.23	0.00	0.00	N

TABLE 8.34. Estimated Values of p_r for Observed Data (continued)

	Group	p_1	p_2	p_3	p_4	p_5	p_6	p_7	Correct Class
50	5	0.22	0.01	0.03	0.00	<u>0.50</u>	0.05	0.20	Y
51	5	<u>0.70</u>	0.00	0.00	0.00	0.30	0.00	0.00	N
52	5	0.00	0.03	0.03	0.14	<u>0.74</u>	0.06	0.00	Y
53	5	0.00	0.00	0.00	0.06	<u>0.91</u>	0.04	0.00	Y
54	5	0.00	0.29	0.14	0.00	<u>0.57</u>	0.00	0.00	Y
55	5	0.00	0.02	0.03	0.01	<u>0.60</u>	0.32	0.00	Y
56	5	0.00	0.00	0.01	0.01	<u>0.85</u>	0.01	0.12	Y
57	5	0.00	0.09	0.12	0.00	<u>0.67</u>	0.11	0.00	Y
58	5	0.00	0.13	<u>0.49</u>	0.08	0.18	0.07	0.06	N
59	5	0.00	<u>0.60</u>	0.22	0.01	0.15	0.00	0.00	N
60	5	0.01	0.32	0.21	0.05	<u>0.41</u>	0.00	0.00	Y
61	6	0.00	0.00	0.00	0.00	0.00	0.09	<u>0.91</u>	N
62	6	0.27	0.00	0.00	0.00	0.00	<u>0.72</u>	0.00	Y
63	6	0.00	0.00	0.06	0.00	0.00	0.45	<u>0.49</u>	N
64	6	0.00	0.00	0.00	0.00	0.00	<u>1.00</u>	0.00	Y
65	6	0.00	0.00	0.06	0.00	0.00	<u>0.57</u>	0.36	Y
66	6	0.00	0.00	0.00	0.00	0.00	<u>1.00</u>	0.00	Y
67	6	0.00	0.00	0.00	0.00	0.00	0.39	<u>0.61</u>	N
68	6	<u>0.55</u>	0.21	0.11	0.02	0.17	0.10	0.00	N
69	6	0.00	0.02	0.04	0.06	<u>0.67</u>	0.20	0.00	N
70	6	0.00	0.00	0.00	0.00	0.00	<u>0.98</u>	0.02	Y
71	6	0.00	0.00	0.01	0.00	0.00	<u>0.79</u>	0.20	Y
72	7	0.00	0.02	0.14	<u>0.38</u>	0.04	0.16	0.26	N
73	7	0.00	0.00	0.00	0.00	0.00	0.09	<u>0.91</u>	Y
74	7	0.00	0.00	0.00	0.00	0.00	0.28	<u>0.72</u>	Y
75	7	0.00	0.01	<u>0.56</u>	0.00	0.00	0.01	0.41	N
76	7	0.00	0.00	0.00	0.00	0.00	0.14	<u>0.86</u>	Y
77	7	0.00	0.00	0.00	0.00	0.00	<u>0.92</u>	0.08	N
78	7	0.00	0.06	<u>0.41</u>	0.00	0.18	0.03	0.32	N
79	7	0.02	0.00	0.00	0.00	0.24	0.09	<u>0.65</u>	Y
80	7	0.00	0.00	0.00	0.00	0.00	0.21	<u>0.78</u>	Y
81	7	0.00	0.00	0.02	0.00	0.00	0.19	<u>0.79</u>	Y

of five categories. The four logit comparisons are shown as

<u>Logit 1</u>

1 vs (2, 3, 4, 5)

<u>Logit 2</u>

2 vs (3, 4, 5)

<u>Logit 3</u>

3 vs (4, 5)

<u>Logit 4</u>

4 vs 5.

TABLE 8.35. Confusion Matrix for Multinomial Logit Model Predictions

Actual Classification	Predicted Classification							
	1	2	3	4	5	6	7	Total
1	6	2	0	0	0	0	1	9
2	2	8	1	1	1	0	0	13
3	0	4	5	0	2	1	0	12
4	0	1	0	11	1	0	0	13
5	1	1	2	0	9	0	0	13
6	1	0	0	0	1	6	3	11
7	0	0	2	1	0	1	6	10
Total	10	16	10	13	14	8	10	81

The logit for the jth equation is given by

$$\ln\left[p_j/\sum_{h>j}^{g}p_h\right], \quad j = 1, 2, \ldots, (g-1).$$

This nested sequence is useful when the categories have been arranged in some meaningful order.

Other Nested Partitions

Any system of *nested partitions* can be used. An alternative sequence for five categories is shown in Figure 8.18. In this case the first logit model compares the group consisting of categories 1 and 2 to the group consisting of categories 3, 4 and 5. The second logit model compares categories 1 and 2. The third logit model compares categories 3 and 4 to category 5 and the fourth logit model compares categories 3 and 4.

Example

To exemplify the use of this single equation approach to the estimation of the simultaneous system, we shall employ a particular *hierarchical system* to estimate a model for the bond data. We also use a backward elimination procedure for each single equation estimation. The hierarchy to be used is

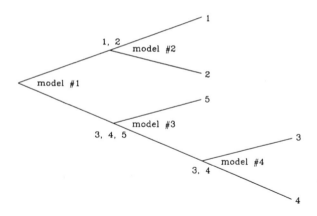

FIGURE 8.18. Example of a Nested Partition of Models

given by the following sequence of six single equations.

No. 1

$$\boxed{[1,2,3,4,5] \quad \text{vs} \quad [6,7]}$$

No. 3   No. 2

$$\boxed{[1,2,3] \quad \text{vs} \quad [4,5]} \qquad \boxed{[6] \quad \text{vs} \quad [7]}$$

No. 5   No. 4

$$\boxed{[1] \quad \text{vs} \quad [2,3]} \qquad \boxed{[4] \quad \text{vs} \quad [5]}$$

No. 6

$$\boxed{[2] \quad \text{vs} \quad [3]}$$

Table 8.36 summarizes the estimated models obtained from a backward elimination procedure applied to each model separately. In the backward elimination process, variables that were not significant at the 0.15 level were eliminated.

It is important to note that only equation 1 uses the entire data set for estimation. Equation 2 uses only the observations corresponding to categories 6 and 7, and equation 3 uses only the observations corresponding to categories 1 to 5. In equation 2, therefore, the probabilities in the logit function are more correctly represented by the conditional probabilities $p[6 \mid [6,7]]$ and $p[7 \mid [6,7]]$ respectively. Having determined $p[6,7]$ from equation 1, equation 2 can be used to determine $p[6 \mid [6,7]]$ and $p[7 \mid [6,7]]$ and hence

$$p[6] = p[6,7]p[6 \mid [6,7]] \qquad p[7] = p[6,7]p[7 \mid [6,7]].$$

TABLE 8.36. Single Equation Estimation by Backward Elimination — Hierarchical System

Logit	INTER-CEPT	LOP-MAR	LFIX-CHAR	LGEAR-RAT	LTD-CAP	LLEVER	LCASH-LTD	LACID-RAT	LCUR-RAT	LREC-TURN	LASS-LTD
1. $\frac{p[1,2,3,4,5]}{p[6,7]}$	27.08 (0.00)	—	—	-52.68 (0.00)	—	29.75 (0.00)	—	-4.82 (0.00)	4.85 (0.02)	—	-45.37 (0.00)
2. $\frac{p[6]}{p[7]}$	-21.26 (0.01)	—	—	34.21 (0.01)	—	-24.88 (0.00)	—				32.84 (0.01)
3. $\frac{p[1,2,3]}{p[4,5]}$	-21.06 (0.00)	2.23 (0.03)	—	32.42 (0.00)	—	-18.17 (0.01)		3.30 (0.01)	-4.34 (0.02)	2.74 (0.01)	32.29 (0.00)
4. $\frac{p[4]}{p[5]}$	355.91 (0.00)	-88.23 (0.00)	-40.33 (0.00)	177.46 (0.00)	—	-149.08 (0.00)	150.51 (0.00)	-117.39 (0.00)	—	-107.45 (0.00)	—
5. $\frac{p[1]}{p[2,3]}$	-1.51 (0.54)	78.16 (0.00)	—	213.16 (0.00)	-469.72 (0.00)	—	-30.18 (0.12)	34.36 (0.00)	34.47 (0.06)	35.62 (0.00)	216.93 (0.00)
6. $\frac{p[2]}{p[3]}$	28.15 (0.02)	3.92 (0.09)	—	9.62 (0.01)	-64.70 (0.01)	—	—	—	—	3.63 (0.06)	—

In a similar fashion $p[1, 2, 3]$ and $p[4, 5]$ can be determined using $p[1, 2, 3, 4, 5]$, $p[1, 2, 3 \mid [1, 2, 3, 4, 5]]$, and $p[4, 5 \mid [1, 2, 3, 4, 5]]$.

8.3.7 OTHER SOURCES OF INFORMATION

Additional discussion of the multinomial logit model is available in Maddala (1983), Fienberg (1980) and McCullagh and Nelder (1989).

8.3.8 THE CONDITIONAL LOGIT MODEL AND CONSUMER CHOICE

In the study of consumer choice the dependent variable in a logistic regression can represent a variety of choices. In such analyses the potential explanatory variables not only vary across consumers but also vary across alternative choices for each consumer. Variables designed to measure consumers' preference for a variety of alternatives are examples. For a particular alternative each consumer assigns a value to each explanatory variable. For another alternative, the same consumer assigns values to the explanatory variables, which are not necessarily the same as for other alternatives. The values assigned to these explanatory variables could represent costs or values of the alternative. A logit model that assumes that the X variables vary by category as well as by individual is given by the conditional logit model

$$p_r/p_s = e^{\mathbf{x}_r'\boldsymbol{\beta}}/e^{\mathbf{x}_s'\boldsymbol{\beta}} = e^{(\mathbf{x}_r'-\mathbf{x}_s')\boldsymbol{\beta}} \quad r, s = 1, 2, \ldots, \ell, \ r \neq s$$
$$p_r = e^{\mathbf{x}_r'\boldsymbol{\beta}}/\sum_{h=1}^{\ell} e^{\mathbf{x}_h'\boldsymbol{\beta}}.$$

Thus, the purpose of the conditional logit model is to obtain estimates of p_{ik}, the probabilities of selection of the kth possible choice by the ith consumer given the values of $\mathbf{x}'$ denoted by $\mathbf{x}_{ik}'$. Note that the parameter vector $\boldsymbol{\beta}$ is assumed to be constant across categories, which contrasts with the multinomial logit model discussed in Section 8.3.5.

A more general model that combines the multinomial logit model and the conditional logit model is given by

$$p_r = e^{\mathbf{x}_r'\boldsymbol{\beta}+\mathbf{Z}'\boldsymbol{\alpha}_r}/\sum_{h=1}^{\ell}[e^{\mathbf{x}_h'\boldsymbol{\beta}+\mathbf{Z}'\boldsymbol{\alpha}_h}], \qquad r = 1, 2, \ldots, \ell,$$

where the vector $\mathbf{Z}$ denotes observations which are specific to the individual but are constant across choice categories as in the multinomial logit model. The constant variables across alternative choices for a given consumer could include age, income and education. Therefore, the conditional probability that the observation is from category $k = r$ given that the observation is from one of the two categories $k = r$ or $k = s$ is assumed to have the same logistic distribution shape as in the binary case. An examination of the ratio for p_r/p_s reveals that this ratio does not depend on any of the other categories. Therefore, regardless of the number of categories, the ratio of the two probabilities remains constant. As mentioned earlier in this section, in the application of this model to consumer choice behavior, this property is referred to as *"independence of irrelevant alternatives"* and in some applications is viewed as a serious weakness of the multinomial logit model.

More generally, the model can be written as

$$p_r = e^{\mathbf{w}_r'\boldsymbol{\gamma}_r} / \sum_{h=1}^{\ell} e^{\mathbf{w}_r'\boldsymbol{\gamma}_r},$$

where $\mathbf{w}_r$ and $\boldsymbol{\gamma}_r$ now combine the previous variables $\mathbf{x}_r$, $\mathbf{Z}$ and the previous parameters $\boldsymbol{\beta}$ and $\boldsymbol{\alpha}_r$.

The maximum likelihood estimators of the vectors $\boldsymbol{\gamma}_r$, $r = 1, 2, \ldots, \ell$, are obtained from the logarithm of the likelihood function using the Newton–Raphson method as in the case of the dichotomous logit model. The observations on the dependent variable are defined in terms of ℓ dummy variables as follows:

$$y_{ik} = \begin{cases} 1 & \text{if observation } i \text{ is in category } k, \quad k = 1, 2, \ldots, \ell \\ 0 & \text{otherwise.} \end{cases}$$

The conditional logit model can be obtained from the multinomial logit model by defining a sequence of stacked vectors. To obtain $\mathbf{x}_r'\boldsymbol{\beta}$ for the conditional logit model from $\mathbf{Z}'\boldsymbol{\alpha}_r$ of the multinomial logit model, we define

$$\mathbf{Z}' = (\mathbf{x}_1', \ldots, \mathbf{x}_r', \ldots, \mathbf{x}_\ell'), \qquad \boldsymbol{\alpha}_r' = (\mathbf{0}', \ldots, \boldsymbol{\beta}', \ldots, \mathbf{0}').$$

Similarly to obtain the multinomial logit model from the conditional logit model, we define

$$\mathbf{x}_r' = (\mathbf{0}', \ldots, \mathbf{Z}', \ldots, \mathbf{0}'), \qquad \boldsymbol{\beta}' = (\boldsymbol{\alpha}_1', \boldsymbol{\alpha}_2', \ldots, \boldsymbol{\alpha}_r', \ldots, \boldsymbol{\alpha}_\ell').$$

Additional discussion of consumer choice models can be found in Ben–Akiva and Lerman (1985) and Maddala (1983).

8.3.9 MULTIVARIATE QUALITATIVE RESPONSE MODELS

Up to this point our discussion of qualitative response models has been restricted to the case of one dependent variable with two or more categories. If there are two qualitative dependent variables, then a single dependent variable can always be constructed using the cross-classification of the two sets of categories. Although this approach is useful for determining the system of linear relationships, it is also of interest to look at certain relationships among the category probabilities of the dependent variable because of the underlying multivariate nature of this variable.

Example

To illustrate this concept we use the panel data introduced in Table 8.25. For the data in Table 8.25 two dependent variables, THISYR and LASTYR, were related to the explanatory variables AGE, EDUC, HUBINC, BLACK, CHILD1, and CHILD2. In this section we estimate a model that combines the two dependent variables into one dependent variable with four categories of (THISYR, LASTYR) given by (0,0), (0,1), (1,0) and (1,1).

For the data in Table 8.25 the two-dimensional contingency table for the 100 observations is given by

		LASTYR	
		0	1
THISYR	0	20	8
	1	15	57

A χ^2 test of independence rejects the independence hypothesis at the 0.000 level. The observed frequencies on the main diagonal are much larger than the expected frequencies (9.8, 46.8) under independence. There is a tendency, therefore, for married women to remain in the same work category in both years. Of interest to us now is whether these probabilities can be related to the set of explanatory variables. In particular we might wish to determine how the interaction is related to the explanatory variables.

Loglinear Models for Dependent Variables

In Chapter 6 the loglinear model was used to represent the variation in the cell frequencies in contingency tables. Using the fact that the cell proportion can be obtained from the cell frequencies by dividing by a constant, the saturated loglinear model for a two dimensional ($r \times c$) table is given by

$$\ln p_{ij} = \mu + \mu_{1(i)} + \mu_{2(j)} + \mu_{12(ij)},$$

where

$$\mu = \frac{1}{rc}\sum_{i=1}^{r}\sum_{j=1}^{c}\ln p_{ij},$$

$$\mu_{1(i)} = \frac{1}{c}\sum_{j=1}^{c}\ln p_{ij} - \mu = \frac{1}{c}\sum_{j=1}^{c}\ln p_{ij} - \frac{1}{rc}\sum_{i=1}^{r}\sum_{j=1}^{c}\ln p_{ij},$$

$$\mu_{2(j)} = \frac{1}{r}\sum_{i=1}^{r}\ln p_{ij} - \mu = \frac{1}{r}\sum_{i=1}^{r}\ln p_{ij} - \frac{1}{rc}\sum_{i=1}^{r}\sum_{j=1}^{c}\ln p_{ij},$$

$$\mu_{12(ij)} \;=\; \ln p_{ij} - \mu_{1(i)} - \mu_{2(j)} - \mu,$$

$$= \ln p_{ij} - \frac{1}{c}\sum_{j=1}^{c}\ln p_{ij} - \frac{1}{r}\sum_{i=1}^{r}\ln p_{ij} + \frac{1}{rc}\sum_{i=1}^{r}\sum_{j=1}^{c}\ln p_{ij}.$$

For a 2×2 table these equations can be simplified to become

$$\mu \;=\; \frac{1}{4}[\ln p_{11} + \ln p_{12} + \ln p_{21} + \ln p_{22}],$$

$$\mu_{1(1)} \;=\; \frac{1}{4}[\ln(p_{11}/p_{21}) + \ln(p_{12}/p_{22})], \qquad \mu_{1(2)} = -\mu_{1(1)},$$

$$\mu_{2(1)} \;=\; \frac{1}{4}[\ln(p_{11}/p_{12}) + \ln(p_{21}/p_{22})], \qquad \mu_{2(2)} = -\mu_{2(1)},$$

$$\mu_{12(11)} \;=\; \frac{1}{4}\ln\left[\frac{p_{11}p_{22}}{p_{21}p_{12}}\right], \qquad \mu_{12(12)} = -\mu_{12(11)} = \mu_{12(21)} = -\mu_{12(22)},$$

$$= \frac{1}{4}\ln\left[\begin{array}{c}\text{ODDS}\\\text{RATIO}\end{array}\right].$$

The row effect parameter $\mu_{1(1)}$ measures the propensity to be in row 1 rather than row 2, whereas $\mu_{2(1)}$ measures the tendency to be in column 1 rather than column 2. The interaction parameter $\mu_{12(11)}$ indicates the tendency for the row effect to differ in each column, or equivalently for the column effect to differ in each row. Alternatively, if the probabilities in the main diagonal are high relative to the off diagonal probabilities, then $\mu_{12(11)}$ will tend to be large.

The two main effects $\mu_{1(1)}$, $\mu_{2(1)}$ and the interaction $\mu_{12(11)}$ can be expressed in terms of logits of the form $\ln[p_{ij}/p_{kl}]$. These parameters can therefore be related to the set of explanatory variables by using a system of logit models to relate the cells of the dependent variable to the explanatory variables.

Relation Between Loglinear Parameters and Logits

If we assume that p_{22} is used as the base cell, for the dependent variable, the system of multinomial logit models used to relate to the explanatory variables is given by

$$\ln[p_{11}/p_{22}] \;=\; \mathbf{x}\boldsymbol{\beta}_A,$$
$$\ln[p_{12}/p_{22}] \;=\; \mathbf{x}\boldsymbol{\beta}_B \quad \text{and}$$
$$\ln[p_{21}/p_{22}] \;=\; \mathbf{x}\boldsymbol{\beta}_C.$$

The effect parameters can therefore be defined as

$$\mu_{1(1)} \;=\; [\mathbf{x}\boldsymbol{\beta}_A + \mathbf{x}\boldsymbol{\beta}_B - \mathbf{x}\boldsymbol{\beta}_C],$$
$$\mu_{2(1)} \;=\; [\mathbf{x}\boldsymbol{\beta}_A - \mathbf{x}\boldsymbol{\beta}_B + \mathbf{x}\boldsymbol{\beta}_C],$$
$$\text{and} \qquad \mu_{12(11)} \;=\; [\mathbf{x}\boldsymbol{\beta}_A - \mathbf{x}\boldsymbol{\beta}_B - \mathbf{x}\boldsymbol{\beta}_C].$$

The reader should note that for convenience the fraction $(1/4)$ has been omitted from these definitions. In a similar fashion, relationships can be established between loglinear model parameters and logit model parameters for $(r \times c)$ contingency tables.

Example

For the data in Table 8.25 a system of three logit models was estimated using the category (THISYR, LASTYR) $= (1,1)$ as the base category. The four probabilities are defined by

		LASTYR	
		0	1
THISYR	0	p_{11}	p_{12}
	1	p_{21}	p_{22}

The estimated logit models are summarized in Table 8.37. The table also contains the model coefficients for the row effect (THISYR), the column effect (LASTYR) and the interaction effect. From the equation for $\mu_{1(1)}$ we can conclude that individuals are less likely to work THISYR when AGE and EDUC are relatively low, when HUBINC is relatively high, when the individual is not BLACK, and if there are preschool children. From the $\mu_{2(1)}$ equation it would appear that all the variables, excluding the intercept, influence the decision to work in the same way. The reader should confirm the similarity of these conclusions to those obtained previously for the models for THISYR and LASTYR using dichotomous response models. From the interaction parameter equation for $\mu_{12(11)}$ we can conclude that individuals are more likely to have the same choice in both periods (positive interaction) if HUBINC is relatively high, if EDUC is relatively low; if they do not have young preschool children, and if they are not BLACK.

A Conditional Probability Approach

In the previous analysis the relationship between the two dependent variables THISYR and LASTYR was studied using a loglinear model that treats the variables as joint covariates. It is also possible to view the two variables in a manner which permits one to be dependent on the other. In other words, given the outcome for LASTYR, what is likely to occur THISYR? For each of the two categories for LASTYR separate logit models can be estimated for the dichotomous response variable THISYR. The two logit models are fit to the two parts of the table separately.

TABLE 8.37. Logistic Regression Model Estimates for 2×2 Contingency Table Parameters

Dependent Variable	INTERCEPT	AGE	HUBINC	EDUC	BLACK	CHILD1	CHILD2
$\ln[p_{11}/p_{22}]$	−0.06	0.03	0.15	−0.36	−1.26	5.58	2.21
	(0.99)	(0.64)	(0.01)	(0.17)	(0.28)	(0.00)	(0.01)
$\ln[p_{12}/p_{22}]$	12.65	−0.22	0.18	−0.76	1.56	−12.64	0.40
	(0.05)	(0.01)	(0.02)	(0.10)	(0.28)	(0.99)	(0.66)
$\ln[p_{21}/p_{22}]$	−9.23	0.11	0.10	0.20	0.64	−13.40	1.51
	(0.01)	(0.09)	(0.10)	(0.16)	(0.44)	(0.99)	(0.09)
$\mu_{1(1)}$ (Row)	21.82	−0.30	0.23	−1.32	−0.34	6.34	1.10
$\mu_{2(1)}$ (Column)	−21.94	0.36	0.07	0.60	-2.18	4.82	3.32
$\mu_{12(11)}$ (Interaction)	-3.48	0.14	-0.13	0.20	−3.46	31.62	0.30

For the case LASTYR $= 0$, the estimated logit model is given by

$$\ln\left[\frac{p[\text{THISYR} = 0 \mid \text{LASTYR} = 0]}{p[\text{THISYR} = 1 \mid \text{LASTYR} = 0]}\right]$$
$$= \underset{(0.00)}{-35.93} + \underset{(0.01)}{0.37} \text{ AGE} - \underset{(0.22)}{0.06} \text{ HUBINC} + \underset{(0.00)}{1.75} \text{ EDUC}$$
$$- \underset{(0.01)}{12.00} \text{ CHILD1} - \underset{(0.86)}{0.28} \text{ CHILD2} + \underset{(0.00)}{7.36} \text{ BLACK}.$$

Similarly, for the case LASTYR $= 1$, the estimated logit model is given by

$$\ln\left[\frac{p[\text{THISYR} = 0 \mid \text{LASTYR} = 1]}{p[\text{THISYR} = 1 \mid \text{LASTYR} = 1]}\right]$$
$$= \underset{(0.06)}{-9.96} + \underset{(0.00)}{0.22} \text{ AGE} - \underset{(0.04)}{0.16} \text{ HUBINC} + \underset{(0.18)}{0.53} \text{ EDUC}$$
$$+ \underset{(0.90)}{5.34} \text{ CHILD1} - \underset{(0.76)}{0.29} \text{ CHILD2} + \underset{(0.57)}{0.82} \text{ BLACK}.$$

From the above two equations, we can conclude that individuals who chose not to work LASTYR are less likely to work THISYR if AGE and EDUC are relatively high, if they are BLACK, and if there are no children under age 2. Individuals who did work LASTYR are more likely to work THISYR if HUBINC is relatively high and AGE is relatively low.

Cited Literature and References

1. Amemiya, Takeshi (1985). *Advanced Econometrics*. Cambridge, MA: Harvard University Press.

2. Andersen, Erling B. (1990). *The Statistical Analysis of Categorical Data*. Berlin: Springer-Verlag.

3. Anderson, T.W. (1984). *An Introduction to Multivariate Statistical Analysis*, Second Edition. New York: John Wiley and Sons.

4. Ben-Akiva, Moshe and Lerman, Steven R. (1985). *Discrete Choice Analysis: Theory and Applications to Travel Demand*. Cambridge, MA: MIT Press.

5. Fienberg, Stephen E. (1980). *The Analysis of Cross Classified Categorical Data*, Second Edition. Cambridge, MA: MIT Press.

6. Fomby, Thomas B., Hill, R. Carter and Johnson, Stanley R. (1984). *Advanced Econometric Methods*. New York: Springer–Verlag.

7. Fox, John (1984). *Linear Statistical Models and Related Methods: With Applications to Social Research*. New York: John Wiley and Sons.

8. Hosmer, David W. and Lemeshow, Stanley (1989). *Applied Logistic Regression*. New York: John Wiley and Sons.

9. Johnson, Richard A. and Wichern, Dean W. (1988). *Applied Multivariate Statistical Analysis*, Second Edition. Englewood Cliffs, NJ: Prentice–Hall.

10. Judge, George G., Griffiths, W.E., Hill, R. Carter, Lütkepohl, Helmut and Lee, Tsoung-Chao (1985). *The Theory and Practice of Econometrics*, Second Edition. New York: John Wiley and Sons.

11. Kryzanowski, W.J. (1988). *Principles of Multivariate Analysis: A User's Perspective*. Oxford: Oxford University Press.

12. Lachenbruch, Peter A. (1975). *Discriminant Analysis*. New York: Hafner Press.

13. Maddala, G.S. (1983). *Limited Dependent and Qualitative Variables in Econometrics*. Cambridge, MA: Cambridge University Press.

14. Mardia, K.V., Kent, J.T. and Bibby, J.M. (1979). *Multivariate Analysis*. London: Academic Press.

15. McCullagh, P. and Nelder, J.A. (1989). *Generalized Linear Models*, Second Edition. London: Chapman and Hall.

16. Morrison, Donald F. (1976). *Multivariate Statistical Methods*, Second Edition. New York: McGraw–Hill.

17. Press, S. James (1972). *Applied Multivariate Analysis*. New York: Holt, Rinehart and Winston.

18. Santner, Thomas J. and Duffy, Dianne E. (1989). *The Statistical Analysis of Discrete Data*. New York: Springer–Verlag.

19. Seber, G.A.F. (1984). *Multivariate Observations*. New York: John Wiley and Sons.

20. Stevens, James (1986). *Applied Multivariate Statistics for the Social Sciences*. Hillsdale, NJ: Lawrence Erlbaum Associates.

21. Tatsuoka, Maurice (1988). *Multivariate Analysis: Techniques for Educational and Psychological Research*, Second Edition. New York: Macmillan Publishing Co.

Exercises for Chapter 8

1. This exercise is based on the Mystery Data in Table V10 in the Data Appendix. The data is a subset of the mystery data in Table 8.11. Table V10 excludes the variables C2, C4, C5, C6 and C7. A comparison of the means across the eight SEXED groups (see Table 8.12) showed that the means on the variables C1, C3, C8, C9 and C10 differed.

   (a) For the five variables use multivariate regression with indicator variables to compare the eight group means for each variable. You will have to use several different indicator variable schemes to obtain a variety of comparisons. Use both dummy coding and effect coding design matrices.

   (b) Use the dummy variable approach in (a) to test the hypothesis that for each education group there is no difference between males and females. Carry out the test separately for each variable and simultaneously for all five variables.

   (c) Repeat the steps in (b) to test the hypothesis that there are no differences among the education categories for each sex.

   (d) Repeat the tests in (b) and (c) using cell parameter coding.

   (e) Repeat the tests in (b) and (c) using the non-full-rank group effects design matrix.

   (f) What can you conclude about the differences among the eight SEXED groups for the five variables?

2. This exercise is based on the R.C.M.P. Officer Data in Table V9 in the Data Appendix. The data pertains to the job satisfaction of police officers in ten different municipal detachments. The four satisfaction factors are to be compared over the ten detachments.

   (a) Perform a MANOVA to compare the means on the four factors over the ten detachments. Perform various multiple comparisons to compare the means.

   (b) Perform a test for parallel profiles.

   (c) Test the hypothesis that detachments 1, 4, 9 and 10 have equal profiles using a multivariate regression approach.

   (d) Test the hypothesis that detachments 2, 3 and 7 have equal profiles using a multivariate regression approach.

   (e) Test the hypothesis that detachments 5, 6 and 8 have equal profiles using a multivariate regression approach.

   (f) Carry out a simultaneous test of all three hypotheses in (c), (d) and (e).

(g) What conclusions can you draw regarding the variation in officer job satisfaction over the ten detachments?

3. This exercise is based on the Shopping Attitude Data in Table V8 in the Data Appendix. The data set contains 200 observations regarding attitudes of female clothing shoppers. In addition to seven attitude variables the respondents age and work status are also included.

(a) Use MANOVA to test the hypothesis that the vector of means on the variables A1 to A7 are the same for the two WORK categories. Also use ANOVA to compare the means for each variable separately. Discuss the results.

(b) For the comparison in (a) perform a test for parallel profiles.

(c) Repeat the tests in (a) and (b) using a multivariate regression approach with indicator variables.

(d) Repeat the analyses in (a), (b) and (c) adding the variable AGE as a covariate. What impact does the addition of AGE have on the earlier results?

(e) Divide the data into two sets of 100 observations each and perform a discriminant analysis of the attitude variables with respect to the two WORK categories for each data set and compare.

(f) Use the results for one data set in (e) to predict the WORK category for the other data set and evaluate the discriminant function as a prediction device.

(g) Repeat the steps in (e) and (f) for the AGE categories.

(h) As in (e) divide the data set into two groups and use logistic regression to relate WORK to the seven attitude variables for each group. Fit two different regressions with and without AGE as an explanatory variable. Discuss the results.

(i) Use the results in one data set in (h) to predict the WORK categories for the other data set and evaluate the logistic regression model as a prediction device. Does the variable AGE make a difference?

(j) Repeat the steps in (h) and (i) by determining multinomial logit models to relate the AGE categories to the seven attitude variables. Evaluate the fitted models as a predictive device by using the fitted model for one set to predict the other set. Fit models with and without the variable WORK as an explanatory variable.

(k) Summarize the analyses and discuss the relationship between shopping attitudes and the variables WORK and AGE.

4. This exercise is based on the Bank Employee Data contained in Tables V11 and V12 in the Data Appendix.

   (a) Using the data in Table V11 use discriminant analysis to relate the variable SEX to the variables LCURRENT, LSTART, EDUC, AGE, SENIOR, and EXPER. Also, examine the one-way ANOVA analyses for each of the six variables. Discuss the relationship between the variable SEX and the explanatory variables.

   (b) Using the discriminant functions obtained in (a) predict the SEX variable in Table V12 and evaluate the model as a prediction tool.

   (c) Repeat steps (a) and (b) by interchanging the roles of Tables V11 and 12.

   (d) Repeat the steps in (a), (b) and (c) using the variable RACE instead of SEX.

   (e) Combine the variables SEX and RACE to form a new variable SEXRACE and repeat the steps in (a), (b) and (c).

   (f) Write a summary discussing the relationship between SEX, RACE and the six variables.

   (g) Repeat the steps in (a), (b) and (c) using the variable JOBCAT instead of SEX.

   (h) Write a summary discussing the relationship between job category and the six variables.

5. This exercise is based on the Bank Employee Data contained in Tables V11 and V12 in the Data Appendix.

   (a) Combine the two tables into one data set of 200 observations.

   (b) Use a single explanatory variable logistic regression model to relate the dependent variable SEX to each of the explanatory variables LCURRENT, LSTART, EDUC, SENIOR, AGE and EXPER. Discuss the set of bivariate relationships studied.

   (c) Carry out a logistic regression relating SEX to all six explanatory variables simultaneously. Discuss the results and compare them to the results in (b). Select a prediction model.

   (d) For the prediction model selected in (c), evaluate the model by constructing a confusion matrix based on the 200 observations.

   (e) Repeat steps (a) through (d) using the dependent variable RACE.

   (f) Determine a new variable SEXRACE with four categories that combine the variables SEX and RACE (SEXRACE = SEX + 2 * RACE). Repeat steps (a) through (d) and discuss the results. Is there a predictable interaction effect between SEX and RACE?

(g) Repeat steps (a) through (e) using probit analysis.

(h) Write a summary outlining the relationship between SEX and RACE and the six variables.

6. This exercise is based on the bank employee data contained in Tables V11 and V12 in the Data Appendix.

   (a) Combine the two tables into one data set of 200 observations. Collapse the categories of the variable JOBCAT by combining categories JOBCAT = 1 and JOBCAT = 2 and also by combining the categories JOBCAT = 4 and JOBCAT = 5. Your new JOBCAT variable should have three categories.

   (b) Use a single explanatory variable multiple logit model to relate the variable JOBCAT to each of the explanatory variables LCURRENT, LSTART, EDUC, SENIOR, AGE and EXPER. Discuss the set of bivariate relationships studied.

   (c) Estimate multiple logit models relating JOBCAT to all six explanatory variables simultaneously. Discuss the results and compare them to the results in (b). Select a prediction model.

   (d) For the prediction model selected in (c) evaluate the model by constructing a confusion matrix based on the 200 observations.

   (e) Repeat steps (a) through (d) using an alternative base case (eg. change from base category 3 to base category 1 for JOBCAT).

7. This exercise is based on the Panel Study Data summarized in Table V13 in the Data Appendix. The first 100 observations in the table are different observations than the ones shown in Table 8.25. The last 100 observations are identical to the observations in Table 8.25.

   (a) Use the first 100 observations to repeat the logistic regression analyses based on Table 8.25 carried out in Chapter 8. Compare your results to the results in the chapter. Select a prediction model for both LASTYR and THISYR.

   (b) Evaluate the selected prediction models in (a) by using the models to predict the outcomes for LASTYR and THISYR in the latter half of Table V13 (also Table 8.25). Discuss your results.

   (c) Use fitted logistic regression models for LASTYR and THISYR given in Chapter 8 to predict the first 100 observations of LASTYR and THISYR in Table V13. Discuss the results.

   (d) Using the first 100 observations in Table V13 repeat the logistic regression analyses carried out in Chapter 8 to predict the interaction between LASTYR and THISYR. Discuss your results and compare to the results obtained in Chapter 8.

(e) Repeat steps (a) through (c) using probit analysis in place of logistic regression.

(f) Write a summary discussing the relationship between LASTYR, THISYR and the six variables.

Questions for Chapter 8

1. The mean vector $\boldsymbol{\mu}(5 \times 1)$ on a random vector $\mathbf{x}(5 \times 1)$ is to be compared across two groups. Denote the mean vectors in the two groups by

$$\boldsymbol{\mu}_1 = \begin{bmatrix} \mu_{11} \\ \mu_{21} \\ \mu_{31} \\ \mu_{41} \\ \mu_{51} \end{bmatrix} \quad \text{and} \quad \boldsymbol{\mu}_2 = \begin{bmatrix} \mu_{12} \\ \mu_{22} \\ \mu_{32} \\ \mu_{42} \\ \mu_{52} \end{bmatrix}.$$

The measurements in $\mathbf{x}$ denote measurements on the same individual under five different circumstances.

(a) Use a comparison of profiles approach to explain what is being tested in the following hypotheses.

   i. $H_0 : \boldsymbol{\mu}_1 = \boldsymbol{\mu}_2$.
   ii. $H_0 : (\mu_{11} - \mu_{12}) = (\mu_{21} - \mu_{22}) = (\mu_{31} - \mu_{32}) = (\mu_{41} - \mu_{42}) = (\mu_{51} - \mu_{52})$.
   iii. $H_0 : \mu_{1j} = \mu_{2j} = \mu_{3j} = \mu_{4j} = \mu_{5j}, \; j = 1, 2$.
   iv. $H_0 : \mathbf{i}'\boldsymbol{\mu}_1 = \mathbf{i}'\boldsymbol{\mu}_2$ where $\mathbf{i}$ a vector of unities.

(b) Assume that you are required to use a multivariate regression program to carry out various hypothesis tests in multivariate analysis of variance. Assume that you are using effect coding for the design matrix with the second group coded -1 and the first group coded $+1$. Give the parameter matrix $\mathbf{B}$ that corresponds to the problem in (a).

(c) For each of the hypotheses outlined in (a) give the matrices $\mathbf{A}$ and $\mathbf{M}$ required for the test $H_0: \mathbf{ABM} = \mathbf{0}$.

2. The design matrix used in the one-way univariate ANOVA model can also be used for the MANOVA model. In ANOVA, the design matrix appears on the right-hand side of a multiple regression whereas in MANOVA the design matrix appears on the right-hand side of a multivariate regression. Using dummy coding the design matrix for a one-way analysis with g groups is given by $\mathbf{Z}$ below. The number of

observations in group j is denoted by n_j, $j = 1, 2, \ldots, g$, $n \sum_{j=1}^{g} n_j$.

$$\mathbf{Z} = \begin{bmatrix}
1 & 1 & 0 & 0 & & 0 \\
1 & 1 & 0 & 0 & & 0 \\
\vdots & \vdots & \vdots & \vdots & \cdots & \vdots \\
1 & 1 & 0 & 0 & & 0 \\
1 & 0 & 1 & 0 & & 0 \\
1 & 0 & 1 & 0 & & 0 \\
\vdots & \vdots & \vdots & \vdots & \cdots & \vdots \\
1 & 0 & 1 & 0 & & 0 \\
1 & 0 & 0 & 1 & & 0 \\
1 & 0 & 0 & 1 & & 0 \\
\vdots & \vdots & \vdots & \vdots & \cdots & \vdots \\
1 & 0 & 0 & 1 & & 0 \\
\vdots & \vdots & \vdots & \vdots & \cdots & \vdots \\
1 & 0 & 0 & 0 & & 1 \\
1 & 0 & 0 & 0 & & 1 \\
\vdots & \vdots & \vdots & \vdots & \cdots & \vdots \\
1 & 0 & 0 & 0 & & 1 \\
1 & 0 & 0 & 0 & & 0 \\
1 & 0 & 0 & 0 & & 0 \\
\vdots & \vdots & \vdots & \vdots & \cdots & \vdots \\
1 & 0 & 0 & 0 & & 0
\end{bmatrix}
\begin{array}{l}
\left.\vphantom{\begin{matrix}1\\1\\ \vdots \\1\end{matrix}}\right\} n_1 \\
\left.\vphantom{\begin{matrix}1\\1\\ \vdots \\1\end{matrix}}\right\} n_2 \\
\left.\vphantom{\begin{matrix}1\\1\\ \vdots \\1\end{matrix}}\right\} n_3 \\
\\
\left.\vphantom{\begin{matrix}1\\1\\ \vdots \\1\end{matrix}}\right\} n_{g-1} \\
\left.\vphantom{\begin{matrix}1\\1\\ \vdots \\1\end{matrix}}\right\} n_g
\end{array}$$

(a) Show that the matrices $\mathbf{Z'Z}$ and $\mathbf{Z'X}$ are given by

$$\mathbf{Z'Z} = \begin{bmatrix}
n & n_1 & n_2 & \cdots & n_{g-1} \\
n_1 & n_1 & 0 & \cdots & 0 \\
n_2 & 0 & n_2 & & \vdots \\
\vdots & \vdots & & \ddots & 0 \\
n_{g-1} & 0 & \cdots & 0 & n_{g-1}
\end{bmatrix} = \begin{bmatrix} n & \mathbf{n'} \\ \mathbf{n} & \mathbf{N} \end{bmatrix},$$

$$\mathbf{Z'X} = \begin{bmatrix}
x_{.1.} & x_{.2.} & x_{.3.} & \cdots & x_{.p.} \\
x_{.11} & x_{.21} & x_{.31} & \cdots & x_{.p1} \\
x_{.12} & x_{.22} & x_{.32} & \cdots & x_{.p2} \\
\vdots & \vdots & \vdots & & \vdots \\
x_{.1(g-1)} & x_{.2(g-1)} & x_{.3(g-1)} & \cdots & x_{.p(g-1)}
\end{bmatrix},$$

where $\mathbf{n'} = (n_1, n_2, \ldots, n_{g-1})$, $\mathbf{N}$ is a diagonal matrix of elements $n_1, n_2, \ldots, n_{g-1}$, and $\mathbf{X}$ is the $(n \times p)$ matrix of elements

x_{ijk} where $i = 1, 2, \ldots, n_k$; $k = 1, 2, \ldots, g$; and $j = 1, 2, \ldots, p$. The order of the rows of $\mathbf{X}$ conform to those of $\mathbf{Z}$ and the dot notation for x_{ijk} is used to denote sums over the dotted subscript.

(b) Show that $(\mathbf{Z}'\mathbf{Z})^{-1}$ can be written as $(\mathbf{Z}'\mathbf{Z})^{-1} = \begin{bmatrix} a & b' \\ b & C \end{bmatrix}$,

where $a = 1/ng$, $b' = -\left[\frac{1}{n_g} \frac{1}{n_g} \cdots \frac{1}{n_g} \right]$,

$$\mathbf{C} = \begin{bmatrix} \frac{1}{n_1} & & & \\ & \frac{1}{n_2} & & \\ & & \ddots & \\ & & & \frac{1}{n_{g-1}} \end{bmatrix} + \frac{1}{n_g} \mathbf{ii'},$$

where $\mathbf{i}$ is a column of unities. HINT: Use the formula for the inverse of a partitioned matrix.

(c) Show that $\widehat{\mathbf{B}} = (\mathbf{Z}'\mathbf{Z})^{-1}\mathbf{Z}'\mathbf{X}$ yields the matrix of estimators

$$\widehat{\mathbf{B}} = \begin{bmatrix} \bar{x}_{\cdot 1g} & \bar{x}_{\cdot 2g} & \cdots & \bar{x}_{\cdot pg} \\ (\bar{x}_{\cdot 11} - \bar{x}_{\cdot 1g}) & (\bar{x}_{\cdot 21} - \bar{x}_{\cdot 2g}) & \cdots & (\bar{x}_{\cdot p1} - \bar{x}_{\cdot pg}) \\ (\bar{x}_{\cdot 12} - \bar{x}_{\cdot 1g}) & (\bar{x}_{\cdot 22} - \bar{x}_{\cdot 2g}) & \cdots & \\ \vdots & & & \\ (\bar{x}_{\cdot 1(g-1)} - \bar{x}_{\cdot 1g}) & (\bar{x}_{\cdot 2(g-1)} - \bar{x}_{\cdot 2g}) & \cdots & (\bar{x}_{\cdot p(g-1)} - \bar{x}_{\cdot pg}) \end{bmatrix}.$$

3. Given a random sample of n_1 observations on $\mathbf{x}(p \times 1)$ from population 1 and n_2 observations from population 2 denote the $n = (n_1 + n_2)$ observations on $\mathbf{x}$ by $\mathbf{X}(n \times p)$. Define $y = 1$ for population 1 and $y = 0$ for population 2 and denote the n y observations by the vector $\mathbf{y}$. Consider the regression model $\mathbf{y} = \mathbf{X}\boldsymbol{\beta}_1 + \mathbf{i}\beta_0 + \mathbf{u}$, where $\mathbf{X}$ does not contain the column of unities $\mathbf{i}$. The least squares estimators of $\boldsymbol{\beta}_1$ and β_0 are given by $\hat{\boldsymbol{\beta}}_1 = \mathbf{S}_{xx}^{-1}\mathbf{S}_{xy}$ and $\hat{\beta}_0 = \bar{y} - \hat{\boldsymbol{\beta}}_1'\bar{\mathbf{x}}$, where $\mathbf{S}_{xx}$ is the sample covariance matrix for $\mathbf{x}$ and $\mathbf{S}_{xy}$ is the sample covariance between y and $\mathbf{x}$, $\bar{\mathbf{x}}$ is the vector of means for $\mathbf{x}$ and $\bar{y}$ is the mean for y.

(a) Show that $\mathbf{S}_{xy} = [\mathbf{X}'\mathbf{y} - n\bar{\mathbf{x}}\bar{y}]/(n-1)$ can be written as $n_1 n_2 (\bar{x}_{\cdot 1} - \bar{x}_{\cdot 2})/n(n-1)$ where $\bar{x}_1$ and $\bar{x}_2$ are the sample means on $\mathbf{x}$ in the two populations. Also show that $\hat{\boldsymbol{\beta}}_1 = n_1 n_2 \mathbf{S}_{xx}^{-1}(\bar{x}_{\cdot 1} - \bar{x}_{\cdot 2})/n(n-1)$ and $\hat{\beta}_0 = n_1/n - n_1 n_2(\bar{x}_{\cdot 1} - \bar{x}_{\cdot 2})'\mathbf{S}_{xx}^{-1}\bar{\mathbf{x}}/n(n-1)$.

(b) Show that for a given observation $\mathbf{x}$, $\hat{y}$ is given by $\hat{y} = \hat{\beta}_0 + \mathbf{x}'\hat{\boldsymbol{\beta}}_1 = n_1/n + n_1 n_2(\mathbf{x} - \bar{\mathbf{x}})'\mathbf{S}_{xx}^{-1}(\bar{\mathbf{x}}_{\cdot 1} - \bar{\mathbf{x}}_{\cdot 2})/n(n-1)$.

(c) Show that if $\mathbf{x} = \bar{\mathbf{x}}_{\cdot 1}$, then $\hat{y}$ is given by

$$\hat{y}_1 = \frac{n_1}{n} + \frac{n_1 n_2}{n(n-1)}(\bar{\mathbf{x}}_{\cdot 1} - \bar{\mathbf{x}})'\mathbf{S}_{xx}^{-1}(\bar{\mathbf{x}}_{\cdot 1} - \bar{\mathbf{x}}_{\cdot 2})$$

and if $\mathbf{x} = \bar{\mathbf{x}}_{\cdot 2}$ then $\hat{y}$ is given by

$$\hat{y}_2 = \frac{n_1}{n} + \frac{n_1 n_2}{n(n-1)}(\bar{\mathbf{x}}_{\cdot 2} - \bar{\mathbf{x}})'\mathbf{S}_{xx}^{-1}(\bar{\mathbf{x}}_{\cdot 1} - \bar{\mathbf{x}}_{\cdot 2}).$$

Also show that

$$(\hat{y}_1 + \hat{y}_2)/2 = y^* = \frac{n_1}{n} + \frac{n_1 n_2 (n_2 - n_1)}{2n^2(n-1)}(\bar{\mathbf{x}}_{\cdot 1} - \bar{\mathbf{x}}_{\cdot 2})'\mathbf{S}_{xx}^{-1}(\bar{\mathbf{x}}_{\cdot 1} - \bar{\mathbf{x}}_{\cdot 2}).$$

(d) Show that a decision rule that places $\mathbf{x}$ in population 1 if $\hat{y} \geq y^*$ is given by the rule which places $\mathbf{x}$ in population 1 if

$$x'\mathbf{S}_{xx}^{-1}(\bar{x}_{\cdot 1} - \bar{x}_{\cdot 2}) \geq \frac{1}{2}(\bar{\mathbf{x}}_{\cdot 1} + \bar{\mathbf{x}}_{\cdot 2})\mathbf{S}_{xx}^{-1}(\bar{\mathbf{x}}_{\cdot 1} - \bar{\mathbf{x}}_{\cdot 2}).$$

(e) Compare the rule given in (d) to the Fisher Criterion given in Section 8.2. Recall that for $\mathbf{x} \sim N(\boldsymbol{\mu}, \boldsymbol{\Sigma})$,

$$f(\mathbf{x}) = [\|2\pi\Sigma\|]^{-p/2} \exp[-1/2(\bar{\mathbf{x}} - \boldsymbol{\mu})'\boldsymbol{\Sigma}^{-1}(\bar{\mathbf{x}} - \boldsymbol{\mu})].$$

4. In discriminant analysis for two groups the minimum total probability of misclassification criterion and the Bayes criterion both assign $\mathbf{x}$ to population 1 if $f_1(\mathbf{x})/f_2(\mathbf{x}) > p_2/p_1$, where p_2 and p_1 are the prior probabilities of membership in groups 2 and 1 respectively.

(a) Show that if the two densities are $\mathbf{x} \sim N(\boldsymbol{\mu}_1, \boldsymbol{\Sigma})$ and $\mathbf{x} \sim N(\boldsymbol{\mu}_2, \boldsymbol{\Sigma})$ then this rule is equivalent to

$$\mathbf{x}'\boldsymbol{\Sigma}^{-1}(\boldsymbol{\mu}_1 - \boldsymbol{\mu}_2) > \frac{1}{2}(\boldsymbol{\mu}_1 + \boldsymbol{\mu}_2)'\boldsymbol{\Sigma}^{-1}(\boldsymbol{\mu}_1 - \boldsymbol{\mu}_2) + \ln(p_2/p_1).$$

(b) Suppose the two densities are univariate Poisson with parameters λ_1 and λ_2 respectively. Show that the above rule suggests that X is assigned to group 1 if

$$X > \Big[\ln(p_2/p_1) + (\lambda_1 - \lambda_2)\Big]/\ln(\lambda_1/\lambda_2).$$

Recall that the Poisson density with parameter λ is given by $f(x) = e^{-\lambda}\lambda^x/x!$.

5. If $\mathbf{A}(p \times p)$ is a symmetric matrix and $\mathbf{x}$ is a nonzero vector then the maximum value of the expression $\mathbf{x}'\mathbf{A}\mathbf{x}/\mathbf{x}'\mathbf{x}$ is λ_1, where λ_1 is the largest eigenvalue of $\mathbf{A}$ and the maximum value is attained at $\mathbf{x} = \mathbf{v}_1$,

where v_1 is the corresponding eigenvector of $\mathbf{A}$. Use this result to show that the maximum value of $\mathbf{b}'\mathbf{G}\mathbf{b}/\mathbf{b}'\mathbf{W}\mathbf{b}$ is λ_1, where λ_1 is the largest eigenvalue of $\mathbf{W}^{-1}\mathbf{G}$ and that the maximum is attained at $\mathbf{b} = \mathbf{v}_1$, where $\mathbf{v}_1$ is the corresponding eigenvector. HINT: Let $\mathbf{b} = \mathbf{W}^{-1/2}\mathbf{x}$ and use the fact that the eigenvalues and eigenvectors of $\mathbf{DC}$ are equal to the eigenvalues and eigenvectors of $\mathbf{CD}$, where $\mathbf{D} = \mathbf{W}^{-1/2}\mathbf{G}$ and $\mathbf{C} = \mathbf{W}^{-1/2}$.

6. A random variable $\mathbf{x}$ is known to have been selected from one of two populations where the populations are $N(\boldsymbol{\mu}_1, \boldsymbol{\Sigma})$ and $N(\boldsymbol{\mu}_2, \boldsymbol{\Sigma})$. Suppose that $\mathbf{x}$ is placed in the population whose squared Mahalanobis distance from $\mathbf{x}$ is the smallest. Show that this criterion is equivalent to the Fisher criterion, which places $\mathbf{x}$ in group 1 if

$$\mathbf{x}'\boldsymbol{\Sigma}^{-1}(\boldsymbol{\mu}_1 - \boldsymbol{\mu}_2) > \frac{1}{2}(\boldsymbol{\mu}_1 + \boldsymbol{\mu}_2)'\boldsymbol{\Sigma}^{-1}(\boldsymbol{\mu}_1 - \boldsymbol{\mu}_2).$$

7. Two bivariate normal populations are defined by $\boldsymbol{\mu}_1 = \begin{bmatrix} \mu_{11} \\ \mu_{21} \end{bmatrix}$, $\boldsymbol{\mu}_2 = \begin{bmatrix} \mu_{12} \\ \mu_{22} \end{bmatrix}$ and common covariance matrix $\boldsymbol{\Sigma} = \begin{bmatrix} \sigma_1^2 & \sigma_1\sigma_2\rho \\ \sigma_1\sigma_2\rho & \sigma_2^2 \end{bmatrix}$.

   (a) Show that the squared Mahalanobis distance between the two populations is given by

   $$d^2 = \frac{(\mu_{11} - \mu_{21})^2\sigma_2^2 + (\mu_{12} - \mu_{22})^2\sigma_1^2 - 2\sigma_1\sigma_2\rho(\mu_{12} - \mu_{22})(\mu_{11} - \mu_{21})}{\sigma_1^2\sigma_2^2(1 - \rho^2)}.$$

   (b) Show that d^2 in (a) can be written as

   $$d^2 = \frac{\Delta_1^2 + \Delta_2^2 - 2\rho\Delta_1\Delta_2}{(1 - \rho^2)}$$

   where $\Delta_1 = \dfrac{(\mu_{11} - \mu_{21})}{\sigma_1}$, $\Delta_2 = \dfrac{(\mu_{21} - \mu_{22})}{\sigma_2}$.

   (c) The larger the value of d^2 the better the ability to classify the unknown $\mathbf{x}$ correctly. For what values of ρ is d^2 larger than the value of d^2 corresponding to $\rho = 0$ given by $d_0^2 = \Delta_1^2 + \Delta_2^2$? What conditions are necessary on Δ_1 and Δ_2 for the discrimination to be improved? Interpret this graphically.

   (d) For what value of ρ is d^2 a maximum? Interpret the answer in terms of the parameters $\boldsymbol{\mu}_1$, $\boldsymbol{\mu}_2$ and ρ.

8. A logistic regression model is given by $\ln[p/(1-p)] = \beta_0 + \beta_1 X$, where X is a dummy variable and p is the probability that $Y = 0$ and $(1-p)$ is the probability that $Y = 1$. Express p as a function of X and show

that the probabilities for the (2×2) table for the conditional density of Y given X are as given below.

		Y	
		0	1
X	0	$e^{\beta_0}/(1+e^{\beta_0})$	$1/(1+e^{\beta_0})$
	1	$e^{\beta_0+\beta_1}/(1+e^{\beta_0+\beta_1})$	$1/(1+e^{\beta_0+\beta_1})$

(a) Explain in words what the quantities in each of the four cells represent.

(b) Let the theoretical joint probabilities be denoted by f_{ij} as shown below and relate the logits $\ln[f_{11}/f_{12}]$ and $\ln[f_{21}/f_{22}]$ to the parameters β_0 and β_1 in (a).

		Y	
		0	1
X	0	f_{11}	f_{12}
	1	f_{21}	f_{22}

(c) Show that $\beta_0 = \ln[f_{11}/f_{12}]$ and $\beta_1 = \ln[(f_{21}/f_{22})/(f_{11}/f_{12})]$ and interpret the parameters β_0 and β_1.

(d) Recall from Chapter 6 that for the loglinear model

$$\ln f_{ij} = \mu + \mu_{1(i)} + \mu_{2(j)} + \mu_{12(ij)} \quad i = 1, 2; \ j = 1, 2,$$

$\mu_{1(1)} = -\mu_{1(2)}; \ \mu_{2(1)} = -\mu_{2(2)}; \ \mu_{12(11)} = -\mu_{12(12)} = \mu_{12(22)} = -\mu_{12(21)}$. Show that $\beta_0 = 2\mu_{2(1)} + 2\mu_{12(11)}, \ \beta_1 = 4\mu_{12(12)}$.

9. Suppose that y_i is distributed as point binomial with density $f(y_i) = p_i^{y_i}(1-p_i)^{(1-y_i)}$.

(a) Show that for the random sample $y_1, y_2, \ldots, y_n$ the likelihood function is given by

$$L = \prod_{i=1}^{n} p_i^{y_i}(1-p_i)^{1-y_i}$$

and the log of the likelihood by

$$\ln L = \sum_{i=1}^{n} [y_i \ln p_i + (1-y_i)\ln(1-p_i)].$$

(b) Assume the x_i' are fixed and $\beta[(c+1) \times 1]$ is a vector of unknown parameters. Define $p_i = e^{x_i'\beta}/[1 + e^{x_i'\beta}]$ and hence $1 - p_i = 1/[1 + e^{x_i'\beta}], \ i = 1, 2, \ldots, n$ and show that $\ln L = \sum_{i=1}^{n} y_i x_i'\beta - \sum_{i=1}^{n} \ln(1 + e^{x_i'\beta})$.

(c) Show that the derivative of $\ln L$ with respect to β is given by $\partial \ln L / \partial \beta = \mathbf{X'y} - \mathbf{X'p}$, where $\mathbf{X}(n \times (c+1))$ is the matrix of fixed X values $\mathbf{x}'_i$, $i = 1, 2, \ldots, n$, $\mathbf{p}(n \times 1)$ is the vector of p_i values $i = 1, 2, \ldots, n$ and $\mathbf{y}(n \times 1)$ is the vector of y_i values $i = 1, 2, \ldots, n$.

(d) Show that the matrix of second derivatives of $\ln L (\partial^2 \ln L / \partial \beta_j \partial \beta_k)_{j,k}$ with respect to the elements of β is given by $-\mathbf{X'WX}$ where $\mathbf{W}$ is the diagonal matrix of elements $p_i(1 - p_i)$, $i = 1, 2, \ldots, n$.

(e) The Newton–Raphson procedure for approximating the maximum of the function $G(\mathbf{y}, \mathbf{X}, \beta) = \ln L$ with respect to β determines a sequence of values β_n such that $G(\mathbf{y}, \mathbf{X}, \beta_{n+1}) > G(\mathbf{y}, \mathbf{X}, \beta_n)$. The method employs the inverse of the Hessian matrix given by

$$\mathbf{H}_n = \left[\frac{\partial^2 G}{\partial \beta_j \partial \beta_k} \right]^{-1}_{\beta = \beta_n}$$

and the vector of first derivatives given by

$$\mathbf{d}_n = \left[\frac{\partial G}{\partial \beta_j} \right]_{\beta = \beta_n}$$

determined from $\beta_{n+1} = \beta_n - \mathbf{H}_n \mathbf{d}_n$. Show that β_{n+1} is given by

$$\beta_{n+1} = \beta_n + (\mathbf{X'WX})^{-1}_{\beta = \beta_n} (\mathbf{X'y} - \mathbf{X'p}_n)$$

where $\mathbf{p}_n$ denotes the vector $\mathbf{p}$ of p_i values evaluated at β_n.

10. Given the observations Y and X below, the logistic regression model $\ln[p_i/(1 - p_i)] = \beta_0 + \beta_1 X$ is to be estimated.

Y	0	0	0	1	1	1
X	1	4	5	7	10	14

(a) Obtain preliminary estimates of p_i using the simple least squares estimators in the model $y_i = \alpha_0 + \alpha_1 x_i + u_i$ given by

$$\hat{\alpha}_0 = \bar{y} - \hat{\alpha}_1 \bar{x}, \quad \hat{\alpha}_1 = \frac{n \sum_{i=1}^{n} xy - \sum_{i=1}^{n} x \sum_{i=1}^{n} y}{n \sum_{i=1}^{n} x^2 - (\sum_{i=1}^{n} x)^2}$$

where $\bar{y} = \sum_{i=1}^{n} y_i / n$ and $\bar{x} = \sum_{i=1}^{n} x_i / n$. Determine the values of $\hat{p}_i = \hat{\alpha}_0 + \hat{\alpha}_1 x_i$. Plot the values of $\hat{p}_i$ vs x_i. Compute the sums $\sum_{i=1}^{n} \hat{p}_i$ and $\sum_{i=1}^{n} x_i \hat{p}_i$ and compare to $\sum_{i=1}^{n} y_i$ and $\sum_{i=1}^{n} x_i y_i$.

(b) Determine the average value of x, $\bar{x}_1$ for $y = 0$ and the average value of x, $\bar{x}_2$ for $y = 1$ and obtain estimates $\bar{p}_1$ and $\bar{p}_2$ at $\bar{x}_1$ and $\bar{x}_2$ respectively using the estimated model obtained in (a). Use the relationships

$$\ln[\bar{p}_j/(1 - \bar{p}_j)] = \tilde{\beta}_0 + \tilde{\beta}_1 \bar{x}_j, \quad j = 1, 2$$

to determine estimates of β_0 and β_1 using $\bar{p}_1$, $\bar{x}_1$, $\bar{p}_2$ and $\bar{x}_2$. Use the estimates of β_0 and β_1 to determine new estimates of p_i, $\tilde{p}_i$, using

$$\tilde{p}_i = e^{\beta_0 + \beta_1 x_i}/[1 + e^{\beta_0 + \beta_1 x_i}], \qquad i = 1, 2, 3, 4, 5, 6,$$

for the x_i data values given in (a). Compare the new estimates $\tilde{p}_i$ to the $\hat{p}_i$ values obtained in (a). Also compute $\sum_{i=1}^{n} \tilde{p}_i$ and $\sum_{i=1}^{n} x_i \tilde{p}_i$ and compare to $\sum_{i=1}^{n} y_i$ and $\sum_{i=1}^{n} x_i y_i$ as in (a). Plot the $\tilde{p}_i$ values against x_i and compare the plot to the plot obtained in (a).

(c) Using the estimates of $\tilde{p}_i$ from (b) determine the diagonal elements $\tilde{p}_i(1 - \tilde{p}_i)$ of the diagonal matrix $\mathbf{W}$ and compute $\mathbf{X'WX}$ where $\mathbf{X}(6 \times 2)$ is the matrix containing a column of unities and the x values given in (a).

(d) Determine new estimators of β_0 and β_1 using the first iteration of the Newton–Raphson procedure

$$\left[\begin{array}{c} \tilde{\beta}_0^* \\ \tilde{\beta}_1^* \end{array} \right] = \left[\begin{array}{c} \tilde{\beta}_0 \\ \tilde{\beta}_1 \end{array} \right] + (\mathbf{X'WX})^{-1}[\mathbf{X'y} - \mathbf{X'\tilde{p}}],$$

where $\tilde{\beta}_0$, $\tilde{\beta}_1$ are the estimates determined in (b) and $\tilde{\mathbf{p}}(6 \times 1)$ is the vector of estimates of $\mathbf{p}(6 \times 1)$ determined in (b).

11. Given the dummy variable $Y = 0$ or 1 and the interval variable X, the linear model $Y = \beta_0 + \beta_1 X + U$ is called a linear probability model. Assume X is fixed and that U is independent of X with $E[U] = 0$.

(a) Let $p_i = P[y_i = 1]$ and $1 - p_i = P[y_i = 0]$ and show that $E[y_i] = p_i = \beta_0 + \beta_1 x_i$.

(b) Since $0 \leq p_i \leq 1$ assume $p_i = \beta_0 + \beta_1 x_i$ when $0 \leq \beta_0 + \beta_1 x_i \leq 1$ and $p_i = 1$ if $\beta_0 + \beta_1 x_i > 1$ and $p_i = 0$ if $\beta_0 + \beta_1 x_i < 0$. Plot p_i as a function of x_i. Can you guarantee that the ordinary least squares estimators of p_i are necessarily in the range $0 \leq p_i \leq 1$.

(c) Show that $E[u_i^2] = p_i(1 - p_i) = (\beta_0 + \beta_1 x_i)(1 - \beta_0 - \beta_1 x_i)$. What standard assumption about linear models does this result contradict?

(d) Suppose you have already obtained the simple least squares estimates $\hat{p}_i$ of the p_i. Use these estimates to obtain a second estimator of p_i that takes into account the result in (c). Show that the estimator is given by $\tilde{\boldsymbol{\beta}} = (\mathbf{X'DX})^{-1}(\mathbf{X'Dy})$, where $\mathbf{D}$ is a diagonal matrix with diagonal elements $1/p_i(1 - p_i)$.

(e) Can you guarantee in (d) that the new estimates of p_i given by $\mathbf{X}\tilde{\boldsymbol{\beta}}$ are in the range $0 \le p_i \le 1$?

12. The logarithm of the likelihood for the multinomial logit model is given by

$$\ln L = \sum_{i=1}^{n}\sum_{j=1}^{g} y_{ij} \ln p_{ij},$$

where

$$p_{ij} = e^{\mathbf{x}'_i\boldsymbol{\beta}_j}/\left[1+\sum_{j=1}^{(g-1)} e^{\mathbf{x}'_i\boldsymbol{\beta}_j}\right], \qquad i = 1, 2, \ldots, n, \ j = 1, 2, \ldots, (g-1),$$

and

$$p_{ig} = 1/\left[1+ \sum_{j=1}^{(g-1)} e^{\mathbf{x}'_i\boldsymbol{\beta}_j}\right], \qquad i = 1, 2, \ldots, n,$$

and $\mathbf{x}_i$ and $\boldsymbol{\beta}_j$ are $(c \times 1)$ vectors.

(a) Show that

$$\frac{\partial p_{ij}}{\partial \boldsymbol{\beta}_j} = p_{ij}(1 - p_{ij})\mathbf{x}_i, \quad j, k = 1, 2, \ldots, (g - 1), \ i = 1, 2, \ldots, n,$$

$$\frac{\partial p_{ij}}{\partial \boldsymbol{\beta}_k} = -p_{ij}p_{ik}\mathbf{x}_i, \qquad j \ne k,$$

$$\frac{\partial p_{ig}}{\partial \boldsymbol{\beta}_j} = -p_{ij}p_{ig}\mathbf{x}_i.$$

(b) Using the expression $\ln L = \sum_{i=1}^{n}\sum_{j=1}^{g} y_{ij} \ln p_{ij}$ and the derivative expressions in (a) show that

$$\frac{\partial \ln L}{\partial \boldsymbol{\beta}_k} = \sum_{i=1}^{n}\left[y_{ik}(1 - p_{ik}) + \sum_{\substack{j=1\\j\ne k}}^{g} y_{ij}(-p_{ik})\right]\mathbf{x}_i$$

$$= \sum_{i=1}^{n}(y_{ik} - p_{ik})\mathbf{x}_i$$

since $\sum_{j=1}^{g} y_{ij} = 1$.

(c) Differentiate the expression in (b) to obtain

$$\frac{\partial^2 \ln L}{\partial\beta_k\partial\beta_k'} = -\sum_{i=1}^{n} p_{ik}(1 - p_{ik})\mathbf{x}_i\mathbf{x}_i'$$

$$\frac{\partial^2 \ln L}{\partial\beta_k\partial\beta_\ell'} = \sum_{i=1}^{n} p_{ik}p_{i\ell}\mathbf{x}_i\mathbf{x}_i'.$$

(d) Let $\boldsymbol{\beta}$ denote the $c(g-1) \times 1$ vector containing the components $\boldsymbol{\beta}_j$, $j = 1, 2, \ldots, (g-1)$. Let $\mathbf{H}^{-1}$ $[c(g-1) \times (c(g-1)]$ denote the matrix with diagonal blocks given by $\partial^2 \ln L/\partial\beta_k\partial\beta_k'$ and off-diagonal blocks by $\partial^2 \ln L/\partial\beta_k\partial\beta_\ell'$. Let $\mathbf{d}$ denote the $c(g-1) \times 1$ vector with $(g-1)$ component vectors $\frac{\partial \ln L}{\partial\beta_k}$, $k = 1, 2, \ldots, (g-1)$. Write an equation for the Newton–Raphson procedure for estimating $\boldsymbol{\beta}$ as in Question 9(e).

13. Recall that the joint density for a random sample $y_1, y_2, \ldots, y_n$ from the point binomial is given by $L = \Pi_{i=1}^{n} p_i^{y_i}(1 - p_i)^{(1-y_i)}$. For a given explanatory variable x_i the probit model assumes that $p_i = F(\beta_0 + \beta_1 x_i)$ where F is the distribution function for the standard normal density

$$p_i = F(z_i) = \int_{-\infty}^{z_i} f(z)dz,$$

and where $f(z)$ is the standard normal density.
Suppose that the explanatory variable x_i is a dummy variable ($x_i = 0$ or 1) and consider the conditional probability table $p[Y \mid X]$ given below

		Y		
		0	1	
X	0	0.20	0.80	1
	1	0.60	0.40	1

Use the fact that

$$p[z \leq z_0] = 0.20 \Rightarrow z_0 = -0.84$$
$$p[z \leq z_0] = 0.60 \Rightarrow z_0 = 0.25$$

to show that $\beta_0 = -0.84$ and $\beta_1 = 1.09$ in the expression $z_i = \beta_0 + \beta_1 x_i$.

14. (a) For the data set given in Question 10 for (x, y), complete 10(a) to obtain estimates $\hat{p}_i$ of p_i.

(b) As in 10(b) compute $\bar{x}_1$ and $\bar{x}_2$ and obtain estimates $\bar{p}_1$ and $\bar{p}_2$ of p_i at $\bar{x}_1$ and $\bar{x}_2$ using the estimated model in 10(a). Now use the probit model to determine z for

$$\int_{-\infty}^{z} f(z)dz = \bar{p}_j, \qquad j = 1, 2,$$

and solve for β_0 and β_1 in the model

$$z_j = \beta_0 + \beta_1 \bar{x}_j, \qquad j = 1, 2.$$

(c) Use the estimates of β_0 and β_1 determined in (b) to generate z_i values for each of the six x_i values and determine the corresponding estimates $\tilde{p}_i$ of p_i using the normal density

$$\int_{-\infty}^{z_i} f(z)dz = \tilde{p}_i.$$

Compare the new estimates of p_i, $\tilde{p}_i$, to the estimates determined from ordinary least squares as in 10(a) and compare $\Sigma\tilde{p}_i$ and $\Sigma x_i \tilde{p}_i$ to Σy_i and $\Sigma x_i y_i$ respectively.

(d) Finish this like 10(c) and 10(d).

(e) Compare the results obtained using the probit model to the estimates obtained for the logit model in 10(d).

9

Principal Components, Factors and Correspondence Analysis

In exploratory studies, researchers often include as many variables as possible to ensure that no relevant variables will be omitted. The resulting data matrices can sometimes be large and difficult to analyze, particularly if the level of correlation among the variables is high. In techniques such as multiple regression and discriminant analysis, variable selection procedures can be employed as a data reduction technique; however this method can result in the loss of one or more important dimensions. An alternative approach is to use all of the variables in $\mathbf{X}$ to obtain a smaller set of new variables that can be used to approximate $\mathbf{X}$. The new variables are called principal components or factors and are designed to carry most of the information in the columns of $\mathbf{X}$. The higher the level of correlation among the columns of $\mathbf{X}$ the fewer the number of new variables required. The techniques of principal components analysis and factor analysis are examples of data reduction techniques.

Principal components analysis and factor analysis operate by replacing the original data matrix $\mathbf{X}$ by an estimate composed of the product of two matrices. The left matrix in the product contains a small number of columns corresponding to the factors or components, whereas the right matrix of the product provides the information that relates the components to the original variables. A scatterplot based on the left matrix is useful for relating the n objects of $\mathbf{X}$ with respect to the new factors. A plot based on the rows of the right matrix can be used to relate the components to the original variables. The decomposition of $\mathbf{X}$ into a product of two matrices is a special case of a matrix approximation procedure called a singular value decomposition. A two-dimensional plot based on this approximation is called a biplot.

A singular value decomposition can also be applied to the $(r \times c)$ matrix formed by an $(r \times c)$ contingency table. This application of the singular value decomposition is called correspondence analysis. A correspondence analysis produces a simultaneous plot locating row and column categories with respect to underlying row and column factors. In correspondence analysis, the underlying row and column factors attach interval scales to the row and column categories, and hence the technique can also be referred to as dual scaling. The extension of the technique to multidimensional contingency tables is called multiple correspondence analysis.

This chapter begins with a discussion of principal components analysis in Section 9.1 followed by an outline of factor analysis in Section 9.2. Section 9.3 describes the singular value decomposition and the biplot. The theory of correspondence analysis is surveyed in Section 9.4.

9.1   Principal Components

Given a set of n observations on p observed variables, the purpose of *principal components analysis* is to determine r new variables, where r is small relative to p. The r new variables called *principal components* must together account for most of the variation in the p original variables. The components are linear transformations of the original variables and are mutually orthogonal. The principal components can be used to provide an approximation for the data matrix $\mathbf{X}$. An example from economics involves a summary of the time series behavior of macro-economic variables for a variety of countries. One component could be representative of the "stock variables" and a second component might represent "flow variables." A second example from zoology involves the measurement of various body parts for various species of birds. The components might represent overall body size, wing span, head size, and so on.

9.1.1   A CLASSIC EXAMPLE

A classic example, which motivates the use of dimension reduction techniques, occurred early in the twentieth century with the attempt to characterize criminals on the basis of a set of body measurements. MacDonell (1902) obtained a correlation matrix relating seven body measurements for a sample of 3000 criminals. The body measurements used were left finger length, left forearm length, left foot length, head length, head breadth, face breadth, and height.

In a principal components analysis outlined in Maxwell (1977), three interesting uncorrelated components were determined from these measurements. The three components together accounted for 84% of the total variation among the seven variables. Each of the components is a linear combination of the original seven variables. The first component which accounts for 54% of the total variance, is a measure of overall size, and the second component which accounts for 21% of the variance represents a contrast between head size and the size of the remainder of the body. The third component contrasts head length and head breadth and carries 9% of the variation. The coefficients for the three components are summarized in Table 9.1. These three components define three uncorrelated measures of human body characteristics.

TABLE 9.1. Summary of Coefficients for Principal Components of Body Measurements

| Variables | Coefficients | | |
	First	Second	Third
Head length	0.538	-0.447	-0.712
Head breadth	0.413	-0.784	0.206
Face breadth	0.575	-0.628	0.309
Left finger length	0.853	0.288	0.056
Left forearm length	0.888	0.339	0.030
Left foot	0.878	0.219	0.048
Height	0.849	0.220	0.005

One might imagine scanning a large group of individuals and noticing that individual differences are characterized in terms of overall body size, head size relative to body size and finally the shape of the head. It is interesting to note also that these three derived dimensions are mutually uncorrelated. Perhaps there are additional components or body characteristics that could be obtained from measurements derived from the feet, thighs, legs and waist. The belief that characteristics of body size and shape could be used to distinguish criminals from the remainder of the population motivated some of the earliest applications of multivariate statistical analysis.

9.1.2 AN AD HOC APPROACH

To provide an initial example, ten observations on each of the variables SMEAN, PMEAN, PERWH and PM2 shown in Table 9.2 are used. These data are derived from a much larger sample of measurements taken on a large number of American cities. The variables measure annual mean of biweekly sulplate readings (SMEAN), annual mean of suspended particulate readings (PMEAN), percent of whites in the population (PERWH), and population density per square mile (PM2). This data is available in Gibbons, McDonald and Gunst (1987). A larger portion of this data set is used later in this section.

The graphs in Figure 9.1 can be used to study the variation in each of the four variables over the ten cities. To understand the principal components analysis problem, the reader should try to imagine replacing the four variables in Figure 9.1 by a single *index variable* in such a way that the index variable accounts for most of the variation in the four variables over the ten cities. In this section an ad hoc approach based on the simple average of the four variables is used to illustrate the problem.

TABLE 9.2. Air Pollution Data

SMEAN	PMEAN	PERWH	PM2
37	108	96.8	49.3
80	112	87.3	52.4
50	244	96.7	22.5
56	78	63.5	22.9
119	92	93.3	32.1
69	125	82.4	26.6
128	114	95.3	93.2
60	99	70.7	19.6
47	76	83.4	25.7
41	81	92.5	51.1

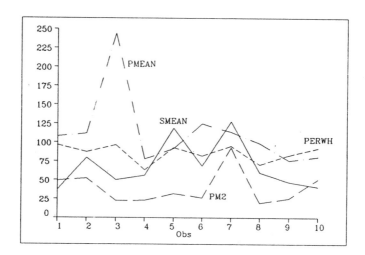

FIGURE 9.1. Variation in Four Variables Over the First Ten Observations

Our objective is to replace the four variables by a single variable, say Z, such that the original observations on the four variables can be approximated by multiplying the single variable Z by appropriate constants a_1, a_2, a_3 and a_4 respectively. In other words, the variation over the ten cities is summarized by the variation in Z. The estimate for each of the original variables is simply a constant multiple of this index. The estimators are given by

$$\text{SMEANH} = a_1 Z$$
$$\text{PMEANH} = a_2 Z$$
$$\text{PERWHH} = a_3 Z$$
$$\text{PM2H} = a_4 Z.$$

TABLE 9.3. Means, Standard Deviations and Component
Coefficients for Air Pollution Data

Variable	No. of Observations	Mean	Standard Deviation	Components Coefficient
PERWH	10	86.19	11.42	1.12
SMEAN	10	68.70	31.63	0.89
PMEAN	10	112.90	48.97	1.47
PM2	10	39.54	22.74	0.51
Grand Mean		76.83		

The variables SMEANH, PMEANH, PERWHH and PM2H are the corresponding estimators of SMEAN, PMEAN, PERWH and PM2 obtained from the index variable Z.

Since the four variables are to be replaced by a single variable, a reasonable candidate for Z would be a simple average of the four variables given by

$$Z = (\text{SMEAN} + \text{PMEAN} + \text{PERWH} + \text{PM2})/4.$$

The variation in Z over the ten observations is therefore an average of the variation in the four variables. To obtain reasonable values of the coefficients a_1, a_2, a_3 and a_4, we use the ratios of the variable averages to the overall mean of the four variables, given by

$$a_1 = \overline{\text{SMEAN}}/\overline{Z},$$
$$a_2 = \overline{\text{PMEAN}}/\overline{Z},$$
$$a_3 = \overline{\text{PERWH}}/\overline{Z},$$
$$a_4 = \overline{\text{PM2}}/\overline{Z},$$

where the bar notation denotes an average taken over the ten observations. Table 9.3 presents the means and standard deviations for the four variables and the values of the coefficients, a_1, a_2, a_3 and a_4.

The approximations to the ten observations for the four variables are shown in Table 9.4. These results are shown graphically in Figure 9.2. In Table 9.4, values of the variables which are approximately two standard deviations from the mean or more, have been underlined. The impact on Z of these more extreme observations can also be seen in the table. The mean of Z is relatively large for observations 3 and 7 and is relatively small for observation 4. The large values of Z correspond to the relatively large value of PMEAN for observation 3 and the relatively large values of SMEAN and PM2 for observation 7. The relatively small value of Z for observation 4 is due to the relatively small value of PERWH.

The variation in Z over the ten observations is also displayed in Figure 9.2. The estimates for each of the variables can be seen to be constant

TABLE 9.4. Estimates of the Four Variables Derived from the Means

OBS	SMEAN	SMEANH	PMEAN	PMEANH	PERWH	PERWHH	PM2	PM2H	ZMEAN
1	37	65.1	108	106.9	96.8	80.9	49.3	37.5	72.8
2	80	74.1	112	121.8	87.3	92.2	52.4	42.7	82.9
3	50	92.4	244	151.7	96.7	114.9	22.5	53.2	103.3
4	56	49.3	78	80.9	63.5	61.3	22.9	28.4	55.1
5	119	75.2	92	123.5	93.3	93.5	32.1	43.3	84.1
6	69	67.7	125	111.3	82.4	84.2	26.6	39.0	75.8
7	128	96.2	114	158.1	95.3	119.7	93.2	55.4	107.7
8	60	55.7	99	91.6	70.7	69.3	19.6	32.1	62.3
9	47	51.9	76	85.2	83.4	64.5	25.7	29.9	58.0
10	41	59.4	81	97.5	92.5	73.8	51.1	34.2	66.4
Means	68.7	68.7	112.9	112.9	86.2	86.2	39.5	39.5	76.8
St.Dev.	31.6	16.1	49.0	26.4	11.4	20.0	22.7	9.2	18.0

proportions of the value of Z (labeled MEAN). The impact of the relatively large and small values on the four estimates can also be seen in this figure. A comparison of the variable values to the estimates shows that some differences are due to relatively large or small values of Z whereas others are due to relatively large or small values of the variable. For observation 1 the difference between SMEAN and SMEANH $(37 - 65.1)$ is due to the relatively low value of SMEAN. For observation 3, the relatively large difference between PM2 and PM2H $(22.5 - 53.2)$ is due primarily to the large value of Z.

From Table 9.4 it can be seen that the standard deviations of the estimates SMEANH, PMEANH, and PM2H are lower than the corresponding standard deviations of the observed values SMEAN, PMEAN and PM2. For PERWHH the reverse is true. The standard deviation of Z is relatively low at 18.0.

Although this ad hoc method is simple and intuitive, the question arises as to whether a better approximation method exists. The principal components approach to the problem is outlined below. This approach attempts to find an index Z that minimizes the Euclidean distance between the original observations and the resulting estimates.

9.1.3 THE PRINCIPAL COMPONENTS APPROACH

Characterizing the First Principal Component

To introduce the principal components method it will be convenient to employ a more general notation. Given a data matrix $\mathbf{X}$ representing n observations on each of p variables, $X_1, X_2, \ldots, X_p$, the purpose of principal

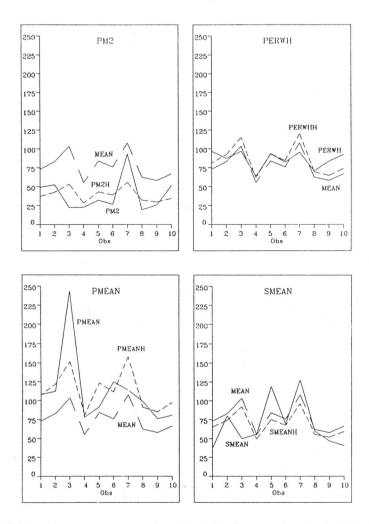

FIGURE 9.2. Approximations to Four Variables Compared to Original Variables and Z

components analysis is to determine a new variable Z_1 that can be used to account for the variation in the p X variables. The result can be used to provide a *matrix approximation* for **X**. The principal component Z_1 is given by a linear combination of the p X variables as

$$Z_1 = v_{11}X_1 + v_{21}X_2 + \ldots + v_{p1}X_p.$$

The approximations for the X variables derived from this principal component are then given by

$$
\begin{aligned}
\widehat{X}_1 &= a_{11}Z_1 \\
\widehat{X}_2 &= a_{21}Z_1 \\
&\vdots \\
\widehat{X}_p &= a_{p1}Z_1.
\end{aligned}
$$

If the principal component Z_1 captures most of the variation in the X variables, then the X approximations $\widehat{X}_j$ should be similar to X_j, $j = 1, 2, \ldots, p$.

The principal components approach to determining Z_1 involves minimizing the sum of squared deviations $\sum_{j=1}^{p} \sum_{i=1}^{n} (\hat{x}_{ij} - x_{ij})^2$, where $z_{i1} = \sum_{j=1}^{p} v_{j1} x_{ij}$ and $\hat{x}_{ij} = a_{j1} z_{i1}$. The x_{ij}, $i = 1, 2, \ldots, n$; $j = 1, 2, \ldots p$, denote the observations in $\mathbf{X}$. In matrix notation an equivalent approach is to determine $(p \times 1)$ vectors $\mathbf{v}_1$ and $\mathbf{a}_1$, where $\mathbf{z}_1$ $(n \times 1) = \mathbf{X}\mathbf{v}_1$ and $\widehat{\mathbf{X}} = \mathbf{z}_1 \mathbf{a}_1'$ such that $tr(\mathbf{X} - \widehat{\mathbf{X}})'(\mathbf{X} - \widehat{\mathbf{X}})$ is minimized. The problem can therefore be stated as $\min tr(\mathbf{X} - \widehat{\mathbf{X}})'(\mathbf{X} - \widehat{\mathbf{X}})$ subject to $\widehat{\mathbf{X}} = \mathbf{z}_1 \mathbf{a}_1'$ and $\mathbf{z}_1 = \mathbf{X}\mathbf{v}_1$. The reader should recall that the trace of a matrix is simply the sum of the diagonal elements.

The Eigenvalue Problem

The solution to this problem involves solving the *eigenvalue–eigenvector problem* given by

$$(\mathbf{X}'\mathbf{X} - \lambda\mathbf{I})\mathbf{v} = 0.$$

Since the magnitude of $\mathbf{v}$ is arbitrary, the customary restriction $\mathbf{v}'\mathbf{v} = 1$ is employed. The solutions to the problem are the *eigenvalues* λ_j, $j = 1, 2, \ldots, s$, and the corresponding *eigenvectors* $\mathbf{v}_j$, $j = 1, 2, \ldots, s$, where the number of solutions s corresponds to the rank of $\mathbf{X}'\mathbf{X}$. The reader who is unfamiliar with eigenvalues and eigenvectors should review these concepts in the Appendix.

The s eigenvectors and corresponding eigenvalues provide s solutions for the desired principal component Z_1. The solution which corresponds to the required minimum employs the largest eigenvalue λ_1 and the corresponding eigenvector $\mathbf{v}_1$. This principal component is called the first principal component. Without loss of generality we assume that the s eigenvalues have been ordered from largest to smallest as $\lambda_1, \lambda_2, \ldots, \lambda_s$ and the corresponding eigenvectors are denoted by $\mathbf{v}_1, \mathbf{v}_2, \ldots, \mathbf{v}_s$. The observations on the principal component Z_1 are therefore given by $\mathbf{z}_1 = \mathbf{X}\mathbf{v}_1$. Given $\mathbf{z}_1$, the vector $\mathbf{a}_1$ in $\widehat{\mathbf{X}} = \mathbf{z}_1 \mathbf{a}_1'$ that minimizes the sum of squared deviations is given by $\mathbf{a}_1 = \mathbf{v}_1$ and hence $\widehat{\mathbf{X}} = \mathbf{z}_1 \mathbf{v}_1'$. The eigenvector $\mathbf{v}_1$ therefore provides both $\mathbf{z}_1$ and $\widehat{\mathbf{X}}$. The quantity $tr(\mathbf{X} - \widehat{\mathbf{X}})'(\mathbf{X} - \widehat{\mathbf{X}})$ has the value

TABLE 9.5. Estimates Using First Principal Component

Obs	SMEAN	SMEANH	PMEAN	PMEANH	PERWH	PERWHH	PM2	PM2H	Z_1
1	37	63.8	108	109.7	96.8	78.5	49.3	36.6	153.7
2	80	70.7	112	121.6	87.3	87.0	52.4	40.5	170.3
3	50	103.6	<u>244</u>	178.3	96.7	127.6	22.5	59.4	249.7
4	56	48.5	78	83.4	<u>63.5</u>	59.7	22.9	27.8	116.8
5	119	70.7	92	121.7	93.3	87.1	32.1	40.6	170.4
6	69	69.0	125	118.8	82.4	85.0	26.6	39.6	166.3
7	<u>128</u>	85.3	114	146.7	95.3	105.0	<u>93.2</u>	48.9	205.4
8	60	56.6	99	97.4	70.7	69.7	19.6	32.5	136.4
9	47	50.8	76	87.5	83.4	62.6	25.7	29.2	122.5
10	41	55.7	81	95.9	92.5	68.6	51.1	32.0	134.3
Mean	68.7	67.5	112.9	116.1	86.2	83.1	39.5	38.7	162.6
Standard Deviation	31.6	16.8	49.0	29.0	11.4	20.7	22.7	9.7	40.6

$tr\mathbf{X}'\mathbf{X} - \lambda_1$. Since $tr\mathbf{X}'\mathbf{X} = \sum_{j=1}^{s} \lambda_j$, where $s = \text{rank}(\mathbf{X}'\mathbf{X})$, $\sum_{j=2}^{s} \lambda_j$ is the required minimum.

Computer Software

The principal component calculations for this section were carried out using SAS PRINCOMP.

Example

For the air pollution data in Table 9.2 the first principal component is given by

$$Z_1 = 0.41 \text{ SMEAN} + 0.71 \text{ PMEAN} + 0.51 \text{ PERWH} + 0.24 \text{ PM2}.$$

In comparison to the ad hoc procedure which employed a simple average, we can see that the first principal component places greater emphasis on PMEAN and less emphasis on PM2. In addition the sum of the weights is 1.87 rather than 1.0 in the case of a simple average. The $\mathbf{X}$ approximations $\mathbf{z}_1\mathbf{a}_1' = \mathbf{z}_1\mathbf{v}_1'$ are summarized in Table 9.5. In Figure 9.3 the approximations are plotted along with the original values of the four variables and the values of the principal component $\mathbf{z}_1$. The estimates for the variables can be seen to be related to the value of Z_1 through multiplication by a constant. The variation in the resulting estimate should be similar to the variation in the original variable.

In Table 9.5, the standard deviation for the first principal component Z_1 is 40.6, which is more than twice the standard deviation (18.0) for the

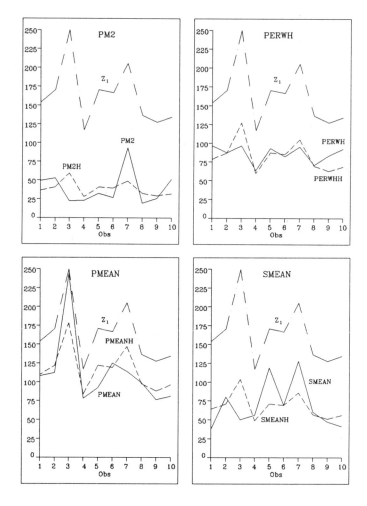

FIGURE 9.3. Approximations to Four Variables Based on the First Principal Component

mean, Z, used in Table 9.4. Most of this increase is due to an overall increase in the magnitude of Z_1 in comparison to Z. The mean of Z_1 (162.6) is a factor (2.12) larger than the mean of Z (76.8). The increase in the standard deviation in the Z_1 variable carries over to provide increases in the standard deviations of the four sets of estimates. We can conclude that the first principal component is more responsive to the variation in the original variables. In particular, the first principal component is much more sensitive to the large value of PMEAN in observation 3. The relatively large coefficient (0.71) for PMEAN results in an index that is more sensitive to the variation in this variable. The sum-of-squared deviations between the

40 original observations and the 40 estimates is 21,809, which is marginally less than the value of 23,419 obtained by using the mean, Z, in the ad hoc approach of Table 9.4.

Since the four variables in this example measure quite different quantities, there is no reason to believe that the variation of all four variables over the ten cities should be related to the variation in a single index variable. The purpose of this example was to illustrate the characteristics of a single principal component. The principal components analysis technique is extended below to obtain additional indices or components to improve the characterization of the variation of the four variables over the ten cities. We shall see that by using additional indices the approximation can be improved.

Generalization to r Principal Components

For the matrix $\mathbf{X'X}$, each of the eigenvalues λ_k and the corresponding eigenvectors $\mathbf{v}_k$, $k = 1, 2, \ldots, r$, where $r \leq s = \text{rank}(\mathbf{X'X})$, can be used to generate components $\mathbf{z}_k = \mathbf{Xv}_k$. The resulting r principal components are given by

$$
\begin{aligned}
Z_1 &= v_{11}X_1 + v_{21}X_2 + \ldots + v_{p1}X_p \\
Z_2 &= v_{12}X_1 + v_{22}X_2 + \ldots + v_{p2}X_p \\
&\vdots \\
Z_r &= v_{1r}X_1 + v_{2r}X_2 + \ldots + v_{pr}X_p.
\end{aligned}
$$

The observations on the r components are defined by $\mathbf{z}_k = \mathbf{Xv}_k$. The entire $(n \times r)$ matrix $\mathbf{Z}$ of observations on the r components is given by $\mathbf{Z} = \mathbf{XV}$, where $\mathbf{V}$ contains r columns corresponding to the first r eigenvectors of $\mathbf{X'X}$.

The X approximations can be improved by using the additional components defined above. For the p X variables, the approximations are given by

$$
\begin{aligned}
\widehat{X}_1 &= v_{11}Z_1 + v_{12}Z_2 + \ldots + v_{1r}Z_r \\
\widehat{X}_2 &= v_{21}Z_1 + v_{22}Z_2 + \ldots + v_{2r}Z_r \\
&\vdots \\
\widehat{X}_p &= v_{p1}Z_1 + v_{p2}Z_2 + \ldots + v_{pr}Z_r.
\end{aligned}
$$

In matrix notation the approximation $\widehat{\mathbf{X}}$ of the matrix $\mathbf{X}$ is given by $\widehat{\mathbf{X}} = \mathbf{ZV'}$, where $\mathbf{Z}$ is the $(n \times r)$ matrix of observations on the first r principal components, and $\mathbf{V}$ $(p \times r)$ is the matrix whose columns are the first r eigenvectors of $\mathbf{X'X}$. For the approximations to $\mathbf{x}_j$, the jth column of $\mathbf{X}$, the equation is given by $\hat{\mathbf{x}}_j = \mathbf{Zv'}_j$, where $\mathbf{v'}_j$ is the jth column of $\mathbf{V'}$.

Since the eigenvectors $\mathbf{v}_k$ are mutually orthogonal, the principal components are also mutually orthogonal. The matrix $\mathbf{Z}'\mathbf{Z} = \Lambda$, where Λ is the diagonal matrix of r eigenvalues λ_k, $k = 1, 2, \ldots, r$. The sum of squares and cross products matrix for the principal components is therefore a diagonal matrix with diagonal elements λ_k, that decline in magnitude. The eigenvalues λ_k are given by $\sum_{i=1}^{n} z_{ik}^2 = \lambda_k$, $k = 1, 2, \ldots, r$. The sum of squares for each principal component is therefore given by the corresponding eigenvalue.

The quantity to be minimized is given by

$$tr(\mathbf{X} - \widehat{\mathbf{X}})'(\mathbf{X} - \widehat{\mathbf{X}}) = tr\mathbf{X}'\mathbf{X} - \sum_{k=1}^{r} \lambda_k$$

and hence if $r = s$ (the rank of $\mathbf{X}'\mathbf{X}$), then this expression has the value zero. The rank of $\mathbf{X}'\mathbf{X}$ is usually $s = p$, where p is the number of variables or columns in $\mathbf{X}$.

Each of the eigenvectors generates a portion of the total variation (or sum of squares) in $\mathbf{X}$ as measured by $tr(\mathbf{X}'\mathbf{X})$. The contribution to $\sum_{k=1}^{s} \lambda_k$ provided by $\mathbf{z}_\ell = \mathbf{Z}\mathbf{v}'_\ell$ is $\mathbf{z}'_\ell\mathbf{z}_\ell = \lambda_\ell$. The proportion of the total variation measured by $tr(\mathbf{X}'\mathbf{X})$, accounted for by the component $\mathbf{z}_\ell$, is given by $\lambda_\ell / \sum_{k=1}^{s} \lambda_k$. The number of components actually used for the approximation of $\mathbf{X}$ can be guided by the measure

$$\sum_{k=1}^{\ell} \lambda_k / \sum_{k=1}^{s} \lambda_k, \quad \text{where} \quad \ell \leq s.$$

In practice, this ratio can be relatively close to 1 even though ℓ is much less than s.

Spectral Decomposition

If the rank of $\mathbf{X}'\mathbf{X}$ is s, then $\mathbf{X}'\mathbf{X}$ can be written in terms of the eigenvalues λ_k and corresponding eigenvectors $\mathbf{v}_k$, $k = 1, 2, \ldots, s$,

$$\mathbf{X}'\mathbf{X} = \sum_{k=1}^{s} \lambda_k \mathbf{v}_k \mathbf{v}'_k,$$

which is called the *spectral decomposition* of $\mathbf{X}'\mathbf{X}$ (see Appendix). Since the magnitudes of the eigenvalues decline exponentially, the terms in the spectral decomposition also decline exponentially. A small number of terms is therefore usually sufficient to approximate $\mathbf{X}'\mathbf{X}$. (Recall that the magnitudes of the elements of $\mathbf{v}_k$ are limited since $\mathbf{v}'_k\mathbf{v}_k = 1$.)

The Full Rank Case

If the rank of $\mathbf{X}$ $(n \times p)$ is $p < n$, then there are p positive eigenvalues $\lambda_1, \lambda_2, \ldots, \lambda_p$ of $\mathbf{X}'\mathbf{X}$ with corresponding eigenvectors $\mathbf{v}_1, \mathbf{v}_2, \ldots, \mathbf{v}_p$. In

TABLE 9.6. Estimates Using First Two Principal Components

OBS	SMEAN	SMEANH	PMEAN	PMEANH	PERWH	PERWHH	PM2	PM2H
1	37	58.9	108	114.6	96.8	77.1	49.3	33.6
2	80	81.3	112	110.9	87.3	90.1	52.4	47.1
3	50	42.2	244	239.4	96.7	109.8	22.5	21.4
4	56	52.9	78	79.0	63.5	61.0	22.9	30.6
5	119	101.0	92	91.6	93.3	95.8	32.1	59.3
6	69	62.9	125	124.9	82.4	83.2	26.6	35.8
7	128	125.9	114	106.2	95.3	116.8	93.2	74.0
8	50	54.2	99	99.8	70.7	69.0	19.6	31.0
9	47	55.5	76	82.9	83.4	64.0	25.7	32.0
10	41	63.4	81	88.3	92.5	70.9	51.1	36.7

this case $\mathbf{X}$ can be written precisely as $\mathbf{X} = \mathbf{ZV'}$, where $\mathbf{V}$ is the matrix whose columns are the eigenvectors $\mathbf{v}_1, \mathbf{v}_2, \ldots, \mathbf{v}_p$ and $\mathbf{VV'} = \mathbf{I}$. The rank of $\mathbf{X'X}$ is p and hence $tr\mathbf{X'X} = \sum_{k=1}^{p} \lambda_k$ and the spectral decomposition is given by $\mathbf{X'X} = \sum_{k=1}^{p} \lambda_k \mathbf{v}_k \mathbf{v}_k'$.

Example

For the air pollution data of Table 9.2, the X matrix approximations provided by the first two principal components Z_1 and Z_2 are shown in Table 9.6 and Figure 9.4. In comparison to Table 9.5 and Figure 9.3, we can see that the use of the first two principal components improves the approximations. The quantity $tr\mathbf{X'X}$ in this case is 301,003 whereas the first two eigenvalues are $\lambda_1 = 279,194$ and $\lambda_2 = 16,261$. The proportion of the total variation in the X values explained by the first two components is therefore 0.98. The sum-of-squared deviations between the 40 observations and the corresponding estimates is $301{,}003 - 279{,}194 - 16{,}261 = 5548$. The addition of the second component results in a reduction in the error sum of squares by 16,261.

In Figure 9.4 it can be seen that the approximation for PMEAN is extremely good even though this variable showed the greatest variance. This result illustrates how the principal component approach is influenced by the variance of the variables. In comparison, the error deviations for PM2 do not seem to have changed between Figures 9.3 and 9.4.

Alternative Characterizations and Geometry

The principal components derived in this section were determined to minimize the sum-of-squared deviations between the actual observations and the approximations based on a small number of components. The principal components can also be obtained using an alternative criterion. Each

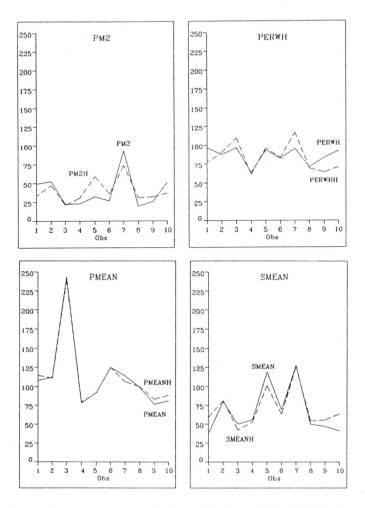

FIGURE 9.4. Approximations to Four Variables Provided by the First Two Principal Components

principal component Z is a linear combination of the p X variables, $Z = \sum_{j=1}^{p} v_j X_j$, such that the sum of the squared coefficients, $\sum_{j=1}^{p} v_j^2$, is unity. The first principal component can be shown to be the unique linear combination that maximizes the sum-of-squared elements $\sum_{i=1}^{n} z_{ik}^2 = \lambda_k$. In other words, the first eigenvalue λ_1 yields the largest possible value of $\sum_{i=1}^{n} z_{ik}^2$ subject to $\mathbf{z}_k = \mathbf{X}\mathbf{v}_k$ and $\mathbf{v}_k'\mathbf{v}_k = 1$. In a similar fashion the second principal component corresponding to λ_2 and $\mathbf{v}_2$ represents the largest possible value of $\sum_{i=1}^{n} z_{ik}^2$ subject to $z_k = \mathbf{X}\mathbf{v}_k$, $\mathbf{v}_k'\mathbf{v}_k = 1$ and $\sum_{i=1}^{n} z_{ik} z_{i1} = 0$, where the z_{i1} denotes values of the first principal component. Finally the rth principal component represents the linear combination of the original X values that maximizes the sum-of-squared elements $\sum_{i=1}^{n} z_{ir}^2$, subject to being orthogonal to the first $(r-1)$ principal components $\mathbf{z}_r'\mathbf{z}_k = 0$, $k = 1, 2, \ldots, (r-1)$ and subject to $\mathbf{z}_r = \mathbf{X}\mathbf{v}_r$, where $\mathbf{v}_r'\mathbf{v}_r = 1$, $k = 1, 2, \ldots, (r-1)$.

As discussed in Chapter 7, geometrically the $\mathbf{X}$ matrix defines n points in a p-dimensional space defined by the p axes $X_1, X_2, \ldots, X_p$. In the full rank case the orthogonal transformation defined by the matrix $\mathbf{V}$ replaces the p X axes by p Z axes which can be viewed as a *rigid rotation* of the X axes in the p-dimensional space. The first principal component Z_1 represents the single axis which most closely approximates the n data points. The estimates $\widehat{\mathbf{X}}_1$ based on Z_1 are the projections of the n observations onto the Z_1 axis. In other words, $\widehat{\mathbf{X}}_1$ minimizes the sum of squared deviations between $\mathbf{X}$ and $\widehat{\mathbf{X}} = \mathbf{Z}_1\mathbf{v}_1'$. Similarly, the second principal component axis Z_2 is an axis orthogonal to Z_1 that most closely approximates the residuals from Z_1. The remaining principal components can similarly be defined as a sequence of mutually orthogonal axes each designed to most closely approximate the observation residuals from the previous components.

In Figure 9.5 two principal components Z_1 and Z_2 are shown corresponding to observations on the two X variables X_1 and X_2. The axis corresponding to Z_1 should appear as the axis that minimizes the sum of perpendicular distances like AM. The angle of rotation relative to the axis X_1 is measured by θ, the angle between Z_1 and X_1. The point $(\hat{x}_{i1}, \hat{x}_{i2})$ denotes the approximation for (x_{i1}, x_{i2}) based on Z_1.

Principal Components and Multivariate Random Variables

For the $(p \times 1)$ multivariate random variable $\mathbf{x}$ with mean vector $\boldsymbol{\mu}$ $(p \times 1)$ and covariance matrix $\boldsymbol{\Sigma}$ $(p \times p)$ of rank p, there is an *orthogonal matrix* $\mathbf{T}$ $(p \times p)$, $\mathbf{T}'\mathbf{T} = \mathbf{I}$, such that $\mathbf{T}'\boldsymbol{\Sigma}\mathbf{T} = \boldsymbol{\Lambda}$ where $\boldsymbol{\Lambda}$ is a diagonal matrix of positive elements $\lambda_1, \lambda_2, \ldots, \lambda_p$ called the eigenvalues of $\boldsymbol{\Sigma}$. The rows of $\mathbf{T}'$ are the eigenvectors of $\boldsymbol{\Sigma}$ and the vector $\mathbf{y} = \mathbf{T}'(\mathbf{x} - \boldsymbol{\mu})$ yields the principal components of $\mathbf{x}$. The covariance matrix for $\mathbf{y}$ is the diagonal matrix $\boldsymbol{\Lambda}$. The p-*dimensional ellipsoid* $\mathbf{x}'\boldsymbol{\Sigma}^{-1}\mathbf{x}$ can be expressed as $\mathbf{y}'\boldsymbol{\Lambda}^{-1}\mathbf{y}$ where variables in $\mathbf{y}$ are defined to be the *principal axes* of the ellipsoid. These principal axes are mutually orthogonal in p-dimensional space. The first principal axis y_1 corresponding to the largest eigenvalue λ_1 defines the direction of maximum variance. The second principal axis y_2 corresponding to the second largest eigenvalue λ_2 defines the direction of maximum variance subject to being orthogonal to the first principal axis. Each succeeding principal axis determines a direction of maximum variance subject to being orthogonal to the previous principal axes.

Example

For the remainder of the discussion of principal components analysis, we shall employ a larger set of 40 observations on each of eleven variables obtained from the same source as in Table 9.2. This (40×11) data matrix

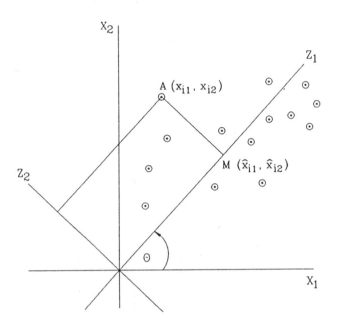

FIGURE 9.5. Geometry of Principal Components

is summarized in Table 9.7 and will be referred to as the air pollution data. The sum of squares and cross products matrix $\mathbf{X'X}$ corresponding to this data is summarized in Table 9.8. The variable TMR in Table 9.7 is used in Section 9.1.5. This data set was also used in Volume I to provide examples for multiple regression.

For the air pollution data the complete set of 11 eigenvalues and eigenvectors for the matrix $\mathbf{X'X}$ are shown in Table 9.9. Each eigenvector column $\mathbf{v}_k$ provides the coefficients for the principal component Z_k, $\mathbf{z}_k = \mathbf{Xv}_k$, $k = 1, 2, \ldots, 11$. Each of the eigenvalues $\lambda_1, \lambda_2, \ldots, \lambda_{11}$ in the table indicates the sum of squares explained by the respective component. The table also gives the cumulative proportion of the total sum of squares explained after adding successive components.

The first eigenvector shows that the first principal component Z_1 is primarily a function of PMAX and SMAX. We can conclude, therefore, that cities with large values of Z_1 should tend to indicate high values of the maximum pollution readings PMAX and SMAX.

The second eigenvector shows that the second principal component Z_2 is strongly positively related to SMAX and strongly negatively related to PMAX. This *bipolar component* therefore contrasts the pollution from SMAX with the pollution from PMAX. Cities with relatively large values of

TABLE 9.7. Air Pollution Data

SMIN	SMEAN	SMAX	PMIN	PMEAN	PMAX
21	37	105	50	108	302
29	80	313	42	112	343
2	50	91	61	244	646
37	56	152	35	78	233
73	119	220	50	92	189
51	69	212	39	125	285
28	128	344	53	114	241
34	60	145	28	99	160
21	47	106	23	76	164
22	41	147	22	81	149
87	229	620	70	160	342
51	124	210	76	135	242
23	69	202	43	100	231
25	123	280	50	156	344
38	60	71	28	60	94
46	66	133	23	106	193
39	65	166	30	83	215
57	228	445	99	221	403
36	126	264	46	143	347
112	153	365	75	215	537
189	273	399	81	175	323
46	92	139	46	112	236
24	165	414	48	148	495
28	75	212	21	70	185
37	162	396	24	77	182
27	79	260	47	121	309
23	54	139	22	102	174
59	81	351	37	144	417
82	100	225	42	86	163
30	42	70	26	62	157
28	77	149	34	81	166
46	72	251	28	74	135
61	81	203	39	121	260
152	194	437	74	198	444
16	68	233	50	124	296
45	125	194	63	145	316
33	51	107	42	101	202
27	58	113	57	125	352
60	228	531	79	162	270
58	73	212	39	111	255

Note: Variable definitions can be found in the introduction to the Data Set V7 given in the Data Appendix.

Z_2 will tend to have relatively high values of the pollution variable SMAX and relatively low values of the pollution variable PMAX.

TABLE 9.7. Air Pollution Data (continued)

PM2	GE65	PERWH	NONPOOR	LPOP	TMR
49.3	70	96.8	89.8	58.0775	664
52.4	89	87.3	78.0	52.0086	929
22.5	47	96.7	84.7	54.1863	621
22.9	48	63.5	67.3	53.3522	825
32.1	98	93.3	86.0	55.3791	1008
26.6	62	82.4	73.6	53.2176	829
93.2	91	95.3	91.3	58.3857	899
19.6	49	70.7	69.0	53.3843	721
25.7	92	83.4	76.1	55.0309	828
51.1	69	92.5	89.7	57.0138	810
122.4	91	84.3	87.0	66.3778	1029
134.8	62	75.1	89.5	63.0144	780
70.2	92	98.2	89.9	61.7086	876
139.3	89	91.2	87.6	68.2883	869
37.9	61	79.1	79.4	53.9185	747
58.6	62	76.6	77.6	56.5840	863
18.6	82	98.8	87.2	55.4654	734
46.5	77	87.6	87.9	55.6367	910
19.2	94	94.9	83.3	55.0730	943
51.7	119	99.6	78.3	53.7020	1400
59.4	95	94.6	86.8	55.3192	964
59.0	65	77.2	79.2	60.0740	823
95.6	76	77.8	85.5	62.3730	978
22.5	113	97.0	85.2	59.1482	1037
46.4	105	97.0	87.7	57.2639	996
50.1	70	98.6	88.2	55.8324	682
34.4	67	93.6	86.9	55.3559	690
43.9	72	94.3	88.0	53.3746	821
42.3	71	94.7	87.0	52.9902	776
21.4	71	61.7	68.9	52.2843	908
24.3	73	94.5	84.9	60.1410	737
3.6	114	99.2	81.9	54.4185	1117
38.9	111	99.7	75.2	55.4029	1282
20.1	122	97.6	77.8	52.7953	1210
55.1	68	93.1	72.8	58.3705	734
146.8	96	87.9	85.1	60.3004	1039
23.5	79	90.0	85.7	52.8047	854
49.8	52	68.2	78.1	53.6184	706
497.7	97	88.0	86.8	70.2917	1046
32.7	101	90.6	84.0	51.1873	978

Note: Variable definitions can be found in the introduction
to the Data Set V7 given in the Data Appendix.

The first two principal components together can be used to define two
orthogonal dimensions, one which measures overall level of pollution and
one that measures a contrast between SMAX and PMAX. The rational

TABLE 9.8. Sample Mean Vector and Sample Sum of Squares and Cross Products Matrix for Air Pollution Data*

	SMIN	SMEAN	SMAX	PMIN	PMEAN	PMAX	PM2	GE65	PERWH	NONPOOR	LPOP	Mean Vector
SMIN	3516.52	6291	13821.1	2539.67	6389.32	13923.4	3094.19	4198.47	4278.62	3938.57	2684.93	47.575
SMEAN	6291	13889.5	31215	5536.82	13888.1	30561.6	8444.56	8838.07	9127.3	8551.22	5899.24	101.975
SMAX	13821.1	31215	74565.4	12539.9	32116.9	72311.6	19531.5	20904	21658.3	20161.9	13903.4	240.650
PMIN	2539.67	5536.82	12539.9	2482.45	6270.12	14043.6	3473.42	3818.27	4096.34	3840.18	2640.89	46.050
PMEAN	6389.32	13888.1	32116.9	6270.12	16634.7	37884.8	8334.27	10003.4	10841.5	10087.6	6926.23	121.425
PMAX	13923.4	30561.6	72311.6	14043.6	37884.8	89119.7	17719.4	22558	24558.6	22811.4	15637.1	274.925
PM2	3094.19	8444.56	19531.5	3473.42	8334.27	17719.4	9844.55	5205.78	5410.73	5220.93	3749.34	61.553
GE65	4198.47	8838.07	20904	3818.27	10003.4	22558	5205.78	7045.35	7947.98	7363.66	4640.28	81.550
PERWH	4278.62	9127.3	21658.3	4096.34	10841.5	24558.6	5410.73	7947.98	7342	7363.66	5029.86	88.565
NONPOOR	3938.57	8551.22	20161.9	3840.18	10087.6	22811.4	5220.93	7363.66	7363.66	6882.94	4708.48	82.723
LPOP	2684.93	5899.24	13903.4	2640.89	6926.23	15637.1	3749.34	4640.28	5029.86	4708.48	3243.43	56.779

*(SSCP matrix elements have been divided by $N - 1 = 39$)

TABLE 9.9. Eigenvalues and Eigenvectors for Sum of Squares and Cross Products Matrix for Air Pollution Data

	V_1	V_2	V_3	V_4	V_5	V_6	V_7	V_8	V_9	V_{10}	V_{11}
SMIN	0.109	0.072	-0.210	0.086	0.631	-0.651	0.242	0.220	-0.022	0.042	-0.004
SMEAN	0.242	0.254	-0.106	-0.051	0.552	0.483	-0.519	0.139	-0.163	-0.093	0.033
SMAX	0.570	0.591	-0.326	-0.218	-0.367	-0.060	0.176	-0.023	0.032	0.022	-0.006
PMIN	0.104	-0.007	0.046	0.035	0.214	0.211	0.169	-0.080	0.919	-0.099	0.080
PMEAN	0.274	-0.161	0.084	0.043	0.285	0.298	0.518	-0.583	-0.298	0.132	-0.072
PMAX	0.626	-0.632	0.150	-0.306	-0.075	-0.146	-0.194	0.163	0.025	-0.025	0.010
PM2	0.149	0.384	0.892	0.042	0.046	-0.169	-0.017	0.025	-0.029	-0.033	-0.012
GE65	0.171	0.008	-0.095	0.472	-0.058	-0.310	-0.519	-0.561	0.125	0.171	-0.102
PERWH	0.182	-0.068	-0.040	0.527	-0.113	0.025	0.126	0.075	-0.142	-0.705	0.362
NONPOOR	0.170	-0.054	0.002	0.487	-0.088	0.201	0.131	0.412	0.024	0.142	-0.689
LPOP	0.117	-0.026	0.031	0.325	-0.064	0.137	0.060	0.260	-0.016	0.644	0.608
Eigenvalues	8,374,002	450,177	165,906	101,049	51,051	13,845	6,942	5,148	2,301	975	312
										Total	9,171,731
Cumulative Proportion	0.913	0.962	0.980	0.991	0.997	0.998	0.999	1.000	1.000	1.000	1.000

for the importance of these two variables can be seen in Table 9.8. The diagonal elements corresponding to SMAX and PMAX account for 70% of the sum of the diagonal elements.

The third principal component, Z_3, is dominated by the variable PM2 and the fourth component, Z_4, is dominated by the variables PERWH, NONPOOR and GE65. Relatively high values of Z_3 suggest high values of PM2, whereas high values of Z_4 suggest relatively high values of PERWH, NONPOOR and GE65.

It is interesting to note that the first four principal components, which together account for 99% of the total sum of squares for the eleven variables, are primarily concerned with measuring variation in the six variables PMAX, SMAX, GE65, NONPOOR, PM2 and PERWH.

Plots of pairs of eigenvectors in a two-dimensional space can also be used to interpret the principal components. The two panels of Figure 9.6 show the relationships between the original variables and the first four components. The first plot relates the variables to Z_1 and Z_2, whereas the second plot relates the variables to Z_3 and Z_4. The points corresponding to the eleven variables can be viewed as the tips of vectors drawn from the origin. Arrows representing the vectors have been omitted to simplify the presentation. In panel (a) it can be seen that PMAX and SMAX dominate the first two components. In panel (b) PM2 dominates the third component, whereas the fourth component represents the variables PERWH, NON-POOR and GE65.

Principal Component Scores

Using the relationships $\mathbf{Z} = \mathbf{XV}$, the values of $\mathbf{Z}$ for a given set of observations $\mathbf{X}$ can be determined using the matrix of eigenvectors $\mathbf{V}$. The $\mathbf{Z}$ values are called the *principal component scores*. In some applications, the *standardized scores* are determined using $\mathbf{Z}^* = \mathbf{Z}\Lambda^{-1/2}$ reflecting the fact that $\mathbf{Z}'\mathbf{Z} = \Lambda$ and hence $\mathbf{Z}^{*'}\mathbf{Z}^* = \mathbf{I}$.

Example Continued

Using the principal components solution given by the eigenvalues and eigenvectors in Table 9.9, the standardized principal component scores for the first four components are summarized in Table 9.10. By examining the scores for the various cities, we can characterize the cities with respect to the various dimensions determined by the principal components. A scatterplot for the first two components is displayed in panel (a) of Figure 9.7. The abbreviation for each of the 40 cities used in the figure is provided in Table 9.10. From the figure we can see that using the Z_1 dimension SC, BA, WH, WI, PH and NY have relatively high values of SMAX and PMAX while MO and GB have relatively low values. Using the Z_2 dimension, we can determine that for PH and NY the values of SMAX are large relative

(a)

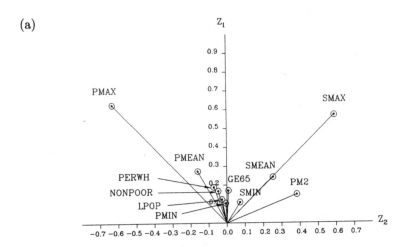

(b)

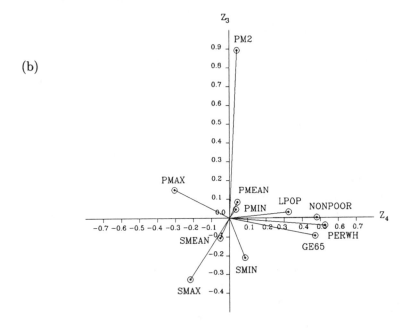

FIGURE 9.6. Relationships Among Principal Components and Original Variables

to PMAX while for AL the reverse is true. From the Z_2 dimension we can see that very few cities contribute to the variance along this dimension.

A second scatterplot is shown in panel (b) of Figure 9.7 relating the cities with respect to the third and fourth components. From this scatterplot we can see that NY is an outlier with respect to Z_3 indicating that NY has a relatively large population density PM2. The fourth dimension does not show any isolated points. From these two scatter plots we can see how principal components can be used to provide spatial representations of the cities with respect to the most important dimensions.

9.1.4   THE VARIOUS FORMS OF $\mathbf{X'X}$ AND PRINCIPAL COMPONENTS

The fact that the magnitudes of the diagonal elements of the sum of squares and cross products matrix (SSCP) influence the nature of the principal components suggests that changes in scales of measurement can influence the principal components solution. For this reason it is more common in practice to use standardized variables. The $\mathbf{X'X}$ matrix based on standardized variables is proportional to a correlation matrix. The covariance matrix can be viewed as a partial step between the SSCP and the correlation matrix. Since the covariance matrix removes the mean of the observations, it corrects the magnitudes of the elements of the SSCP for the overall level. It does not, however, correct for differences in the variances among the variables.

As outlined above the error sum of squares that results from the approximation of $\mathbf{X}$ by the first r principal components is given above by $\left(tr\mathbf{X'X} - \sum_{j=1}^{r} \lambda_j\right)$, or equivalently $\left(\sum_{j=1}^{p} \sum_{i=1}^{n} x_{ij}^2 - \sum_{j=1}^{r} \lambda_j\right)$. In addition, if all components are used, the error sum of squares is zero and hence $\sum_{j=1}^{p} \sum_{i=1}^{n} x_{ij}^2 = \sum_{j=1}^{p} \lambda_j$. The ratio $\sum_{j=1}^{r} \lambda_j / \sum_{j=1}^{p} \lambda_j$ therefore was used above to measure the proportion of the total sum of squares of the variables that is accounted for by the first r principal components. If the matrix $\mathbf{X'X}$ is a covariance matrix, then $tr\mathbf{X'X} = \sum_{j=1}^{p} s_{x_j}^2$ denotes the sum of the variances. If $\mathbf{X'X}$ is a correlation matrix, then $tr\mathbf{X'X} = p$ since the diagonal elements of a correlation matrix are unity. In general, therefore, each principal component Z_ℓ accounts for a portion $\lambda_\ell / \sum_{j=1}^{p} \lambda_j$ of the total variation in the Xs. The variation is measured in one of three ways depending on which type of SSCP matrix is used for $\mathbf{X'X}$. The example below will be used to demonstrate how the principal components solution is affected by the form of $\mathbf{X'X}$.

Example

We now study the principal components solutions for both the covariance matrix and the correlation matrix for the air pollution data. The covariance and correlation matrices are shown in Tables 9.11 and 9.12.

(a)

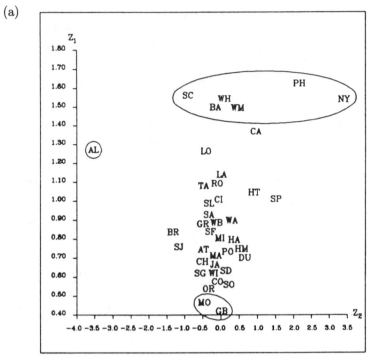

(b)

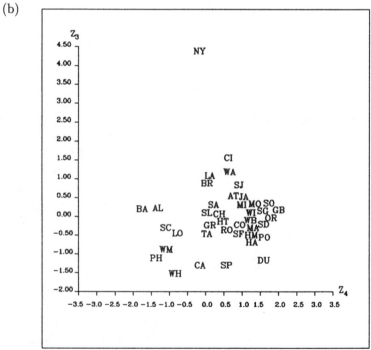

FIGURE 9.7. Scatterplot of Component Scores on the First Four Principal Components

TABLE 9.10. Data for Plot of Component Scores

City	Abbreviation	Z_1	Z_2	Z_3	Z_4
San Jose	SJ	0.763	-1.200	0.758	0.778
Roanoke	RO	1.098	-0.169	-0.266	-0.462
Albuquerque	AL	1.276	-3.580	1.507	-1.465
Charleston	CH	0.681	-0.496	-0.078	0.148
Harrisburg	HA	0.799	0.295	-0.672	1.113
Greenville	GR	0.880	-0.503	-0.243	0.019
Hartford	HT	1.050	0.842	-0.167	0.281
Columbus	CO	0.589	-0.146	-0.236	0.719
Orlando	OR	0.548	-0.376	0.011	1.444
Sacramento	SO	0.586	-0.007	0.160	1.415
Philadelphia	PH	1.629	2.087	-1.188	-1.540
Washington	WA	0.903	0.231	1.106	0.493
Minneapolis	MI	0.814	-0.085	0.290	1.009
Los Angeles	LA	1.145	-0.024	1.099	-0.026
Greensboro	GB	0.404	-0.049	0.146	1.724
Jacksonville	JA	0.651	-0.263	0.368	0.905
Madison	MA	0.714	-0.343	-0.337	1.060
Wilmington	WM	1.501	0.378	-1.000	-1.353
Tacoma	TA	1.085	-0.522	-0.530	-0.113
Scranton	SC	1.566	-0.960	-0.370	-1.258
Canton	CA	1.372	0.904	-1.378	-0.311
Atlanta	AT	0.742	-0.434	0.423	0.688
Baltimore	BA	1.504	-0.210	0.127	-1.914
Portland	PO	0.735	0.139	-0.628	1.287
Springfield, MA	SP	1.015	1.454	-1.376	0.411
Salt Lake	SL	0.993	-0.301	-0.065	0.038
Wichita	WI	0.623	-0.257	0.033	1.247
Lorain	LO	1.268	-0.461	-0.441	-0.985
Hamilton	HM	0.751	0.479	-0.592	1.039
Montgomery	MO	0.471	-0.532	0.135	1.150
San Diego	SD	0.627	-0.103	-0.251	1.296
Duluth	DU	0.715	0.580	-1.244	1.417
Wilkes Barre	WB	0.872	-0.329	-0.225	0.882
Wheeling	WH	1.539	0.020	-1.598	-1.025
San Antonio	SA	0.931	-0.383	0.174	0.026
Cincinnati	CI	1.004	-0.266	1.483	0.503
Saginaw	SG	0.624	-0.633	0.078	1.285
Baton Rouge	BR	0.841	-1.396	0.855	-0.152
New York	NY	1.544	3.336	4.309	-0.338
Springfield, OH	SF	0.858	-0.277	-0.338	0.734

The principal components solution based on the covariance matrix and correlation matrix are shown in Tables 9.13 and 9.14 respectively. Each table contains the eigenvalues and eigenvectors. For the covariance matrix solution, the first principal component is dominated by SMAX and PMAX, and the second component represents a contrast between SMAX

TABLE 9.11. Sample Covariance Matrix for Air Pollution Data

	SMIN	SMEAN	SMAX	PMIN	PMEAN	PMAX	PM2	GE65	PERWH	NONPOOR	LPOP
SMIN	1285.28	1476.45	2432.98	357.791	628.237	865.531	170.087	326.906	66.810	3.125	-16.737
SMEAN	1476.45	3580.13	6845.89	862.437	1544.37	2590.95	2223.33	535.399	98.340	118.562	112.03
SMAX	2432.98	6845.89	17080	1495.33	2970.26	6308.64	4839.94	1311.81	353.982	261.24	245.71
PMIN	357.791	862.437	1495.33	371.126	695.901	1418.82	655.308	64.510	18.386	31.601	26.897
PMEAN	628.237	1544.37	2970.26	695.901	1939.17	4617.49	882.321	103.837	89.777	44.083	32.683
PMAX	865.531	2590.95	6308.64	1418.82	4617.49	13883	817.568	141.401	215.231	70.648	27.859
PM2	170.087	2223.33	4839.94	655.308	882.321	817.568	6211.12	190.95	-41.707	132.465	260.988
GE65	326.906	535.399	1311.81	64.510	103.837	141.401	190.95	405.074	122.586	27.728	10.232
PERWH	66.810	98.340	353.982	18.386	89.777	215.231	-41.707	122.586	106.894	38.301	1.282
NONPOOR	3.125	118.562	261.244	31.601	44.083	70.648	132.465	27.728	38.301	40.956	11.891
LPOP	-16.737	112.03	245.712	26.897	32.6827	27.8592	260.988	10.2316	1.282	11.891	20.099

TABLE 9.12. Sample Correlation Matrix for Air Pollution Data

	SMIN	SMEAN	SMAX	PMIN	PMEAN	PMAX	PM2	GE65	PERWH	NONPOOR	LPOP
SMIN	1.000	0.688	0.519	0.518	0.398	0.205	0.060	0.453	0.180	0.014	-0.104
SMEAN	0.688	1.000	0.876	0.748	0.586	0.368	0.472	0.445	0.159	0.310	0.418
SMAX	0.519	0.876	1.000	0.594	0.516	0.410	0.470	0.499	0.262	0.312	0.419
PMIN	0.518	0.748	0.594	1.000	0.820	0.625	0.432	0.166	0.092	0.256	0.311
PMEAN	0.398	0.586	0.516	0.820	1.000	0.890	0.254	0.117	0.197	0.156	0.166
PMAX	0.205	0.368	0.410	0.625	0.890	1.000	0.088	0.060	0.177	0.094	0.053
PM2	0.060	0.472	0.470	0.432	0.254	0.088	1.000	0.120	0.051	0.263	0.739
GE65	0.453	0.445	0.499	0.166	0.117	0.060	0.120	1.000	0.589	0.215	0.113
PERWH	0.180	0.159	0.262	0.092	0.197	0.177	0.051	0.589	1.000	0.579	0.028
NONPOOR	0.014	0.310	0.312	0.256	0.156	0.094	0.263	0.215	0.579	1.000	0.415
LPOP	-0.104	0.418	0.419	0.311	0.166	0.053	0.739	0.113	0.028	0.415	1.000

and PMAX. From the diagonal elements of the covariance matrix in Table 9.11, we can see that together the variances of PMAX and SMAX represent about 50% of the trace of the covariance matrix. As a result, the first principal component is dominated by these two variables. Thus even after removing the mean of the variables, PMAX and SMAX still dominate the first two principal components.

For the correlation matrix, the first eigenvector is positively related to all eleven variables. The first component therefore measures overall air pollution and is also related positively to the demographic variables PERWH, NONPOOR, GE65, LPOP and PM2. The first component is therefore an index measuring the impact of all eleven variables. Since in a correlation matrix all variables account for the same proportion of the variance, the weights for the first component tend to be equal. The second component seems to contrast the nonpollution variables NONPOOR, PM2 and LPOP with the pollution variables SMIN, PMEAN and PMAX.

Interpretation Using Correlations

A useful way of interpreting the principal components is to examine the correlations between the principal components and the original variables. The correlations are shown in Table 9.15 for the correlation matrix solution. As can be seen from Table 9.15 the correlations display the same pattern of variation as the eigenvectors in Table 9.14. The difference in magnitudes between comparable columns is a constant of proportionality equal to $\sqrt{\lambda_j}$, where λ_j is the eigenvalue corresponding to the eigenvector $\mathbf{v}_j$ in column j. The matrix of correlations is therefore given by $\mathbf{V}\mathbf{\Lambda}^{1/2}$.

Example

The squares of the correlation coefficients in Table 9.15 describe the portion of the total variance in the X variable that is explained by that component. For the variable SMAX the first component accounts for 74.5% $[(0.863)^2 = 0.745]$ of the total variation in SMAX. [Recall that each X variable has a variance of 1 for a correlation matrix.] For the variable PMAX the first two components account for 64.7% $[(0.606)^2 = 0.367$ and $(0.529)^2 = 0.280]$ of the variation. These two portions can be added together because the principal components are mutually uncorrelated.

Standardized Principal Components

We have seen that the principal components solution for the matrix $\mathbf{X}$ is not the same for the three types of $\mathbf{X}'\mathbf{X}$ matrices. In practice the most frequent starting point for principal components analysis is the correlation matrix and hence the X variables are standardized. In such cases it is customary also to standardize the resulting principal components. Since the variances of the principal components $Z_1, Z_2, \ldots, Z_p$ are given by the

TABLE 9.13. Eigenvalues and Eigenvectors for Covariance Matrix

	V_1	V_2	V_3	V_4	V_5	V_6	V_7	V_8	V_9	V_{10}	V_{11}
SMIN	0.111	0.054	-0.210	0.640	0.519	-0.491	0.049	0.106	-0.025	0.063	0.045
SMEAN	0.311	0.179	-0.107	0.539	-0.438	0.326	-0.483	0.033	-0.164	-0.094	-0.028
SMAX	0.729	0.413	-0.329	-0.388	0.016	-0.127	0.146	-0.014	0.033	0.014	0.002
PMIN	0.089	-0.024	0.046	0.218	-0.177	0.081	0.209	0.100	0.916	-0.100	0.032
PMEAN	0.209	-0.207	0.083	0.297	-0.203	0.211	0.758	-0.248	-0.292	0.112	-0.013
PMAX	0.502	-0.795	0.144	-0.097	0.112	-0.061	-0.256	0.056	0.024	-0.016	0.002
PM2	0.231	0.340	0.892	0.051	0.158	-0.061	-0.033	0.015	-0.029	-0.017	-0.028
GE65	0.048	0.056	-0.091	0.027	0.611	0.655	-0.106	-0.367	0.136	0.138	-0.033
PERWH	0.015	-0.003	-0.035	-0.025	0.230	0.344	0.213	0.671	-0.164	-0.542	0.103
NONPOOR	0.011	0.010	0.007	-0.012	-0.018	0.166	0.025	0.571	0.008	0.764	-0.247
LPOP	0.011	0.017	0.035	-0.015	-0.038	0.060	-0.034	0.052	-0.015	0.258	0.960
Eigenvalues	27,390.8	10,842.2	4,362.6	1,358.3	399.7	254.2	170.5	67.0	60.3	11.1	6.2
											Total 44,922.92
Cumulative Proportion	0.610	0.851	0.948	0.978	0.987	0.993	0.997	0.998	1.000	1.000	1.000

TABLE 9.14. Eigenvalues and Eigenvectors for Correlation Matrix

	V_1	V_2	V_3	V_4	V_5	V_6	V_7	V_8	V_9	V_{10}	V_{11}
SMIN	0.278	-0.254	0.255	-0.452	-0.354	0.287	0.032	0.455	0.342	0.217	-0.089
SMEAN	0.419	0.006	-0.012	-0.284	-0.192	-0.257	0.065	-0.042	-0.133	-0.565	0.546
SMAX	0.399	0.071	0.036	-0.197	0.115	-0.634	-0.365	0.015	-0.182	0.318	-0.337
PMIN	0.391	-0.192	-0.186	0.049	-0.233	0.329	0.213	-0.445	-0.447	0.407	0.050
PMEAN	0.358	-0.348	-0.139	0.291	0.100	0.135	0.040	0.105	-0.029	-0.526	-0.572
PMAX	0.280	-0.396	-0.116	0.429	0.310	-0.162	-0.014	0.078	0.432	0.262	0.428
PM2	0.249	0.381	-0.383	-0.123	0.193	0.465	-0.578	-0.045	0.178	-0.045	0.082
GE65	0.225	0.153	0.507	-0.221	0.516	0.132	0.286	-0.432	0.238	-0.048	-0.093
PERWH	0.168	0.175	0.563	0.387	0.101	0.218	-0.199	0.380	-0.444	0.020	0.187
NONPOOR	0.201	0.408	0.176	0.436	-0.571	-0.115	-0.01	-0.245	0.397	-0.017	-0.121
LPOP	0.219	0.500	-0.339	0.033	0.157	-0.047	0.599	0.428	-0.026	0.125	-0.046
Eigenvalues	4.69	1.79	1.71	1.23	0.57	0.36	0.23	0.18	0.14	0.07	0.03
										Total	11.000
Cumulative Proportion	0.426	0.589	0.744	0.856	0.908	0.941	0.961	0.978	0.991	0.997	1.000

corresponding eigenvalues $\lambda_1, \lambda_2, \ldots, \lambda_p$; the *standardized principal components* are given by $Z_1^* = Z_1/\sqrt{\lambda_1}$, $Z_2^* = Z_2/\sqrt{\lambda_2}, \ldots, Z_p^* = Z_p/\sqrt{\lambda_p}$. From the relationships $\mathbf{X} = \mathbf{Z}\mathbf{V}'$ and $\mathbf{Z} = \mathbf{X}\mathbf{V}$, we have $\mathbf{X} = \mathbf{Z}^* \Lambda^{1/2} \mathbf{V}'$ and the standardized components are given by $\mathbf{Z}^* = \mathbf{X}\mathbf{V}\Lambda^{-1/2}$. We denote $\mathbf{V}\Lambda^{1/2}$ as $\mathbf{V}^*$ and write $\mathbf{X} = \mathbf{Z}^*\mathbf{V}^{*\prime}$ to relate the original variables to the standardized components. The elements of the matrix $\mathbf{V}^* = \mathbf{V}\Lambda^{-1/2}$ are usually referred to as *standardized scoring coefficients* and are used to obtain the component scores from the original data. The matrix of standardized scoring coefficients for the correlation matrix of the example is shown in Table 9.16.

The correlation matrix between $\mathbf{X}$ and $\mathbf{Z}^*$ is given by $\mathbf{X}'\mathbf{Z}^* = \mathbf{V}\Lambda^{1/2} = \mathbf{V}^*$, and hence the new coefficients relating $\mathbf{Z}^*$ to $\mathbf{X}$ are equivalent to the correlations between the Xs and the standardized components. These correlations are also equal to the correlations between the Xs and the components. In our earlier discussion of the principal components analysis example, the correlations between the X variables and the components were useful for interpretation. It is common to use the correlation matrix $\mathbf{V}^*$ to interpret the principal components. The $\mathbf{V}^*$ matrix corresponding to the pollution data correlation matrix is equivalent to the matrix in Table 9.15.

Communality or Variance Explained

If only the first r principal components are to be retained, it is of interest to determine how much of the variation in each individual X variable is explained by the approximation. If all components are used, the expression for the variance of x_j is obtained from $\mathbf{X}'\mathbf{X} = \mathbf{V}\Lambda\mathbf{V}' = \mathbf{V}^*\mathbf{V}^{*\prime}$, and hence $x_j'x_j = v_j^{*\prime}v_j^* = (v_j'\Lambda^{1/2})(\Lambda^{1/2}v_j)$ where v_j denotes the jth column of $\mathbf{V}'$. The expression for $x_j'x_j$ can be recognized as the sum of squares in the ith row of the correlation matrix $\mathbf{V}\Lambda^{1/2}$, which is the matrix of correlations between X and the principal components. The sum of squares of the elements of the jth row of $\mathbf{V}^*$ therefore gives the variance of X_j. If only the first r components are used, then the sum of squares of the jth row denotes the part of the variance of X_j explained by the first r components. In the case of the correlation matrix, the variances of the Xs are unity, and hence the variance explained by the first r components is also the proportion of the total variance. This proportion is referred to as the *communality* of the variable X_j.

Example

Using only the first four components, the communalities for the eleven variables can be obtained by summing the squared correlations in each row for the first four columns in Table 9.15. These communalities are SMIN (0.84), SMEAN (0.92), SMAX (0.80), PMIN (0.84), PMEAN (0.96), PMAX (0.90), PM2 (0.82), PERWH (0.91), NONPOOR (0.77), GE65 (0.78), and

TABLE 9.15. Correlations Between Variables and Principal Components from Correlation Matrix

	Z_1	Z_2	Z_3	Z_4	Z_5	Z_6	Z_7	Z_8	Z_9	Z_{10}	Z_{11}
SMIN	0.602	-0.340	0.334	-0.502	-0.269	0.171	0.015	0.196	0.128	0.056	-0.017
SMEAN	0.906	0.008	-0.016	-0.315	-0.146	-0.153	0.031	-0.018	-0.050	-0.145	0.102
SMAX	0.863	0.095	0.047	-0.218	0.088	-0.378	-0.174	0.007	-0.068	0.082	-0.063
PMIN	0.847	-0.257	-0.243	0.054	-0.176	0.196	0.102	-0.191	-0.167	0.104	0.009
PMEAN	0.776	-0.466	-0.181	0.323	0.076	0.080	0.019	0.045	-0.011	-0.135	-0.107
PMAX	0.606	-0.529	-0.152	0.476	0.235	-0.096	-0.007	0.034	0.161	0.067	0.080
PM2	0.539	0.510	-0.500	-0.137	0.147	0.277	-0.277	-0.019	0.067	-0.012	0.015
GE65	0.488	0.205	0.662	-0.245	0.392	0.079	0.137	-0.186	0.089	-0.012	-0.017
PERWH	0.364	0.234	0.736	0.429	0.076	0.130	-0.095	0.163	-0.166	0.005	0.035
NONPOOR	0.436	0.546	0.230	0.484	-0.433	-0.069	-0.005	-0.105	0.148	-0.004	-0.023
LPOP	0.473	0.668	-0.443	0.036	0.119	-0.028	0.286	0.184	-0.010	0.032	-0.009

TABLE 9.16. Standardized Scoring Coefficients

	Z_1	Z_2	Z_3	Z_4	Z_5	Z_6	Z_7	Z_8	Z_9	Z_{10}	Z_{11}
SMIN	0.128	-0.190	0.195	-0.407	-0.466	0.481	0.067	1.060	0.916	0.845	-0.478
SMEAN	0.193	0.004	-0.009	-0.255	-0.252	-0.431	0.136	-0.097	-0.356	-2.203	2.926
SMAX	0.184	0.053	0.027	-0.177	0.152	-1.063	-0.762	0.035	-0.488	1.237	-1.804
PMIN	0.180	-0.143	-0.141	0.043	-0.306	0.551	0.445	-1.034	-1.197	1.586	0.268
PMEAN	0.165	-0.260	-0.106	0.262	0.132	0.226	0.082	0.243	-0.078	-2.050	-3.071
PMAX	0.129	-0.296	-0.088	0.387	0.408	-0.271	-0.029	0.181	1.156	1.022	2.295
PM2	0.114	0.285	-0.292	-0.111	0.254	0.780	-1.209	-0.104	0.477	-0.176	-0.439
PERWH	0.077	0.131	0.430	0.348	0.132	0.365	-0.417	0.883	-1.190	0.079	1.002
NONPOOR	0.092	0.305	0.134	0.393	-0.752	-0.192	-0.023	-0.570	1.064	-0.064	-0.647
GE65	0.104	0.114	0.387	-0.199	0.680	0.221	0.598	-1.005	0.637	-0.185	-0.498
LPOP	0.100	0.373	-0.259	0.029	0.206	-0.078	1.251	0.995	-0.068	0.488	-0.247

LPOP (0.87). Thus 96% of the variation in PMEAN is explained by the first four components, whereas for NONPOOR this percentage is 77.

How Many Principal Components?

Recall that one of the objectives in principal components analysis was to replace the set of p original variables with a small subset of r principal components. The assumption was that because of the covariance relationships among the variables a small value of r would usually be sufficient to retain most of the variation. The sum of squared deviations between the original matrix $\mathbf{X}$ and the estimated values based on the first r components is given by $tr\mathbf{X}'\mathbf{X} - \sum_{j=1}^{r} \lambda_j = \sum_{j=r+1}^{p} \lambda_j$ and hence the proportion of the total sum of squares accounted for by the first r components is given by $\sum_{j=1}^{r} \lambda_j / \sum_{j=1}^{p} \lambda_j$. Some cut-off proportion, therefore, can be used to determine the number of components to retain.

Average Criterion

Since the total variation is given by $\sum_{j=1}^{p} \lambda_j$, where λ_j is the variance of Z_j, a possible rule of thumb is to retain those components whose variance exceeds the average $\bar{\lambda} = \sum_{j=1}^{p} \lambda_j / p$. In other words, retain Z_j if $\lambda_j > \bar{\lambda}$. For correlation matrices, $\sum_{j=1}^{p} \lambda_j = p$ and hence $\bar{\lambda} = 1$. This criterion becomes the *eigenvalue-one-criterion*, which is commonly used in factor analysis and will be discussed in Section 9.2.

Example

For the SSCP matrix $\mathbf{X}'\mathbf{X}$ of the example, the average eigenvalue criterion requires that eigenvalues above 833,794 be retained, and hence only the first component representing 91% of the variance would be retained. For the covariance matrix the criterion suggests that eigenvalues above 4084 correspond to factors that should be retained. Thus the first three components representing 95% of the variance should be retained. For the correlation matrix the eigenvalue-one-criterion suggests the retention of four components. The first four components account for 86% of the variation for the correlation matrix.

Geometric Mean Criterion

An alternative criterion based on the eigenvalues is the generalized variance. Since $|\mathbf{X}'\mathbf{X}| = \Pi_{j=1}^{p} \lambda_j$, we have that $|\mathbf{X}'\mathbf{X}|^{1/p} = [\Pi_{j=1}^{p} \lambda_j]^{1/p} =$ the geometric mean, $\bar{\lambda}_m$, of the eigenvalues. The *average generalized variance* is given by the geometric mean of the eigenvalues, $\bar{\lambda}_m$ and hence the criterion retain Z_j if $\lambda_j > \bar{\lambda}_m$. Recall that the geometric mean is useful for averaging a set of numbers containing a few extremes.

Example

For SSCP matrices and covariance matrices, the geometric mean provides a more useful criterion than the one based on the sum. For the SSCP matrix, the geometric mean of the eigenvalues is 23,247 and hence the first five factors should be retained. These factors jointly account for 99.7% of the variation. For the covariance matrix, the geometric mean of the eigenvalues is 347 and hence the first five components should be retained. These five factors jointly account for 98.7% of the variation. All components corresponding to the later eigenvalues are ignored.

A Test for Equality of Eigenvalues in Covariance Matrices

Since the eigenvalues decline in a geometric fashion, it can often be argued that the last $(p-r)$ eigenvalues are primarily due to "noise". In such cases it is of interest to test the null hypothesis that the last $(p-r)$ eigenvalues are equal. Under the assumption that the X observations have been sampled from a multivariate normal distribution, the test statistic is given by

$$[n - (2p + 11)/6]\left[(p - r)\ln\bar{\lambda}_{p-r} - \sum_{j=r+1}^{p}\ln\lambda_j\right]$$

where λ_j, $j = 1, 2, \ldots, p$ are the eigenvalues of the covariance matrix and where $\bar{\lambda}_{p-r} = \sum_{j=r+1}^{p}\lambda_j/(p-r)$. If the null hypothesis is true, this statistic has a χ^2 distribution with $\frac{1}{2}(p-r+2)(p-r-1)$ degrees of freedom. A special case of this hypothesis was discussed as a test of sphericity in Chapter 7. Later in this chapter, in factor analysis, the scree test will also be concerned with the equality of the latter eigenvalues of the correlation matrix.

A Cross Validation Approach

For large data sets the data can be divided into g mutually exclusive subsets. A principal components solution is determined using all data excluding one of the groups. The principal components solution is used to predict the observations in the omitted group. The goodness of fit is evaluated using

$$T_j(r) = tr[\mathbf{X}(j) - \widehat{\mathbf{X}}(j)]'[\mathbf{X}(j) - \widehat{\mathbf{X}}(j)],$$

where j denotes the groups omitted. For each value r of the number of components each group is omitted once and is predicted by the remaining group. The total measure of error $T(r) = \sum_{j=1}^{g}T_j(r)$ over the g groups is determined. As the number of principal components r increases, the total error $T(r)$ decreases. When the relative change in total error as measured by $[T(r) - T(r-1)]/T(r-1)$ is considered small it is not necessary to add additional principal components.

Other approaches to cross validation based on the likelihood function will be introduced in the section on factor analysis.

Should All the Variables be Retained?

If only a small number of components are required to retain most of the variation among the variables, it may be possible to eliminate some of the variables without affecting the components. Depending on the application, it may be useful to use principal components analysis to determine if some variables can be discarded. If a principal component has negligible variation and is dominated by a particular variable, then it may be possible to eliminate this variable.

Example

For the SSCP matrix the variation accounted for by the last few principal components is negligible. The last component is dominated by LPOP and NONPOOR (see Table 9.9) whereas the second last component is dominated by LPOP and PERWH. The variables LPOP, NONPOOR and PERWH, therefore, are much less important when it comes to accounting for the variation among the raw X variables. For the covariance matrix, the last two components are negligible and also correspond to LPOP, NONPOOR and PERWH. For the correlation matrix, there is not a single dominant variable for the latter components. For the variable SMEAN, components 10 and 11 of the correlation matrix explain a portion $(0.565)^2 + (0.546)^2 = 0.617$ of the variance, and since these components are negligible, perhaps SMEAN could be deleted. A similar argument could also be applied to PMEAN, which has the portion $(0.526)^2 + (0.573)^2 = 0.605$ accounted for by the last two components. We see below in the application of principal components analysis to multiple regression that negligible components can sometimes be useful and therefore should not always be discarded. As we see below variables that dominate the latter principal components may indicate outliers.

9.1.5   PRINCIPAL COMPONENTS, MULTIPLE REGRESSION AND SUPPLEMENTARY POINTS

Multiple Regression

A useful alternative to a variable selection procedure in multiple regression is provided by the use of a subset of the principal components obtained from the set of all explanatory variables. By using a small subset of the components as explanatory variables, the number of explanatory variables can often be reduced considerably. A second advantage of the components is that they are mutually uncorrelated, and hence the variation in the dependent variable explained by each component can be determined independently of the other components. Since the components are mutually uncorrelated, the presence of any one component does not affect the regression coefficients of the other components.

Example

To illustrate the application of this technique the air pollution data given in Table 9.7 is be used. The dependent variable TMR (total mortality rate) will be the dependent variable. The explanatory variables are the eleven standardized principal components derived from the correlation matrix as defined in Table 9.16. A forward stepwise regression yields the equation

$$\text{TMR} = 891.575 + \underset{(.000)}{86.355} \ Z_1^* + \underset{(.000)}{66.646} \ Z_3^* - \underset{(.000)}{70.767} \ Z_4^*$$

$$+ \underset{(.000)}{78.198} \ Z_5^* + \underset{(.054)}{22.371} \ Z_7^* - \underset{(.029)}{25.608} \ Z_8^* + \underset{(.031)}{25.149} \ Z_9^*,$$

which has an R^2 of 0.862. The remaining components were not significant at the 0.10 level. The contributions to R^2 from each of the above components are

$$Z_1^*(0.259), \quad Z_3^*(0.154), \quad Z_4^*(0.174), \quad Z_5 * (0.212),$$
$$Z_7^*(0.017), \quad Z_8^*(0.023), \quad Z_9^*(0.022).$$

From this equation we can see that not all of the major components, but three of the minor components, are related to TMR. The components Z_7^*, Z_8^* and Z_9^* account for the proportions 0.021, 0.017 and 0.013 of the total variance respectively. The components Z_2^*, Z_6^*, Z_{10}^* and Z_{11}^* were not included in the regression. Examining the eigenvectors in Table 9.14, we can conclude that the seventh component measures a contrast between LPOP and PM2, and the eighth component measures a contrast between the variables GE65 and PMIN and the variables SMIN and LPOP. The ninth component contrasts NONPOOR and PMAX with PMIN and PERWH. This example illustrates that if the underlying purpose of the principal components analysis is a multiple regression, it may not be wise to drop the minor principal components before the regression.

If the relationships between the standardized principal components and the original variables are taken into account, the multiple regression can be expressed in terms of the original variables. Using the matrix of standardized scoring coefficients given in Table 9.16 of Section 9.1.5, substitution for Z_1^*, Z_3^*, Z_4^*, Z_5^*, Z_7^*, Z_8^* and Z_9^* in the regression equation was carried out. Table 9.17 shows how the seven components contribute to the coefficients of the eleven variables. The last column of this table shows the sums of the rows that are the regression coefficients for the original variables. The numbers in the columns for each component are obtained by multiplying the corresponding column of scoring coefficients by the regression coefficient. The resulting equation is given by

$$\text{TMR} = 891.575 + 13.846 \, \text{SMIN} + 10.976 \, \text{SMEAN} + 11.922 \, \text{SMAX}$$

$$-14.604 \, \text{PMIN} - 7.357 \, \text{PMEAN} + 33.572 \, \text{PMAX} + 5.825 \, \text{PM2}$$

$$-40.787 \, \text{PERWH} - 28.781 \, \text{NONPOOR} + 157.327 \, \text{GE65} + 6.279 \, \text{LPOP}.$$

TABLE 9.17. Calculation of Standardized Regression Coefficients from Equation Based on Principal Components

Variable	Contributions from Components							Sum
	Z_1^*	Z_3^*	Z_4^*	Z_5^*	Z_7^*	Z_8^*	Z_9^*	
SMIN	11.079	13.011	28.869	-36.517	1.504	-27.149	23.049	13.846
SMEAN	16.685	-.623	18.100	-19.772	3.064	2.490	-8.967	10.976
SMAX	15.894	1.830	12.559	11.892	-17.065	-0.913	-12.276	11.922
PMIN	15.595	-9.461	-3.110	-23.969	9.967	26.495	-30.121	-14.604
PMEAN	14.289	-7.070	-18.566	10.329	1.851	-6.229	-1.968	-7.357
PMAX	11.163	-5.922	-27.392	31.950	-0.669	-4.654	29.095	33.572
PM2	9.938	-19.519	7.855	19.933	-27.052	2.673	12.006	5.825
PERWH	6.699	28.707	-24.674	10.378	-9.329	-22.627	-29.941	-40.787
NONPOOR	8.022	8.967	-27.835	-58.810	-0.521	14.619	26.776	-28.781
GE65	8.987	25.834	14.122	53.216	13.382	25.753	16.034	157.327
LPOP	8.717	-17.294	-2.096	16.176	28.001	-25.493	-1.732	6.279

Since the correlation matrix has been used, the explanatory variables are standardized, and hence the magnitudes of the regression coefficients can be compared. The most important variables are therefore GE65, PERWH, PMAX and NONPOOR.

The multiple regression of TMR on all eleven explanatory variables yields the standardized regression coefficients: SMIN (26.134), SMEAN (–31.564), SMAX (23.928), PMIN (–14.764), PMEAN (47.177), PMAX (–2.036), PM2 (4.921), PERWH (–53.116), NONPOOR (–24.437), GE65 (166.136) and LPOP (4.073). Comparing these coefficients to those obtained from the principal components, we can see that the major differences are in the coefficients of the pollution variables SMIN, SMEAN, SMAX, PMIN, PMEAN and PMAX. This is to be expected because these six variables represent two groups of three variables that are mutually correlated. The average correlation coefficient among the three S variables is 0.69 and among the three P variables is 0.78. The coefficients for the demographic variables GE65, NONPOOR, PERWH and LPOP are remarkably similar in both regressions.

Supplementary Dimensions and Points

The variable TMR in the above analysis can be viewed as a *supplementary dimension* in that it was not used to determine the original principal components. The regression of TMR onto the 11 components is the projection of a new vector (TMR) onto the space generated by the original eleven variables. The resultant regression estimator $\widehat{TMR}$ is a supplementary dimension which can also be plotted in a manner similar to the plots in Figure 9.6.

In a similar fashion, a new city that was not used in the original data matrix can also be located in the space generated by the principal components. Using the vector of X values for the new city, say x_ℓ $(p \times 1)$, the principal component scores are given by z_ℓ $(p \times 1)$ where $z_\ell' = x_\ell' V$. The coordinates of the new city with respect to the p principal components are given by the elements of z_ℓ. The point corresponding to the new city is called a *supplementary point*. This point could be plotted as in Figure 9.7.

9.1.6 OUTLIERS AND ROBUST PRINCIPAL COMPONENTS ANALYSIS

Identification of Outliers

Principal components analysis can be used to identify *multivariate outliers*. For each of the original observations the principal component scores can be examined for outliers using techniques available for univariate and bivariate analysis. By comparing standardized principal component scores to the standard normal distribution, upper and lower extremes for each component are easily identified.

Since the first few unstandardized principal components have large variances, these components tend to be strongly related to variables that have relatively large variances and covariances. Observations which are outliers with respect to the first few components usually correspond therefore to outliers on one or more of the original variables. These outliers are therefore usually detectable by studying the frequency distributions of the original variables. The last few unstandardized principal components represent linear functions of the original variables with minimal variance. These latter components are sensitive to observations that are inconsistent with the correlation structure of the data but are not outliers with respect to the original variables. These outliers are usually detectable using a bivariate scatter plot. In Figure 9.8 Point A corresponds to an observation that is an outlier in both the X_1 domain and the X_2 domain. It is not, however, an outlier with respect to the *correlation structure*. This observation would therefore appear as an outlier on one of the first few principal components. Observation B in the figure is not an outlier with respect to the domains of the two variables but is inconsistent with the correlation structure. This observation should appear as a large value for one of the last principal components but should not appear as an outlier on the first few principal components.

If the number of original variables is large, there may be outliers that are not detectable using univariate or bivariate analyses. This point was discussed in Chapter 7 in connection with the study of outliers in multivariate distributions. Large values of observations on the minor components can reflect such multivariate outliers. Scatterplots of the standardized minor principal components can also be used to identify potential outliers.

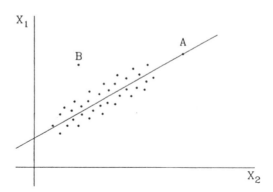

FIGURE 9.8. Scatterplot Showing Two Types of Outliers

In addition to scatterplots, it is also possible to identify outliers using functions of the component scores. For each observation i, the sum of squares $\sum_{j=1}^{p} z_{ij}^{*2}$ represents the sum of squares of the standardized component values. This sum is equivalent to the *Mahalanobis distance* of observation i from the mean for the sample. In Chapter 7 this distance was used for the identification of outliers.

An examination of the partial sum $\sum_{j=\ell}^{p} z_{ij}^{*2}$, which is the sum of squares for the last $(p - \ell + 1)$ components, is useful to determine how much of the variation in observation i is distributed over these latter components. If the last few components contain most of the variation in a particular observation, this could be an indication that this observation is an outlier with respect to the structure correlation. If the variation in an observation is dominated by the latter components, this is an indication that this observation is different from the majority of the data. By examining values of the ratio $\sum_{j=\ell}^{p} z_{ij}^{*2} / \sum_{j=1}^{p} z_{ij}^{*2}$ for each observation i, the relative importance of the last $(p - \ell + 1)$ components can be determined. The complete sum $\sum_{j=1}^{p} z_{ij}^{*2}$ over all standardized components is equivalent to the Mahalanobis distance of the observation from the centre of the data. An alternative statistic based on the same principle is to determine $\max_{\ell \leq j \leq p} |z_{ij}^{*}|$ for each observation $i = 1, 2, \ldots, n$.

Example

For the air pollution data discussed in this chapter, the principal component analysis of the correlation matrix was summarized in Table 9.14. The standardized values of the principal components for the forty observations are shown in Table 9.18. Component values that exceed 2 in absolute value have been highlighted with an asterisk. Observations that have relatively large values of principal components are New York (components 1, 2, 3, 6 and 7), Canton (1 and 5), Baltimore (6, 9 and 11), Wilmington (8 and 9), Albuquerque (4), Philadelphia (6), Los Angeles (7), Springfield (6), San Antonio (9), and Lorain (7). The sums of squares of these principal component values are summarized in Table 9.19. This value is equivalent to the Mahalanobis distance of the observation from the mean for all 40 observations. From this column one can determine that the cities farthest from the centre of the data are New York (34.709), Wilmington (26.183), Canton (24.314) and Albuquerque (23.103). All four of these cities have large values of at least one of the principal components.

For the eleventh component, the largest absolute value corresponds to Baltimore. From the eigenvector corresponding to this component, we observe that this component contrasts PMEAN with SMEAN and PMAX. This would suggest that the scatterplots relating PMEAN with SMEAN and PMEAN with PMAX could show an outlier for Baltimore. An ordinary least squares fit for the regression of PMAX on PMEAN yields a studentized residual of 2.931 for Baltimore. The Cook's D value for this observation is 0.153. These two statistics together indicate that Baltimore is an outlier with respect to the correlation structure between PMAX and PMEAN similar to B in Figure 9.8.

The sum of squares for the last four principal components (INDEX 1) and the sum of squares for the last seven components (INDEX 2) are also shown in Table 9.19. The ratios of these two indices to the Mahalanobis distance are shown in the columns INDEX 1R, and INDEX 2R respectively. Using the values of INDEX 1R we observe that for the cities San Antonio (0.851), Jacksonville (0.798) and Springfield (0.729) the value of INDEX 1R is relatively high. For San Antonio, most of this sum of squares is due to principal component 9 (−3.285), which is a contrast between PMAX and NONPOOR and the variables PMIN and PERWH. Scatterplots relating PMAX and NONPOOR to PMIN and PERWH should reveal an anomaly. The ordinary least squares fit of NONPOOR on PERWH shows a studentized residual of (2.22) for this city whereas the Cook's D value is only 0.076. This indicates that San Antonio is located well above the OLS line but is in the middle of the range of values of PERWH. In other words, PERWH is average but NONPOOR is large relative to PERWH.

The city of Jacksonville has moderately large values of all of the last four components but none of these values is larger than 1.5 in absolute value. An examination of the eigenvector for the last four components seems to suggest that the relationship among the six variables SMIN, SMEAN, SMAX, PMIN, PMAX and PMEAN is different for Jacksonville than for other cities. The Jacksonville values are compared to the average values for these variables in Table 9.20.

Relative to the average for the S variables, Jacksonville has a relatively low SMEAN value even though its SMIN and SMAX values are similar to the average. For the P variables, Jacksonville has relatively low PMIN and

TABLE 9.18. Standardized Principal Component Scores
For Air Pollution Data

	Z_1^*	Z_2^*	Z_3^*	Z_4^*	Z_5^*	Z_6^*
San Jose	-0.363	0.502	0.059	1.635	-0.642	1.004
Roanoke	-0.194	-0.588	0.220	-0.060	1.232	-0.828
Albuquerque	0.261	-1.948	-0.946	3.703*	0.513	0.690
Charleston	-1.342	-1.208	-1.524	-1.413	0.202	-0.647
Harrisburg	0.042	0.305	1.011	-0.552	-0.804	0.529
Greenville	-0.587	-1.073	-0.435	-0.369	0.273	-0.132
Hartford	0.465	1.021	0.335	0.458	-0.410	-0.682
Columbus	-1.337	-0.915	-1.121	-1.175	0.054	-0.402
Orlando	-1.143	0.203	0.279	-0.613	1.136	0.304
Sacramento	-0.879	1.120	0.314	0.671	-0.910	-0.064
Philadelphia	1.912	0.604	-0.732	-0.998	-0.239	-2.485*
Washington	0.479	0.576	-1.647	0.043	-1.829	0.859
Minneapolis	-0.052	1.337	0.406	0.877	0.135	0.365
Los Angeles	0.786	1.278	-0.881	0.836	1.026	-0.045
Greensboro	-1.331	0.211	-0.270	-0.787	-1.054	0.493
Jacksonville	-0.911	-0.103	-0.717	-0.617	-0.147	0.053
Madison	-0.609	0.423	1.080	0.549	-0.558	0.104
Wilmington	1.736	-1.202	-0.549	0.555	-1.393	-1.081
Tacoma	0.248	-0.352	0.604	0.536	0.610	-0.620
Scranton	1.609	-1.864	1.035	0.086	1.572	1.083
Canton	2.001*	-1.151	1.042	-1.684	-2.630*	0.784
Atlanta	-0.414	-0.020	-1.023	-0.391	-0.361	0.363
Baltimore	0.829	-0.029	-1.293	0.500	0.770	-2.782*
Portland	-0.466	1.266	1.285	-0.086	1.131	-0.466
Springfield, MA	0.144	1.247	1.241	-0.584	0.249	-2.236*
Salt Lake	-0.018	0.227	0.272	1.217	0.496	-0.355
Wichita	-0.843	0.596	0.353	0.720	-0.692	-0.064
Lorain	0.252	-0.517	0.422	1.005	-0.311	-1.240
Hamilton	-0.258	0.168	0.817	-0.373	-1.905	0.410
Montgomery	-1.616	-0.822	-0.970	-1.663	0.631	0.058
San Diego	-0.610	0.942	0.218	0.321	-0.487	-0.016
Duluth	-0.490	0.650	1.753	-0.598	0.853	-0.101
Wilkes Barre	-0.094	-0.242	1.169	-0.435	1.815	1.213
Wheeling	1.737	-1.924	1.514	-1.139	0.595	0.517
San Antonio	-0.317	-0.319	-0.568	0.305	1.382	0.179
Cincinnati	0.599	0.416	-0.460	0.191	0.464	1.537
Saginaw	-0.703	0.029	0.479	0.478	-0.724	0.832
Baton Rouge	-0.655	-1.195	-1.567	0.212	-0.492	0.265
New York	2.391*	2.568*	-2.271*	-1.220	1.294	2.168*
Springfield, OH	-0.257	-0.278	1.064	-0.141	0.147	0.432

PMAX values; however, its PMEAN value is much closer to average. The
profile for Jacksonville on these six variables seems to be different than the
means for the other cities. For Springfield the reason for the moderately

TABLE 9.18. Standardized Principal Component
Scores For Air Pollution Data (continued)

	Z_7^*	Z_8^*	Z_9^*	Z_{10}^*	Z_{11}^*
San Jose	0.501	-0.007	0.027	1.933	0.798
Roanoke	-1.537	-1.329	-0.063	1.221	0.483
Albuquerque	-0.335	0.863	0.300	-1.669	0.390
Charleston	-0.227	0.227	-0.236	1.335	1.110
Harrisburg	0.612	-0.770	-0.047	0.492	0.844
Greenville	-0.817	1.080	-0.662	0.278	-0.602
Hartford	-0.483	-1.406	-0.675	0.165	-0.300
Columbus	-0.494	0.934	-1.164	-1.096	-1.060
Orlando	0.625	-0.630	0.209	-0.803	0.018
Sacramento	-0.572	0.598	0.633	-0.311	-1.400
Philadelphia	0.943	0.925	-0.341	1.351	-1.874
Washington	1.516	-1.018	0.268	1.051	-1.157
Minneapolis	1.143	0.038	-0.098	1.057	-0.429
Los Angeles	2.032*	1.265	0.560	-0.636	-0.849
Greensboro	-0.263	0.000	-0.119	-0.592	0.880
Jacksonville	-0.087	1.380	1.123	-1.498	-1.255
Madison	-0.774	0.117	0.129	0.164	1.097
Wilmington	0.289	-2.517*	-2.729*	-1.755	-1.022
Tacoma	0.204	-0.596	-0.323	-1.482	1.010
Scranton	0.266	0.018	0.653	0.591	-0.396
Canton	0.396	1.599	0.627	-0.932	1.883
Atlanta	1.494	0.739	0.174	-0.033	0.255
Baltimore	0.426	-0.064	2.013*	0.071	2.158*
Portland	1.261	-0.114	0.697	-0.092	0.262
Springfield, MA	-0.652	-0.339	-0.170	-1.655	1.533
Salt Lake	-1.048	0.184	-0.850	0.988	0.283
Wichita	-0.761	0.856	0.087	-1.691	-1.207
Lorain	-2.087*	0.995	1.782	1.271	-1.287
Hamilton	-1.319	0.720	-0.221	0.813	0.419
Montgomery	0.480	-1.327	1.007	-0.566	-0.391
San Diego	1.123	1.228	-1.127	0.087	1.009
Duluth	0.165	-0.589	-0.716	0.336	-1.140
Wilkes Barre	0.453	0.649	-0.815	-0.166	0.015
Wheeling	0.387	0.535	0.406	0.544	-0.903
San Antonio	-0.007	1.265	-3.285*	1.269	0.806
Cincinnati	0.868	-1.164	0.670	-0.665	0.814
Saginaw	-0.147	-1.240	0.082	-0.091	-0.596
Baton Rouge	0.104	-1.559	0.899	1.171	0.627
New York	-2.995*	-0.072	-0.019	-0.507	0.357
Springfield, OH	-0.684	-1.477	1.312	0.052	-1.186

large values of the components 8, 9 and 11 would appear to be relatively
large values of SMAX and GE65.

TABLE 9.19. Sums of Squares and Partial Sums of Squares of Standardized Principal Components

City	Total Sum of Squares Mahalanobis Distance	INDEX 2	INDEX 2R	INDEX 1	INDEX 1R
San Jose	9.112	6.050	0.663	4.375	0.480
Roanoke	8.504	8.067	0.948	3.498	0.411
Albuquerque	23.103	4.628	0.200	3.775	0.163
Charleston	11.220	3.636	0.324	3.125	0.278
Harrisburg	4.276	2.853	0.667	1.551	0.362
Greenville	4.632	2.807	0.606	2.046	0.441
Hartford	5.004	3.421	0.683	2.552	0.509
Columbus	10.229	4.966	0.485	4.556	0.445
Orlando	4.664	2.862	0.613	1.087	0.233
Sacramentp	6.555	3.977	0.606	2.816	0.429
Philadelphia	18.993	13.435	0.707	6.312	0.332
Washington	13.217	9.938	0.751	3.552	0.268
Minneapolis	5.500	2.774	0.504	1.313	0.238
Los Angeles	11.958	8.226	0.687	3.039	0.254
Greensboro	5.076	2.564	0.505	1.140	0.224
Jacksonville	8.759	7.002	0.801	6.990	0.798
Madison	4.203	2.185	0.519	1.262	0.300
Wilmington	26.182	21.112	0.806	17.916	0.684
Tacoma	5.317	4.478	0.842	3.678	0.691
Scranton	11.798	4.653	0.394	0.934	0.079
Canton	24.314	15.059	0.619	7.368	0.303
Atlanta	4.513	3.140	0.695	0.643	0.142
Baltimore	19.846	17.234	0.868	8.718	0.439
Portland	7.147	3.667	0.513	0.576	0.080
Springfield, MA	14.186	10.726	0.756	5.238	0.369
Salt Lake	4.893	3.285	0.671	1.814	0.370
Wichita	7.834	6.123	0.781	5.060	0.645
Lorain	14.957	13.436	0.898	7.444	0.497
Hamilton	7.849	6.946	0.884	1.404	0.178
Montgomert	10.882	3.884	0.356	3.251	0.298
San Diego	6.718	5.305	0.789	3.806	0.566
Duluth	7.138	3.040	0.425	2.274	0.318
Wilkes Barre	7.712	6.087	0.789	1.115	0.144
Wheeling	12.650	2.336	0.184	1.564	0.123
San Antonio	17.219	16.600	0.964	14.656	0.851
Cincinnati	7.030	6.248	0.888	2.913	0.414
Saginaw	4.102	3.147	0.767	1.908	0.465
Baton Rouge	9.690	5.329	0.549	5.006	0.516
New York	34.709	15.742	0.453	0.391	0.011
Springfield, OH	7.289	5.991	0.822	5.315	0.729

TABLE 9.20. Comparison of Jacksonville to Other Cities

	Jacksonville	Mean for All 40 Cities
SMIN	46	47.6
SMEAN	66	101.9
SMAX	133	130.7
PMIN	23	46.1
PMEAN	106	121.4
PMAX	193	274.9

Influence

The measurement of the *influence* that an observation has on the estimators of regression coefficients can be obtained from closed form expressions (see Chapter 3 of Volume I). For principal components analysis, however, there are no simple expressions available for the measurement of influence. For observations believed to be outliers or that have a major impact, the principal components analysis should be determined for a reduced sample excluding these observations. By comparing the solution for the complete sample to the solution from the reduced sample, we can determine the impact of the omitted observations.

Robust Principal Components Analysis

As outlined above, outliers can sometimes bring about large increases in variances, covariances and correlations. The relative magnitude of these measures of variation and covariation has an important impact on the principal components solution particularly for the first few components. For this reason, if outliers are believed to be present, it is of value to begin a principal components analysis with a *robust estimator* of the covariance matrix or correlation matrix. A comparison of the robust solution with the conventional solution should be made to determine the effects of the outliers. The iterative procedure outlined in Chapter 7 for robust estimation of covariance and correlation matrices could also be applied to obtain an iterative procedure for principal components analysis.

Rank Correlation and Robust Principal Components Analysis

The original data matrix $\mathbf{X}$ can be transformed to a matrix of column ranks $\mathbf{X}_R$ before performing a principal components analysis. This procedure is useful when the original variables have very different scales of measurement and/or when there are outliers present.

9.1.7 OTHER SOURCES OF INFORMATION

Extensive discussion of principal components analysis can be found in Jolliffe (1986). This topic is also outlined in Seber (1984) and Jackson (1990).

9.2 The Exploratory Factor Analysis Model

Like principal components analysis, the essential purpose of *factor analysis* is to describe the variation among many variables in terms of a few underlying but unobservable random variables called factors. Unlike principal components analysis, the underlying model in factor analysis specifies a small number of *common factors*. All the covariances or correlations are explained by the common factors. Any portion of the variance unexplained by the common factors is assigned to residual error terms which are called *unique factors*. The unique factors are assumed to be mutually uncorrelated. The factor analysis model therefore assumes that the covariance matrix or correlation matrix can be divided into two parts. The first part of the matrix is generated by the common factors and the second part is generated by the errors or unique factors. The error portion of the matrix is diagonal. Although principal components analysis is primarily concerned with explaining the variance of the variables, factor analysis is concerned with explaining the covariance. Factor analysis can also be viewed as a statistical procedure for grouping variables into subsets such that the variables within each set are mutually highly correlated, whereas at the same time variables in different subsets are relatively uncorrelated.

Much of the early development of factor analysis was done near the beginning of the twentieth century by psychologists seeking a better understanding of the dimensions of human intelligence. Later in this section an example application of factor analysis is provided based on a study of job related stress for police officers. A random sample of officers was asked to respond to a variety of questions regarding the amount of stress they experience from various work situations. Analysis of the correlation matrix obtained from the sample yields several underlying components of stress such as workload stress, stress due to risk of injury, stress from organizational factors, and stress from uncooperative people and unpleasant duties. Thus factor analysis allows the many real life situations contributing to stress to be summarized by a small number of underlying factors. These factors can then be used to study the relationship between stress and other variables such as type of neighborhood, type of supervision, type of police duty and police officer personality characteristics.

9.2.1   THE FACTOR ANALYSIS MODEL AND ESTIMATION

The Model

The model known as *common factor analysis* is composed of three sets of variables: a set of p observed variables $X_1, X_2, \ldots, X_p$ with mean vector $\boldsymbol{\mu}$ ($p \times 1$) and covariance matrix $\boldsymbol{\Sigma}$ ($p \times p$); a set of r unobserved variables called common factors $F_1, F_2, \ldots, F_r$, where $r \ll p$, and a set of p unique but unobserved factors $U_1, U_2, \ldots, U_p$. The model is given by the p equations

$$
\begin{aligned}
(X_1 - \mu_1) &= a_{11}F_1 + a_{12}F_2 + \ldots + a_{1r}F_r + U_1 \\
(X_2 - \mu_2) &= a_{21}F_1 + a_{22}F_2 + \ldots + a_{2r}F_r + U_2 \\
&\ \ \vdots \\
(X_p - \mu_p) &= a_{p1}F_1 + a_{p2}F_2 + \ldots + a_{pr}F_r + U_p
\end{aligned}
$$

or equivalently in matrix notation

$$(\mathbf{x} - \boldsymbol{\mu}) = \mathbf{A}\mathbf{f} + \mathbf{u},$$

where $(\mathbf{x} - \boldsymbol{\mu})$ is the $p \times 1$ vector of elements $X_i - \mu_i$, $i = 1, 2, \ldots, p$;

$\mathbf{f}$ is the $r \times 1$ vector of linearly independent common factors, F_j, $j = 1, 2, \ldots, r$;

$\mathbf{A}$ is the $p \times r$ *factor pattern matrix* (consisting of the unknown *factor loadings*) a_{ij}, $i = 1, 2, \ldots, p$; $j = 1, 2, \ldots, r$;

and $\mathbf{u}$ is the $p \times 1$ vector of unique factors U_i, $i = 1, 2, \ldots, p$.

The factors $F_1, F_2, \ldots, F_r$ are common to all p X variables, whereas the error or unique factor U_i is unique to X_i. The common factors are assumed to have mean 0 and variance 1 and are mutually uncorrelated. The unique factors are assumed to have mean 0 and variance $\sigma_{u_i}^2$, $i = 1, 2, \ldots, p$. In addition, it is assumed that all of the common factors are uncorrelated with the unique factors. Given these assumptions, we may express the X covariance matrix $\boldsymbol{\Sigma}$ in the form

$$\boldsymbol{\Sigma} = \mathbf{A}\mathbf{A}' + \boldsymbol{\Psi}$$

where $E[(\mathbf{x} - \boldsymbol{\mu})(\mathbf{x} - \boldsymbol{\mu})'] = \boldsymbol{\Sigma}$ the ($p \times p$) covariance matrix;

$$
\begin{aligned}
E[\mathbf{f}] &= E[\mathbf{u}] = 0; \\
E[\mathbf{f}\mathbf{f}'] &= \mathbf{I}, \text{ a } (r \times r) \text{ identity matrix;} \\
E[\mathbf{u}\mathbf{u}'] &= \boldsymbol{\Psi}, \text{ a } (p \times p) \text{ diagonal matrix with diagonal elements } \sigma_{u_i}^2, \\
& \qquad i = 1, 2, \ldots, p; \\
E[\mathbf{u}\mathbf{f}'] &= 0, \text{ no correlation between unique factors} \\
& \qquad \text{and common factors;}
\end{aligned}
$$

and where the *factor structure matrix*, which is equivalent to the factor pattern matrix, is given by $\mathrm{Cov}(\mathbf{x}, \mathbf{f}) = \mathbf{A}$. If the X variables are standardized, the elements of $\mathbf{A}$ represent correlations between the X variables and the factors.

The variance of each variable X_i can be written as

$$\sigma_i^2 = \sum_{j=1}^{r} a_{ij}^2 + \sigma_{u_i}^2,$$

and hence the variance is divided into two parts. The first part $\sum_{j=1}^{r} a_{ij}^2$ is the variance explained by the common factors and is usually referred to as the *communality*. The second term $\sigma_{u_i}^2$ is called the *unique variance* or *specific variance*. The entire covariance between X_i and X_k is given by $\sum_{j=1}^{r} a_{ij} a_{kj}$.

Factor Analysis Using the Correlation Matrix

Usually in practice, the X variables are assumed to be standardized and hence $\boldsymbol{\mu} = 0$ and $\boldsymbol{\Sigma}$ is a correlation matrix. In that case, $\sigma_i^2 = 1$, and $\sum_{j=1}^{r} a_{ij}^2$ represents the proportion of the X_i variance explained by the common factors. The factor loadings in this case are correlations between the factors and the X variables. For the remainder of the discussion of factor analysis, we assume that the variables are standardized and hence $\boldsymbol{\mu} = 0$ and $\boldsymbol{\Sigma}$ is the correlation matrix $\boldsymbol{\rho}$. The equations for the factor analysis model, therefore, become

$$\mathbf{x} = \mathbf{Af} + \mathbf{u}$$
$$\text{and} \quad \boldsymbol{\rho} = \mathbf{AA}' + \boldsymbol{\Psi}.$$

Indeterminacy

A major problem in the estimation of the factor model is the fact that the relationship $\boldsymbol{\rho} = \mathbf{AA}' + \boldsymbol{\Psi}$ is *indeterminate*. The indeterminacy arises from several sources. The number of parameters to be estimated are the pr elements of $\mathbf{A}$ and the p diagonal elements of $\boldsymbol{\Psi}$ yielding a total of $p(r+1)$ parameters. The correlation matrix contains $p(p+1)/2$ unique elements and hence $p(p+1)/2$ should exceed $p(r+1)$. Even if this condition is satisfied, the elements of $\mathbf{A}$ are only unique up to an *orthogonal transformation*, $\mathbf{B} = \mathbf{AT}$, since for any *orthogonal matrix* $\mathbf{T}$, $\mathbf{BB}' = \mathbf{ATT}'\mathbf{A}' = \mathbf{AA}'$ since $\mathbf{TT}' = \mathbf{I}$. Such orthogonal transformations or *rigid rotations* are studied later in Section 9.2.2. In addition, if r is prespecified, it may not be possible to determine r linearly independent factors from a given correlation matrix. The indeterminacy problem, however, has not deterred the use of factor analysis in social science research. This indeterminacy has permitted the examination of a variety of solutions for the purpose of selecting the most useful. Factor analysis is used in general to obtain a relatively small number of factors from a large number of variables in such a way that the communalities are close to 1 and the factors are easily interpreted. It should therefore be viewed as a data reduction technique rather than a method for deriving True factors.

Estimation of the Factor Model Using Principal Components

Given an observed data matrix $\mathbf{X}$ $(n \times p)$, the factor analysis model can be expressed as

$$\mathbf{X} = \mathbf{F}\mathbf{A}' + \mathbf{U},$$

where

$\mathbf{F}(n \times r)$ is the unobserved matrix of values of the r common factors for the n observational units;

$\mathbf{A}'$ is the $(r \times p)$ unknown factor pattern or loading matrix; and

$\mathbf{U}$ is the $(n \times p)$ matrix of unobserved errors or values of unique factors for the n observational units.

Thus unlike a multiple linear regression model the entire right-hand side of the model is unobserved.

For a given value of r, principal components analysis can be used to estimate the matrices $\mathbf{A}$ and $\boldsymbol{\Psi}$ in the relation $\boldsymbol{\rho} = \mathbf{A}\mathbf{A}' + \boldsymbol{\Psi}$. From Section 9.1 the principal components analysis model can be written as $\mathbf{X} = \mathbf{Z}\mathbf{V}'$, where $\mathbf{V}$ is the matrix of eigenvectors of $\mathbf{X}'\mathbf{X}$. By writing $\mathbf{X} = (\mathbf{Z}\boldsymbol{\Lambda}^{-1/2})(\boldsymbol{\Lambda}^{1/2}\mathbf{V}')$, we can obtain an expression for $\mathbf{X}$ in terms of new components $\mathbf{Z}\boldsymbol{\Lambda}^{-1/2}$, which have unit variances. Therefore, we may let $\widehat{\mathbf{F}} = \mathbf{Z}\boldsymbol{\Lambda}^{-1/2}$ and $\widehat{\mathbf{A}}' = \boldsymbol{\Lambda}^{1/2}\mathbf{V}'$ to estimate the factor model. The principal component equations are given by $X_i = v_{i1}Z_1 + v_{i2}Z_2 + \ldots + v_{ip}Z_p$, $i = 1, 2, \ldots, p$, and in terms of standardized variables are given by

$$X_i = (v_{i1}\sqrt{\lambda_1})\left(\frac{Z_1}{\sqrt{\lambda_1}}\right) + (v_{i2}\sqrt{\lambda_2})\left(\frac{Z_2}{\sqrt{\lambda_2}}\right) + \ldots + (v_{ip}\sqrt{\lambda_p})\left(\frac{Z_p}{\sqrt{\lambda_p}}\right)$$

$$i = 1, 2, \ldots, p.$$

We now have a model that conforms to a common factor analysis model with $r = p$ factors.

Since the number of factors r is usually considerably less than p, we can modify the above by retaining only the first r components. Partition the matrices $\mathbf{Z}$, $\mathbf{V}'$, and $\boldsymbol{\Lambda}$ as follows:

$$\mathbf{Z} = (\mathbf{Z}_1\ \mathbf{Z}_2), \quad \mathbf{V}' = \begin{bmatrix} \mathbf{V}'_1 \\ \mathbf{V}'_2 \end{bmatrix} \quad \text{and} \quad \boldsymbol{\Lambda} = \begin{bmatrix} \boldsymbol{\Lambda}_1 & 0 \\ 0 & \boldsymbol{\Lambda}_2 \end{bmatrix},$$

where $\mathbf{Z}_1$ is $n \times r$, $\mathbf{V}'_1$ is $r \times p$ and $\boldsymbol{\Lambda}_1$ is $r \times r$. The remaining matrices are $\mathbf{Z}_2$ $[n \times (p-r)]$, $\mathbf{V}'_2$ $(p-r) \times p$ and $\boldsymbol{\Lambda}_2(p-r) \times (p-r)$.

The model can now be written

$$
\begin{aligned}
\mathbf{X} &= \mathbf{Z}_1 \mathbf{V}_1' + \mathbf{Z}_2 \mathbf{V}_2' \\
&= (\mathbf{Z}_1 \Lambda_1^{-1/2})(\Lambda_1^{1/2} \mathbf{V}_1') + \mathbf{Z}_2 \mathbf{V}_2' \\
&= \widehat{\mathbf{F}} \widehat{\mathbf{A}}' + \widehat{\mathbf{U}}
\end{aligned}
$$

where $\widehat{\mathbf{F}} = \mathbf{Z}_1 \Lambda_1^{-1/2}$, $\widehat{\mathbf{A}}' = \Lambda_1^{1/2} \mathbf{V}_1'$ and $\widehat{\mathbf{U}} = \mathbf{Z}_2 \mathbf{V}_2'$.

The principal component analysis solution is only an approximation since the resulting $\widehat{\mathbf{U}}' \widehat{\mathbf{U}}$ will not in general be a diagonal matrix. It is hoped that the off-diagonal elements of $\widehat{\mathbf{U}}' \widehat{\mathbf{U}}$ are negligible. This assumption characterizes the difference between principal component analysis and factor analysis. Principal component analysis is concerned with capturing the bulk of the variance whereas factor analysis is concerned with explaining all the covariance. The last $(p - r)$ terms in the principal component model equations are dropped and replaced by the error term $\mathbf{U}$.

As we see later, one advantage of the principal components method of estimating the factor analysis model is that the matrix of the estimated factor scores $\widehat{\mathbf{F}}$ can be obtained directly from $\mathbf{Z}_1$ and Λ_1 by computing $\widehat{\mathbf{F}} = \mathbf{X} \mathbf{V}_1 \Lambda_1^{-1/2}$. The factor score coefficient matrix is therefore given by $\mathbf{V}_1 \Lambda_1^{-1/2}$. With other factor solution methods, it is not always possible to obtain the factor scores in such a simple fashion, and in some cases regression analysis must be used to obtain an estimator $\widehat{\mathbf{F}}$ of the factor scores $\mathbf{F}$.

Example

For the air pollution data example of Section 9.1, Table 9.15 shows the correlations between the components and the original variables. These correlations are the required factor loadings $\mathbf{A}' = \Lambda^{1/2} \mathbf{V}$. This matrix of factor loadings is often called the factor pattern matrix. A four-factor solution is provided by the first four columns of this table. From the eigenvalues shown earlier in Table 9.14 we can conclude that the eigenvalue-one-criterion yields four factors. These four factors account for 85.6% of the variation. The correlation matrix in Table 9.21 shows the correlations among the X residuals after removing the four factors. The diagonal elements of this matrix represent the specific variances for each variable. Subtracting these variances from one yields the communalities. Recall that the off-diagonal elements of this matrix are assumed to be zero. For the most part the correlations are less than 0.10 in absolute value.

Estimation of the Common Factor Model

As shown above for a given r, principal components analysis can be used to estimate the common factor model. An alternative tradition in psychol-

TABLE 9.21. Correlation Matrix for Factor Model Residuals (Diagonal elements are specific variances)

	SMIN	SMEAN	SMAX	PMIN	PMEAN	PMAX	PM2	PERWH	NONPOOR	GE65	LPOP
SMIN	0.159	-0.006	-0.092	0.029	-0.004	-0.050	0.007	0.010	0.102	-0.115	0.004
SMEAN	-0.006	0.079	0.024	-0.003	-0.014	-0.029	-0.072	-0.025	0.066	-0.065	-0.012
SMAX	-0.092	0.024	0.196	-0.089	-0.030	0.047	-0.050	-0.015	-0.020	-0.026	-0.023
PMIN	0.029	-0.003	-0.089	0.155	-0.017	-0.086	-0.008	-0.000	0.057	-0.020	-0.027
PMEAN	-0.004	-0.014	-0.030	-0.017	0.044	-0.007	0.026	0.019	-0.041	0.032	0.017
PMAX	-0.050	-0.029	0.047	-0.086	-0.007	0.102	0.020	-0.012	-0.076	0.089	0.034
PM2	0.007	-0.072	-0.050	-0.008	0.026	0.020	0.180	0.059	-0.069	0.050	-0.074
PERWH	0.010	-0.025	-0.015	-0.000	0.019	-0.012	0.059	0.087	-0.084	-0.018	0.009
NONPOOR	0.102	0.066	-0.020	0.057	-0.041	-0.076	-0.069	-0.084	0.225	-0.142	-0.071
GE65	-0.115	-0.065	-0.026	-0.020	0.032	0.089	0.050	-0.018	-0.142	0.221	0.048
LPOP	0.004	-0.012	-0.023	-0.027	0.017	0.034	-0.074	0.009	-0.071	0.048	0.131

ogy involves estimation of the communalities and r before determining the factors.

Estimation of the model begins with the sample correlation matrix $\mathbf{R}$. For the sample data we write $\mathbf{R} = \widehat{\mathbf{A}}\widehat{\mathbf{A}}' + \widehat{\boldsymbol{\Psi}}$, where $\widehat{\mathbf{A}}\widehat{\mathbf{A}}'$ and $\widehat{\boldsymbol{\Psi}}$ are estimates of $\mathbf{A}\mathbf{A}'$ and $\boldsymbol{\Psi}$. The first step in the common factor model is to determine an estimate $\widehat{\boldsymbol{\Psi}}$ of $\boldsymbol{\Psi}$. Given $\widehat{\boldsymbol{\Psi}}$, the second step is to determine the number of factors r and the matrix estimate $\widehat{\mathbf{A}}$. Since $\widehat{\boldsymbol{\Psi}}$ is a diagonal matrix, the matrix $(\mathbf{R} - \widehat{\boldsymbol{\Psi}})$ is a correlation matrix with the 1s in the diagonal replaced by estimates of the communalities.

The estimation of $\boldsymbol{\Psi}$ must take into account that $\mathbf{A}\mathbf{A}' = (\rho - \boldsymbol{\Psi})$ is *positive definite*. Given $\mathbf{R}$ and an estimator $\widehat{\boldsymbol{\Psi}}$ of $\boldsymbol{\Psi}$, if the elements of $\widehat{\boldsymbol{\Psi}}$ are too large, the matrix $\widehat{\mathbf{A}}\widehat{\mathbf{A}}' = \mathbf{R} - \widehat{\boldsymbol{\Psi}}$ could become *negative definite*. A common approach is to minimize r, the rank of $\widehat{\mathbf{A}}\widehat{\mathbf{A}}'$, and at the same time insure that $\widehat{\mathbf{A}}\widehat{\mathbf{A}}'$ is positive definite. Thus, if the elements of $\widehat{\boldsymbol{\Psi}}$ are too small, the rank of $\widehat{\mathbf{A}}\widehat{\mathbf{A}}'$ may be too large, whereas if the diagonal elements of $\widehat{\boldsymbol{\Psi}}$ are too large, the matrix $\widehat{\mathbf{A}}\widehat{\mathbf{A}}'$ could become negative definite. One estimate of $\boldsymbol{\Psi}$ is given by $\widehat{\boldsymbol{\Psi}}_1$, where $\widehat{\boldsymbol{\Psi}}_1$ is a diagonal matrix of diagonal elements $d_{1i} = (1 - R_i^2)$, $i = 1, 2, \ldots, p$, and where R_i is the largest correlation between X_i and the remaining variables $X_1, \ldots, X_p$. An alternative estimate of $\boldsymbol{\Psi}$ is given by $\widehat{\boldsymbol{\Psi}}_2$, where $\widehat{\boldsymbol{\Psi}}_2$ is a diagonal matrix of elements d_{2i}, $i = 1, 2, \ldots, p$, with $d_{2i} = (1 - R_i^2)$ and where R_i^2 is the square of the multiple correlation between X_i and the remaining X variables. A commonly used method known as the *principal factor method* uses either of these two methods to determine $\widehat{\mathbf{A}}\widehat{\mathbf{A}}'$.

Determination of the Number of Factors

The most important step in the estimation of the factor analysis model is the estimation of r, the number of factors. If r is too large, some of the residual or error factors will be mixed in with the common factors, and if r is too small, important common factors will be omitted. Three lower bounds derived by Guttman for the value of r are available. The most commonly used lower bound is where r is at least as large as the number of eigenvalues in $\mathbf{R}$ that exceed 1 (*eigenvalue 1 criterion*). This is the most commonly used criterion. This criterion was introduced in Section 9.1 for principal component analysis. The eigenvalue of 1 is the arithmetic mean of the eigenvalues of a correlation matrix. The value of 1 is also the variance of each of the X variables, and hence the eigenvalue-one-criterion suggests that a factor be retained if it explains at least as much as a single variable.

Two additional criteria involving lower bounds are based on the estimation of the communalities using $(\mathbf{R} - \widehat{\boldsymbol{\Psi}})$. The number of positive eigenvalues of $(\mathbf{R} - \widehat{\boldsymbol{\Psi}}_1)$ and the number of positive eigenvalues of $(\mathbf{R} - \widehat{\boldsymbol{\Psi}}_2)$ are lower bounds to the value of $\mathbf{r}$. The estimators $\widehat{\boldsymbol{\Psi}}_1$ and $\widehat{\boldsymbol{\Psi}}_2$ of $\widehat{\boldsymbol{\Psi}}$ were defined

above. For the air pollution data discussed in Section 9.1, the eigenvalue
one criterion would suggest $r = 4$ factors.

A Useful Preliminary Test

In Chapter 7 a test for zero correlation was introduced for sample correlation matrices. If the true correlation matrix is diagonal there are no
common factors and hence $r = 0$. This test should be carried out first to
ensure that the factor analysis being carried out is meaningful. Chance
correlations can be used to generate chance factors.

Scree Test

A popular alternative approach to the determination of the number of factors is the *scree test*. This procedure employs a graph of the eigenvalues
(vertical) versus the eigenvalue number (horizontal). Since the eigenvalues
are ordered from largest to smallest, this graph yields a downward sloping and usually exponential shaped curve. The typical shape of a *scree
graph* consists of two parts: a rapidly downward sloping first part with an
exponential shape followed by a second part which is almost a horizontal
line. The typical shape is therefore similar to a scythe or hockey stick. The
almost horizontal part is viewed as random variation around a constant
ordered from largest to smallest. This part is referred to as the *scree*, since
it resembles a scree of rock debris at the foot of a mountain. These small
eigenvalues correspond to the unique factors or error terms, which are not
required to explain the correlations among the variables. The large eigenvalues represent variation explained by the common factors. The correct
number of factors r corresponds to the eigenvalue number to the immediate
left of the beginning of the scree called the elbow.

Example

A plot of the eigenvalues for the air pollution data correlation matrix is
shown in Figure 9.9. From this plot it would appear that the scree begins
at the fifth component and hence four factors should be used. In Figure 9.9
an almost horizontal line could be fitted by eye to the last seven eigenvalues.
In this case the scree test and the eigenvalue-one-criterion agree on $r = 4$
factors. The two criteria, however, do not always agree.

The Broken Stick Model

An interesting way to examine the distribution of eigenvalues is by the
broken stick model. Suppose we have a stick of length one unit which is
broken at random into p pieces. The expected length of the kth longest segment is given by $(1/p) \sum_{j=k}^{p} 1/j = \Delta_k$. The average of the eigenvalues for
a correlation matrix is unity and hence the eigenvalues can be compared to
the expected lengths for the broken stick segments. If the proportion of the

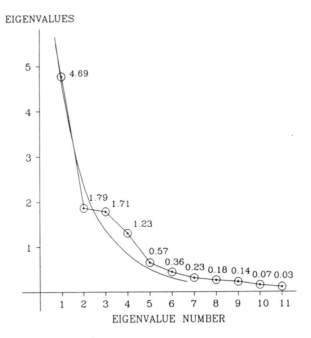

EIGENVALUES

FIGURE 9.9. Eigenvalues from Air Pollution Data Correlation Matrix

variance accounted for by the kth largest eigenvalue exceeds this amount, then that component should be retained. All components corresponding to eigenvalues below the expected length are discarded. This model is discussed in Jolliffe (1986).

Example

For 11 variables the values of Δ_k are given by:

k	1	2	3	4	5	6	7	8	9	10	11
Δ_k	0.275	0.184	0.138	0.108	0.085	0.067	0.052	0.039	0.027	0.017	0.008

For the air pollution example, the proportion of variance accounted for by the eleven eigenvalues are 0.426, 0.163, 0.155, 0.112, 0.052, 0.032, 0.021, 0.017, 0.013, 0.006, 0.003, which suggests that after the first eigenvalue the remaining eigenvalues may simply be random variation comparable to the broken stick model. In Figure 9.9 it could also be argued that there is only one factor and that the scree begins at eigenvalue 2.

Equal Correlation Structure and the Number of Factors

In Chapter 7 a procedure was presented for testing the null hypothesis of equal off-diagonal elements in a correlation matrix. It can be shown that for a $(p \times p)$ correlation matrix with equal off-diagonal elements given by ρ, the first eigenvalue is given by $[1 + \rho(p - 1)]$. The remaining eigenvalues are equal and are given by $(1 - \rho)$. This correlation structure is therefore consistent with a single factor model. If the hypothesis of zero correlation is rejected this test for equal correlation structure provides a useful follow-up test. For the air pollution example the critical χ^2 value for the equal correlation structure is 208 with 54 degrees of freedom. The p-value for the test is therefore less than 0.0000.

Principal Factor Approach

The principal component analysis solution to the factor model discussed above begins with a specified r and simultaneously determines estimates of the communalities and the elements of the factor pattern matrix $\mathbf{A}$ using the first r eigenvectors of $\mathbf{R}$. A variation of this approach is to estimate both the communalities and r using the second or third lower bound approach outlined above. This method is commonly called the principal factor approach. The eigenvectors of $\widehat{\mathbf{A}}\widehat{\mathbf{A}}' = (\mathbf{R} - \widehat{\mathbf{\Psi}})$ corresponding to the positive eigenvalues of $\mathbf{A}\mathbf{A}'$ are then used to determine $\widehat{\mathbf{A}}_1$ as in the case of principal component analysis. If desired, the estimate $\widehat{\mathbf{A}}_1$ of $\mathbf{A}$ can then be used to determine new communalities $\widehat{\mathbf{A}}_1\widehat{\mathbf{A}}_1'$, which are then used to obtain new eigenvectors. The process can be iterated until the change in the communality estimates is minimal.

Computer Software

The calculations required for the factor analysis examples in this section were performed using SAS PROC FACTOR.

Example

Using the third lower bound or squared multiple correlation R_i^2 to estimate the communality for X_i, a principal factor approach was applied to the air pollution data correlation matrix. The factor pattern matrix and eigenvalues for the first four factors are given in Table 9.22. The prior communality estimates (squared multiple correlations) and final communalities are also shown in Table 9.22. A comparison of this factor pattern matrix to the first four factors for the principal component solution shows that the two solutions are very similar. The communality estimates based on the first four principal components are also shown in the table. The principal component communalities are in general marginally higher than the principal factor squared multiple correlations. When the principal component communalities are higher than the principal factor communalities, this is often due to artificially high communalities in the principal component method. Using prior communalities of unity can result in some of the error variation being included in the common factors.

9.2.2 FACTOR ROTATION

As suggested in Section 9.2.1, there is no unique solution to the factor analysis model. In this section factor rotation will be introduced as a means of obtaining factors that are more easily interpretable.

The Theory of Rigid Rotation

In the factor analysis model presented in Section 9.2.1 given by

$$\mathbf{X} = \mathbf{FA}' + \mathbf{U},$$

the factors $\mathbf{F}$ are not unique. By introducing an *orthogonal transformation* matrix $\mathbf{T}$, $\mathbf{T}'\mathbf{T} = \mathbf{TT}' = \mathbf{I}$, new factors $\mathbf{G} = \mathbf{FT}$ may be defined that also satisfy the factor analysis model. The factor loading matrix becomes $\mathbf{B}' = \mathbf{T}'\mathbf{A}'$ and the model is given by

$$
\begin{aligned}
\mathbf{X} &= \mathbf{FTT}'\mathbf{A}' + \mathbf{U} \\
&= \mathbf{GB}' + \mathbf{U}.
\end{aligned}
$$

The orthogonal transformation is a *rigid rotation* of the r axes in a p-dimensional space. Since there are infinitely many factor solutions that yield the same correlation matrix, the question arises as to whether there is an optimum set of factors.

The most common criterion applied to factor rotation is known as *simple structure*, first advocated by Thurstone in his pioneering work in factor analysis. In his characterization of simple structure, Thurstone used five criteria to describe the numerical properties of the factor loading matrix $\mathbf{A}$. The essence of these criteria is that the observed variables should fall into mutually exclusive categories in such a way that the variables in a given category exhibit loadings that are high on the same single factor, moderate to low on a very few factors and negligible on the remaining factors. Simple structure does not allow for the possibility of a "general factor" which was the subject of much debate among psychometricians in the first half of this century.

Since the description of simple structure is qualitative, it is necessary to determine a quantitative description that can be used generally to determine an objective rather than a subjective solution. In order to define such quantitative criteria, the notation given in Figure 9.10 is required. This figure illustrates the factor loading matrix for the new factors $G_1, G_2, \ldots, G_r$.

TABLE 9.22. Factor Analysis of Air Pollution Data Correlation Matrix Using Principal Factor Method

	F_1	F_2	F_3	F_4	Prior Communality Estimates	Final Communality Estimates	Communality Estimates From First Four Principal Components
SMIN	0.580	0.166	-0.376	0.469	0.724	0.725	0.840
SMEAN	0.909	-0.074	0.016	0.331	0.928	0.942	0.920
SMAX	0.851	-0.157	-0.026	0.195	0.854	0.788	0.804
PMIN	0.835	0.265	0.171	-0.004	0.827	0.797	0.845
PMEAN	0.786	0.508	0.068	-0.265	0.928	0.950	0.956
PMAX	0.610	0.569	0.022	-0.401	0.878	0.857	0.897
PM2	0.503	-0.359	0.530	0.051	0.642	0.666	0.820
PERWH	0.345	-0.334	-0.619	-0.429	0.741	0.798	0.913
NONPOOR	0.403	-0.482	-0.084	-0.435	0.638	0.591	0.774
GE65	0.455	-0.332	-0.523	0.135	0.641	0.609	0.779
LPOP	0.443	-0.499	0.528	-0.104	0.680	0.735	0.868
Eigenvalues	4.512	1.542	1.401	1.004			

	G_1	G_2	G_3	...	G_r	c_i^2
X_1	b_{11}	b_{12}	b_{13}	...	b_{1r}	$\sum_{j=1}^{r} b_{1j}^2$
X_2	b_{21}	b_{22}				$\sum_{j=1}^{r} b_{2j}^2$
$\vdots$	$\vdots$	$\vdots$	$\vdots$		$\vdots$	$\vdots$
X_p	b_{p1}	b_{p2}	b_{p3}	...	b_{pr}	$\sum_{j=1}^{r} b_{pj}^2$

FIGURE 9.10. Factor Loading Matrix Based on New Factors

Varimax

Simple structure suggests that we design a $\mathbf{B}'$ matrix in such a way that only one of the elements $\{b_{ij}\}$ is close to 1 in each row and that most elements in a row are close to zero. The communality of the variable X_i is given by $c_i^2 = \sum_{j=1}^{r} b_{ji}^2$. Since not all variables will have the same communality, each row of $\mathbf{B}'$ can be normalized by dividing through by c_i.

The method of rotation, called *normalized varimax rotation* is commonly used to achieve simple structure. This method maximizes the quantity

$$\sum_{j=1}^{r} \left\{ \frac{\sum_{i=1}^{p} (b_{ij}^2/c_i^2)^2}{p} - \left[\frac{\sum_{i=1}^{p} (b_{ij}^2/c_i^2)}{p} \right]^2 \right\}$$

which is the sum of the variances of each b_{ij}^2/c_i^2 over each j. For each factor j, the variance of b_{ij}^2/c_i^2 is maximized, which tends to force the quantity to zero or 1 since $|b_{ij}/c_i| < 1$. The dividing by c_i^2 normalizes the values of b_{ij}^2 taking into account the differences in the communalities. (The reader should replace (b_{ij}^2/c_i^2) by X_i in the above expression to recognize the conventional expression for a sample variance.)

The varimax method is by far the most commonly used method of rotation. It is ideal for producing orthogonal factors that approach the simple structure objective. Beyond this it is also commonly used as a starting point for oblique methods of rotation to be discussed later in this section.

The process of computing the coefficients for the rotation is an iterative procedure. Factors are typically rotated in pairs until the quantity defined above is maximized. Graphically, this can be viewed as a rigid rotation of perpendicular axes in such a way that the new axes tend to pass through the data points. In other words, one of the two coordinates or loadings is forced to zero or 1. The geometry of rotation will be discussed later in this section.

Example

For the air pollution data correlation matrix introduced in Section 9.1, the first four principal components were rotated using varimax rotation. Table 9.23 shows the factor loadings for the rotated factors. For each variable, the largest factor loading is underlined. By concentrating on factor loadings that exceed 0.5, the four factors can be seen to represent the four clusters of variables given by (1) PMIN, PMEAN, PMAX; (2) SMIN, SMEAN, SMAX, GE65; (3) PM2, LPOP; (4) PERWH, NONPOOR, GE65. The first factor therefore measures the particulate levels, and the second factor measures the sulfate levels as well as the fraction of the population over age 65. (For some reason cities with high sulphate readings also tend to have a larger proportion of aged people.) The third factor measures population level and density; the fourth factor decreases with increases in percent white, percent above poverty level, and percent over age 65. Since without loss of generality the factor loading signs can all be changed by multiplying by -1, it would be more convenient to change all the signs for factor four. With the sign change, the fourth factor now represents a demographic variable strongly related to percent white and percent above poverty and also related to the percent of the population over age 65.

The communalities column in Table 9.23 indicates that the variances in the original X variables explained by the four factors varies from 0.77 for NONPOOR to 0.96 for PMEAN. These communalities are identical to the variances explained by the first four unrotated principal components in Table 9.15. The communalities are therefore not changed by rotation. The variances explained by each factor are shown at the bottom of Table 9.23. These individual variances are not the same as the variances explained by the first four principal components in Table 9.14 (see eigenvalues at the bottom of the table). The total variance explained by the first four principal components (85.6%) given in Table 9.14 does not change after rotation.

For comparison purposes, the varimax rotated solution for five factors is shown in Table 9.24. The first three factors remain essentially unchanged from the four factor solution except that the second factor now has a higher correlation with PMIN and a lower correlation with GE65. The fourth factor measures a contrast between GE65 and PERWH and the fifth factor represents PERWH and NONPOOR. The lowest communality is now .81 whereas most communalities are above 0.88. The sum of the variances explained is 9.99, which is 90.8% of the total of 11. The five-component solution seems to provide a very good summary of the variation carried by the 11 variables.

A varimax rotation was also carried out for the four factor solution obtained using the principal factor method shown in Table 9.22. The factor loadings are shown in Table 9.25. The factor pattern obtained is very similar to the rotated principal components solution in Table 9.23. In addition, the five factor varimax rotated solution was also obtained using the principal

TABLE 9.23. Factor Loading Matrix After Varimax Rotation of First Four Principal Components

	F_1	F_2	F_3	F_4	Communalities
SMIN	0.236	0.877	-0.116	0.042	0.84
SMEAN	0.406	0.744	0.444	-0.066	0.92
SMAX	0.358	0.669	0.441	-0.186	0.80
PMIN	0.757	0.397	0.337	0.008	0.84
PMEAN	0.943	0.223	0.106	-0.071	0.96
PMAX	0.940	0.038	-0.042	-0.102	0.90
PM2	0.110	0.145	0.886	0.037	0.82
PERWH	0.087	0.197	-0.130	-0.922	0.91
NONPOOR	0.117	-0.044	0.400	-0.774	0.77
GE65	-0.117	0.704	-0.001	-0.520	0.78
LPOP	0.044	-0.007	0.919	-0.147	0.87
Variance Explained by Rotated Components	2.743	2.537	2.338	1.798	

TABLE 9.24. Factor Loading Matrix After Varimax Rotation of First Five Principal Components

	F_1	F_2	F_3	F_4	F_5	Communalities
SMIN	0.129	0.913	-0.154	0.196	-0.029	.91
SMEAN	0.313	0.796	0.410	0.168	0.115	.94
SMAX	0.329	0.601	0.455	0.357	0.086	.81
PMIN	0.664	0.567	0.288	-0.098	0.143	.88
PMEAN	0.923	0.303	0.113	0.028	0.070	.96
PMAX	0.971	0.056	-0.001	0.080	0.011	.95
PM2	0.097	0.154	0.899	-0.006	0.018	.84
PERWH	0.149	-0.008	-0.134	0.690	0.634	.92
NONPOOR	0.048	0.076	0.262	0.094	0.936	.96
GE65	-0.030	0.336	0.092	0.898	0.059	.93
LPOP	0.044	-0.004	0.916	0.020	0.204	.88
Variance Explained by Rotated Components	2.495	2.386	2.236	1.503	1.371	

factor method. The resultant factor loadings are shown in Table 9.26. Once again the five-factor rotated solution for the principal factors is similar to the rotated five-factor solution for the principal components.

TABLE 9.25. Varimax Rotated Factors For Four Factor Principal Factor Analysis of Air Pollution Correlation Matrix

	F_1	F_2	F_3	F_4	Prior Communality Estimates	Final Communality Estimates
SMIN	0.213	0.816	-0.109	-0.023	0.724	0.750
SMEAN	0.360	0.776	0.449	-0.085	0.928	0.950
SMAX	0.323	0.664	0.441	-0.214	0.854	0.822
PMIN	0.703	0.427	0.345	0.003	0.827	0.841
PMEAN	0.932	0.247	0.122	-0.071	0.928	0.950
PMAX	0.918	0.062	-0.019	-0.096	0.878	0.894
PM2	0.110	0.147	0.794	0.018	0.642	0.672
PERWH	0.089	0.154	-0.091	-0.870	0.741	0.798
NONPOOR	0.097	-0.013	0.393	-0.653	0.638	0.667
GE65	-0.069	0.578	0.012	-0.518	0.641	0.664
LPOP	0.047	0.007	0.846	-0.125	0.680	0.741
Variance Explained	2.524	2.340	2.056	1.537		

Other Rotation Methods

Although the varimax method of rotation is by far the most commonly used, there are several other methods commonly available in statistics computer software. A brief outline appears below.

Quartimax Criterion

A second method of rotation, *quartimax*, maximizes the sum of the variances of the b_{ij}^2 over the entire loading matrix. Hence the quantity

$$\frac{\sum_{j=1}^{r} \sum_{i=1}^{p} (b_{ij}^2)^2}{pr} - \left[\frac{\sum_{j=1}^{r} \sum_{i=1}^{p} (b_{ij}^2)}{pr} \right]^2$$

is maximized. This is equivalent to maximizing the quantity $\sum_{j=1}^{r} \sum_{i=1}^{p} b_{ij}^4$. This method of rotation usually produces a general factor since the variance is computed over the entire matrix, not just in each column as in the case of varimax rotation. After the first general factor the loadings on the remaining factors tend to be lower than for varimax rotation.

Orthomax

A general class of orthogonal rotation criteria can be constructed using a weighted average of the raw varimax and quartimax criteria. The *orthomax*

TABLE 9.26. Varimax Rotated Factors for Five Factor Principal Factor Analysis of Air Pollution Correlation Matrix

	F_1	F_2	F_3	F_4	F_5	Prior Communality Estimates	Final Communality Estimates
SMIN	0.164	0.823	-0.111	-0.019	0.178	0.724	0.756
SMEAN	0.310	0.784	0.448	-0.104	0.162	0.928	0.964
SMAX	0.326	0.564	0.472	-0.121	0.399	0.854	0.869
PMIN	0.638	0.545	0.323	-0.103	-0.141	0.827	0.853
PMEAN	0.912	0.306	0.126	-0.089	-0.007	0.928	0.957
PMAX	0.937	0.061	0.005	-0.044	0.093	0.878	0.898
PM2	0.099	0.140	0.801	-0.011	0.008	0.642	0.693
PERWH	0.109	0.069	-0.112	-0.776	0.407	0.741	0.802
NONPOOR	0.060	0.051	0.332	-0.741	-0.025	0.638	0.678
GE65	-0.032	0.391	0.042	-0.333	0.629	0.641	0.682
LPOP	0.043	-0.004	0.844	-0.158	0.012	0.680	0.743
Variance Explained	2.377	2.188	2.037	1.335	0.809		

criterion maximizes the quantity

$$\sum_{j=1}^{r}\left[\sum_{i=1}^{p}b_{ij}^4 - \frac{\gamma}{p}\left(\sum_{i=1}^{p}b_{ij}^2\right)^2\right]$$

where $0 \leq \gamma \leq 1$. For $\gamma = 0$, this criterion becomes the quartimax, but for $\gamma = 1$ this criterion becomes the raw varimax. For $\gamma = 0.5$ the criterion is sometimes referred to as the *biquartimax* and for $\gamma = r/2$ it is equivalent to the *equamax* criterion.

Example

For comparison purposes four factor rotations using the criteria quartimax, orthomax with $\gamma = 0.5$ (biquartimax) and $\gamma = r/2$ (equamax) are compared in Table 9.27. For each variable the highest factor loading is underlined. A comparison of the three methods reveals essentially no difference among the solutions. In every case the highest loading for any variable appears on the same factor. A comparison of these three solutions to the varimax rotated solution in Table 9.23 also shows that for this example there is virtually no difference between the varimax solution and the solutions in Table 9.27.

TABLE 9.27. Comparison of Factor Loadings for Various Methods of Rotation

	Quartimax				Biquartimax				Equamax			
	F_1	F_2	F_3	F_4	F_1	F_2	F_3	F_4	F_1	F_2	F_3	F_4
SMIN	0.241	0.872	-0.126	-0.071	0.239	0.875	-0.120	-0.057	0.230	0.880	-0.110	-0.017
SMEAN	0.415	0.747	0.434	0.037	0.411	0.745	0.439	0.052	0.397	0.743	0.450	0.091
SMAX	0.367	0.676	0.433	0.160	0.363	0.671	0.437	0.174	0.350	0.663	0.445	0.209
PMIN	0.763	0.394	0.326	-0.025	0.761	0.395	0.331	-0.016	0.752	0.401	0.344	0.006
PMEAN	0.946	0.219	0.096	0.062	0.945	0.220	0.101	0.068	0.941	0.227	0.113	0.080
PMAX	0.940	0.034	-0.050	0.100	0.940	0.035	-0.047	0.102	0.940	0.041	-0.036	0.104
PM2	0.119	0.151	0.883	-0.048	0.116	0.147	0.885	-0.042	0.103	0.143	0.888	-0.027
PERWH	0.089	0.226	-0.127	0.915	0.087	0.218	-0.013	0.919	0.085	0.171	-0.135	0.926
NONPOOR	0.122	-0.015	0.404	0.772	0.119	-0.030	0.401	0.774	0.114	-0.067	0.395	0.775
GE65	-0.110	0.722	-0.003	0.496	-0.113	0.713	-0.001	0.508	-0.122	0.687	-0.002	0.540
LPOP	0.053	0.005	0.920	0.140	0.049	-0.002	0.919	0.145	0.038	-0.016	0.918	0.152

Oblique Rotation

In some applications it is preferable to permit a minor amount of correlation among factors. Rotation methods that permit the factors to be correlated are called *oblique*. There does not seem to be one single popular method for oblique rotation, and the use of nonorthogonal rotation requires considerable expertise. As this topic is not discussed here, the interested reader is referred to more specialized texts on factor analysis, such as Harman (1976) and Gorsuch (1983).

Procrustes Rotation

If a hypothesized factor pattern is to be tested, or if two factor analysis solutions are to be compared, a *procrustes rotation* can be used. In this technique the rotation is carried out to obtain a solution that is as close as possible to some hypothesized factor pattern or to a previous solution. The previous solution may arise from the same data but a different technique, or may be from a different sample on the same variables. Beginning with an orthogonal solution, the researcher may choose to reduce the low loadings to zero and then use the procrustes method to obtain a new solution, which most closely approximates the new target. Interested readers should consult the two factor analysis textbooks listed in the previous paragraph.

The Geometry of Factor Analysis

The term rigid rotation, used to label the orthogonal transformation process described above, is derived from the geometry associated with the transformation. The p X variables can be used to generate a p-dimensional vector space and the r factors can be viewed as linear transformations of the p X variables. Since the variables and factors have mean 0 and variance 1, the correlation between any pair of vectors is equivalent to the cosine of the angle between them, and the length of any vector is 1.

In panel (a) of Figure 9.11 the relationship between X_i and F_j is shown. The correlation between the two variables is $\cos\theta$ and the projection of X_i onto F_j is given by $\widetilde{X}_{ij}$, which lies along F_j with $\widetilde{X}_{ij} = X_i \cos\theta = a_{ij}$. Recall that a_{ij} is the factor loading for X_i on F_j. In panel (b) of Figure 9.11 the projection of X_i onto the plane of F_j and F_k is shown. The projection onto this plane is given by the vector $\widetilde{X}_{ijk}$, which makes an angle of θ with F_j and $(90° - \theta)$ with F_k. The coordinates of $\widetilde{X}_{ijk}$ are given by (a_{ij}, a_{ik}) where $a_{ij} = X_i \cos\theta$ and $a_{ik} = X_i \sin\theta$. The coordinates of $\widetilde{X}_{ijk}$ are therefore the factor loadings, and hence the factor loadings a_{ij} and a_{ik} are the coordinates of X_i in the projection of X_i onto the plane formed by F_j and F_k.

The characterization of the factors F_j and F_k relative to X_i depend on the magnitudes of a_{ij} and a_{ik} respectively. The closer these correlations are to 1 in absolute value the more important is the variable X_i in the

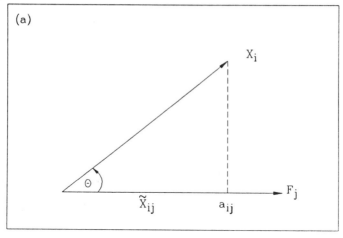

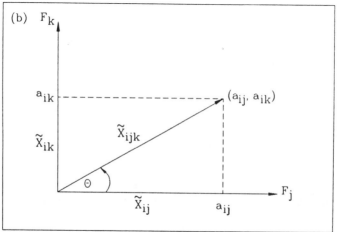

FIGURE 9.11. Geometry of Factor Analysis

characterization of these two factors. In the extreme case that $a_{ij} = 1$, F_j is considered to be equivalent to X_i, whereas if $a_{ij} = 0$, then F_j is unrelated to X_i.

For any given pair of factors F_j and F_k, the entire set of loadings for all p random variables can be plotted in the two-dimensional space generated by F_j and F_k. Figure 9.12 shows an example consisting of 14 such variables. The coordinates of the 14 variables with respect to the two factors F_j and F_k indicate the magnitudes of the correlations of the variables with the two factors. From the figure, it would appear that variables 1, 7, 9 and 13 move in the same direction as F_k and F_j, whereas variables 5 and 10 move in a direction opposite to F_j and F_k. For variables 2, 6 and 12 the direction is positive for F_k and negative for F_j and for variables 8, 11 and 14 the

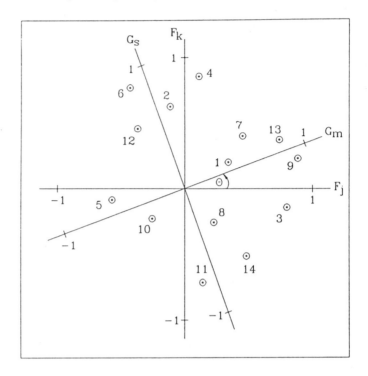

FIGURE 9.12. Geometry of Factor Rotation

opposite is true. Variable 3 is strongly related to F_j and very weakly related to F_k; for variable 4 the reverse is true. It can be seen from this figure that a rigid rotation of the factors through an angle θ to obtain new factors G_m and G_s will yield correlations (loadings) with respect to the new factors that have a tendency to be close to zero on one factor and close to one on the other factor. Exceptions to this are the variables 3 and 4, which are now more equally correlated with both new factors.

Example

Figures 9.13 and 9.14 contain plots showing the loadings for the first two factors of the four-factor principal component solutions before and after rotation. In Figure 9.13 the loadings plotted are before rotation whereas in Figure 9.14 the loadings plotted are after rotation. In Figure 9.13 many of the loadings are in the middle of the range (0,1) and hence many of the points are in the middle of one of the four quadrants. In Figure 9.13 the first factor shows moderate to high loadings for all variables, but in Figure 9.14 the first factor displays high loadings for only three variables. Similar results can be seen for the second factor.

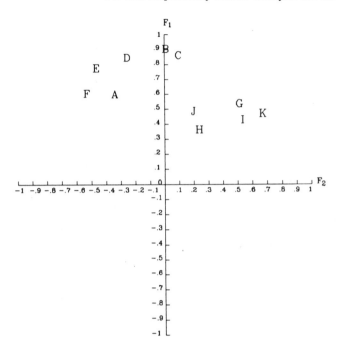

FIGURE 9.13. Unrotated Factor Loadings for Factors 1 and 2

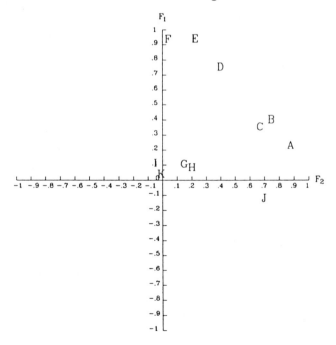

FIGURE 9.14. Rotated Factor Loadings for Factors 1 and 2

9.2.3 FACTOR SCORES

Factor analysis is often used as a preliminary data reduction step forming a part of a larger analysis. In later analyses the factors obtained are often used in place of the original variables from which they were derived. An example of such a process is provided by the multiple regression example in Section 9.1.6 which employed the principal components as explanatory variables. In order to use the derived factors, observations on the factors must be generated for the n rows of the original X matrix. These factor values are called *factor scores*.

If the unrotated factors have been obtained using principal components or the principal factor method, then the unrotated factor score can be obtained using the eigenvectors and eigenvalues of $\widehat{\mathbf{A}}\widehat{\mathbf{A}}'$. As outlined in 9.2.4 the factors can be expressed in terms of $\mathbf{X}$ and the eigenvectors and eigenvalues of $\widehat{\mathbf{A}}\widehat{\mathbf{A}}'$. The rotated factor scores can then be obtained using the orthogonal transformation matrix $\mathbf{T}$ used for the rotation.

If some other technique is used to obtain $\widehat{\mathbf{A}}$ and $\widehat{\boldsymbol{\Psi}}$, then weighted least squares is often used to obtain the estimated factor scores, $\widehat{\mathbf{F}}$. Given $\widehat{\mathbf{A}}$ and $\widehat{\boldsymbol{\Psi}}$, $\widehat{\mathbf{F}}$ is given by $\widehat{\mathbf{F}} = (\widehat{\mathbf{A}}'\widehat{\boldsymbol{\Psi}}^{-1}\widehat{\mathbf{A}})^{-1}\widehat{\mathbf{A}}'\widehat{\boldsymbol{\Psi}}^{-1}\mathbf{X}$. The matrix required to obtain $\widehat{\mathbf{F}}$ from $\mathbf{X}$ is often labeled the *factor score coefficient matrix*.

Factor Score Example

Table 9.28 summarizes the factor scores for the 40 cities on the first four factors of the varimax rotated, five factor principal component solution summarized in Table 9.24. A scatterplot of the 40 cities based on the first two factors is shown in Figure 9.15. The abbreviations for the 40 cities are also shown in Table 9.28. From Figure 9.15 we can see that AL has a large positive value of F_1 and a small positive value of F_2. The city of Albuquerque therefore has relatively high particulate readings and relatively low sulphate readings. For Canton (CA) the value of F_1 is near zero and F_2 is relatively large and hence Canton has relatively high sulphate readings. The cities of Greensboro (GB) and Scranton (SC) are near zero on F_2 but are at opposite ends of the scale on F_1. A second scatterplot in Figure 9.15 plots the cities with respect to the third and fourth factors. From the second scatterplot we can conclude that NY is an outlier with respect to the dimension of the third factor, whereas WA is unusual with respect to the fourth factor. Since F_3 is a population size and density factor NY is not an unexpected outlier. F_4 is an indicator of a large proportion of white people and of people over age 65. Washington therefore has a relatively large nonwhite population with relatively few over age 65.

(a)

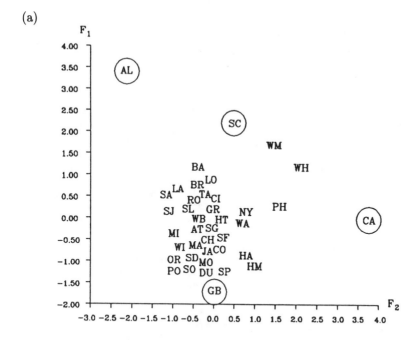

(b)

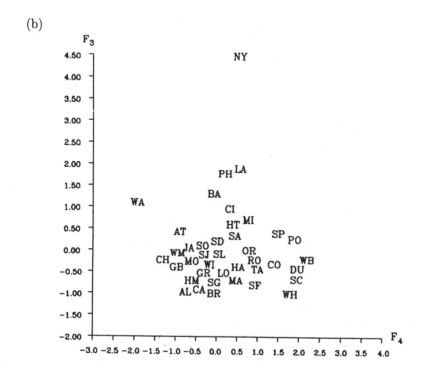

FIGURE 9.15. Scatterplots for Cities on First Four Rotated Factors

TABLE 9.28. Data for Plot of Factor Scores

City	Abbreviation	F_1	F_2	F_3	F_4
San Jose	SJ	0.215	-1.036	-0.198	-0.447
Roanoke	RO	0.408	-0.460	-0.304	0.756
Albuquerque	AL	3.493	-2.082	-0.943	-0.846
Charleston	CH	-0.435	-0.019	-0.309	-1.480
Harrisburg	HA	-0.843	0.817	-0.470	0.313
Greenville	GR	0.229	-0.089	-0.608	-0.487
Hartford	HT	-0.231	0.017	0.536	0.283
Columbus	CO	-0.595	-0.094	-0.402	1.242
Orlando	OR	-0.889	-0.805	-0.187	0.651
Sacramento	SO	-1.039	-0.726	-0.109	-0.407
Philadelphia	PH	0.336	1.618	1.697	0.034
Washington	WA	-0.056	0.673	1.085	-2.056
Minneapolis	MI	-0.373	-0.845	0.556	0.567
Los Angeles	LA	0.603	-0.905	1.814	0.437
Greensboro	GB	-1.453	0.058	-0.437	-1.133
Jacksonville	JA	-0.649	-0.197	-0.022	-0.837
Madison	MA	-0.661	-0.371	-0.784	0.298
Wilmington	WM	1.751	1.477	-0.113	-1.031
Tacoma	TA	0.614	-0.275	-0.384	0.778
Scranton	SC	2.099	0.784	-0.756	1.789
Canton	CA	0.062	3.801	-0.884	-0.476
Atlanta	AT	-0.274	-0.011	0.343	-1.044
Baltimore	BA	1.180	-0.364	1.224	-0.223
Portland	PO	-1.050	-0.782	0.177	1.728
Springfield, MA	SP	-1.142	0.273	0.309	1.341
Salt Lake	SL	0.327	-0.544	-0.266	-0.142
Wichita	WI	-0.664	-0.712	-0.415	-0.338
Lorain	LO	0.789	-0.150	-0.655	0.021
Hamilton	HM	-1.045	1.007	-0.855	-0.627
Montgomery	MO	-0.965	-0.209	-0.358	-0.811
San Diego	SD	-0.856	-0.513	0.082	-0.153
Duluth	DU	-1.160	-0.154	-0.505	1.786
Wilkes Barre	WB	-0.009	-0.344	-0.388	1.902
Wheeling	WH	1.236	2.133	-1.091	1.608
San Antonio	SA	0.594	-1.009	0.254	0.292
Cincinnati	CI	0.416	-0.124	0.876	0.201
Saginaw	SG	-0.433	-0.326	-0.776	-0.327
Baton Rouge	BR	0.650	-0.309	-0.326	-1.834
New York	NY	0.095	0.703	4.436	0.427
Springfield, OH	SF	-0.275	0.099	-0.841	0.724

Regression Example

To provide an example, the five rotated factors of Table 9.24 are used as explanatory variables in a multiple regression model with dependent variable TMR. This example follows the example in Section 9.1.5 where principal components were used as explanatory variables. Using the factor scores obtained from the five rotated principal components shown in Table 9.24, the estimated multiple regression relationship with TMR is given by

$$TMR = 981.575 \quad + \quad \underset{(.135)}{19.752} \; F_1 + \underset{(.000)}{73.957} \; F_2 + \underset{(.145)}{19.237} \; F_3$$
$$+ \quad \underset{(.000)}{121.250} \; F_4 - \underset{(.001)}{46.918} \; F_5.$$

The contributions to R^2 were F_1 (0.014), F_2 (0.190), F_3 (0.013), F_4 (0.511) and F_5 (0.076) and hence the overall R^2 value was 0.804. Since the factors are standardized, we can compare the magnitudes of the coefficients. The factor F_4, which reflects the negative of NONPOOR and PERWH, was the most important, accounting for 51% of the variation in TMR. The first and third factors F_1 and F_3 seem to contribute little to the explanation of TMR. The first factor measures PMIN, PMEAN and PMAX whereas the third factor measures PM2 and LPOP. The second factor F_2 accounts for 19% of the variation in TMR and is primarily a measure of SMIN, SMEAN, SMAX and PMIN. The fifth factor F_5 accounts for 7.6% of the variation in TMR and is a measure of PERWH and GE65.

For comparison, a multiple regression of TMR on PMIN, SMAX, GE65, NONPOOR and PERWH produced an R^2 of 0.87. The reader may recall from the results of Section 9.1.5 that some of the minor principal components were related to TMR. Since the factor analysis only retained the first five factors some of the information from the later components has been omitted.

9.2.4 THE MAXIMUM LIKELIHOOD ESTIMATION METHOD

The Maximum Likelihood Approach

Under the assumption of multivariate normality, the maximum likelihood estimators for $\mathbf{A}$ and $\boldsymbol{\Psi}$ in the factor analysis model are obtained by maximizing

$$-\frac{1}{2}(n-1)[\ln|\boldsymbol{\Sigma}| - tr\,\boldsymbol{\Sigma}^{-1}\mathbf{S}]$$

subject to $\boldsymbol{\Sigma} = \mathbf{A}\mathbf{A}' + \boldsymbol{\Psi}$. To get a unique solution it is customary to require also that $\mathbf{A}'\boldsymbol{\Psi}^{-1}\mathbf{A}$ be diagonal. Since maximum likelihood estimators are scale invariant, it does not matter whether $\boldsymbol{\Sigma}$ is a covariance matrix or a correlation matrix. If $\boldsymbol{\Sigma}$ is a covariance matrix, the correlation matrix is given by $\rho = \mathbf{D}^{-1}\boldsymbol{\Sigma}\mathbf{D}^{-1}$, where $\mathbf{D}$ is the diagonal matrix of standard deviations.

$$\rho = \mathbf{D}^{-1}\mathbf{A}\mathbf{A}'\mathbf{D}^{-1} + \mathbf{D}^{-1}\boldsymbol{\Psi}\mathbf{D}^{-1} = \mathbf{A}^*\mathbf{A}^{*\prime} + \boldsymbol{\Psi}^*.$$

The two solutions can therefore be related using the maximum likelihood estimators of the standard deviations.

The equation for the maximum likelihood estimators does not have an analytical solution and hence must be solved iteratively. Unfortunately the solution does not necessarily converge. Improper solutions with negative diagonal elements of $\widehat{\boldsymbol{\Psi}}$ can occur and hence additional constraints requiring positive diagonal elements for $\widehat{\boldsymbol{\Psi}}$ must be imposed. Cases of negative residual variances (and hence communalities that exceed 1) are commonly called *Heywood cases*.

Goodness of Fit

An advantage of the maximum likelihood approach is that a goodness of fit test can be carried out for the model $\boldsymbol{\Sigma} = \mathbf{A}\mathbf{A}' + \boldsymbol{\Psi}$. Given the maximum likelihood solution, $\widehat{\mathbf{A}}$ and $\widehat{\boldsymbol{\Psi}}$, the maximum likelihood estimator of $\boldsymbol{\Sigma}$ under the model is given by $\widehat{\boldsymbol{\Sigma}} = \widehat{\mathbf{A}}\widehat{\mathbf{A}}' + \widehat{\boldsymbol{\Psi}}$. A test of goodness of fit is carried out by comparing $\widehat{\boldsymbol{\Sigma}}$ to $\dfrac{(n-1)}{n}\mathbf{S}$ which is the maximum likelihood estimator of $\boldsymbol{\Sigma}$ under no restrictions. The test statistic, including Bartlett's correction, is given by

$$\left[n - \left(\frac{2p + 11}{6}\right) - \frac{2r}{3}\right] \ln \left|\frac{\widehat{\boldsymbol{\Sigma}}}{\frac{(n-1)}{n}\mathbf{S}}\right|$$

which in large samples has a χ^2 distribution with $\frac{1}{2}[(p - r)^2 - (p + r)]$ degrees of freedom if the model is correct. This test is related to the test for sphericity discussed in Chapter 7. In the test introduced here the null hypothesis is that the residual covariance matrix is diagonal after removing the r common factors defined by $\widehat{\mathbf{A}}$.

This test statistic is only valid if $\mathbf{A}\mathbf{A}'$ has full rank r. If r is too large, the test statistic is not reliable. A common approach to maximum likelihood factor analysis is to begin with the solution $r = 1$ and then to continue increasing r by increments of 1 until the goodness of fit criterion is satisfied or until the degrees of freedom are negative.

Example

For the air pollution data covariance matrix, a maximum likelihood factor analysis was carried out. The squared multiple correlation was used to estimate the communalities. For values of r (the number of factors) varying from 1 to 6, the maximum likelihood solution yielded the χ^2 goodness of fit values and p-values summarized in Table 9.29. From the p-values shown in the table it would appear that a five-factor solution provides an adequate fit for the data.

The factor pattern matrix for the five-factor varimax rotated solution is given in Table 9.30. The first two factors appear to represent the PMIN, PMEAN, PMAX group and the SMIN, SMEAN, SMAX group respectively. The third factor is dominated by PM2 and LPOP whereas the fifth factor is dominated by SMAX. The fourth factor represents a demographic factor consisting of the variables PERWH and NONPOOR and the variable GE65. In comparison to the five-factor solution derived from the principal

TABLE 9.29. Chi-Square Goodness of Fit Test for Number of Factors

No. of Factors	1	2	3	4	5	6
CHI Square	190.712	108.486	72.430	28.571	13.270	3.849
d.f.	44	34	25	17	10	4
p value	0.0001	0.0001	0.0001	0.0387	0.2089	0.4268

factor method shown in Table 9.26, the first three factors are very similar. The variances explained by the three factors in each case are also similar. For the maximum likelihood approach, the fourth factor represents a demographic factor consisting of PERWH, NONPOOR and GE65 whereas for the principal factor method the fourth factor represents only two demographic variables PERWH and NONPOOR. The fifth factor for the principal factor method primarily represents GE65, but for the maximum likelihood method the fifth factor represents SMAX. The fifth factor for the principal factor method does however have a weak correlation with PERWH. A comparison of the final communality estimates in Tables 9.26 and 9.30 also indicates some differences between the two solutions.

Cross Validation

A disadvantage of the χ^2 goodness of fit test is that it tends to overestimate the number of factors, r. A useful procedure is to divide the sample into two halves and to obtain the maximum likelihood solution for the two samples separately. Denoting the two sets of sample estimators by $\widehat{\Sigma}_1$, $\widehat{\Sigma}_2$, $(n_1 - 1)\mathbf{S}_1/n_1$, and $(n_2 - 1)\mathbf{S}_2/n_2$ respectively, the goodness of fit can be evaluated by determining the likelihood statistics

$$\ln|\widehat{\Sigma}_1| - \ln\left|\mathbf{S}_2\frac{(n_2 - 1)}{n_2}\right| + tr\left[\frac{(n_2 - 1)}{n_2}\,\mathbf{S}_2\widehat{\Sigma}_1^{-1}\right] - p \quad \text{and}$$

$$\ln|\widehat{\Sigma}_2| - \ln\left|\mathbf{S}_1\frac{(n_1 - 1)}{n_1}\right| + tr\left[\frac{(n_1 - 1)}{n_1}\mathbf{S}_1\widehat{\Sigma}_2^{-1}\right] - p.$$

The value of r that minimizes these statistics is an estimate of the required number of factors.

In the absence of normality the models can be evaluated using the discrepancy functions

1. Ordinary Least Squares

$$\sum_{j<k}\sum(s_{jk} - \hat{\sigma}_{jk})^2, \qquad \text{where } s_{jk} \text{ and } \hat{\sigma}_{jk} \text{ are elements of } \frac{(n-1)}{n}\mathbf{S} \text{ and } \widehat{\Sigma} \text{ respectively.}$$

TABLE 9.30. Varimax Rotated Factors For Five Factor Maximum Likelihood Factor Analysis For Air Pollution Data

	F_1	F_2	F_3	F_4	F_5	Prior Communality Estimates	Final Communality Estimates
SMIN	0.179	0.868	-0.118	-0.114	0.036	0.724	0.814
SMEAN	0.327	0.759	0.436	-0.121	0.262	0.928	0.956
SMAX	0.312	0.532	0.401	-0.222	0.640	0.854	1.000
PMIN	0.668	0.508	0.350	-0.021	-0.047	0.827	0.830
PMEAN	0.950	0.270	0.134	-0.079	0.007	0.928	1.000
PMAX	0.923	0.033	-0.012	-0.053	0.151	0.878	0.880
PM2	0.123	0.117	0.785	0.004	0.086	0.642	0.653
PERWH	0.127	0.039	-0.089	-0.987	0.029	0.741	1.000
NONPOOR	0.046	0.035	0.411	-0.618	-0.035	0.638	0.555
GE65	-0.048	0.417	0.019	-0.581	0.243	0.641	0.573
LPOP	0.046	-0.025	0.898	-0.103	0.056	0.680	0.822
Variance Explained	2.476	2.1378	2.105	1.790	0.576		

2. Generalized Least Squares

$$\frac{1}{2}tr\left\{\left[\frac{(n-1)}{n}\mathbf{S} - \widehat{\boldsymbol{\Sigma}}\right]\left[\mathbf{S}\frac{(n-1)}{n}\right]^{-1}\right\}^{2}.$$

Akaike and Schwartz Criteria

In small samples, dividing the sample in half may not be advantageous. In such circumstances two other goodness of fit criteria that can be used are the *Akaike criterion* and the *Schwartz criterion*. Denoting the value of the log likelihood statistic above by L, the Akaike goodness of fit criterion is given by $(L+2q)$ where $q = p(r+1)-r(r-1)/2$ for the r factor model. The Schwartz criterion is given by $(L+q\ln n^*)$ where n^* is Bartlett's correction given by $n^* = [n - (2p + 4r + 11)/6]$. The number of factors r is chosen to minimize the value of the criterion. These two criteria seek to insure that decreases in L due to fitting additional factors are sufficiently large to justify the additional factors. A major difficulty with the χ^2 goodness of fit test is that in small samples many competing models are equally satisfactory. Large samples, however, tend to reject all models. The difficulty with large samples is that more complex models are obtained that attempt to explain residuals that are negligible for practical purposes. These goodness of fit criteria tend to guard against the overfitting of large models to explain negligible residuals. Application of cross validation techniques to the determination of factors in capital asset pricing models is discussed in Conway and Reinganum (1988) and Jobson (1988).

Example

The Akaike and Schwartz criteria were used to evaluate the maximum likelihood factor analyses for various numbers of factors for the air pollution data. For each value of r = number of factors, the values of the criteria are given in the Table 9.31. Using the Schwartz criterion the minimum value occurs at four factors whereas the Akaike criterion is minimized at five factors. Since the Schwartz criterion suggests that the four-factor solution should be sufficient, the four-factor maximum likelihood solution is presented in Table 9.32. A comparison of this four-factor solution with the principal four-factor solution reveals that the two sets of factors are remarkably similar. It would seem that in stretching to extract the unnecessary fifth factor the two methods obtained different results.

9.2.5 Results From A Simulation Study

In a simulation study carried out by Hakistan, Rogers and Cattell (1982), the eigenvalue-one, scree test and maximum likelihood χ^2 test procedures for estimating the number of factors were compared. For a correlation matrix derived from a factor model consisting of r factors and p variables,

TABLE 9.31. Values for Akaike and Schwartz Criteria

No. of Factors	1	2	3	4	5	6
AKAIKE	269.47	194.84	171.14	133.90	129.03	129.05
SCHWARTZ	153.31	124.44	120.19	108.33	111.80	116.88

TABLE 9.32. Varimax Rotated Factors for Four-Factor Maximum Likelihood Factor Analysis of Air Pollution Data

	F_1	F_2	F_3	F_4	Final Communality Estimates
SMIN	0.199	0.788	-0.158	-0.106	0.697
SMEAN	0.309	0.858	0.393	-0.120	1.000
SMAX	0.269	0.710	0.394	-0.237	0.788
PMIN	0.683	0.491	0.293	-0.010	0.793
PMEAN	0.951	0.278	0.115	-0.074	1.000
PMAX	0.904	0.092	0.006	-0.057	0.829
PM2	0.130	0.151	0.770	0.002	0.632
PERWH	-0.129	-0.044	0.095	0.986	1.000
NONPOOR	-0.046	-0.086	-0.377	0.614	0.528
GE65	-0.056	0.447	0.020	-0.587	0.547
LPOP	0.044	-0.028	0.933	-0.111	0.886
Variance Explained	2.435	2.421	2.049	1.800	

sample correlation matrices were generated. A variety of factor models were employed in order to provide variation with respect to the following:

1. Ratio of r/p.

2. Complexity of the factors in terms of degree of departure of factor loadings from the extremes 0 or 1.

3. Sample size n.

4. Level of communalities.

5. Number of variables, p.

6. Existence of minor factors.

The results of the study indicate that the estimate of the number of major factors is more reliable with low values of r/p, less complex factors, large sample size, high communalities, large numbers of variables, and no minor factors.

For the eigenvalue-one criterion, if the value of r/p was low, the estimated number of factors tended to be correct unless the communalities were low, in which case the estimated number of factors was too large. Complex factors and high values of r/p tended to produce underestimates of the number of factors especially for high communalities. Larger samples generally produce better results than smaller samples. The addition of minor factors did not in general reduce the reliability of the estimates.

For the scree test, high communalities yielded more reliable estimates, whereas low communalities tended to produce large overestimates of the number of factors. Overestimates also tended to be produced with a large number of factors, p, with increased factor complexity, and with lower sample size. The ratio r/p, however, did not seem to have an impact. The existence of minor factors tended in general to increase the magnitude and frequency of the overestimates.

For the likelihood ratio test, the estimates were more precise with high communalities. For low communalities the estimates tended to be low. Underestimation of the number of factors was more frequent with larger p and with large values of r/p. Factorial complexity did not seem to have an effect on the reliability. The degree of error was markedly reduced by large samples. The presence of minor factors tended to result in the overestimation of the number of major factors; was much more pronounced in large samples than in small ones, and when the ratio r/p was low. The degree of overestimation in the presence of minor factors was larger than the degree of underestimation when minor factors were not present.

A comparison of the three criteria showed that the eigenvalue one criterion was less influenced by the presence of minor factors than the other two. In addition, the magnitude of overestimation for likelihood ratio and scree is much larger in this case than when no minor factors are present.

In the absence of minor factors, high communalities in general yielded reliable estimates for the scree and likelihood ratio tests. For the eigenvalue-one-criterion it was also necessary to have low values of r/p in order to obtain reliable results. When communalities were low, the likelihood ratio test tended to underestimate and the scree test tended to overestimate. The performance of the eigenvalue one criterion, however, also depended on the ratio r/p. For small values of r/p, the eigenvalue one criterion tends to overestimate the number of factors whereas for high values of r/p the tendency is to underestimate.

9.2.6   A Second Example

In a survey of R.C.M.P. (Royal Canadian Mounted Police) officers, responses were obtained for eighteen stress items. The 18 stress variables are measures of stress due to (1) insufficient resources, (2) unclear job responsibilities, (3) personality conflicts, (4) investigation where there is serious injury or fatality, (5) dealing with obnoxious or intoxicated people, (6) having to use firearms, (7) notifying relatives about death or serious injury, (8) tolerating verbal abuse in public, (9) unsuccessful attempts to solve a series of offences, (10) lack of availability of ambulances, doctors, and so on, (11) poor presentation of a case by the prosecutor resulting in dismissal of the charge, (12) heavy workload, (13) not getting along with unit commander, (14) many frivolous complaints lodged against officers by the public, (15) engaging in high speed chases, (16) becoming involved in physical violence with an offender, (17) investigating domestic quarrels, (18) having to break up fights or quarrels in bars and cocktail lounges. Additional background on this survey is available in Jobson and Schneck (1982).

Each of the 18 stress variables represents a composite scale derived from two questions. Each officer was asked how stressful (1 = very little to 5 = very much) he or she found each situation and also how often (1 = never to 5 = always) this type of situation occurred on the unit. The composite scale was determined by subtracting 1 from the frequency scale and then multiplying it by the stress scale to obtain an index of stress (0 to 20) for each officer on each of the 18 types of stress.

The responses from 56 officers for the eighteen stress items are displayed in Table 9.33. The means, standard deviations, and correlation matrix are shown in Table 9.34. The variables are labeled STR1 through STR18. The eigenvalues of the correlation matrix are also shown in Table 9.34.

Using an eigenvalue-one-criterion, the number of factors should be 5. The fifth eigenvalue, however, is only 1.026. The maximum likelihood goodness of fit statistics are shown in Table 9.35. The χ^2 test of goodness of fit test suggests three or at most four factors and the Akaike criterion also suggests four factors. The Schwartz criterion, however, seems to suggest that two factors are sufficient.

The scree plot for the eigenvalues of the correlation matrix is shown in Figure 9.16. The scree part of the plot seems to have two parts with two eigenvalues to the left of the longest scree. The next three eigenvalues to the right may represent sampling error and/or several minor factors. Depending on the choice of scree, a two-factor or six-factor solution is suggested. Using the eigenvalue-one-criterion a five-factor solution is suggested. For simplicity a four-factor solution using various approaches is now presented. Table 9.36 contains the factor pattern matrix for the vector varimax rotated solution for the principal component, principal factor, and maximum likelihood methods. A comparison of the factor loadings reveals the solutions to be very similar. The first factor could be labeled organizational stress since the highest loadings are for organizational characteristics such as lack of resources, unclear job responsibility, personality conflicts, and poor presentation of cases by prosecutors. The second factor is labeled stress from unpleasant duties such as use of firearms, investigation and notification of

TABLE 9.33. R.C.M.P. Stress Data

STR1	STR2	STR3	STR4	STR5	STR6	STR7	STR8	STR9
4	3	3	4	3	4	6	6	3
12	3	8	8	6	5	5	9	4
12	8	15	1	3	2	8	4	12
8	4	5	1	9	0	2	9	4
5	5	6	15	12	3	4	3	10
12	2	3	6	4	10	9	15	15
6	3	6	4	9	3	6	2	3
6	8	5	2	6	3	1	4	8
6	6	4	3	4	4	4	4	4
2	3	0	6	4	0	6	4	3
12	2	1	3	6	1	6	6	4
6	8	5	2	9	0	2	3	4
6	2	2	2	6	1	1	6	1
2	0	6	3	4	1	6	9	4
2	6	2	6	6	5	6	6	4
3	3	2	4	3	2	2	3	3
8	4	4	8	9	5	6	8	5
2	4	3	6	6	4	4	6	4
3	3	4	2	6	0	4	9	6
2	3	6	4	4	2	4	4	2
2	6	2	2	9	2	4	6	4
3	3	3	2	6	3	8	6	8
8	4	8	3	4	2	6	3	10
4	3	8	3	4	5	12	6	8
6	3	6	2	3	0	2	2	4
6	4	8	12	9	5	10	9	8
6	6	4	6	4	5	6	2	8
6	8	8	6	12	5	8	8	8
4	6	9	20	1	6	9	6	2
20	20	12	9	16	10	12	12	16
9	8	9	4	4	8	6	8	6
6	3	3	3	3	2	1	2	6
9	4	4	9	6	4	6	9	3
12	4	3	3	12	12	10	6	8
9	4	4	8	4	3	4	12	8
12	8	10	6	12	4	4	10	8
15	4	2	3	10	0	2	2	6
15	2	1	2	3	1	3	3	6
8	8	6	6	20	4	12	3	12
12	6	3	2	9	5	2	3	10
12	4	4	8	12	5	4	3	8
12	6	12	4	6	8	3	9	12
15	2	4	6	20	6	4	12	8
12	1	6	3	8	4	6	6	8
12	12	15	2	3	4	6	12	10
6	12	15	2	3	1	2	4	6
2	3	4	1	6	0	1	3	8
9	2	4	2	2	2	3	2	3
12	9	1	4	6	3	2	9	12
6	4	4	6	4	4	3	4	8
12	8	3	4	12	0	4	8	8
6	3	8	2	6	0	8	9	3
6	6	8	4	6	4	3	4	6
8	3	8	3	4	0	3	6	4
0	0	4	6	4	3	9	9	0
1	1	6	2	3	1	6	4	6

TABLE 9.33. R.C.M.P. Stress Data (continued)

STR10	STR11	STR12	STR13	STR14	STR15	STR16	STR17	STR18
4	6	3	3	6	8	8	8	6
6	5	15	5	3	10	4	10	4
10	8	6	4	8	2	2	2	1
4	15	9	3	10	3	6	4	5
5	10	8	4	10	8	8	8	12
5	10	20	4	5	10	8	15	15
3	12	2	3	2	2	4	9	4
10	15	6	8	4	1	3	6	2
4	4	6	0	2	8	8	2	8
0	2	0	0	2	4	6	4	4
3	2	2	0	2	6	9	4	6
4	10	4	0	0	2	8	3	4
2	8	2	2	1	1	4	1	6
3	9	6	1	6	6	1	2	6
3	4	3	0	8	10	8	10	8
3	8	2	0	3	2	2	2	4
4	8	4	4	6	6	8	9	8
5	4	4	0	3	3	4	6	6
3	6	4	0	4	2	6	6	9
3	2	3	0	2	2	2	4	4
4	6	3	0	3	6	6	4	4
2	2	3	3	0	3	6	9	6
4	8	3	3	6	4	2	4	2
12	12	3	0	0	10	12	9	4
2	6	4	1	1	2	2	3	2
4	9	4	5	4	5	8	12	12
4	9	3	0	3	3	4	4	6
8	12	8	5	10	10	8	9	8
2	2	12	6	4	2	3	6	2
20	15	15	15	12	12	20	16	16
12	6	8	5	6	4	6	6	8
3	3	1	2	12	2	2	2	2
6	6	9	8	9	6	6	9	6
1	3	15	4	2	4	12	12	4
0	4	8	6	4	6	6	6	4
5	10	10	15	6	10	15	12	8
2	1	6	0	4	2	4	9	9
0	2	6	0	9	2	2	3	6
6	8	6	4	12	9	9	6	4
5	1	12	6	4	5	10	8	4
4	3	8	4	6	8	8	8	8
9	8	12	8	12	8	12	6	15
4	4	10	4	8	4	4	9	9
2	4	12	6	6	4	6	6	6
6	15	9	8	10	5	2	6	2
2	9	4	15	6	4	4	2	4
8	8	3	12	10	3	6	9	3
4	6	4	9	9	4	4	9	6
12	9	6	8	12	3	6	9	8
6	8	3	3	3	3	3	3	4
12	12	8	12	12	6	8	4	12
3	6	6	4	8	6	3	6	4
2	6	6	5	9	4	8	6	2
0	8	3	0	6	6	4	1	2
5	6	0	0	9	6	6	9	15
4	6	4	4	2	4	1	4	4

TABLE 9.34. R.C.M.P. Stress Data – Correlations, Means and Standard Deviations

	STR1	STR2	STR3	STR4	STR5	STR6	STR7	STR8	STR9	STR10	STR11	STR12	STR13	STR14	STR15	STR16	STR17	STR18
STR1	1.00	0.39	0.19	0.01	0.41	0.36	0.04	0.29	0.56	0.32	0.07	0.64	0.41	0.34	0.22	0.36	0.33	0.27
STR2	0.39	1.00	0.53	0.11	0.29	0.29	0.13	0.16	0.51	0.62	0.48	0.28	0.58	0.35	0.26	0.45	0.19	0.13
STR3	0.19	0.53	1.00	0.07	-0.03	0.19	0.26	0.20	0.32	0.33	0.44	0.27	0.48	0.24	0.20	0.09	0.02	-0.06
STR4	0.01	0.11	0.07	1.00	0.18	0.39	0.35	0.20	0.04	0.04	-0.05	0.28	0.12	0.09	0.30	0.20	0.36	0.30
STR5	0.41	0.29	-0.03	0.18	1.00	0.27	0.21	0.17	0.37	0.25	0.16	0.32	0.23	0.28	0.34	0.53	0.42	0.35
STR6	0.36	0.29	0.19	0.39	0.27	1.00	0.51	0.39	0.47	0.35	0.04	0.65	0.23	0.08	0.46	0.54	0.63	0.39
STR7	0.04	0.13	0.26	0.35	0.21	0.51	1.00	0.33	0.26	0.25	0.10	0.24	-0.00	0.00	0.49	0.37	0.47	0.24
STR8	0.29	0.16	0.20	0.20	0.17	0.39	0.33	1.00	0.26	0.28	0.28	0.53	0.29	0.24	0.45	0.31	0.46	0.49
STR9	0.56	0.51	0.32	0.04	0.37	0.47	0.26	0.26	1.00	0.54	0.34	0.53	0.45	0.38	0.33	0.43	0.40	0.30
STR10	0.32	0.62	0.33	0.04	0.25	0.35	0.25	0.28	0.54	1.00	0.55	0.27	0.48	0.40	0.35	0.45	0.35	0.37
STR11	0.07	0.48	0.44	-0.05	0.16	0.04	0.10	0.28	0.34	0.55	1.00	0.11	0.39	0.29	0.20	0.19	0.12	0.15
STR12	0.64	0.28	0.27	0.28	0.32	0.65	0.24	0.53	0.53	0.27	0.11	1.00	0.44	0.23	0.41	0.40	0.52	0.32
STR13	0.41	0.58	0.48	0.12	0.23	0.23	-0.00	0.29	0.45	0.48	0.39	0.44	1.00	0.48	0.24	0.37	0.38	0.18
STR14	0.34	0.35	0.24	0.09	0.28	0.08	0.00	0.24	0.38	0.40	0.29	0.23	0.48	1.00	0.27	0.17	0.17	0.31
STR15	0.22	0.26	0.20	0.30	0.34	0.46	0.49	0.45	0.33	0.35	0.20	0.41	0.24	0.27	1.00	0.66	0.52	0.48
STR16	0.36	0.45	0.09	0.20	0.53	0.54	0.37	0.31	0.43	0.45	0.19	0.40	0.37	0.17	0.66	1.00	0.61	0.53
STR17	0.33	0.19	0.02	0.36	0.42	0.63	0.47	0.46	0.40	0.35	0.12	0.52	0.38	0.17	0.52	0.61	1.00	0.53
STR18	0.27	0.13	-0.06	0.30	0.35	0.39	0.24	0.49	0.30	0.37	0.15	0.32	0.18	0.31	0.48	0.53	0.53	1.00
Mean	7.53	4.82	5.51	4.64	3.32	6.69	5.10	6.10	6.46	4.80	6.98	6.17	4.12	5.82	5.01	6.01	6.44	6.12
Std Dev	4.35	3.35	3.52	3.48	2.72	4.12	2.90	3.18	3.40	3.68	3.71	4.17	4.05	3.58	2.82	3.62	3.44	3.65
Eigen-values	6.61	2.36	1.40	1.22	1.02	0.88	0.67	0.64	0.59	0.50	0.39	0.35	0.30	0.29	0.23	0.18	0.15	0.12

TABLE 9.35. Goodness of Fit Statistics For Maximum Likelihood Solution R.C.M.P. Stress Data

Factors	χ^2	Prob.	Akaike	Schwartz
1	238.258	0.0001	352.893	212.903
2	159.642	0.0064	296.889	202.116
3	119.376	0.1151	282.803	211.276
4	86.873	0.4836	274.921	222.525
5	65.218	0.7300	277.462	237.973
6	49.470	0.8320	284.725	254.769

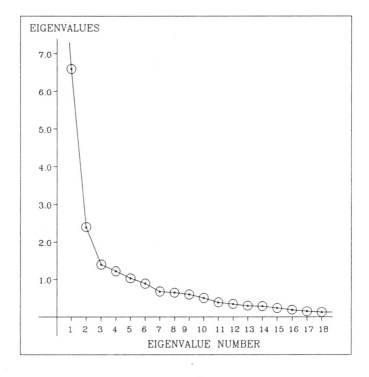

FIGURE 9.16. Scree Plot for Eigenvalues of Stress Stress Data Correlation Matrix

relatives of a serious injury or a fatality, verbal abuse from the public, and high-speed chases. The third factor reflects workload stress and insufficient resources. The fourth factor is labeled stress due to risk of injury. This factor is related to stress from dealing with obnoxious and intoxicated people, physical violence, domestic quarrels and fights in bars and cocktail lounges.

The final communalities are similar for the three methods with the principal component communalities generally larger. The final communalities

TABLE 9.36. Varimax Rotated Factor Pattern Matrix for the Four-Factor Solution of the R.C.M.P. Stress Data

| | Factor 1 | | | Factor 2 | | | Factor 3 | | | Factor 4 | | | Communalities | | | |
| | | | | | | | | | | | | | SMC | Final | | |
	P.C.	P.F.	MLE	P.C.	P.F.	MLE	P.C.	P.F.	MLE	P.C.	P.F.	MLE	Prior	P.C.	P.F.	MLE
STR1	0.178	0.221	0.309	-0.005	0.040	0.016	0.835	0.725	0.618	0.257	0.260	0.279	0.609	0.797	0.644	0.556
STR2	0.764	0.749	0.787	0.091	0.091	0.087	0.277	0.226	0.133	0.082	0.122	0.148	0.673	0.676	0.637	0.668
STR3	0.707	0.666	0.705	0.259	0.219	0.264	0.186	0.141	0.120	-0.439	-0.342	-0.433	0.590	0.796	0.630	0.770
STR4	-0.071	-0.038	-0.032	0.626	0.508	0.465	0.058	0.067	0.119	0.047	0.077	0.087	0.385	0.402	0.270	0.239
STR5	0.119	0.140	0.188	0.098	0.156	0.160	0.283	0.258	0.203	0.665	0.539	0.530	0.450	0.547	0.402	0.384
STR6	0.060	0.090	0.118	0.715	0.699	0.640	-0.454	-0.404	-0.412	0.143	0.151	0.238	0.694	0.742	0.685	0.651
STR7	0.128	0.123	0.102	0.810	0.738	0.800	-0.096	-0.082	-0.087	0.049	0.059	0.053	0.582	0.686	0.571	0.662
STR8	0.212	0.203	0.164	0.507	0.474	0.427	0.235	0.244	0.366	0.217	0.196	0.158	0.528	0.406	0.364	0.369
STR9	0.474	0.474	0.506	0.173	0.199	0.204	0.545	0.467	0.382	0.249	0.254	0.277	0.575	0.615	0.548	0.522
STR10	0.731	0.684	0.661	0.174	0.182	0.191	0.101	0.101	0.079	0.328	0.320	0.324	0.621	0.684	0.615	0.586
STR11	0.806	0.696	0.635	0.040	0.044	0.064	-0.145	-0.078	-0.015	0.107	0.099	0.036	0.511	0.685	0.503	0.409
STR12	0.115	0.147	0.153	0.441	0.447	0.375	0.782	0.741	0.908	0.096	0.097	0.102	0.738	0.831	0.782	1.000
STR13	0.632	0.626	0.650	0.047	0.052	0.029	0.462	0.415	0.349	0.104	0.135	0.150	0.673	0.627	0.586	0.569
STR14	0.498	0.449	0.467	-0.117	-0.050	-0.051	0.245	0.239	0.169	0.369	0.297	0.240	0.449	0.459	0.350	0.308
STR15	0.250	0.237	0.226	0.614	0.589	0.586	0.036	0.053	0.132	0.439	0.403	0.355	0.612	0.634	0.570	0.540
STR16	0.258	0.261	0.309	0.443	0.449	0.439	0.196	0.167	0.142	0.615	0.614	0.609	0.740	0.681	0.676	0.681
STR17	0.065	0.078	0.100	0.614	0.607	0.566	0.298	0.282	0.264	0.474	0.451	0.502	0.685	0.696	0.659	0.653
STR18	0.089	0.085	0.098	0.374	0.378	0.342	0.078	0.120	0.139	0.702	0.605	0.570	0.571	0.648	0.532	0.472
Variance Explained	3.416	3.074	3.232	3.269	2.944	2.756	2.516	2.039	2.039	2.412	1.966	2.013				

are also similar to the prior communalities obtained from the squared multiple correlations. The weakest communalities are for STR4, STR8, and STR14 around 40%. The majority of the principal component communalities are over 60%. The principal component method accounts for 64% of the variance whereas the principal factor and maximum likelihood methods account for 56% of the variance.

From the scree plot it is apparent that a two factor model may also be sufficient to account for the correlation among the stress items. The Schwartz goodness of fit criterion also indicated two factors. The varimax rotated factor pattern matrix for the maximum likelihood solution for two factors is shown in Table 9.37. From the factor loadings in Table 9.37, it would appear that the first factor represents stress related to the unpleasant aspects of police work, and the second factor is related to organizational stress. In comparison to the four factor solution discussed above, it would appear that the second and fourth factors have combined to form the unpleasant duties stress factor. This factor also includes the stress from a heavy workload which was part of factor three. The insufficient resources part of factor three is added to the organizational stress factor to make the organizational stress factor in the two-factor model. The total variation explained by the two-factor solution is 44%. It would appear, therefore, that separate factors for workload stress and stress due to risk of injury are of only minor importance. The communalities and factor loadings in Table 9.37 for STR1, STR3, STR5 and STR14 are quite low suggesting that there may be other factors which have not been identified.

9.2.7 OTHER SOURCES OF INFORMATION

Extensive discussions of factor analysis can be found in Gorsuch (1983), Mulaik (1972), Jolliffe (1986) and Lawley and Maxwell (1971).

9.3 Singular Value Decomposition and Matrix Approximation

In Section 9.1, it was demonstrated that the least squares approximation to the $(n \times p)$ data matrix $\mathbf{X}$ could be accomplished by replacing the p columns of $\mathbf{X}$ by a smaller number of columns $r < p$ derived from the principal components of the matrix $\mathbf{X}'\mathbf{X}$. The matrix approximation is denoted by $\widehat{\mathbf{X}} = \mathbf{Z}_1 \mathbf{V}'_1$ where $\mathbf{Z}_1$ $(n \times r)$ denotes the observations or scores on the first r principal components and $\mathbf{V}_1$ $(p \times r)$ denotes the matrix whose columns consist of the first r eigenvectors of $\mathbf{X}'\mathbf{X}$. The r principal components are linear combinations of the X variables and are given by $\mathbf{Z}_1 = \mathbf{X}\mathbf{V}_1$. The r principal components can be viewed as the unobservable underlying dimensions that generate the observable X variables. This

TABLE 9.37. Varimax Rotated Factor Pattern Matrix for the Two Factor Maximum Likelihood Solution of the R.C.M.P. Stress Data

	Factor 1	Factor 2	Final Communalities
STR1	0.378	0.429	0.327
STR2	0.147	0.804	0.669
STR3	0.001	0.604	0.365
STR4	0.447	-0.035	0.201
STR5	0.461	0.244	0.272
STR6	0.746	0.179	0.589
STR7	0.558	0.044	0.313
STR8	0.533	0.209	0.327
STR9	0.412	0.586	0.513
STR10	0.299	0.685	0.559
STR11	0.019	0.617	0.382
STR12	0.616	0.316	0.479
STR13	0.230	0.684	0.521
STR14	0.150	0.497	0.270
STR15	0.656	0.216	0.477
STR16	0.684	0.328	0.576
STR17	0.816	0.133	0.683
STR18	0.626	0.122	0.407
Variance Explained	4.399	3.531	7.930

principal component analysis is related to a more general form of matrix approximation based on the *singular value decomposition* of **X**.

In this section the theory of singular value decomposition is used to obtain matrix approximations to **X** that can be characterized in terms of both column labels and row labels. The biplot will be introduced as a graphical method for interpreting a matrix approximation based on two dimensions.

9.3.1 SINGULAR VALUE DECOMPOSITION AND PRINCIPAL COMPONENTS

As outlined in Section 3.2 of the Appendix, the $(n \times p)$ data matrix **X** of full rank $p < n$ can be expressed in the form

$$\mathbf{X} = \mathbf{UDV}', \tag{9.1}$$

where **D** $(p \times p)$ is a diagonal matrix of positive diagonal elements $\alpha_1, \alpha_2, \ldots,$ α_p arranged in descending order called the *singular values* of **X**, and **U**

$(n \times p)$ and $\mathbf{V}$ $(p \times p)$ are matrices whose columns contain the *left and right singular vectors* of $\mathbf{X}$ respectively. The singular vectors are denoted by $\mathbf{u}_1, \mathbf{u}_2, \ldots, \mathbf{u}_p$ and $\mathbf{v}_1, \mathbf{v}_2, \ldots, \mathbf{v}_p$ for $\mathbf{U}$ and $\mathbf{V}$ respectively and the vectors in each set are mutually orthogonal, hence $\mathbf{U}'\mathbf{U} = \mathbf{V}'\mathbf{V} = \mathbf{I}$.

A least squares approximation to $\mathbf{X}$ of dimension $r < p$ is provided by the matrix $\widehat{\mathbf{X}}$, where

$$\widehat{\mathbf{X}} = \sum_{j=1}^{r} \alpha_j \mathbf{u}_j \mathbf{v}_j',$$

and hence $\mathbf{Y} = \widehat{\mathbf{X}}$ minimizes the expression $tr(\mathbf{X} - \mathbf{Y})(\mathbf{X} - \mathbf{Y})'$ subject to the rank of $\mathbf{Y}$ being less than or equal to r.

The singular value decomposition can be related to principal component analysis, since for $\mathbf{X}$ given by (9.1) $\mathbf{X}'\mathbf{X} = \mathbf{VD}^2\mathbf{V}'$, and hence the right singular vectors of $\mathbf{X}$ are the eigenvectors of $\mathbf{X}'\mathbf{X}$ and the eigenvalues of $\mathbf{X}'\mathbf{X}$ are the squares of the singular values of $\mathbf{X}$. In a similar fashion for the matrix $\mathbf{XX}' = \mathbf{UD}^2\mathbf{U}'$, it can be seen that the left singular vectors of $\mathbf{X}$ are the eigenvectors of $\mathbf{XX}'$ and that the eigenvalues of $\mathbf{XX}'$ are the squares of the singular values of $\mathbf{X}$. The eigenvalues of $\mathbf{X}'\mathbf{X}$ and $\mathbf{XX}'$ are therefore equivalent.

For the principal components of $\mathbf{X}'\mathbf{X}$ given by $\mathbf{Z} = \mathbf{XV}$, it can be seen that $\mathbf{Z} = \mathbf{UD}$ and hence that the principal components of $\mathbf{X}'\mathbf{X}$ are simply a scaled version of the left singular vectors of $\mathbf{X}$. The singular value decomposition of $\mathbf{X}$ can therefore be expressed as $\mathbf{X} = \mathbf{ZV}'$ as outlined in Section 9.1

From the symmetry present it can be seen that principal components can also be defined for $\mathbf{XX}'$ say $\mathbf{W} = \mathbf{X}'\mathbf{U}$ and hence $\mathbf{W} = \mathbf{VD}$. The principal components of $\mathbf{XX}'$ are obtained by scaling the right singular vectors of $\mathbf{X}$ and the eigenvectors of $\mathbf{XX}'$ are the left singular vectors of $\mathbf{X}$. We can conclude therefore that the principal components of $\mathbf{XX}'$ are related to the eigenvectors of $\mathbf{X}'\mathbf{X}$, and the eigenvectors of $\mathbf{XX}'$ are related to the principal components of $\mathbf{X}'\mathbf{X}$. The principal components of $\mathbf{XX}'$ can be used to derive underlying dimensions for the rows of $\mathbf{X}$ in a similar fashion to the principal components of $\mathbf{X}'\mathbf{X}$ being used to generate column dimensions for $\mathbf{X}$ as discussed in Section 9.1. The use of the components of $\mathbf{XX}'$ to define row types is often referred to as, *Q-type factor analysis* in contrast to, *R-type factor analysis* based on $\mathbf{X}'\mathbf{X}$ discussed in Sections 9.1 and 9.2.

9.3.2 BIPLOTS AND MATRIX APPROXIMATION

As suggested above the elements of a singular value decomposition of a matrix can be used to provide a least squares approximation to the matrix of a given dimension. By using the singular value decomposition format, portions of $\mathbf{U}$, $\mathbf{V}$ and $\mathbf{D}$ can be used to write $\widehat{\mathbf{X}} = \mathbf{U}_1\mathbf{D}_1\mathbf{V}_1'$ where $\mathbf{U}_1(n \times r)$, $\mathbf{D}_1(r \times r)$ and $\mathbf{V}_1(p \times r)$ represent the first r columns of $\mathbf{U}$ and $\mathbf{V}$ and the corresponding portion of $\mathbf{D}$. The columns of $\mathbf{V}_1$ provide information about the first r column or variable components of $\mathbf{X}$, and the columns of $\mathbf{U}_1$ provide information about the first r row or object components of $\mathbf{X}$. The term *biplot* has the prefix "bi" to refer to the simultaneous consideration of both column dimensions and row dimensions. The suffix plot refers to the graphical presentation of this information which will be outlined below.

Constructing Biplots

A biplot is used to provide a two-dimensional representation for a data matrix $\mathbf{X}$. Only two dimensions are usually employed to keep the presentation simple. It is assumed that a singular value decomposition approximation for $\mathbf{X}$ based on $r = 2$ dimensions is adequate. This of course should be evaluated by examining the magnitudes of the singular values beyond $r = 2$. The sum of these remaining residual singular values should ideally represent only a small portion of $tr\mathbf{D}$.

A singular value decomposition approximation for $\mathbf{X}$ based on two dimensions is given by $\widehat{\mathbf{X}} = \mathbf{U}_1\mathbf{D}_1\mathbf{V}_1'$, where the rows of $\mathbf{V}_1'(2 \times p)$ are the eigenvectors of $\mathbf{X}'\mathbf{X}$ and the columns of $\mathbf{U}_1(n \times 2)$ are the eigenvectors of $\mathbf{XX}'$. There are several ways of employing the three elements of the right-hand side of the equation for $\widehat{\mathbf{X}}$. We begin with the most common form which is called the *principal components plot*.

The Principal Components Biplot

In principal components analysis an approximation for $\mathbf{X}'\mathbf{X}$ was given by $\widehat{\mathbf{X}}'\widehat{\mathbf{X}} = \mathbf{V}_1\mathbf{D}_1^2\mathbf{V}_1' = \mathbf{V}_1\boldsymbol{\Lambda}_1\mathbf{V}_1'$, where $\boldsymbol{\Lambda}_1$ denotes the (2×2) diagonal matrix consisting of the first two eigenvalues in the diagonal, and the two columns of $\mathbf{V}_1$ are the principal component loadings for the first two principal components of $\mathbf{X}'\mathbf{X}$. The approximation $\widehat{\mathbf{X}} = \mathbf{U}_1\mathbf{D}_1\mathbf{V}_1' = \mathbf{Z}_1\mathbf{V}_1'$ therefore represents the product of the principal component scores (two columns of $\mathbf{Z}_1$) and the corresponding loadings (two rows of $\mathbf{V}_1'$).

It was demonstrated in Section 9.1 that the principal component scores for the n objects could be plotted as a scatterplot using the two principal components as axes. In addition it was also shown that the loadings of the principal components (eigenvectors) could be plotted as rays drawn from the origin on a graph that contains principal components as axes. In the plot of object scores the objects can be related to the dimensions represented by the two components. In the plot of principal component loadings the relation between the new dimensions and the original variables could be observed. The principal components biplot simply combines the above two plots into one. By combining the two plots the relationships between the objects and the variables can be related through the principal components.

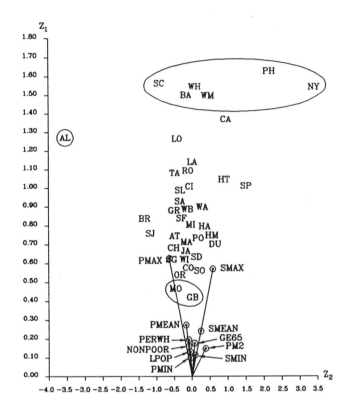

FIGURE 9.17. Principal Components Biplot for Air Pollution Data

Example

In panel (a) of each of Figures 9.6 and 9.7 in Section 9.1 plots of principal component scores and principal component loadings are illustrated for the first two principal components derived from the air pollution data. Combining these two plots yields the biplot shown in Figure 9.17. From the biplot we can see that the first component is strongly related to SMAX and PMAX, and that the cities of Philadelphia, New York, Wheeling, Baltimore, Wichita and Scranton have relatively high scores along this dimension. The second component represents a contrast between SMAX and PMAX. New York therefore seems to have high values of SMAX relative to PMAX and the reverse is true for Albuquerque.

Covariance Biplot

If the variables in the columns of $\mathbf{X}$ have been mean corrected then the matrix $\mathbf{X'X}$ is a scalar multiple of the covariance matrix [say $\mathbf{X'X} = (n -$

1)S]. In the singular value decomposition $\mathbf{X} = \mathbf{UDV}'$, we may write $\mathbf{X} = \mathbf{GH}$, where $\mathbf{H} = \mathbf{DV}'/(n-1)$ and $\mathbf{G} = \mathbf{U}(n-1)^{1/2}$. It follows then that $\mathbf{HH}' = \mathbf{VD}^2\mathbf{V}'/(n-1) = \mathbf{V}\mathit{\Lambda}\mathbf{V}'/(n-1) = \mathbf{X}'\mathbf{X}/(n-1) = \mathbf{S}$ and $\mathbf{GG}' = \widehat{\mathbf{X}}'\mathbf{S}^{-1}\widehat{\mathbf{X}}$. Denoting the rows of $\mathbf{G}$ by $\mathbf{g}_i$, $i = 1, 2, \ldots, n$, the distance between the points represented by $\mathbf{g}_i$ and $\mathbf{g}_\ell$ is the Mahalanobis distance between the observations $\hat{\mathbf{x}}_i$ and $\hat{\mathbf{x}}_\ell$ (the ith and ℓth rows of $\widehat{\mathbf{X}}$). Denoting the columns of $\mathbf{H}$ and $\mathbf{h}_j$, $j = 1, 2, \ldots, p$, the sample covariance between the variables $\hat{\mathbf{x}}_j$ and $\hat{\mathbf{x}}_k$ (columns of $\widehat{\mathbf{X}}$) is the cross-product between the vectors $\mathbf{h}_j$ and $\mathbf{h}_k$. The sample variance of $\hat{\mathbf{x}}_j$ is illustrated by the squared length of $\mathbf{h}_j$. Because the vectors $\mathbf{h}_j$ derived from the columns of $\mathbf{H}$ describe the covariance matrix, the plot is called a *covariance biplot*. The vectors $\mathbf{h}_j$ are plotted as rays, whereas the vectors $\mathbf{g}_i$ are plotted as points. A plot that only contains the rays corresponding to $\mathbf{h}_j$ is sometimes called an *h-plot*.

Symmetric Biplot

In a *symmetric biplot*, the singular value decomposition approximation to $\mathbf{X}$ is expressed as $\widehat{\mathbf{X}} = \mathbf{GH}'$ where $\mathbf{G} = \mathbf{UD}^{1/2}$ and $\mathbf{H}' = \mathbf{D}^{1/2}\mathbf{V}'$. In this case, the two columns of $\mathbf{G}$ are taken as coordinates for n points in a two-dimensional space and similarly the two rows of $\mathbf{H}'$ are taken as coordinates for p points in a two-dimensional space. Since each element of $\widehat{\mathbf{X}}$ is given by the product of a row of $\mathbf{G}$ and a column of $\mathbf{H}'$, then

$$\hat{x}_{ij} = g_{i1}h_{1j} + g_{i2}h_{2j} = \mathbf{g}_i\mathbf{h}_j,$$

where $\mathbf{g}_i$ denotes the ith row of $\mathbf{G}$ and $\mathbf{h}_j$ denotes the jth column of $\mathbf{H}$. The magnitude of $\hat{x}_{ij}$ therefore depends on the magnitude of the scalar product between $\mathbf{g}_i$ and $\mathbf{h}_j$. In Figure 9.18 the point corresponding to g_i is shown as a vector, and the columns of $\mathbf{H}$ are shown as points. In the figure we can see that $\hat{x}_{ij}$ is relatively large, whereas $\hat{x}_{ik}$ is relatively small. The point $\hat{x}_{im}$ is relatively large but negative.

9.3.3 OTHER SOURCES OF INFORMATION

Additional discussion of biplots can be found in Jackson (1990) and Seber (1984).

9.4 Correspondence Analysis

Correspondence analysis is a technique that uses singular value decomposition to analyze a matrix of nonnegative data. The technique simultaneously characterizes the relationship among the rows and also among the columns of the data matrix. The outcome of a correspondence analysis is a pair of *bivariate plots*. One bivariate plot is based on the first two principal axes derived from the row profiles, and the second plot is based on the first two

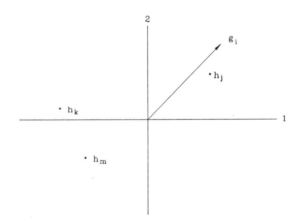

FIGURE 9.18. Symmetric Biplot

principal axes obtained from the column profiles. Points representing the row categories are plotted using the row principal axes and points representing the column categories are plotted using the column principal axes. The spatial relationships among the two sets of categories can then be studied using the two bivariate plots. By using the same pair of axes to denote both pairs of principal axes the two bivariate plots can be superimposed on one another. With both plots appearing on the same axes the spatial relationship between the row categories and column categories can also be related.

The theory of correspondence analysis is based on the generalized singular value decomposition of matrices, which is outlined in the Appendix. Readers unfamiliar with such decompositions should review Section 3 in the Appendix.

The discussion of correspondence analysis in this chapter begins with the study of two-dimensional contingency tables first discussed in Chapter 6. The technique is then extended to the two-dimensional product multinomial scheme introduced in Section 6.3 for the weighted least squares methodology. Finally, the multiple correspondence analysis technique is introduced for the study of higher dimensional contingency tables. The SAS computer software procedure CORRESP will be used throughout this section to perform the necessary data analysis.

TABLE 9.38. Correspondence Matrix of Observed Cell Densities for an $(r \times c)$ Contingency Table

		Columns					
		1	2	3	...	c	Row Masses
	1	o_{11}	o_{12}	o_{13}	...	o_{1c}	$o_{1.}$
	2	o_{21}	o_{22}	o_{23}	...	o_{2c}	$o_{2.}$
Rows	3	o_{31}	o_{32}	o_{33}	...	o_{3c}	$o_{3.}$
	$\vdots$	$\vdots$	$\vdots$	$\vdots$			
	r	o_{r1}	o_{r2}	o_{r3}	...	o_{rc}	$o_{r.}$
	Column Masses	$o_{.1}$	$o_{.2}$	$o_{.3}$	...	$o_{.c}$	1

9.4.1 CORRESPONDENCE ANALYSIS FOR TWO-DIMENSIONAL CONTINGENCY TABLES

Some Notation

In this section correspondence analysis is outlined as a technique that can be used to study interaction in a two-dimensional contingency table. In Chapter 6, the study of the n observation, $(r \times c)$ contingency table began with the cell frequencies n_{ij}, $i = 1, 2, \ldots, r$, $j = 1, 2, \ldots, c$, as displayed in Table 6.5. For the purposes of the discussion of correspondence analysis this table is converted to a table of observed *cell proportions* or *cell densities*. This notation is introduced in Table 9.38. The cell density for cell (i, j) is denoted by $o_{ij} = n_{ij}/n$, where n_{ij} denotes the sample frequency in cell (i, j). The row and column marginal densities are given by $o_{i.} = n_{i.}/n$ and $o_{.j} = n_{.j}/n$ respectively where $n_{i.}$ and $n_{.j}$ are the row and column marginal frequencies respectively.

Computer Software

The calculations in this section were performed using the SAS PROC CORRESP.

Example

The first contingency table to be used as an example in this section is illustrated in Table 9.39. The data was obtained from the files of a student-run legal advice service for the poor. This table examines the relationship be-

TABLE 9.39. Contingency Table for Criminal Charge Data

| Convicted | Sex | Charge | | | | | Totals |
		Impaired Driving	Theft Under $1000	Mischief	Possession of Narcotics	Other	
No	Male	8	11	5	7	12	43
	Female	5	15	3	1	6	30
Yes	Male	105	32	11	23	37	208
	Female	32	57	6	2	25	122
Totals		150	115	25	33	80	403

TABLE 9.40. Correspondence Matrix for Criminal Charge Data

Convicted	Sex	Impaired Driving	Theft Under $1000	Mischief	Possession of Narcotics	Other	Row Mass
No	Male	2.0	2.7	1.2	1.7	3.0	10.7
	Female	1.2	3.7	0.7	0.2	1.5	7.4
Yes	Male	26.1	7.9	2.7	5.7	9.2	51.6
	Female	7.9	14.1	1.5	0.5	6.2	30.3
Column Mass		37.2	28.6	6.2	8.1	19.9	100.0

tween the type of criminal charge and the eventual outcome of the case for both males and females. The corresponding matrix of cell densities and row and column marginal densities are shown in Table 9.40. The numbers are given as percentages and hence represent $100o_{ij}$. The column of row masses on the right presents the row marginals as percents $100o_{i.}$, and the row of column masses (last row) displays the column marginals, $100o_{.j}$. The majority of the clients were convicted (81.9%); a total of 51.6% of the sample were convicted males and 30.3% of the sample were convicted females. The two most common offences were impaired driving (37.2%) and theft under $1000 (28.6%). The most common offence for males was impaired driving (28.1% of the sample) and the most common female offence was theft under $1000 (17.8% of the sample).

Correspondence Matrix and Row and Column Masses

The $(r \times c)$ matrix of cell densities as shown in Table 9.38 is denoted by $\mathbf{O}$ and is called the *correspondence matrix*. The $(r \times 1)$ vector of row marginals $o_{i.}$, $i = 1, 2, \ldots, r$, is denoted by $\mathbf{r}$ and similarly the $(c \times 1)$ vector of column marginals $o_{.j}$, $j = 1, 2, \ldots, c$, is denoted by $\mathbf{c}$. These *row and column marginal vectors* can be written as $\mathbf{r} = \mathbf{O}e_c$ and $\mathbf{c} = \mathbf{O}'e_r$

TABLE 9.41. Matrix $\mathbf{R}$ for Row Profiles

		Columns					
		1	2	3	$\ldots$	c	Totals
Rows	1	$n_{11}/n_1.$	$n_{12}/n_1.$	$n_{13}/n_1.$	$\ldots$	$n_{1c}/n_1.$	1
	2	$n_{21}/n_2.$	$n_{22}/n_2.$	$n_{23}/n_2.$	$\ldots$	$n_{2c}/n_2.$	1
	3	$n_{31}/n_3.$	$n_{32}/n_3.$	$n_{33}/n_3.$	$\ldots$	$n_{3c}/n_3.$	1
	$\vdots$	$\vdots$	$\vdots$	$\vdots$	$\vdots$	$\vdots$	$\vdots$
	r	$n_{r1}/n_r.$	$n_{r2}/n_r.$	$n_{r3}/n_r.$	$\ldots$	$n_{rc}/n_r.$	1
	Column Mass	$n._1/n$	$n._2/n$	$n._3/n$	$\ldots$	$n._c/n$	1

where $\mathbf{e}_c$ $(c \times 1)$ and $\mathbf{e}_r$ $(r \times 1)$ are vectors of unities. The vectors $\mathbf{r}$ and $\mathbf{c}$ are also referred to respectively as *row and column masses*. Diagonal matrices constructed from the row and column masses are denoted by $\mathbf{D}_r$ $(r \times r)$ and $\mathbf{D}_c$ $(c \times c)$ respectively. The diagonal elements of $\mathbf{D}_r$ are the elements of $\mathbf{r}$ and the diagonal elements of $\mathbf{D}_c$ are the elements of $\mathbf{c}$.

Row and Column Profiles

Beginning with the table of cell frequencies n_{ij} for each row i, the $(c \times 1)$ vector of row conditional densities is determined from $n_{ij}/n_i.$, $j = 1, 2, \ldots, c$, and is denoted by $\mathbf{r}_i$. These row conditional densities are called *row profiles*. The complete set of r row profiles will be denoted by the $(r \times c)$ matrix $\mathbf{R}$ with rows given by $\mathbf{r}_i$, $i = 1, 2, \ldots, r$. Similarly the vector of column conditional densities $n_{ij}/n._j$, $i = 1, 2, \ldots, r$, for column j is denoted by the $(r \times 1)$ vector $\mathbf{c}_j$ and will be referred to as the *column profile* for column j. The complete set of column profiles is denoted by the $(r \times c)$ matrix $\mathbf{C}$ with columns given by $\mathbf{c}_j$, $j = 1, 2, \ldots, c$. The matrices $\mathbf{R}$ and $\mathbf{C}$ are illustrated in Tables 9.41 and 9.42 respectively. The reader should recall from Chapter 6 that the row and column profiles can be compared to the column and row marginal densities or masses to judge the departure from independence ($\mathbf{r}_i$ is compared to $\mathbf{c}$ and $\mathbf{c}_j$ is compared to $\mathbf{r}$).

Example

For the criminal charge data the row profile matrix $\mathbf{R}$ and the column profile matrix $\mathbf{C}$ are summarized in Tables 9.43 and 9.44. Figures 9.19 and 9.20 provide a graphical representation of the profiles. The row profiles

TABLE 9.42. Matrix $\mathbf{C}$ for Column Profiles

		Columns					
		1	2	3	...	c	Row Mass
Rows	1	$n_{11}/n_{.1}$	$n_{12}/n_{.2}$	$n_{13}/n_{.3}$	...	$n_{1c}/n_{.c}$	$n_{1.}/n$
	2	$n_{21}/n_{.1}$	$n_{22}/n_{.2}$	$n_{23}/n_{.3}$	...	$n_{2c}/n_{.c}$	$n_{2.}/n$
	3	$n_{31}/n_{.1}$	$n_{32}/n_{.2}$	$n_{33}/n_{.3}$	...	$n_{3c}/n_{.c}$	$n_{3.}/n$
	$\vdots$	$\vdots$	$\vdots$	$\vdots$		$\vdots$	$\vdots$
	r	$n_{r1}/n_{.1}$	$n_{r2}/n_{.2}$	$n_{r3}/n_{.3}$	...	$n_{rc}/n_{.c}$	$n_{r.}/n$
	Totals	1	1	1	...	1	1

in Table 9.43 and Figure 9.19 compare the four sex/conviction categories. The two female profiles (no and yes) are quite similar to each other, but the two male profiles are different from each other. For the column profiles in Table 9.44 and Figure 9.20 the impaired driving and possession of narcotics profiles are similar to each other. Also the mischief and other profiles are similar. The profile for the theft under $1000 is quite different from the other four column profiles. Since theft under $1000 is the only offence dominated by females we shall see that this provides a partial explanation for this different column profile.

Departure from Independence

The purpose of correspondence analysis in the study of contingency tables is usually to study the departure of the observed cell frequencies from the cell frequencies expected under independence. Although it is possible to compare the observed cell frequencies to expected frequencies from other models, the independence model is the most commonly used base for comparisons.

   In Chapter 6 it was outlined that under the independence assumption the theoretical row profiles for each row should be equal to the column marginals and equivalently the true column profiles for each column should be equal to the row marginals. For the sample correspondence matrix therefore the matrix differences $(\mathbf{R} - \mathbf{e}_r\mathbf{c}')$ and $(\mathbf{C} - \mathbf{re}_c')$ measure the degree of departure or deviation from independence in the sample. Equivalently, under independence the cross product of the sample row and column marginal vectors or masses should be approximately equal to the correspondence ma-

TABLE 9.43. Row Profiles for Criminal Charge Data

Convicted	Sex	Impaired Driving	Theft Under $1000	Mischief	Possession of Narcotics	Other	Total
No	Male	0.186	0.256	0.116	0.163	0.279	1.000
	Female	0.167	0.500	0.100	0.033	0.200	1.000
Yes	Male	0.505	0.154	0.053	0.111	0.178	1.000
	Female	0.262	0.467	0.049	0.016	0.205	1.000
Column Mass		0.372	0.286	0.062	0.081	0.199	1.000

TABLE 9.44. Column Profiles for Criminal Charge Data

Convicted	Sex	Impaired Driving	Theft Under $1000	Mischief	Possession of Narcotics	Other	Row Mass
No	Male	0.053	0.096	0.200	0.212	0.150	0.107
	Female	0.033	0.130	0.120	0.030	0.075	0.074
Yes	Male	0.700	0.278	0.440	0.697	0.463	0.516
	Female	0.214	0.496	0.240	0.061	0.312	0.303
		1.000	1.000	1.000	1.000	1.000	1.000

trix $\mathbf{O}$ of observed cell densities. The matrix difference $(\mathbf{O} - \mathbf{rc}')$ is also therefore a measure of the deviation from independence.

Averaging the Profiles

The r row profiles given by $\mathbf{r}_i$, $i = 1, 2, \ldots, r$, have been obtained from samples of size $n_{i\cdot}$, $i = 1, 2, \ldots, r$. Each profile $\mathbf{r}_i$ therefore is representative of a proportion $n_{i\cdot}/n$ of the data. A weighted average of the r profiles can be used to obtain an *average row profile* given by $\sum_{i=1}^{r} \mathbf{r}_i(n_{i\cdot}/n) = \mathbf{c}$, which is the $(c \times 1)$ vector of column masses. Similarly a weighted average of the c column profiles is given by $\sum_{j=1}^{c} \mathbf{c}_j(n_{\cdot j}/n) = \mathbf{r}$, which is the $(r \times 1)$ vector of row masses and can be called on *average column profile*. These *average profiles* are shown in Tables 9.41 and 9.42 as column and row masses respectively. The matrices of differences $(\mathbf{R} - \mathbf{e}_r\mathbf{c}')$ and $(\mathbf{C} - \mathbf{e}_c\mathbf{r}')$ defined above as departures from independence also therefore represent *profile deviations* from the average profiles.

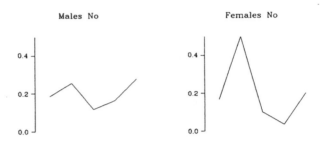

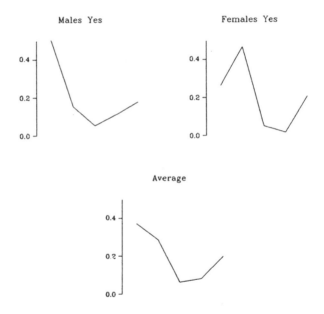

FIGURE 9.19. Row Profiles for Criminal Charge Data

Example

The average row and column profiles for the criminal charge data are shown in Figures 9.19 and 9.20. The differences between these averages and the individual row and column profiles represent the departure from independence. Figures 9.21 and 9.22 show these profile differences. These differences can also be obtained by subtraction using Tables 9.43 and 9.44. The results are summarized in Tables 9.45 and 9.46.

From the row profile deviations, it would appear that the male convicted category has more convictions than expected for impaired driving and fewer than expected for theft under $1000. For females, it would appear

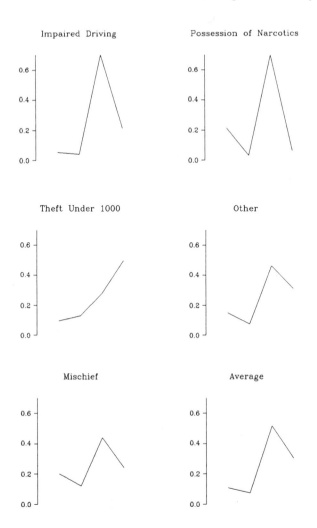

FIGURE 9.20. Column Profiles for Criminal Charge Data

that more females are charged with theft under $1000 than would be expected under independence, whereas fewer than expected are charged with impaired driving. From the column profile deviations it, would appear that for impaired driving an unusually large number of males are convicted but for theft under $1000 an unusually large number of females are convicted. The number of males convicted of theft under $1000 is much less than expected under independence. For the possession of narcotics many more males are charged and convicted than expected, whereas for females the reverse is true. For the mischief charge there tend to be fewer convictions than expected for both sexes.

TABLE 9.45. Row Profile Deviations from Independence

Convicted	Sex	Impaired Driving	Theft Under $1000	Mischief	Possession of Narcotics	Other
No	Male	−0.186	−0.030	0.054	0.082	0.080
	Female	−0.205	0.124	0.038	−0.048	0.001
Yes	Male	0.133	−0.132	−0.009	0.030	−0.021
	Female	−0.110	0.181	−0.013	−0.065	0.006

TABLE 9.46. Column Profile Deviations from Independence

Convicted	Sex	Impaired Driving	Theft Under $1000	Mischief	Possession of Narcotics	Other
No	Male	−0.054	−0.011	0.093	0.105	0.043
	Female	−0.041	0.056	0.046	−0.044	0.001
Yes	Male	0.284	−0.238	−0.076	0.181	−0.053
	Female	−0.089	0.193	−0.063	−0.242	0.009

Relationship to Pearson Chi-Square Statistic

The Pearson Chi-square statistic for testing independence was given in Section 6.1 as

$$G^2 = \sum_{i=1}^{r} \sum_{j=1}^{c} \frac{(n_{ij} - n_{i.}n_{.j}/n)^2}{n_{i.}n_{.j}/n}.$$

This expression can be written alternatively in the forms

$$G^2 = \sum_{i=1}^{r} n_{i.} \left[\sum_{j=1}^{c} \left(\frac{n_{ij}}{n_{i.}} - n_{.j}/n \right)^2 / (n_{.j}/n) \right] \qquad (9.2)$$

or

$$G^2 = \sum_{j=1}^{c} n_{.j} \left[\sum_{i=1}^{r} \left(\frac{n_{ij}}{n_{.j}} - n_{i.}/n \right)^2 / (n_{i.}/n) \right]. \qquad (9.3)$$

In (9.2) for each row i, the square of each row profile deviation in column j is divided by the column j marginal and the results are summed over all c columns. For each row i, the sum therefore yields a weighted average of the squared profile deviations over the columns. Since the column weights are the inverse of the column masses or marginals, large deviations that occur in columns with low column mass are given greater weight in the average. Finally, the weighted average for the squared row profile deviations is multiplied by the row mass to obtain a total squared deviation

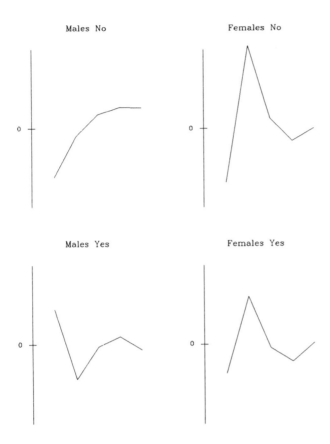

FIGURE 9.21. Row Profile Deviations from Independence for Criminal Charge Data

for the row. The sum of the resulting row total squared deviations yields the Pearson Chi-square statistic. A similar description can be given to the column oriented version in (9.3).

Example

Table 9.47 shows the cell contributions to the total Chi-square statistic. From the table we can see that males convicted of impaired driving and males and females convicted of theft under $1000 make the largest individual contributions to χ^2. For the row profiles the convicted males and convicted females yield the largest χ^2 contribution whereas for column profiles the largest contributions to χ^2 come from the impaired driving and theft under $1000 charges.

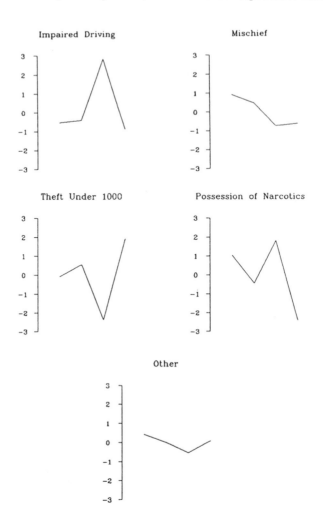

FIGURE 9.22. Column Profile Deviations from Independence for Criminal Charge Data

Total Inertia

The two versions of the Pearson Chi-square statistic given by (9.2) and (9.3) can also be expressed in the forms

$$G^2 = \sum_{i=1}^{r} n_{i\cdot} (\mathbf{r}_i - \mathbf{c})' \mathbf{D}_c^{-1} (\mathbf{r}_i - \mathbf{c}) \tag{9.4}$$

TABLE 9.47. Contributions to Chi-Square Statistic for Criminal Charge Data

Convicted	Sex	Impaired Driving	Theft Under $1000	Mischief	Possession of Narcotics	Other	Totals
No	Male	4.00	0.13	2.04	3.44	1.41	11.02
	Female	3.41	4.84	0.70	0.86	0.00	9.81
Yes	Male	9.83	12.61	0.28	2.09	0.44	25.25
	Female	3.96	14.14	0.32	6.39	0.03	24.84
Totals		21.20	31.72	3.34	12.78	1.88	70.92

and

$$G^2 = \sum_{j=1}^{c} n_{\cdot j}(\mathbf{c}_j - \mathbf{r})'\mathbf{D}_r^{-1}(\mathbf{c}_j - \mathbf{r}) \tag{9.5}$$

respectively. The statistic G^2/n is called the *total inertia* and the two statistics shown in (9.4) and (9.5) when divided by n represent the totals of inertia for the row points and column points respectively. Thus the total inertia can be viewed as a measure of the magnitude of the total row squared deviations or equivalently the magnitude of the column squared deviations. Later we shall see that singular value decomposition can be used to allocate this total inertia to various dimensions. The total inertia, G^2/n, may also be expressed in the form

$$tr[\mathbf{D}_r^{-1}(\mathbf{O} - \mathbf{rc}')\mathbf{D}_c^{-1}(\mathbf{O} - \mathbf{rc}')']. \tag{9.6}$$

Example

For the criminal charge data the value of the Pearson Chi-square statistic is 70.917 which has a p-value of 0.000. The total inertia given by (9.6) represents the magnitude of the departure from independence that needs to be explained. A generalized singular value decomposition of this matrix of deviations $(\mathbf{O} - \mathbf{rc}')$ is used to decompose the deviations by separately generating components for row profile deviations and column profile deviations respectively.

Generalized Singular Value Decomposition

The theory of the *generalized singular value decomposition* outlined in the Appendix can be employed to obtain a relationship between the total inertia given by (9.6) and an approximation to the matrix of deviations $(\mathbf{O} - \mathbf{rc}')$. For the matrix $(\mathbf{O} - \mathbf{rc}')$, the generalized singular value decomposition subject to the conditions $\mathbf{A}'\mathbf{D}_r^{-1}\mathbf{A} = \mathbf{I}$, $\mathbf{B}'\mathbf{D}_c^{-1}\mathbf{B} = \mathbf{I}$ is given by

$$(\mathbf{O} - \mathbf{rc'}) = \mathbf{AD}_{\mu}\mathbf{B'} = \sum_{k=1}^{K} \mu_k \mathbf{a}_k \mathbf{b}'_k,$$

where the columns of $\mathbf{A}(r \times K)$ and $\mathbf{B}(c \times K)$ are denoted by $\mathbf{a}_k$ and $\mathbf{b}_k$ respectively and $\mu_1, \mu_2, \ldots, \mu_k$ are the diagonal elements of the diagonal matrix $\mathbf{D}_{\mu}(K \times K)$. The dimension K is the rank of the matrix being decomposed which in this application is $\min[(r-1), (c-1)]$.

The vectors $\mathbf{a}_k$, $k = 1, 2, \ldots, K$, are called the *principal axes* of the columns of $(\mathbf{O} - \mathbf{rc'})$; the vectors $\mathbf{b}_k$, $k = 1, 2, \ldots, K$, are called the principal axes of the rows of $(\mathbf{O} - \mathbf{rc'})$. The diagonal elements $\mu_1, \mu_2, \ldots, \mu_K$ of $\mathbf{D}_{\mu}$ are called the *singular values* of $(\mathbf{O} - \mathbf{rc'})$. The total inertia can therefore be written as

$$tr[\mathbf{D}_r^{-1}(\mathbf{O} - \mathbf{rc'})\mathbf{D}_c^{-1}(\mathbf{O} - \mathbf{rc'})] = \sum_{k=1}^{K} \mu_k^2$$

which is the sum of the squares of the singular values.

Example

The total inertia for the criminal charge contingency table is $70.917/403 = 0.17957$. This total can be allocated to the three dimensions ($K = 3$) as $\mu_1^2 = 0.14191$, $\mu_2^2 = 0.03286$ and $\mu_3^2 = 0.00120$. The three percentages of the total inertia are 80.65%, 18.67% and 0.68%. The number of dimensions is three since the contingency table contains four rows and four columns. The first dimension of the generalized singular value decomposition therefore accounts for a large majority of the deviation from independence as measured by the Pearson Chi-square statistic.

Coordinates for Row and Column Profiles

For the generalized singular value decomposition of $(\mathbf{O} - \mathbf{rc'})$ given by $\mathbf{AD}_{\mu}\mathbf{B'}$ the columns of the matrices $\mathbf{A}$ and $\mathbf{B}$ provide the principal axes for the columns and rows of $(\mathbf{O} - \mathbf{rc'})$ respectively. Each row of $(\mathbf{O} - \mathbf{rc'})$ can be expressed as a linear combination of the rows of $\mathbf{B'}$ (columns of $\mathbf{B}$), and hence the coordinates for the rows of $(\mathbf{O} - \mathbf{rc'})$ in the space generated by the rows of $\mathbf{B'}$ are given by $\mathbf{AD}_{\mu}$. The coordinates for the ith row of $(\mathbf{O} - \mathbf{rc'})$ are given by the ith row of $\mathbf{AD}_{\mu}$. Similarly the coordinates for the columns of $(\mathbf{O} - \mathbf{rc'})$ with respect to the space generated by the columns of $\mathbf{A}$ are provided by the columns of $\mathbf{D}_{\mu}\mathbf{B'}$.

To obtain coordinates for the row and column profile deviations, the relationships

$$(\mathbf{R} - \mathbf{e}_r \mathbf{c'}) = \mathbf{D}_r^{-1}(\mathbf{O} - \mathbf{rc'})$$

and

$$(\mathbf{C} - \mathbf{re}'_c) = \mathbf{D}_c^{-1}(\mathbf{O} - \mathbf{rc'})$$

TABLE 9.48. Coordinates for Row
Profiles on Row Principal Axes

		Row Principal Axes (Columns of **B**)				
		1	2	3	...	K
	1	v_{11}	v_{12}	v_{13}	...	v_{1K}
	2	v_{21}	v_{22}	v_{23}	...	v_{2K}
Rows	3	v_{31}	v_{32}	v_{33}	...	v_{3K}
	$\vdots$	$\vdots$	$\vdots$	$\vdots$		$\vdots$
	r	v_{r1}	v_{r2}	v_{r3}	...	v_{rK}

TABLE 9.49. Coordinates for Column
Profiles on Column Principal Axes

		Column Principal Axes (Columns of **A**)				
		1	2	3	...	K
	1	w_{11}	w_{12}	w_{13}	...	w_{1K}
	2	w_{21}	w_{22}	w_{23}	...	w_{2K}
Columns	3	w_{31}	w_{32}	w_{33}	...	w_{3K}
	$\vdots$	$\vdots$	$\vdots$	$\vdots$		$\vdots$
	c	w_{c1}	w_{c2}	w_{c3}	...	w_{cK}

can be used. The required coordinates for the row and column profile deviations are therefore given by

$$\mathbf{V}(r \times K) = \mathbf{D}_r^{-1} \mathbf{A} \mathbf{D}_\mu = \mathbf{D}_r^{-1}(\mathbf{O} - \mathbf{rc}')\mathbf{D}_c^{-1}\mathbf{B}$$

and

$$\mathbf{W}(c \times K) = \mathbf{D}_c^{-1} \mathbf{B} \mathbf{D}_\mu = \mathbf{D}_c^{-1}(\mathbf{O} - \mathbf{rc}')'\mathbf{D}_r^{-1}\mathbf{A}$$

respectively. Since $\mathbf{r}'\mathbf{V} = \mathbf{0}$ and $\mathbf{c}'\mathbf{W} = \mathbf{0}$ the coordinates of the row and column profile deviations are related to each other by the equations

$$\mathbf{V}\mathbf{D}_\mu = \mathbf{D}_r^{-1}\mathbf{O}\mathbf{W} \quad \text{and} \quad \mathbf{W}\mathbf{D}_\mu = \mathbf{D}_c^{-1}\mathbf{O}'\mathbf{V}. \tag{9.7}$$

The coordinates for the r row profile deviations are given by the elements V_{ik}, $i = 1, 2, \ldots, r$, $k = 1, 2, \ldots, k$, of $\mathbf{V}$ and are shown in Table 9.48. Similarly the coordinates for the c column profile deviations are given by the elements w_{jk}, $j = 1, 2 \ldots, c$, $k = 1, 2, \ldots, k$, of $\mathbf{W}$ and are shown in Table 9.49.

Each row of $\mathbf{V}$ in Table 9.48 provides the coordinates for a row profile deviation with respect to the K principal axes given by the columns of $\mathbf{B}$. Each column of $\mathbf{V}$ provides the coordinates for the r profile deviations

TABLE 9.50. Coordinates for Row Profiles on Row Principal Axes for Criminal Charge Data

Row Profile		Principal Axes		
		1	2	3
No	Males	0.04	0.50	−0.03
No	Females	0.55	0.09	0.11
Yes	Males	−0.35	−0.05	0.01
Yes	Females	0.44	−0.11	−0.03

TABLE 9.51. Coordinates for Column Profiles on Column Principal Axes for Criminal Charge Data

Column Profile	Principal Axes		
	1	2	3
Impaired Driving	−0.34	−0.16	0.01
Theft Under $1000	0.52	−0.06	0.00
Mischief	0.08	0.34	0.11
Possession of Narcotics	−0.50	0.37	−0.01
Other	0.07	0.13	−0.05

with respect to a particular principal axis or column of **B**. Similarly the coordinates for the column profiles are given by Table 9.49. The coordinates for the first two principal axes (v_{i1}, v_{i2}), $i = 1, 2, \ldots, r$, can be used to locate the r profile deviations in a two-dimensional space defined by the two principal axes. For the column profile deviations the coordinates for the first two principal axes are given by (w_{j1}, w_{j2}), $j = 1, 2, \ldots, c$.

Example

For the criminal charge data the coordinates for the row and column profile deviations on their respective dimensions are shown in Tables 9.50 and 9.51. For the row profiles it would appear that the first dimension reflects a contrast between females charged and males convicted. The second row dimension is primarily a measure of males charged but not convicted. For the column profiles the first dimension represents a contrast between theft under $1000 and the crimes of narcotics possession and impaired driving. The second dimension for column profile deviations seems to reflect a contrast between the three charges mischief, narcotics possession and other offences with the charge impaired.

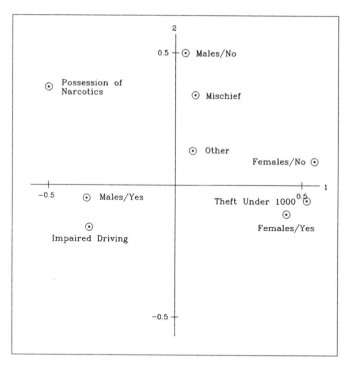

FIGURE 9.23. Correspondence Analysis Plot for Criminal Charge Data

Figure 9.23 shows the above results graphically for the first two dimensions. From the plot the relationships between the row and column dimensions can be studied. The contrast between charged females and convicted males seems to be related to a contrast between theft under $1000 and the two crimes, narcotics possession and impaired driving. This relationship is consistent with the conclusions that were drawn earlier on comparisons of the profiles. Relating the two second dimensions, it would appear that the not convicted males category is positively related to the charges of mischief, narcotics possession and other, and negatively related to impaired driving. In other words, in comparison to the relationships expected under independence, males tend to have higher rates of conviction for impaired driving and lower rates of conviction for the remaining criminal charges.

The generalized singular value decomposition of the matrix of deviations $(\mathbf{O} - \mathbf{rc}') = \mathbf{A}\mathbf{D}_\mu\mathbf{B}'$ can be approximated by the first row and column dimensions. Using only these first dimensions, the cell values are given by $\mathbf{a}_1\mathbf{b}_1'/\mu_1$ where $\mathbf{a}_1$ and $\mathbf{b}_1$ are the first columns of $\mathbf{A}$ and $\mathbf{B}$ respectively and μ_1 is the first diagonal element of $\mathbf{D}_\mu$. Table 9.52 compares the deviations from independence derived from $\mathbf{O} - \mathbf{rc}'$ (in percents) to the approximations obtained from the first row and column dimensions (shown in brackets). By comparing the nonbracketed figures to the bracketed figures in Table 9.52

TABLE 9.52. Deviations from Independence (Percents) Compared to Deviation Explained by First Row and Column Dimensions (Shown in Brackets)

Conviction	Sex	Impaired Driving	Theft Under $1000	Mischief	Possession of Narcotics	Other	Totals
No	Male	−1.98	−0.32	0.58	0.86	0.86	0.00
		(−0.16)	(0.19)	(0.00)	(−0.05)	(0.02)	(0.00)
	Female	−1.53	1.60	0.28	-0.36	0.01	0.00
		(−1.37)	(1.62)	(0.05)	(−0.44)	(0.14)	(0.00)
Yes	Male	6.84	−6.79	−0.47	1.48	−1.06	0.00
		(5.94)	(−7.00)	(−0.22)	(1.90)	(−0.62)	(0.00)
	Female	−3.33	5.51	−0.39	−1.98	0.19	0.00
		(−4.41)	(5.29)	(0.17)	(−1.41)	(0.46)	(0.00)
Totals		0.00	0.00	0.00	0.00	0.00	
		(0.00)	(0.00)	(0.00)	(0.00)	(0.00)	

it can be seen that for females and convicted males the approximation provided by the first dimensions is quite good. The largest deviations remaining appear to be for nonconvicted males which is explained by the second dimension. The reader should recall that the first dimension accounts for 80.65% and the second dimension accounts for a remaining 18.67%.

The results obtained here can be related to loglinear models discussed in Chapter 6. The results obtained by fitting a three-dimensional loglinear model to Table 9.39 yields significant interactions between sex and charge and between charge and convict. The interaction between sex and convict was not significant. From the loglinear analysis, one could conclude that after controlling for the charge there is no relation between the sex of the individual and the likelihood of conviction. The reader is left to explore further the relationships among the categories of this table.

Partial Contributions to Total Inertia

A weighted average of the coordinates in a column of $\mathbf{V}$ given by $\mathbf{r'V}$ yields the zero vector, and hence the weighted average of each column of $\mathbf{V}$ is zero. A weighted average of the squares of the elements in the columns of $\mathbf{V}$ is given by $\mathbf{V'D}_r\mathbf{V} = \mathbf{D}_\mu^2$. Thus using the row marginals as weights the coordinates with respect to each principal axis have zero mean and variance μ_k^2, $k = 1, 2, \ldots, K$.

Since the squares of the singular values are given by μ_k^2 and since $\sum_{k=1}^K \mu_k^2$ represents the total inertia we can conclude that the weighted variances of the columns of $\mathbf{V}$ indicate the contribution of each principal axis to the total inertia. The total inertia can be expressed as $\sum_{k=1}^K \mu_k^2 = \sum_{k=1}^K \sum_{i=1}^r o_i \cdot V_{ik}^2$, and hence $o_i \cdot \sum_{k=1}^K V_{ik}^2$ represents the contribution made by the ith row

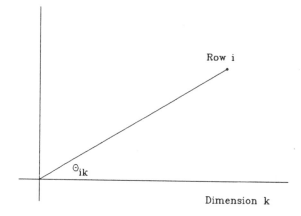

FIGURE 9.24. Squared Cosines for Allocation of Total Inertia

profile deviation to the total inertia. In a similar manner the elements of
W can be used to relate the total inertia to the column profile deviations
and their principal axes.

Squared Cosines

For each row profile deviation the inertia $o_i. \sum_{k=1}^{K} V_{ik}^2$ can be allocated to
the K dimensions as $o_i. V_{i1}^2, o_i. V_{i2}^2, \ldots, o_i. V_{ik}^2$. The proportions

$$o_i. V_{ik}^2 \Big/ o_i. \sum_{k=1}^{K} V_{ik}^2 = V_{ik}^2 \Big/ \sum_{k=1}^{K} V_{ik}^2$$

therefore represent an allocation of the row profile deviation to the kth dimension. This ratio represents the square of the cosine of the angle between
the dimension k and a ray drawn from the origin to the point representing the row profile (see Figure 9.24.). Denoting this angle by θ_{ik} we have
$\cos^2 \theta_{ik} = V_{ik}^2 / \sum_{i=1}^{K} V_{ik}^2$. The closer this value is to 1 the closer the ith row
point is to the kth axis and hence the more important that point is to the
kth dimension.

Example

Tables 9.53 and 9.54 summarize the partial contributions to total inertia
and squared cosines for the row profile and column profile dimensions respectively. The tables also contain the total inertia for each profile and the
corresponding total mass.

For the row profiles, we can see from Table 9.53 that the points corresponding to females and convicted males are close to dimension 1 and
nonconvicted males are close to dimension 2. For the column profiles in

TABLE 9.53. Partial Contributions to Inertia and Squared Cosines (in brackets) for Row Profiles

		Row Mass	Row Total Inertia	Dimensions 1	Dimensions 2	Dimensions 3
Male	No	0.11	0.16	0.00 (0.01)	0.82 (0.99)	0.07 (0.00)
Female	No	0.07	0.14	0.16 (0.94)	0.02 (0.02)	0.75 (0.04)
Male	Yes	0.52	0.35	0.43 (0.98)	0.04 (0.02)	0.01 (0.00)
Female	Yes	0.30	0.35	0.41 (0.93)	0.12 (0.06)	0.17 (0.00)

TABLE 9.54. Partial Contributions to Inertia and Squared Cosines (in brackets) for Column Profiles

	Column Mass	Column Total Inertia	Dimensions 1	Dimensions 2	Dimensions 3
Impaired Driving	0.37	0.30	0.30 (0.81)	0.30 (0.19)	0.02 (0.00)
Theft Under $1000	0.29	0.45	0.55 (0.99)	0.03 (0.01)	0.00 (0.00)
Mischief	0.06	0.05	0.00 (0.04)	0.22 (0.87)	0.57 (0.08)
Possession of Narcotics	0.08	0.18	0.14 (0.64)	0.35 (0.36)	0.01 (0.00)
Other	0.20	0.03	0.01 (0.19)	0.10 (0.71)	0.39 (0.10)

Table 9.54, impaired driving and theft under $1000 are close to dimension 1 and mischief and other are close to dimension 2. The offence possession of narcotics, however, is in between dimensions 1 and 2.

Principle of Distributional Equivalence

A useful property of the *Chi-square metric* in correspondence analysis is that if two row profiles are identical they may be replaced by a single row profile that is the sum of the two profiles. This collapsing of the two rows will not affect the geometry of the column profiles. Similarly a row profile can be subdivided into two or more rows without affecting the geometry of the column profiles. If two row profiles are identical they occupy identical positions in the row space. This property guarantees the invariance of the solution with respect to the coding of the original variables.

Generalized Least Squares Approximation

The objective of correspondence analysis is to replace the matrix of deviations $(\mathbf{O} - \mathbf{rc}')$ by an approximation say $\mathbf{X}(r \times c)$ based on a small number of the principal axes of the generalized singular value decomposition. Most commonly $\mathbf{X}$ is given by the first two terms of $\sum_{k=1}^{K} \mathbf{a}_k \mathbf{b}_k' \mu_k$ and hence is a rank two approximation to $(\mathbf{O} - \mathbf{rc'})$. It can be shown that minimizing the expression

$$tr[\mathbf{D}_r^{-1}(\mathbf{O} - \mathbf{rc'} - \mathbf{X})\mathbf{D}_c^{-1}(\mathbf{O} - \mathbf{rc'} - \mathbf{X})']$$

among all matrices $\mathbf{X}$ of rank two or less yields $\mathbf{X} = \sum_{k=1}^{2} a_k \mathbf{b}_k' \mu_k$. Thus the generalized singular value decomposition can be used to provide an approximation to the deviations $(\mathbf{O} - \mathbf{rc'})$. This approximation maximizes the proportion of the total inertia that can be allocated to two dimensions.

In terms of the row and column profile deviation coordinates provided in Tables 9.48 and 9.49, only the first two columns are required. The first column of Table 9.48 provides the row profile coordinate locations along the first row principal axis, whereas the second column of Table 9.48 provides the row profile coordinate locations along the second row principal axis. Similarly the first two columns of Table 9.49 provide the coordinates for the column profiles along the first two column principal axes.

Relationship to Generalized Singular Value Decomposition of $\mathbf{O}$

The discussion so far has been concerned with the generalized singular value decomposition of the matrix of deviations $(\mathbf{O} - \mathbf{rc'})$. It is also possible to express the solution in terms of a generalized singular value decomposition of the correspondence matrix $\mathbf{O}$. The generalized singular value decomposition of $\mathbf{O}$ is given by $\mathbf{O} = \mathbf{A}^* \mathbf{D}_\mu^* \mathbf{B}^{*'}$, where $\mathbf{A}^{*'} \mathbf{D}_r^{-1} \mathbf{A}^* = \mathbf{B}^{*'} \mathbf{D}_c^{-1} \mathbf{B}^* = \mathbf{I}$ and where $\mathbf{A}^* = [\mathbf{r}, \mathbf{A}]$, $\mathbf{B}^* = [\mathbf{c}, \mathbf{B}]$ and $\mathbf{D}_\mu^* = \begin{bmatrix} 1 & 0 \\ 0 & \mathbf{D}_\mu \end{bmatrix}$. Therefore in comparison to the decomposition of $(\mathbf{O} - \mathbf{rc'})$ the decomposition of $\mathbf{O}$ simply adds one dimension without changing the other singular values and principal axes. The dimensions added are simply the row and column average profiles or marginal densities. The matrices of coordinates for row and column profiles are now given by $\mathbf{V}^* = \mathbf{D}_r^{-1} \mathbf{A}^* \mathbf{D}_\mu^*$ and $\mathbf{W}^* = \mathbf{D}_c^{-1} \mathbf{B}^* \mathbf{D}_\mu^*$ respectively.

Row and Column Profile Deviations and Eigenvectors

The relations (9.7) between the coordinate vectors for the row and column profiles deviations can be used to obtain eigenequations. (The reader may wish to review the relation between singular value decompositions and symmetric decompositions outlined in the Appendix.) These eigenequations will reveal the relationship between the coordinate vectors and the eigenvalues and eigenvectors of the two symmetric matrices $\mathbf{D}_c^{-1/2} \mathbf{O}' \mathbf{D}_r^{-1} \mathbf{O} \mathbf{D}_c^{-1/2}$ and $\mathbf{D}_r^{-1/2} \mathbf{O} \mathbf{D}_c^{-1} \mathbf{O}' \mathbf{D}_r^{-1/2}$. These relationships will be useful in the next section.

Using the relations given by (9.7) the following equations can be obtained:

$$D_r^{-1}OD_c^{-1}O'V = VD_\mu^2 \qquad (9.8)$$

$$D_c^{-1}O'D_r^{-1}OW = WD_\mu^2, \qquad (9.9)$$

where $V'D_rV = D_\mu^2$ and $W'D_cW = D_\mu^2$.

By introducing the notation $H = D_r^{-1/2}OD_c^{-1/2}$, $U = D_c^{1/2}WD_\mu^{-1}$ and $X = D_r^{1/2}VD_\mu^{-1}$, the equations (9.8) and (9.9) can be expressed as

$$HH'X = XD_\mu^2 \quad \text{and} \quad H'HU = UD_\mu^2.$$

Since $H'H$ and HH' are symmetric matrices we can conclude that the diagonal elements of D_μ^2 are eigenvalues of the two matrices $H'H$ and HH'. In addition the columns of X are the eigenvectors of HH', whereas the columns of U are the eigenvectors of $H'H$. The matrices X and U are related by $XD_\mu = HU$. These relationships will be employed in the next section to relate the correspondence analysis of contingency tables to the correspondence analysis of frequency response tables.

Correspondence Analysis for Multidimensional Tables

Although correspondence analysis is designed to analyze a two-dimensional contingency table it can also be used to study larger dimensional tables. In the example studied in this section the contingency table analyzed was three-dimensional. A two-dimensional table was derived by constructing a single dimension from two dimensions using the cross-product of the two sets of categories. Since correspondence analysis is designed to study the departure from independence, in the example presented the departure being studied is relative to a partial independence model. The independence model fitted is equivalent to a partial independence model for the underlying three-dimensional table. The fitted interaction term corresponds to the two variables combined to form the one dimension. The correspondence analysis in this case therefore reflects the departures from the partial independence model. For the three-dimensional table additional analyses could be carried out for the other two possible partial independence models. For higher dimensional tables, a variety of models can be studied by choosing a variety of possible combinations of variables to be formed into the two final dimensions. In the next section, an alternative approach to the study of multidimensional contingency tables known as multiple correspondence analysis will be introduced.

9.4.2   OTHER SOURCES OF INFORMATION

The outline of theory provided in this section parallels the discussion in Greenacre (1984). Alternative outlines of this topic are available in Lebart, Morineau and Warwick (1984) and Andersen (1990).

9.4.3   CORRESPONDENCE ANALYSIS AND FREQUENCY RESPONSE TABLES

An alternative application of correspondence analysis which yields similar results to those outlined in Section 9.4.1, is the *frequency response table* or *product multinomial sample* as studied in Section 6.3. In the weighted least squares methodology, multidimensional contingency tables are structured as two-dimensional tables with all categorical variables being grouped into two mutually exclusive categories: response variables and factor or explanatory variables. The observations are assumed to be derived from independent samples from each group. The two sets of categories will be labelled response categories for the response variables and subpopulations or groups for the factors or explanatory variables.

Following the notation of Table 6.47 the r rows of the frequency response table denote the subpopulations and the c columns denote the response categories. A sample size n will be assumed to have been selected by randomly selecting $n_1., n_2., \ldots, n_r., \sum_{i=1}^{r} n_i. = n$, from the r groups respectively. The sample size obtained in the jth response category of the ith group is denoted by n_{ij}, $i = 1, 2, \ldots, r$, $j = 1, 2, \ldots, c$.

Example

For the driver injury data used in Chapter 6, the frequency response table illustrated in Table 6.49 is employed in this section to provide an example. Table 9.55 reproduces the frequencies from Table 6.49 and also presents the cell densities o_{ij} as percentages. In this section the variation in injury level response over the driver condition-seatbelt groups is studied. The objective is to define an underlying dimension or relationship among the injury level categories that maximizes the variation among the four driver condition-seatbelt groups. A possible result might be that the injury level characteristic that is most important in distinguishing the four groups is a contrast between the two highest injury level categories and the two lowest injury level categories. Since the row categories are derived from the cross classification of two categorical variables the interaction among these two variables is also being fitted. The deviations being studied in this case are the deviations from a partial independence model that assumes an interaction between driver condition and seatbelt usage.

TABLE 9.55. Driver Injury Level Response to Driver Condition/Seatbelt Usage, Cell Frequencies and Cell Densities (Percentage)

Group		Driver Injury Level				
Driver Condition	Seatbelt Usage	None	Minimal	Minor	Major/Fatal	Totals
Normal	Yes	12,500	604	344	38	13,486
		(14.406)	(0.696)	(0.396)	(0.044)	(15.542)
	No	61,971	3,519	2,272	237	67,999
		(71.421)	(4.056)	(2.618)	(0.273)	(78.368)
Been	Yes	313	43	15	4	375
Drinking		(0.361)	(0.050)	(0.017)	(0.005)	(0.432)
	No	3,992	481	370	66	4,909
		(4.601)	(0.554)	(0.426)	(0.076)	(5.658)
Totals		78,776	4,647	3,001	345	86,769
		(90.788)	(5.356)	(3.459)	(0.398)	100.00

A Dual Scaling Approach

The derivation of a correspondence analysis for the frequency response table presented in this section employs a *dual scaling* approach. The objective will be to determine axes or scales to characterize the response categories and subpopulations so that the variation in response pattern across the groups can be studied. The approach is based on a one-way analysis of variance study of the response variation across groups.

Review of One-way ANOVA Notation

In one-way analysis of variance random samples of $n_i.$ individuals are selected from groups $i = 1, 2, \ldots, r$. For each individual j a response is measured yielding the observation y_{ij}, $i = 1, 2, \ldots, r$, $j = 1, 2, \ldots, c$. For the r groups the means are denoted by $\bar{y}_{i.}$ where $\bar{y}_{i.} = \sum_{j=1}^{c} y_{ij}/n_{i.}$. The grand mean is denoted by $\bar{y} = \sum_{i=1}^{r} \sum_{j=1}^{n_i.} y_{ij}/n$. The sums of squares used to measure variation within groups, among groups, and the total are given by $SSW = \sum_{i=1}^{r} \sum_{j=1}^{n_i.} (y_{ij} - \bar{y}_{i.})^2$, $SSA = \sum_{i=1}^{r} n_{i.} (\bar{y}_{i.} - \bar{y}_{..})^2$ and $SST = \sum_{i=1}^{r} \sum_{j=1}^{n_i.} (y_{ij} - \bar{y}_{..})^2$ respectively. The variation in the group means $\bar{y}_{i.}$ over the r groups is considered large if the ratio $(n-r)SSA/SSW/(r-1)$ is large. In one-way analysis of variance the assumption of independence, homogeneity of variance and normality lead to an F distribution with $(r-1)$ and $(n-r)$ degrees of freedom if the true group means are equal.

Scaling the Response Categories

In the context of the frequency response table, we assume that there is an underlying scale of measurement, say Z, for the response variable and that

TABLE 9.56. Response Frequency Table With Scaled Values Included

Groups	Response Categories					Group Means
	1	2	3	...	c	
1	n_{11} values of z_1	n_{12} values of z_2	n_{13} values of z_3	...	n_{1c} values of z_c	$t_1 = \sum_{j=1}^{c} n_{1j} z_j / n_1.$
2	n_{21} values of z_1	n_{22} values of z_2	n_{23} values of z_3	...	n_{2c} values of z_c	$t_2 = \sum_{j=1}^{c} n_{2j} z_j / n_2.$
$\vdots$						
r	n_{r1} values of z_1	n_{r2} values of z_2	n_{r3} values of z_3	...	n_{rc} values of z_c	$t_r = \sum_{j=1}^{c} n_{rj} z_j / n_r.$
Column Means	z_1	z_2	z_3	...	z_c	Grand Mean $\bar{t} = \bar{z} = \sum_{i=1}^{r} \sum_{j=1}^{c} n_{ij} z_j / n$

for the c response categories the values of Z are $z_1, z_2, \ldots, z_c$ (note that these z values are not ordered). Each individual in a given response category is assumed to have the same value of Z. Thus for the jth response category the sum of the hypothetical observations in group i is $n_{ij} z_j$, and the mean for group i is given by $t_i = \sum_{j=1}^{c} n_{ij} z_j / n_i. = \sum_{j=1}^{c} o_{ij} z_j / r_i$, where $r_i = n_i./n$. The grand mean is given by $\bar{t} = \bar{z} = \sum_{i=1}^{r} \sum_{j=1}^{c} n_{ij} z_j / n = \sum_{i=1}^{r} \sum_{j=1}^{c} o_{ij} z_i$. These quantities are shown in Table 9.56.

The $(r \times 1)$ vector of group means $\mathbf{t}$ with elements t_i, $i = 1, 2, \ldots, r$, is given by $\mathbf{t} = \mathbf{D}_r^{-1} \mathbf{O} \mathbf{z}$, where $\mathbf{D}_r$ is the diagonal matrix with diagonal elements equal to the row marginals $n_i./n$, $i = 1, 2, \ldots, r$, and where $\mathbf{z}$ $(c \times 1)$ is vector of values z_j corresponding to the response categories $j = 1, 2, \ldots, c$. The grand mean is given by $\bar{t} = \bar{z} = \mathbf{c}' \mathbf{z}$ where $\mathbf{c}$ $(c \times 1)$ is a vector of column marginals $o_{.j} = n_{.j}/n$, $j = 1, 2, \ldots, c$.

To maximize the ratio $(n-r)SSA/SSW(r-1)$ it is sufficient to maximize the ratio $\eta^2 = SSA/SST$ since $SSA + SSW = SST$. The two sums of squares can be written as

$$SSA = \sum_{i=1}^{r} n_i. (t_i - \bar{t})^2 = n \mathbf{z}' \mathbf{O}' \mathbf{D}_r^{-1} \mathbf{O} \mathbf{z} - n \mathbf{z}' \mathbf{c} \mathbf{c}' \mathbf{z}$$

and

$$SST = \sum_{i=1}^{r} \sum_{j=1}^{c} n_{ij} (z_j - \bar{t})^2 = n \mathbf{z}' \mathbf{D}_c \mathbf{z} - n \mathbf{z}' \mathbf{c} \mathbf{c}' \mathbf{z}.$$

Since these sums of squares are defined relative to the grand mean $\bar{t}$, there is no loss of generality in assuming that $\bar{t} = 0$. The ratio to maximize therefore becomes $\mathbf{z}'\mathbf{O}'\mathbf{D}_r^{-1}\mathbf{Oz}/\mathbf{z}'\mathbf{D}_c\mathbf{z}$.

Since multiplication of $\mathbf{z}$ by an arbitrary constant does not change the value of the above ratio, the scale can be restricted by assuming $\mathbf{z}'\mathbf{D}_c\mathbf{z} = 1$. Maximizing the numerator $\mathbf{z}'\mathbf{O}'\mathbf{D}_r^{-1}\mathbf{Oz}$ subject to the condition $\mathbf{z}'\mathbf{D}_c\mathbf{z} = 1$ yields the equation

$$(\mathbf{D}_c^{-1}\mathbf{O}'\mathbf{D}_r^{-1}\mathbf{O} - \eta^2\mathbf{I})\mathbf{z} = \mathbf{0} \tag{9.10}$$

where η^2 is the value of the ratio to be maximized.

An examination of equation (9.9) in the previous section and (9.10) above reveals that there is a relationship between $\mathbf{z}$ and $\mathbf{w}$ and between η^2 and the elements of $\mathbf{D}_\mu^2$. Choosing η^2 to be the largest diagonal element in $\mathbf{D}_\mu^2$ and letting $\mathbf{w}_1$ be the corresponding column of $\mathbf{W}$ we have $\mathbf{z} = \mathbf{w}_1$.

In a similar fashion the eigenvector of $\mathbf{V}$ in (9.8) corresponding to η^2 the largest eigenvalue, is given by $\mathbf{v}_1$. Using (9.7) we have $\mathbf{v}_1 = \mathbf{D}_r^{-1}\mathbf{Ow}_1/\eta = \mathbf{D}_r^{-1}\mathbf{Oz}/\eta$ and hence $\mathbf{t} = \mathbf{v}_1\eta$, where $\mathbf{t}$ is the vector of group means. Thus the group means or row scores can be obtained by multiplying the row profile coordinates on the first principal axis by the first singular value.

From these results, we can conclude that a correspondence analysis of the matrix of profile deviations discussed in the previous section and the scaling of the response categories to maximize the variation among group responses introduced in this section yields the same solution. The coordinate vector which locates the column profile deviations along their first principal axis in Section 9.4.1 is also the coordinate vector that describes the location of the response factors along the scale that maximizes the variation among group means.

The coordinate vector that locates the row profile deviations along their first principal axis is also the vector that describes the magnitudes of the group means $t_1, t_2, \ldots, t_r$ scaled by η^2.

Example

For the accident data of Table 9.55 the Pearson Chi-square statistic is 682.372 with 9 degrees of freedom. The evidence against homogeneity of response across the four driver condition/seatbelt groups is therefore quite strong. The correspondence analysis procedure in this case results in 97.89% (0.00770) of the total inertia (0.00786) being allocated to the first dimension. The second dimension accounts for almost all of the remaining 2% (0.00014). The three singular values were 0.0877, 0.0119 and 0.0049. The row profile and column profile coordinates are summarized in Table 9.57. Since the second dimension is relatively minor only the first dimension is plotted in Figure 9.25. From the profile coordinates in Table 9.57 and the plot in Figure 9.25, it is clear that the severity of injury is strongly related to the presence of one or both of the seatbelt and been drinking factors.

TABLE 9.57. Row Profile and Column Profile Coordinates for Accident Data

Driver Condition	Seatbelt Usage	Row Profile Coordinates		Driver Injury Level	Column Profile Coordinates	
		1	2		1	2
Normal	Yes	−0.07	0.01	None	−0.03	−0.00
Normal	No	−0.01	−0.01	Minimal	0.22	0.04
Been Drinking	Yes	0.24	0.17	Minor	0.30	0.05
Been Drinking	No	0.34	−0.01	Major/Fatal	0.59	0.02

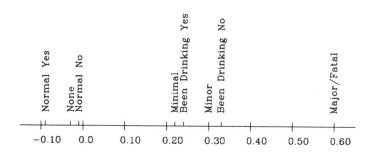

FIGURE 9.25. Correspondence Analysis for Accident Data

The scale values for the four injury levels are given by the column coordinates: None (-0.03), Minimal (0.22), Minor (0.30) and Major/Fatal (0.59). The mean scale values for the four groups can be obtained from the row profile coordinates by multiplying by the first singular value which is (0.0877). These group means are given by −0.006 (Normal Yes), −0.001 (Normal No), 0.021 (Been Drinking Yes) and 0.030 (Been Drinking No). The column profile scores have been selected so as to maximize the variance among the four row means.

Some Alternative Approaches to Correspondence Analysis

The dual scaling approach to correspondence analysis described above is only one of several alternative approaches to obtaining the same decomposition of the two-dimensional contingency table. Other approaches have been referred to as reciprocal averaging, bivariate correlation, linear regression and canonical correlation. These various approaches are outlined in Greenacre (1984) and Nishisato (1980). The reciprocal averaging method has been employed in ecology and will not be discussed here. A brief description of the bivariate correlation, simultaneous linear regression and canonical correlation approaches is provided below.

Bivariate Correlation

Given the two-dimensional $(r \times c)$ contingency table, the objective of the bivariate correlation approach is to assign a $(r \times 1)$ vector of scale values, say $\mathbf{y}$, to the row categories and a $(c \times 1)$ vector of scale values, say $\mathbf{z}$, to the column categories such that the correlation η between the two newly created variables is maximized. The solution to this problem is provided by the principal coordinates corresponding to the first singular vectors $\mathbf{v}_1$ and $\mathbf{w}_1$ for the row and column profile deviations and the first singular value μ_1 as outlined in Section 9.4.1. The vectors of optimal scale values are $\mathbf{y} = \mathbf{v}_1$ and $\mathbf{z} = \mathbf{w}_1$ and the maximum correlation in $\eta = \mu_1$. This solution is also equivalent to the dual scaling solution outlined above. The vector $\mathbf{z}$ represents the optimal scale values and the vector $\mathbf{t} = \mathbf{y}\eta$ denotes the mean scores for the groups.

Simultaneous Linear Regression

Suppose we have an $(r \times c)$ contingency table with cell densities o_{ij}, $i = 1, 2, \ldots, r$, $j = 1, 2, \ldots, c$.

Suppose also we have available vectors of scores $\mathbf{y}$ $(r \times 1)$ for the row categories and $\mathbf{z}$ $(c \times 1)$ for the column categories. Let $\mathbf{x}$ $(c \times 1)$ denote the average values for the column categories based on the $\mathbf{y}$ scores and similarly let $\mathbf{t}$ $(r \times 1)$ denote the average values for the row categories based on $\mathbf{z}$ scores. Suppose the average values for the row categories $\mathbf{t}$ are plotted against the row category scores $\mathbf{y}$. Would the result be a straight line? Similarly would the relation between the assigned column scores $\mathbf{z}$ and the derived column scores $\mathbf{x}$ be linear? Since the answers to these questions depends on the original selection of scores $\mathbf{y}$ and $\mathbf{z}$ can they be chosen so that the above two plots are linear?

The solution is to select the $\mathbf{y}$ and $\mathbf{z}$ score vectors corresponding to the first principal coordinate vectors. The resulting average values derived from $\mathbf{y}$ and $\mathbf{z}$ are $\mathbf{x} = \eta\mathbf{z}$ and $\mathbf{t} = \eta\mathbf{y}$, and hence the plot of $\mathbf{x}$ against $\mathbf{z}$ and $\mathbf{t}$ against $\mathbf{y}$ lie on the same straight line through the origin.

Example

For the accident data example discussed above, the necessary scales are the coordinates for the row and column profiles on the first singular vectors as shown in Table 9.57. The correlation coefficient between these two vectors based on the frequencies in the contingency table is the first singular value 0.0877. Table 9.58 shows the score vectors and the derived average scores for both the row and column categories. A plot of the linear relationship is given in Figure 9.26.

TABLE 9.58. Relationship Between Assigned Scores and Derived Means for Row and Column Categories

Row Categories		Assigned Row	Row Means Derived From	Column Categories Driver Injury	Assigned Column	Column Means Derived From
Driver Condition	Seatbelt Usage	Scores	Column Scores	Level	Scores	Row Scores
Normal	Yes	−0.07	−0.006	None	−0.03	−0.003
Normal	No	−0.01	−0.001	Minimal	0.22	0.019
Been Drinking	Yes	0.24	.021	Minor	0.30	0.026
Been Drinking	No	0.34	0.030	Major/Fatal	0.59	0.052

Note: The derived columns are obtained from the scores columns by multiplying by $\eta = 0.0877$.

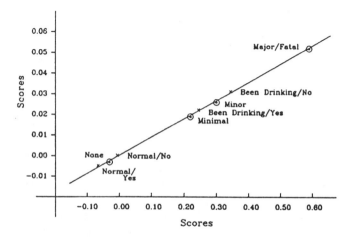

FIGURE 9.26. Linear Relationship Between Assigned Scores (Horizontal) and Derived Scores (Vertical)

Canonical Correlation

Canonical correlation analysis introduced in Chapter 7 can also be applied to two-dimensional contingency tables. In canonical correlation observations on two sets of variables summarized in two data matrices $\mathbf{X}(n \times q)$ and $\mathbf{Y}(n \times s)$ are used to determine a pair of linear relationships one for $\mathbf{X}$ and one for $\mathbf{Y}$. The relationships denoted by $\mathbf{w} = \mathbf{Xb}$ and $\mathbf{z} = \mathbf{Ya}$ are determined so that the correlation coefficient r_{zw} between the $(n \times 1)$ vectors $\mathbf{z}$ and $\mathbf{w}$ is maximized. The $(n \times 1)$ vectors $\mathbf{z}$ and $\mathbf{w}$ denote observations on the underlying variable Z and W. The coefficient vectors $\mathbf{a}$ and $\mathbf{b}$ are $(s \times 1)$ and $(q \times 1)$ respectively. These coefficient vectors are the eigenvectors obtained by solving the eigenequations

$$(\mathbf{R}_{xx}^{-1}\mathbf{R}_{xy}\mathbf{R}_{yy}^{-1}\mathbf{R}_{yx} - \lambda\mathbf{I})\mathbf{b} = \mathbf{0}$$
$$(\mathbf{R}_{yy}^{-1}\mathbf{R}_{yx}\mathbf{R}_{xx}^{-1}\mathbf{R}_{xy} - \lambda\mathbf{I})\mathbf{a} = \mathbf{0},$$

where the correlation matrix derived from $\mathbf{X}$ and $\mathbf{Y}$ is denoted by $\mathbf{R} = \begin{bmatrix} \mathbf{R}_{xx} & \mathbf{R}_{xy} \\ \mathbf{R}_{yx} & \mathbf{R}_{yy} \end{bmatrix}$. The vectors $\mathbf{b}$ and $\mathbf{a}$ are the eigenvectors corresponding to the largest common eigenvalue λ and $\sqrt{\lambda}$ is the correlation r_{zw} between $\mathbf{z}$ and $\mathbf{w}$. In the context of correspondence analysis the variables Z and W correspond to the underlying scales derived for the row and column categories.

For each of the $(r+c)$ categories in the $(r \times c)$ contingency table dummy variables are defined. The n observations for the row dummies are denoted by $\mathbf{Z}_1(n \times r)$ and for the column dummies the observations are denoted by $\mathbf{Z}_2(n \times c)$. The entire matrix of observations on both sets of dummy variables is denoted by $\mathbf{Z} = (\mathbf{Z}_1\mathbf{Z}_2)$. The matrix $(\mathbf{Z}'\mathbf{Z}/n)$ can be written as $\begin{bmatrix} \mathbf{D}_r & \mathbf{O} \\ \mathbf{O} & \mathbf{D}_c \end{bmatrix}$, where $\mathbf{O}(r \times c)$, $\mathbf{D}_r$ and $\mathbf{D}_c$ are defined above.

The sample covariance matrices for the two sets of variables are given by $[\mathbf{D}_r - \mathbf{rr}']$ and $[\mathbf{D}_c - \mathbf{cc}']$ respectively. The sample covariance matrix between the two sets of variables is given by $[\mathbf{O} - \mathbf{rc}']$. The vectors $\mathbf{r}$ and $\mathbf{c}$ are the mean vectors for the row and column dummies respectively. The two sets of dummy variables are not linearly independent since only $(r-1)$ and $(c-1)$ dummy variables are sufficient to identify the r and c categories respectively. As a result the covariance matrices for $\mathbf{Z}_1$ and $\mathbf{Z}_2$ have ranks of $(r-1)$ and $(c-1)$ respectively. To eliminate this problem of singular matrices the covariance matrices will be replaced by $\mathbf{D}_r$, $\mathbf{D}_c$ and $\mathbf{O}$ respectively.

The result of this change simply causes the first eigenvalue/eigenvector solution to be trivial. It does not affect the remaining eigenvalue-eigenvectors solutions. An alternative procedure, which obtains full rank matrices by eliminating one dummy variable from each set, yields the same solutions as the nontrivial solutions obtained using $\mathbf{D}_r$, $\mathbf{D}_c$ and $\mathbf{O}$.

As in the development of the theory of canonical correlation in Section 7.5.2 the eigenvalue-eigenvector equations are given by

$$(\mathbf{D}_r^{-1}\mathbf{O}\mathbf{D}_c^{-1}\mathbf{O}' - \lambda_y\mathbf{I})\mathbf{y} = \mathbf{0}$$
$$(\mathbf{D}_c^{-1}\mathbf{O}'\mathbf{D}_r^{-1}\mathbf{O} - \lambda_z\mathbf{I})\mathbf{z} = \mathbf{0},$$

where $\mathbf{y}'\mathbf{D}_r\mathbf{y} = \mathbf{z}'\mathbf{D}_c\mathbf{z} = 1$, the canonical variables are $\mathbf{x} = \mathbf{Z}_2\mathbf{z}$ and $\mathbf{t} = \mathbf{Z}_1\mathbf{y}$ and the means satisfy $\bar{t} = \mathbf{e}'\mathbf{t}/n = \bar{x} = \mathbf{e}'\mathbf{x}/n = 0$. The vectors $\mathbf{y}$ and $\mathbf{z}$ are the required canonical weights. The solution therefore involves determining the eigenvalues λ_y, λ_z and corresponding eigenvectors $\mathbf{y}$ and $\mathbf{z}$ of the two matrices $\mathbf{D}_r^{-1}\mathbf{O}\mathbf{D}_c^{-1}\mathbf{O}'$ and $\mathbf{D}_c^{-1}\mathbf{O}'\mathbf{D}_r^{-1}\mathbf{O}$. As in the case of canonical correlation analysis discussed above $\lambda_y = \lambda_z = \lambda$ say. The common eigenvalue λ represents the square of the correlation between the derived scores $\mathbf{t}$ and $\mathbf{x}$.

The required eigenvectors and eigenvalues can be obtained from a correspondence analysis as outlined in Section 9.4.1. The vectors $\mathbf{y}$ and $\mathbf{z}$ correspond to the principal coordinates for the row and column axes in the generalized singular value decomposition of the matrix $(\mathbf{O}-\mathbf{rc}')$. The corresponding canonical correlations $\sqrt{\lambda}$ are given by the singular values of this decomposition. The number of canonical variables that can be obtained is $\min[(r-1),(c-1)]$. The mean values of the new canonical variables evaluated in each of the row categories and column categories are given by $\sqrt{\lambda}\mathbf{z}$ for the row categories and $\sqrt{\lambda}\mathbf{y}$ for the column categories.

Example

For the criminal charge data presented in Table 9.39, the correspondence analysis results of Tables 9.50 and 9.51 provide the vectors of canonical weights. For the first pair of canonical variables, the weights are males not convicted (0.04), females not convicted (0.55), males convicted (-0.35), females convicted (0.44) for the row catgories and impaired driving (-0.34), theft under 1000 (0.52), mischief (0.08), possession of narcotics (-0.50), other (0.07) for the column categories. The first canonical variable for the row categories represents a contrast between all females and convicted males. The first column canonical variable represents a contrast between the charge theft under 1000 and the charges of impaired driving and possession of narcotics. The correlation between these two canonical variables is 0.377 which is the first singular value of the decomposition. Multiplying the principal coordinate vectors above by 0.377 yields the mean scores for the four row categories and the five column categories. These averages are given by (0.015, 0.207, –0.132, 0.166) and (–0.128, 0.196, 0.030, –0.189, 0.026). Thus for the canonical variable representing charges, convicted males have the lowest value (–0.132) and nonconvicted females have the highest value (0.207). Convicted females also have a relatively high value (0.166), whereas for nonconvicted males the value is close to zero (0.015). For the canonical variable representing sex-conviction status the charges of impaired driving and possession of narcotics have negative values (–0.128 and –0.189), whereas the charge of theft has the highest positive value at (0.196). The values corresponding to the charges mischief and other are quite small.

9.4.4 OTHER SOURCES OF INFORMATION

The dual scaling approach to correspondence analysis discussed here is similar to the presentation in Greenacre (1984) and Nishisato (1980). A discussion of the bivariate correlation and simultaneous linear regression is available in Nishisato (1980). The canonical correlation approach is discussed in Greenacre (1984).

9.4.5 CORRESPONDENCE ANALYSIS IN MULTIDIMENSIONAL TABLES

Although singular value decompositions can be used to approximate two-dimensional tables, there does not seem to be a simple extension of such a procedure to approximate tables of higher dimensions. One approach that has been used for multidimensional tables is called multiple correspondence analysis. In this procedure singular value decomposition is used to simultaneously approximate all possible two-dimensional subtables that can be derived from the multidimensional table. An outline of this procedure is provided below.

Multiple Correspondence Analysis and Burt Matrices

Multiple correspondence analysis begins by constructing dummy indicator variables for each category of each variable. A data matrix $\mathbf{Z}$ is then constructed with rows corresponding to observations on subjects and columns corresponding to dummy variables. For each subject, row unities appear in one dummy column for each variable, whereas the remaining dummy columns in that row contain zeroes. The rows of the resulting data matrix can be clustered so that identical rows are combined to produce frequencies in each row representing the number of occasions that the particular row type occurred in the data matrix. The resulting matrix $\mathbf{Z}^*$ therefore has rows that correspond to the individual cells of the multidimensional table. The data matrix $\mathbf{Z}^*$ becomes the starting point for a singular value decomposition on the column profiles. Dimensions for the row profiles of $\mathbf{Z}^*$ can also be obtained from the decomposition of $\mathbf{Z}^*$. The analysis for row profiles will not be discussed here.

An equivalent procedure to the analysis of $\mathbf{Z}^*$ is to first compute the matrix $\mathbf{B} = \mathbf{Z}'\mathbf{Z}$ using the matrix $\mathbf{Z}$ defined above. The new matrix $\mathbf{B}$ is called a *Burt matrix*. The matrix $\mathbf{B}$ contains block diagonal matrices reflecting frequencies for variable categories. The set of off-diagonal blocks in $\mathbf{B}$ represent the set of all two-dimensional contingency tables that can be constructed by studying two variables at a time. The matrix $\mathbf{B}$ is square symmetric and hence only row or column profiles need to be approximated. The singular value decomposition of the rows and columns of $\mathbf{B}$ yield the same results as the singular value decomposition of the columns of $\mathbf{Z}$ discussed above. The rank of the matrix $\mathbf{B}$ is equal to the number of columns of $\mathbf{Z}$ less the number of underlying variables used to generate $\mathbf{Z}$. Equivalently, the rank is given by rank $= \sum_{j=1}^{p}(k_j - 1)$, where $k_j =$ number of categories for variable j and p is the number of variables.

Although the number of possible dimensions is given by $\sum_{j=1}^{p}(k_j - 1)$ these dimensions are not all "interesting". Some of the dimensions are required for standardizing and centering and are considered to be artificial. As a rule of thumb only those dimensions with principal inertias that exceed $1/p$ are considered to be relevant. For a multidimensional contingency table with only one observation per cell, the principal inertias for all dimensions would necessarily be $1/p$ and hence $1/p$ is a useful baseline. This rationale was suggested by Greenacre (1984).

The multiple correspondence analysis measures the departure from independence for all two-dimensional tables that can be derived from the multidimensional table. For a three-dimensional table, the expected frequencies are determined for the three possible two-dimensional tables obtained by collapsing the table over the third variable. The diagonal blocks of the Burt matrix are also fitted using the independence model and hence the off-diagonal cells of these blocks are obtained from the products of the diagonal elements. The model for the expected frequency in the jth diagonal block is given by $n_{ii}n_{\ell\ell}/n$, $i, \ell = 1, 2, \ldots, k_j$, where $n = \sum_{i=1}^{k_j} n_{ii}$ and n_{ii}, $n_{\ell\ell}$ denote the diagonal elements in block j.

In Chapter 6 an outline of loglinear models illustrated that fitting a two-way interaction involved fitting the observed frequencies in the corresponding two-dimensional collapsed table. Multiple correspondence analysis therefore focuses on the difference between the independence model and the model which includes all two-way interaction terms. The dimensions with the largest inertias will represent the most important two-way interactions.

Example

The $4 \times 2 \times 2$ contingency table for the accident data introduced in Section 6.2 is used as an example. The Burt matrix resulting from this table is shown in Table 9.59. The various submatrices have been outlined in the table. The three block-diagonal matrices can be seen to contain the total frequencies for the various categories of the variables. In addition, the off-diagonal blocks show the three two-dimensional contingency tables each of which appears twice.

The multiple correspondence analysis of the Burt matrix yields five dimensions $(3 + 1 + 1)$. Two of the dimensions yield principal inertias that exceed the $\frac{1}{3}$ criterion (there are three variables). The two principal inertias are 0.373 and 0.334. Since the second dimension has a principal inertia that is barely above $\frac{1}{3}$ we can ignore it for the remainder of the analysis. The coordinates of the categories on the first dimension are summarized in Table 9.60 and are plotted in Figure 9.27. As expected based on previous results, higher levels of injury are associated with the category been drinking and with seatbelt nonusage.

TABLE 9.59. Burt Matrix for Three Dimensional Contingency Table for Accident Data

| | Injury Level | | | | Driver Condition | | Seatbelt Usage | |
	None	Minimal	Minor	Major/Fatal	Normal	Been Drinking	Yes	No
None	78776	0	0	0	74471	4305	12813	65963
Minimal	0	4647	0	0	4123	524	647	4000
Minor	0	0	3001	0	2616	385	2642	359
Major/Fatal	0	0	0	345	275	70	42	303
Normal	74471	4123	2616	275	81485	0	13486	67999
Been Drinking	4305	524	385	70	0	5284	375	4909
Yes	12813	647	2642	42	13486	375	13861	0
No	65963	4000	359	803	67999	4909	0	72908

TABLE 9.60. Coordinates on First Column Dimension for Accident Data

| | | | CATEGORY | | | | |
None	Minimal	Minor	Major/Fatal	Normal	Been Drinking	Yes	No
−0.19	1.52	2.13	3.96	−0.18	2.76	−1.14	0.22

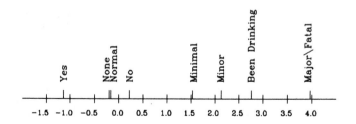

FIGURE 9.27. Multiple Correspondence Analysis for Accident Data

A Second Example

For the charge data introduced in Table 9.39 a $5 \times 2 \times 2$ three-dimensional contingency table relating charge, conviction and sex was constructed and analyzed in Burt matrix form. Although there are a total of six dimensions only two of the dimensions produced principal inertias that exceeded $\frac{1}{3}$. These inertias were 0.463 and 0.386 respectively for the first two dimensions. The coordinates on the first two dimensions are shown in Table 9.61 and a corresponding plot appears in Figure 9.28. The first dimension places high positive values on the charge theft and high negative values on nar-

TABLE 9.61. Column Coordinates for First Two Dimensions for Criminal Charge Data

| | Column Dimensions | |
Category	1	2
Impaired	−0.75	−0.64
Theft	1.13	−0.31
Mischief	0.23	1.64
Narcotics	−1.06	1.65
Other	0.14	0.47
No	0.69	1.47
Yes	−0.15	−0.33
Male	−0.61	0.20
Female	1.00	−0.33

cotics and impaired. This dimension also assigns high positive values to no conviction and female and a moderately high negative value to male. The first dimension suggests that females are more associated with theft and that males are more associated with impaired driving and possession of narcotics. In addition it would appear that no conviction is more associated with theft and females than with males, impaired driving and narcotics possession. The second dimension assigns high positive values to no conviction, mischief and narcotics and high negative values to impaired. The second dimension indicates that the charges of mischief and narcotics possession are less likely to result in conviction than the charge of impaired driving.

9.4.6   OTHER SOURCES OF INFORMATION

Multiple correspondence analysis is discussed more extensively in Greenacre (1984) and Lebart, Morineau and Warwick (1984). A brief discussion is also available in Andersen (1990). Extensive references to the research literature can be found in these texts.

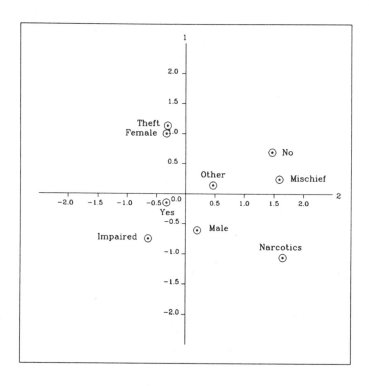

FIGURE 9.28. Multiple Correspondence Analysis for Criminal Charge Data

Cited Literature and References

1. Andersen, Erling B. (1990). *The Statistical Analysis of Categorical Data*. Berlin: Springer–Verlag.

2. Bernstein, Ira H. (1987). *Applied Multivariate Analysis*. New York: Springer–Verlag.

3. Conway, Delores A. and Reinganum, Marc R. (1988). "Stable Factors in Security Returns: Identification Using Cross-Validation," *Journal of Business and Economic Statistics* 6, 1–15.

4. Everitt, B.S. (1984). *An Introduction to Latent Variable Models*. London: Chapman and Hall.

5. Gibbons, D.I., McDonald, G.C. and Gunst, R.F. (1987). "The Complementary Use of Regression Diagnostics and Robust Estimation," *Naval Research Logistics* 34, 109–131.

6. Gifi, A. (1990). *Nonlinear Multivariate Analysis*. New York: John Wiley and Sons.

7. Gorsuch, R.L. (1983). *Factor Analysis*, Second Edition. Hillsdale, NJ: Lawrence Erlbaum.

8. Greenacre, M.J. (1984). *Theory and Applications of Correspondence Analysis*. New York: Academic Press.

9. Hakistan, A.R., Rogers, W.T. and Cattell, R.B. (1982). "The Behavior of Number-of-Factors Rules with Simulated Data," *Multivariate Behavioral Research* 17, 193–219.

10. Harman, H.H. (1976). *Modern Factor Analysis*, Third Edition. Chicago: University of Chicago Press.

11. Hawkins, D.M. (1980). *Identification of Outliers*. London: Chapman and Hall.

12. Jackson, J. Edward (1990). *A User's Guide to Principal Components*. New York: John Wiley and Sons.

13. Jambu, Michel (1991). *Exploratory and Multivariate Data Analysis*. New York: Academic Press.

14. Jobson, J.D. (1988). Comment on "Stable Factors in Security Returns: Identification Using Cross-Validation," by Conway and Reinganum, *Journal of Business and Economic Statistics* 6, 16–20.

15. Jobson, J.D. and Schneck, Rodney (1982). "Constituent Views of Organizational Effectiveness: Evidence From Police Organizations." *Academy of Management Journal* 25, 25–46.

16. Johnson, Richard A. and Wichern, Dean W. (1988). *Applied Multivariate Statistical Analysis*, Second Edition. Englewood Cliffs, NJ: Prentice–Hall.

17. Jolliffe, I.T. (1986). *Principal Component Analysis*. New York: Springer-Verlag.

18. Lawley, D.N. and Maxwell, A.E. (1971). *Factor Analysis as a Statistical Method*, Second Edition. London: Butterworth and Company.

19. Lebart, L, Morineau, A. and Warwick, K.M. (1984). *Multivariate Descriptive Statistical Analysis*. New York: John Wiley and Sons.

20. MacDonnell, W.R. (1902). "On Criminal Anthropometry and the Identification of Criminals," *Biometrika* 1, 177–227.

21. Mardia, K.V., Kent, J.T. and Bibby, J.M. (1979). *Multivariate Analysis*. London: Academic Press.

22. Maxwell, A.E. (1977). *Multivariate Analysis in Behavioral Research*, London: Chapman and Hall.

23. Morrison, Donald F. (1976). *Multivariate Statistical Methods*, Second Edition. New York: McGraw–Hill.

24. Mulaik, Stanley A. (1972). *The Foundations of Factor Analysis*. New York: McGraw-Hill.

25. Nishisato, S. (1980). *Analysis of Categorical Data, Dual Scaling and Its Application*. Toronto: University of Toronto Press.

26. Seber, G.A.F. (1984). *Multivariate Observations*. New York: John Wiley and Sons.

27. Stevens, James (1986). *Applied Multivariate Statistics for the Social Sciences*. Hillsdale, NJ: Lawrence Erlbaum Associates.

28. Theil, Henri (1971). *Principles of Econometrics*. New York: John Wiley and Sons.

Exercises for Chapter 9

1. This exercise is based on the Air Pollution Data in Table V7 in the Data Appendix.

   (a) Carry out a principal components analysis of the covariance matrix for the eleven variables excluding TMR. Interpret all of the components whose eigenvalues exceed the geometric mean of the eigenvalues. How important is each of these components in explaining the variation in the underlying variables?

   (b) Repeat the steps in (a) using the correlation matrix. Interpret all components whose eigenvalues exceed one. How important is each component in explaining the variation in the underlying variables?

   (c) Obtain the standardized principal component scores for all of components generated in (a) and regress TMR on those scores. Discuss the results. Allocate the variation in TMR to each component.

   (d) Repeat (c) for the principal components obtained in (b).

   (e) Use principal components analysis to identify and characterize outliers as illustrated in Chapter 9 using the results in (b). Discuss the results.

2. This exercise is based on the Financial Accounting Data in Table V6 in the Data Appendix.

   (a) Carry out a principal components analysis of the covariance matrix for the 12 variables excluding RETCAP. Interpret all the components whose eigenvalues exceed the geometric mean of the eigenvalues. How important is each of these components in explaining the variation in the underlying variables.

   (b) Repeat the steps in (a) using the correlation matrix. Interpret all components whose eigenvalues exceed one. How important is each component in explaining the variation in the underlying variables.

   (c) Obtain the standardized principal component scores for all of the components generated in (a) and regress RETCAP on these scores. Discuss the results. Allocate the variation in RETCAP to each component.

   (d) Repeat (c) for the principal components obtained in (b).

   (e) Use principal components analysis to identify and characterize outliers as illustrated in Chapter 9 using the results in (b). Discuss the results.

3. This exercise is based on the Shopping Attitude Data in Table V8 in the Data Appendix.

    (a) Beginning with a principal components analysis of the variables A1 to A18 retain four factors and carry out a varimax rotation. Obtain a scree test plot and justify the four factor solution. Interpret the rotated factors. Comment on the communalities of the variables. Examine plots of the rotated factor patterns.

    (b) Examine the varimax rotated four-factor solution derived from the squared multiple correlation approach to the principal factor method and compare the factors to (a).

    (c) Examine the maximum likelihood factors after varimax rotation and compare to the factors in (a) and (b). Carry out a χ^2 test of goodness of fit for the number of factors from one factor through six factors and discuss the results. Also determine the cross validation test statistics.

    (d) Carry out an alternative method of rotation and repeat the analysis in (a). Compare the results to the previous factor solutions.

    (e) Select one of the four factor solutions determined in (a) through (d) and obtain the factor scores for the four rotated factors. Relate the factors to the variables WORK and AGE using one-way ANOVA. Discuss the differences in means for the four factors over the categories of WORK and AGE.

    (f) Use a two-way ANOVA to relate the factors to the WORK and AGE categories. Is there any interaction? Compare the results to (e).

    (g) Use regression analysis to relate AGE as an interval variable to the four factors. Discuss the results.

    (h) Write an overall summary discussing the shopping attitude factors and their relation to WORK and AGE.

4. This exercise is based on the Air Pollution Data in Table V7 in the Data Appendix.

    (a) Beginning with the principal components of the correlation matrix use the eigenvalue-one-criterion and varimax rotation to obtain a factor analysis solution. Examine the scree plot and factor pattern plots. Discuss the results.

    (b) Repeat the analysis in (a) beginning with the squared multiple correlation approach to the principal factor approach. Discuss the results and compare to (a).

    (c) Examine various alternative solutions in (a) and (b) by altering the number of factors criterion and/or the varimax rotation criterion.

(d) Use the maximum likelihood method with varimax rotation to obtain a factor analysis for the correlation matrix. Use the χ^2 statistic to examine the significance of various factors. Also examine the various cross-validation statistics. Discuss the results. Compare the solution to the results in (a), (b) and (c).

(e) Write an overall summary discussing the factors that seem to be present in the data.

5. This exercise is based on the Financial Accounting Data in Table V6 in the Data Appendix.

(a) Beginning with the principal components of the correlation matrix use the eigenvalue-one-criterion and varimax rotation to obtain a factor analysis solution. Examine the scree plot and factor pattern plots. Discuss the results.

(b) Repeat the analysis in (a) beginning with the squared multiple correlation approach to the principal factor approach. Discuss the results and compare to (a).

(c) Examine various alternative solutions in (a) and (b) by altering the number of factors criterion and/or the varimax rotation criterion.

(d) Use the maximum likelihood method with varimax rotation to obtain a factor analysis for the correlation matrix. Use the χ^2 statistic to examine the significance of various factors. Also examine the various cross-validation statistics. Discuss the results. Compare the solution to the results in (a), (b) and (c).

(e) Write an overall summary discussing the factors that seem to be present in the data.

6. This exercise is based on the Air Pollution Data in Table V7 in the Data Appendix. For the principal component analysis of the correlation matrix construct a biplot. Interpret the plot.

7. This exercise is based on the Financial Accounting Data in Table V6 in the Data Appendix. For the principal component analysis of the correlation matrix construct a biplot. Interpret the plot.

8. This exercise is based on the Bus Data of Table V1 in the Data Appendix.

(a) Construct three two-dimensional tables relating the variable AT-TEND to all possible pairs of the variables SEX, DAY and GARAGE by combining the categories in pairs. Use correspondence analysis to study the three tables and discuss the results. Give a dual scaling interpretation for your results.

(b) Repeat the analyses in (a) using the three three-dimensional tables and multiple correspondence analysis.

(c) Combine your results from (a) and (b) and write an overall conclusion.

9. This exercise is based on the Accident Data in Table V2 of the Data Appendix. For some analyses it may be necessary to combine categories because of the cells with zero frequency.

(a) Use correspondence analysis to study the two-dimensional table relating DRIVER INJURY LEVEL to the variable which combines POINT OF IMPACT and DRIVER CONDITION. Repeat the procedure to relate DRIVER INJURY LEVEL to the variable which combines SEATBELT and POINT OF IMPACT. Discuss the results. Give a dual scaling interpretation for your results in each case.

(b) Use multiple correspondence analysis to analyze the two three-dimensional tables of (a):

   i. DRIVER INJURY LEVEL vs POINT OF IMPACT vs DRIVER CONDITION, and

   ii. DRIVER INJURY LEVEL vs POINT OF IMPACT vs SEATBELT.

Discuss your results.

(c) Use correspondence analysis to relate INJURY LEVEL to all three variables by combining the latter three variables into one variable. Discuss your results.

(d) Use multiple correspondence analysis to analyze the four-dimensional table relating the four variables in Table V2. Discuss your results.

(e) Combine your results from (a), (b) and (c) and write an overall conclusion.

10. This exercise is based on the Accident Data in Table V3 of the Data Appendix.

(a) Use correspondence analysis to study the two-dimensional table relating DRIVER INJURY LEVEL to the variable which combines SPEED LIMIT and DRIVER CONDITION. Repeat the procedure to relate DRIVER INJURY LEVEL to the variable which combines SEATBELT and SPEED LIMIT. Discuss the results. Give a dual scaling interpretation for your results in each case.

(b) Use multiple correspondence analysis to analyze the two three-dimensional tables of (a):

i. DRIVER INJURY LEVEL vs SPEED LIMIT vs DRIVER CONDITION, and

ii. DRIVER INJURY LEVEL vs SPEED LIMIT vs SEAT-BELT.

Discuss your results.

(c) Use correspondence analysis to relate INJURY LEVEL to all three variables by combining the latter three variables into one variable. Discuss your results.

(d) Use multiple correspondence analysis to analyze the four-dimensional table relating the four variables in Table V2. Discuss your results.

(e) Combine your results from (a), (b) and (c) and write an overall conclusion.

Questions for Chapter 9

1. Given a data matrix $\mathbf{X}(n \times p)$, let $\mathbf{z}(n \times 1)$ be an index variable and let $\mathbf{a}(p \times 1)$ be a vector of constants so that $\widehat{\mathbf{X}} = \mathbf{za}'$ provides an approximation for $\mathbf{X}$. Assume that $\mathbf{a}$ is scaled so that $\mathbf{a}'\mathbf{a} = 1$.

   (a) Show that $\sum_{i=1}^{n}\sum_{j=1}^{p}(x_{ij} - \hat{x}_{ij})^2$ can be written as $tr[(\mathbf{X} - \widehat{\mathbf{X}})'(\mathbf{X} - \widehat{\mathbf{X}})]$.

   (b) Substitute $\widehat{\mathbf{X}} = \mathbf{za}'$ and show that

   $$tr[(\mathbf{X} - \widehat{\mathbf{X}})'(\mathbf{X} - \widehat{\mathbf{X}})] = tr\ \mathbf{X}'\mathbf{X} - tr\ \mathbf{az}'\mathbf{X} - tr\ \mathbf{X}'\mathbf{za}'$$
   $$+tr\ \mathbf{az}'\mathbf{za}' = tr\ \mathbf{X}'\mathbf{X} - 2\mathbf{z}'\mathbf{Xa} + \mathbf{z}'\mathbf{z}$$

   using properties of the trace and also that $\mathbf{a}'\mathbf{a} = 1$.

   (c) Differentiate the expression on the right-hand side in (b) with respect to $\mathbf{z}$, holding $\mathbf{a}$ fixed, and hence show that $\mathbf{Z} = \mathbf{Xa}$ when the derivative is zero.

   (d) Using the expression for $tr[(\mathbf{X} - \widehat{\mathbf{X}})'(\mathbf{X} - \widehat{\mathbf{X}})]$ given in (b) substitute $\mathbf{Xa}$ for $\mathbf{z}$ to obtain

   $$tr[(\mathbf{X} - \widehat{\mathbf{X}})'(\mathbf{X} - \widehat{\mathbf{X}})] = tr\ \mathbf{X}'\mathbf{X} - \mathbf{a}'\mathbf{X}'\mathbf{Xa}.$$

   (e) Minimize the expression in (d) with respect to $\mathbf{a}$ subject to $\mathbf{a}'\mathbf{a} = 1$. Use a Lagrange multiplier λ to include this condition and show that setting the derivative equal to zero yields the expression $(\mathbf{X}'\mathbf{X} - \lambda\mathbf{I})\mathbf{a} = 0$ and hence that λ is an eigenvalue of $\mathbf{X}'\mathbf{X}$ with corresponding eigenvector $\mathbf{a}$.

2. Given the $(n \times p)$ data matrix $\mathbf{X}$ let $\mathbf{z}(n \times 1)$ be given by $\mathbf{z} = \mathbf{Xv}$ where $\mathbf{v}(p \times 1)$ is a vector of constants with $\mathbf{v}'\mathbf{v} = 1$.

   (a) Show that the vector $\mathbf{v}$ that maximizes $\mathbf{z}'\mathbf{z}$ subject to $\mathbf{v}'\mathbf{v} = 1$ is obtained from the equation $(\mathbf{X}'\mathbf{X} - \lambda\mathbf{I})\mathbf{v} = 0$. Use a Lagrange multiplier λ on the condition $\mathbf{v}'\mathbf{v} = 1$, and maximize the function $\mathbf{v}'\mathbf{X}'\mathbf{Xv} - \lambda(\mathbf{v}'\mathbf{v} - 1)$ with respect to $\mathbf{v}$.

   (b) Show that the largest eigenvalue λ_1 that is obtained in (a) is the one that maximizes $\mathbf{z}'\mathbf{z}$.

   (c) Let $\mathbf{z}_2 = \mathbf{Xv}_2$ be a second linear transformation of $\mathbf{X}$ so that $\mathbf{z}_2'\mathbf{z}_1 = 0$. Show that this implies that $\mathbf{v}_2'\mathbf{v}_1 = 0$.

   (d) Determine $\mathbf{z}_2$ in (c) so that $\mathbf{z}_2'\mathbf{z}_2$ is maximized subject to $\mathbf{v}_2'\mathbf{v}_2 = 1$ and subject to $\mathbf{v}_2'\mathbf{v}_1 = 0$. Use Lagrangian multipliers λ_2 for $\mathbf{v}_2'\mathbf{v}_2 = 1$ and θ for $\mathbf{v}_2'\mathbf{v}_1 = 0$. Maximize the function $\mathbf{v}_2'\mathbf{X}'\mathbf{Xv}_2 - \lambda_2(\mathbf{v}_2'\mathbf{v}_2 - 1) - \theta(\mathbf{v}_1'\mathbf{v}_2)$ with respect to $\mathbf{v}_2$.

(e) Show the general result that the kth principal component $z_k = Xv_k$ maximizes $z'_k z_k$ subject to $v'_k v_k = 1$ and $v'_k v_j = 0$, $j \neq k$, $j = 1, 2, \ldots, k - 1$.

3. Given the eigenvalues $\lambda_1, \lambda_2, \ldots, \lambda_p$ and corresponding eigenvectors $v_1, v_2, \ldots, v_p$ of $X'X$ where $v'_j v_j = 1$ and $v'_j v_k = 0$, $j \neq k$, $j, k = 1, 2, \ldots, p$, let V denote the $(p \times p)$ matrix whose columns are the eigenvectors and let $\Lambda(p \times p)$ be the diagonal matrix whose diagonal elements are the eigenvalues $\lambda_1, \lambda_2, \ldots, \lambda_p$. Denote the corresponding principal components of X by z_j, $j = 1, 2, \ldots, p$, where $z_j = Xv_j$ and let $Z(n \times p)$ denote the matrix whose columns are the vectors z_j, $j = 1, 2, \ldots, p$. Use these properties and the fact that $(X'X - \lambda_j I)V_j = 0$ to show the results summarized below.

(a) $V'V = I$, $V' = V^{-1}$, $VV' = I$.

(b) $Z = XV$.

(c) $Z'Z = \Lambda$, $\quad (Z'Z)^{-1} = \Lambda^{-1}$.

(d) $X = ZV'$.

(e) $X'X = V\Lambda V'$ and $X'XV = V\Lambda$, $\quad (X'X)^{-1} = V\Lambda^{-1}V'$.

(f) Let Z be partitioned into $Z_1(n \times r)$ and $Z_2[n \times (p - r)]$ and let $V_1(p \times r)$ and $V_2[p \times (p-r)]$ be the corresponding partitions of V. Let $\Lambda_1(r \times r)$ and $\Lambda_2((p - r) \times (p - r))$ denote corresponding diagonal matrices formed from the diagonal matrix Λ where

$$\Lambda = \begin{bmatrix} \Lambda_1 & 0 \\ 0 & \Lambda_2 \end{bmatrix}. \text{ Show that}$$

$$X = Z_1 V'_1 + Z_2 V'_2$$

and

$$X'X = V_1 \Lambda_1 V'_1 + V_2 \Lambda_2 V'_2.$$

4. In Question 2 the eigenvalues and eigenvectors were determined for the data matrix $X'X$. Let $\text{Cov}(x) = \Sigma$ for the random variable $x(p \times 1)$ and define $z_j = x'v_j$. Repeat the steps in question 2 by maximizing $V(z_j)$ subject to $v'_j v_j = 1$ and also $\text{Cov}(z_j, z_k) = 0$, $j \neq k$. Use Σ in place of $X'X$ and $v'_j \Sigma v_k$ in place of $\text{Cov}(z_j, z_k)$.

5. For the eigenvalues $\lambda_1, \lambda_2, \ldots, \lambda_p$ and corresponding eigenvectors $v_1, v_2, \ldots, v_p$ determined from $(\Sigma - \lambda_j I)v_j = 0$ with $z_j = x'v_j$ show the following results.

(a) $\lambda_j = v'_j \Sigma v_j$, $j = 1, 2, \ldots, p$.

(b) $V(z_j) = \lambda_j$ where $z_j = x'v_j$, $j = 1, 2, \ldots, p$.

(c) $\text{Cov}(z) = \Lambda$ where Λ is a diagonal matrix with diagonal elements λ_j, and z is the $(p \times 1)$ vector of elements z_j, $j = 1, 2, \ldots, p$.

(d) $\Lambda = V'\Sigma V$ where $V(p \times p)$ contains the eigenvectors v_j, $j = 1, 2, \ldots, p$, as columns.

(e) $\Sigma = V\Lambda V' = \sum_{j=1}^{p} \lambda_j v_j v_j'$ (The spectral decomposition of Σ).

(f) $\Sigma^{-1} = V\Lambda^{-1}V'$.

(g) Let $\mu_X = E[x]$, $\mu_Z = E[z]$, $z' = x'V$ and show that $\mu_Z' = \mu_X'V$.

(h) Use the results of (f) and (g) to show that

$$(x - \mu_X)'\Sigma^{-1}(x - \mu_X) = (z - \mu_Z)'\Lambda^{-1}(z - \mu_Z)$$

and hence that the principal components define the principal axes of the ellipsoid of constant probability for the multivariate normal.

(i) If the random variable x in (h) has been standardized to have variance 1, show that if $\Sigma = \rho$ (the correlation matrix) the expression in (h) can be written in terms of principal components as in Section 9.1.6.

6. Let $\lambda_1, \lambda_2, \ldots, \lambda_p$ denote the eigenvalues of $X'X$ and let $v_1, v_2, \ldots, v_p$ denote the corresponding eigenvectors. Show that the eigenvalues and eigenvectors of $(X'X)^{-1}$ are given by $1/\lambda_1, 1/\lambda_2, \ldots, 1/\lambda_p$ and $v_1, v_2, \ldots, v_p$ respectively.

7. Consider the multiple linear regression model given by $y = X\beta + u$ and let $Z = XV$ denote the corresponding principal components derived from $X'X$. Let γ be defined such that $X\beta = Z\gamma$ and hence $y = Z\gamma + u$.

(a) Show that $\gamma = V'\beta$ and $\beta = V\gamma$.

(b) Given that $\hat{\beta} = (X'X)^{-1}X'y$ and $\hat{\gamma} = (Z'Z)^{-1}Z'y$, show that $\hat{\beta} = V\hat{\gamma}$ and $\hat{\gamma} = V'\hat{\beta}$.

(c) Show that the covariance matrix of $\hat{\beta}$ given by $\sigma^2(X'X)^{-1}$ can be written as $\sigma^2 \sum_{j=1}^{p} v_j v_j'/\lambda_j$, which is the spectral decomposition of $\sigma^2(X'X)^{-1}$. Explain how multicollinearity in X shows up as relatively large terms in the above spectral decomposition.

(d) Given that the regression sum of squares for $\hat{\gamma}$ is given by $\hat{\gamma}'Z'y$ show that this sum of squares can be separately allocated to each component z_j, $j = 1, 2, \ldots, p$.

8. The equal correlation–equal variance matrix is given by

$$\Sigma(p \times p) = \sigma^2 \begin{bmatrix} 1 & \rho & \rho & & \rho \\ \rho & 1 & & & \\ \vdots & & 1 & & \\ \vdots & & & \ddots & \rho \\ \rho & & & & 1 \end{bmatrix}.$$

(a) Show that $\Sigma = \sigma^2 \rho \mathbf{ii}' + \sigma^2(1-\rho)\mathbf{I}$ where $\mathbf{i}(p \times 1)$ is a vector of unities.

(b) Given that the eigenvalue λ, must satisfy the characteristic equation $|\Sigma - \lambda \mathbf{I}| = 0$ use the fact that $|c\mathbf{I} + b\mathbf{df}'| = c + b\mathbf{d}'\mathbf{f}$ for $\mathbf{I}(p \times p)$, $\mathbf{d}(p \times 1)$ and $\mathbf{f}(p \times 1)$ to show that $\lambda = \sigma^2[1 + \rho(p-1)]$ satisfies the characteristic equation.

(c) Using λ derived in (b) show that the eigenvector $\mathbf{v}$ that satisfies $(\Sigma - \lambda \mathbf{I})\mathbf{v} = 0$ is given by $\mathbf{v} = \mathbf{i}/\sqrt{p}$.

(d) Let λ_1 and $\mathbf{v}_1$ denote the eigenvalue and eigenvector determined in (b) and (c). Show that all remaining eigenvalues are equal to $\lambda_j = \sigma^2(1-\rho)$, $j = 2, \ldots, p$, and all remaining eigenvectors are orthogonal to the vector $\mathbf{i}$. [HINT: see Appendix, Section 1.3, item 12 for properties of determinants.]

9. Two pairs of orthogonal axes X_1, X_2 and Z_1, Z_2 are related by the equations

$$\begin{aligned} Z_1 &= X_1 \cos \phi + X_2 \sin \phi \\ Z_2 &= -X_1 \sin \phi + X_2 \cos \phi, \end{aligned}$$

where ϕ represents the angle of rotation between the X_1, X_2 axes and the Z_1, Z_2 axes.

(a) Given the covariance matrix $\Sigma = \begin{bmatrix} 5 & \sqrt{3} \\ \sqrt{3} & 3 \end{bmatrix}$ of $\begin{bmatrix} X_1 \\ X_2 \end{bmatrix}$, show that the eigenvectors and eigenvalues of Σ are given by $\lambda_1 = 6$, $\lambda_2 = 2$ and

$$\mathbf{v}_1 = \begin{bmatrix} \sqrt{3}/2 \\ 1/2 \end{bmatrix}, \quad \mathbf{v}_2 = \begin{bmatrix} -1/2 \\ \sqrt{3}/2 \end{bmatrix}.$$

Give the equations relating the components Z_1 and Z_2 to X_1 and X_2.

(b) Show that the angle ϕ of rotation from X_1 to Z_1 is $30°$.

10. The factor analysis model requires that the correlation matrix be expressible as $\rho = \mathbf{AA}' + \Psi$, where Ψ is a diagonal matrix, $\rho(p \times p)$, $\Psi(p \times p)$ and $\mathbf{A}(p \times r)$.

(a) Assume $r = 1$ and $p = 3$ and denote the elements of $\mathbf{A}$, ρ and Ψ by

$$\mathbf{A} = \begin{bmatrix} a_1 \\ a_2 \\ a_3 \end{bmatrix}, \quad \rho = \begin{bmatrix} 1 & \rho_{12} & \rho_{13} \\ \rho_{12} & 1 & \rho_{23} \\ \rho_{13} & \rho_{23} & 1 \end{bmatrix},$$

$$\Psi = \begin{bmatrix} \sigma_{u_1}^2 & 0 & 0 \\ 0 & \sigma_{u_2}^2 & 0 \\ 0 & 0 & \sigma_{u_3}^2 \end{bmatrix}.$$

Show that the model yields the six equations given by

$$1 = a_1^2 + \sigma_{u_1}^2 \qquad \rho_{12} = a_1 a_2$$

$$1 = a_2^2 + \sigma_{u_2}^2 \qquad \rho_{13} = a_1 a_3$$

$$1 = a_3^2 + \sigma_{u_3}^2 \qquad \rho_{23} = a_2 a_3.$$

(b) Assume ρ is the correlation matrix

$$\begin{bmatrix} 1 & 0.8 & 0.6 \\ 0.8 & 1 & 0.2 \\ 0.6 & 0.2 & 1 \end{bmatrix}$$

and show that

$$a_1 = \sqrt{12/5}$$
$$a_2 = \sqrt{4/15}$$
$$a_3 = \sqrt{3/20}.$$

(c) What does the solution for a_1 imply about $\sigma_{u_1}^2$ and how does this violate the conditions of the model? Is this called a Heywood case?

(d) Determine a condition on the parameters σ_{12}, σ_{13} and σ_{23} that would eliminate any Heywood cases. If the partial correlation between variables 2 and 3 given variable 1 is given by $(\rho_{23} - \rho_{12}\rho_{13})/[(1-\rho_{12}^2)(1-\rho_{13}^2)]^{1/2}$ what does your condition say about this partial correlation coefficient? Can you draw a general conclusion about when a one factor model can be used in place of three variables?

(e) For the model in (a) there were precisely six unknowns and six equations. Assuming Σ known in general and $\mathbf{A}$ and Ψ given above, how many independent equations usually result and how many unknowns must be solved? Give a minimum condition on the relationship between p and r in order for there to be a sufficient number of equations.

11. Given the estimated factor model

$$\mathbf{R} = \widehat{\mathbf{A}}\widehat{\mathbf{A}}' + \widehat{\boldsymbol{\Psi}} \quad \text{and} \quad \mathbf{X} = \widehat{\mathbf{A}}\widehat{\mathbf{F}} + \widehat{\mathbf{U}},$$

where $E[\mathbf{U}\mathbf{U}'] = \boldsymbol{\Psi}$ and $\widehat{\boldsymbol{\Psi}} = \widehat{\mathbf{U}}\widehat{\mathbf{U}}'$, by referring to the weighted least squares estimator in multiple regression, explain why the estimates of the factor scores given by

$$\widehat{\mathbf{F}} = (\widehat{\mathbf{A}}'\widehat{\boldsymbol{\Psi}}^{-1}\widehat{\mathbf{A}})^{-1} \, (\widehat{\mathbf{A}}'\widehat{\boldsymbol{\Psi}}^{-1}\mathbf{X})$$

are called weighted least squares.

12. Given $\boldsymbol{\rho} = \begin{bmatrix} 1.00 & 0.63 & 0.45 \\ 0.63 & 1.00 & 0.35 \\ 0.45 & 0.35 & 1.00 \end{bmatrix}$.

(a) Show that the standardized variables X_1, X_2 and X_3 can be generated by the single factor model

$$\begin{aligned} X_1 &= 0.9 \, F_1 + U_1 \\ X_2 &= 0.7 \, F_1 + U_2 \\ X_3 &= 0.5 \, F_1 + U_3, \end{aligned}$$

where $V(F_i) = 1$, $\operatorname{Cov}(U_i, F_i) = 0$, $i = 1, 2, 3$ and

$$\boldsymbol{\Psi} = \begin{bmatrix} 0.19 & 0 & 0 \\ 0 & 0.51 & 0 \\ 0 & 0 & 0.75 \end{bmatrix}$$

by showing that $\boldsymbol{\rho} = \mathbf{A}\mathbf{A}' + \boldsymbol{\Psi}$ where $\mathbf{A} = \begin{bmatrix} 0.9 \\ 0.7 \\ 0.5 \end{bmatrix}$.

(b) The eigenvalues and eigenvectors of the correlation matrix $\boldsymbol{\rho}$ above are

$$\boldsymbol{\lambda} = \begin{bmatrix} 1.96 \\ 0.68 \\ 0.36 \end{bmatrix} \quad \text{and} \quad \mathbf{V} = \begin{bmatrix} 0.625 & -0.219 & 0.749 \\ 0.593 & -0.491 & -0.638 \\ 0.507 & 0.843 & -0.177 \end{bmatrix}$$

respectively. Determine the factor loadings for a one-factor model based on this principal component solution and compare the loadings to the solution in (a).

(c) Compare the communalities for the one factor models obtained in (a) and (b).

Johnson and Wichern (1988).

13. Assume that the three variables X_1, X_2 and f have mean 0 and variance 1 and let $\mathrm{Corr}(X_1, X_2) = r_{12}$, $\mathrm{Corr}(X_1, f) = \mathrm{Corr}(X_2, f) = r_f$.

(a) Show that if the angle between f and X_1 and between f and X_2 is θ, where $r_{X_1 f} = r_{X_2 f} = \cos \theta$ then the largest possible angle between X_1 and X_2 is 2θ and hence the minimum value of $r_{12} = \cos 2\theta$. Show that if $r_{X_1 f} = r_{X_2 f} = 0.707$ then the minimum value of r_{12} is 0. HINT: Show that the angle between X_1 and X_2 is maximized if X_1, X_2 and f are in the same plane with X_1 and X_2 on opposite sides of f. From Mulaik (1972).

(b) Suppose X_1 and X_2 satisfy the factor model

$$X_1 = 0.707f + U_1 \qquad \mathrm{Corr}(U_1, U_2) = 0,$$
$$X_2 = 0.707f + U_2 \qquad \mathrm{Corr}(f, U_2) = 0,$$
$$\mathrm{Corr}(f, U_1) = 0.$$

Show that $\mathrm{Corr}(X_1, X_2) = 0.50$ and that the partial correlation between X_1 and X_2 controlling for f is 0. What condition given in the model ensures that this partial correlation is zero? Explain.

(c) Determine the communality and specific variance for each of the variables X_1 and X_2 in (b).

14. (a) Given the correlation matrix for X_1 and X_2

$$\rho = \begin{bmatrix} 1 & r_{12} \\ r_{12} & 1 \end{bmatrix},$$

show that the eigenvalues and eigenvectors are given by

$$\lambda = \begin{bmatrix} 1 + r_{12} \\ 1 - r_{12} \end{bmatrix}, \qquad \mathbf{V} = \begin{bmatrix} \sqrt{2}/2 & -\sqrt{2}/2 \\ \sqrt{2}/2 & \sqrt{2}/2 \end{bmatrix}.$$

(b) Show that, if the first standardized principal component is used as the factor f in the one-factor model

$$X_1 = a_1 f + U_1$$
$$X_2 = a_2 f + U_2,$$

then $\mathrm{Cov}(U_1, U_2) = 0$ is only satisfied if $r_{12} = 1$.

(c) What is the answer in (b) if f is not constrained to be the first principal component?

(d) Suppose that X_1 and X_2 are both dependent on another factor, say g. How does this assumption affect the previous assumption that $\mathrm{Cov}(U_1, U_2) = 0$?

15. Given an $(n \times p)$ matrix $\mathbf{X}$ of rank p, denote the singular value decomposition by $\mathbf{X} = \mathbf{UDV'}$, where $\mathbf{U}$, $\mathbf{D}$, and $\mathbf{V}$ are defined in Section 9.3.

   (a) Show that $\mathbf{X'X} = \mathbf{V'\Lambda V}$ and $\mathbf{XX'} = \mathbf{U\Lambda U'}$, where $\Lambda = \mathbf{D}^2$ and relate $\mathbf{V}$, Λ and $\mathbf{U}$ to the principal components $\mathbf{Z}$ of $\mathbf{X'X}$ and $\mathbf{W}$ of $\mathbf{XX'}$.

   (b) Let $\hat{\mathbf{X}} = \mathbf{U_1 D_1 V'_1}$ where $\hat{\mathbf{X}}$ is based on the first two diagonal elements of $\mathbf{D}$ (denote by $\mathbf{D_1}$). Explain why a biplot that plots $\mathbf{U_1 D_1}$ and $\mathbf{V_1}$ simultaneously in two dimensions is called a principal components plot.

   (c) Assume that the columns of $\mathbf{X}$ are mean-corrected and denote the singular value decomposition of $\mathbf{X}$ by $\mathbf{GH}$ where $\mathbf{H} = \mathbf{DV'}/ (n-1)^{1/2}$ and $\mathbf{G} = (n-1)^{1/2}\mathbf{U}$. Show that $\mathbf{HH'} = \mathbf{VD^2V'} = \mathbf{S}$ where $\mathbf{S}$ is the covariance matrix for the variables in $\mathbf{X}$ and $\mathbf{GG'} = \mathbf{XS^{-1}X'}$.

   (d) Use the information in (c) to provide an interpretation for the columns of $\mathbf{H}$ and the rows of $\mathbf{G}$, and hence interpret the biplot that plots the first two columns of $\mathbf{G}$ and the first two rows of $\mathbf{H}$.

16. Show that the Pearson Chi-square statistic for testing independence in a two-dimensional contingency table can be written in the forms shown in equations (9.2) to (9.6) given in Section 9.4.1.

17. Review the theory of singular value decompositions given in the Appendix and show that the generalized singular value decomposition given by

$$\mathbf{O} = \mathbf{A^* D^*_\mu B^{*'}}, \quad \text{where} \quad \mathbf{A^{*'} D_r^{-1} A^*} = \mathbf{B^{*'} D_c^{-1} B^*} = \mathbf{I},$$

can be expressed as a singular value decomposition given by

$$\mathbf{F} = \mathbf{D_r^{-1/2} O D_c^{-1/2}} = \mathbf{L D^*_\mu M}, \quad \text{where} \quad \mathbf{L'L} = \mathbf{M'M} = \mathbf{I}$$

by defining $\mathbf{L}$ and $\mathbf{M}$ appropriately.

18. Given the singular value decomposition of $\mathbf{F} = \mathbf{D_r^{-1/2} O D_c^{-1/2}} = \mathbf{L D^*_\mu M}$ where $\mathbf{L'L} = \mathbf{M'M} = \mathbf{I}$, review the theory of matrix decompositions in the Appendix and show that spectral decompositions of $\mathbf{FF'}$ and $\mathbf{F'F}$ are given by

$$\mathbf{FF'} = \mathbf{L D^{*2}_\mu L'} \quad \text{and} \quad \mathbf{F'F} = \mathbf{M' D^{*2}_\mu M}.$$

In each case give the eigenvalues and eigenvectors. What is the relationship between $\mathbf{L}$ and $\mathbf{M}$?

19. Given that the generalized singular value decomposition of $(O - rc')$ is given by $AD_\mu B'$ where $A'D_r^{-1}A = B'D_c^{-1}B = I$ and that the coordinates of row and column profiles are given by $F = D_r^{-1}(O - rc')D_c^{-1}B$ and $G = D_c^{-1}(O - rc')'D_r^{-1}A$ show that $F = D_r^{-1}AD_\mu$, $G = D_c^{-1}BD_\mu$ and $GD_\mu = D_c^{-1}O'F$, $FD_\mu = D_r^{-1}OG$. HINT: Show that $r'D_r^{-1}(O - rc') = 0'$ and $(O - rc')D_c^{-1}c = 0$.

20. Given the definitions of SSA and SST in Section 9.4.3 derive the expressions

$$
\begin{aligned}
SSA &= nz'O'D_r^{-1}Oz - nz'cc'z \\
SST &= nz'D_cz - nz'cc'z.
\end{aligned}
$$

21. Define a Lagrangian expression to maximize the quantity $z'O'D_r^{-1}Oz$ subject to $z'D_cz = 1$, and show that finding the maximum involves solving the eigenvalue problem given by $(D_c^{-1}O'D_r^{-1}O - \eta^2 I)z = 0$. Show also that η^2 is the required maximum. [HINT: η^2 is the required Lagrange multiplier].

22. Assume you have observations on three categorical variables in the form of a $2 \times 2 \times 2$ three-dimensional contingency table. Construct a Z matrix of indicator variables to be used in a multiple correspondence analysis. Construct the Z^* and Burt matrices as outlined in Section 9.4.5, and hence confirm that the nonzero observations in the rows of Z^* are cell frequencies and that the Burt matrix consists of block diagonal elements of frequencies and off-diagonal blocks which are two-dimensional contingency tables.

23. Assume that correspondence analysis is to be applied to a three-dimensional contingency table after constructing a single dimension from the cross classification of two of the original three dimensions. Explain why a correspondence analysis in this case represents the study of the residuals from a partial independence model introduced in Chapter 6.

10

Cluster Analysis and Multidimensional Scaling

This chapter continues the discussion of data reduction techniques begun in Chapter 9. In Chapter 9 the focus was on reducing the number of variables or columns of the data matrix $\mathbf{X}$. Chapter 10 begins by focusing on the reduction of the number of rows of $\mathbf{X}$. Since the rows of $\mathbf{X}$ represent observational units, the approach is to combine the units into groups of relatively homogeneous units called clusters. For this approach to data reduction, the various techniques available are commonly called cluster analysis.

In principal components analysis, the starting point is a sum of squares and cross products matrix of the form $\mathbf{X'X}$, which is usually a covariance matrix or correlation matrix. These matrices are used to measure the degree of similarity in variation among pairs of variables over the observational units. Variables that are similar tend to have relatively large off-diagonal elements in $\mathbf{X'X}$ whereas variables that are not related tend to have relatively small off-diagonal elements.

In cluster analysis, the starting point is a proximity matrix that measures the similarity of the observational units over the variables. The proximity matrix can often be obtained from $\mathbf{X}$ using the matrix $\mathbf{XX'}$, which is a sum of squares and cross products matrix for observational units. Off-diagonal elements of $\mathbf{XX'}$ that are relatively large correspond to objects that are similar. As in the case of principal components analysis, the form of $\mathbf{XX'}$ can vary depending on whether the X variables have been mean-centered or standardized.

In some situations, it is difficult to obtain precise measurements in the form of a data matrix $\mathbf{X}$ however, it is possible to obtain a proximity matrix that provides information about the degree of similarity among the observational units. Under the assumption that the measures of proximity are ordinal measures and that they actually represent some underlying interval scale measures, it is possible to determine interval dimensions that are consistent with the given proximities. Such techniques are commonly referred to as multidimensional scaling. Multidimensional scaling is used to determine a low-dimensional graphical representation of the relationships among the objects given by the proximity matrix. In multidimensional scaling the given matrix of proximities is used to generate a matrix of the form $\mathbf{XX'}$. This sum of squares and cross products matrix is then used to generate the underlying dimensions.

This chapter begins with an outline of how proximity matrices can be derived from data matrices. The section on proximity matrices also discusses the measurement of proximity between groups of observational units. Section 10.2 then outlines a variety of approaches to cluster analysis. The discussion of cluster analysis uses the various measures of proximity outlined in Section 10.1. Multidimensional scaling is presented in Section 10.3.

10.1   Proximity Matrices Derived from Data Matrices

The multivariate data matrix $\mathbf{X}$ $(n \times p)$, consists of observations obtained from the measurement of n subjects or objects with respect to p features or characteristics. The p columns of $\mathbf{X}$ are usually referred to as variables whereas the n rows are commonly called the profiles or patterns of the observational units. The term *profile* has been used previously in this text to refer to row or column densities for contingency tables (Chapters 6 and 9) and to refer to the components of a multivariate mean vector (Chapters 7 and 8). A profile is simply a vector of measurements whose elements are to be compared. In this chapter the profiles are the n $(1 \times p)$ vectors that constitute $\mathbf{X}$.

A *proximity matrix* is an $(n \times n)$ matrix that summarizes the degree of *similarity* or *dissimilarity* among all possible pairs of profiles in $\mathbf{X}$. This matrix is denoted by $\mathbf{P}$ with elements p_{rs}, $r, s = 1, 2, \ldots, n$. The element p_{rs} denotes the *proximity measure* between observational units r and s. The matrix $\mathbf{XX'}$ is an example of a proximity matrix. A variety of measures of proximity are introduced in this section. Proximity measures are also introduced for relating two groups of observational units in Section 10.1.2. The proximity measures introduced are employed in Sections 10.2 and 10.3 in the study of cluster analysis and multidimensional scaling.

To provide examples throughout this section, the air pollution data introduced in Chapter 9 are used. Table 10.1 shows the observations for 10 of the 40 cities introduced in Table 9.7. Table 10.1 also contains the standardized data obtained by subtracting the variable means and dividing by the variable standard deviations. Because the 11 variables use a variety of measurement scales it is sometimes preferable to standardize the variables to obtain a common scale. In the next section we will be concerned with the measurement of proximity among these 10 cities.

TABLE 10.1. Air Pollution Data for Ten Cities*

City	SMEAN	PMEAN	SMAX	PMAX	SMIN	PMIN	PM2	PERWH	NON-POOR	GE65	LPOP
Albuquerque	50	244	91	646	2	61	22.5	96.7	84.7	47	54.2
	-0.92	1.98	-1.08	2.07	-1.20	0.71	-0.86	0.76	0.31	-1.21	-0.69
Atlanta	91	112	139	236	46	46	59.0	77.2	79.2	65	60.1
	0.17	-0.43	-0.56	-0.41	0.30	-0.07	-0.03	-0.66	-0.38	-0.37	0.44
Columbus	60	99	145	160	34	28	19.6	70.7	69.0	49	53.4
	-0.66	-0.67	-0.49	-0.87	-0.11	-1.02	-0.93	-1.13	-1.64	-1.11	-0.84
Los Angeles	123	156	280	344	25	50	139.3	91.2	87.6	89	68.3
	1.03	0.37	0.98	0.24	-0.42	0.14	1.80	0.36	0.67	0.73	2.00
Minneapolis	69	100	202	231	23	43	70.2	98.2	89.9	92	61.7
	-0.41	-0.65	0.13	-0.44	-0.49	-0.23	0.23	0.87	0.96	0.87	0.75
Montgomery	42	72	70	157	30	26	21.4	61.7	68.9	71	52.3
	-1.14	-1.16	-1.31	-0.89	-0.25	-1.12	-0.89	-1.78	-1.66	-0.10	-1.05
Salt Lake	79	121	260	309	27	47	50.1	98.6	88.2	70	55.8
	-0.15	-0.27	0.76	0.03	-0.35	-0.02	-0.23	0.89	0.74	-0.14	-0.37
Scranton	153	215	365	537	112	75	51.7	99.6	78.3	119	53.7
	1.83	1.45	1.91	1.41	2.53	1.45	-0.20	0.97	-0.49	2.12	- 0.78
Washington	124	135	210	242	51	76	134.8	75.1	89.5	62	63.0
	1.06	-0.01	0.22	-0.37	0.47	1.50	1.70	-0.81	0.91	-0.51	1.00
Wichita	54	102	139	174	23	22	34.4	93.6	86.9	67	55.4
	-0.82	-0.61	-0.56	-0.78	-0.49	-1.33	-0.59	0.53	0.58	- 0.28	-0.46

*First row – original data; second row – standardized data.

10.1.1   THE MEASUREMENT OF PROXIMITY BETWEEN OBJECTS

Proximity measures usually reflect the degree of similarity or the degree of dissimilarity. As two objects become more similar, the value of a *similarity measure* increases whereas the corresponding *dissimilarity measure* declines in value. An example of a similarity measure between two objects is a correlation coefficient between the objects based on the p measurements. An example of a dissimilarity measure based on the same p observations is the Euclidean distance between the two objects. The two types of proximity measures are defined more generally by the properties summarized below.

Similarity

Given two objects r and s, the proximity measure p_{rs} is a measure of similarity if p_{rs} satisfies the following:

1. $0 \leq p_{rs} \leq 1$ for all objects r, s;

2. $p_{rs} = 1$ if and only if r and s are identical;

3. $p_{rs} = p_{sr}$.

The most common measure of similarity is the Pearson correlation coefficient. Since a correlation coefficient has the range $[-1, 1]$, it is customary to use either the absolute value of the coefficient or to add 1.0 to the value of the coefficient and then divide the result by 2. In either case the revised coefficient lies in the required range. Similarity proximity measures are sometimes called *Q-type* or *correlation-type* measures.

Dissimilarity

A proximity measure p_{rs} is a measure of dissimilarity if p_{rs} satisfies the following:

1. $p_{rs} \geq 0$ for all objects r, s;

2. $p_{rs} = 0$ if objects r and s are identical;

3. $p_{rs} = p_{sr}$.

The most commonly used measure of dissimilarity is the Euclidean distance. An alternative measure of dissimilarity is the Mahalanobis distance between two observations introduced in Chapter 7. Dissimilarity measures are commonly referred to as *distance-type* measures.

An outline of dissimilarity measures is provided next, followed by a discussion of similarity measures. The relationship between correlation-type measures and distance-type measures is discussed in connection with profiles. The section ends with a discussion of proximity measures for categorical data.

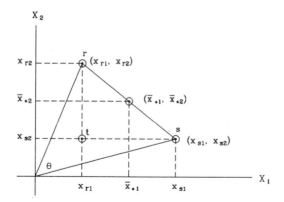

FIGURE 10.1. Two-Dimensional Representation of Proximity Between Two Points

Euclidean Distance

The rth and sth rows of the data matrix $\mathbf{X}$ will be denoted by $(x_{r1}, x_{r2}, \ldots, x_{rp})$ and $(x_{s1}, x_{s2}, \ldots, x_{sp})$ respectively. These two rows correspond to the observations on two objects for all p variables. Geometrically, the two profiles can be viewed as the coordinates of two points in a p-dimensional space. A convenient measure of dissimilarity between the objects r and s can be obtained from the *Euclidean distance* between the two points. This distance is denoted by d_{rs} where

$$d_{rs}^2 = \sum_{j=1}^{p} (x_{rj} - x_{sj})^2.$$

The quantity d_{rs}^2 will be referred to as the *squared Euclidean distance*.

For two $(p = 2)$ dimensions the distance between objects r and s can be represented as shown in Figure 10.1. In this case the square of the Euclidean distance is given by

$$d_{rs}^2 = (x_{r1} - x_{s1})^2 + (x_{r2} - x_{s2})^2.$$

The quantities can be related to the sides of a triangle formed by the points r, s and t in Figure 10.1.

Using Mean-Centered Variables

It can be seen from the expression for d_{rs}^2 that the Euclidean distance between r and s would be unaffected if the variables x_1 and x_2 were each *mean-corrected* or *mean-centered*. The mean $(x_{r1} + x_{s1})/2$ would be subtracted from both x_{r1} and x_{s1} and similarly the mean $(x_{r2} + x_{s2})/2$ would

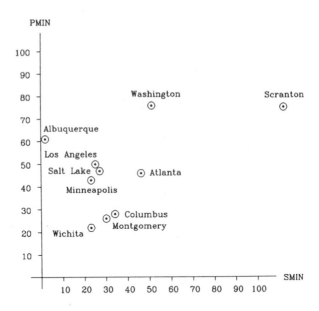

FIGURE 10.2. Scatterplot for Ten Cities

be subtracted from x_{r2} and x_{s2}. The resulting value of d_{rs}^2 therefore would remain unchanged. This result extends to the case of p variables in a similar fashion.

Example

A scatterplot is used in Figure 10.2 to show the variation among the ten cities of Table 10.1 with respect to the two pollution measures SMIN and PMIN. With respect to these two measures we can see from the figure that Washington and Scranton are furthest away from the remaining eight cities. The city of Scranton is quite far away in that it is twice as far from the group in comparison to Washington. The cities of Los Angeles, Salt Lake and Minneapolis are quite close to each other as are the cities of Wichita, Columbus and Montgomery.

To study the relationships among the ten cities with respect to all 11 dimensions, a proximity matrix of squared Euclidean distances can be used. Table 10.2 shows this Euclidean distance proximity matrix based on the measurements given in Table 10.1. From the table it would appear that Albuquerque is quite distant from all other nine cities. Scranton also tends to be distant from the other nine cities. The two closest cities are Columbus and Wichita. Another similar pair is Los Angeles and Salt Lake. Excluding Albuquerque and Scranton the remaining seven cities all seem to be close to Minneapolis.

TABLE 10.2. Squared Euclidean Distances

	Albuquerque	Atlanta	Columbus	Los Angeles	Minneapolis	Montgomery	Salt Lake	Scranton	Washington	Wichita
Albuquerque	0	193,771	263,285	142,932	210,794	273,274	160,230	116,789	198,304	247,787
Atlanta	193,771	0	9,409	36,260	6,573	17,516	21,218	164,839	8,334	7,378
Columbus	263,285	9,409	0	62,840	14,390	7,276	39,205	227,779	19,866	1,826
Los Angeles	142,932	36,260	62,840	0	26,019	95,407	5,487	58,472	18,138	57,928
Minneapolis	210,794	6,573	14,390	26,019	0	29,436	10,947	150,703	8,997	9,865
Montgomery	273,274	17,516	7,276	95,407	29,436	0	65,995	278,069	41,439	7,695
Salt Lake	160,230	21,218	39,205	5,487	10,947	65,995	0	87,837	11,531	34,775
Scranton	116,789	164,839	227,779	58,472	150,703	278,069	87,837	0	126,360	219,261
Washington	198,304	8,334	19,866	18,138	8,997	41,439	11,531	126,360	0	19,787
Wichita	247,787	7,378	1,826	57,928	9,865	7,695	34,775	219,261	19,787	0

Euclidean Distance in Matrix Form

Given an $(n \times p)$ data matrix $\mathbf{X}$ with $(1 \times p)$ row vectors $\mathbf{x}'_1, \mathbf{x}'_2, \ldots, \mathbf{x}'_n$, the square of the Euclidean distance, d^2_{rs}, between objects r and s can be written as

$$d^2_{rs} = (\mathbf{x}_r - \mathbf{x}_s)'(\mathbf{x}_r - \mathbf{x}_s), \qquad r, s = 1, 2, \ldots, n.$$

The $(n \times n)$ matrix of d^2_{rs} values is often called a *squared Euclidean distance matrix*.

If the p X variables are mean-centered the data matrix is denoted by $\mathbf{X}^*$ as defined in Section 7.1.1 of Chapter 7. Since removing the means from the p variables does not affect the distance between two points, the squared Euclidean distance between objects r and s is given by

$$d^2_{rs} = (\mathbf{x}^*_r - \mathbf{x}^*_s)'(\mathbf{x}^*_r - \mathbf{x}^*_s), \qquad r, s = 1, 2, \ldots, n,$$

where $\mathbf{x}^{*\prime}_r$ and $\mathbf{x}^{*\prime}_s$ are the corresponding rows of $\mathbf{X}^*$.

Standardized Euclidean Distance

A disadvantage of the Euclidean distance as a measure of proximity is its sensitivity to the scales of measurement. It is possible for one or a few variables to dominate the distance measure because of large differences in scale. In general, if the scales of measurement are not common for all p variables, it is preferable to use a weighted distance given by $\sum_{j=1}^{p} w_j (x_{rj} - x_{sj})^2$ where the weights w_j reflect the importance of the variables $j = 1, 2, \ldots, p$.

A special case of the *weighted Euclidean distance* is the *standardized Euclidean distance*. The standardized Euclidean distance is given by

$$d^2_{rs} = \sum_{j=1}^{p} \left(\frac{1}{s^2_j}\right)(x_{rj} - x_{sj})^2 = (\mathbf{x}_r - \mathbf{x}_s)'\mathbf{D}^{-1}(\mathbf{x}_r - \mathbf{x}_s),$$

where s^2_j, $j = 1, 2 \ldots, p$ denotes the variance of the variable X_j over the n objects and $\mathbf{D}$ is the diagonal matrix with diagonal elements given by s^2_j, $j = 1, 2, \ldots, p$. The $(p \times 1)$ vectors $\mathbf{x}_r$ and $\mathbf{x}_s$ denote the observations on the two profiles.

Computer Software

The calculations in this section were performed using the SPSSX program PROXIMITIES.

Example

The proximity matrix of squared Euclidean distances based on the standardized observations in Table 10.1 is shown in Table 10.3. In this case the two closest cities are Columbus and Montgomery. The cities of Salt Lake, Wichita and Minneapolis are quite close to each other and similarly Atlanta, Washington and Los Angeles are relatively close. As observed previously, Albuquerque and Scranton are distant from each other and from the remaining eight cities.

Mahalanobis Distance and Multivariate Distance

An extension of the system of weights that also takes into account the covariances among the variables is given by the *Mahalanobis distance* $(\mathbf{x}_r - \mathbf{x}_s)'\mathbf{S}^{-1}(\mathbf{x}_r - \mathbf{x}_s)$ introduced in Section 7.1.1. This distance is called a multivariate measure of distance since it takes into account the covariance structure among the p variables. If the original variables are first transformed to principal components before computing the Euclidean distance, then the Euclidean distances based on all the principal components are equivalent to the Mahalanobis distances. The Mahalanobis distance may not be useful as a measure of distance in some applications because it removes the correlation effects. These correlation effects may be important elements in distinguishing between objects and hence should not be removed.

Euclidean Distance and the Centroid

The Euclidean distance between two profiles, d_{rs}, can be related to the *centroid* between the two objects. Denoting the mean on variable j by $\bar{x}_{\cdot j}$, $[\bar{x}_{\cdot j} = (x_{rj} + x_{sj})/2]$, the centroid is given by $(\bar{x}_{\cdot 1}, \bar{x}_{\cdot 2}, \ldots, \bar{x}_{\cdot p})$. The squared Euclidean distance can be written as

$$d_{rs}^2 = 2\Big[\sum_{j=1}^{p}(x_{rj} - \bar{x}_{\cdot j})^2 + \sum_{j=1}^{p}(x_{sj} - \bar{x}_{\cdot j})^2\Big]. \tag{10.1}$$

Each of the two parts on the right-hand side of (10.1) represents the square of a distance between a profile and the centroid of the two profiles. The sum represents the sum of squared deviations of the two profiles from their centroid. Thus the *variation of the two profiles around their centroid* is proportional to the square of the distance between them. This *variation between the two profiles* can also be characterized as the *variation within the group* formed by joining the two profiles. This characterization will be used in the discussion of cluster analysis in Section 10.2 For the two-variable case Figure 10.1 shows the centroid as the midpoint of the line joining points r and s.

Manhattan or City Block Metric

An alternative distance-type metric is the *Manhattan* or *city-block metric*, which is based on the absolute values of the differences among the coordi-

TABLE 10.3. Squared Euclidean Distances Using Standardized Data

	Albuquerque	Atlanta	Columbus	Los Angeles	Minneapolis	Montgomery	Salt Lake	Scranton	Washington	Wichita
Albuquerque	0.0	33.5	44.2	30.2	38.6	55.6	22.3	34.3	32.4	32.6
Atlanta	33.5	0.0	10.9	12.9	10.0	14.5	9.0	35.8	7.7	9.1
Columbus	44.2	10.9	0.0	35.9	32.6	2.8	23.4	57.4	25.0	14.4
Los Angeles	30.2	12.9	35.9	0.0	8.7	43.1	10.7	28.6	9.4	17.8
Minneapolis	38.6	10.0	32.6	8.7	0.0	36.8	4.8	35.1	15.1	8.5
Montgomery	55.6	14.5	2.8	43.1	36.8	0.0	30.9	64.6	31.0	19.2
Salt Lake	22.3	9.0	23.4	10.7	4.8	30.9	0.0	26.8	12.3	5.1
Scranton	34.3	35.8	57.4	28.6	35.1	64.6	26.8	0.0	35.7	45.1
Washington	32.4	7.7	25.0	9.4	15.1	31.0	12.3	35.7	0.0	17.4
Wichita	32.6	9.1	14.4	17.8	8.5	19.2	5.1	45.1	17.4	0.0

nates. This metric is given by

$$b_{rs} = \sum_{j=1}^{p} |x_{rj} - x_{sj}|.$$

For the two-dimensional case in Figure 10.1, the distance b_{rs} represents the sum of the distances from points r to t and from points s to t. With the city block metric, a constant difference between each of the p coordinates in the amount a has the same effect on total distance as changing the difference between one set of coordinates by the amount pa. This is not true for the Euclidean distance metric where the distance in the second case would be larger than in the first case. The city-block metric therefore is much less sensitive to outliers.

Minkowski Metrics

The Euclidean distance and city-block metrics are special cases of the *Minkowski metric* which is given by

$$m_{rs} = \left[\sum_{j=1}^{p} |x_{rj} - x_{rs}|^{\lambda} \right]^{1/\lambda}.$$

The Euclidean and city-block distances correspond to $\lambda = 2$ and $\lambda = 1$ respectively. In general the larger the value of λ the greater the emphasis given to differences in coordinates on a given variable. In addition to the three properties of a dissimilarity measure given above, all Minkowski metrics also satisfy the following:

4. $p_{rs} = 0$ only if $\mathbf{x}_r = \mathbf{x}_s$;

5. $p_{rs} \leq p_{rm} + p_{ms}$ for all points r, s and m.

Distance Measures Averaged over Variables

Some users of distance measures choose to divide the distance measure by the number of variables in the expression. For example the squared Euclidean distance and Euclidean distance are sometimes given by

$$\frac{1}{p} \sum_{j=1}^{p} (x_{rj} - x_{sj})^2$$

and

$$\frac{1}{p} \left[\sum_{j=1}^{p} (x_{rj} - x_{sj})^2 \right]^{1/2},$$

respectively. This modification does not affect the results when measuring proximity since all such measures are divided by the same amount. Some software packages use these versions of the distance and squared distance measures.

Correlation Type Measures of Similarity

An alternative approach, to the measurement of proximity between two points r and s in a p-dimensional space, is to use the angle between the two $(p \times 1)$ vectors of observations $\mathbf{x}_r$ and $\mathbf{x}_s$. In Figure 10.1 the profiles of objects r and s are shown as points in two-dimensional space. The two points can be viewed as tips of vectors drawn from the origin with an angle θ between the two vectors. A useful measure of similarity is the cosine of the angle θ.

In general the cosine of the angle between the vectors $\mathbf{x}_r$ and $\mathbf{x}_s$ is given by

$$c_{rs} = \sum_{j=1}^{p} x_{rj} x_{sj} \Big/ \sqrt{\sum_{j=1}^{p} x_{rj}^2 \sum_{j=1}^{p} x_{sj}^2}.$$

It can be seen from Figure 10.1 that c_{rs} does not depend on the lengths of the two vectors, and hence proportional changes in the coordinates $\mathbf{x}_r$ and/or $\mathbf{x}_s$ will not change c_{rs}. The quantities $\sum_{j=1}^{p} x_{rj}^2$ and $\sum_{j=1}^{p} x_{sj}^2$ are the squared lengths of the vectors. This *cosine coefficient* is also sometimes referred to as the *congruency coefficient*.

The profiles $\mathbf{x}_r$ and $\mathbf{x}_s$ can be mean-centered to yield $(\mathbf{x}_r - \bar{x}_{r\cdot}\mathbf{e})$ and $(\mathbf{x}_s - \bar{x}_{s\cdot}\mathbf{e})$, where $\mathbf{e}$ $(p \times 1)$ is a vector of unities and $\bar{x}_{r\cdot} = \sum_{j=1}^{p} x_{rj}/p$ and $\bar{x}_{s\cdot} = \sum_{j=1}^{p} x_{sj}/p$ are the means for profiles r and s respectively. The cosine of the angle between the mean-centered vectors is equivalent to the Pearson correlation between the vectors $\mathbf{x}_r$ and $\mathbf{x}_s$. The resulting *correlation coefficient* similarity measure is given by

$$
\begin{aligned}
q_{rs} &= \frac{(\mathbf{x}_r - \bar{x}_{r\cdot}\mathbf{e})'(\mathbf{x}_s - \bar{x}_{s\cdot}\mathbf{e})}{\sqrt{(\mathbf{x}_r - \bar{x}_{r\cdot}\mathbf{e})'(\mathbf{x}_r - \bar{x}_{r\cdot}\mathbf{e})(\mathbf{x}_s - \bar{x}_{s\cdot}\mathbf{e})'(\mathbf{x}_s - \bar{x}_{s\cdot}\mathbf{e})}} \\
&= \sum_{j=1}^{p}(x_{rj} - \bar{x}_{r\cdot})(x_{sj} - \bar{x}_{s\cdot}) \Big/ \sqrt{\sum_{j=1}^{p}(x_{rj} - \bar{x}_{r\cdot})^2 \sum(x_{sj} - \bar{x}_{s\cdot})^2}.
\end{aligned}
$$

The measures c_{rs} and q_{rs} are often called *Q-type* measures of similarity. A disadvantage of the Q-type measure of similarity is its sensitivity to the direction of the scale of measurement for each of the variables. If some items are measured in the positive direction (i.e., honest, sociable) and others are measured in the negative direction (i.e., tardy, sloppy) the number of positive and negative type measures will affect the overall profile means and hence the value of q_{rs}. In such cases some scales could be reversed so that the directions are all the same.

As discussed earlier, since correlation type measures have the range $[-1, 1]$, it is customary to use either the absolute value of the coefficient or to add 1.0 to the coefficient and divide the resulting sum by 2.0.

Example

For the ten-city data of Table 10.1 the cosine and correlation measures of similarity are shown in Tables 10.4 and 10.5 respectively. The cosine measure of proximity emphasizes the similarity in pattern of variation over the variables rather than the absolute level. From Table 10.4 we can conclude that Albuquerque continues to show a weak similarity with the other nine cities. The proximities among the other nine cities however show cosine values close to 1. In comparison to the Euclidean distances in Tables 10.2 and 10.3, Scranton now shows a strong similarity with the other eight cities. The cosine of the angle between Los Angeles and Scranton is very close to 1.

The correlation proximity matrix is given in Table 10.5. For this matrix Albuquerque shows a high correlation with Atlanta. Thus after correcting for the differences in the overall levels these two profiles exhibit similar patterns of variation over the eleven variables. The strongest correlation in the matrix is between Los Angeles and Salt Lake (0.99), whereas the weakest correlations are between Albuquerque and each of the cities Columbus, Minneapolis and Wichita. In addition to Albuquerque, the city of Montgomery also appears to be somewhat different from the other eight cities when correlation is used to measure similarity.

Similarity Matrices

The similarity measures derived from Q-type proximities can be summarized in an $(n \times n)$ sum of squares and cross-products matrix for objects rather than variables. For the data matrix $\mathbf{X}$ the matrix $\mathbf{XX'}$ $(n \times n)$ is a raw sum of squares and cross-products matrix for the n objects. For the cosine coefficient, the similarity matrix is derived from the matrix $\mathbf{X}^+$ where $\mathbf{X}^+$ is the standardized $\mathbf{X}$ matrix given below.

$$\mathbf{X}^+ = \begin{bmatrix} \dfrac{x_{11}}{\left(\sum\limits_{j=1}^{p} x_{1j}^2\right)^{1/2}} & \dfrac{x_{12}}{\left(\sum\limits_{j=1}^{p} x_{1j}^2\right)^{1/2}} & \cdots & \dfrac{x_{1p}}{\left(\sum\limits_{j=1}^{p} x_{1j}^2\right)^{1/2}} \\[3em] \dfrac{x_{21}}{\left(\sum\limits_{j=1}^{p} x_{2j}^2\right)^{1/2}} & \dfrac{x_{22}}{\left(\sum\limits_{j=1}^{p} x_{2j}^2\right)^{1/2}} & \cdots & \dfrac{x_{2p}}{\left(\sum\limits_{j=1}^{p} x_{2j}^2\right)^{1/2}} \\[3em] \vdots & \vdots & \cdots & \vdots \\[1em] \dfrac{x_{n1}}{\left(\sum\limits_{j=1}^{p} x_{nj}^2\right)^{1/2}} & \dfrac{x_{n2}}{\left(\sum\limits_{j=1}^{p} x_{nj}^2\right)^{1/2}} & \cdots & \dfrac{x_{np}}{\left(\sum\limits_{j=1}^{p} x_{nj}^2\right)^{1/2}} \end{bmatrix}.$$

TABLE 10.4. Cosine Measures of Similarity

	Albuquerque	Atlanta	Columbus	Los Angeles	Minneapolis	Montgomery	Salt Lake	Scranton	Washington	Wichita
Albuquerque	1.00	0.88	0.82	0.86	0.82	0.88	0.86	0.89	0.82	0.83
Atlanta	0.88	1.00	0.98	0.98	0.98	0.93	0.97	0.97	0.98	0.98
Columbus	0.82	0.98	1.00	0.98	0.98	0.96	0.98	0.96	0.99	0.99
Los Angeles	0.86	0.98	0.98	1.00	0.98	0.94	0.99	0.99	0.99	0.97
Minneapolis	0.82	0.98	0.98	0.98	1.00	0.96	0.99	0.96	0.97	0.99
Montgomery	0.88	0.93	0.96	0.94	0.96	1.00	0.94	0.93	0.94	0.97
Salt Lake	0.86	0.97	0.98	0.99	0.99	0.94	1.00	0.98	0.98	0.97
Scranton	0.89	0.97	0.96	0.99	0.96	0.93	0.98	1.00	0.97	0.94
Washington	0.82	0.98	0.99	0.99	0.97	0.94	0.98	0.97	1.00	0.97
Wichita	0.83	0.98	0.99	0.97	0.99	0.97	0.97	0.94	0.97	1.00

TABLE 10.5. Correlation Measures of Similarity

	Albuquerque	Atlanta	Columbus	Los Angeles	Minneapolis	Montgomery	Salt Lake	Scranton	Washington	Wichita
Albuquerque	1.00	0.92	0.76	0.80	0.75	0.91	0.78	0.84	0.76	0.78
Atlanta	0.92	1.00	0.92	0.96	0.92	0.91	0.95	0.96	0.93	0.92
Columbus	0.76	0.92	1.00	0.97	0.94	0.84	0.97	0.92	0.96	0.97
Los Angeles	0.80	0.96	0.97	1.00	0.96	0.84	0.99	0.97	0.98	0.93
Minneapolis	0.75	0.92	0.94	0.96	1.00	0.84	0.98	0.91	0.90	0.95
Montgomery	0.91	0.91	0.84	0.84	0.84	1.00	0.84	0.84	0.78	0.90
Salt Lake	0.78	0.95	0.97	0.99	0.98	0.84	1.00	0.96	0.96	0.94
Scranton	0.84	0.96	0.92	0.97	0.91	0.84	0.96	1.00	0.95	0.87
Washington	0.76	0.93	0.96	0.98	0.90	0.78	0.96	0.95	1.00	0.88
Wichita	0.78	0.92	0.97	0.93	0.95	0.90	0.94	0.87	0.88	1.00

For the correlation coefficient q_{rs}, the data matrix is given by $\mathbf{X}^{++}$, which contains mean-centered and standardized measurements as shown below.

$$\mathbf{X}^{++} = \begin{bmatrix} \dfrac{(x_{11} - \bar{x}_{1\cdot})}{s_1} & \dfrac{(x_{12} - \bar{x}_{1\cdot})}{s_1} & \cdots & \dfrac{(x_{1p} - \bar{x}_{1\cdot})}{s_1} \\[2mm] \dfrac{(x_{21} - \bar{x}_{2\cdot})}{s_2} & \dfrac{(x_{22} - \bar{x}_{2\cdot})}{s_2} & \cdots & \dfrac{(x_{2p} - \bar{x}_{2\cdot})}{s_2} \\[2mm] \vdots & \vdots & & \vdots \\[2mm] \dfrac{(x_{n1} - \bar{x}_{n\cdot})}{s_n} & \dfrac{(x_{n2} - \bar{x}_{n\cdot})}{s_n} & \cdots & \dfrac{(x_{np} - \bar{x}_{n\cdot})}{s_n} \end{bmatrix}.$$

The quantities $\bar{x}_r$ and s_r are given by $\bar{x}_{r\cdot} = \sum_{j=1}^{p} x_{rj}/p$ and $s_r = \sum_{j=1}^{p}(x_{rj} -\bar{x}_{r\cdot})^2/p$, $r = 1, 2, \ldots, n$.

A third form of $\mathbf{X}$ matrix involves *mean-centered profiles* that have not been standardized. In this case the $\mathbf{X}\mathbf{X}'$ matrix is proportional to a covariance matrix among the profiles.

Double Mean-Centered

If the observations are first of all mean-centered by variables and then mean-centered by row, the resulting observations are given by $(x_{ij} - \bar{x}_{i\cdot} - \bar{x}_{\cdot j} + \bar{x}_{\cdot\cdot})$ where

$$\bar{x}_{i\cdot} = \sum_{j=1}^{p} x_{ij}/p,$$

$$\bar{x}_{\cdot j} = \sum_{i=1}^{n} x_{ij}/n,$$

and

$$\bar{x}_{\cdot\cdot} = \sum_{i=1}^{n}\sum_{j=1}^{p} x_{ij}/np.$$

In this case the elements of the revised data matrix have no row effects and no column effects and are said to be *double mean-centered*. This transformation will be used in multidimensional scaling.

Profile Shape, Scatter and Level

When the variables all have the same scale of measurement (or are standardized) an alternative geometric representation that can be used to compare two or more profiles is the *profile plot*. An example of two profile plots is shown in Figure 10.3. The p observations corresponding to a particular profile are plotted in order of the columns of $\mathbf{X}$. The variation among the observations and the level or magnitude of the observations can be quickly observed from the profile plot. In addition the profiles corresponding to two or more objects can be compared using a profile plot.

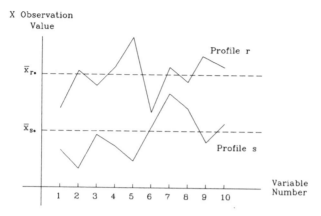

FIGURE 10.3. Profile Comparison

In Figure 10.3, the squared Euclidean distance d_{rs}^2 between objects r and s can be viewed as the sum of squared distances between the two values shown for each variable. When comparing the two profiles it is possible to compare the overall *levels*, the *scatter* and also the *shape*.

The level of the rth profile is given by the mean of the p observations, $\bar{x}_{r\cdot} = \sum_{j=1}^{p} x_{rj}/p$. The scatter of the rth profile is given by $v_r^2 = \sum_{j=1}^{p}(x_{rj} - \bar{x}_{r\cdot})^2$, which represents the variation of the profile observations around the mean level $\bar{x}_{r\cdot}$. In Figure 10.3 the profile levels $\bar{x}_{r\cdot}$ and $\bar{x}_{s\cdot}$ are shown for the two profiles. The two profile scatters v_r^2 and v_s^2 measure the variation of the profiles around the levels $\bar{x}_{r\cdot}$ and $\bar{x}_{s\cdot}$ respectively.

The shape of a profile is measured relative to a second profile by computing the Pearson correlation coefficient q_{rs} between the two profiles. The closer that q_{rs} is to 1, the greater the tendency for the two profiles to display the same shape or pattern.

The square of the Euclidean distance between the two profiles d_{rs}^2 can be expressed in terms of the measures of level, scatter and shape. The relationship is given by

$$d_{rs}^2 = (v_r - v_s)^2 + p(\bar{x}_{r\cdot} - \bar{x}_{s\cdot})^2 + 2v_r v_s(1 - q_{rs}). \qquad (10.2)$$

This expression shows that the Euclidean distance consists of three components reflecting differences in scatter, level and shape in that order. The component $(\bar{x}_{r\cdot} - \bar{x}_{s\cdot})^2$ reflects differences in level whereas the component $(v_r - v_s)^2$ reflects differences in scatter. Finally the component $(1 - q_{rs})$ reflects differences in shape.

In some applications the measurement of scatter and shape are combined and referred to as shape. In this context, profiles are compared with respect to level using $p(\bar{x}_{r\cdot} - \bar{x}_{s\cdot})^2$ and with respect to shape using $d_{rs}^2 - p(\bar{x}_{r\cdot} -$

$\bar{x}_{s\cdot})^2 = [v_r^2 + v_s^2 - 2v_r v_s q_{rs}]$. In this case, the shape coefficient represents the squared Euclidean distance between the mean-corrected profiles given by

$$d_{rs}^{*\,2} = \sum_{j=1}^{p} (x_{rj}^* - x_{sj}^*)^2,$$

where $x_{rj}^* = (x_{rj} - \bar{x}_{r\cdot})$ and $x_{sj}^* = (x_{sj} - \bar{x}_{s\cdot})$.

If the profiles are mean-centered the level component disappears. If in addition the profiles are standardized with respect to scatter the squared Euclidean distance becomes $d_{rs}^2 = 2(1 - q_{rs})$, which simply reflects the difference in shape. Consequently for *standardized profiles* there is a simple relationship between Euclidean distance and the correlation coefficient.

Example

The profiles for the ten cities with respect to the 11 variables are plotted in Figure 10.4. Since the variables should all have the same scale of measurement the standardized data from Table 10.1 has been used to produce the four plots. The first profile plot in panel (a) compares the cities of Albuquerque and Scranton. These two cities appear to be quite distinct with respect to SMEAN, SMAX, SMIN and GE65. Scranton has relatively high values for all six pollution variables, and Albuquerque has relatively high values for the three P variables (PMAX, PMEAN, PMIN) and relatively low values for the three S variables (SMAX, SMEAN, SMIN). The two cities are quite similar with respect to the demographic variables PM2, LPOP, PERWH and NONPOOR. In Scranton, the proportion of individuals over age 65 is relatively high whereas for Albuquerque this proportion is relatively low.

The second panel (b) compares the profiles for Columbus and Montgomery. The two profiles are quite similar and all observations are negative. The two cities appear to have relatively low values for all 11 variables.

The third panel (c) compares the profiles for Minneapolis, Salt Lake and Wichita and the fourth panel (d) compares the profiles for Los Angeles, Washington and Atlanta. In both cases the three cities in the plot appear to have a lot in common. From the plots in panel (c) Minneapolis appears to lie between Salt Lake and Wichita on the six pollution variables and for the demographic variables Salt Lake lies between Wichita and Minneapolis. In panel (d) the three profiles cross frequently. A comparison of panels (c) and (d) indicates that the three cities in panel (c) tend to have lower values than the three cities in panel (d). The squared Euclidean distances in Table 10.3 however indicate that the six cities have a lot in common. The largest differences appear to be between Wichita and each of Los Angeles and Washington.

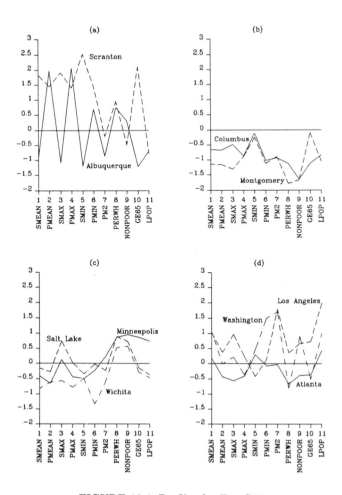

FIGURE 10.4. Profiles for Ten Cities

Some Relationships between Similarity and Euclidean Distance

The above discussion of profile plots has illustrated the relationship between the Euclidean distance d_{rs}^2 between two profiles and the differences in level, scatter and shape. It is also of interest to relate the Euclidean distance to the elements of the matrix $\mathbf{XX'}$. The quantity d_{rs}^2 is given by

$$d_{rs}^2 = \sum_{j=1}^{p}(x_{rj} - x_{sj})^2 = \sum_{j=1}^{p}x_{rj}^2 + \sum_{j=1}^{p}x_{sj}^2 - 2\sum_{j=1}^{p}x_{rj}x_{sj}$$

and hence the three components of d_{rs}^2 can be obtained from the elements of $\mathbf{XX'}$.

If the measurements for each profile are standardized so that $\sum_{j=1}^{p} x_{rj}^2 = \sum_{j=1}^{p} x_{sj}^2 = 1$, the Euclidean distance becomes $d_{rs}^2 = 2(1 - \sum_{j=1}^{p} x_{rj} x_{sj}) = 2(1 - c_{rs})$. If in addition the measurements for each profile are standardized so that $\sum_{j=1}^{p} x_{rj} = \sum_{j=1}^{p} x_{sj} = 0$, then $d_{rs}^2 = 2(1 - q_{rs})$.

The relationship between d_{rs}^2 and the cosine measure of similarity can also be obtained from the cosine law in trigonometry. The squared Euclidean distance d_{rs}^2 can be expressed as

$$d_{rs}^2 = d_r^2 + d_s^2 - 2d_r d_s \cos\theta,$$

where d_r^2 and d_s^2 denote the squared lengths of the vectors $\mathbf{x}_r$ and $\mathbf{x}_s$ given by $d_r^2 = \sum_{j=1}^{p} x_{rj}^2$, $d_s^2 = \sum_{j=1}^{p} x_{sj}^2$ and where $\cos\theta$ is the cosine coefficient $c_{rs} = \sum_{j=1}^{p} x_{rj} x_{sj} / d_r d_s$.

If $d_r^2 = d_s^2 = 1$ then $d_{rs}^2 = 2(1 - \cos\theta)$ and if $\sum_{j=1}^{p} x_{rj} = \sum_{j=1}^{p} x_{sj} = 0$, then $d_{rs}^2 = 2(1 - q_{rs})$ where q_{rs} is the correlation between the profiles.

From the above relationships between d_{rs}^2 and the correlation type measures c_{rs} and q_{rs} a Euclidean distance matrix can be obtained from a similarity matrix. Q-type similarity measures are often converted to Euclidean distance measures to employ computer algorithms for cluster analysis.

Proximity Measures for Categorical Data

Up to this point in the discussion of proximity measurement between rows in data matrices, it has been assumed that the values of the variables were measured on an interval scale. In this section the concept of proximity is extended to categorical variables. Assume that all p variables are categorical with the number of categories for each variable equal to $k_1, k_2, \ldots, k_p$ respectively. Assume that the columns of the $\mathbf{X}$ matrix are now composed of dummy variables representing the categories of the p variables. The total number of dummy variables required is $K = \sum_{j=1}^{p} k_j$ and hence for n observations the $\mathbf{X}$ matrix is now $(n \times K)$. Each row of the $\mathbf{X}$ matrix will contain precisely p unities and $(K - p)$ zeros. There will be one unity corresponding to each variable.

To measure the proximity between the rows of $\mathbf{X}$ the matrix $\mathbf{XX}'$ can be used. The diagonal elements of this matrix are all equal to p whereas the off-diagonal elements are equal to f_{rs}, where f_{rs} is the number of variables in which the two rows have the same categories (e.g., unities in common).

The cosine coefficient of similarity between rows r and s in this case is given by

$$c_{rs} = \sum_{j=1}^{K} x_{rj} x_{sj} \Big/ \sqrt{\sum_{j=1}^{K} x_{rj}^2 \sum_{j=1}^{K} x_{sj}^2}$$

$$= f_{rs}/p,$$

which is simply the proportion of the p variables in which the two rows matched.

The correlation coefficient between the rows r and s can also be obtained from $\mathbf{X}$. Since each row has the same sum (p) and the sum of squares (p) the means and variances of the rows are equal to p/K and $(p/K)(1-p/K)$ respectively. The correlation between rows r and s in this case is therefore given by

$$q_{rs} = \left[\frac{f_{rs}}{K} - \left(\frac{p}{K}\right)^2\right] \bigg/ \frac{p}{K}\left(1 - \frac{p}{K}\right).$$

Since the means and variances of the rows are all equal, the Euclidean distance d_{rs} between rows r and s is given by

$$d_{rs}^2 = 2(p - f_{rs}).$$

Example

Assume the data matrix $\mathbf{X}$ is constructed from observations made on three categorical variables and that the variables have 2, 3 and 4 levels respectively. Observations on the three categorical variables can be used to create a three-dimensional contingency table with dimensions $2 \times 3 \times 4$. The total number of observations n is distributed over the 24 cells of the contingency table. The $\mathbf{X}$ matrix consists of $2 + 3 + 4 = 9$ dummy variables with each row consisting of 3 unities and 6 zeroes. One of each of the 3 unities corresponds to each of the 3 variables. The n rows of $\mathbf{X}$ will consist of only 24 types depending on the location of the 3 unities in each row. A comparison of any 2 rows may yield 3, 2, 1 or 0 unities in common. If there are 3 unities in common the rows are identical. If there are 2 unities or 1 unity in common, then the rows agree with respect to two or one of the variables respectively. If there is no agreement between rows the two objects have different categories for all three variables. For the cosine measure of similarity the possible values are 1, 2/3, 1/3 or 0 since $f_{rs} = 3, 2, 1$ or 0 and $p = 3$. For the correlation coefficient measure the possible values are 1, 1/2, 0 and $-1/2$ corresponding to the values of $frs = 3, 2, 1,$ and 0 respectively. Finally for the Euclidean distance measure, the measures of dissimilarity are given by 0, 2, 4, and 6 corresponding to the values of $frs = 3, 2, 1$ and 0 respectively. The number of cases for each proximity measure will depend on the various cell frequencies.

Matching Coefficients for Binary Variables

In the special case that the p categorical variables each contain only two categories (binary), an alternative approach can be used to measure proximity between subjects. Each variable for each subject is coded either 0 or 1 to indicate which of the two attributes is present. A (2×2) table can be used to describe the proximity in terms of the four possible categories for each of p variables as shown in Figure 10.5.

In Figure 10.5, a represents the number of variables in which both subjects were coded 0, and d represents the number of variables in which both

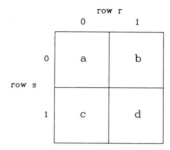

FIGURE 10.5. Relationship Between Two Binary Variables

subjects were coded 1. The sum $(c + b)$ represents the number of variables in which the subjects had different coding. The total number of variables observed is $p = (a + b + c + d)$.

Two commonly used measures of similarity in this case are the *simple matching coefficient* and *Jaccard's coefficient*. For the simple matching coefficient the similarity measure is given by $(a + d)/(a + b + c + d)$ which measures the proportion of variables in which the subjects have the same coding.

For Jaccard's coefficient the similarity measure is given by $d/(b+c+d)$. In this case the number of variables in which both subjects were coded 0 has been omitted. If the purpose of the measure of similarity is to indicate how similar the subjects are with respect to attributes present (coded 1) and to ignore the impact of attributes absent (coded 0), then Jaccard's coefficient is more appropriate. By excluding variables in which neither subject has the attribute, similarity is only measured with respect to attributes in common. If two subjects are both missing a large number of attributes it may not be desirable to say they are similar. For example in numerical taxonomy the Jaccard coefficient is often preferred. Since a fish and a bird have few attributes in common one would not want to say that the two species are similar.

If the dummy variable $\mathbf{X}$ matrix is used for binary variables (2 columns for each X variable) as in the case of categorical data discussed above, the quantities K, p and f_{rs} are related to a, b, c and d in Figure 10.5 by $p = (a + b + c + d)$, $K = 2p$ and $f_{rs} = (a + d)$. The measures of proximity introduced for categorical data are now given by $c_{rs} = (a+d)/(a+b+c+d)$, $q_{rs} = [(a+d) - (b+c)]/(a+b+c+d)$ and $d_{rs}^2 = (b+c)$. The c_{rs} measure is therefore equivalent to the simple matching coefficient and d_{rs}^2 is obtained by subtracting the matching coefficient from 1. The correlation coefficient q_{rs} in this case is commonly called the *Hamann coefficient*.

If in the case of binary variables an $\mathbf{X}$ matrix is formed by using 0–1 coding (1 column per variable), a 0 appears in a column when the subject does not have the attribute and a 1 appears when the attribute is present. In this case the $\mathbf{XX'}$ matrix yields cosine and correlation measures given by

$$c_{rs} = d/[(b+d)(c+d)]^{1/2}$$

and

$$q_{rs} = (ad - bc)/[(a+b)(a+c)(b+d)(c+d)]^{1/2}.$$

The c_{rs} coefficient is usually called the *Ochiai coefficient* and the q_{rs} is referred to as the *phi coefficient*.

There are many other *binary similarity coefficients* available. A large list of such coefficients can be found in the Proximities chapter of the SPSSX User's Guide.

Example

Table 10.6 shows the available grounds for divorce in 1982 for twenty selected U.S. states. For each of the nine categories a state is coded 1 if that ground is available and is coded 0 otherwise. A comparison of the rows of the table allows one to conclude that the states of Washington, Montana, Oregon and Nebraska have identical responses with only one available ground "marriage breakdown". The states of Rhode Island, Massachusetts and New Hampshire also have identical rows. For these three states there are eight grounds available for divorce. Finally, the two midwestern states North Dakota and Oklahoma have identical rows. It is interesting to note that identical states seem to be located in the same geographical area in all three cases.

To determine the similarity among all twenty states, similarity matrices were constructed using both the simple matching coefficient and Jaccard's coefficient. These two similarity matrices are displayed in Table 10.7. From the simple matching coefficients in the lower triangle of the table, we can observe that there are exactly nine possible values of $(a+d)$ which is the number of matches. Thus states with eight out of nine matches have the coefficient 0.889, and states with seven out of nine matches have the coefficient 0.778. From this table of similarities we can conclude that Florida is quite similar to the four western states of Nebraska, Montana, Oregon and Washington with a similarity of 0.889. Other examples of strong similarities are provided by Louisiana and South Dakota, Maine with each of North Dakota and Oklahoma and finally Texas with West Virginia.

The upper triangle of Table 10.7 shows the similarities as measured by the Jaccard coefficient. For this coefficient, similarity is measured using only the grounds present not the grounds not present. Of course the states that are identical have the coefficient value 1.000 as above. The largest coefficient value less than 1 appears to be Maine with each of North Dakota and Oklahoma with a coefficient of 0.875. For these three states, seven of the eight grounds that are available in North Dakota and Oklahoma

TABLE 10.6. Grounds for Divorce in Twenty Selected U.S. States in 1982*

State	Abbre- viations	Marriage Breakdown	Cruelty	Desertion	Non- Support	Alcohol/Drug Addiction	Felony	Impotency	Insanity	A Period of Separation
Florida	FL	1	0	0	0	0	0	0	1	0
Louisiana	LA	0	1	1	1	1	1	0	0	1
Maine	ME	1	1	1	1	1	0	1	1	0
Maryland	MD	0	1	1	0	0	1	1	1	1
Massachusetts	MA	1	1	1	1	1	1	1	0	1
Montana	MT	1	0	0	0	0	0	0	0	0
Nebraska	NE	1	0	0	0	0	0	0	0	0
New Hampshire	NH	1	1	1	1	1	1	1	0	1
New York	NY	0	1	1	0	0	1	0	0	1
North Dakota	ND	1	1	1	1	1	1	1	1	0
Oklahoma	OK	1	1	1	1	1	1	1	1	0
Oregon	OR	1	0	0	0	0	0	0	0	0
Rhode Island	RI	1	1	1	1	1	1	1	0	1
South Carolina	SC	0	1	1	0	1	0	0	0	1
South Dakota	SD	0	1	1	1	1	1	0	0	0
Texas	TX	1	1	1	0	0	1	0	1	1
Vermont	VT	0	1	1	1	0	1	0	1	1
Virginia	VA	0	1	1	0	0	0	0	0	1
Washington	WA	1	0	0	0	0	0	0	0	0
West Virginia	WV	1	1	1	0	1	1	0	1	1

*Source: The World Almanac 1983 published for the Boston Herald American by Newspaper Enterprise Association Inc., New York.

TABLE 10.7. Similarity Matrix for Twenty U.S. States – Grounds for Divorce, Lower Triangle; Simple Matching Coefficient, Upper Triangle; Jaccard Coefficient

	FL	LA	ME	MD	MA	MT	NE	NH	NY	ND	OK	OR	RI	SC	SD	TX	VT	VA	WA	WV
FL	X	0.000	0.286	0.143	0.111	0.500	0.500	0.111	0.000	0.250	0.250	0.500	0.111	0.000	0.000	0.333	0.143	0.000	0.500	0.286
LA	0.111	X	0.444	0.500	0.750	0.000	0.000	0.750	0.667	0.556	0.556	0.000	0.750	0.667	0.833	0.500	0.714	0.667	0.000	0.625
ME	0.444	0.444	X	0.444	0.667	0.143	0.143	0.667	0.222	0.875	0.875	0.143	0.667	0.375	0.500	0.444	0.444	0.222	0.143	0.556
MD	0.333	0.556	0.444	X	0.556	0.000	0.000	0.556	0.667	0.556	0.556	0.000	0.556	0.429	0.375	0.714	0.714	0.667	0.000	0.625
MA	0.111	0.778	0.667	0.556	X	0.125	0.125	1.000	0.500	0.778	0.778	0.125	1.000	0.500	0.625	0.556	0.556	0.500	0.125	0.667
MT	0.889	0.222	0.333	0.222	0.222	X	1.000	0.125	0.000	0.125	0.125	1.000	0.125	0.000	0.000	0.167	0.000	0.000	1.000	0.143
NE	0.889	0.222	0.333	0.444	0.222	0.778	X	0.125	0.000	0.125	0.125	1.000	0.125	0.000	0.000	0.167	0.000	0.000	1.000	0.143
NH	0.111	0.778	0.667	0.556	1.000	0.222	0.222	X	0.500	0.778	0.778	0.125	1.000	0.500	0.625	0.556	0.556	0.500	0.125	0.667
NY	0.333	0.778	0.222	0.778	0.556	0.444	0.444	0.556	X	0.333	0.333	1.000	0.500	0.600	0.500	0.667	0.667	1.000	0.000	0.571
ND	0.333	0.556	0.889	0.556	0.778	0.222	0.222	0.778	0.333	X	1.000	0.125	0.778	0.333	0.625	0.556	0.556	0.333	0.125	0.667
OK	0.333	0.556	0.889	0.556	0.778	0.222	0.222	0.778	0.333	1.000	X	0.125	0.778	0.333	0.625	0.556	0.556	0.333	0.125	0.667
OR	0.889	0.222	0.333	0.222	0.222	1.000	1.000	0.222	0.444	0.222	0.222	X	0.125	0.000	0.000	0.167	0.000	0.000	1.000	0.143
RI	0.111	0.778	0.667	0.556	1.000	0.222	0.222	1.000	0.556	0.778	0.778	0.222	X	0.500	0.625	0.556	0.556	0.500	0.125	0.667
SC	0.333	0.778	0.444	0.556	0.556	0.444	0.444	0.556	0.778	0.333	0.333	0.444	0.556	X	0.500	0.556	0.429	0.500	0.000	0.571
SD	0.222	0.889	0.556	0.444	0.667	0.333	0.333	0.667	0.667	0.667	0.667	0.333	0.667	0.667	X	0.375	0.571	0.600	0.000	0.500
TX	0.556	0.556	0.444	0.778	0.556	0.444	0.444	0.556	0.778	0.556	0.556	0.444	0.556	0.556	0.444	X	0.714	0.500	0.167	0.857
VT	0.333	0.778	0.444	0.778	0.556	0.222	0.222	0.556	0.778	0.556	0.556	0.222	0.556	0.556	0.667	0.778	X	0.667	0.167	0.625
VA	0.333	0.778	0.222	0.778	0.556	0.444	0.444	0.556	1.000	0.333	0.333	0.444	0.556	0.778	0.667	0.778	0.778	X	0.000	0.571
WA	0.889	0.222	0.333	0.222	0.222	1.000	1.000	0.222	0.444	0.222	0.222	1.000	0.222	0.444	0.333	0.444	0.222	0.444	X	0.143
WV	0.444	0.667	0.556	0.667	0.667	0.333	0.333	0.667	0.667	0.667	0.667	0.333	0.667	0.667	0.556	0.889	0.667	0.667	0.333	X

are also available in Maine. All of the grounds available in Maine are also available in North Dakota and Oklahoma. An interesting comparison between the two similarity measures used in Table 10.7 is provided by the similarity measure between Florida and the four western states, Nebraska, Oregon, Montana and Washington. For the four western states, there is only one ground available "marriage breakdown", whereas for Florida there are two grounds available: "marriage breakdown" and "insanity". The simple matching coefficient between Florida and each of the western states was 0.889 whereas for the Jaccard coefficient the similarity measure is only 0.500. The Jaccard coefficient simply indicates that only one of the two grounds available in Florida or Montana is available in both Florida and Montana. The fact that these two states are almost identical with respect to grounds not available does not influence the similarity measure.

Mixtures of Categorical and Interval Scales Variables

If the data matrix $\mathbf{X}$ contains a mixture of dummy variables and interval scaled variables it is difficult to combine the variables to determine a measure of similarity. One approach would be to standardize the variables or columns of $\mathbf{X}$ before computing the matrix $\mathbf{XX}'$.

A second alternative would be to compute separate measures of proximity for the categorical and interval variables and then to combine the two proximity measures using appropriate weights. A third alternative would be to convert the interval variables into categorical variables by constructing classes and then treating all the variables as categorical.

10.1.2   THE MEASUREMENT OF PROXIMITY BETWEEN GROUPS

In the previous section techniques were introduced for measuring the proximity between objects or rows of a data matrix. In this section a variety of approaches are introduced for the measurement of *proximity between two groups* of objects. The purpose of studying the measurement of proximity among groups will be demonstrated in Section 10.2 with the outline of cluster analysis.

Assume that two groups of objects denoted by s and r contain n_s and n_r objects respectively. The observations on the p variables for the n_r objects in group r are denoted by x_{rjm}, $j = 1, 2, \ldots, p$, $m = 1, 2, \ldots, n_r$, and similarly for group s the observations are denoted by x_{sjm}, $j = 1, 2, \ldots, p$; $m = 1, 2, \ldots, n_s$. Figure 10.6 illustrates the notation for $p = 2$ variables. In the figure group r contains $n_r = 6$ objects and group s contains $n_s = 7$ objects. The mth object in group r has the coordinates (x_{r1m}, x_{r2m}) whereas the mth object in group s has the coordinates (x_{s1m}, x_{s2m}).

A variety of possible approaches to the measurement of proximity between two groups are introduced in this section. The function $p_{rs}(j, k)$ is

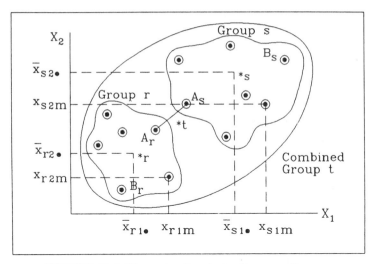

FIGURE 10.6. Measuring Proximity Between Groups

used to denote the measure of proximity between the jth object of group r and the kth object of group s.

Single Linkage or Nearest Neighbor

The *single linkage* or *nearest neighbor* measure of proximity between two groups is based on the strongest measure of proximity between objects in the two groups. Thus although there can be many objects involved, the proximity measure used is based on only one pair of objects. If a dissimilarity measure such as Euclidean distance is being used to measure proximity, then the single linkage approach uses the smallest possible Euclidean distance measure between objects in the two groups. For a similarity measure such as a correlation coefficient the single linkage measure will be based on the maximum possible correlation between objects in the two groups. In Figure 10.6 the distance between Ar and As represents the smallest Euclidean distance between points in the two groups.

Complete Linkage or Furthest Neighbor

The *complete linkage* or *furthest neighbor* measure of proximity between two groups is derived from the weakest link between objects in the two groups. The complete linkage measure is therefore the opposite of the single linkage measure. For a dissimilarity type measure such as Euclidean distance, the largest possible distance between objects in the two groups is used to represent the proximity between the groups. For a similarity measure such as the correlation coefficient the smallest possible value over all possible pairs is used to measure the proximity. In Figure 10.6 the distance between B_r and B_s appears to be the largest distance between objects in the two groups.

Average Linkage

As an alternative to using the proximity measure based on only one of the possible pairs of objects, an average can be determined over all possible pairs of objects. If there are n_r and n_s objects in the two groups r and s respectively, there are a total of $(n_r)(n_s)$ paired measures of proximity. The *average linkage measure* is given by the average of these $(n_r)(n_s)$ measures given by $\sum_{j=1}^{n_r} \sum_{k=1}^{n_s} p_{rs}(j, k)/n_r n_s$. The average linkage is sometimes referred to as UPGMA which stands for "unweighted pair group method using averages."

Example

To illustrate the application of these between-group measures of proximity the ten city data will be used. Various measures of proximity between the two groups $\boxed{1}$ [Columbus, Montgomery] and $\boxed{2}$ [Salt Lake, Wichita] are determined. The squared Euclidean distances in Table 10.8 were obtained from Table 10.3.

The single linkage measure of proximity between groups $\boxed{1}$ and $\boxed{2}$ is the minimum of the distances between elements in the two groups given by

$$p_{12} = \min(23.4,\ 14.4,\ 30.9,\ 19.2) = 14.4.$$

The complete linkage measure is the maximum of these distances

$$p_{12} = \max(23.4,\ 14.4,\ 30.9,\ 19.2) = 30.9.$$

The average linkage between the two groups is given by

$$p_{12} = [23.4 + 14.4 + 30.9 + 19.2]/4 = 21.98.$$

An Algorithm for Updating the Proximity Measures

When objects or groups of objects are combined to form a new group, it is necessary to be able to revise or update the proximity measures between the new group and the remaining groups in the data set. A useful algorithm that can be used to revise the matrix of proximity measures is given by

$$p_{tu} = \alpha_r p_{ru} + \alpha_s p_{su} + \beta p_{rs} + \gamma |p_{ru} - p_{su}| \tag{10.3}$$

where

1. t is the new reference for the group resulting from the combining of groups r and s.

TABLE 10.8. Squared Euclidean Distances Between Cities

| | | 1 | | 2 | |
		Columbus	Montgomery	Salt Lake	Wichita
1	Columbus	0.0	2.8	23.4	14.4
	Montgomery	2.8	0.0	30.9	19.2
2	SaltLake	23.4	30.9	0.0	5.1
	Wichita	14.4	19.2	5.1	0.0

2. u is the reference for some other group other than r or s.

3. α_r, α_s, β and γ are coefficients that depend on the proximity measure being used.

Assuming the proximity measure being used is a dissimilarity measure, the complete linkage method employs $\alpha_r = \alpha_s = \frac{1}{2}$, $\beta = 0$ and $\gamma = \frac{1}{2}$ whereas for the single linkage method the values are $\alpha_r = \alpha_s = \frac{1}{2}$, $\beta = 0$ and $\gamma = -\frac{1}{2}$. If the proximity measure is a similarity measure the values of γ are interchanged for the complete and single linkage methods. For the average linkage method, $\alpha_r = n_r/(n_r + n_s)$, $\alpha_s = n_s/(n_r + n_s)$, $\beta = 0$ and $\gamma = 0$.

Using the relationship (10.3) allows the proximity matrix to be updated after groups r and s are combined to form the new group t. It is important to note that the updated proximity matrix does not require the original data matrix. The original proximity matrix among the objects is the only link required to the original data.

Example

The previous example is used to illustrate the use of (10.3) for the single linkage method. Groups 1 and 2 each contain two objects and hence there are two steps required to proceed from the original proximity matrix with all groups consisting of one object. Formula (10.3) must be applied twice. If group 1 is formed first, then the new measure of proximity between each member of 2 and group 1 is given by

$$\frac{1}{2}(23.4) + \frac{1}{2}(30.9) - \frac{1}{2}|23.4 - 30.9| = 23.4$$

for Salt Lake and

$$\frac{1}{2}(14.4) + \frac{1}{2}(19.2) - \frac{1}{2}|14.4 - 19.2| = 14.4$$

for Wichita. When Wichita and Salt Lake are combined using single linkage the proximity measure between group $\boxed{1}$ and the new group $\boxed{2}$ is given by

$$\frac{1}{2}(23.4) + \frac{1}{2}(14.4) - \frac{1}{2}|23.4 - 14.4| = 14.4.$$

This result is equivalent to the result obtained earlier for the single linkage proximity measure between groups $\boxed{1}$ and $\boxed{2}$. The same result would be obtained if Wichita and Salt Lake were combined first followed by Columbus and Montgomery.

A similar procedure can be carried out to illustrate the complete linkage and average linkage methods. The reader is invited to try these two cases using (10.3).

Distance Between Centroids

The *Euclidean distance between group centroids* can also be used to measure proximity between two groups of objects. If the coordinates for the centroids in groups r and s are given respectively by $(\bar{x}_{r1.}, \bar{x}_{r2.}, \ldots, \bar{x}_{rp.})$ and $(\bar{x}_{s1.}, \bar{x}_{s2.}, \ldots, \bar{x}_{sp.})$, the square of the Euclidean distance between the two centroids is given by

$$d_{rs}^2 = \sum_{j=1}^{p} (\bar{x}_{rj.} - \bar{x}_{sj.})^2.$$

This measure is referred to as the distance between centroids.

For the two-variable case displayed in Figure 10.6 the two centroids are denoted by $*r$ and $*s$. The Euclidean distance between the two points is the required proximity measure between the groups r and s.

Incremental Sums of Squares

An alternative measure of proximity based on the Euclidean distance between centroids uses the fact that there are a total of $(n_r)(n_s)$ distances between the two groups. A measure of the total distances between the two groups is given by $n_r n_s d_{rs}^2$. Since there are a total of $(n_r + n_s)$ objects an average is provided by $n_r n_s d_{rs}^2 / (n_r + n_s)$. This measure of average distance is equivalent to the *change in within group sum of squares*, or *incremental sum of squares*, resulting from the combining of groups r and s. This concept is illustrated below.

For the rth group the within group sum of squares is given by

$$SSW_r = \sum_{m=1}^{n_r} \sum_{j=1}^{p} (x_{rjm} - \bar{x}_{rj.})^2.$$

Similarly for group s the within group sum of squares is given by

$$SSW_s = \sum_{m=1}^{n_s} \sum_{j=1}^{p} (x_{sjm} - \bar{x}_{sj\cdot})^2.$$

If groups r and s are combined to form a new group t, a new centroid $(\bar{x}_{t1\cdot}, \bar{x}_{t2\cdot}, \ldots, \bar{x}_{tp\cdot})$ is obtained and the new within group sum of squares for group t is given by

$$SSW_t = \sum_{m=1}^{n_r+n_s} \sum_{j=1}^{p} (x_{tjm} - \bar{x}_{tj\cdot})^2.$$

The *increase in the total within group sum of squares* as a result of joining groups r and s is given by $SSW_t - (SSW_r + SSW_s)$. This increase in total within group sum of squares is equivalent to the total distance $n_r n_s d_{rs}^2/(n_r + n_s)$. This incremental sum of squares is commonly used as a measure of proximity between two groups. The reader should recall from Section 10.1.1 that for $n_r = n_s = 1$ the squared Euclidean distance between the points and the centroid were related to the Euclidean distance between the points (see (10.1)). In Figure 10.6 the centroid for group t is given by $*t$ which lies along the line joining $*r$ and $*s$.

Relationship to Analysis of Variance

The total within group sum of squares for the combined group t can be viewed as a total sum of squares as in ANOVA. The subgroups r and s yield a total within group sum of squares and a total between group sum of squares. As in the analysis of variance, the total sum of squares contains the two components, the within sum of squares and the between sum of squares. The *total sum of squares* is equivalent to SSW_t defined above and the *within group sum of squares* is $(SSW_r + SSW_s)$ defined above. The difference between the two is the *between sum of squares* given by

$$
\begin{aligned}
SSG_t &= \sum_{j=1}^{p} [n_r (\bar{x}_{rj\cdot} - \bar{x}_{tj\cdot})^2 + n_s (\bar{x}_{sj\cdot} - \bar{x}_{tj\cdot})^2] \\
&= \frac{n_r n_s}{(n_r + n_s)} \sum_{j=1}^{p} (\bar{x}_{rj\cdot} - \bar{x}_{sj\cdot})^2 = \frac{n_r n_s}{(n_r + n_s)} d_{rs}^2.
\end{aligned}
$$

This between sum of squares is the incremental sum of squares used to measure the proximity between groups r and s.

Algorithms for Determining Proximity Measures Based on Centroids and Sums of Squares

For both the *centroid method* and the *incremental sum of squares method*, the proximity measure can be obtained in a sequential fashion using the algorithms outlined below. The sequence involves a series of steps in which a measure of proximity is determined between two objects or groups of objects at each step. The process begins with a proximity matrix based on the squared Euclidean distances among the original objects. At a given point in the sequence, the proximity measures among the groups r, s and u are denoted by p_{rs}, p_{ru} and p_{su}. Group r and s are combined to form a new group denoted by t, and a new proximity measure p_{tu} is required relating t and u.

For the incremental sum of squares method, the process begins with proximity measures given by $p_{rs} = \frac{1}{2}d_{rs}^2$ where d_{rs}^2 denotes the square of the Euclidean distance between objects r and s. After combining objects r and s to form the new group t, the incremental sum of squares proximity measure between t and u is given by

$$p_{tu} = \left(\frac{1}{n_t + n_u}\right)\left[(n_u + n_r)p_{ru} + (n_u + n_s)p_{su} - n_u p_{rs}\right], \qquad (10.4)$$

where $n_t = (n_r + n_s)$.

For the centroid method the process begins with proximity measures given by $p_{rs} = d_{rs}^2$. After combining objects r and s to form group t the proximity measure between t and u is given by

$$p_{tu} = \left(\frac{n_r}{n_r + n_s}\right)p_{ru} + \left(\frac{n_s}{n_r + n_s}\right)p_{su} - \frac{n_r n_s}{(n_r + n_s)^2}p_{rs}. \qquad (10.5)$$

The application of these algorithms is illustrated in the example below.

The above two algorithms are special cases of (10.3) defined above. For the incremental sum of squares, $\alpha_r = (n_r + n_u)/(n_t + n_u)$, $\alpha_s = (n_s + n_u)/(n_t + n_u)$, $\beta = -n_u/(n_t + n_u)$ and $\gamma = 0$. For the centroid method, $\alpha_r = n_r/(n_r + n_s)$, $\alpha_s = n_s/(n_r + n_s)$, $\beta = -n_r n_s/(n_r + n_s)^2$ and $\gamma = 0$. Although these two algorithms can be viewed as special cases of (10.3) it is important to keep in mind that the original proximity matrix is assumed to be squared Euclidean distance.

The incremental sum of squares measure is also referred to as *Ward's method*, and the *centroid method* is also referred to as UPGMC which is an abbreviation for the expression "unweighted pair group method using centroids."

Example

In this example we determine the proximity measure between Wichita and the [Columbus, Montgomery] group and then the proximity measure between Salt Lake and the [Columbus, Montgomery] group. These two proximity measures are then used to determine the measure between the groups [Wichita, Salt Lake] and [Columbus, Montgomery]. The two algorithms presented above are used repeatedly.

For the centroid method, the distance between Wichita and [Columbus, Montgomery] applying (10.5) and the information in Table 10.8 is given by

$$\frac{1}{2}(14.4) + \frac{1}{2}(19.2) - \frac{1}{4}(2.8) = 16.1.$$

In (10.5), r corresponds to Columbus, s corresponds to Montgomery and u corresponds to Wichita. This proximity measure is based on the three squared Euclidean distance measures among the three cities. Similarly for the distance between Salt Lake and [Columbus, Montgomery] the measure is given by

$$\frac{1}{2}(23.4) + \frac{1}{2}(30.9) - \frac{1}{4}(2.8) = 26.45.$$

In (10.5) r corresponds to Columbus, s corresponds to Montgomery and u corresponds to Salt Lake.

Finally the distance between the centroids of the two groups can now be obtained from (10.5) and is given by

$$\frac{1}{2}(16.1) + \frac{1}{2}(26.45) - \frac{1}{4}(5.1) = 20.075.$$

In (10.5) r corresponds to Wichita, s corresponds to Salt Lake and u corresponds to the group containing Columbus and Montgomery.

For the incremental sum of squares or Ward's method a similar series of steps is required. The squared Euclidean distances can be converted to the incremental sum of squares measure by dividing the squared distances by 2. Using (10.4) and Table 10.8 the incremental sum of squares that would result from combining Wichita with Columbus and Minneapolis is given by

$$\frac{1}{3}[2(14.4) + 2(19.2) - (2.8)] = 21.47.$$

Similarly

$$\frac{1}{3}[2(23.4) + 2(30.9) - (2.8)] = 35.27$$

determines the incremental sum of squares that would result from combining Salt Lake with Columbus and Montgomery. Finally the incremental sum of squares that would result from combining the two groups is determined by applying (10.4) and is given by

$$\frac{1}{4}[(3)(21.47) + (3)(35.27) - (2)(5.1)] = 40.0.$$

Ultrametric Inequality

The general algorithm given by (10.3) provides a sequential method for updating proximity measures between groups as groups expand in size and hence become less numerous. Usually the proximity measure is such that the distance between the two groups being combined (groups r and s) is smaller than the distance between the new combined group t and any other group u. If this is true then the distance-type proximity measure is said to satisfy the *ultrametric inequality*. This inequality implies that no distance measure in the new matrix can be smaller than the smallest distance measure in the previous matrix. The purpose of identifying this property is that it guarantees that as groups are combined the distance-type proximity measures will be monotonically nondecreasing.

If the parameters in (10.3) satisfy the conditions $(\alpha_r + \alpha_s + \beta) \geq 1$ and $\gamma \geq \max(-\alpha_r, -\alpha_s)$, then the proximity measure will have the ultrametric property. For the measures discussed in this section only the centroid method does not satisfy this property. For the centroid method $(\alpha_r + \alpha_s + \beta) = n_r n_s / (n_r + n_s)^2$ which is not ≥ 1. For this reason the centroid method is seldom used.

Sums of Squares Derived from MANOVA Matrices

Given g groups of objects with group sizes $n_1, n_2, \ldots, n_g$; where each object is measured on a p-dimensional variable $\mathbf{x}$ $(p \times 1)$, MANOVA notation introduced in Chapter 8 can be used to characterize differences between the g groups. In the discussion of MANOVA, three sum of squares and cross-product matrices were used to measure variation and were denoted by $\mathbf{T}$, $\mathbf{W}$ and $\mathbf{G}$. The three matrices were labeled total sum of squares, within sum of squares and among sum of squares respectively and satisfied the relation $\mathbf{T} = \mathbf{W} + \mathbf{G}$.

The matrices $\mathbf{T}$, $\mathbf{W}$ and $\mathbf{G}$ are $p \times p$ matrices with jth diagonal elements given by $t_{jj} = \sum_{k=1}^{g} \sum_{i=1}^{n_k} (x_{ijk} - \bar{x}_{.j.})^2$, $w_{jj} = \sum_{k=1}^{g} \sum_{i=1}^{n_k} (x_{ijk} - \bar{x}_{.jk})^2$ and $g_{jj} = \sum_{k=1}^{g} \sum_{i=1}^{n_k} (\bar{x}_{.jk} - \bar{x}_{.j.})^2 = \sum_{k=1}^{g} n_k (\bar{x}_{.jk} - \bar{x}_{.j.})^2$ respectively. For each of these three elements, the first or inside summation represents the sum of squares over the observations in a group for variable j, whereas the outer sum determines the sum over all groups. Each of the diagonal elements therefore respectively represents the total sum of squares, within group sum of squares and among group sum of squares over all g groups for one variable.

To determine the three sums of squares over all p variables the sums of the diagonal elements of these matrices are required. The three sums of squares are given by

$$
tr\mathbf{T} = \sum_{j=1}^{p} t_{jj} = \sum_{j=1}^{p} \sum_{k=1}^{g} \sum_{i=1}^{n_k} (x_{ijk} - \bar{x}_{.j.})^2,
$$

$$
tr\mathbf{W} = \sum_{j=1}^{p} w_{jj} = \sum_{j=1}^{p} \sum_{k=1}^{g} \sum_{i=1}^{n_k} (x_{ijk} - \bar{x}_{.jk})^2,
$$

$$trG = \sum_{j=1}^{p} g_{jj} = \sum_{j=1}^{p} \sum_{k=1}^{g} \sum_{i=1}^{n_k} (\bar{x}_{.jk} - \bar{x}_{.j.})^2$$

$$= \sum_{j=1}^{p} \sum_{k=1}^{g} n_k (\bar{x}_{.jk} - \bar{x}_{.j.})^2.$$

By reordering the sequences of summation signs these quantities can be viewed as sums over g groups of sums of squares determined over p variables. Denoting the sums of squares for group k by SSW_k, SST_k and SSG_k we have $trT = \sum_{k=1}^{g} SST_k$, $trW = \sum_{k=1}^{g} SSW_k$ and $trG = \sum_{k=1}^{g} SSG_k$, where

$$SSW_k = \sum_{j=1}^{p} \sum_{i=1}^{n_k} (x_{ijk} - \bar{x}_{.jk})^2,$$

$$SSG_k = \sum_{j=1}^{p} n_k (\bar{x}_{.jk} - \bar{x}_{.j.})^2$$

and

$$SST_k = \sum_{j=1}^{p} \sum_{i=1}^{n_k} (x_{ijk} - \bar{x}_{.j.})^2.$$

The previous discussion on the measurement of proximity between groups using the incremental sum of squares method illustrated the relationship between the three sums of squares. It was shown that when groups r and s are joined to form a new group t that

$$SSW_t = \frac{n_r n_s}{(n_r + n_s)} \sum_{j=1}^{p} (\bar{x}_{.jr} - \bar{x}_{.js})^2 + SSW_r + SSW_t.$$

Therefore since $SST_t = SST_r + SST_s$ remains fixed, the effect on the among group sum of squares is

$$SSG_t = SSG_r + SSG_s - \frac{n_r n_s}{(n_r + n_s)} \sum_{j=1}^{p} (\bar{x}_{.jr} - \bar{x}_{.js})^2.$$

For the jth diagonal elements of the matrices W and G the elements w_{jj} increase by $n_r n_s (\bar{x}_{.jr} - \bar{x}_{.js})^2/(n_r + n_s)$ while the elements g_{jj} decrease by this same amount. The net effect on trW and trG of the combining of groups r and s is therefore an increase in trW of $n_r n_s \sum_{j=1}^{p} (\bar{x}_{.jr} - \bar{x}_{.js})^2/(n_r + n_s)$ and a corresponding decrease in trG of this same amount.

A Multivariate Measure of Proximity

Assume that the $(p \times 1)$ observation vectors x_{ik}, $k = 1, 2, \ldots, g$, represent random samples from g multivariate populations with $(p \times 1)$ mean vectors

$\boldsymbol{\mu}_k$ and common $(p \times p)$ covariance matrix $\boldsymbol{\Sigma}$. Denoting the sample within-group covariance matrices by $\mathbf{S}_1, \mathbf{S}_2, \ldots, \mathbf{S}_g$ the pooled estimate of $\boldsymbol{\Sigma}$ is given by $\overline{\mathbf{S}} = \sum_{k=1}^{g}(n_k - 1)\mathbf{S}_k/(n - g)$ and $\mathbf{W} = (n - g)\overline{\mathbf{S}}$ (see Section 8.1). For any two groups r and s, a measure of proximity between the two groups that takes into account the covariances is given by the squared Mahalanobis distance $(n-g)(\bar{\mathbf{x}}_{.r} - \bar{\mathbf{x}}_{.s})'\mathbf{W}^{-1}(\bar{\mathbf{x}}_{.r} - \bar{\mathbf{x}}_{.s})$, where $\bar{\mathbf{x}}_{.r}$ and $\bar{\mathbf{x}}_{.s}$ are the sample mean vectors for groups r and s respectively.

10.2   Cluster Analysis

The object of *cluster analysis* is to partition a set of objects into groups or clusters in such a way that the profiles of objects in the same cluster are very similar, whereas the profiles of objects in different clusters are quite distinct. Since the concept of clusters is closely linked to the concept of proximity between objects and groups of objects, the techniques discussed in Section 10.1 play an important role in cluster identification.

In some applications of cluster analysis, the objects are believed to belong to a few natural groups whereas in other cases the quest is simply to find a convenient grouping. The first case is sometimes called *classification*; the latter application is often referred to as *dissection*. Other terms commonly used for cluster-analysis-type procedures include *pattern recognition* and *numerical taxonomy*. Both the development of clustering techniques and the application of such techniques have appeared in many different fields of study. Engineering, zoology, medicine, linguistics, anthropology, psychology and marketing are just some of the fields of application.

Like principal component analysis, cluster analysis can be viewed as a data reduction technique. Instead of reducing the number of variables or columns required to characterize $\mathbf{X}$ as in principal components analysis, cluster analysis reduces the number of distinct objects or rows of $\mathbf{X}$ by creating groups of objects called clusters.

Cluster analysis techniques can be classified into five major types; hierarchical, partitioning, Q-sort, density and clumping. In the *hierarchical* approach the process proceeds sequentially such that at each step only one object or group of objects changes group membership and the groups at each step are nested with respect to previous groups. Thus once an object has been assigned to a group it is never removed from the group later on in the clustering process. The hierarchical method produces a complete sequence of cluster solutions beginning with n clusters (one for each object) and ending with one cluster containing all n objects. In some applications a set of nested clusters is the required solution whereas in other applications only one of the cluster solutions in the hierarchy is selected as the solution.

The *partitioning method* begins with a given number of clusters, say g, as the objective and then partitions the objects to obtain the required g clusters. In contrast to the hierarchical method, partitioning techniques permit objects to change group membership throughout the cluster formation process. The partitioning method usually begins with an initial solution, after which reallocation occurs according to some optimality criterion.

The *Q-sort* methods include a variety of techniques that are similar to factor analysis. These methods usually begin with the $\mathbf{XX'}$ matrix and are concerned with grouping objects together whose off-diagonal elements in $\mathbf{XX'}$ are relatively large.

The *density* or *mode* seeking procedure assumes that the objects will be allocated in space in such a way that there are several dense areas with regions in between that are very sparse. This method assumes the existence of natural clusters.

Finally, the *clumping method* unlike the above three techniques permits the clusters to overlap. Another term used for this type of cluster analysis is *fuzzy clustering* because the clusters are allowed to overlap.

In this section the hierarchical method is emphasized. A variety of algorithms for performing a hierarchical cluster analysis will be outlined. Techniques for evaluating the cluster solution will also be introduced.

While cluster structure can vary across many differing research applications, for the most part the techniques outlined here assume that the quest is to find natural clusters. *Natural clusters* are assumed to satisfy the properties of *external isolation* and *internal cohesion*. External isolation suggests that points in one cluster should be separated from points in another cluster by an empty area of space. Internal cohesion requires that points within a cluster should be close together. This characterization of the "natural cluster" therefore does not allow cluster overlap.

10.2.1   HIERARCHICAL METHODS

The most common approach to cluster analysis is the hierarchical method. The method proceeds sequentially yielding a nested arrangement of objects in groups. The hierarchical process can be represented conveniently using a *tree diagram* as illustrated in Figure 10.7. This figure illustrates the hierarchical clustering process for a sample of five objects.

To begin with on the left of Figure 10.7, there are five objects that can be viewed as five clusters each containing a single object. At step 1 in the process (moving one step to the right) objects 1 and 2 are joined to form a group. Similarly at step 2 objects 4 and 5 are joined to form a group. After step 2 is complete, there are now three clusters, two each containing two objects and one cluster containing only object 3. At step 3, object 3 joins the cluster containing objects 1 and 2, and finally in step 4 all objects are joined to form a single cluster of five objects.

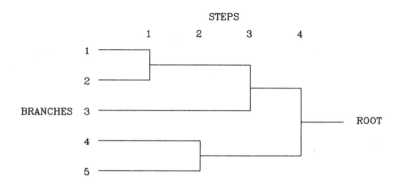

FIGURE 10.7. Tree Diagram for Hierarchical Clustering

Agglomerative versus Divisive Processes

At each step in a hierarchical clustering process, proceeding from left to right, two groups are joined together. Once the groups are joined together they are never separated later in the process. This hierarchical process is called *agglomerative* because, as the process moves sequentially from n clusters to one cluster, the sizes of the clusters increase and the number of clusters decrease. The agglomerative process moves from the branches of the tree on the left to the root of the tree on the right. A hierarchical process that moves in the reverse order is called *divisive*. The divisive process begins with all objects in one cluster on the right of Figure 10.7 and moves to the left. This process therefore moves from the root of the tree to the branches of the tree. The divisive approach is not commonly used and will not be discussed in this chapter.

At each step of the hierarchical process, the value of an *objective function* or *clustering criterion* must be computed to determine which two groups are to be joined. The objective function is usually based on a measure of proximity between groups. The methods outlined in Section 10.1.2 for measuring proximity between groups are used in the hierarchical process to determine which groups are to be joined at each step.

Beginning with the proximity matrix for the n objects the two closest objects are joined in step 1 to form a group. Before selecting the next pair of objects or clusters to be joined, the proximity matrix must be revised to reflect the proximities between the new cluster and the remaining objects. The formula given by (10.3) in Section 10.1.2 can be used to revise the proximities between the new cluster and the remaining objects. After revising the proximity matrix a selection of the two closest clusters to be joined in step 2 can be made. This process continues until all objects are contained in one cluster. After each step of the process the proximity matrix is revised to reflect the relationships among the groups that exist at that time. The revised proximity matrix is then used to determine the groups to be joined at the next step.

The agglomerative hierarchical clustering process does not provide a single cluster solution. In fact each step of the process is a cluster solution. The determination of the appropriate number of clusters involves selecting one of the steps of the hierarchical process using a second *optimality criterion*. A variety of optimality criteria are outlined in Section 10.2.2. The example below illustrates the hierarchical process using the ten-city example introduced in Section 10.1.

Computer Software

The cluster analysis calculations in this section have been carried out using the SPSSX program CLUSTER, the SAS program CLUSTER and Wishart's CLUSTAN software package.

Example

The five methods introduced in Section 10.1.2 for measuring group proximity are employed here to obtain a five cluster solution for the ten city data introduced in Table 10.1. In each case the process begins with the standardized Euclidean distance matrix given in Table 10.3. At each step in the process the revised proximities are obtained using formula (10.3). The optimality criterion value is the closest proximity value among groups at that stage of the process. The results are summarized below. The criterion value shown represents the optimality criterion value or proximity value as determined by that method.

Single Linkage Method

Step	Clusters Combined	Criterion Value
1	[Columbus] & [Montgomery]	2.8
2	[Salt Lake] & [Minneapolis]	4.8
3	[Wichita] & [Salt Lake, Minneapolis]	5.1
4	[Washington] & [Atlanta]	7.7
5	[Los Angeles] & [Wichita, Salt Lake, Minneapolis]	8.7

Five-Cluster Solution:
1. Albuquerque
2. Scranton
3. Columbus, Montgomery
4. Washington, Atlanta
5. Los Angeles, Wichita, Salt Lake, Minneapolis

Complete Linkage Method

Step	Clusters Combined	Criterion Value
1	[Columbus] & [Montgomery]	2.8
2	[Salt Lake] & [Minneapolis]	4.8
3	[Washington] & [Atlanta]	7.7
4	[Wichita] & [Salt Lake, Minneapolis]	8.5
5	[Los Angeles] & [Washington, Atlanta]	12.9

Five-Cluster Solution: 1. Albuquerque
2. Scranton
3. Columbus, Montgomery
4. Washington, Atlanta, Los Angeles
5. Wichita, Salt Lake, Minneapolis

Average Linkage Method

Step	Clusters Combined	Criterion Value
1	[Columbus] & [Montgomery]	2.8
2	[Salt Lake] & [Minneapolis]	4.8
3	[Wichita] & [Salt Lake, Minneapolis]	6.8
4	[Washington] & [Atlanta]	7.7
5	[Los Angeles] & [Washington, Atlanta]	11.1

Five-Cluster Solution: Same as Complete Linkage Method

Ward's Method

Step	Clusters Combined	Criterion Value
1	[Columbus] & [Montgomery]	1.4
2	[Salt Lake] & [Minneapolis]	2.4
3	[Wichita] & [Salt Lake, Minneapolis]	3.7
4	[Washington] & [Atlanta]	3.8
5	[Los Angeles] & [Washington, Atlanta]	6.2

Five-Cluster Solution: Same as Complete Linkage and Average Linkage Methods

Centroid Method

Step	Clusters Combined	Criterion Value
1	[Columbus] & [Montgomery]	2.8
2	[Salt Lake] & [Minneapolis]	4.8
3	[Wichita] & [Salt Lake, Minneapolis]	5.6
4	[Washington] & [Atlanta]	7.7
5	[Los Angeles] & [Wichita, Salt Lake, Minneapolis]	8.2

Five-Cluster Solution: Same as Single Linkage

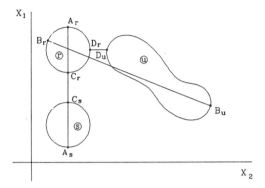

FIGURE 10.8. Single Linkage versus Complete Linkage Using the Distance Measure

From the above results it can be seen that there are only two different five cluster solutions. The difference between the two solutions involves the location of Los Angeles. In the single linkage and centroid methods Los Angeles belongs to the group that also includes Wichita, Salt Lake and Minneapolis, whereas in the other three methods Los Angeles belongs to a group containing Washington and Atlanta. A discussion of the differences among the hierarchical cluster solutions obtained from the five methods is given below. A larger example involving all 40 cities is illustrated in Section 10.2.3. An example based on binary data is illustrated in Section 10.2.2.

Comparison of Group Proximity Measures

Assuming that the squared Euclidean distance is the underlying proximity measure, it will be useful to compare the various group proximity measures in the context of a hierarchical clustering process. The single linkage and complete linkage measures are compared in Figure 10.8. Using the single linkage measure, groups $\boxed{r}$ and $\boxed{u}$ are closer than groups $\boxed{r}$ and $\boxed{s}$. The distance between $\boxed{r}$ and $\boxed{u}$ is the distance from Dr to Du, and the distance from $\boxed{r}$ to $\boxed{s}$ is the distance from Cr to Cs. For the complete linkage measure, groups $\boxed{r}$ and $\boxed{s}$ are closer since the distance from Br to Bu is greater than the distance from Ar to As. As a result of this difference it is easy to imagine how single linkage clustering leads to *chain-like clusters*, whereas complete linkage clustering leads to *compact clusters*. As can be seen from Figure 10.8, it is possible in single linkage clustering for an object in one cluster to be closer to an object in another cluster than to some objects in its own cluster.

A second interesting comparison between single and complete linkage is the impact of cluster size on the proximity measure. Imagine two distinct clusters that are growing in size within a confined space. The single linkage measure between the two clusters will remain constant, whereas the complete linkage measure will increase. Imagine also an *isolated point* or outlier and its proximity measure to a cluster growing in size. The single linkage measure of proximity to the outlier will remain fixed, but the complete linkage measure will tend to increase. In a hierarchical process, the proximity measure increases as clusters increase in size. Since the complete linkage measure is based on the weakest link, an isolated point more quickly becomes relatively close to an existing cluster with the complete linkage method than with the single linkage method. With the single linkage method outliers tend to remain as isolated points until very late in the hierarchical process. The single linkage method is said to be *space conserving*, whereas the complete linkage method is called *space diluting* or space filling.

Both the single and complete linkage measures employ only a single proximity measure to represent group proximity and hence are very susceptible to extreme observations. In a single linkage hierarchical process, a single outlier lying between two clusters can result in the eventual joining of the two groups. In the case of a complete linkage process, small changes in the location of particular points or errors can have a substantial impact on the hierarchical solution.

The average linkage, centroid and Ward's methods are usually preferable to the single linkage and complete linkage methods because of their relative insensitivity to extremes or outliers. Depending on the types of clusters expected this property could also be a disadvantage.

The average linkage measure is determined by averaging the proximities between all pairs of objects (one object from each group). This averaging process has some interesting properties. In Figure 10.9 two groups are shown, A with one point and B with two points (B_L and B_U). The average distance in the one dimension is

$$\bar{d} = [(X_1 - X_2 - h)^2 + (X_1 - X_2 + h)^2]/2 = (X_1 - X_2)^2 + h^2.$$

The average linkage therefore between A and B increases as B_L and B_U move away from the center at B. Thus the average distance between point A and group B increases as the distance between points in B increase. The average linkage between two groups based on squared Euclidean distance therefore grows as the two groups become less compact.

A second interesting property of the average linkage method can also be illustrated using Figure 10.9. If B represents a single point group and A represents a single point group the average squared Euclidean distance is given by $(X_1 - X_2)^2$. In comparison to the previous example, A is closer to group B in this case than it was when B was considered to be a group of two points. Thus as the size of the group increases, the average linkage measure increases unless all points in the group are located at the centroid.

The behavior of the average linkage method of hierarchical clustering in the presence of outliers can be explained using the above mentioned

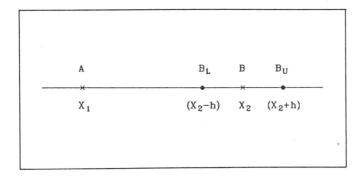

FIGURE 10.9. Distance Between Clusters Using Average Linkage

properties. Initially although the beginning clusters are small and compact, outliers tend to remain isolated. As clusters grow in size the average proximity measure between clusters grows until a point is reached where outliers have proximity measures of a similar magnitude. Since the size and compactness of the clusters influences the average linkage measure, outliers are most likely to link with other outliers and least likely to link with less compact and/or large clusters. In addition to lying between the single linkage and complete linkage methods with respect to space dilution, the average linkage method is also characterized by its tendency to form nonconformist groups of outliers near the end of the hierarchical process.

The centroid method and Ward's method are easily compared since the proximity measures are both based on d_{rs}^2, the squared Euclidean distance between group centroids. Since the coefficient of d_{rs}^2 is 1 for the centroid method and $n_r n_s/(n_r + n_s)$ for Ward's method it is only necessary to determine the impact that the cluster sizes n_r and n_s have on the Ward's measure.

The coefficient $n_r n_s/(n_r + n_s)$ can also be written in other forms such as $n_s/(1 + n_s/n_r)$ and $1/(1/n_r + 1/n_s)$. From these expressions we can conclude that as n_r and n_s get larger in size so does the coefficient $n_r n_s/(n_r + n_s)$ and also as n_r increases relative to n_s the coefficient increases. Given two clusters whose centroids are a distance d_{rs}^2 apart we can conclude that Ward's method would yield a larger measure of proximity between them as the cluster sizes increase and as the cluster sizes become less equal. We can conclude therefore that in comparison to the centroid method, Ward's method has a greater tendency to form equal size and/or smaller clusters. The centroid method therefore tends to join clusters whose centroids are close, whereas in Ward's method smaller clusters reach out to other more distant smaller clusters. Like single linkage the centroid method is space contracting, whereas Ward's method like the complete linkage method is space diluting.

It is also interesting to examine the impact that outliers might have on the hierarchical clustering process and also to compare these processes to the average linkage method discussed above. Consider the proximity measure between a single outlier and a group with n_s objects. The coefficient of d_{rs}^2 in this case is given by $n_s/(1+n_s)$. When n_s is small this coefficient is also relatively small and as n_s increases the coefficient approaches 1. Since the coefficient would be relatively small early in the hierarchical process there would be a tendency for isolated points to be more compatible with small clusters at the early stages of the process. Two isolated points would have a coefficient of $\frac{1}{2}$ and hence are more likely to join together early in the process. Two large clusters would have a coefficient of $n_r n_s/(n_r + n_s)$. When compared to a coefficient of $\frac{1}{2}$ for two single point clusters it can be seen that a large distance between outliers can eventually be overcome as $n_r n_s/(n_r + n_s)$ increases relative to $\frac{1}{2}$. When the average linkage method is compared to Ward's method it is not uncommon to find that in Ward's method the outliers are grouped much earlier in the hierarchical process. An example illustrating the impact of outliers on the average linkage and Ward's methods is provided below.

Some Multivariate Approaches to Hierarchical Clustering

In Section 10.1, the incremental sum of squares method for measuring proximity between groups was shown to be related to changes in $tr\mathbf{W}$ and $tr\mathbf{G}$, where $\mathbf{W}$ and $\mathbf{G}$ are the within and the among group sums of squares matrices respectively. After groups r and s are joined at step ℓ, the new matrices $\mathbf{W}_\ell$ and $\mathbf{G}_\ell$ can be expressed in terms of these matrices at the previous step $(\ell - 1)$ as

$$\mathbf{W}_\ell = \mathbf{W}_{\ell-1} + \frac{n_r n_s}{(n_r + n_s)}(\bar{\mathbf{x}}_r - \bar{\mathbf{x}}_s)(\bar{\mathbf{x}}_r - \bar{\mathbf{x}}_s)'$$

and

$$\mathbf{G}_\ell = \mathbf{G}_{\ell-1} - \frac{n_r n_s}{(n_r + n_s)}(\bar{\mathbf{x}}_r - \bar{\mathbf{x}}_s)(\bar{\mathbf{x}}_r - \bar{\mathbf{x}}_s)'.$$

Each step in Ward's algorithm attempts to minimize the increase in $tr\mathbf{W}_\ell$ and minimize the decrease in $tr\mathbf{G}_\ell$. Because this criterion does not take into account the off-diagonal elements of $\mathbf{W}$ and $\mathbf{G}$ it tends to produce *spherical shaped clusters*. This method is optimal therefore if the underlying covariance matrix $\mathbf{\Sigma}$, is spherical, namely $\mathbf{\Sigma} = \sigma^2\mathbf{I}$.

An alternative criterion, which takes into account the covariances among the variables, minimizes $|\mathbf{W}|$ instead of $tr\mathbf{W}$. When groups r and s are joined the revised value of $|\mathbf{W}|$ is given by

$$|\mathbf{W}_\ell| = |\mathbf{W}_{\ell-1}|\left(\frac{n_r n_s}{n_r + n_s}\right)(\bar{\mathbf{x}}_r - \bar{\mathbf{x}}_s)'\mathbf{W}_{k-1}^{-1}(\bar{\mathbf{x}}_r - \bar{\mathbf{x}}_s).$$

Therefore at each step the two groups selected to be joined must minimize a function of the Mahalanobis distance between the group centroids. Recall that the Mahalanobis distance is equivalent to the Euclidean distance with principal components used in place of the X variables.

In comparison to the criterion $tr\mathbf{W}$ the criterion $|\mathbf{W}|$ takes into account correlation effects. The use of the latter criterion would therefore tend to generate *elliptical shaped clusters*. Both criteria tend to produce clusters of the same shape because of the assumption of homogeneity of covariance matrices across groups. An alternative criterion, which permits greater variation in cluster shape, is to minimize $\prod_{k=1}^{g} |\mathbf{W}_k|^{n_k}$ where $\mathbf{W}_k$ and n_k correspond to cluster group k.

An alternative multivariate criterion for cluster choice is related to discriminant analysis and MANOVA. The criterion seeks to choose clusters that maximize $tr\mathbf{GW}^{-1}$. Since this criterion tends to maximize the largest eigenvalue of $\mathbf{GW}^{-1}$ clusters obtained tend to be elongated.

An Example with Outliers

In the above discussion comparing the various measures of group proximity in a hierarchical clustering process, it was suggested that outliers could have a strong impact on the clustering process and that the various methods would respond differently to such extreme observations. The example presented here will compare the average linkage method to Ward's method. From the earlier discussion, we would expect the outliers to have less impact on the Ward's solution than on the average linkage solution. In addition we would expect outliers to remain isolated for a longer period in the average linkage process than in the case of Ward's method.

Given that there are unusual observations there are two ways of determining the impact of the outliers on the cluster solution. The variable on which the outlier is measured can be removed for all objects (a column of $\mathbf{X}$) or the object with the unusual value can be removed from the analysis (a row of $\mathbf{X}$). After the clusters have been determined with respect to the reduced $\mathbf{X}$ matrix the omitted observations can be used to calculate values for each cluster in the case of omitted variables. In the case of omitted objects the location of the objects can be determined relative to the existing clusters.

The World Data in Table 10.9 contains observations on 12 variables for 25 countries. The variables are intended to represent various dimensions of economic and physical well-being. To identify outliers standardized values were computed for each country for each of the 12 variables. An examination of the standardized values (not shown) revealed three observations that exceed an absolute value of 3. The Hong Kong population density (POP) showed a standardized value of 4.78, the inflation rate (INFLAT) for Argentina showed a value of 4.19, and the number of hospital beds per capita (BEDS) in Japan showed a standardized value of 3.35. These three observations are treated as outliers in the following cluster analyses.

For the purposes of comparison to the outlier detection methods of Chapter 7, Table 10.10 shows the Mahalanobis distances from the centroid for each country for the full **X** matrix, the **X** matrix omitting three variables, POP, INFLAT and BEDS and for the **X** matrix omitting three countries Argentina, Hong Kong and Japan. These three **X** matrices are used in the example. Using the Mahalanobis distance measure we can conclude that Hong Kong and Argentina are farthest from the centroid when all variables are used. Japan is the third most distant country from the centroid, although it is much closer to the centroid than Argentina or Hong Kong. The country of Israel has approximately the same distance from the centroid as Japan. Israel, however, did not have any standardized values above 1.7 (not shown). After removing the three variables POP, INFLAT and BEDS, Argentina, Israel and Syria yield the largest distances from the centroid. When Argentina, Japan and Hong Kong were removed from the data the variation among the Mahalanobis distances seemed much less extreme.

For each of the three data matrices two hierarchical cluster analyses were performed using the average linkage method and Ward's method. Figures 10.10, 10.11, and 10.12 show tree diagrams which compare the various solutions from the eight-cluster stage to the single cluster final step. The numbers in brackets report the value of the proximity measure at that step. Within each figure the two methods can be compared for a given data matrix. For the full data matrix Figure 10.10 illustrates the tendency for Ward's method to determine more equal sized clusters. In contrast Figure 10.10 illustrates the tendency for the average linkage method to form some large clusters early with outliers remaining more isolated until later in the process. Even though Argentina is regarded as an outlier, it is close enough to Chile and Mexico that it forms a cluster at stage 7 in the average linkage method. For Hong Kong, however, the closest countries are Australia, Japan, Canada and United States at values of 31.15, 35.94, 33.15 and 35.54 respectively. Hong Kong therefore tends to remain isolated until very late in the process. The squared Euclidean distances from Israel to Canada, the United States, Australia and Japan are 12.38, 13.84, 9.65 and 17.10 respectively. Israel is therefore an isolated point as was also seen in Table 10.10. Israel joins these four countries at stage 7 in Ward's method but not until stage 4 in the average linkage method.

In Figure 10.11 the two analyses for the matrix with three variables missing (POP, INFLAT and BEDS) are compared. Now that there are no isolated points the two solutions are almost identical. The order in which some groups are joined differs slightly but the membership in the groups is the same.

For the data matrix omitting the countries of Argentina, Hong Kong and Japan Figure 10.12 shows the results of the two cluster analyses. In this

TABLE 10.9. Data Matrix for World

Numerical Code	Country	WORK	RET	EXPEC	POP	DWELL	BIRTH	BEDS	LITERACY	INFLAT	INFANT	GNP	SCHOOL
1	Canada	681	104	802	3	28	148	670	980	40	79	1415	103
2	United States	664	119	786	26	27	157	531	995	36	105	1677	94
3	Mexico	555	36	738	41	52	341	89	903	577	53	218	53
4	Brazil	593	43	680	16	38	306	406	778	2269	706	167	36
5	Chile	627	58	757	16	47	216	344	911	307	195	145	67
6	Argentina	609	86	750	11	35	243	531	939	6721	353	214	71
7	Ecuador	545	37	717	33	50	368	183	802	280	695	116	55
8	Algeria	507	37	642	9	60	389	266	447	105	880	256	51
9	Nigeria	494	24	520	92	40	504	92	424	55	1142	95	29
10	Kenya	457	31	547	35	59	539	134	592	130	802	31	21
11	South Africa	581	41	571	26	40	387	457	782	162	833	217	30
12	Mozambique	531	32	482	17	49	451	89	272	198	1156	17	7
13	India	581	28	578	234	61	329	4	408	56	1049	29	39
14	Pakistan	525	28	514	130	50	433	54	262	58	1056	34	16
15	Zaire	513	25	540	13	59	451	310	612	238	1073	16	23
16	Colombia	591	38	660	25	49	292	170	853	240	500	135	50
17	Japan	681	104	819	320	30	119	1291	990	20	55	1125	96
18	Indonesia	577	36	572	87	41	321	442	673	47	844	52	41
19	Philippines	552	34	658	184	56	333	152	833	231	506	57	65
20	Australia	663	101	798	2	28	157	636	990	67	100	1170	96
21	Egypt	557	39	620	48	53	375	50	435	133	451	65	66
22	Syria	491	28	672	56	64	235	114	400	173	480	161	60
23	Israel	587	88	770	204	33	465	629	918	3046	119	656	80
24	Zambia	486	23	550	9	49	481	209	657	374	883	37	17
25	Hong Kong	693	76	792	5246	41	140	442	773	34	75	608	72

1985 Data: WORK -   % of population in working age category 15–64 ×10
RET -   % of population in retired age caggory 65 and over ×10
EXPEC -   female life expectancy in years ×10
POP -   population density per square mile
DWELL -   population per dwelling ×10
BIRTH -   birth rate per 100,000

BEDS -   number of hospital beds per 100,000
LITERACY - % literacy rate ×10
INFLAT -   % inflation rate ×10
GNP -   GNP per capita in U.S. dollars ÷10
SCHOOL -   secondary school enrollment ratio ×100
           (enrolled/eligible)
INFANT -   deaths per 100 live births

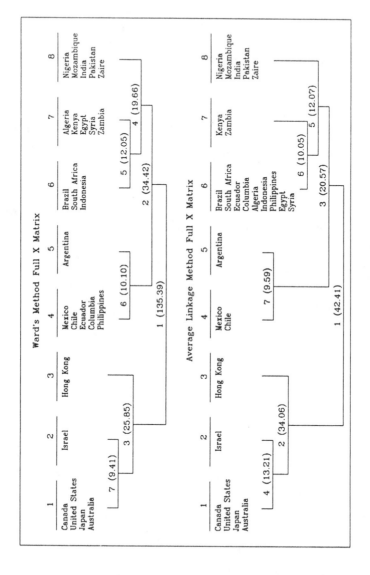

FIGURE 10.10. Tree Diagrams for Full **X** Matrix – World Data

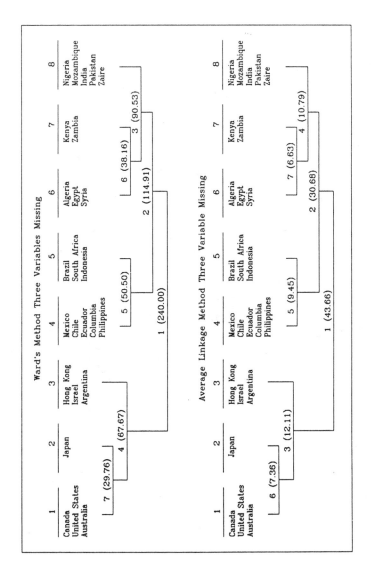

FIGURE 10.11. Tree Diagrams with Three Variables Omitted – World Data

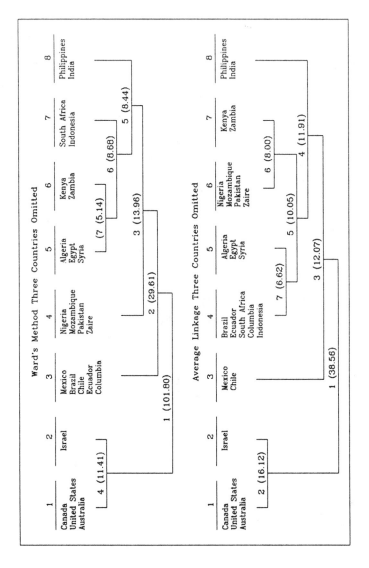

FIGURE 10.12. Tree Diagrams with Three Countries Omitted – World Data

TABLE 10.10. Squared Mahalanobis Distances for
Three Data Matrices – World Data

Country	All Observations	Omitting Variables	Omitting Countries
Canada	7.980	6.612	9.823
United States	14.196	10.602	13.876
Mexico	14.315	11.626	16.194
Brazil	10.983	8.244	15.881
Chile	13.466	4.675	12.613
Argentina	20.833	16.817	–
Ecuador	13.306	9.547	12.482
Algeria	8.572	8.074	10.458
Nigeria	13.001	11.103	13.125
Kenya	11.737	9.739	10.267
South Africa	5.547	4.892	6.894
Mozambique	9.428	7.968	9.190
India	12.600	10.310	14.545
Pakistan	6.437	5.712	9.012
Zaire	8.404	6.477	9.185
Columbia	4.184	3.531	5.708
Japan	16.959	3.514	–
Indonesia	7.577	6.464	9.828
Phillipines	8.810	7.704	10.932
Australia	5.735	3.701	6.003
Egypt	12.667	12.290	16.195
Syria	16.546	15.944	16.528
Israel	16.086	15.166	18.340
Zambia	5.672	5.091	4.922
Hong Kong	22.954	10.195	–

case there is one difference between the two solutions at the eight-cluster level. Ward's method puts South Africa and Indonesia in a separate group, whereas the average linkage method places Mexico and Chile in a separate group. For Ward's method Brazil, Ecuador and Columbia are with Mexico and Chile, whereas in the case of the average linkage method these three countries are placed with South Africa and Indonesia.

It is interesting also to compare the results of the **X** matrix with three variables missing (Figure 10.11) to the results for the full **X** matrix (Figure 10.10). After omitting the three variables contributing the outliers, the countries of Japan, Argentina and Hong Kong moved much closer to Canada, United States and Australia. In the case of Argentina, the change in distances resulted in its moving away from its South American neighbors to form a Hong Kong–Argentina–Israel cluster at an early stage in the process. Other than this major difference the Ward's solution for the remaining countries produced similar results. For the average linkage method however, the tendency for isolated points to remain isolated caused the two solutions to be less similar.

For the three-country omitted matrix Japan, Argentina and Hong Kong have been omitted. In comparing this case to the full X matrix case there is perhaps one puzzling difference. For the full X matrix India is clustered with Nigeria, Mozambique, Zaire and Pakistan, whereas the Philippines is clustered with Brazil, Ecuador, South Africa, Columbia and Indonesia. In the three-country omitted case India joins with the Philippines to form a separate cluster of two. The reason for this change is the impact on the standardization of the variables when the outlier countries are removed. An examination of the Mahalanobis distances in Table 10.10 also shows some changes in location relative to the centroid as a result of the changes in the data.

10.2.2 ASSESSING THE HIERARCHICAL SOLUTION AND CLUSTER CHOICE

As outlined above the hierarchical method of cluster analysis produces a nested sequence of cluster solutions ranging from n (the total number of ojects) to 1. The cluster solutions that are selected from this hierarchy for later use depends on the particular application. In some cases a nested hierarchy within a range of solutions may be used to summarize the relationships among various subgroups. An example would be groups of plants or animals in numerical taxonomy. Alternatively, a particular cluster solution (only one step in the hierarchy) may be selected to be used as a convenient grouping for further analyses. Depending on the application, the hierarchical solution may require additional study before the particular choice of solution (or solutions) is made. In this section techniques are presented for assessing the quality of the hierarchical solution and for choosing an appropriate cluster solution.

Dendograms and Derived Proximities

In hierarchical clustering a tree diagram such as Figure 10.7 can be used to keep track of the sequential clustering process. In such tree diagrams it is also of interest to indicate the proximity value at each step. This *derived proximity measure* indicates the degree of similarity of the two clusters that have just been joined at that step. When these proximity values are included with the tree, the tree diagram is usually called a dendogram. A *dendogram* therefore contains a derived proximity scale that shows the value of the proximity measure at each step of the hierarchical process. Figures 10.10, 10.11 and 10.12 in Section 10.2.1 are examples of *partial dendograms* since only the final stages of the process were shown. The proximity values monotonically increase if the hierarchical clustering process satisfies the ultrametric inequality introduced in Section 10.1.2. A table which summarizes the information in the dendogram is called an *agglomeration schedule*.

For the $(n-1)$ steps in the sequential process the order of the steps forms a one-to-one relation with the proximity measures in the dendogram. The number of the step at which two given objects first appear together in the same cluster is called the *partition rank*. The set of corresponding derived proximities in the dendogram can be used to obtain a new proximity matrix called the *derived proximity matrix*. When two groups of objects are joined, all possible pairs of objects derived from objects in opposite groups are assigned the same derived proximity value. Since there are only $(n - 1)$ unique proximity values in the dendogram, the derived proximity matrix that contains a total of $n(n - 1)/2$ values must therefore have many values in common.

Cophenetic Correlation and Cluster Validity

As a measure of cluster validity it is sometimes of interest to compare the derived proximity matrix to the original proximity matrix. The most common method of comparison is to compute a Pearson correlation between the original values and the derived values. The resulting correlation is called a *cophenetic correlation*. The magnitude of this correlation should be very close to 1 for a high-quality solution. This measure can also be used to compare alternative cluster solutions obtained from different algorithms.

Stress

An alternative method of comparing the two sets of proximities in the case of Euclidean distances is to compute the *stress* measure

$$\sum_{i<j}^{n}\sum^{n}(p_{ij} - \hat{p}_{ij})^2 / \sum_{i<j}\sum \hat{p}_{ij}^2,$$

where p_{ij} denotes the original proximity and $\hat{p}_{ij}$ the derived proximity. This measure is commonly used in multidimensional scaling to evaluate scaling solutions. This topic is outlined in Section 10.3.

Alternative Derived Proximities Based on Centroids

An alternative approach to deriving proximities from the results of the hierarchical process would be to return to each of the steps and compute the distance between the centroids of the clusters joined at each step. These distances could then be used as the derived proximity measures. In this case, the centroid method is not being used as a criterion for forming the clusters but is used as the measure for the derived proximity. In this case, however, the derived proximities would not satisfy the ultrametric inequality.

TABLE 10.11. Agglomeration Schedule for Ward's and Average Linkage Methods: Ten-City Data

Step	Cluster Groups Combined	Proximity Values Ward's	Proximity Values Average Linkage	Change in Proximity Values Ward's	Change in Proximity Values Average Linkage
1	(3) (6)	1.42	2.84		
2	(5) (7)	2.41	4.81	0.99	1.97
3	(5,7) (10)	3.72	6.78	1.31	1.97
4	(9) (2)	3.84	7.68	0.12	0.90
5	(2,9) (4)	6.12	11.10	2.28	3.42
6	(2,4,9) (5,7,10)	10.29	12.22	4.17	1.12
7	(1) (8)	17.17	26.47	6.88	4.25
8	(1,8) (2,4,5,7,9,10)	30.13	34.34	12.96	7.87
9	(3,6) (1,2,4,5,7,8,9,10)	38.08	38.66	9.95	4.32

Note: For pairs of individuals Ward's measure is one-half the squared Euclidean distance.

Example

For the ten-city example introduced above the squared Euclidean distance matrix for the standardized data was given in Table 10.3. Using this proximity matrix both the Ward's and the average linkage methods were used to produce a hierarchical cluster analysis. The agglomeration schedules for the two methods are shown in Table 10.11. Because the two procedures produced an identical sequence of clusters it is possible to compare the agglomeration schedules with respect to the proximity values. The dendogram for the two processes is shown in Figure 10.13. Since the two procedures produced an identical sequence of clusters the same tree diagram can be used for both procedures.

A comparison of the derived proximities to the original proximities is shown in Tables 10.12 and 10.13. In Table 10.12, the upper right triangle contains the partition rank, while the lower left triangle contains the three proximity measures. The first proximity measure is derived from Ward's method, the second proximity measure is derived from the average linkage method and the third is the original squared Euclidean distance measure given in Table 10.3. (The reader should recall that Ward's measure is one-half the squared Euclidean distance when comparing individual objects.) The means and standard deviations for the proximities are given at the bottom of Table 10.12. Comparing the original proximities to the derived proximities shows several unusual cases. The twelfth ranked original proximity (out of a total of 45) between Atlanta and Columbus is 10.9 though these two cities are not joined until the last step (see Table 10.13). The two pairs, Atlanta and Montgomery and Columbus and Wichita also represent pairs that are joined much later than their original proximities would suggest. The cophenetic correlation between the original proximities and the

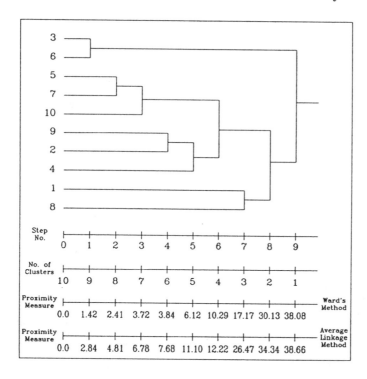

FIGURE 10.13. Dendogram for Hierarchical Cluster Analysis of Ten Cities —
Ward's Method and Average Linkage Method

derived proximities is given by 0.72 and 0.74 for Ward's method and the
average linkage method respectively.

Choosing the Number of Clusters

In the hierarchical clustering process, a sequence of cluster solutions is
obtained with an "ideal" solution appearing for each possible number of
clusters from n to 1. A second step of the cluster analysis is often to select an optimal number of clusters. To assist with the determination of the
appropriate solution an optimality criterion is usually used. As the number of clusters g declines from n to 1 the cluster solution is evaluated by
computing one or more available optimality criteria. At the completion of
the hierarchical process the optimality criteria are studied to determine the
optimum value of g.

The simplest approach to cluster choice uses the value of the group proximity measure for the two groups joined at each step. As the process moves
from step 1 to step $(n-1)$ the value of the group proximity measure, say s,
will increase (for dissimilarity measures). If n is large the increases in s will
be small initially but will tend to grow exponentially. Thus as g decreases

TABLE 10.12. Comparison of Original Proximity Matrix to Derived Proximities: Ten-City Data

	Albuquerque 1	Atlanta 2	Columbus 3	Los Angeles 4	Minneapolis 5	Montgomery 6	Salt Lake 7	Scranton 8	Washington 9	Wichita 10
Albuquerque 1	X	[8]	[9]	[8]	[8]	[9]	[8]	[7]	[8]	[8]
Atlanta 2	30.13 / 34.34 / 33.5	X	[9]	[5]	[6]	[9]	[6]	[8]	[4]	[6]
Columbus 3	38.08 / 38.66 / 44.2	38.08 / 38.66 / 10.9	X	[9]	[9]	[1]	[9]	[9]	[9]	[9]
Los Angeles 4	30.13 / 34.34 / 30.2	6.12 / 11.10 / 12.9	38.08 / 38.66 / 35.9	X	[6]	[9]	[6]	[8]	[5]	[6]
Minneapolis 5	30.13 / 34.34 / 38.6	10.29 / 12.22 / 10.00	38.08 / 38.66 / 32.6	10.29 / 12.22 / 8.7	X	[9]	[2]	[8]	[6]	[3]
Montgomery 6	38.08 / 38.66 / 55.6	38.66 / 38.66 / 14.5	1.42 / 2.84 / 2.8	38.08 / 38.66 / 43.1	38.08 / 38.66 / 36.8	X	[9]	[9]	[6]	[3]
Salt Lake 7	30.13 / 34.34 / 22.3	10.29 / 12.22 / 9.0	38.08 / 38.66 / 23.4	10.29 / 12.22 / 10.7	2.41 / 4.81 / 4.8	38.08 / 38.66 / 30.9	X	[8]	[8]	[8]
Scranton 8	17.17 / 26.47 / 34.3	30.13 / 34.34 / 35.8	38.08 / 38.66 / 57.4	30.13 / 34.34 / 28.6	30.13 / 34.34 / 35.1	38.08 / 38.66 / 64.6	30.13 / 34.34 / 26.8	X	[8]	[8]
Washington 9	30.13 / 34.34 / 32.4	3.84 / 7.68 / 7.7	38.08 / 38.66 / 25.0	6.12 / 11.10 / 9.4	10.29 / 12.22 / 15.1	38.08 / 38.66 / 31.0	10.29 / 12.22 / 12.3	30.13 / 34.34 / 35.7	X	[6]
Wichita 10	30.13 / 34.34 / 32.6	10.29 / 12.22 / 9.1	38.08 / 38.66 / 14.4	10.29 / 12.22 / 17.8	3.72 / 6.78 / 8.5	38.08 / 38.66 / 19.2	3.72 / 6.78 / 5.1	30.13 / 34.34 / 45.1	10.29 / 12.22 / 17.4	X

	Average Proximity	Standard Deviation
Ward's	24.62	13.64
Average Linkage	27.07	13.11
Original	25.15	15.20

Note: The lower left triangle compares the derived proximities to the original proximities. The order of entries in each cell is Ward's, average linkage and original proximities. The upper right triangle shows the partition rank for each pair.

TABLE 10.13. Comparison of Original Proximity Matrix to Derived Proximities: Ten-City Data

Original Pair	Original Proximity	Original Rank	Derived Proximities Wards	Derived Proximities Average Linkage	Partition Rank	Original Pair	Original Proximity	Rank	Derived Proximities Wards	Derived Proximities Average Linkage	Partition Rank
(3,6)	2.8	1	1.42	2.84	1	(7,8)	26.8	24	30.13	34.34	8
(5,7)	4.8	2	2.41	4.81	2	(4,8)	28.6	25	30.13	34.34	8
(7,10)	5.1	3	3.72	6.78	3	(1,4)	30.2	26	30.13	34.34	8
(2,9)	7.7	4	3.84	7.68	4	(6,7)	30.9	27	38.08	38.66	9
(5,10)	8.5	5	3.72	6.78	3	(6,9)	31.0	28	38.08	38.66	9
(4,5)	8.7	6	10.29	12.22	6	(1,9)	32.4	29	30.13	34.34	8
(2,7)	9.0	7	10.29	12.22	6	(1,10)	32.6	30.5	30.13	34.34	8
(2,10)	9.1	8	10.29	12.22	6	(3,5)	32.6	30.5	38.08	38.66	9
(4,9)	9.4	9	6.12	11.10	5	(1,2)	33.5	32	30.13	34.34	8
(2,5)	10.0	10	10.29	12.22	6	(1,8)	34.3	33	17.17	26.47	7
(4,7)	10.7	11	10.29	12.22	6	(5,8)	35.1	34	30.13	34.34	8
(2,3)	10.9	12	38.08	38.66	9	(8,9)	35.7	35	30.13	34.34	8
(7,9)	12.3	13	10.29	12.22	6	(2,8)	35.8	36	30.13	34.34	8
(2,4)	12.9	14	6.12	11.10	5	(3,4)	35.9	37	38.08	38.66	9
(3,10)	14.4	15	38.08	38.66	9	(5,6)	36.8	38	38.08	38.66	9
(2,6)	14.5	16	38.08	38.66	9	(1,5)	38.6	39	30.13	34.34	8
(5,9)	15.1	17	10.29	12.22	6	(4,6)	43.1	40	38.08	38.66	9
(9,10)	17.4	18	10.29	12.22	6	(1,3)	44.2	41	38.08	38.66	9
(4,10)	17.8	19	10.29	12.22	6	(8,10)	45.1	42	30.13	34.34	8
(6,10)	19.2	20	38.08	38.66	9	(1,6)	55.6	43	38.08	38.66	9
(1,7)	22.3	21	30.13	34.34	8	(3,8)	57.4	44	38.08	38.66	9
(3,7)	23.4	22	38.08	38.66	9	(6,8)	64.6	45	38.08	38.08	9
(3,9)	25.0	23	38.08	38.66	9						

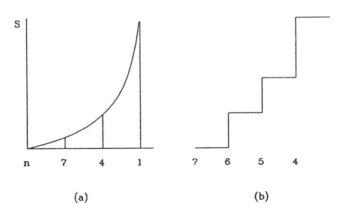

FIGURE 10.14. Plotting the Clustering Criterion

from n to 1 the value of the group proximity measure should behave as in Figure 10.14(a). One approach to the selection of an appropriate value of g would be to examine the behavior of s in a neighborhood of the expected g. If a large change in s occurs at some value of g then the solution $(g + 1)$ immediately prior to this step should be chosen. Figure 10.14(b) illustrates this concept; the function seems to have a much larger jump at step 4 than the two prior steps and hence the five-cluster solution should be selected. An alternative graphical approach involves plotting the changes in S, ΔS, as a function of the number of clusters. Initially the curve ΔS should rise slowly but eventually should rise rapidly when distant clusters are combined. An elbow or dramatic change in the slope of ΔS would be indicative of the appropriate end to the clustering process.

Example

For the ten-city example discussed above, the proximity values for the nine steps for both Ward's method and the average linkage method are summarized in Table 10.11. The changes in the values of these two criteria are also shown in the table. For Ward's incremental sum of squares method the criterion value increases more rapidly beginning with the four-cluster solution. For the average linkage method the criterion value is relatively large at the five-cluster stage and after the four-cluster stage. The formation of the four-cluster stage did not require an average linkage value much greater than for the five-cluster solution. It would appear that a four- or five-cluster solution is reasonable. A summary of the four groups for the four-cluster solution is given below. It is interesting to note that in the original proximity matrix Scranton is equidistant from both Albuquerque and Salt Lake and that Albuquerque is closer to Salt Lake than it is to Scranton. Since the three-cluster solution involves the joining together of Scranton and Albuquerque the fact that Salt Lake is not part of this cluster seems inappropriate. Leaving the two cities as single object clusters therefore seems more justified.

Group 1	Group 2	Group 3	Group 4
Columbus	Washington	Albuquerque	Scranton
Montgomery	Minneapolis		
	Los Angeles		
	Atlanta		
	Salt Lake		
	Wichita		

A Binary Data Example

For the divorce data introduced in Table 10.7 cluster analyses were carried out using the two similarity matrices summarized in Table 10.8. Both the single linkage and complete linkage criteria are applied to both the simple matching coefficient and Jaccard proximity matrices. The derived proximity results for the 19 stages of the four hierarchical analyses are summarized in Table 10.14. Figure 10.15 plots the behavior of the similarity measures over the 19 stages of the hierarchical process. Panel (a) of the figure compares the single linkage (SL) and complete linkage (CL) results for the simple matching coefficient; panel (b) compares the two clustering criteria for Jaccard's coefficient.

For the simple matching coefficient the two clustering criteria provide identical results up to the seven-cluster stage. Between six clusters and four clusters, the difference between the two proximities is 0.778 versus 0.667 in all three cases. Below four clusters the difference between the two proximities is quite large. Using the simple matching coefficient a four cluster solution may be acceptable.

For the Jaccard coefficient the two clustering criteria yield identical derived proximities up to ten clusters. Between nine clusters and seven clusters the difference between the two similarities is always less than 0.100. At six and five clusters the difference jumps to 0.143 and 0.214, while after four clusters the difference between the two coefficients is approximately 0.3. Using the Jaccard coefficient possible solutions range between seven and four depending on the purpose. Certainly with seven clusters the differences between the two proximities is small and the overall degree of similarity is relatively high.

Tables 10.15 and 10.16 show the clustering results for the simple matching coefficient and the Jaccard coefficient respectively. Each table compares the single linkage results to the complete linkage results. For the simple matching coefficient, the single linkage clusters and complete linkage clusters are quite similar. At the five-cluster solution for the single linkage

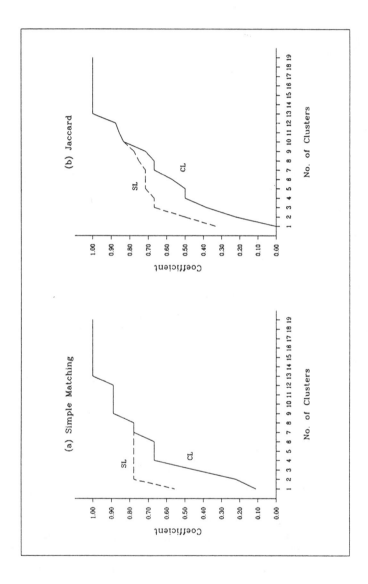

FIGURE 10.15. Behavior of Derived Proximities Single Linkage (SL) versus Complete Linkage (CL) Simple Matching Coefficient and Jaccard Coefficient

TABLE 10.14. Cluster Analysis Derived Proximities for Divorce
Data

		Value of Similarity Measure			
	Number	Simple Matching Coefficient		Jaccard's Coefficient	
Stage	of	Single	Complete	Single	Complete
	Clusters	Linkage	Linkage	Linkage	Linkage
1	19	1.000	1.000	1.000	1.000
2	18	1.000	1.000	1.000	1.000
3	17	1.000	1.000	1.000	1.000
4	16	1.000	1.000	1.000	1.000
5	15	1.000	1.000	1.000	1.000
6	14	1.000	1.000	1.000	1.000
7	13	1.000	1.000	1.000	1.000
8	12	0.889	0.889	0.875	0.875
9	11	0.889	0.889	0.857	0.857
10	10	0.889	0.889	0.833	0.833
11	9	0.889	0.889	0.778	0.714
12	8	0.778	0.778	0.750	0.667
13	7	0.778	0.778	0.714	0.667
14	6	0.778	0.667	0.714	0.571
15	5	0.778	0.667	0.714	0.500
16	4	0.778	0.667	0.667	0.500
17	3	0.778	0.444	0.667	0.375
18	2	0.778	0.222	0.500	0.222
19	1	0.556	0.111	0.333	0.000

method, New York, South Carolina and Vermont are joined with Mary-
land, Texas, and Virginia, but in the complete linkage method the first
three states mentioned are joined with Louisiana and South Dakota.

For the Jaccard coefficient in Table 10.16 there is greater variation between single linkage and complete linkage than in the case of simple matching. For the state of Florida, which is somewhat isolated with respect to
the other states, the single linkage process does not link it with other states
until the two-cluster stage is reached. In comparison the complete linkage
method links Florida at the four-cluster stage. This is an example of the
difference in behavior expected of the two methods with respect to isolated
points. Another common difference between the two linkage methods is in
the size of the clusters. At the five-cluster stage under single linkage the
clusters have sizes 12, 4, 2, 1, 1, whereas for the complete linkage method
the cluster sizes are 6, 6, 4, 3, 1. As expected the complete linkage clusters
have less variance in size.

Test Statistics for Number of Clusters

At each step of the hierarchical process, a derived proximity measure is
available indicating the group proximity measure corresponding to the two

TABLE 10.15. Cluster Membership Based on Simple Matching Coefficient
Single Linkage (SL) vs Complete Linkage (CL)

State	12 SL	12 CL	11 SL	11 CL	10 SL	10 CL	9 SL	9 CL	8 SL	8 CL	7 SL	7 CL	6 SL	6 CL	5 SL	5 CL	4 SL	4 CL	3 SL	3 CL	2 SL	2 CL
FL	1	1	1	1	1	1	1	1	1	1	1	1	1	1	1	1	1	1	1	1	1	1
LA	2	2	2	2	2	2	2	2	2	2	2	2	2	2	2	2	2	2	2	2	2	2
ME	3	3	3	3	3	3	3	3	3	3	3	3	3	3	3	3	3	3	3	3	3	2
MD	4	4	4	4	4	4	4	4	4	4	4	4	4	4	4	4	4	3	4	2	2	2
MA	5	5	5	5	5	5	5	5	5	5	5	5	5	5	5	5	2	3	2	3	2	2
MT	6	5	6	6	6	6	1	1	1	1	1	1	1	1	1	1	1	1	1	1	1	1
NE	6	6	6	6	6	6	6	6	6	6	6	6	1	1	1	1	1	1	1	1	1	1
NH	5	5	5	5	5	5	5	5	5	5	5	5	5	5	5	5	2	3	2	3	2	2
NY	7	7	7	7	7	7	7	7	6	6	6	6	6	6	4	2	4	2	2	2	2	2
ND	8	8	8	8	3	3	3	3	3	3	3	3	3	3	3	3	3	3	3	3	3	2
OK	8	8	8	8	3	3	3	3	3	3	3	3	3	3	3	3	3	3	3	3	3	2
OR	6	6	6	6	6	6	1	1	1	1	1	1	1	1	1	1	1	1	1	1	1	1
RI	5	5	5	5	5	5	5	5	5	5	5	5	5	5	5	5	2	3	2	3	2	2
SC	9	9	9	9	8	8	7	7	7	7	7	7	6	6	4	2	4	2	2	2	2	2
SD	10	10	10	10	2	2	2	2	2	2	2	2	2	2	2	2	2	2	2	2	2	2
TX	11	11	10	10	9	9	8	8	8	8	2	7	2	4	2	4	2	4	2	2	2	2
VT	12	12	11	11	10	10	9	9	8	8	6	6	6	6	4	2	4	2	2	2	2	2
VA	7	7	7	7	7	7	6	6	6	6	6	6	6	6	4	2	4	2	2	2	2	2
WA	6	6	6	6	6	6	1	1	1	1	1	1	1	1	1	1	1	1	1	1	1	1
WV	11	11	10	10	9	9	8	8	8	8	6	7	6	6	4	4	4	4	2	2	2	2

Number of Clusters

TABLE 10.16. Cluster Membership Based on Jaccard's Coefficient
Single Linkage (SL) vs Complete Linkage (CL)

State	12 SL	12 CL	11 SL	11 CL	10 SL	10 CL	9 SL	9 CL	8 SL	8 CL	7 SL	7 CL	6 SL	6 CL	5 SL	5 CL	4 SL	4 CL	3 SL	3 CL	2 SL	2 CL
FL	1	1	1	1	1	1	1	1	1	1	1	1	1	1	1	1	1	1	1	1	1	1
LA	2	2	2	2	2	2	2	2	2	2	2	2	2	2	2	2	2	2	2	2	2	2
ME	3	3	3	3	3	3	3	3	2	3	2	3	2	3	2	3	2	3	2	3	2	2
MD	4	4	4	4	4	4	4	4	3	4	3	4	3	4	2	4	2	4	2	2	2	2
MA	5	5	5	5	5	5	5	5	3	5	3	3	3	3	2	3	2	3	2	3	2	2
MT	6	6	6	6	6	6	3	6	4	6	4	5	4	5	3	5	3	3	3	1	1	1
NE	6	6	6	6	6	6	5	6	4	6	4	5	4	4	3	5	3	3	3	1	1	1
NH	5	5	5	5	5	5	5	5	2	5	2	3	2	3	2	3	2	3	2	3	2	2
NY	7	7	7	7	7	7	6	7	5	4	5	5	5	4	4	4	4	4	2	2	2	2
ND	3	3	3	3	3	3	3	3	2	3	2	3	2	3	2	3	2	3	2	3	2	2
OK	3	3	3	3	3	3	3	3	2	3	2	3	2	3	2	3	2	3	2	3	2	2
OR	6	6	6	6	6	6	5	6	4	6	4	5	4	4	3	5	3	3	3	1	1	1
RI	5	5	5	5	6	5	3	5	2	5	2	3	2	2	2	2	2	2	2	2	2	2
SC	8	8	8	8	8	8	7	8	6	7	6	6	6	6	5	5	2	3	2	2	2	2
SD	9	9	9	9	9	9	8	9	7	8	7	7	6	6	5	5	2	2	2	2	2	2
TX	10	10	10	10	10	10	9	9	8	8	7	7	4	4	4	4	4	4	2	2	2	2
VT	11	11	11	11	10	10	9	9	8	8	7	7	5	5	4	4	4	4	3	2	2	2
VA	7	7	7	7	7	7	6	7	5	4	5	5	5	4	4	5	3	3	2	2	2	2
WA	6	6	6	6	6	6	5	6	4	6	4	5	4	3	3	5	3	1	3	1	1	1
WV	12	12	10	10	9	9	8	9	7	8	7	7	3	4	2	4	2	4	2	2	2	2

clusters joined at that step. If the algorithm employed satisfies the utrametric inequality the derived measures increase monotonically through the entire process. The Clustan software package (see Wishart 1987) employs these derived proximity measures to develop two test statistics for the optimal number of clusters. The statistics are also outlined in Mojena (1977).

The first statistic is called the *upper tail rule* and is based on the assumption that if in truth there are no clusters, the derived proximity measures are simply a set of order statistics corresponding to samples from some underlying probability distribution. If the underlying derived proximity measures can be assumed to be normally distributed, the measured values obtained from the hierarchical process can be treated as a sample from a normal distribution. Denoting the derived measures by $s_1, s_2, \ldots, s_{n-1}$ corresponding to $1, 2, \ldots, (n-1)$ clusters, the sample mean $\bar{s} = \sum_{j=1}^{n-1} s_j$, and the sample standard deviation $v = \left[\sum_{j=1}^{n-1}(s_j - \bar{s})^2/(n-2)\right]^{1/2}$ can be used to derive a test statistic. Standardized values of the observed proximity measures are given by $(s_j - \bar{s})/v$. If this test statistic is large relative to a standardized normal statistic then it can be concluded that the cluster formed at step j is nonoptimal. The value of the corresponding proximity measure s_j at step j in this case is deemed to be too large. This first rule is conservative since if s_j is relatively large the value of $\bar{s}$ and v will also be too large and hence the standardized value will be too small.

A second test statistic called the *moving average rule* employs a moving average fitted to a line obtained by the ordered proximity values $s_1, s_2, \ldots, s_{n-1}$. Assuming that there are no clusters the ordered s_j values are expected to be approximately linear with some slope b_j. Using an r point moving average based on the r points prior to s_{j+1}, the fitted value of s_{j+1}, $\hat{s}_{j+1}$ is given by $(\bar{s}_j + L_j + b_j)$, where b_j is the moving average least squares slope of the line and $L_j = (r-1)b_j/2$. Denoting an estimate of the standard deviation of $\hat{s}_{j+1}$ by v_j a test statistic is given by $(s_{j+1} - \hat{s}_{j+1})/v_j$. Once again large values of this statistic relative to a standard normal statistic would indicate that the $(j+1)$ grouping is not optimal. In comparison to the upper tail rule, this statistic has the advantage that the quantities used to derive $\hat{s}_{j+1}$ do not depend on s_{j+1} and the higher values.

Example

For the ten-city example the Clustan package was used to compute the standardized values of the proximity measures for the upper tail rule described above. The standardized values for both the average linkage method and Ward's method are shown below.

Stage	No. of Clusters							
	9	8	7	6	5	4	3	2
Average Linkage	−0.80	−0.58	−0.56	−0.52	−0.18	0.59	0.95	2.06
Ward's		−0.82	−0.68	−0.64	−0.61	−0.06	0.62	1.34	1.75

Since the data set contained only ten cities, the relatively large values of the proximity measures corresponding to the latter four stages caused the standard deviation to be large relative to the mean. As a result the standardized values were relatively low.

For the moving average rule based on a four-point moving average the standardized deviations from the forecasts are given below.

| | No. of Clusters | | | | |
Stage	6	5	4	3	2
Average Linkage	−0.80	1.79	3.88	0.31	0.83
Ward's	−0.72	5.22	2.24	0.72	−0.25

The moving average deviations at five and four clusters appear to be relatively large. Since the moving averages do not include values beyond the forecast stage the standardized deviations tend to be higher than for the upper tail rule. Either four or five clusters could be justified using the moving average rule.

Some ANOVA Type Statistics

Regardless of which criterion is used to carry out the hierarchical process (average linkage, complete linkage, Ward's, single linkage, etc.), if the original proximity matrix is squared Euclidean distance, then the sums of squares matrices T, W and G can be used to construct a variety of measures to assist with cluster choice. As outlined in Section 10.1.2 the quantities $tr T$, $tr W$ and $tr G$ measure total sum of squares, within sum of squares and among sum of squares respectively. At each step of the hierarchical process $tr W$ increases by $(SSW_t - SSW_r - SSW_s)$ whereas $tr G$ decreases by the same amount. The total sum of squares remains fixed over the entire process.

Pseudo-F, Pseudo-t^2 and Beale's F Ratio

Three F-type statistics that are sometimes used for cluster choice are derived from the changes in the sums of squares described above. Two statistics produced by the SAS procedure CLUSTER are called pseudo-F and pseudo-t^2. The *pseudo-F statistic* is given by

$$F^* = [tr G/(g-1)]/[tr W/(n-g)]$$

and has also been termed the *variance ratio criterion*. Under the assumption of multivariate normality with spherical covariance matrix, this statistic is the conventional one-way ANOVA statistic for testing equality of cluster means. With this rather strong assumption, F^* has an F distribution with $p(g-1)$ and $p(n-g)$ degrees of freedom if the cluster mean vectors are equal. This statistic could be compared to a tabular F using an appropriate Bonferroni p-value to assess the significance of the clusters.

An alternative use of the F^* statistic is to simply monitor the behavior of F^* over the various stages of the process. Initially as g decreases, F^* will decline as $tr\mathbf{W}$ gradually increases and $tr\mathbf{G}$ gradually decreases. The gradual almost monotonic decline in F^* that occurs as like objects are joined, will eventually end, as largely dissimilar objects or clusters are joined. At some point in the process a sudden relatively large decline in F^* should occur if the joining of two clusters result in a large change in $tr\mathbf{W}$ and $tr\mathbf{G}$. The value of g immediately prior to this point should then be considered as a possible optimum value of g. This statistic should perform well if there are a few quite distinct spherical-shaped clusters.

A second statistic that is similar in principal to F^* is the *pseudo-t^2* statistic given by

$$t^{*^2} = \frac{[SSW_t - SSW_r - SSW_s](n_r + n_s - 2)}{[SSW_r + SSW_s]}.$$

The numerator sum of squares in t^{*^2} measures the incremental sum of squares resulting from the joining of clusters r and s to form a new cluster t. The denominator sum of squares is the sum of the within sum of squares for the two clusters being joined. The two cluster sizes are n_r and n_s respectively. Under the assumption of multivariate normality with spherical covariance matrix, the pseudo-t^2 statistic has an F-distribution with p and $p(n_r + n_s - 2)$ degrees of freedom if the two clusters being joined are not distinct. As above this statistic could be compared to a Bonferroni p-value to perform an approximate test of cluster significance. As in the case of F^*, t^{*^2} can be used to monitor the hierarchical process. A relatively large value of t^{*^2} at g clusters would suggest $(g + 1)$ as a possible cluster choice.

Monitoring the pseudo-t^2 statistic is equivalent to monitoring the statistic $[SSW_r + SSW_s]/SSW_t$ over the hierarchical process. A sudden decline in this amount would also be indicative of the joining of two very distinct clusters.

A third F-type statistic is commonly referred to as *Beale's F-ratio* and is given by

$$F' = \frac{[tr\mathbf{W}_1 - tr\mathbf{W}_2]}{tr\mathbf{W}_2} \bigg/ \left[\left(\frac{n - g_1}{n - g_2}\right)\left(\frac{g_2}{g_1}\right)^{2/p} - 1\right],$$

where $\mathbf{W}_1$ and $\mathbf{W}_2$ denote the matrix $\mathbf{W}$ corresponding to g_1 and g_2 clusters respectively and where $g_2 > g_1$. If the g_2 solution is significantly better than the g_1 solution F' will be relatively large. F' can be compared to an F statistic for $p(g_2 - g_1)$ and $p(n - g_2)$ degrees of freedom. This is only an approximate F as in the two previous statistics outlined above.

If the two cluster solutions g_2 and g_1 are consecutive $[g_2 = (g_1 + 1)]$ then F' is given by

$$F' = \frac{[SSW_t - SSW_r - SSW_s]}{tr\mathbf{W}_2} \bigg/ \left[\left(\frac{n - g_1}{n - g_1 - 1}\right)\left(\frac{g_1 + 1}{g_1}\right)^{2/p} - 1\right].$$

In this case the increase in the within group sum of squares is compared to the total within group sum of squares prior to this point in the process. The ratio should therefore be a more reliable measure of change in $tr\mathbf{W}$ than the t^{*^2} statistic, when the hierarchical process is governed by the value of $[SSW_t - SSW_r - SSW_s]$ as in Ward's method.

R^2-Type Measures

A measure of the partition of the total sum of squares $tr\mathbf{T}$, between $tr\mathbf{W}$ and $tr\mathbf{G}$, is given by $R_g^2 = tr\mathbf{G}/tr\mathbf{T}$. This ratio indicates the proportion of the total variation among the objects that is accounted for by variation among cluster groups. As the number of clusters declines R_g^2 also declines. A sudden decrease in R_g^2 would indicate the joining of two clusters that are quite distinct. Another statistic related to R_g^2 is called the *semipartial R^2* and is given by $\Delta R^2 = R_g^2 - R_{(g-1)}^2$. The semipartial R^2 statistic computes the ratio of $[SSW_t - SSW_r - SSW_s]$ to $tr\mathbf{T}$. Both of these statistics are produced by the SAS procedure CLUSTER. Since the numerator of ΔR^2 represents the incremental sum of squares, this quantity can be monitored throughout the process even though Ward's method may not be the hierarchical clustering criterion being used. If the average linkage method is being used to choose clusters the semipartial R^2 provides information using an alternative criterion. The ΔR^2 statistic is also useful for comparing two or more alternative heirarchical solutions based on different criteria.

Example

For the ten-city data the values of the various ANOVA-type statistics are shown in Table 10.17 for Ward's method. Examining the results for F^* and t^{*^2} does not reveal an obvious solution to the number of clusters problem. These results are typical of the results this author has encountered with a variety of data sets using these criteria. For the statistic F' the value seems to increase at four clusters and again at three clusters. The R^2 statistics exhibit behavior already observed in connection with the behavior of Ward's method. Large changes in R^2 begin to occur around the four-cluster solution. When it comes to making inferences there is no "easy to use, works every time statistic." Cluster analysis is an exploratory data analysis technique requiring judgement and cross-validation techniques, as well as the comparison of a variety of solutions derived from alternative clustering criteria.

TABLE 10.17. ANOVA-Type Statistics for Ten-City Data

Stage	No. of Clusters	Ward's Incremental Sum of Squares	Cumulative Incremental Sum of Squares	R_g^2	ΔR_g^2	F^*	t^{*2}	F'
1	9	1.42	1.42	0.987	0.013	9.49	–	–
2	8	2.41	3.83	0.966	0.021	8.13	–	1.63
3	7	3.72	7.55	0.933	0.067	6.96	1.54	1.81
4	6	3.84	11.39	0.899	0.101	7.12	–	1.38
5	5	6.12	17.51	0.845	0.155	6.81	1.59	1.84
6	4	10.29	27.80	0.754	0.246	6.13	4.18	2.36
7	3	17.17	44.97	0.603	0.397	5.32	–	2.65
8	2	30.13	75.10	0.336	0.664	4.05	6.58	2.94
9	1	<u>38.08</u>	113.18	0.000	0.336	–	9.66	1.84
Total		113.18						

Correlation Type Measures of Cluster Quality

Correlation type measures of *cluster quality* are based on a comparison of the original proximity matrix and the cluster group location of each object. The measures are based on the principal that objects that are in the same cluster at any step should have closer original proximity measures than objects that are in different clusters. At each step of the clustering process, all pairs of objects are assigned a new or derived proximity value based on whether or not the pair are both in the same cluster group. Pairs in which objects belong to the same cluster are assigned the value 0, whereas those in which objects are in different clusters are assigned the value 1. All pairs coded 0 are called *within pairs* and all pairs coded 1 are called *between pairs*. Correlation coefficients between the original proximities and the assigned values can be used to determine cluster quality.

Point-Biserial Correlation

The Pearson correlation coefficient between the original $n(n-1)/2$ proximities and the corresponding assigned values (0 or 1) is called the *point-biserial correlation*. The correlation can also be determined using the expression

$$r_b = (\bar{d}_b - \bar{d}_w)(n_b n_w / n_d^2)^{1/2} / s_d,$$

where the subscripts b and w correspond to the groups of pairs coded 1 (between pairs) and coded 0 (within pairs) respectively. The means of the original proximities for the two groups are denoted by $\bar{d}_b$ and $\bar{d}_w$. The number of pairs in each of the two groups is denoted by n_b and n_w. The total number of pairs $n(n-1)/2 = (n_b + n_w)$ is denoted by n_d and the standard deviation of the original proximities is denoted by s_d. A relatively high value of this correlation coefficient (close to 1) would indicate that pairs coded 1 tend to have low proximity values (relatively dissimilar) whereas pairs coded 0 tend to have high proximity values (relatively similar).

Gamma and G(+)

An alternative measure of correlation between the original proximities and the assigned numerical values can be obtained using a *coefficient of concordance*. The two groups of pairs defined above, within and between, are compared, one from each group, to yield a total of $(n_w)(n_b)$ comparisons. All comparisons in which the original proximity of the within pair, exceeded (more similar) the original proximity of the between pair, are classified as *concordant* whereas in the opposite case the comparison is classified as *discordant*. The total number of comparisons is therefore divided into two groups containing $S(+)$ concordant and $S(-)$ discordant comparisons, hence $S(+) + S(-) = n_w n_b$. The gamma coefficient of concordance is given by

$$\gamma = [S(+) - S(-)]/[S(+) + S(-)].$$

A value of γ close to 1 therefore indicates a close agreement between original proximities and cluster group membership. Kendall's coefficient of concordance between the original proximities and the coded 0–1 values is equivalent to the gamma coefficient if all ties are eliminated from the calculation.

An alternative measure of agreement is given by $G(+) = S(-)/[S(+) + S(-)]$. For this coefficient, a value close to 0 is indicative of close agreement between cluster group membership and the original proximities.

Example

For the ten-city example Table 10.18 shows the original proximity values for each pair as well as the 0–1 coding for each of the nine sets of clusters. These coding columns can be used to judge the quality of various cluster solutions. The point-biserial, gamma and $G(+)$ correlations for the latter four cluster solutions are summarized below. The correlations for the four-cluster solution are optimum in all three cases. The four-cluster solution appears to be the best when it comes to preserving the order established by the original proximities. This solution is consistent with the solution selected above based on the values of Ward's and the average linkage criteria.

	No. of Clusters			
	5	4	3	2
Point-Biserial	0.51	0.74	0.71	0.37
Gamma	0.92	0.95	0.89	0.49
$G(+)$	0.04	0.02	0.06	0.25

TABLE 10.18. Coding for Correlation Comparison
of Original Proximities and Coded Pairs

Pair	Original Proximity	No. of Clusters in Solution							
		9	8	7	6	5	4	3	2
(3,6)	2.8	0	0	0	0	0	0	0	0
(5,7)	4.8	1	0	0	0	0	0	0	0
(7,10)	5.1	1	1	0	0	0	0	0	0
(2,9)	7.7	1	1	1	0	0	0	0	0
(5,10)	8.5	1	1	0	0	0	0	0	0
(4,5)	8.7	1	1	1	1	1	0	0	0
(2,7)	9.0	1	1	1	1	1	0	0	0
(2,10)	9.1	1	1	1	1	1	0	0	0
(4,9)	9.4	1	1	1	1	0	0	0	0
(2,5)	10.0	1	1	1	1	1	0	0	0
(4,7)	10.7	1	1	1	1	1	0	0	0
(2,3)	10.9	1	1	1	1	1	1	1	1
(7,9)	12.3	1	1	1	1	1	0	0	0
(2,4)	12.9	1	1	1	1	0	0	0	0
(3,10)	14.4	1	1	1	1	1	1	1	1
(2,6)	14.5	1	1	1	1	1	1	1	1
(5,9)	15.1	1	1	1	1	1	0	0	0
(9,10)	17.4	1	1	1	1	1	0	0	0
(4,10)	17.8	1	1	1	1	1	0	0	0
(6,10)	19.2	1	1	1	1	1	1	1	1
(1,7)	22.3	1	1	1	1	1	1	1	0
(3,7)	23.4	1	1	1	1	1	1	1	1
(3,9)	25.0	1	1	1	1	1	1	1	1
(7,8)	26.8	1	1	1	1	1	1	1	0
(4,8)	28.6	1	1	1	1	1	1	1	0
(1,4)	30.2	1	1	1	1	1	1	1	0
(6,7)	30.9	1	1	1	1	1	1	1	1
(6,9)	31.0	1	1	1	1	1	1	1	1
(1,9)	32.4	1	1	1	1	1	1	1	0
(1,10)	32.6	1	1	1	1	1	1	1	0
(3,5)	32.6	1	1	1	1	1	1	1	1
(1,2)	33.5	1	1	1	1	1	1	1	0
(1,8)	34.3	1	1	1	1	1	1	0	0
(5,8)	35.1	1	1	1	1	1	1	1	0
(8,9)	35.7	1	1	1	1	1	1	1	0
(2,8)	35.8	1	1	1	1	1	1	1	0
(3,4)	35.9	1	1	1	1	1	1	1	1
(5,6)	36.8	1	1	1	1	1	1	1	1
(1,5)	38.6	1	1	1	1	1	1	1	0
(4,6)	43.1	1	1	1	1	1	1	1	1
(1,3)	44.2	1	1	1	1	1	1	1	1
(8,10)	45.1	1	1	1	1	1	1	1	0
(1,6)	55.6	1	1	1	1	1	1	1	1
(3,8)	57.4	1	1	1	1	1	1	1	1
(6,8)	64.6	1	1	1	1	1	1	1	1

10.2.3   COMBINING HIERARCHICAL CLUSTER ANALYSIS WITH OTHER MULTIVARIATE METHODS

Interpretation of the Cluster Solution

After a cluster analysis solution has been obtained, it is usually of interest to characterize the clusters with respect to the variables used to derive them. As a result of the cluster analysis, a new categorical variable denoting the cluster group membership can be added to the original data matrix $\mathbf{X}$. The task of characterizing the clusters therefore consists of relating the new categorical variable to the original p variables in $\mathbf{X}$. Each of the p variables can be related to the cluster variable individually using ANOVA or simultaneously using MANOVA. Since there are usually a large number of variables a principal components analysis could be used initially to reduce the number of variables. Alternatively, a discriminant analysis based on the original p variables can be used to characterize the differences among the cluster groups. These techniques will now be demonstrated in the example below.

Example

The air pollution data in Table 9.7 contains observations for 40 cities. So far in this chapter the analysis has been based on only 10 of these cities. A cluster analysis will now be carried out using the entire (40 × 11) data matrix in Table 9.7. Using Ward's incremental sum of squares method a hierarchical cluster analysis was carried out. An examination of the behavior of the incremental sum of squares over the 39 stages revealed a steady exponential rise without any sudden or major jumps. The moving average rule based on a 16-point moving average indicated that three or five clusters might be appropriate. For convenience a three-cluster solution will be used for this example. The three groups of cities are summarized below.

Group 1		Group 2	Group 3
San Jose	Portland	Roanoke	Philadelphia
Albuquerque	Springfield	Charleston	Wilmington
Harrisburg	Salt Lake	Greenville	Scranton
Hartford	Wichita	Columbus	New York
Orlando	Lorain	Jacksonville	Wheeling
Sacramento	Hamilton	Atlanta	Canton
Washington	San Diego	Montgomery	Baltimore
Minneapolis	Duluth	Baton Rouge	
Los Angeles	Wilkes Barre	San Antonio	
Greensboro	Cincinnati		
Madison	Saginaw		
Tacoma	Springfield		

TABLE 10.19. ANOVA Results for Three
Cluster Groups on Eleven Air Pollution
Variables

	Group 1	Group 2	Group 3
SMEAN	−0.30	−0.61*	1.81*
PMEAN	−0.27	−0.35*	1.39*
PMAX	−0.24	−0.21	1.08*
SMIN	−0.27	−0.35*	1.39*
SMAX	−0.28	-0.56*	1.67*
PMIN	−0.29	−0.39*	1.51*
PERWH	0.43*	−1.25*	0.13
NONPOOR	0.45*	−1.39*	0.25
GE65	0.13	−0.93*	0.75*
PM2	−0.18	−0.04	0.68*
LPOP	−0.01	−0.45*	0.61*

For comparison purposes a three-cluster solution was also generated using the average linkage method. The outcome in this case resulted in a
single city cluster for Albuquerque, a second cluster identical to the cluster
in group 3 of Ward's and a third cluster which contained all the cities in
groups 1 and 2 of the Ward's solution (except Albuquerque). This provides
another example of the tendency for the average linkage method to leave
outliers isolated until late in the process.

ANOVA

Using a one-way analysis of variance, a comparison of means over the three
groups (determined using Ward's method) was carried out for each of the
eleven variables. The results from this analysis are shown in Table 10.19.
Since the variables have been standardized the overall mean in each case is
zero and the standard deviation is always 1. A 95% confidence interval for
the mean in each case is $(−0.32, 0.32)$. Values in the table that are outside
of this interval have been marked with an *. An examination of the values in
the table reveals that the group 3 means tend to be large particularly for the
pollution variables. In addition the means for PM2, GE65 and LPOP are
also significantly large. In contrast to group 3, group 2 shows means that are
significantly negative in most cases. The means for PERWH, NONPOOR
and GE65 are particularly large and negative. For group 1 the means for
PERWH and NONPOOR are significantly positive, whereas the remainder
of the means are not significantly different from zero. The means for the
air pollution variables however are all negative and very close to the lower
boundary of −0.32.

MANOVA and Discriminant Analysis

A multivariate analysis of variance shows that the three (11×1) mean vectors are significantly different. A discriminant analysis can therefore be used to find two linear combinations of the eleven variables that best characterize the group differences. In this case, both discriminant functions obtained were highly significant and are given by

$$
\begin{aligned}
y_1 = \ & 0.12 \ \text{SMEAN} + 0.71 \ \text{SMAX} + 0.45 \ \text{SMIN} + 0.30 \ \text{PMEAN} \\
& +0.08 \ \text{PMAX} + 0.08 \ \text{PMIN} + 0.56 \ \text{PM2} + 0.00 \ \text{LPOP} \\
& -0.36 \ \text{GE65} - 0.44 \ \text{PERWH} - 0.89 \ \text{NONPOOR}
\end{aligned}
$$

and

$$
\begin{aligned}
y_2 = \ & 0.51 \ \text{SMEAN} + 0.31 \ \text{PMEAN} + 0.15 \ \text{PMAX} - 0.17 \ \text{SMAX} \\
& +0.11 \ \text{SMIN} - 0.20 \ \text{PMIN} + 0.06 \ \text{PERWH} + 0.75 \ \text{NONPOOR} \\
& +0.05 \ \text{LPOP} + 0.51 \ \text{GE65} - 0.26 \ \text{PM2}.
\end{aligned}
$$

The first discriminant function contrasts the six pollution variables and population density with the three demographic variables GE65, PERWH and NONPOOR. It would appear that cities with large values of y_1 tend to have relatively large values of the pollution variables and relatively small values of the three demographic variables. The means on y_1 for the three groups were -1.7, 1.8 and 3.6 indicating therefore that groups 1 and 3 differ the most on the y_1 dimension with group 3 having the largest values of y_1 and group 1 having the smallest values.

For the second discriminant function the negative coefficients seem to be minor, whereas the positive coefficients are dominated by NONPOOR, GE65 and SMEAN. The correlation between y_2 and the 11 variables is positive in all cases with the highest correlation occurring between y_2 and each of SMEAN, SMAX, NONPOOR, PERWH, PMIN, GE65, PMEAN and SMIN. Cities with large values of y_2 therefore tend to have higher values of the pollution variables and the three demographic variables. The means for groups 1, 2 and 3 were 0.35, -2.77 and 2.35 respectively. The y_2 dimension therefore seems to separate cluster groups 2 and 3 with group 3 having high values of y_2 and group 1 having low values of y_2. Combining the results from the two discriminant functions, it would appear that group 3 has relatively high values of both the pollution variables and the demographic variables with the pollution variables being higher in a relative sense than the demographic variables. Group 1 appears to have moderate values of the 11 variables with the pollution variables being smaller relatively than the demographic variables. Group 2 appears to have relatively low values of all variables with the demographic variables being smaller than the pollution variables. The discriminant function results and the ANOVA results correspond very well overall, as they should.

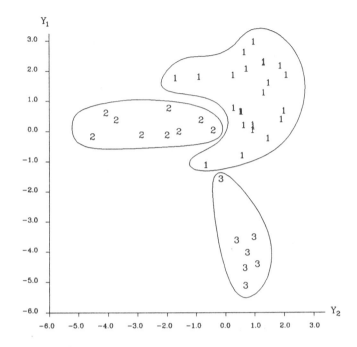

FIGURE 10.16. Discriminant Analysis of Cluster Groups

A scatterplot of the 40 cities with respect to the two discriminant functions is shown in Figure 10.16. The cluster group number is used as a plotting symbol. For the most part the three groups of cities are well separated with respect to the two dimensions.

Principal Components Analysis and Factor Analysis

Since there are 11 variables involved in the analysis, it may also be useful to use principal components analysis or factor analysis to reduce the number of variables before comparing the groups. A factor analysis of the correlation matrix for the eleven variables was carried out in Chapter 9 using the principal component method. Table 9.23 shows the loadings for the four factors obtained from a varimax rotation. Using the factor scores on the four factors, the means on the four factors were compared across the three groups using one-way analysis of variance. The means are summarized in Table 10.20.

With the exception of factor 3 the means were significantly different over the three cluster groups. From the factor loadings in Table 9.23, factor 1 represents the pollution variables PMIN, PMEAN and PMAX, whereas factor 2 represents the pollution variables SMIN, SMEAN and SMAX, as well as the variable GE65. The last factor, factor 4, is negatively corre-

TABLE 10.20. Cluster Group
Mean on Four Factors

	Cluster Group Means		
Factor	1	2	3
1	-0.27	-0.14	1.10
2	-0.27	-0.36	1.39
3	-0.07	-0.31	0.64
4	-0.56	1.40	0.13

lated with PERWH and NONPOOR. From these results we can conclude that cluster group 3 has relatively high values for the two sets of pollution variables, whereas group 2 has relatively low values of PERWH and NON-POOR. Group 1 appears to have relatively high values of PERWH and NONPOOR. These results are comparable to the ANOVA results obtained above using the original 11 variables.

Principal Components Analysis Prior to Cluster Analysis

In applications involving a relatively large number of variables, it is sometimes advantageous to first perform a principal components analysis on the data matrix X before the hierarchical cluster analysis. By keeping only a small portion of the principal components the number of variables used for the cluster analysis can be reduced substantially. This practice, however, must be carried out with extreme caution because of the weighting and scaling effects that principal components analyses can introduce.

In Chapter 9, it was demonstrated that the form of the $X'X$ matrix used for principal component analysis is usually one of a correlation matrix, covariance matrix or raw sum of squares and cross-products matrix. Depending on which of these three forms is used the nature of the principal components vary. In addition, it has been suggested in Section 10.1 that standardization of the variables is usually an important preliminary step before performing a cluster analysis. In general, the correlation matrix should be used to generate the principal components to ensure that certain variables do not dominate the solution because of scale differences. This would be consistent with using standardized variables in cluster analysis since in both cases the beginning data matrix X is the same.

A second consideration of extreme importance in cluster analysis is the effective weight given to each of the variables. A data matrix X may contain a few groups of highly correlated variables and hence may only represent a few underlying dimensions. In a cluster analysis each standardized variable in X is usually given the same weight. Thus if some dimensions are overrepresented by sets of highly correlated variables, the resulting cluster analyses will give greater weight to these overrepresented dimensions. A preliminary principal components analysis of the correlation matrix can be useful for extracting the underlying dimensions before the cluster analysis is carried out. It is also important to keep in mind that if only the most important components are retained for the cluster analysis, outlier effects carried by the latter components will be lost. This may or may not be advantageous. An important issue remaining is whether the resulting principal component scores should be standardized prior to the cluster analysis. This issue is discussed next.

In Chapter 9 it was demonstrated that a small number of principal components can be used to provide an approximation to the data matrix $\mathbf{X}$. Recall that if the matrix of unstandardized principal component scores is given by $\mathbf{Z}(n \times p)$ then $\mathbf{X} = \mathbf{ZV}$ where $\mathbf{V'V} = \mathbf{VV'} = \mathbf{I}$ and $\mathbf{V}$ is the matrix of eigenvectors of $\mathbf{X'X}$. The matrix $\mathbf{XX'}$ therefore can be written as $\mathbf{ZZ'}$. If $\mathbf{Z} = [\mathbf{Z_1 Z_2}]$ is a partition of $\mathbf{Z}$ then $\mathbf{XX'} = \mathbf{Z_1 Z_1'} + \mathbf{Z_2 Z_2'}$ and hence if $\mathbf{Z_1}$ provides a close approximation to $\mathbf{X}$ then $\mathbf{Z_1 Z_1'}$ approximates $\mathbf{XX'}$. If, however, the principal component scores are standardized, then $\mathbf{X} = \mathbf{Z^* V^*}$, where $\mathbf{Z^*} = \mathbf{Z} \Lambda^{-1/2}$, $\mathbf{V^*} = \Lambda^{1/2} \mathbf{V}$ and $\mathbf{Z'Z} = \Lambda$. In this case, $\mathbf{XX'} = \mathbf{Z^* V^* V^{*'} Z^{*'}}$, and hence using $\mathbf{Z^* Z^{*'}}$ to approximate $\mathbf{XX'}$ involves in effect a changing of the scale since $\mathbf{Z^* Z^{*'}} = \mathbf{XV'} \Lambda^{-1} \mathbf{VX'}$. Proximity measures based on $\mathbf{Z^* Z^{*'}}$, which are concerned with scale such as Euclidean distance will yield different results than the same proximity measures based on $\mathbf{ZZ'}$. If principal components are used prior to a cluster analysis, it would therefore seem "safer" to use the unstandardized principal component scores for the cluster analysis. If the proximity measure being used however is a correlation type measure, then the scaling of the principal component scores prior to clustering should not affect the proximities. The use of principal components analysis as a preliminary step to cluster analysis is demonstrated next.

Example

For the data matrix of 40 cities used in the previous example a principal components analysis of the correlation matrix was carried out. The unstandardized principal component scores for the first four components were then used as the input for a Ward's hierarchical cluster analysis. The results of the principal components analysis can be viewed in Tables 9.14 and 9.15. These two tables can be used to interpret the four components. The first four components represent the eigenvalues that exceed one and together account for 85.6% of the variation.

A three-cluster solution produced clusters very similar to those obtained in the previous example. In this case the cities of Orlando and Greensboro moved from group 1 to group 2, whereas the cities of Washington, Los Angeles and Cincinnati moved to group 3 from group 1. The reason for these slight changes is the differences in weights attached to the various dimensions by the principal components. A comparison of the means for

TABLE 10.21. Comparison of Cluster Group Means on the Four
Principal Components

| Principal | Cluster Group | | | Overall Standard Deviation |
Component	1	2	3	of Component
1	−0.482	−1.939	3.050	2.165
2	0.457	−0.709	−0.089	1.337
3	0.822	−0.915	−0.555	1.307
4	0.549	−0.663	−0.314	1.109
Group Size	19	11	10	

the principal component scores across the three cluster groups is given in
Table 10.21. The magnitudes of the means for each principal component
reflect the differing standard deviations for the four components. The standard deviations of the components are the square roots of the eigenvalues
in Table 9.14. The interpretation of the differences in means is left as an
exercise for the reader. It would appear, however, that the use of principal
components analysis has reduced slightly the impact of the six pollution
variables while increasing the relative importance of the five demographic
variables.

10.2.4 OTHER CLUSTERING METHODS

Although the hierarchical method is the most commonly used technique
for clustering there are other approaches. The remaining methods are commonly classified into the four categories: partitioning methods, Q-sort meth-
ods, density methods, and clumping methods. Of these methods, the partitioning methods are more commonly used and are most similar to the
hierarchical approach. These methods are briefly outlined next.

Partitioning Methods

Partitioning methods usually begin with a preselected or target number of
groups or clusters, say g. An initial allocation of the objects into g groups
is carried out followed by a determination of the proximity between each
object and each group. Objects are then placed in the groups to which
they are closest. The proximities are recalculated each time objects change
groups. The process of reassigning objects to new groups continues until
all objects belong to the closest group. The method consists of two major phases, a choice of the initial g clusters and the reallocation process.
Partitioning methods do not usually use the initial proximity matrix but
instead employ the original data matrix $\mathbf{X}$. This is a contrast to the hierarchical method, which sequentially updates the proximity matrix without
ever returning to $\mathbf{X}$. The difference lies in the fact that in the hierarchical

method objects cannot change groups, whereas in the partitioning method this is not the case.

The k-Means Algorithm

A commonly used partitioning method is the *k-means algorithm* which measures the proximity between groups using the Euclidean distance between group centroids. In this method the quest is for k clusters. Using the notation employed here the method would be labeled a g-means algorithm. Beginning with the initial selection of g groups objects are reassigned until they are located in the group with the nearest centroid. As objects are reassigned the group centroids must be revised. For a given level of g, equilibrium is reached when all objects are located in the group whose centroid is closest.

Selecting the Initial Partition

The initial group phase can begin with either a preselected set of g clusters or, with a preselected set of g *seed points* or objects that are then used to locate the remaining objects. The initial set of g clusters may be available from another clustering solution, from a previous study, or may be dictated by some underlying hypothesis.

The number of possible initial solutions is large if n is large relative to g. The number of possible partitions of n objects into g groups is given by

$$\frac{1}{g!} \sum_{i=1}^{g} \binom{g}{i} (-1)^{g-i} i^n,$$

which is of the order $g^n/g!$ when n is large. To evaluate all possible partitions would therefore be prohibitive if n is large. In practice therefore a few well-chosen initial solutions are required. Several different starting configurations should be used to ensure the validity of the final solution.

In the absence of a beginning set of g clusters the initial stage begins with a set of g objects or seed points around which g clusters are formed. Proximity measures are computed between each of the $(n - g)$ remaining objects. Other criteria that could be used include $tr\mathbf{W}$, $|\mathbf{W}|$ and $\mathbf{G}^{-1}\mathbf{W}$. In addition a procedure suggested in the Clustan procedure RELOCATE is based on a split of the squared Euclidean distance into two components. The two components are the size component based on differences between profile means and the shape component based on the variances of the two profiles and the covariance between them. An example using the RELOCATE procedure is given below.

An Example Using RELOCATE

The RELOCATE procedure is a generalization of the *k*-means algorithm. Given a particular partition of the n objects into g groups, the RELOCATE procedure uses a given criterion to measure the proximity between each object and g groups. Each object is then placed into the closest group. The process is carried out sequentially so that after each object is placed, the criterion is recomputed for placing the next object. The process continues until all objects are located in their closest group. It is therefore possible for an object to move more than once before equilibrium is reached. A hierarchical clustering process can be carried out by joining the two closest clusters at each equilibrium. The initial partition can be based on a random allocation or can be based on a prior cluster analysis solution.

For the purposes of testing cluster validity, the Clustan manual suggests that the outcome of three alternative two-step procedures be compared. Initially a random allocation is made to g' clusters where g' is larger than the expected solution. One first-step solution consists of reallocating the initial solution using the size component whereas a second first-step solution consists of reallocating the initial solution using the shape component. The third first-step solution consists of the initial random allocation with no relocation. The second step of the recommended process produces a heirarchical solution by using the Euclidean distance measure to reduce the number of clusters and at each step to reallocate the objects. The procedure is carried out for each of the three first-step solutions and produces three sets of hierarchical solutions each ranging from g' clusters to 1 cluster.

Using the air pollution data for the 40 cities discussed above three initial ten-group solutions were generated. Then, using the squared Euclidean distance criterion, the RELOCATE procedure was applied to join clusters and to relocate objects. At the five-cluster stage all three methods produced the same solution. The solutions of course then remained the same for the later stages of the process. For the five-cluster stage the cluster groups are given by

1		2	3	4
San Jose	Salt Lake	Charleston	Hartford	Wilmington
Roanoke	Wichita	Greenville	Philadelphia	Scranton
Harrisburg	Lorain	Columbus	Washington	Canton
Sacramento	Hamilton	Orlando	Los Angeles	Wheeling
Minneapolis	San Diego	Greensboro	Baltimore	
Madison	Duluth	Jacksonville	Cincinnati	5
Tacoma	Wilkes Barre	Atlanta	New York	Albuquerque
Portland	Saginaw	Montgomery		
Springfield, OH	Springfield, MA	San Antonio		
		Baton Rouge		

The three-cluster solution derived from this RELOCATE procedure was compared to the three-group solution obtained in Section 10.2.4 using

Ward's. The three cluster solution derived from the RELOCATE procedure is identical in terms of the three cluster sizes 24, 9 and 7. There are some differences in group membership however since in comparison to the Ward's solution, Roanoke and San Antonio move to group 1, whereas Orlando and Greensboro move to group 2.

Classification Typologies and Q-Sort Methods

The *classification typology* approach to clustering objects has been used extensively in psychology and psychiatry for the classification of human personalities and various psychological disorders. The approach is based on the properties of a spectral decomposition of the $(n \times n)$ matrix $\mathbf{XX}'$. Since the number of variables p is usually much less than the number of individuals n, the rank of $\mathbf{XX}'$ is usually p. The spectral decomposition therefore yields a set of p components which may be viewed as a set of "pure" types. The classification process then involves assigning the n individuals to the closest "pure" type.

This method may be viewed as a factor analysis carried out on the n rows of $\mathbf{X}$ rather than the p columns of $\mathbf{X}$ and hence is also referred to as a Q-sort method rather than the conventional R-sort method of factor analysis. For a data matrix $\mathbf{X}(n \times p)$ with appropriate standardization, the matrix $\mathbf{X}'\mathbf{X}$ is called an R-type matrix, whereas the matrix $\mathbf{XX}'$ is called a Q-type matrix. The R-type matrix summarizes the correlations among the columns or variables of $\mathbf{X}$, while the Q-type matrix summarizes the correlations among the rows or profiles of $\mathbf{X}$.

The eigenvectors of $\mathbf{XX}'$ provide vectors of coefficients that can be used to obtain the pure types as a linear combination of the n objects. A rotation of the initial solution is usually carried out in an attempt to obtain pure types that depend on only a small number of individuals. Ideally any one individual should be primarily a determinant of one and only one "pure" type. Thus after rotation each individual should load highly on only one factor or ideal type. An extensive discussion of this approach to cluster analysis is available in Overall and Klett (1972).

Density Methods

In applications where *natural clusters* are expected, methods are used that search for regions of high density commonly called modes. Natural clustering usually suggests that there should be many points in space that are very close to other points and that these clusters are separated by areas with very few points. Single linkage approaches are usually used in this category. A popular technique is called *mode analysis*. This method determines dense points which are used to define the initial clusters. A radius r and a number of points k is selected initially. Around each point or object a sphere of radius r is determined and the number of points contained in the sphere is then calculated. All points with at least k other points contained in the sphere are called dense points. The initial clusters are defined by the dense points in such a way that if a dense point belongs to more than one cluster the relevant clusters are combined. Clusters are also combined if the distance between them is less than a threshold value c, which is the average of the $2k$ smallest distances between the original n points. Any point separated from all dense points by at least r forms its own cluster. After the initial solution has been determined r can be increased and the process repeated.

The *taxmap method* also uses a single linkage approach. Beginning with the proximity matrix, the two closest individuals are selected to form the first cluster. A new proximity matrix is computed relating the cluster to the other points. The closest point to the cluster is again determined. The average measure of proximity among the three is computed and compared to the measure of proximity between the first two. The difference between the two measures is referred to as the measure of discontinuity. If this measure is larger than some preassigned value, then the individual is not added to the cluster and a new cluster is initiated with the rejected point. The process is now repeated for the new cluster.

Clumping Techniques or Fuzzy Clustering

Clumping techniques usually begin with a proximity matrix. Beginning with a preselected level of proximity p all points are joined if their level of proximity is at least as large as p. A cluster is formed by finding the largest possible subset in which all points are joined to all other points in the set. If one cluster has points that are joined to at least k points in another cluster, the two clusters are combined. The integer k is preselected. In this method the clusters are allowed to overlap to a certain extent.

10.2.5   CLUSTER VALIDITY AND CLUSTER ANALYSIS METHODOLOGY

The discussion of cluster analysis presented in this chapter suggests that the methodology is exploratory. The outcome of the analysis depends to a large extent on the technique selected, on the variables selected, and on the underlying cluster structure — if indeed there is one. The following quotation from Milligan (1981) sets the stage for the discussion presented in this section.

> An inherent problem in the use of a clustering algorithm in practice is the difficulty of validating the resulting data partition. This is a particularly serious issue since virtually any clustering algorithm will produce partitions for any data set, even random noise data which contains no cluster structure. Thus, an applied researcher is often left in a quandry as to whether the obtained

clustering of a real life data set actually represents significant cluster structure or an arbitrary partition of random data.

Cluster Validity

To be valid therefore, a cluster solution should not be a structure that could have occurred by chance by random sampling from a homogeneous population. The structure must be unusual to be valid. The underlying randomness assumption could be expressed in several ways. The location of the n points in Euclidean space is random, or the assignment of n points to g clusters is random, or finally, the rank order of the observed proximities is random. Tests for randomness usually appeal to one of these concepts of randomness. An extensive discussion of this topic is provided in Jain and Dubes (1988).

The *validity* of a cluster structure can be examined in several ways. *External criteria* measure cluster solutions against a priori information regarding structure. Evaluation of clustering algorithms using samples from known clusters is an example of an external evaluation of cluster methodology. *Internal criteria* are used to evaluate a cluster solution relative to the underlying sample data matrix and the corresponding proximity matrix. With internal criteria the issue is the goodness of fit of the cluster solution relative to the original proximity matrix. This issue was explored in Section 10.2.2. A third criterion for the evaluation of a cluster solution is *replicability* which involves the use of cross-validation procedures. Comparison of results from split-half samples would be an example of a replicability evaluation. Finally, a comparison of cluster solutions obtained from alternative clustering algorithms applied to the same data matrix, constitute what is usually referred to as the application of *relative criteria*. In this case indices of agreement can be computed between alternative cluster solutions.

Monte Carlo Studies

A large variety of Monte Carlo experiments have been carried out over the past two decades for the purpose of evaluating cluster analysis methodology. The studies carried out have had a variety of purposes. To illustrate the variety of experiments, four studies carried out by Milligan (1980) and (1981), Milligan and Cooper (1985) and (1988) are summarized below. The purposes of the four studies were to evaluate clustering algorithms, rules for cluster choice, criteria for measuring cluster internal validity and variable standardization procedures. These four reports are used here for illustration because they are all based on the same underlying design.

The Underlying Cluster Population

The experimental design for the experiments consisted of 36 cells derived from a $4 \times 3 \times 3$ design. The first design factor was the number of clusters (2, 3, 4 or 5) and the second factor was the number of variables (4, 6 or 8). The third factor represented the distribution of points across clusters and consisted of three levels: (a) equal distribution, (b) one cluster with 10% and remaining clusters equal and (c) one cluster with 60% and remaining clusters equal.

The points for the clusters were selected randomly from a truncated multivariate normal with diagonal covariance matrix. The cluster means and variances were selected to guarantee complete separation on one dimension and to ensure that the clusters satisfied the properties of external isolation and internal cohesion. The construction of cluster boundaries with respect to the various dimensions was somewhat complex and will not be described here. Each of the 36 cells was used to generate samples of 50 points each. Three replications were used to yield 108 data sets of 50 points each.

Evaluation of Clustering Algorithms – Milligan (1980)

Ten different types of *error perturbation* were used in the 108 data sets to yield a total of 1080 data sets. The types of error were addition of outliers (2 levels), measurement error (2 levels), addition of random noise (2 levels), addition of random noise dimensions (2 levels), use of a correlation type proximity measure (2 different measures), standardization of the variables and finally, the original unperturbed data set.

Fifteen clustering algorithms were evaluated including single linkage, complete linkage, average linkage, centroid method, Ward's method and the k-means partitioning method. The k-means method was evaluated with random starts and with preliminary solutions based on the other methods. The best all-around method seemed to be the average linkage method, although the single linkage method performed best in the presence of outliers. The k-means algorithm following an average linkage method start also performed well.

Evaluation of Internal Criterion Measures – Milligan (1981)

Using the same 108 data sets described above, 432 sets were constructed using four different error characteristics; no error, random noise, measurement error and random noise dimensions. For each of the 432 data sets four hierarchical algorithms (single linkage, complete linkage, average linkage and Ward's) were employed. Thirty internal criterion measures were evaluated. A set of six measures were judged to be outstanding. Three of these six measures were the point-biserial correlation, gamma and $G(+)$, which are described in Section 10.2.2.

Cluster Choice – Milligan and Cooper (1985)

Using the 108 data sets described above four clustering methods (single linkage, complete linkage, average linkage and Ward's) were applied to yield a total of 432 sets of hierarchical solutions. Each set of solutions was evaluated using a total of 30 cluster choice criteria.

Some of the best performing choice criteria were gamma, Beale's F' ratio, pseudo-F, $G(+)$, Mojena's average rule, point-biserial correlation and the statistic $[SSW_r + SSW_s]/SSW_t$, which is equivalent to using the statistic pseudo-t^2. All of the above were in the top 9 out of 30 criteria studied. The pseudo-F statistic showed the best results by correctly identifying the number of clusters in 390 out of 432 whereas the Mojena procedure was correct on 289 out of 432 cases and was rated ninth out of 30.

Interestingly, procedures that have been designed on the basis of multivariate normal mixtures for cluster structure tended to perform poorly. The procedures in this category included $|\mathbf{T}|/|\mathbf{W}|$, $tr\,\mathbf{W}^{-1}\mathbf{G}$, $tr\,\mathbf{W}$, $\log(|\mathbf{T}|/|\mathbf{W}|)$ and $g^2|\mathbf{W}|$. This result may be due to the particular design of clusters used in the study.

Variable Standardization Procedures – Milligan and Cooper (1988)

In a Monte Carlo evaluation of various standardization procedures, the following six standardization procedures were compared:

(a) $Z = (X - \overline{X})/S$

(b) $Z = X/\max(x)$

(c) $Z = X/[\max(x) - \min(x)]$

(d) $Z = X/\Sigma X$

(e) $Z = \mathrm{Rank}(X)$

(f) no standardization, $Z = X$.

The 108 basic data sets described above were perturbed with respect to cluster separation (2 levels), cluster variance (2 levels) and error conditions (4 levels). Four different hierarchical methods, single linkage, complete linkage, average linkage and Ward's were used to perform the analyses. Cluster recovery was evaluated at levels g, $(g + 3)$, $(g + 6)$ and $(g + 9)$ where g is the correct number of clusters.

The results of the Monte Carlo study indicated that division by the range [method (c)] was usually the best method of standardization. The rank transformation [method (e)] performed very poorly.

TABLE 10.22. Comparison of Locations of Pairs of
Points Between True Solution and Derived Solution

| | | True Solution | |
		Pair in Same Cluster	Pair Not in Same Cluster
Derived Solution	Pair in Same Cluster	a	b
	Pair Not in Same Cluster	c	d

On the Measurement of Cluster Recovery and External Measurement Criteria

Over the decade spanned by the four Monte Carlo studies outlined above, improvements have been made with respect to the measurement of external validity. Based on a number of studies, the recommended measure of external validity is the adjusted Rand index developed by Hubert and Arabie (1985). This index, which is used to compare a derived cluster solution to a true cluster solution, is given by

$$Ra = (a + d - n_c)/(a + b + c + d - n_c),$$

where n_c is an adjustment to correct for chance agreement. The parameters a, b, c and d are defined by Table 10.22. The adjustment for chance agreement, n_c, is given by

$$n_c = [n(n^2 + 1) - (n + 1)\Sigma n_i^2. - (n + 1)\Sigma n_{.j}^2 + 2\Sigma\Sigma n_i^2. n_{.j}^2/n]/[2(n - 1)],$$

where n_{ij} denotes the number of points in cluster i for the derived solution which are also in cluster j of the true solution. The marginals are given by $n_i. = \sum_j n_{ij}$ and $n_{.j} = \sum_i n_{ij}$ and the grand total by $n = \sum_i \sum_j n_{ij}$. The index was used in Milligan and Cooper (1988) but was not used in their prior studies. See Milligan and Cooper (1986) for a comparison to other external criteria.

A second important consideration in Monte Carlo studies is at what number of clusters the derived and true solutions should be compared. In most studies the known true solution of g clusters is compared to the g clusters derived solution. It has been argued (see Edelbrock 1979) that due to outliers, a point in the tree may be reached where all points except the outliers are correctly classified. Requiring that the derived solution have the same number of clusters as the true solution, forces the comparison calculation to be made at a point in the tree that does not represent the ideal solution given the outliers. Thus, algorithms that tend to combine outliers into clusters at an early stage such as Ward's method would tend to appear superior. This in fact seems to be the case, since many early studies found Ward's criterion to be superior. To combat this difficulty Milligan and Cooper (1988) compared solutions at levels g, $(g+3)$, $(g+6)$ and $(g+9)$ where g is the true number of clusters. They found that at level g and $(g+3)$ Ward's method was superior to the average linkage method, whereas at levels $(g+6)$ and $(g+9)$ the average linkage method was superior to Ward's method. Generally the single linkage and complete linkage methods were inferior to the average linkage and Ward's methods. The single linkage method tended to behave like the average linkage method in terms of its cluster recovery pattern over the four measurement levels g, $(g+3)$, $(g+6)$ and $(g+9)$; however, its index was always lower. Similarly the cluster recovery pattern for the complete linkage method was similar to the pattern for Ward's method, but the index was at lower levels.

10.2.6 OTHER SOURCES OF INFORMATION

Summaries of cluster analysis can be found in Anderberg (1973), Everitt (1974), Gordon (1981), Jain and Dubes (1988), Lorr (1983), Mardia, Kent and Bibby (1979) and Seber (1984).

10.3 Multidimensional Scaling

Multidimensional scaling (usually abbreviated MDS) is a body of techniques that uses proximities between objects to produce a spatial representation of the objects. The proximity matrix is usually a dissimilarity matrix. The derived spatial representation consists of a geometric configuration of points on a map, each point corresponding to one of the objects. The greater the similarity among the objects the closer the objects will be on the map. A proximity matrix consisting of distances among cities for instance can be used to construct a spatial representation preserving the given intercity distances. Unlike the map of cities, however, in many applications the measures of proximity used to relate the objects are not based on direct measurement. Instead the proximity measures are based on perceptions of similarity derived from human judgements. An example would be the perceived degree of similarity among automobile brands by consumers. A two-dimensional representation in this case may contain dimensions that appear to represent size and luxury.

Multidimensional scaling which is based on measured proximities is usually referred to as metric MDS, whereas nonmetric MDS is used to characterize the technique when the proximities are based on judgement. In metric MDS the spatial representation attempts to preserve the distances among the objects, whereas in nonmetric MDS the spatial representation only preserves the rank order among the dissimilarities. Thus in nonmetric

MDS if objects (A and B) are perceived to be closer than (A and C) and (B and C), then the spatial configuration will preserve these similarity rankings. In the derived geometric configuration the distance between points (A and B) will be less than the distances between (A and C) and (B and C). In the case of the perceived similarities among automobiles, A and B may represent compact cars produced by two North American companies, whereas automobile C may represent a compact car produced in Europe.

Once the dimensions or scales have been determined the second step of the analysis involves the interpretation of the results. Scatterplots showing the location of the objects with respect to the derived dimensions are useful for providing a graphical representation of the dissimilarity relationships. If measurements of object characteristics are available that are believed to be contributing to the perceived dissimilarities, other analyses can be carried out to assist with interpretation. The derived dimensions can be related to the measured characteristics using other multivariate techniques. In the study of automobiles, characteristics such as fuel economy, size, style and luxury, can be related to the derived scales. Unexplained differences that still remain may lead the researcher to look for other unknown factors.

If the dissimilarity matrix is based on an average of the dissimilarity matrices of a sample of individuals, then individual differences can also be studied. Once the derived dimensions have been determined for the entire sample, a set of coordinates can be determined to locate the objects for each individual. The differences among individuals is handled by the assigning of individual weights to the various dimensions. The rationale for the individual difference scaling approach is that individual differences are attributable to the differences in importance that individuals attach to the various common scales. The underlying scales are assumed to be constant across individuals.

Like cluster analysis, multidimensional scaling is an exploratory data analysis technique. Cluster analysis seeks to classify objects into groups using dissimilarity measures derived from observed measurements; however, multidimensional scaling seeks to determine the underlying dimensions that contribute to the perceived differences among the objects. As in the case of principal components and factor analyses, multidimensional scaling is concerned with understanding the underlying dimensions that contribute to differences among objects. Factor analysis uses measures obtained from objects along known dimensions, whereas multidimensional scaling uses overall measures of dissimilarity of the objects to derive underlying dimensions.

This section begins with a discussion of metric multidimensional scaling and also introduces the fundamental theorem of multidimensional scaling. Although the metric method is not commonly used, it provides the motivation for the techniques used in the more popular nonmetric multidimensional scaling, which is discussed in the second section. A brief outline of other types of nonmetric multidimensional scaling is then provided in the last section.

10.3.1 METRIC MULTIDIMENSIONAL SCALING

Metric multidimensional scaling begins with an $(n \times n)$ proximity matrix **D** of dissimilarities, δ_{rs}, $r, s = 1, 2, \ldots, n$, which assigns a measure of dissimilarity to all possible pairs of n objects. The diagonal elements of **D** are therefore zero. The objective of metric MDS is to define a set of p underlying dimensions defined by measures $X_1, X_2, \ldots, X_p$, such that

1. the coordinates of the n objects along the p derived dimensions yield a Euclidean distance matrix, and

2. the elements of the Euclidean distance matrix are equivalent to, or closely approximate, the elements δ_{rs} of **D**.

In contrast to most other multivariate techniques in multidimensional scaling, the $(n \times p)$ **X** matrix of observations is derived from the given matrix **D** of dissimilarities.

Constructing a Positive Semidefinite Matrix Based on **D**

A dissimilarity matrix **D** with zeroes in the main diagonal is not positive semidefinite. A positive definite matrix $\mathbf{A}(n \times n)$, however, can be constructed based on the elements δ_{rs} of **D**. The elements a_{rs} of the new matrix **A** can be determined using the relationship

$$a_{rs} = -\frac{1}{2}[\delta_{rs}^2 - \delta_{r.}^2 - \delta_{.s}^2 + \delta_{..}^2], \quad r, s = 1, 2, \ldots, n, \qquad (10.6)$$

where

$$\delta_{r.}^2 = \frac{1}{n} \sum_{s=1}^{n} \delta_{rs}^2,$$

$$\delta_{.s}^2 = \frac{1}{n} \sum_{r=1}^{n} \delta_{rs}^2, \text{ and}$$

$$\delta_{..}^2 = \frac{1}{n^2} \sum_{r=1}^{n} \sum_{s=1}^{n} \delta_{rs}^2.$$

In matrix notation the relationship is given by

$$\mathbf{A} = -\frac{1}{2}[\mathbf{I}_n - \frac{1}{n}\mathbf{i}_n\mathbf{i}_n']\mathbf{D}^2[\mathbf{I}_n - \frac{1}{n}\mathbf{i}_n\mathbf{i}_n'],$$

where $\mathbf{I}_n$ is an $(n \times n)$ identity matrix, $\mathbf{i}_n$ is an $(n \times 1)$ vector of unities, and $\mathbf{D}^2$ is the matrix whose elements are the squares of the elements of **D**.

The matrix $[\mathbf{I}_n - \frac{1}{n}\mathbf{i}_n\mathbf{i}_n']$ is called a *centering matrix*. The matrix $\mathbf{A}$ has been derived from the matrix $\mathbf{D}^2$ by centering the rows and columns of $\mathbf{D}^2$ (double mean centering). The rows and columns of $\mathbf{A}$ therefore sum to zero, and hence the rank of $\mathbf{A}$ is at most $(n-1)$.

The Fundamental Theorem of MDS

The matrix $\mathbf{D}$ $(n \times n)$ of dissimilarities δ_{rs}, $r, s = 1, 2, \ldots, n$ is said to be *Euclidean* if there exists a dimension p and a set of n points given by the $(1 \times p)$ vectors $(\mathbf{x}_1', \mathbf{x}_2', \ldots, \mathbf{x}_n')$ such that

$$\delta_{rs}^2 = (\mathbf{x}_r - \mathbf{x}_s)'(\mathbf{x}_r - \mathbf{x}_s), \qquad r, s = 1, 2, \ldots, n.$$

In other words if the X observations were known, $\mathbf{D}$ would be the *Euclidean distance matrix* derived from $\mathbf{X}(n \times p) = (\mathbf{x}_1', \mathbf{x}_2', \ldots, \mathbf{x}_n')$ as outlined in Section 10.1.1.

The *fundamental theorem of MDS* states that the given dissimilarity matrix $\mathbf{D}$ is Euclidean if and only if the matrix $\mathbf{A}$ defined by (10.6) is positive semidefinite. This fundamental theorem provides the key to obtaining the MDS solution for a given dissimilarity matrix $\mathbf{D}$. If the matrix $\mathbf{D}$ is Euclidean, then the matrix $\mathbf{A}$ can be written as $\mathbf{A} = \mathbf{X}^*\mathbf{X}^{*\prime}$, where $\mathbf{X}$ is the $(n \times p)$ matrix consisting of the coordinates of the n points in p-dimensional space and $\mathbf{X}^*$ is the matrix of mean centered columns of $\mathbf{X}$. There is no loss of generality therefore in assuming that the p X variables each have mean zero. We shall assume for the remainder of this section that the p X variables have mean zero.

The MDS Solution

Given a dissimilarity matrix $\mathbf{D}$ the matrix $\mathbf{A}$ is constructed using (10.6). The eigenvectors $\mathbf{v}_1, \mathbf{v}_2, \ldots, \mathbf{v}_n$ and corresponding eigenvalues $\lambda_1, \lambda_2, \ldots, \lambda_n$ of the matrix $\mathbf{A}$ are then used to obtain underlying measures $\mathbf{x}_1, \mathbf{x}_2, \ldots, \mathbf{x}_n$. If $\mathbf{A}$ is positive semidefinite of rank p, then p of the eigenvalues of $\mathbf{A}$ will be positive and the remaining $(n-p)$ eigenvalues will be zero. The matrix $\mathbf{A}$ is expressible as $\mathbf{A} = \mathbf{V}\mathbf{\Lambda}\mathbf{V}'$, where $\mathbf{V}$ is the matrix of eigenvectors of $\mathbf{A}$ and $\mathbf{\Lambda}$ is the diagonal matrix of eigenvalues.

The number of positive eigenvalues permit the determination of p whereas the absence of any negative eigenvalues supports the positive semidefinite claim. Thus given the matrix $\mathbf{D}$ the rank of $\mathbf{A}$ is not specified in advance but is determined from the number of positive eigenvalues.

For the p nonzero eigenvalues the X coordinates can be defined by

$$\mathbf{x}_j = \mathbf{v}_j\sqrt{\lambda_j}, \qquad j = 1, 2, \ldots, p,$$

where it is assumed that $\mathbf{v}_j'\mathbf{v}_j = 1$. Equivalently

$$\mathbf{X}(n \times p) = \mathbf{V}\mathbf{\Lambda}^{1/2}.$$

The rows of $\mathbf{X}$, given by $\mathbf{x}_1', \mathbf{x}_2', \ldots, \mathbf{x}_n'$ have the property that

$$\delta_{rs}^2 = (\mathbf{x}_r - \mathbf{x}_s)'(\mathbf{x}_r - \mathbf{x}_s), \qquad r, s = 1, 2, \ldots, n,$$

and hence the dissimilarity relationships in $\mathbf{D}$ are preserved by the scaling solution given by $\mathbf{X}$. The p X variables or dimensions (scales) have mean zero and are only unique up to a constant.

An Approximate Solution

Usually in practice the objective is to obtain a small number of dimensions say $k \ll p$ such that the derived dissimilarity relationships are approximately equal to the original matrix $\mathbf{D}$. A common approach is to retain the first r eigenvectors and corresponding eigenvalues so that $\mathbf{A}$ is approximated by $\widehat{\mathbf{A}} = \mathbf{V}_0 \mathbf{\Lambda}_0 \mathbf{V}_0'$, where $\mathbf{\Lambda}_0$ and $\mathbf{V}_0$ denote submatrices of $\mathbf{\Lambda}$ and $\mathbf{V}$ corresponding to the r largest eigenvalues and corresponding eigenvectors respectively. The corresponding scale values are given by $\mathbf{X}^{(0)} = \mathbf{V}_0 \mathbf{\Lambda}_0^{1/2}$ and the resulting distances are given by

$$d_{rs}^{(0)^2} = (\mathbf{x}_r^{(0)} - \mathbf{x}_s^{(0)})'(\mathbf{x}_r^{(0)} - \mathbf{x}_s^{(0)}).$$

If the first k eigenvalues account for most of the variation in $\mathbf{A}$ then the approximations of the δ_{rs}^2 by the $d_{rs}^{(0)^2}$ should be good.

A useful measure of goodness of fit is based on the square of the Pearson correlation RSQ between the δ_{rs} and $d_{rs}^{(0)}$, $r, s = 1, 2, \ldots, n$. This value should be close to 1 to assure a reasonable fit. Other goodness of fit measures called STRESS and SSTRESS will be introduced in the next section in connection with nonmetric MDS.

Computer Software

The calculations in this section were carried out using the SPSSX programs PROXIMITIES and ALSCAL.

Example

This example is based on the ten-city data discussed in Sections 10.1 and 10.2. The ten cities represent a subset of the 40 cities studied in the Air Pollution Data of Chapter 9. The multidimensional scaling analysis begins with the squared Euclidean distance matrix, based on standardized data, introduced in Table 10.3. The objective is to use metric multidimensional scaling to derive a spatial representation for the ten cities. The original observations will then be used to provide an interpretation for the derived dimensions.

The dissimilarity matrix of Table 10.3 is repeated in the lower left of the matrix of Table 10.23. The diagonal elements of 0.00 have been omitted. The upper right portion of the matrix in Table 10.23 contains the scaling matrix $\mathbf{A}$ derived according to (10.6). The eigenvalues and eigenvectors of $\mathbf{A}$ are used to derive the spatial representation for the ten cities. The coordinates for the first four dimensions are summarized in Table 10.24. Figure 10.17 (a) and (b) show plots of the ten cities with respect to the first two and last two dimensions respectively. As indicated in Table 10.24, the first four eigenvalues represent 93.2% of the sum of the diagonal elements of $\mathbf{A}$ and hence account for most of the variation.

In multidimensional scaling, the objectives include minimizing the number of dimensions while at the same time preserving the dissimilarity relationships of the original dissimilarity matrix. Often only two dimensions are used. Table 10.25 summarizes the original Euclidean distances δ_{rs} and compares them to the Euclidean distances based on only the first two dimensions $[d_{rs}(2)]$ and then based on the first four dimensions $[d_{rs}(4)]$. The original dissimilarities are shown in ascending order. Derived distances that do not preserve the ranking of the original dissimilarities are coded with an asterisk. The derived dissimilarities $d_{rs}(4)$ show fewer asterisks than the dissimilarities $d_{rs}(2)$. The relationship between the original distance and the two sets of derived distances is shown in panels (a) and (b) of Figure 10.18. Comparing panels (a) and (b) of this figure shows that the four-dimensional solution provides a much better approximation than the two-dimensional case. Table 10.25 also shows the deviations between the original distances and the distances based on two dimensions, $[\delta_{rs} - d_{rs}(2)]$. Examination of these deviations reveals some large differences suggesting that the spatial representation for some cities (in two dimensions) is inaccurate. A comparison of the original distances to the four-dimensional distances shows that the two sets of distances are quite close. The correlation coefficients between δ_{rs} and each of $d_{rs}(2)$ and $d_{rs}(4)$ are 0.936 and 0.997 respectively.

As in the case of principal components analysis in Chapter 9, it can be seen here from Table 10.23 that dimensions are sensitive to the magnitudes of the diagonal elements of $\mathbf{A}$. Cities corresponding to large diagonal elements such as Scranton, Albuquerque, Columbus and Montgomery therefore dominate the first dimension. An examination of the differences $[\delta_{rs} - d_{rs}(2)]$ reveals that a number of distances involving Washington are too small. In addition a few distances involving either Scranton or Albuquerque are also too small. An examination of Figure 10.17(b) which plots the cities with respect to the third and fourth dimensions confirms, that the cities of Washington, Scranton and Albuquerque are furthest from the origin. In panel (a) of Figure 10.17 arrows have been attached to the cities indicating the direction the corresponding points should move so that the derived distances could approach the original distances.

To provide some interpretation for the derived dimensions, a correlation matrix was generated to relate the four dimensions to the 11 variables in the air pollution data set. The results are shown in Table 10.26. The variables SSUM and PSUM represent the sums of the three S variables and

TABLE 10.23. Dissimilarity Matrix for Ten Cities – Lower Left Excluding Diagonal / Beginning Scaling Matrix – Upper Right Including Diagonal

	1 Albuquerque	2 Atlanta	3 Columbus	4 Los Angeles	5 Minneapolis	6 Montgomery	7 Salt Lake	8 Scranton	9 Washington	10 Wichita
1. Albuquerque	21.052	-4.713	-4.903	-0.368	-4.923	-8.008	0.982	5.887	-2.033	-2.973
2. Atlanta	33.5	3.022	2.732	-0.733	0.362	3.527	-1.383	-3.878	1.302	-0.238
3. Columbus	44.2	10.9	13.342	-7.073	-5.778	14.537	-3.423	-9.518	-2.188	2.272
4. Los Angeles	30.2	12.9	35.9	8.412	3.707	-8.078	0.462	2.417	3.147	-1.893
5. Minneapolis	38.6	10.0	32.6	8.7	7.702	-5.283	3.057	-1.188	-0.058	2.402
6. Montgomery	55.6	14.5	2.8	43.1	36.8	18.533	-4.578	-10.523	-2.593	2.467
7. Salt Lake	22.3	9.0	23.4	10.7	4.8	30.9	3.212	0.717	-0.903	1.857
8. Scranton	34.3	35.8	57.4	28.6	35.1	64.6	26.8	25.022	-1.698	-7.238
9. Washington	32.4	7.7	25.0	9.4	15.1	31.0	12.3	35.7	7.282	-2.258
10. Wichita	32.6	9.1	14.4	17.8	8.5	19.2	5.1	45.1	17.4	5.602

(a)

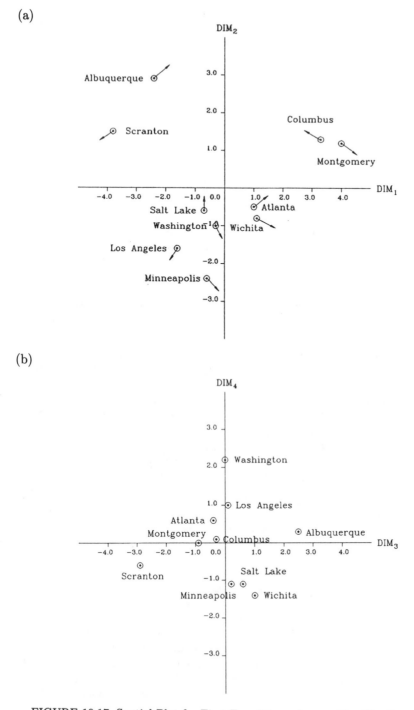

(b)

FIGURE 10.17. Spatial Plot for First Four Dimensions — Ten-City Data

TABLE 10.24. Coordinates for First Four Dimensions — Ten-City Data

City	X_1	X_2	X_3	X_4	
1	−2.4	2.9	2.5	0.3	
2	1.0	−0.5	−0.4	0.6	
3	3.3	1.3	−0.3	0.1	
4	−1.6	−1.6	0.1	1.0	
5	−0.6	−2.4	0.2	−1.1	
6	4.0	1.2	−0.9	0.0	
7	−0.7	−0.6	0.6	−1.1	
8	−3.8	1.5	−2.9	−0.6	
9	−0.3	−1.0	0.0	2.2	
10	1.10	−0.8	1.0	−1.4	
Eigenvalues	52.25	24.83	17.26	11.18	Total 105.52

Sum of Diagonal Elements of Matrix 113.18

First Four Dimensions Represent 93.2% of Variation

three P variables respectively. Because of the small number of observations (10) large correlations are required to feel confident about the relationships. It would appear that DIM1 is strongly related to the S variables, whereas DIM2 is strongly related to the P variables. Dimensions 3 and 4 show somewhat weaker correlations although DIM3 seems to have a negative correlation with the S variables. The five demographic variables may also be important although the correlations are somewhat weak. A comparison of the various profiles shown earlier in Figure 10.4 indicates that Albuquerque and Scranton are quite different with respect to the S variables. Albuquerque has low values and Scranton has high values. The third dimension, DIM3, confirms this result. Note that in Figure 10.17 Albuquerque and Scranton differ along this dimension. For the city of Washington, the profiles shown in Figure 10.4 indicate that Washington differs from Los Angeles, Minneapolis and Wichita with respect to PERWH and GE65 and from Atlanta with respect to NONPOOR and PM2. These differences could explain why for DIM4 in Figure 10.17 Washington is different than these other cities.

*Metric Multidimensional Scaling Beginning with **D***

In general the original dissimilarity matrix **D** is given and the underlying **X** matrix is not available. The MDS derived dimensions are then interpreted using other information. For this example we have behaved as if the underlying **X** matrix was not available when generating the dimensions. The original **X** matrix was then used to relate the derived dimensions to the original variables. This example has been used for convenience and also to allow comparisons to other techniques such as cluster analysis and princi-

TABLE 10.25. Comparison of Original Distances to Derived Distances Based on Two and Four Dimensions

City Pair	δ_{rs}	$d_{rs}(2)$	$d_{rs}(4)$	$\delta_{rs} - d_{rs}(2)$
Columbus-Montgomery	1.67	0.63	0.89	1.04
Minneapolis-Salt Lake	2.19	1.73	1.76	0.46
Salt Lake-Wichita	2.26	1.87	1.92	0.39
Atlanta-Washington	2.77	1.38*	2.19	1.40
Minneapolis-Wichita	2.92	2.37	2.53	0.55
Los Angeles-Minneapolis	2.95	1.18*	2.43*	1.77
Atlanta-Salt Lake	3.00	1.76*	2.63	1.24
Atlanta-Wichita	3.02	0.32*	2.47*	2.70
Los Angeles-Washington	3.07	1.48*	1.90*	1.58
Atlanta-Minneapolis	3.16	2.47	3.03	0.69
Los Angeles-Salt Lake	3.27	1.30*	2.55*	1.97
Atlanta-Columbus	3.30	2.95	3.00*	0.35
Salt Lake-Washington	3.51	0.63*	3.42	2.87
Atlanta-Los Angeles	3.59	2.85*	2.93*	0.75
Columbus-Wichita	3.79	3.03	3.65	0.76
Atlanta-Montgomery	3.81	3.42	3.51*	0.39
Minneapolis-Washington	3.89	1.38*	3.55*	2.51
Washington-Wichita	4.17	1.41*	4.05	2.76
Los Angeles-Wichita	4.22	2.85*	3.87*	1.37
Montgomery-Wichita	4.38	3.46	4.20	0.92
Albuquerque-Salt Lake	4.72	3.95	4.59	0.77
Columbus-Salt Lake	4.84	4.49	4.74	0.34
Columbus-Washington	5.18	4.28*	4.79	0.90
Salt Lake-Scranton	5.35	3.70*	5.11	1.65
Los Angeles-Scranton	5.50	3.82*	5.13	1.67
Albuquerque-Los Angeles	5.56	4.66	5.31	0.90
Montgomery-Salt Lake	5.57	5.05	5.38	0.52
Montgomery-Washington	5.69	4.77*	5.36*	0.92
Albuquerque-Washington	5.71	4.51*	5.53	1.20
Albuquerque-Wichita	5.71	5.11	5.60	0.60
Columbus-Minneapolis	5.71	5.41	5.56*	0.30
Albuquerque-Atlanta	5.79	4.87*	5.71	0.92
Albuquerque-Scranton	5.86	2.00*	5.83	3.86
Minneapolis-Scranton	5.92	4.96*	5.87	0.96
Scranton-Washington	5.97	4.31*	5.90	1.66
Atlanta-Scranton	5.98	5.20*	5.87*	0.79
Columbus-Los Angeles	5.99	5.74	5.83*	0.26
Minneapolis-Montgomery	6.07	5.82	6.02	0.24
Albuquerque-Minneapolis	6.21	5.59*	6.20	0.63
Los Angeles-Montgomery	6.57	6.25	6.42	0.32
Albuquerque-Columbus	6.65	5.90*	6.57	0.75
Scranton-Wichita	6.72	5.38*	6.69	1.34
Albuquerque-Montgomery	7.46	6.58	7.43	0.88
Columbus-Scranton	7.58	7.06	7.54	0.51
Montgomery-Scranton	8.04	7.72	7.99	0.32

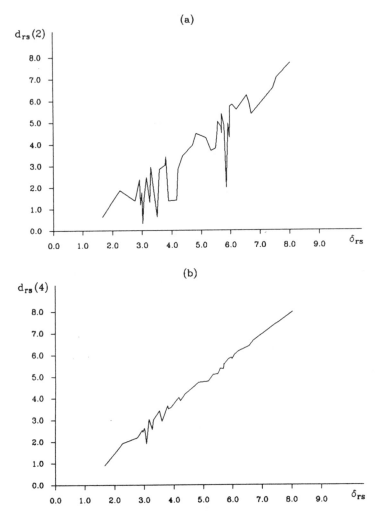

FIGURE 10.18. Comparison of Original Euclidean Distances and Derived Distances

pal components analysis. In particular in Chapter 9, it has already been demonstrated that spatial relationships can be obtained using the first few principal components or factors.

Relation to Cluster Analysis

In Section 10.2, it was outlined how beginning with a proximity matrix the objects can be clustered in a hierarchical fashion and that the results can be used to obtain derived proximities among the objects. These derived

TABLE 10.26. Correlation Matrix Relating
Dimensions to Air Pollution Variables

	DIM1	DIM2	DIM3	DIM4
SMEAN	0.54	0.06	−0.36	−0.25
SMAX	0.26	0.02	−0.64	−0.08
SMIN	0.75	0.13	−0.37	−0.14
SSUM	0.45	0.05	−0.56	−0.14
PMEAN	0.06	0.64	0.26	0.00
PMAX	0.07	0.65	0.26	−0.08
PMIN	0.34	0.58	−0.05	−0.20
PSUM	−0.01	0.66	0.25	−0.07
GE65	0.34	−0.26	−0.32	−0.10
PERWH	−0.39	0.22	−0.23	0.38
NONPOOR	0.36	0.08	−0.17	0.32
PM2	0.21	−0.25	−0.15	−0.22
LPOP	−0.07	−0.45	−0.01	−0.17

proximities could then be related to the original proximities to evaluate
the cluster procedure.

A cluster analysis concentrates on accurately fitting the small dissimilarities or proximities. At the early stages of the hierarchical process the
group proximities are "close" to the original proximities. As the clusters
grow in size however the group proximities are much less comparable to
the original proximities. Thus a hierarchical cluster analysis does not tend
to provide reliable proximities at the large end of the scale. In contrast,
a scaling procedure such as principal components tends to concentrate on
the large dissimilarities and does a very poor job of fitting the small dissimilarities. It is often useful therefore to combine the results from a cluster
analysis and an MDS analysis on the spatial plot of the objects. The results
of the hierarchical clustering process can be shown on the plot hence con-
firming the proximity of various objects. This is illustrated for the ten-city
data below.

Example

Figure 10.19 is a reproduction of Figure 10.18(a) with the names of the
cities omitted. Curves have been drawn on the figure to indicate the order
of the clustering process and the values of Ward's criterion in each case
has also been shown. The information used to complete the plot can be
obtained from Table 10.9 and Figure 10.10. Figure 10.19 shows that in
the first step of the clustering process Columbus and Montgomery were
joined at a criterion value of 1.42 and at step 7 of the hierarchical process
Albuquerque and Scranton were joined at a criterion value of 17.17. We can
see from the figure that Albuquerque and Scranton appear closer on the

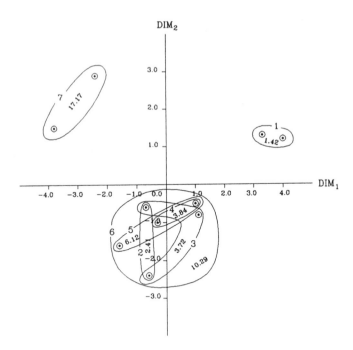

FIGURE 10.19. Spatial Plot of First Two Dimensions with Cluster Information

plot than their proximity would justify. For the six cities in the main cluster some of the plotted distances are inconsistent with the proximities derived from the clustering. These results tend to confirm that it is dangerous to judge the relative distances on the MDS plot for objects that are relatively close. For the large distances, however, it appears safe to conclude that there are three or four clusters depending on whether Albuquerque and Scranton are combined.

Improving the Solution

It is possible at this stage to improve the two-dimensional representation of the ten cities by revising the coordinates for the ten objects. One approach would be to use a numerical approximation procedure to revise the coordinates to bring the derived distances closer to the original distances. This is usually carried out by determining revised coordinates in the first two dimensions so that $\sum_{r<s} \sum [d_{rs}(2) - \delta_{rs}]^2$ is minimized. Numerical procedures such as Newton–Raphson or steepest descent are usually used to obtain revised coordinates and revised values of $d_{rs}(2)$, say $\hat{d}_{rs}(2)$ such that the above sum of squares is minimized. In some instances an iterative procedure is carried out that revises both $d_{rs}(2)$ and the coordinates in a

series of alternating steps. This iterative procedure will be discussed in the next section in connection with nonmetric scaling.

Using Similarities

If the proximity matrix is a similarity matrix (e.g., a correlation matrix) $\mathbf{C}$ with elements c_{rs} satisfying $c_{rr} = 1$, $c_{rs} = c_{sr}$, $0 \leq c_{rs} \leq 1$, $r, s = 1, 2, \ldots, n$, the matrix can be transformed to a dissimilarity matrix using the expression $\delta_{rs}^2 = (2 - 2c_{rs})$. In Section 10.1 this was the relationship obtained between squared Euclidean distances and correlations for standardized variables.

Relation to Principal Components Analysis

As illustrated in Chapter 9, if the matrix $\mathbf{X}(n \times p)$ is known it can often be approximated by a small number of principal components $k \ll p$. The components can be used to produce plots showing the relationships among the n objects in $\mathbf{X}$. Such plots were illustrated in Chapter 9. In the context of metric scaling as discussed in this section $\mathbf{X}$ is unknown. Only the matrix of dissimilarities $\mathbf{D}$ is available. If $\mathbf{X}$ is known then there is no need to generate the matrix $\mathbf{A}$ to determine $\mathbf{X}$.

Metric MDS and Principal Coordinates Analysis

Principal coordinates analysis uses a given similarity matrix $\mathbf{S}$ $(n \times n)$ to derive a spatial representation for the n objects. ($\mathbf{S}$ might be a covariance matrix among the n objects or other $\mathbf{XX}'$ type matrix). Denoting the elements of $\mathbf{S}$ by s_{rs}, $r, s = 1, 2, \ldots, n$, a new matrix $\mathbf{C}$ is obtained by computing

$$c_{rs} = s_{rs} - \bar{s}_{r\cdot} - \bar{s}_{\cdot s} + \bar{s}_{\cdot\cdot},$$

where $\bar{s}_{r\cdot} = \frac{1}{n}\sum_{s=1}^{n} s_{rs}$, $\bar{s}_{\cdot s} = \frac{1}{n}\sum_{r=1}^{n} s_{rs}$ and $\bar{s}_{\cdot\cdot} = \frac{1}{n^2}\sum_{r=1}^{n}\sum_{s=1}^{n} s_{rs}$. Equivalently $\mathbf{C} = (\mathbf{I} - \mathbf{i}_n\mathbf{i}_n')\mathbf{S}(\mathbf{I} - \mathbf{i}_n\mathbf{i}_n')$ as in the case of (10.6). Principal components analysis is then applied to $\mathbf{C}$ to determine coordinates for the n objects. Setting $\mathbf{S} = -\frac{1}{2}\mathbf{D}^2$ where $\mathbf{D}^2$ is a matrix of squared Euclidean distances results in the metric MDS approach outlined above.

An Alternative Derivation for $\mathbf{A}$

Given the $(n \times n)$ matrix $\mathbf{D}$ of dissimilarities, a spatial configuration for the n objects can be obtained from the positive semidefinite matrix $\mathbf{A}$ as outlined above in (10.6). The elements of $\mathbf{A}$ were defined as deviations from the means of the rows and columns of $\mathbf{D}$ and hence the row and column means of the elements of $\mathbf{A}$ are zero. An alternative approach to obtaining a spatial configuration is to define the elements of $\mathbf{A}$ using an alternative reference point. In Section 10.1, the cosine law was used to relate squared

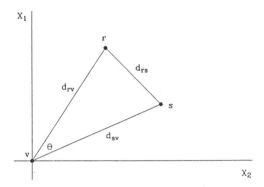

FIGURE 10.20. Cosine Law Relationship Between Three Points

Euclidean distance to the cosine measure of similarity. This approach can also be used to obtain the positive semi-definite matrix $\mathbf{A}$.

A particular reference point v is selected from the set of n points in p-dimensional space. For each pair of points, say r and s, the squared Euclidean distances relating the three points are given by d_{rs}^2, d_{rv}^2 and d_{sv}^2 respectively. Denoting by θ the angle between the two vectors joining v to r and s respectively, the cosine law yields the relationship

$$d_{rs}^2 = d_{rv}^2 + d_{sv}^2 - 2d_{rv}\, d_{sv}\, \cos\theta.$$

Figure 10.20 illustrates the relationship for two dimensions X_1 and X_2. The $(n \times n)$ matrix of elements a_{rs}, $r, s = 1, 2, \ldots, n$, is given by

$$a_{rs} = (d_{rv}^2 + d_{sv}^2 - d_{rs}^2) = 2d_{rv}\, d_{sv}\, \cos\theta$$

and can be used to obtain a positive semidefinite matrix $\mathbf{A}$. The matrix $\mathbf{A}$ can be used to obtain a spatial configuration for the n points as outlined above.

The Additive Constant Problem

If the given dissimilarity matrix $\mathbf{D}$ $(n \times n)$ for n objects is Euclidean, then there exists an integer p such that a p-dimensional configuration can be determined for the n objects. In some applications the dissimilarities are estimated in such a way that the matrix $\mathbf{D}$ may not be Euclidean. The dissimilarities may be valid as interval scaled distances, but they may not be acceptable as ratio scaled distances. Interval distances are correct up to a constant c, but the origin is undefined (e.g., Celcius and Fahrenheit temperatures are interval scales but not ratio scales since the zero point is arbitrary).

The true distances δ_{rs} can be related to the dissimilarity values d_{rs} by the equation $\delta_{rs} = d_{rs} + c$, $r, s = 1, 2, \ldots, n$. If the value of c is sufficiently large the observed dissimilarities may not be Euclidean. To illustrate this, recall that Euclidean distances must satisfy the triangle inequality given by

$$\delta_{rs} \le \delta_{ru} + \delta_{su}.$$

By subtracting a constant c from all three terms in this inequality a point can be reached such that $d_{rs} > d_{ru} + d_{su}$.

If the dissimilarity matrix $\mathbf{D}$ is not Euclidean, the matrix $\mathbf{A}$ derived according to (10.6) will no longer be positive semidefinite and hence will yield at least one negative eigenvalue. If $\mathbf{A}$ is not positive semidefinite it will not be possible to derive p dimensions to reproduce $\mathbf{D}$. It is still possible however to derive dimensions corresponding to the positive eigenvalues of $\mathbf{A}$ and hence to approximate $\mathbf{D}$. If the negative eigenvalues are relatively small the approximation based on the positive eigenvalues should be adequate.

An alternative approach is to determine a constant c that can be added to all the off-diagonal elements of $\mathbf{D}$ to ensure that the matrix is Euclidean. If c is sufficiently large, $\mathbf{D}$ will be Euclidean, however what is required is the smallest value of c to guarantee that $\mathbf{D}$ is Euclidean. As c increases the dimension p required to approximate $\mathbf{D}$ also increases. The objective in MDS is to minimize p and hence c should be as small as possible. Some MDS computer routines determine an approximate value of c although it is possible to determine c precisely (see Cailliez 1983).

Application of Metric Scaling

If we are provided with the exact map distances between the major cities of Europe we could use metric scaling to produce a map that reproduced the between-city distances exactly. We could not however produce a map that showed the cities located correctly with respect to their true location in N–S and E–W coordinates. The derived locations of the cities would have to be moved around in N–S and E–W directions as well as perhaps rotated to obtain the correct orientation. If however we could place two of the cities in their correct location, then the remaining cities would automatically be correctly located using the between city distances. There are only $(n-2)$ independent dimensions available from the n cities.

If instead of the exact map distances, we were provided with the schedule of a major airline, which gave the required flying time between the major cities, the map locations could only be approximated. The flying times are also based on factors such as stopovers and general flying conditions. The matrix of flying times however would still be Euclidean. The derived coordinates in two dimensions would simply reflect some error due to the nondistance factors.

In experiments dealing with humans the dissimilarities are often based on judgements that are subject to measurement error. The dissimilarities obtained are therefore only approximations and in addition the nature of the underlying dimensions is somewhat vague. In this case the MDS solution is used to derive insights about relationships among the objects as perceived by the subjects in the experiment. The analysis does not however yield models that can be used to make precise individual predictions.

In addition to the problems associated with the measurement of true dissimilarity, MDS procedures also are intended to minimize the number of derived dimensions. With so many levels of approximation inherent in the MDS process the attempt to reproduce the dissimilarities precisely does not seem justified. Nonmetric scaling discussed in the next section seeks only to preserve an ordinal relationship between the original dissimilarities and the derived distances. If two objects A and B are perceived to be more similar than objects C and D, then the derived distances should also reflect this relative relationship. If the differences in perceived similarity are only slight, scaling procedures cannot be relied upon to preserve the differences. Only large differences in perceived similarity are in general preserved by the scaling process.

10.3.2   Nonmetric Multidimensional Scaling

In *nonmetric MDS* a matrix **D** of dissimilarities δ_{rs} is often derived from human responses to questionnaires or experimental procedures. The respondents or subjects are usually asked to make comparisons among sets of objects or stimuli. The purpose of the *nonmetric scaling analysis* is to obtain insights into the nature of the perceived dissimilarities. Such analyses have been used to measure attitudes and preferences in law, political science and sociology; to make cross-cultural comparisons in anthropology; to study human perceptions in psychology and linguistics; and to evaluate product designs and advertising campaigns in marketing. The fields of psychology and marketing have provided much of the research literature on techniques of scaling and the required experimental designs. Review papers such as Carroll and Arabie (1980) and Young (1984) provide excellent summaries of the research literature.

In nonmetric multidimensional scaling much effort must be devoted to the design of experimental procedures to measure dissimilarity. A partial list of the types of procedures is given below.

1. *Paired comparisons* — subjects are asked to compare all possible pairs of a set of objects and to rate the degree of similarity.

2. *Partitions* — subjects are asked to divide the set of objects into a small number of mutually exclusive categories.

3. *Rankings* — subjects are asked to rank the objects with respect to a specified criterion.

4. *Triadic comparisons* — subjects are asked to rank the degree of similarity among three possible pairings of sets of three objects.

5. *Tetrads* — Subjects are asked to compare all possible pairs of objects and to indicate the most similar and/or the most dissimilar pairs.

Additional discussion on the measurement of dissimilarities in experimental situations can be found in Coxon (1982) and Green, Carmone and Smith (1989).

The techniques of design will not be discussed in this text. For additional background, interested readers should consult the research literature of the particular discipline as well as the references on MDS provided here. Although MDS can be applied to nonsymmetric matrices the discussion presented here will be restricted to symmetric matrices.

Ordinal Scaling

In nonmetric MDS, the given dissimilarities δ_{rs} are used to generate a set of derived distances d_{rs}, which are approximately related to the given dissimilarities δ_{rs} by a monotonic increasing function f. In this case we write

$$d_{rs} \approx f(\delta_{rs}),$$

where f is a function with the property that

$$\delta_{rs} < \delta_{uv} \Leftrightarrow f(\delta_{rs}) < f(\delta_{uv}).$$

The rank correlation between the δ_{rs} and $f(\delta_{rs})$ is therefore unity, whereas the rank correlation between δ_{rs} and the d_{rs} is close to 1. A plot of the d_{rs} versus δ_{rs} should be very close to a monotonically increasing function. Because only the rank order is important the scaling is ordinal and is commonly called nonmetric scaling.

The most common approach to determining the elements d_{rs} and the underlying configuration $x_1, x_2, \ldots, x_p$ is an iterative process commonly referred to as the *Shepard–Kruskal algorithm*.

Shepard–Kruskal Algorithm

The Shepard–Kruskal algorithm for nonmetric MDS is illustrated in Figure 10.21. After determining the dissimilarity matrix $\mathbf{D}$ and the corresponding scaling matrix $\mathbf{A}$ using (10.6) an iterative process begins that successively revises the dissimilarities and object coordinates until an adequate fit is achieved. The objective of the iterative process is to obtain a spatial representation in a given dimension such that the Euclidean distances among the objects are monotonicially related to the original dissimilarities.

The iterative part of the process contains four steps. The first step or *initial phase* selects the dimension p and determines the initial configuration $\mathbf{X}^{(0)}$ and resulting distances $d_{rs}^{(0)}$. The second step or *nonmetric phase* then uses monotone regression to relate the $d_{rs}^{(0)}$ to the δ_{rs}. The estimated regression produces a new set of pseudo-dissimilarities $\hat{d}_{rs}^{(0)}$ called *disparities* that are monotonically related to the δ_{rs}. The third step of the process

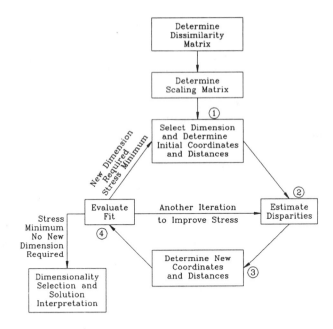

FIGURE 10.21. Shepard–Kruskal Algorithm for Nonmetric Scaling

called the *metric phase* revises the spatial configuration to obtain $\mathbf{X}^{(1)}$ in order to obtain new distances $d_{rs}^{(1)}$ which are more closely related to the disparities $\hat{d}_{rs}^{(0)}$ generated in step two. The fourth step is the *evaluation phase* which determines the goodness of fit of the distances $d_{rs}^{(1)}$ and the disparities $\hat{d}_{rs}^{(0)}$. If the fit is not adequate steps 2 and 3 are repeated. For the repetition of step 2, the distances $d_{rs}^{(1)}$ are related to the original dissimilarities δ_{rs} using monotone regression to generate new disparities $\hat{d}_{rs}^{(1)}$. A new step 3 is then carried out to determine a new spatial representation $\mathbf{X}^{(2)}$ and new distances $d_{rs}^{(2)}$. The evaluation phase then compares the $\hat{d}_{rs}^{(1)}$ to the $d_{rs}^{(2)}$. Finally after obtaining solutions over a range of dimensions a solution dimensionality is selected. This solution is then interpreted. This step will be referred to as the *selection and interpretation phase*. A more detailed presentation of the techniques involved in steps 2, 3, 4 and 5 is given below.

The Nonmetric Phase and Monotone Regression

In the nonmetric phase, disparities $\hat{d}_{rs}^{(0)}$ are determined from the distances $d_{rs}^{(0)}$ in such a way that the $\hat{d}_{rs}^{(0)}$ are monotonically related to the original dissimilarities δ_{rs}. The $\hat{d}_{rs}^{(0)}$ are the result of a regression of the $d_{rs}^{(0)}$ on the δ_{rs} subject to the condition that the fitted relationship is monotonic.

This regression is therefore called *monotone regression*. A useful successive approximation method for obtaining the regression estimates is called the *pool-adjacent violators* algorithm which is outlined next.

The Pool Adjacent Violators Algorithm

This approach to determining the disparities $\hat{d}_{rs}^{(0)}$ from the $d_{rs}^{(0)}$ and δ_{rs} begins by ranking the δ_{rs} values from lowest to highest before comparing them to the corresponding $d_{rs}^{(0)}$ values. Beginning with the lowest ranked value of δ_{rs}, the adjacent $d_{rs}^{(0)}$ values are compared for each δ_{rs} to determine if they are monotonically related to the δ_{rs}. As long as $d_{rs}^{(0)} < d_{uv}^{(0)}$ when $\delta_{rs} < \delta_{uv}$, then $\hat{d}_{rs}^{(0)} = d_{rs}^{(0)}$. Whenever a block of consecutive values of $d_{rs}^{(0)}$ are encountered that violate the required monotonicity property the $d_{rs}^{(0)}$ values are averaged together with the most recent nonviolator $d_{rs}^{(0)}$ value to obtain an estimator $\hat{d}_{rs}^{(0)}$. This value of $\hat{d}_{rs}^{(0)}$ is then assigned to all points in the particular block. This procedure is illustrated by the following example.

Example

δ_{rs}	1	2	3	4	5	6	7	8	9	10
$d_{rs}^{(0)}$	10	8	11	5	13	11	9	14	6	16
		9		8		11		10		
		8.5				10.6				
$\hat{d}_{rs}^{(0)}$	8.5	8.5	8.5	8.5	10.6	10.6	10.6	10.6	10.6	16

In this example the *blocks of violators* that are underlined above are averaged to obtain initial estimates of disparities. The new estimates are then checked for monotonicity. If blocks of violators remain the averaging process continues. In the above example two steps are required to obtain a monotonic set of disparities. In this case the resulting disparities are constant for periods of four and five consecutive dissimilarities.

Figure 10.22(a) plots the relationships between each of $d_{rs}^{(0)}$ and $\hat{d}_{rs}^{(0)}$ and the original dissimilarities δ_{rs}. The plot in Figure 10.22(b) applies to the rank image method to be discussed below. These plots are commonly referred to as *Shepard diagrams*. The behavior exhibited by the plot for $d_{rs}^{(0)}$ is not unlike the behavior exhibited in the distances obtained for the two-dimensional solution for the example in Section 10.3.1. The reader should note how the plot for $\hat{d}_{rs}^{(0)}$ is monotonically nondecreasing. Figure 10.23(a) shows the relationship between $\hat{d}_{rs}^{(0)}$ and $d_{rs}^{(0)}$. This plot is usually referred to as an *image diagram*. Figure 10.23(a) pertains to the $d_{rs}^{(0)}$ and $\hat{d}_{rs}^{(0)}$ values shown in Figure 10.22(a). Figure 10.23(b) relates to Figure 10.22(b) and is discussed below.

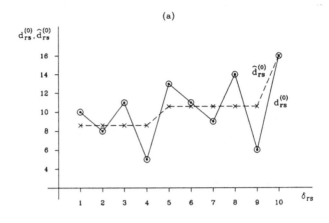

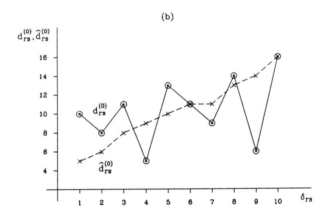

FIGURE 10.22. Shepard Diagrams for Example

Ties and Types of Monotonicity

The requirement that $\hat{d}_{rs}^{(0)} \leq \hat{d}_{uv}^{(0)}$ if $\delta_{rs} < \delta_{uv}$ is called *weak monotonicity* whereas the requirement that $\hat{d}_{rs}^{(0)} < \hat{d}_{uv}^{(0)}$ if $\delta_{rs} < \delta_{uv}$ is called *strong monotonicity*. Note that the disparities determined in the example above satisfy the weak monotonicity requirement but not the requirement of strong monotonicity. The weak monotonicity requirement results in flat/horizontal regions in the Shepard diagram relating $\hat{d}_{rs}^{(0)}$ to δ_{rs}. Strong monotonicity does not allow such flat regions.

To obtain disparities that satisfy the strong monotonicity requirement the *Guttman rank image approach* is usually used. In this method the distances $d_{rs}^{(0)}$ are simply ranked and then relabelled to obtain the disparities

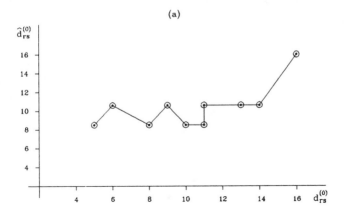

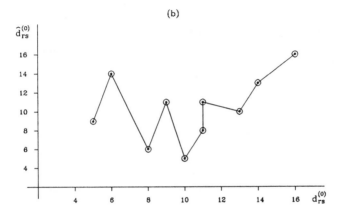

FIGURE 10.23. Image Diagrams for Example

$\hat{d}_{rs}^{(0)}$. The following example illustrates this approach for the data of the previous example.

Example

δ_{rs}	1	2	3	4	5	6	7	8	9	10
$d_{rs}^{(0)}$	10	8	11	5	13	11	9	14	6	16
$\hat{d}_{rs}^{(0)}$	5	6	8	9	10	11	11	13	14	16

The Shepard diagrams relating the distances and disparities to the observed dissimilarities appears in Figure 10.22(b). In comparison to panel (a) the strong monotonicity requirement in this example produces an almost linear relationship. The image diagram for this example is shown in panel (b) of Figure 10.23. In comparison to the image diagram of Figure 10.23(a), panel (b) of this figure has much more scatter indicating a greater variation between $\hat{d}_{rs}^{(0)}$ and $d_{rs}^{(0)}$.

Ties in the Original Dissimilarities

If there are ties encountered in the original dissimilarity data so that $\delta_{rs} = \delta_{uv}$, then there is no guarantee that $d_{rs}^{(0)} = d_{uv}^{(0)}$. There are two approaches to the estimation of $d_{rs}^{(0)}$ when two or more δ_{rs} are tied. The *primary approach* permits the $d_{rs}^{(0)}$ to be unequal, even though the δ_{rs} are tied, while the *secondary approach* requires that $d_{rs}^{(0)} = d_{uv}^{(0)}$ when $\delta_{rs} = \delta_{uv}$.

In general the primary approach is commonly used. The secondary approach demands a level of precision that is often not justified given the reliability of the original dissimilarities. The primary approach permits a step in $d_{rs}^{(0)}$ when δ_{rs} is constant, whereas the secondary approach demands a constant $d_{rs}^{(0)}$ when δ_{rs} is contant. The primary approach therefore permits vertical changes in $d_{rs}^{(0)}$ while δ_{rs} is constant. The combination of weak monotonicity and the primary approach to ties tends to yield a Shepard diagram with many steps. At the opposite extreme, strong monotonicity with a secondary approach to ties tends to yield a more smooth curve for a Shepard diagram.

The Metric Phase

At the completion of the nonmetric phase, the set of disparities $\hat{d}_{rs}^{(0)}$ and distances $d_{rs}^{(0)}$ have been generated using the original dissimilarities δ_{rs}. If the two sets of derived measures are close the scaling is considered to be good. A measure that is commonly used to determine goodness of fit is called STRESS and is given by

$$\text{STRESS} = \sum_{r<s}\sum [\hat{d}_{rs}^{(0)} - d_{rs}^{(0)}]^2 / \sum_{r<s}\sum d_{rs}^{(0)^2}, \qquad (10.7)$$

which is the sum-of-squared deviations normalized by $\sum_{r<s}\sum d_{rs}^{(0)^2}$. Alternative normalizing denominators that are also used are $\sum_{r<s}\sum \hat{d}_{rs}^{(0)^2}$ and $\sum_{r<s}\sum [d_{rs}^{(0)} - \bar{d}_{..}^{(0)}]^2$ where $\bar{d}_{..}^{(0)}$ denotes the average value of $d_{rs}^{(0)}$.

An alternative measure which is used in the ALSCAL procedure (to be outlined later) is given by

$$\text{SSTRESS} = \left\{ \sum_{r<s}\sum [\hat{d}_{rs}^{(0)^2} - d_{rs}^{(0)^2}]^2 / \sum\sum \hat{d}_{rs}^{(0)^4} \right\}^{1/2}.$$

Note that the denominator of STRESS is determined from $d_{rs}^{(0)}$, whereas the denominator of SSTRESS employs $\hat{d}_{rs}^{(0)}$.

The two goodness of fit measures STRESS and SSTRESS tend to place greater emphasis on deviations corresponding to the larger dissimilarities.

This is particularly true for SSTRESS which uses the squares of the d_{rs}. For this reason the relative positions of objects that are close together should not be used to derive conclusions. Objects close together in the spatial representation should simply be viewed as a cluster of similar objects. An alternative goodness of fit measure based on relative differences is given by

$$\text{STRESSR} = \sum_{r<s}\sum(1 - \hat{d}_{rs}^{(0)}/d_{rs}^{(0)})^2.$$

We have already seen that a useful diagram for comparing the values of $\hat{d}_{rs}^{(0)}$ to the values of $d_{rs}^{(0)}$ is the image diagram. Figure 10.23 illustrated the image diagram for the examples discussed above.

In the metric phase a new configuration $\mathbf{X}^{(1)}$ is sought to replace $\mathbf{X}^{(0)}$. The new distances derived from the $\mathbf{X}^{(1)}$ matrix, $d_{rs}^{(1)}$, must minimize the STRESS measure. In other words the $\mathbf{X}^{(1)}$ is selected in such a way that STRESS is minimized. In this step the disparities $\hat{d}_{rs}^{(0)}$ are held constant and the distances $d_{rs}^{(1)}$ change with $\mathbf{X}^{(1)}$. This procedure requires a numerical approximation procedure such as the method of steepest descent or the Newton–Raphson method. These numerical procedures yield expressions for $\mathbf{X}^{(1)}$ in terms of $\mathbf{X}^{(0)}$, $\hat{d}_{rs}^{(0)}$ and $d_{rs}^{(0)}$. These procedures are not outlined here. For the steepest descent method the interested reader is referred to MDS texts such as Coxon (1982) or Davidson (1983). For the Newton–Raphson procedure the reader should consult the ALSCAL program description in Schiffman, Reynolds and Young (1981).

The Evaluation Phase

In the evaluation phase the value of the goodness of fit measures obtained over all previous iterations are compared for the given dimensionality. When the changes in the goodness of fit measure as a result of the last iteration are sufficiently small, the procedure is usually terminated. At this stage the optimal fit has been obtained for a given dimension, say p. At this stage the analyst hopes that the minimum achieved is a global minimum for the given level of p.

Selection and Interpretation Phase

After the minimum stress solution has been determined for a range of dimensionalities a selection of the solution dimensionality is made. A plot of the minimum stress value as a function of the fitted dimensionality usually yields a downward sloping exponential shaped curve. One approach to selecting the appropriate dimensionality is to look for an elbow in the plot of the stress values. In some cases the curve may decline steeply until a certain dimension is reached after which the function becomes relatively flat. The dimension corresponding to the point of the change in slope is then selected as the appropriate dimension. The shape of the curve is based on the property that initially as the dimensionality increases toward the true value, the stress value decreases because of the increasing flexability for fitting presented by the additional dimensions. Eventually, however, as the dimensionality goes beyond the true value, the stress value changes very little since the increased flexability is now fitting noise.

The overall magnitude of the stress value depends on the stress formula used and the criteria used to perform monotone regression. Unless otherwise indicated it is assumed that the weak monotonicity criterion with a primary approach to ties will be used. In addition the stress will be evaluated according to (10.7).

Ideally the stress value of the solution should be less than 0.10 and preferably as low as 0.05; however, the value of this index is also influenced by the number of points being fitted and the amount of error or noise in the data. The larger the number of points to be fitted, the greater the number of squared differences in the stress computation and hence the greater the stress value. A common rule of thumb is to require that the number of points should exceed $(4p - 1)$ where p is the dimensionality. Also, as the error variance increases the true configuration is less distinct and hence more difficult to fit. A portion of the squared differences between fitted distances and disparities is due to the error in the original dissimilarities.

Monte Carlo Studies of the Stress Function

A number of Monte Carlo studies have been carried out to study the impact that the three factors, number of points, dimensionality and error, have on the value of the stress function. Using various known configurations with a variety of true dimensions the variation in stress values have been studied and related to the above three factors. Tables 10.27 and 10.28 show the results from two such studies. Table 10.27 is derived from a study by Wagenaar and Padmos (1971) which shows the relationship between stress and various levels of error and fitted dimensionality. This table is based on $n = 10$ points or stimuli. The second table, Table 10.28 is derived from a study by Spence and Graef (1974). In this second table the number of points was 36. A study of the tables allows one to see the impact that error and fitted dimensionality have on stress at each true dimension level. A comparison of the two tables provides some information about the impact of the number of points on the stress values. Another study illustrating the impact of error and dimensionality on stress is outlined in Young (1970).

In practice the two tables of stress values can be used to assist in selecting the appropriate dimensionality. For each level of error and true dimensionality there is a pattern of stress values corresponding to the fitted dimensionalities from 1 to 5. By finding a stress value pattern similar to the one obtained in an actual MDS analysis the researcher can guess at the true dimensionality and the magnitude of error. This approach is illustrated in the example discussed below.

TABLE 10.27. Stress Values for Simulated Data for Ten Objects

True Dimension	Error Level						Fitted Dimension
	0.00	0.05	0.10	0.20	0.40	∞	
1	0.000	0.006	0.040	0.105	0.216	0.408	1
	0.000	0.004	0.019	0.038	0.099	0.197	2
	0.000	0.000	0.009	0.017	0.054	0.106	3
	0.000	0.000	0.003	0.008	0.024	0.053	4
	0.000	0.000	0.002	0.002	0.007	0.025	5
2	0.214	0.194	0.204	0.275	0.319	0.408	1
	0.000	0.008	0.037	0.087	0.131	0.197	2
	0.000	0.005	0.018	0.038	0.066	0.106	3
	0.000	0.001	0.012	0.015	0.025	0.053	4
	0.000	0.000	0.002	0.006	0.010	0.025	5
3	0.264	0.282	0.278	0.297	0.361	0.408	1
	0.070	0.079	0.092	0.122	0.166	0.197	2
	0.000	0.006	0.026	0.058	0.094	0.106	3
	0.000	0.001	0.009	0.017	0.056	0.053	4
	0.000	0.000	0.000	0.005	0.024	0.025	5

Source: Wagenaar and Padmos (1971)

TABLE 10.28. Stress Values for Simulated Data for 36 Objects

True Dimension	Error Level					Fitted Dimension
	0.0000	0.0625	0.1225	0.2500	∞	
1	0.000	0.096	0.192	0.331	0.529	1
	0.000	0.089	0.164	0.257	0.349	2
	0.000	0.081	0.140	0.203	0.259	3
	0.000	0.074	0.121	0.167	0.205	4
	0.000	0.069	0.107	0.142	0.169	5
2	0.334	0.342	0.360	0.438	0.529	1
	0.000	0.078	0.144	0.255	0.349	2
	0.000	0.071	0.123	0.200	0.259	3
	0.000	0.064	0.106	0.162	0.205	4
	0.000	0.060	0.096	0.138	0.169	5
3	0.413	0.403	0.428	0.470	0.529	1
	0.197	0.196	0.228	0.289	0.349	2
	0.000	0.063	0.121	0.203	0.259	3
	0.000	0.057	0.104	0.163	0.205	4
	0.000	0.053	0.091	0.137	0.169	5
4	0.387	0.447	0.467	0.502	0.529	1
	0.204	0.233	0.255	0.312	0.349	2
	0.097	0.125	0.160	0.221	0.259	3
	0.000	0.060	0.111	0.172	0.205	4
	0.000	0.055	0.097	0.143	0.169	5

Source: Spence and Graef (1974)

It is also useful to examine the behavior of the stress function when there is no spatial configuration (i.e., random data). Monte Carlo studies have also been carried out to examine this behavior. In the two tables discussed above the error column ∞ refers to the no-configuration case. An examination of the stress values in this case illustrates quite clearly that increasing the fitted dimensionality reduces the stress quite dramatically. Also, by comparing the two tables we can see that increasing the number of points from 10 to 36 also increases the overall levels of stress in the error $= \infty$ case. In light of these results it is of interest to consider the null hypothesis of no spatial configuration for a given MDS analysis. Since there are no statistical models available for testing this hypothesis a Monte Carlo study will be used to provide some guidance.

In a study by Spence and Ogilvie (1973) the behavior of the stress function was studied for populations with no spatial configuration (random data). Plots showing the stress value as a function of fitted dimensionality were presented for $n = 12, 18, 26, 36$ and 48 points. Corresponding to $n = 18$ the stress values were approximately 0.468, 0.288, 0.198, 0.144 and 0.108, for dimensionalities 1 through 5 respectively. Since these values are based on means the standard deviations of the stress values should also be considered. Based on a plot of the standard deviations given in their paper, Spence and Ogilvie suggest using the three standard deviation lower limit as a critical region for rejection of the null hypothesis. Corresponding to $n = 18$ the approximate standard deviations obtained from their graph are 0.015, 0.009, 0.008, 0.008 and 0.007 for dimensions 1 through 5 respectively. Using the above means for $n = 18$ and applying the three standard deviation criterion results in the lower limits of 0.423, 0.261, 0.174, 0.120 and 0.087 for the stress values. These lower limits can provide some guidance to the researcher regarding the acceptability of the no-true-spatial-configuration hypothesis.

In addition to the study of the stress values as a function of dimensionality, cross validation can also be used to provide some guidance. If the data can be subdivided so that two or more separate MDS analyses can be carried out for the same stimuli the stress value functions can be compared. This process will be illustrated in the example discussed below.

Any fitted solution should be examined carefully for large differences between the fitted distances and the disparities. Generally at the lower levels of d_{rs}, there will be many small differences since the fitting procedure concentrates on fitting the large dissimilarities. If there are any large differences at the larger end of the d_{rs} scale this could be due to an outlier or error. A study of the impact of outliers on MDS is provided by Spence and Lewandowsky (1989).

The example below will be used to illustrate the selection and interpretation phase of MDS. The example will also illustrate the use of cross validation.

The ALSCAL Algorithm

The statistical computer software packages SPSSX and SAS employ the ALSCAL algorithm created by Takane, Young and de Leeuw (1977) and Young, Takane and Lewyckyj (1978). A description of the algorithm is provided in Schiffman, Reynolds and Young (1981). This algorithm has been used to perform the MDS analyses for the example discussed below.

The ALSCAL program begins by estimating an additive constant c to ensure that the triangle inequality holds for all triples among the original dissimilarities δ_{rs}. The scaling matrix is then computed according to (10.6). The initial configuration is derived using the metric scaling procedure for the prescribed number of dimensions. Kruskal's least squares monotone regression with weak monotonicity is used to generate disparities. The estimated disparities are then normalized to minimize the SSTRESS expression described above and then a Newton–Raphson procedure is used to obtain a new configuration, which minimizes SSTRESS. Iteration at each dimensionality continues until the change in SSTRESS on a given iteration is sufficiently small. The algorithm also determines STRESS and RSQ for the final solution at each dimensionality.

Example

A total of 182 Royal Canadian Mounted Police Officers (RCMP) working in 18 municipal detachments in Alberta, Canada, were asked a number of questions regarding the amount of stress they experienced due to a variety of sources. (See example at the end of the factor analysis section of Chapter 9 for a description of the various stress items.) The responses obtained were used to construct the Euclidean distance matrix shown in the lower left triangle of Table 10.29. Using the ALSCAL procedure in SPSSX, nonmetric MDS solutions were generated for dimensions 1 through 5 yielding the STRESS values of 0.351, 0.134, 0.068, 0.044 and 0.029 respectively. The corresponding RSQ values were 0.646, 0.900, 0.964, 0.982 and 0.991 respectively. Using the "elbow" approach it would appear that a dimensionality of 3 is appropriate. Examining the stress values in Tables 10.27 and 10.28, it would appear that the pattern exhibited by these STRESS values lies somewhere between the error level 0.40, true dimension level 3, values for 10 points in Table 10.27, and the error level 0.0625, true dimension level 3, values for 36 points in Table 10.28. Since in this case there are 18 points it would appear that the three-dimensional solution is reasonable.

For the three-dimensional solution the coordinates are summarized in Table 10.30. Figure 10.24 shows plots locating the 18 detachments with respect to the three dimensions. A Shepard diagram illustrating the relationship between the final disparities and the original dissimilarities is given in Figure 10.25. The final disparities are also summarized in the upper right triangle of Table 10.29.

TABLE 10.29. Original Distances (lower left) and Fitted Disparities (upper right) — RCMP Data

	A	B	C	D	E	F	G	H	I	J	K	L	M	N	O	P	Q	R
A	X	3.20	1.66	1.66	1.88	2.09	2.46	2.80	2.85	1.38	1.88	2.67	2.63	3.20	3.00	2.09	1.26	2.80
B	11.53	X	1.47	4.33	3.90	0.70	3.90	2.67	2.58	2.46	2.46	0.70	3.65	2.67	0.70	1.26	2.67	1.26
C	7.45	7.10	X	3.40	2.85	0.70	3.00	2.05	2.05	1.78	2.09	1.26	2.64	2.32	1.26	1.26	2.09	1.66
D	7.72	14.53	11.82	X	2.85	3.66	2.09	4.48	4.33	1.66	2.09	3.82	3.57	4.25	4.37	3.57	1.26	4.03
E	7.99	13.46	10.88	10.90	X	3.00	2.09	2.67	2.09	2.09	2.63	3.40	1.66	2.09	3.20	3.20	2.09	2.80
F	8.68	3.95	5.06	12.75	11.36	X	3.24	2.05	2.09	2.05	1.88	0.86	3.00	2.09	0.88	0.86	2.09	0.94
G	9.93	13.36	10.99	8.88	9.41	11.73	X	3.00	3.00	2.09	2.32	3.67	2.09	2.63	3.57	3.57	1.66	3.00
H	10.59	10.44	8.43	14.87	10.47	8.34	11.02	X	1.88	3.20	3.54	2.85	2.46	2.09	2.09	2.09	3.15	2.09
I	10.81	10.01	8.33	14.81	8.51	8.72	11.12	8.30	X	2.46	2.85	2.64	1.88	0.70	1.47	2.85	2.85	1.26
J	7.06	9.88	7.78	7.50	9.15	8.49	8.78	11.73	9.78	X	0.94	2.09	2.46	2.58	2.46	2.09	0.70	2.09
K	8.08	9.74	9.13	8.49	10.04	8.31	9.45	12.08	10.82	5.56	X	2.09	3.00	2.46	2.64	2.67	1.26	2.09
L	10.32	4.96	7.04	13.04	11.78	5.16	13.03	10.86	10.25	9.34	9.07	X	3.57	2.81	1.26	1.26	2.46	1.66
M	10.10	12.73	10.24	12.08	7.72	11.11	9.40	9.94	8.32	9.93	11.45	12.34	X	1.66	2.80	3.20	2.58	2.32
N	11.50	10.41	9.43	14.43	9.31	9.02	10.10	8.53	5.05	9.98	9.67	10.67	7.51	X	1.88	3.40	3.00	1.26
O	11.03	4.73	6.94	14.87	11.51	5.31	12.21	9.12	7.10	9.58	10.29	6.78	10.54	8.09	X	1.66	2.80	0.86
P	8.84	5.96	6.30	12.24	11.70	5.13	12.11	9.21	10.93	9.24	10.51	6.54	11.51	11.94	7.21	X	2.05	2.09
Q	6.80	10.30	8.66	5.93	8.65	8.90	7.38	11.50	10.89	4.64	6.54	9.68	9.97	11.38	10.57	8.36	X	2.46
R	10.56	6.48	7.50	13.66	10.56	5.54	11.01	8.71	6.85	8.87	8.60	7.65	9.45	5.90	5.30	8.86	9.80	X

(a)

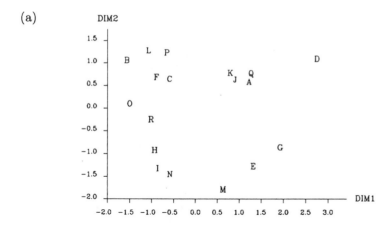

(b)

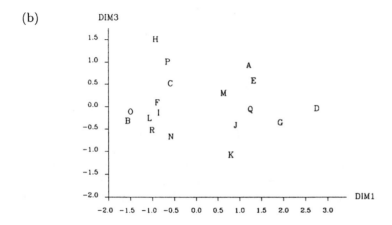

(c)

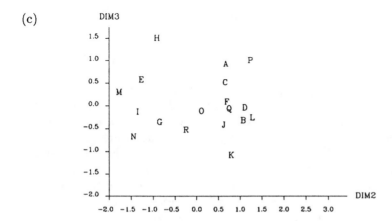

FIGURE 10.24. Spatial Plots for RCMP Detachments

TABLE 10.30. Coordinates for Scaling Dimensions – RCMP Data

Detachment	DIM1	DIM2	DIM3
A	1.15	0.61	0.88
B	−1.63	1.00	−0.37
C	−0.65	0.59	0.46
D	2.69	1.04	−0.09
E	1.26	−1.33	0.53
F	−0.95	0.63	0.04
G	1.86	−0.92	−0.41
H	−1.00	−0.98	1.45
I	−0.89	−1.38	−0.19
J	0.86	0.58	−0.47
K	0.73	0.73	−1.13
L	−1.13	1.22	−0.31
M	0.55	-1.85	0.25
N	−0.65	−1.51	−0.73
O	−1.57	0.04	−0.17
P	−0.72	1.17	0.95
Q	1.18	0.68	−0.11
R	−1.08	−0.31	−0.58

Using Ward's criterion a hierarchical clustering was carried out using the original dissimilarities in Table 10.29. The sequence of clusters obtained is shown graphically in Figure 10.26. This figure illustrates how the hierchical clustering takes place and relates the clusters to the spatial configuration provided by the first two scaling dimensions. The changes in the total within group sum of squares for the first 14 steps of the process are summarized below.

1	2	3	4	5	6	7
7.8	10.8	12.8	14.0	14.5	19.9	21.0

8	9	10	11	12	13	14
25.2	29.8	29.9	34.4	36.0	47.6	49.1

The remaining three steps in the process yielded very large values of the change in the within total sum of squares. From Figure 10.26 it would appear that the four distinct clusters make sense with respect to the first two dimensions. An examination of the two plots in Figure 10.24 that contain the third dimension seem to suggest that detachments H and K are at the two extremes of this third dimension. In addition detachments A and P appear to have relatively large positive values of this dimension whereas detachment N has a relatively large negative value.

To aid in the interpretation of the spatial relationships the three derived dimensions were related to several other sets of variables. One set of vari-

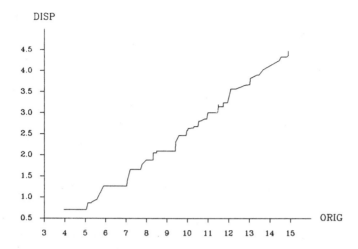

FIGURE 10.25. Shepard Diagram for Scaling of RCMP Data – Three-Dimensional Solution

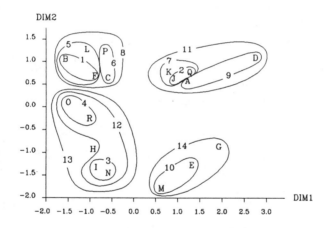

FIGURE 10.26. Spatial Plot for RCMP Detachments with Hierarchical Clusters

ables consisted of opinions of residents of the various communities with respect to the severity of the crime situation in the community, the feelings of lack of safety, the quality of performance of the RCMP, the degree of perceived interaction between RCMP and the community, and the behavior of the RCMP while carrying out their duties. In general communities with high values on DIM1 tended to view the crime situation as serious, were more fearful for their personal safety, and tended to have negative views regarding the behavior, performance and interaction of the RCMP. A second set of variables consisted of responses from the same individuals with respect to age, education, income, length of residency and family size. No relationships were found between these variables and the three dimensions.

A third set of variables consisted of responses from the RCMP officers regarding their satisfaction with various aspects of their job and working conditions. Relating these variables to the three dimensions shows that DIM2 tended to be large if the officers were happy with the commanding officer, their coworkers and the working conditions. Using the first two dimensions, we can conclude that detachments in the south-east quadrant of panel (a) of Figure 10.26 tend to have high crime rates, a low regard for the RCMP, and the police officers are not happy with their working conditions or with the commanding officers or coworkers. In a similar fashion the other three quandrants of the figure can be characterized. The third dimension derived from the scaling did not seem to be related to any of the variables discussed above. Another useful characterization of the dimensions is provided by other community characteristics. Detachments that are within 30 minutes of a major city (bedroom suburbs) tended to have low values of DIM1. Communities that are further away from the major city and experiencing high rates of growth tended to have high values of DIM1 while the relatively stagnant communities showed lower values of DIM1. Additional background on this police study is available in Jobson and Schneck (1982).

For the purposes of cross validation, the responses of the 182 police officers were divided randomly into two groups so that approximately half of the respondents in each detachment were in different groups. Separate MDS solutions were obtained for each of the two groups. The STRESS values and RSQ values for the two groups are given below.

			Dimension				
			1	2	3	4	5
Group	1	STRESS	0.333	0.156	0.102	0.057	0.028
	2	STRESS	0.285	0.154	0.088	0.048	0.029
Group	1	RSQ	0.669	0.868	0.923	0.969	0.991
	2	RSQ	0.739	0.866	0.941	0.978	0.990

Although the two sets of STRESS values exhibit the same behavior, the STRESS value at three dimensions seems to be a little higher than for the combined data seen earlier. Perhaps the reduction in sample size has increased the error component sufficiently. The fourth dimension may be required to fit some of this noise.

To compare the two sets of coordinates for the three-dimensional solutions a canonical correlation analysis was carried out to relate the two sets of dimensions. The canonical correlations obtained for the first two sets of functions were quite high, and these two dimensions were strongly related to the original dimensions DIM1 and DIM2. The third canonical functions were not correlated nor were they correlated with DIM3. It would appear that the third dimension in all three cases contains a large noise component. These results appear to be consistent with the two factors obtained in the factor analysis of the stress data in Table 9.37. The two factors obtained were related to the unpleasant aspects of police work and organizational stress. These two factors appear to be related to the two scaling dimensions obtained from MDS. A correlation analysis relating the two stress factors to the MDS dimensions showed that DIM1 is strongly and positively related to both stress factors, whereas DIM2 is related to a contrast between the two stress factors. If DIM2 is large, stress due to unpleasant duties is high relative to stress due to organizational matters.

Other Examples

The above example was derived from survey data. Examples in which the dissimilarities are derived from experiments designed to measure preferences for products as in marketing can be found in Schiffman, Reynolds and Young (1981) and Green, Carmone and Smith (1989). Other examples can be found in Romney, Shepard and Nerlove (1972).

10.3.3   OTHER SCALING MODELS

In this section a brief outline of other types of scaling models is provided. In the previous two sections, the beginning data matrix was assumed to be an $(n \times n)$ symmetric matrix of dissimilarities among n objects derived from observations received from a group of respondents. The output of the multidimensional scaling analysis was a spatial representation for the n objects based on the "averaged" dissimilarity ratings of the group. Alternative MDS models have been developed to analyze data matrices other than the single symmetric matrix of dissimilarities. Two such variations are discussed here. *Individual difference models* are designed to simultaneously determine a spatial representation for a group and to determine how the representation for each individual differs from the average. *Preference models* or *unfolding models* are designed to determine spatial representations for objects based on preference rankings obtained from individuals. These two types of MDS models are outlined below.

Individual Difference Models

Assume that an $(n \times n)$ dissimilarity matrix $\mathbf{D}$ is obtained from the responses supplied by a group of m individuals, and assume that $\mathbf{X}$ $(n \times p)$ denotes the matrix of coordinates for the underlying spatial configuration. For each object j the coordinate vector is given by $\mathbf{x}_j$ $(p \times 1)$ where $\mathbf{x}_j$ locates the object in the p-dimensional space.

Individual difference scaling models assume that for each individual the matrix of coordinates can differ from $\mathbf{X}$ and hence $\mathbf{X}_k$, $k = 1, 2, \ldots, m$ denotes the coordinate matrix for individual k. The matrix $\mathbf{X}_k$ is assumed to be related to $\mathbf{X}$ by some function g_k, $\mathbf{X}_k = g_k(\mathbf{X})$, $k = 1, 2, \ldots, m$.

The simplest and most commonly used model relating $\mathbf{X}_k$ to $\mathbf{X}$ is the weighted Euclidean model, which assumes $\mathbf{X}_k = \mathbf{X}\mathbf{W}_k$ where $\mathbf{W}_k$ $(p \times p)$ is a diagonal matrix of nonnegative weights for the p coordinates. Thus each individual is permitted to apply a different set of weights to the group coordinates $\mathbf{X}$. The input to the analysis is a set of m dissimilarity matrices, one for each individual. The output of the analysis consists of the group spatial configuration given by $\mathbf{X}$ and a set of weight matrices for the m individuals. The matrix $\mathbf{X}$ and the weight matrix $\mathbf{W}_k$ can be combined to determine a spatial representation for the n objects for each individual k.

A more general model than the weighted Euclidean model permits the individual coordinate matrix to be given by $\mathbf{X}_k = \mathbf{X}\mathbf{W}_k\mathbf{T}_k$, where $\mathbf{T}_k$ is a linear transformation matrix. In this case the spatial configuration for each individual consists of both a rotation of the group dimensions as well as a stretching or shrinking of the coordinates. This model of individual scaling is sometimes referred to as a three-mode model. The solution to the MDS analysis in this case consists of three sets of matrices, $\mathbf{X}$, $\mathbf{W}_k$ and $\mathbf{T}_k$, $k = 1, 2, \ldots, m$.

Preference Models and Multidimensional Unfolding

In this context the beginning data matrix consists of the rank order preferences of m individuals to a set of n objects. In this case the input matrix is asymmetric $(n \times m)$. The analysis begins with a dissimilarity matrix determined from the ranking by each individual k of each object i denoted by δ_{ik}. A *nonmetric multidimensional unfolding* model simultaneously fits a spatial configuration for both subjects and objects. As in the case of Section 10.3.2 the spatial configuration is determined so that the Euclidean distances d_{ik} are monotonically related to the initial ranks δ_{ik}. In this case, there are coordinates $\mathbf{X}$ $(n \times p)$ defining the location of the objects and coordinates $\mathbf{Y}$ $(m \times p)$ defining the location of the individuals with respect to the same p axes. The derived distances are given by $d_{ik} = \sum_{j=1}^{p} (x_{ij} - y_{kj})^2$.

10.3.4   OTHER SOURCES OF INFORMATION

Surveys of the various types of multidimensional scaling can be found in Schiffman, Reynolds and Young (1981), Coxon (1982), Davidson (1983), Kruskal and Wish (1978) and Green, Carmone and Smith (1989). An interesting survey of examples is provided in Young (1987).

Cited Literature and References

1. Anderberg, M.R. (1973). *Cluster Analysis for Applications*. New York: Academic Press.

2. Cailliez, Francis (1983). "The Analytical Solution of the Additive Constant Problem," *Psychometrika* 48, 305–308.

3. Carroll, J.D. and Arabie, P. (1980). "Multidimensional Scaling" in *Annual Review of Psychology*, M.R. Rosenzwerg and L.W. Porter (eds.). Palo Alto, CA: Annual Reviews.

4. Coxon, A.P.M. (1982). *The User's Guide to Multidimensional Scaling*. Exter, NH: Heinemann Educational Books.

5. Davidson, M.L. (1983). *Multidimensional Scaling*. New York: John Wiley and Sons.

6. Edelbrock, Craig (1979). "Mixture Model Tests of Hierarchical Clustering Algorithms: The Problem of Classifying Everybody," *Multivariate Behavioral Research* 14, 367–384.

7. Everitt, B.S. (1974). *Cluster Analysis*. London: Heinemann.

8. Gifi, A. (1990). *Nonlinear Multivariate Analysis*. New York: John Wiley and Sons.

9. Gordon, A.D. (1981). *Classification*. London: Chapman and Hall.

10. Green, Paul E., Carmone, Frank J. and Smith, Scott M. (1989). *Multidimensional Scaling: Concepts and Applications*. Boston: Allyn and Bacon.

11. Hartigan, J.A. (1975). *Clustering Algorithms*. New York: John Wiley and Sons.

12. Hubert, L. and Arabie, P. (1985). "Comparing Partitions," *Journal of Classification* 2, 193–218.

13. Jain, A.K. and Dubes, R.C. (1988). *Algorithms for Clustering Data*. Englewood Cliffs, NJ: Prentice-Hall.

14. Jambu, Michel (1991). *Exploratory and Multivariate Data Analysis*. New York: Academic Press.

15. Jobson, J.D. and Schneck, Rodney (1982). "Constituent Views of Organizational Effectiveness: Evidence from Police Organizations," *Academy of Management Journal* 25, 25–46.

16. Kruskal, J.B. and Wish, M. (1978). *Multidimensional Scaling*. Beverly Hills, CA: Sage Publications.

17. Lorr, Maurice (1983). *Cluster Analysis for Social Scientists.* San Francisco: Jossey-Bass.

18. Mardia, K.V., Kent, J.T. and Bibby, J.M. (1979). *Multivariate Analysis.* London: Academic Press.

19. Milligan, Glen W. (1980). "An Examination of the Effect of Six Types of Error Perturbation of Fifteen Clustering Algorithms," *Psychometrika* 45, 325–342.

20. Milligan, Glen W. (1981). "A Monte Carlo Study of Thirty Internal Criterion Measures for Cluster Analysis," *Psychometrika* 46, 187–199.

21. Milligan, Glen W. and Cooper, Martha C. (1985). "An Examination of Procedures for Determining the Number of Clusters in a Data Set," *Psychometrika* 50, 159–179.

22. Milligan, Glen W. and Cooper, Martha C. (1986). "A Study of the Comparability of External Criteria for Hierarchical Cluster Analysis," *Multivariate Behavioral Research* 21, 441–458.

23. Milligan, Glen W. and Cooper, Martha C. (1988). "A Study of Standardization of Variables in Cluster Analysis," *Journal of Classification* 5, 181–204.

24. Mojena, R. (1977). "Hierarchical Grouping Methods and Stopping Rules: An Evaluation," *The Computer Journal* 20, 359–363.

25. Overall, John E. and Klett, C. James (1972). *Applied Multivariate Analysis.* New York: McGraw-Hill.

26. Romney, A.K., Shepard, R.N. and Nerlove, S.B. (eds.) (1972). *Multidimensional Scaling: Theory and Applications in the Behavioral Sciences, Volume II: Applications.* New York: Seminar Press.

27. Schiffman, S.S., Reynolds, M.L. and Young, F.W. (1981). *Introduction to Multidimensional Scaling: Theory, Methods and Applications.* London and New York: Academic Press.

28. Seber, G.A.F. (1984). *Multivariate Observations.* New York: John Wiley and Sons.

29. Shepard, R.N., Romney, A.K. and Nerlove, S.B. (eds.) (1972). *Multidimensional Scaling: Theory and Applications in the Behavioral Sciences, Volume I: Theory.* New York: Seminar Press.

30. Spence, Ian and Ogilvie, John (1973). "A Table of Expected Stress Values for Random Rankings in Nonmetric Multidimensional Scaling," *Multivariate Behavioral Research* 8, 511–517.

31. Spence, Ian and Graef, Jed (1974). "The Determination of the Underlying Dimensionality of an Empirically Obtained Matrix of Proximities," *Multivariate Behavioral Research* 9, 331–341.

32. Spence, Ian and Lewandowsky, Stephan (1989). "Robust Multidimensional Scaling," *Psychometrika* 54, 501–513.

33. Takane, Y., Young, F.W. and de Leeuw, J. (1977). "Nonmetric Individual Differences Scaling: An Alternating Least Squares Method with Optimal Scaling Features," *Psychometrika* 42, 3–27.

34. Waganaar, W.A. and Padmos, P. (1971). "Quantitative Interpretation of Stress in Kruskal's Multidimensional Scaling Technique," *British Journal of Mathematical and Statistical Psychology* 24, 101–110.

35. Wishart, David (1987). *Clustan User Manual*, 4th Edition. Edinburgh: University of Saint Andrews.

36. Young, Forest W. (1970). "Nonmetric Multidimensional Scaling: Recovery of Metric Information," *Psychometrika* 35, 455–473.

37. Young, F.W., Takane, Y. and Lewyckyj, R. (1978), "Three Notes on Alscal," *Psychometrika* 43, 433–435.

38. Young, F.W. (1984). "Scaling" in *Annual Review of Psychology*, M.R. Rosenzwerg and L.W. Porter (eds.). Palo Alto, CA: Annual Reviews.

39. Young, F.W. (1987). *Multidimensional Scaling: History, Theory and Applications*, (R.M. Hamer, ed.). Hillsdale, NJ: Lawrence Erlbaum Associates.

Exercises for Chapter 10

1. This exercise is based on the U.S. Crime Data contained in Table V17 in the Data Appendix.

   (a) Obtain proximity matrices for the fifteen states using both squared Euclidean distances and the cosine coefficient. Compare the proximity matrices. Does the change in proximity measure have an impact on the relationships among the states? Why might the proximity relationships change? Which measure would you recommend?

   (b) Standardize the data in Table V17 so that for each type of crime the mean over the 15 states is 0 and the standard deviation is 1. Determine proximity matrices using squared Euclidean distance and the correlation coefficient. Compare the proximity matrices. Does the change in proximity measure have an impact on the relationships among the states? Compare these two proximity matrices to the two matrices obtained in (a).

   (c) Using the covariance matrix derived from Table V17 carry out a principal components analysis. Obtain the unstandardized scores and the standardized scores for the first two components. Determine a proximity matrix relating the 15 states using the squared Euclidean distance method. Determine two such matrices, one based on the first two unstandardized components and one based on the standardized components. Compare the results to (a) and (b).

   (d) Using the two sets of principal component scores in (c) obtain two plots for the 15 states based on the first two components. Discuss the results by interpreting the first two components.

   (e) Repeat the analyses in (c) and (d) using the correlation matrix from Table V17. Compare the results to (a), (b), (c) and (d).

   (f) Write an overall summary discussing the proximity characteristics of the 15 states.

2. This exercise is based on the Air Pollution Data in Table V21 contained in the Data Appendix.

   (a) Obtain proximity matrices for the twelve cities using both squared Euclidean distances and the cosine coefficient. Compare the proximity matrices. Does the change in proximity measure have an impact on the relationships among the cities? Why might the proximity relationships change? Which measure would you recommend?

(b) Standardize the data in Table V21 so that for each variable the mean is zero and the standard deviation is 1. Determine proximity matrices using squared Euclidean distance and the correlation coefficient. Compare the two proximity matrices. Does the change in proximity measure have an impact on the relationships among the cities? Compare these two proximity matrices to the two matrices obtained in (a).

(c) Using the covariance matrix derived from Table V21 carry out a principal components analysis. Obtain the unstandardized scores and the standardized scores for the first two components. Determine a proximity matrix relating the 12 cities using the squared Euclidean distance method. Determine two such matrices, one based on the first two unstandardized components and one based on the standardized components. Compare the results to (a) and (b).

(d) Using the two sets of principal component scores in (c) obtain two plots for the 12 cities based on the first two components. Discuss the results by interpreting the first two components.

(e) Repeat the analyses in (c) and (d) using the correlation matrix from Table V21. Compare the results to (a), (b), (c) and (d).

(f) Write an overall summary discussing the proximity characteristics of the 12 cities.

3. This exercise is based on the U.S. Divorce Data contained in Table V15 in the Data Appendix.

(a) Use the simple matching coefficient and the Jaccard coefficient to determine proximity matrices relating the twenty states with respect to the available grounds for divorce. Compare the proximity measures and discuss the differences between the two coefficients.

(b) The columns of the data matrix formed by Table V15 represent binary indicator variables. Use the squared Euclidean distance, cosine coefficient and correlation measures to determine proximity measures. Compare the three proximity matrices. Also compare the proximity matrices to the matrices obtained in (a).

(c) Create nine new indicator variables by transforming the binary variables in Table V17 using

$$\text{NEWVAR} = -1 * \text{OLDVAR} + 1.$$

Using the new data matrix containing old and new variables (18 altogether), determine proximity measures using squared Euclidean distance, cosine coefficient and correlation coefficient measures. Compare the three sets of proximity measures. Also compare these measures to those obtained in (a) and (b).

(d) Write a summary discussion outlining the properties of the various proximity measures for categorical data used in this exercise.

4. This exercise is based on the Automobile Data in Table V18 in the Data Appendix.

   (a) Determine a proximity matrix for the 20 automobiles using the correlation, squared Euclidean distance and cosine measures of similarity. Discuss the similarity relationships found.

   (b) Define dummy variables for the categories defined by the ranked variables (e.g., ENGSIZE has five categories and hence requires five dummy variables). Also, define a new dummy variable for the (FOR $= 0$) category using NONFOR $= -1 * FOR + 1$. Using the new data matrix containing the entire set of 19 dummy variables determine proximity matrices using the measures correlation, cosine and squared Euclidean distance. Discuss the relationships obtained. Compare the results to (a).

   (c) Write a summary discussing the differences among the various measures. Comment on the problem of combining categorical and ordinal variables.

5. This exercise in based on the U.S. Crime Data contained in Table V16 in the Data Appendix.

   (a) Use Ward's algorithm to carry out a hierarchical cluster analysis for the data in Table V16. Standardize the data first. Discuss the results and choose a particular solution.

   (b) Use other statistical methods to characterize the clusters from your solution in (a) using the data in Table V16. Compare the clusters.

   (c) Repeat steps (a) and (b) using the average linkage method instead of Ward's method. Compare the solution obtained using Ward's to the solution obtained using the average linkage method.

   (d) Use a partitioning method such as the k-means algorithm to obtain solutions in the neighborhood of the number of clusters selected in (a) and (c). Compare the partitioning method solution to the solutions obtained in (a) and (c).

   (e) Use principal components analysis applied to the standardized-data in Table V16 to obtain a plot of the states with respect to the first two principal components. Compare the plot locations of the states to the clusters obtained in (a), (c) and (d).

(f) Provide an overall summary regarding the similarity of the states with respect to the crime rates.

6. This exercise is based on the Air Pollution Data contained in Table V7 in the Data Appendix.

   (a) Use Ward's algorithm to carry out a hierarchical cluster analysis for the data in Table V7. Standardize the data first. Discuss the results and choose a particular solution.

   (b) Use other statistical methods to characterize the clusters from your solution in (a) using the data in Table V7. Compare the clusters.

   (c) Repeat steps (a) and (b) using the average linkage method instead of Ward's method. Compare the solution obtained using Ward's to the solution using the average linkage method.

   (d) Use a partitioning method such as the k-means algorithm to obtain solutions in the neighborhood of the solutions in (a) and (c). Compare the partitioning method solution to the solutions in (a) and (c).

   (e) Use principal components analysis applied to the standardized data in Table V21 to obtain a plot of the cities with respect to the first two principal components. Compare the plot locations of the cities to the clusters obtained in (a), (c) and (d).

   (f) Provide an overall summary regarding the similarity of the cities with respect to the variables.

7. This exercise is based on the U.S. Divorce Data contained in Table V14 in the Data Appendix.

   (a) Select two different measures of proximity (e.g., simple matching coefficient and Jaccard coefficient) and carry out a hierarchical cluster analysis using the single linkage method. Choose a particular solution in each case. Discuss the results including a comparison of the two solutions.

   (b) Repeat the steps in (a) using the same proximity measures but replace the single linkage method by the complete linkage method. Compare the results in (a) and (b).

   (c) For the data matrix in Table V14 define nine new variables by transforming the old variables using

   $$\text{NEWVAR} = -1 * \text{OLDVAR} + 1.$$

   Using the data matrix containing all 18 variables (OLD and NEW) determine principal components. Using the scores based on the first two principal components plot the location of the states.

(d) Provide an overall summary discussion regarding the similarity among the states with respect to available grounds for divorce.

8. This exercise is based on the U.S. Crime Data in Table V17.

(a) Use a multidimensional scaling program to obtain a spatial representation for the 15 states based on a Euclidean distance matrix derived from the data in Table V17. Discuss the results.

(b) Obtain the principal components for the data in Table V17 and interpret the components. Obtain a scatterplot for the 15 states based on the first two components. How does this spatial representation compare to the one obtained in (a).

(c) Use cluster analysis to relate the 15 states and compare the results to the results in (a) and (b).

(d) Use other multivariate techniques to relate the original data to the scaling dimensions.

(e) Provide an overall summary discussion regarding the relationships among the 15 states.

9. This exercise is based on the Air Pollution Data in Table V21.

(a) Use a multidimensional scaling program to obtain a spatial representation for the ten cities based on a Euclidean distance matrix derived from the data in Table V21. Discuss the results.

(b) Obtain the principal components for the data in Table V21 and interpret the components. Obtain a scatterplot for the 12 cities based on the first two components. How does the spatial representation compare to the one in (a).

(c) Use cluster analysis to relate the 12 cities and compare the results to the results in (a) and (b).

(d) Use other multivariate techniques to relate the original data to the scaling dimensions.

(e) Provide an overall summary discussion regarding the relationships among the ten cities.

10. This exercise is based on the U.S. Divorce Data in Table V15.

(a) Obtain a dissimilarity matrix for the 20 states by determining the Euclidean distance derived from the matrix of nine binary variables in Table V15. Use multidimensional scaling to obtain a spatial representation for the 20 states. Choose a solution and interpret the derived spatial configuration.

(b) Carry out a hierarchical cluster analysis using the same dissimilarity matrix as in (a). Compare the cluster results to the results in (a).

(c) Repeat the steps in (a) and (b) using a dissimilarity matrix derived from the Euclidean distance based on the matrix of dummy variables (two for each of the grounds variables. Discuss the solution. Compare your solution to the ones in (a) and (b).

11. This exercise is based on the Automobile Data in Table V18.

(a) Obtain a dissimilarity matrix for the 20 automobiles by determining a Euclidean distance matrix based on the variables in Table V18. Use multidimensional scaling to obtain a spatial representation for the 20 automobiles. Choose a solution and interpret the derived spatial configuration.

(b) Carry out a hierarchical cluster analysis using the dissimilarity matrix in (a). Compare the cluster results to the scaling results in (a).

(c) Repeat the steps in (a) and (b) using a dissimilarity matrix based on the matrix of dummy variables (one dummy variable for each category of each variable). Discuss the solution and compare your solution to (a).

12. This exercise is based on the Cola Similarity Data in Table V19.

(a) Use nonmetric multidimensional scaling to obtain a spatial configuration for the dissimilarity matrix given in Table V19. Obtain the STRESS values for dimensions 1 through 5 and compare them to the simulation values shown in Table 10.27. Select the appropriate dimension.

(b) Examine the spatial representation for the chosen solution in (a) and interpret the result.

13. This exercise is based on the Car Similarity Data in Table V20.

(a) Use nonmetric multidimensional scaling to obtain a spatial configuration for the dissimilarity matrix given in Table V20. Obtain the STRESS values for the dimensions 1 through 5 and compare them to the simulation values shown in Tables 10.27 and 10.28. Select the appropriate dimension.

(b) Examine the spatial representation for the chosen solution in (a) and interpret the result.

Questions for Chapter 10

1. Let $\mathbf{X}$ $(n \times p)$ denote a data matrix of n observations on each of p variables and let x_{ij} denote the observation in row i and column j. If the observations in a row of $\mathbf{X}$ denote the coordinates of an object in a p-dimensional space, the Euclidean distance between the objects in rows r and s is given by d_{rs}, where

$$d_{rs}^2 = \sum_{j=1}^{p}(x_{rj} - x_{sj})^2.$$

   (a) Denote the rth row of $\mathbf{X}$ by $\mathbf{x}_r'$ and show that

$$d_{rs}^2 = (\mathbf{x}_r - \mathbf{x}_s)'(\mathbf{x}_r - \mathbf{x}_s).$$

   (b) Let $\bar{\mathbf{x}} = \sum_{r=1}^{n} \mathbf{x}_r/n$ denote the mean of the row vectors and let $\mathbf{x}_r^* = (\mathbf{x}_r - \bar{\mathbf{x}})$ and $\mathbf{x}_s^* = (\mathbf{x}_s - \bar{\mathbf{x}})$. Show that

$$d_{rs}^2 = (\mathbf{x}_r^* - \mathbf{x}_s^*)'(\mathbf{x}_r^* - \mathbf{x}_s^*).$$

2. Let $\bar{\mathbf{x}}_{rs}$ denote the mean of $\mathbf{x}_r$ and $\mathbf{x}_s$, hence $\bar{\mathbf{x}}_{rs} = (\mathbf{x}_r + \mathbf{x}_s)/2$. Show that

$$d_{rs}^2 = 2(\mathbf{x}_r - \bar{\mathbf{x}}_{rs})'(\mathbf{x}_r - \bar{\mathbf{x}}_{rs}) + 2(\mathbf{x}_s - \bar{\mathbf{x}}_{rs})'(\mathbf{x}_s - \bar{\mathbf{x}}_{rs}),$$

   where d_{rs} is the Euclidean distance between $\mathbf{x}_r$ and $\mathbf{x}_s$. Give an interpretation for the right-hand side of this expression.

3. For the $(n \times p)$ matrices $\mathbf{X}^+$ and $\mathbf{X}^{++}$ defined in Section 10.1, show that the elements of $\mathbf{X}^+\mathbf{X}^{+\prime}$ and $\mathbf{X}^{++}\mathbf{X}^{++\prime}$ are given by the cosine coefficient c_{rs} and the correlation coefficient q_{rs} also defined in Section 10.1.

4. Each row of the data matrix $\mathbf{X}$ is called a profile.

   (a) Show that the Euclidean distance between the rth and sth profile given by d_{rs} can be written as

$$d_{rs}^2 = (v_r - v_s)^2 + p(\bar{x}_r. - \bar{x}_s.)^2 + 2v_r v_s(1 - q_{rs})$$

   where $v_r^2 = \sum_{j=1}^{p}(x_{rj} - \bar{x}_r.)^2$, $\bar{x}_r. = \sum_{j=1}^{p} x_{rj}/p$ (similarly for v_s^2 and $\bar{x}_s.$) and q_{rs} is the correlation coefficient between rows r and s.

   (b) The squared Euclidean distance is composed of three components: level, scatter and shape. Define these terms with reference to the three terms in d_{rs}^2 and explain what the terms actually measure.

(c) Use the concepts in (b) to help explain the differences between d_{rs} and $2(1 - q_{rs})$ as measures of proximity. When might one measure be preferable to the other?

5. Assume that the **X** matrix contains a total of K dummy variables corresponding to p categorical variables. Let k_j denote the number of categories for the jth variable, $K = \sum_{j=1}^{p} k_j$.

   (a) Show that the cosine coefficient is given by $c_{rs} = f_{rs}/p$, where f_{rs} is the number of variables in which the rth and sth rows have the same categories (unities in common).

   (b) Show that the correlation coefficient is given by

   $$q_{rs} = \left[\frac{f_{rs}}{K} - \left(\frac{p}{K}\right)^2\right] \Big/ \frac{p}{K}\left(1 - \frac{p}{K}\right).$$

   (c) Show that the square of the Euclidean distance between rows r and s is given by $d_{rs}^2 = 2(p - f_{rs})$.

6. If the categorical variables are binary (only two categories), an alternative method of coding is to use only one dummy variable for each variable so that **X** has p dummy variables corresponding to the p classification variables. Let a = number of variables in which r and s both are coded 0, and d = number of variables in which both are coded 1. Let c = number of variables in which r is coded 0 and s is coded 1, and let b = number of variables in which r is coded 1 and s is coded 0.

   (a) Show that the cosine coefficient c_{rs} based on **X** is given by $d/[(b+d)(c+d)]^{1/2}$.

   (b) Show that the correlation coefficient q_{rs} based on **X** is given by $(ad - bc)/[(a+b)(a+c)(b+d)(c+d)]^{1/2}$.

7. For any group proximity measure that satisfies the ultrametric inequality, (10.3) can be used to update the proximity measures after two groups are joined.

   (a) Show that for the complete linkage method $\alpha_r = \alpha_s = \frac{1}{2}$, $\beta = 0$ and $\gamma = \frac{1}{2}$ assuming a dissimilarity measure.

   (b) Show that for the single linkage method $\alpha_r = \alpha_s = \frac{1}{2}$, $\beta = 0$ and $\gamma = -\frac{1}{2}$ assuming a dissimilarity measure.

   (c) Show that for the average linkage method $\alpha_r = n_r/(n_r + n_s)$, $\alpha_s = n_s/(n_r + n_s)$, $\beta = 0$, and $\gamma = 0$ assuming a dissimilarity measure.

8. Let $(x_{r1m}, x_{r2m}, \ldots, x_{rpm})$ denote observations on the p variables for the mth object in group r. Similarly let $(x_{s1m}, x_{s2m}, \ldots, x_{spm})$ denote observations on the p variables for the mth object in group s. Let the means for each group on the p variables be denoted by $(\bar{x}_{r1\cdot}, \bar{x}_{r2\cdot}, \ldots, \bar{x}_{rp\cdot})$ and $(\bar{x}_{s1\cdot}, \bar{x}_{s2\cdot}, \ldots, \bar{x}_{sp\cdot})$ for groups r and s respectively. The within group sums of squares for the two groups are given by

$$SSW_r = \sum_{m=1}^{n_r} \sum_{j=1}^{p} (x_{rjm} - \bar{x}_{rj\cdot})^2 \quad \text{and}$$

$$SSW_s = \sum_{m=1}^{n_s} \sum_{j=1}^{p} (x_{sjm} - \bar{x}_{sj\cdot})^2,$$

where n_r and n_s denote the number of objects in groups r and s respectively. When groups r and s are joined to form group t the new centroid is given by $(\bar{x}_{t1\cdot}, \bar{x}_{t2\cdot}, \ldots, \bar{x}_{tp\cdot})$ and the within group sum of squares is given by

$$SSW_t = \sum_{m=1}^{(n_r+n_s)} \sum_{j=1}^{p} (x_{tjm} - \bar{x}_{tj\cdot})^2.$$

Denote the squared distance between the centroids of groups r and s by d_{rs}^2 where

$$d_{rs}^2 = \sum_{j=1}^{p} (\bar{x}_{rj\cdot} - \bar{x}_{sj\cdot})^2.$$

(a) Show that $n_r n_s d_{rs}^2/(n_r + n_s)$ is equivalent to the incremental sum of squares given by $SSW_t - (SSW_r + SSW_s)$.

(b) In an analysis of variance with two groups r and s the within group sum of squares is given by $(SSW_r + SSW_s)$ as defined above. Show that SSW_t corresponds to the total sum of squares and $n_r n_s d_{rs}^2/(n_r + n_s)$ corresponds to the between group sum of squares. (HINT: $\bar{x}_{tj\cdot}$, $j = 1, 2, \ldots, p$ is the grand mean for the two groups.)

(c) The incremental sum of squares measure of proximity between two groups r and s is given by $s_{rs} = n_r n_s d_{rs}^2/(n_r + n_s)$ where d_{rs}^2 is defined above. Let s_{tu}, s_{ru}, s_{su} denote the proximities between groups t and u, r and u and s and u respectively. If t corresponds to the combination of groups r and s show that

$$s_{tu} = \frac{1}{(n_t + n_u)} [(n_u + n_r)s_{ru} + (n_u + n_s)s_{su} - n_u s_{rs}].$$

(d) Show that the expression for s_{tu} given in (c) can be obtained from (10.3) with $\alpha_r = (n_r + n_u)/(n_t + n_u)$, $\alpha_s = (n_s + n_u)/(n_t + n_u)$, $\beta = -n_u/(n_t + n_u)$ and $\gamma = 0$.

9. Let g denote the number of groups of objects and let $\mathbf{T}$, $\mathbf{W}$ and $\mathbf{G}$ denote the total, within and among group sums of squares matrices as defined in Section 10.1.2. Show that the incremental sum of squares resulting from the combining of groups r and s is given by an increase in $tr\mathbf{W}$ and a corresponding decrease in $tr\mathbf{G}$ of $n_r n_s d_{rs}^2/(n_r + n_s)$.

10. The quality of a cluster analysis can be determined by comparing the group membership of the objects to the original proximities. Objects that are in the same cluster group should in general be closer as measured by the original proximities than objects that are not in the same group. Correlation measures can be used to measure the cluster quality of a given solution. Pairs in which the objects are in the same group are coded 0, and pairs that are not in the same group are coded 1. This newly created dummy variable defined over the $n(n-1)/2$ pairs can be correlated with the original proximities to measure cluster quality.

  (a) Show that the Pearson correlation between the two sets of values yields the point-biserial correlation as given in Section 10.2.2.

  (b) Show that if all ties are eliminated, Kendall's coefficient of concordance is given by γ defined in Section 10.2.2.

11. Let $\mathbf{D}$ $(n \times n)$ denote a dissimilarity matrix with elements δ_{rs}, $r, s = 1, 2, \ldots, n$. Assume that $\mathbf{D}$ is Euclidean and hence that there exists a set of p $(n \times 1)$ vectors $\mathbf{x}_1, \mathbf{x}_2, \ldots, \mathbf{x}_n$ such that

$$\delta_{rs}^2 = (\mathbf{x}_r - \mathbf{x}_s)'(\mathbf{x}_r - \mathbf{x}_s), \qquad r, s = 1, 2, \ldots, n.$$

  (a) Show that $\delta_{rs}^2 = \mathbf{x}_r'\mathbf{x}_r + \mathbf{x}_s'\mathbf{x}_s - 2\mathbf{x}_r'\mathbf{x}_s$ and hence that $a_{rs} = -\frac{1}{2}\delta_{rs}^2 = -\frac{1}{2}\mathbf{x}_r'\mathbf{x}_r - \frac{1}{2}\mathbf{x}_s'\mathbf{x}_s + \mathbf{x}_r'\mathbf{x}_s$.

  (b) Define $b_{rs} = a_{rs} - \bar{a}_r - \bar{a}_s + \bar{a}_{..}$ where $\bar{a}_r = \frac{1}{n}\sum_{s=1}^{n} a_{rs}$, $\bar{a}_s = \frac{1}{n}\sum_{r=1}^{n} a_{rs}$ and $\bar{a}_{..} = \frac{1}{n^2}\sum_{r=1}^{n}\sum_{s=1}^{n} a_{rs}$; and show that

$$\begin{aligned} b_{rs} &= \mathbf{x}_r'\mathbf{x}_r - \mathbf{x}_r'\bar{\mathbf{x}} - \bar{\mathbf{x}}'\mathbf{x}_s + \bar{\mathbf{x}}'\bar{\mathbf{x}} \\ &= (\mathbf{x}_r - \bar{\mathbf{x}})'(\mathbf{x}_s - \bar{\mathbf{x}}) = \mathbf{x}_r^{*\prime}\mathbf{x}_s^{*}, \end{aligned}$$

  where $\bar{\mathbf{x}} = \frac{1}{n}\sum_{r=1}^{n} \mathbf{x}_r$.

  (c) Let $\mathbf{B}$ denote the matrix of elements b_{rs}, $r, s = 1, 2, \ldots, n$. Show that $\mathbf{B} = \mathbf{X}^*\mathbf{X}^{*\prime}$, where $\mathbf{X}^*$ $(n \times p)$ is the matrix of row deviations $(\mathbf{x}_r - \bar{\mathbf{x}})'$, $r = 1, 2, \ldots, n$ so that $\mathbf{B}$ has the form $\mathbf{B} = \mathbf{Z}\mathbf{Z}'$.

(d) Show that you could use principal components to determine two underlying vectors z_1 and z_2 so that $\widehat{B} = \begin{bmatrix} z_1' \\ z_2' \end{bmatrix} [z_1 \ z_2]$. What properties does this approximation to B have?

12. Given an $(n \times p)$ matrix X, a rigid notation of the columns of X is given by $Z = XV$, where $V' = V^{-1}$ (V is orthogonal).

(a) Show that the squared Euclidean distance between rows r and s, d_{rs}^2, is not affected by rigid rotations and hence that

$$d_{rs}^2 = (x_r - x_s)'(x_r - x_s) = (z_r - z_s)'(z_r - z_s).$$

(b) Assume X $(n \times p)$ is derived from a matrix D of dissimilarities δ_{rs}, $r, s = 1, 2, \ldots, n$. Given the results from (a) and Question 1 what conclusions can you draw regarding the uniqueness of X?

13. Suppose that the $(n \times p)$ data matrix X $(n \times p)$ contains p variables that have been standardized to have mean 0 and variance 1.

(a) Show that the matrix $X'X$ is a correlation matrix.

(b) Show that the principal components of X given by $Z = XV$ satisfy $ZZ' = XX'$.

(c) Show that the squared Euclidean distances δ_{rs}^2 are given by $\delta_{rs}^2 = (x_r - x_s)'(x_r - x_s) = (z_r - z_s)'(z_r - z_s)$, where x_r, x_s denote columns of X and z_r, z_s denote the corresponding columns of Z, $r, s = 1, 2, \ldots, n$.

(d) Show that δ_{rs}^2 in (c) can be written as $\delta_{rs}^2 = \sum_{j=1}^{p}(z_{rj} - z_{sj})^2$. What does each term of this sum represent?

(e) Show that $\sum_{r=1}^{n}\sum_{s=1}^{n}\delta_{rs}^2 = 2n\sum_{j=1}^{p}\sum_{r=1}^{n}z_{rj}^2$. (HINT: The means of the components are necessarily zero.)

(f) Show that $\sum_{r=1}^{n}\sum_{s=1}^{n}\delta_{rs}^2/n^2 = 2p/n$.

Appendix

1 Matrix Algebra

1.1 MATRICES

Matrix

A *matrix* of order $(n \times p)$ is a rectangular array of elements consisting of n rows and p columns. A matrix is denoted by a boldface letter, say $\mathbf{A}$, where

$$\mathbf{A} = \begin{bmatrix} a_{11} & a_{12} & \cdots & a_{1p} \\ a_{21} & a_{22} & \cdots & a_{2p} \\ a_{31} & a_{32} & \cdots & a_{3p} \\ \vdots & & & \vdots \\ a_{n1} & a_{n2} & \cdots & a_{np} \end{bmatrix}.$$

The elements are denoted by a_{ij}, $i = 1, 2, \ldots, n$, $j = 1, 2, \ldots, p$, where the first subscript i refers to the row location, and the second subscript j refers to the column location of the element. The matrix is also sometimes denoted by $((a_{ij}))$.

Example of a Matrix

The matrix $\mathbf{B} = \begin{bmatrix} 1 & 9 & -2 \\ 4 & 6 & 3 \end{bmatrix}$ has 2 rows and 3 columns and is a (2×3) matrix, whereas the matrix $\mathbf{C} = \begin{bmatrix} -6 & 2 \\ 4 & -5 \\ -3 & 8 \end{bmatrix}$ has 3 rows and 2 columns and is a (3×2) matrix.

Transpose of a Matrix

The *transpose* of the matrix $\mathbf{A}$ $(n \times p)$ is the matrix $\mathbf{B}$ $(p \times n)$ obtained by interchanging rows and columns so that

$$b_{ij} = a_{ji}, \quad i = 1, 2, \ldots, p; \ j = 1, 2, \ldots, n.$$

The transpose of $\mathbf{A}$ is usually denoted by $\mathbf{B} = \mathbf{A}'$. Some additional properties of a matrix transpose are

1. $\mathbf{A}' = \mathbf{B}'$ if and only if $\mathbf{A} = \mathbf{B}$,

2. $(\mathbf{A}')' = \mathbf{A}$.

Example of a Matrix Transpose

The transpose of the matrices $\mathbf{B}$ and $\mathbf{C}$ defined in the above example are given by

$$\mathbf{B}' = \begin{bmatrix} 1 & 4 \\ 9 & 6 \\ -2 & 3 \end{bmatrix} \quad \text{and} \quad \mathbf{C}' = \begin{bmatrix} -6 & 4 & -3 \\ 2 & -5 & 8 \end{bmatrix}.$$

Exercise — Matrix Transpose

Using the matrices $\mathbf{B}$ and $\mathbf{C}$ defined above verify that $(\mathbf{B}')' = \mathbf{B}$ and $(\mathbf{C}')' = \mathbf{C}$.

Row Vector and Column Vector

A *row vector* is a matrix with only one row and is denoted by a lower case letter with boldface type

$$\mathbf{b} = [b_1, b_2, \ldots, b_p].$$

A *column vector* is a matrix with only one column and is also denoted by a lower case boldface letter

$$\mathbf{a} = \begin{bmatrix} a_1 \\ a_2 \\ \vdots \\ a_n \end{bmatrix}.$$

Example of Row and Column Vectors

The (4×1) matrix $\mathbf{d} = \begin{bmatrix} 2 \\ -4 \\ 4 \\ -6 \end{bmatrix}$ is a column vector, but the (1×5) matrix $\mathbf{f} = \begin{bmatrix} 3 & 5 & -7 & 4 & -2 \end{bmatrix}$ is a row vector.

Square Matrix

A matrix is *square* if the number of rows n is equal to the number of columns p ($n = p$). A *square matrix* with m rows and columns is said to have *order* m.

Symmetric Matrix

A square matrix $\mathbf{A}$ of order m is *symmetric*, if the transpose of $\mathbf{A}$ is equal to $\mathbf{A}$.

$$\mathbf{A} = \mathbf{A}' \quad \text{if } \mathbf{A} \text{ symmetric.}$$

Diagonal Elements

The elements a_{ii}, $i = 1, 2, \ldots, m$, are called the *diagonal* elements of the square matrix $\mathbf{A}$ with elements a_{ij}, $i = 1, 2, \ldots, m$, $j = 1, 2, \ldots, m$.

Trace of a Matrix

The sum of the diagonal elements of $\mathbf{A}$ is called the *trace* of $\mathbf{A}$ and is denoted by $tr(\mathbf{A})$. The trace of the $(m \times m)$ matrix is given by $tr(\mathbf{A}) = \sum_{i=1}^{m} a_{ii}$.

Example — Square, Symmetric, Diagonal Elements, Trace

The matrix $\mathbf{H} = \begin{bmatrix} 8 & 6 & -4 \\ 6 & 2 & -1 \\ -4 & -1 & 7 \end{bmatrix}$ is a (3×3) or square matrix of order 3.

The matrix is symmetric since $\mathbf{H}' = \mathbf{H}$. The diagonal elements of $\mathbf{H}$ are the elements 8, 2 and 7. The trace of $\mathbf{H} = tr\mathbf{H} = 17$ which is the sum of the diagonal elements.

Exercise — Symmetric Matrix, Trace

Verify that the matrix $\mathbf{A} = \begin{bmatrix} a & b & c \\ b & d & e \\ c & e & f \end{bmatrix}$ is symmetric and that $tr(\mathbf{A}) = (a + d + f)$.

Null or Zero Matrix

The *null or zero matrix* denoted by $\mathbf{0}$ is the matrix whose elements are all zero.

Identity Matrix

The *identity matrix* of order m is the square matrix whose diagonal elements are all unity, and whose off-diagonal elements are all zero. The identity matrix of order m is usually denoted by $\mathbf{I}_m$ and is given by

$$\mathbf{I}_m = \begin{bmatrix} 1 & 0 & 0 & \cdots & 0 \\ 0 & 1 & 0 & \cdots & 0 \\ 0 & 0 & 1 & & \vdots \\ \vdots & \vdots & & \ddots & 0 \\ 0 & 0 & \cdots & 0 & 1 \end{bmatrix}.$$

Diagonal Matrix

A *diagonal matrix* is a matrix whose off-diagonal elements are all zero. The identity matrix is a special case of a diagonal matrix.

Submatrix

A *submatrix* of a matrix $\mathbf{A}$ is a matrix obtained from $\mathbf{A}$ by deleting some rows and columns of $\mathbf{A}$.

Example of Submatrix

The (2×2) matrix $\mathbf{C}^* = \begin{bmatrix} 4 & -4 \\ 2 & 1 \end{bmatrix}$ is a submatrix of the matrix $\mathbf{C} = \begin{bmatrix} 9 & -6 & 3 \\ 4 & -4 & 5 \\ 2 & 1 & -5 \end{bmatrix}$. $\mathbf{C}^*$ is obtained from $\mathbf{C}$ by deleting the first row and the third column.

1.2   MATRIX OPERATIONS

Equality of Matrices

Two matrices $\mathbf{A}$ and $\mathbf{B}$ are *equal*, if and only if each element of $\mathbf{A}$ is equal to the corresponding element of $\mathbf{B}$: $a_{ij} = b_{ij}$, $i = 1, 2, \ldots, n$, $j = 1, 2, \ldots, p$.

Addition of Matrices

The *addition* of two matrices $\mathbf{A}$ and $\mathbf{B}$ is carried out by adding together corresponding elements. The two matrices must have the same order. The sum is given by $\mathbf{C}$ where

$$\mathbf{C} = \mathbf{A} + \mathbf{B}$$

and where $c_{ij} = a_{ij} + b_{ij}$, $i = 1, 2, \ldots, n$, $j = 1, 2, \ldots, p$.

Additive Inverse

A matrix $\mathbf{B}$ is the *additive inverse* of a matrix $\mathbf{A}$, if the matrices sum to the null matrix $\mathbf{0}$

$$\mathbf{B} + \mathbf{A} = \mathbf{0},$$

where $b_{ij} + a_{ij} = 0$ or $b_{ij} = -a_{ij}$, $i = 1, 2, \ldots, n$, $j = 1, 2, \ldots, p$. This additive inverse is denoted by $\mathbf{B} = -\mathbf{A}$. If $\mathbf{C} = \mathbf{A} + \mathbf{B}$ then $\mathbf{C}' = \mathbf{A}' + \mathbf{B}'$.

Example

The sum of the matrices $\mathbf{A}$ and $\mathbf{B}$ given below is denoted by $\mathbf{C}$.

$$\mathbf{A} = \begin{bmatrix} 4 & 3 \\ -6 & 9 \\ 2 & -4 \end{bmatrix} \quad \mathbf{B} = \begin{bmatrix} 2 & -5 \\ -2 & -2 \\ 4 & 6 \end{bmatrix}$$

$$\mathbf{C} = \begin{bmatrix} 4+2 & 3-5 \\ -6-2 & 9-2 \\ 2+4 & -4+6 \end{bmatrix} = \begin{bmatrix} 6 & -2 \\ -8 & 7 \\ 6 & 2 \end{bmatrix}.$$

The matrix $\mathbf{D} = \begin{bmatrix} -4 & -3 \\ 6 & -9 \\ -2 & 4 \end{bmatrix}$ is the additive inverse of the matrix $\mathbf{A}$

since $\mathbf{A} + \mathbf{D} = \begin{bmatrix} 0 & 0 \\ 0 & 0 \\ 0 & 0 \end{bmatrix}.$

Exercise

Verify that $\mathbf{C}' = \mathbf{A}' + \mathbf{B}'$ using the matrices $\mathbf{A}$, $\mathbf{B}$ and $\mathbf{C}$ in the previous example.

Scalar Multiplication of a Matrix

The *scalar multiplication* of a matrix $\mathbf{A}$ by a scalar k is carried out by multiplying each element of $\mathbf{A}$ by k. This scalar product is denoted by $k\,\mathbf{A}$ and the elements by ka_{ij}, $i = 1, 2, \ldots, n$, $j = 1, 2, \ldots, p$.

Product of Two Matrices

The *product* of two matrices $\mathbf{A}$ $(n \times p) = (a_{ij})$ and $\mathbf{B}$ $(p \times m) = (b_{jk})$ is denoted by $\mathbf{C} = \mathbf{AB}$, if the number of columns (p) of $\mathbf{A}$ is equal to the number of rows (p) of $\mathbf{B}$, and if the elements of $\mathbf{C} = (c_{ij})$ are given by

$$c_{ik} = \sum_{j=1}^{p} a_{ij} b_{jk}.$$

The order of the product matrix $\mathbf{C}$ is $(n \times m)$.

Example

The matrix $\mathbf{A} = \begin{bmatrix} 4 & 3 \\ -6 & 9 \\ 2 & -4 \end{bmatrix}$ when multiplied by the scalar 3 yields the

matrix $3\mathbf{A} = \begin{bmatrix} 12 & 9 \\ -18 & 27 \\ 6 & -12 \end{bmatrix}$.

The product of the two matrices $\mathbf{A}$ and $\mathbf{B}$, where $\mathbf{A}$ is given above and $\mathbf{B} = \begin{bmatrix} 0 & 6 & 2 \\ 1 & -2 & 3 \end{bmatrix}$, is given by

$$\mathbf{C} = \mathbf{AB} = \begin{bmatrix} (4)(0) + (3)(1) & (4)(6) + (3)(-2) & (4)(2) + (3)(3) \\ (-6)(0) + (9)(1) & (-6)(6) + (9)(-2) & (-6)(2) + (9)(3) \\ (2)(0) + (-4)(1) & (2)(6) + (-4)(-2) & (2)(2) + (-4)(3) \end{bmatrix}$$

$$= \begin{bmatrix} 3 & 18 & 17 \\ 9 & -54 & 15 \\ -4 & 20 & -8 \end{bmatrix}.$$

Some additional properties are:

1. $\mathbf{A} + \mathbf{B} = \mathbf{B} + \mathbf{A}$.

2. $(\mathbf{A} + \mathbf{B}) + \mathbf{C} = \mathbf{A} + (\mathbf{B} + \mathbf{C})$.

3. $a(\mathbf{A} + \mathbf{B}) = a\mathbf{A} + a\mathbf{B}$.

4. $(a + b)\mathbf{A} = a\mathbf{A} + b\mathbf{A}$.

5. $(\mathbf{AB})\mathbf{C} = \mathbf{A}(\mathbf{BC})$.

6. $(\mathbf{A} + \mathbf{B})\mathbf{C} = \mathbf{AC} + \mathbf{BC}$.

7. In general $\mathbf{AB} \neq \mathbf{BA}$ even if the dimensions conform for multiplication.

8. For square matrices $\mathbf{A}$ and $\mathbf{B}$

$$tr(\mathbf{A} + \mathbf{B}) = tr\mathbf{A} + tr\mathbf{B} \quad \text{and} \quad tr(\mathbf{AB}) = tr(\mathbf{BA}).$$

Exercise

Given $\mathbf{A} = \begin{bmatrix} 1 & -1 \\ 2 & 1 \end{bmatrix}$, $\mathbf{B} = \begin{bmatrix} 2 & 0 \\ 1 & -1 \end{bmatrix}$ and $\mathbf{C} = \begin{bmatrix} -1 & 2 \\ 0 & 3 \end{bmatrix}$ verify the properties 1 through 8 given above.

Multiplicative Inverse

The *multiplicative inverse* of the square matrix $\mathbf{A}$ $(m \times m)$ is the matrix $\mathbf{B}$ $(m \times m)$ satisfying the equation $\mathbf{AB} = \mathbf{BA} = \mathbf{I}_m$, where $\mathbf{I}_m$ is the identity matrix of order m. The multiplicative inverse of $\mathbf{A}$ is usually denoted by $\mathbf{B} = \mathbf{A}^{-1}$. If $\mathbf{A}$ has an inverse it is unique.

Some additional properties are:

1. If $\mathbf{C} = \mathbf{AB}$ then $\mathbf{C}^{-1} = \mathbf{B}^{-1}\mathbf{A}^{-1}$

2. $(\mathbf{A}')^{-1} = (\mathbf{A}^{-1})'$

3. If $\mathbf{C} = \mathbf{AB}$ then $\mathbf{C}' = \mathbf{B}'\mathbf{A}'$.

4. $(k\mathbf{A})^{-1} = (1/k)\mathbf{A}^{-1}$.

Example

The inverse of the matrix $\mathbf{A} = \begin{bmatrix} 4 & -2 \\ -2 & 2 \end{bmatrix}$ is given by $\mathbf{A}^{-1} = \begin{bmatrix} \frac{1}{2} & \frac{1}{2} \\ \frac{1}{2} & 1 \end{bmatrix}$

since $\mathbf{AA}^{-1} = \begin{bmatrix} 4 & -2 \\ -2 & 2 \end{bmatrix}\begin{bmatrix} \frac{1}{2} & \frac{1}{2} \\ \frac{1}{2} & 1 \end{bmatrix} = \begin{bmatrix} 1 & 0 \\ 0 & 1 \end{bmatrix}$.

A Useful Result

Given a symmetric nonsingular matrix $\mathbf{A}$ $(p \times p)$ and matrices $\mathbf{B}$ and $\mathbf{C}$ of order $(p \times q)$, the inverse of the matrix $[\mathbf{A} + \mathbf{BC}']$ is given by

$$[\mathbf{A} + \mathbf{BC}']^{-1} = \mathbf{A}^{-1} - \mathbf{A}^{-1}\mathbf{B}[\mathbf{I} + \mathbf{C}'\mathbf{A}^{-1}\mathbf{B}]^{-1}\mathbf{C}'\mathbf{A}^{-1}.$$

An important special case of this result is when $\mathbf{B}$ and $\mathbf{C}$ are $(p \times 1)$ vectors, say $\mathbf{b}$ and $\mathbf{c}$. The inverse of the matrix $[\mathbf{A} + \mathbf{bc}']$ is given by

$$[\mathbf{A} + \mathbf{bc}']^{-1} = \mathbf{A}^{-1} - \frac{\mathbf{A}^{-1}\mathbf{bc}'\mathbf{A}^{-1}}{1 + \mathbf{c}'\mathbf{A}^{-1}\mathbf{b}}.$$

Exercise

(a) Verify the above two results by using matrix multiplication.

(b) The equicorrelation matrix has the form $\mathbf{B} = (1 - \rho)\mathbf{I} + \rho\mathbf{ii}'$, where ρ is a constant correlation coefficient and $\mathbf{i}$ $(n \times 1)$ is a vector of unities. Show by using the above result that the matrix $\mathbf{B}^{-1}$ is given by

$$\mathbf{B}^{-1} = \frac{1}{(1 - \rho)}\mathbf{I} - \frac{\rho\mathbf{ii}'}{(1 - \rho)^2 + \rho(1 - \rho)n}.$$

Idempotent Matrix

An *idempotent* matrix $\mathbf{A}$ is a matrix that has the property $\mathbf{AA} = \mathbf{A}$. If $\mathbf{A}$ is idempotent and has full rank then $\mathbf{A}$ is an identity matrix.

Example

The matrix $\mathbf{B} = \begin{bmatrix} 1 & -1 \\ 0 & 0 \end{bmatrix}$ is idempotent since $\begin{bmatrix} 1 & -1 \\ 0 & 0 \end{bmatrix}\begin{bmatrix} 1 & -1 \\ 0 & 0 \end{bmatrix} = \begin{bmatrix} 1 & -1 \\ 0 & 0 \end{bmatrix}$.

Exercise

Show that the mean centering matrix $\left[\mathbf{I} - \frac{1}{n}\mathbf{ii}'\right]$ is idempotent, where $\mathbf{I}$ is an $(n \times n)$ identity matrix and $\mathbf{i}$ $(n \times 1)$ is a vector of unities.

Kronecker Product or Direct Product

If $\mathbf{A}(n \times m)$ and $\mathbf{B}(p \times q)$, the *Kronecker product* of $\mathbf{A}$ and $\mathbf{B}$ is denoted by $\mathbf{A} \otimes \mathbf{B}$ and is given by the matrix

$$\begin{bmatrix} a_{11}\mathbf{B} & a_{12}\mathbf{B} & \cdots & a_{1m}\mathbf{B} \\ a_{21}\mathbf{B} & a_{22}\mathbf{B} & \cdots & a_{2m}\mathbf{B} \\ \vdots & & & \\ a_{n1}\mathbf{B} & a_{n2}\mathbf{B} & \cdots & a_{nm}\mathbf{B} \end{bmatrix}.$$

Example

Given $\mathbf{A} = \begin{bmatrix} 2 & -1 \\ 3 & 4 \end{bmatrix}$ $\mathbf{B} = \begin{bmatrix} 3 \\ 5 \end{bmatrix}$ then

$$\mathbf{A} \otimes \mathbf{B} = \begin{bmatrix} 2\begin{bmatrix}3\\5\end{bmatrix} & -1\begin{bmatrix}3\\5\end{bmatrix} \\ 3\begin{bmatrix}3\\5\end{bmatrix} & 4\begin{bmatrix}3\\5\end{bmatrix} \end{bmatrix} = \begin{bmatrix} 6 & -3 \\ 10 & -5 \\ 9 & 12 \\ 15 & 20 \end{bmatrix} \quad \text{and}$$

$$\mathbf{B} \otimes \mathbf{A} = \begin{bmatrix} 3\begin{bmatrix}2 & -1\\3 & 4\end{bmatrix} \\ 5\begin{bmatrix}2 & -1\\3 & 4\end{bmatrix} \end{bmatrix} = \begin{bmatrix} 6 & -3 \\ 9 & 12 \\ 10 & -5 \\ 15 & 20 \end{bmatrix}.$$

Some properties of Kronecker products are as follows:

1. $(\mathbf{A} \otimes \mathbf{B}) \otimes \mathbf{C} = \mathbf{A} \otimes (\mathbf{B} \otimes \mathbf{C})$.

2. $(\mathbf{A} + \mathbf{B}) \otimes \mathbf{C} = (\mathbf{A} \otimes \mathbf{C}) + (\mathbf{B} \otimes \mathbf{C})$.

3. $(\mathbf{A} \otimes \mathbf{B})(\mathbf{C} \otimes \mathbf{D}) = \mathbf{A}\mathbf{C} \otimes \mathbf{B}\mathbf{D}$.

4. $(\mathbf{A} \otimes \mathbf{B})' = \mathbf{A}' \otimes \mathbf{B}'$.

5. $tr(\mathbf{A} \otimes \mathbf{B}) = tr(\mathbf{A})tr(\mathbf{B})$.

6. For vectors $\mathbf{a}$ and $\mathbf{b}$

$$\mathbf{a}' \otimes \mathbf{b} = \mathbf{b} \otimes \mathbf{a}' = \mathbf{ba}'.$$

7. In general $\mathbf{A} \otimes \mathbf{B} \neq \mathbf{B} \otimes \mathbf{A}$ as demonstrated in the above example.

Exercise

(a) Use the matrices $\mathbf{A}$ and $\mathbf{B}$ defined in the example above to show that $(\mathbf{A} \otimes \mathbf{B})' = (\mathbf{A}' \otimes \mathbf{B}')$.

(b) Use the matrices $\mathbf{A} = \begin{bmatrix} 1 & -1 \\ 3 & 2 \end{bmatrix}$ and $\mathbf{B} = \begin{bmatrix} -1 & 2 \\ 5 & 3 \end{bmatrix}$ to verify that $tr(\mathbf{A} \otimes \mathbf{B}) = tr(\mathbf{A})tr(\mathbf{B})$.

1.3 DETERMINANTS AND RANK

Determinant

The *determinant* of a square matrix $\mathbf{A}$ is a scalar quantity denoted by $|\mathbf{A}|$ and is given by

$$|\mathbf{A}| = \sum_{k=1}^{n!}(-1)^{j(k)}a_{11(k)}a_{22(k)} \cdots a_{nn(k)}.$$

The determinant represents the sum of $n!$ terms, each term consisting of the product of n elements of $\mathbf{A}$. For each term of the summation, the first subscripts are the integers 1 to n in their natural order, and the second subscripts represent a particular permutation of the integers 1 to n. The power $j(k)$ is 1 or 2 depending on whether the second subscripts represent an odd or an even number of interchanges with the integers in their natural order.

For the 2×2 matrix

$$\mathbf{A} = \begin{bmatrix} a_{11} & a_{12} \\ a_{21} & a_{22} \end{bmatrix}$$

the determinant is given by

$$|\mathbf{A}| = \underset{\text{(0 interchanges)}}{a_{11}a_{22}} - \underset{\text{(1 interchange)}}{a_{12}a_{21}}.$$

For the 3×3 matrix

$$\mathbf{B} = \begin{bmatrix} b_{11} & b_{12} & b_{13} \\ b_{21} & b_{22} & b_{23} \\ b_{31} & b_{32} & b_{33} \end{bmatrix}$$

the determinant is given by

$$|\mathbf{B}| = \underset{(0 \text{ interchanges})}{b_{11}b_{22}b_{33}} - \underset{(1 \text{ interchange})}{b_{11}b_{23}b_{32}} + \underset{(2 \text{ interchanges})}{b_{12}b_{23}b_{31}}$$

$$- \underset{(1 \text{ interchange})}{b_{12}b_{21}b_{33}} + \underset{(2 \text{ interchanges})}{b_{13}b_{21}b_{32}} - \underset{(1 \text{ interchange})}{b_{13}b_{22}b_{31}} .$$

NOTE: $n! = 3! = 6$ terms.

Nonsingular

If $|\mathbf{A}| \neq 0$ then $\mathbf{A}$ is said to be *nonsingular*.

The determinant of the matrix $\mathbf{A}$ $(n \times n)$ can be evaluated in terms of the determinants of submatrices of $\mathbf{A}$. The determinant of the submatrix of $\mathbf{A}$ obtained after deleting the jth row and kth column of $\mathbf{A}$ is called the *minor* of a_{jk} and is denoted by $|A^{jk}|$. The *cofactor* of a_{jk} is the quantity $(-1)^{j+k}|A^{jk}|$. The determinant of $\mathbf{A}$ can be expressed by $|\mathbf{A}| = \sum_{j=1}^{n} a_{jk}(-1)^{j+k}|A^{jk}|$ for any k and is called the cofactor expansion of $\mathbf{A}$.

Example

The determinant of the matrix $\mathbf{B} = \begin{bmatrix} 1 & 3 & -2 \\ 4 & 5 & 1 \\ -3 & -4 & 7 \end{bmatrix}$ can be determined

by expanding about the first row obtaining

$$1(-1)^2 \begin{vmatrix} 5 & 1 \\ -4 & 7 \end{vmatrix} + 3(-1)^3 \begin{vmatrix} 4 & 1 \\ -3 & 7 \end{vmatrix} + (-2)(-1)^4 \begin{vmatrix} 4 & 5 \\ -3 & -4 \end{vmatrix} = -52.$$

Equivalently expanding about the third column

$$-2(-1)^4 \begin{vmatrix} 4 & 5 \\ -3 & -4 \end{vmatrix} + 1(-1)^5 \begin{vmatrix} 1 & 3 \\ -3 & -4 \end{vmatrix} + 7(-1)^6 \begin{vmatrix} 1 & 3 \\ 4 & 5 \end{vmatrix} = -52.$$

Exercise

Verify the value of the determinant in the example by expanding about the second row.

Some useful properties of the determinant are:

1. $|\mathbf{A}| = |\mathbf{A}|'$.

2. If each element of a row (or column) of $\mathbf{A}$ is multiplied by the scalar k then the determinant of the new matrix is $k|\mathbf{A}|$.

3. $|k\mathbf{A}| = |\mathbf{A}|k^p$ if $\mathbf{A}$ is $(p \times p)$.

4. If each element of a row (or column) of $\mathbf{A}$ is zero then $|\mathbf{A}| = 0$.

5. If two rows (or columns) of $\mathbf{A}$ are identical then $|\mathbf{A}| = 0$.

6. The determinant of a matrix remains unchanged if the elements of one row (or column) are multipled by a scalar k and the results added to a second row (or column).

7. The determinant of the product $\mathbf{AB}$ of the square matrices $\mathbf{A}$ and $\mathbf{B}$ is given by $|\mathbf{AB}| = |\mathbf{A}||\mathbf{B}|$.

8. If $\mathbf{A}^{-1}$ exists then $|\mathbf{A}^{-1}| = |\mathbf{A}|^{-1}$.

9. $\mathbf{A}^{-1}$ exists if and only if $|\mathbf{A}| \neq 0$.

10. $|\mathbf{A} \otimes \mathbf{B}| = |\mathbf{A}|^n |\mathbf{B}|^m$ where $\mathbf{A}(n \times n)$ and $\mathbf{B}(m \times m)$.

11. $(\mathbf{A} \otimes \mathbf{B})^{-1} = \mathbf{A}^{-1} \times \mathbf{B}^{-1}$.

12. If $\mathbf{A}(p \times p)$ is non-singular, $\mathbf{B}(p \times m)$ and $\mathbf{C}(m \times p)$ then $|\mathbf{A} + \mathbf{BC}| = |\mathbf{A}|^{-1}|\mathbf{I}_p + \mathbf{A}^{-1}\mathbf{BC}| = |\mathbf{A}^{-1}||\mathbf{I}_m + \mathbf{CA}^{-1}\mathbf{B}|$ where $\mathbf{I}_p(p \times p)$ and $\mathbf{I}_m(m \times m)$ are identity matrices.

Relation Between Inverse and Determinant

The inverse of the matrix $\mathbf{A}$ is given by $\mathbf{A}^{-1} = \frac{1}{|\mathbf{A}|}\mathbf{A}^*$, where $\mathbf{A}^*$ is the transpose of the matrix of cofactors of $\mathbf{A}$. The matrix $\mathbf{A}^*$ is called the *adjoint* of $\mathbf{A}$.

Example

For the (2×2) matrix $\mathbf{A} = \begin{bmatrix} 9 & -2 \\ -4 & 6 \end{bmatrix}$ the determinant is given by $54 - 8 = 46$. The matrix of cofactors is given by $\begin{bmatrix} 6 & 4 \\ 2 & 9 \end{bmatrix}$ and hence the adjoint matrix is $\mathbf{A}^* = \begin{bmatrix} 6 & 2 \\ 4 & 9 \end{bmatrix}$. The inverse of $\mathbf{A}$ is therefore given by $\mathbf{A}^{-1} = \frac{1}{46}\begin{bmatrix} 6 & 2 \\ 4 & 9 \end{bmatrix}$.

Exercise

(a) Verify that $|\mathbf{A}^{-1}| = |\mathbf{A}|^{-1}$ using the matrix $\mathbf{A}$ in the previous example.

(b) Verify the inverse of the matrix $\mathbf{A}$ in the example above using the adjoint and determinant of $\mathbf{A}^{-1}$ to get $\mathbf{A}$.

(c) Use the properties above to show that the determinant of the equi-correlation matrix is given by $|\rho\mathbf{I} + (1-\rho)\mathbf{ii}'| = (1-\rho)^{n-1}[1 + \rho(n-1)]$, where $\mathbf{i}$ $(n \times 1)$ is a vector of unities.

Rank of a Matrix

The *rank* of a matrix $\mathbf{A}$, rank $(\mathbf{A})$, is the order of the largest nonsingular submatrix of $\mathbf{A}$. If $\mathbf{A}$ is nonsingular then $\mathbf{A}$ is said to have *full rank*.

Example

Given the matrix $\mathbf{B} = \begin{bmatrix} 6 & 4 \\ 3 & 2 \\ -1 & 4 \end{bmatrix}$ the rank is the order of the largest

nonsingular submatrix of $\mathbf{B}$. Since $\mathbf{B}$ is (3×2) the rank cannot exceed 2 since a (2×2) matrix is the largest square submatrix of $\mathbf{B}$. The three possible submatrices are

$$\begin{bmatrix} 6 & 4 \\ 3 & 2 \end{bmatrix}, \quad \begin{bmatrix} 3 & 2 \\ -1 & 4 \end{bmatrix} \quad \text{and} \quad \begin{bmatrix} 6 & 4 \\ -1 & 4 \end{bmatrix}.$$

The first of these submatrices has determinant zero and hence is singular. The remaining two submatrices are nonsingular. The rank of the matrix $\mathbf{B}$ is therefore 2.

Exercise

(a) Verify that the rank of the matrix $\mathbf{A}$ given by $\mathbf{A} = \begin{bmatrix} 2 & 3 & -4 \\ 1 & 4 & 2 \\ 6 & 11 & 0 \end{bmatrix}$

is 2 and determine all (2×2) matrices that have rank 2.

(b) Using the matrix $\mathbf{B}$ in the above example and $\mathbf{A}$ in (a) verify that rank $(\mathbf{BA}) = 2$.

The following properties are useful.

1. If $\mathbf{A}$ and $\mathbf{B}$ are nonsingular matrices with the appropriate dimensions and if $\mathbf{C}$ is arbitrary of appropriate dimension, then

    (a) rank $(\mathbf{AB}) \le \min[\text{rank}\,(\mathbf{A}), \text{rank}\,(\mathbf{B})]$
    (b) rank $(\mathbf{AC}) = \text{rank}\,(\mathbf{C})$
    (c) rank $(\mathbf{CA}) = \text{rank}\,(\mathbf{C})$
    (d) rank $(\mathbf{ACB}) = \text{rank}\,(\mathbf{C})$.

2. The rank of an idempotent matrix is equal to its trace.

1.4 QUADRATIC FORMS AND POSITIVE DEFINITE MATRICES

Quadratic Form

Given a symmetric matrix $\mathbf{A}$ $(n \times n)$ and an $(n \times 1)$ vector $\mathbf{x}$, a *quadratic form* is the scalar obtained from the product

$$\mathbf{x}'\mathbf{A}\mathbf{x} = \sum_{i=1}^{n} \sum_{j=1}^{n} a_{ij} x_i x_j.$$

The matrix $\mathbf{A}$ must be square, but need not be symmetric, since for any square matrix $\mathbf{B}$ the quadratic form $\mathbf{x}'\mathbf{B}\mathbf{x}$ can be written equivalently in terms of a symmetric matrix $\mathbf{A}$. Thus

$$\mathbf{x}'\mathbf{A}\mathbf{x} = \mathbf{x}'\mathbf{B}\mathbf{x}.$$

There is no loss of generality therefore, in assuming the matrix $\mathbf{A}$ in a quadratic form is symmetric.

Exercise

Given $\mathbf{x} = \begin{bmatrix} a \\ -2a \\ -a \end{bmatrix}$ and $\mathbf{A} = \begin{bmatrix} 1 & 3 & 0 \\ 3 & 2 & -1 \\ 0 & -1 & 4 \end{bmatrix}$ determine the value of $\mathbf{x}'\mathbf{A}\mathbf{x}$ and solve for a in the equation $\mathbf{x}'\mathbf{A}\mathbf{x} = 10$.

Congruent Matrix

A square matrix $\mathbf{B}$ is *congruent* to a square matrix $\mathbf{A}$ if there exists a nonsingular matrix $\mathbf{P}$ such that $\mathbf{A} = \mathbf{P}'\mathbf{B}\mathbf{P}$. By defining the linear transformation $\mathbf{y} = \mathbf{P}\mathbf{x}$ the quadratic form $\mathbf{y}'\mathbf{B}\mathbf{y}$ is equivalent to $\mathbf{x}'\mathbf{A}\mathbf{x}$. If $\mathbf{A}$ is of rank r, there exists a nonsingular matrix $\mathbf{P}$ such that the congruent matrix $\mathbf{B}$ is diagonal. In this case $\mathbf{B}$ has r nonzero diagonal elements and the remaining $(n-r)$ diagonal elements are zero, therefore $\mathbf{y}'\mathbf{B}\mathbf{y} = \sum_{i=1}^{r} b_{ii} y_i^2$.

Positive Definite

A real symmetric matrix $\mathbf{A}$ $(n \times n)$ is *positive definite* if the quadratic form $\mathbf{x}'\mathbf{A}\mathbf{x}$ is positive for all $(n \times 1)$ vectors $\mathbf{x}$. If $\mathbf{A}$ is positive definite and nonsingular, then $\mathbf{A}$ is congruent to an identity matrix in that there exists a nonsingular matrix $\mathbf{P}$ such that

$$\mathbf{x}'\mathbf{A}\mathbf{x} = \mathbf{y}'\mathbf{y}, \quad \text{where } \mathbf{y} = \mathbf{P}\mathbf{x}.$$

Positive Semidefinite, Negative Definite, Nonnegative Definite

The matrix $\mathbf{A}$ is *positive semidefinite* if $\mathbf{x}'\mathbf{A}\mathbf{x} \geq 0$ for all $\mathbf{x}$, and $\mathbf{x}'\mathbf{A}\mathbf{x} = 0$ for at least one $\mathbf{x}$. It is *negative definite* if $\mathbf{x}'\mathbf{A}\mathbf{x} < 0$ for all $\mathbf{x}$, and *negative semidefinite* if $\mathbf{x}'\mathbf{A}\mathbf{x} \leq 0$ for all $\mathbf{x}$ and $\mathbf{x}'\mathbf{A}\mathbf{x} = 0$ for at least one $\mathbf{x}$. The matrix $\mathbf{A}$ is *nonnegative definite* if it is not *negative definite*.

Exercise

Suppose that $\mathbf{A} = \mathbf{a}\mathbf{a}'$ where $\mathbf{a}(p \times 1)$ and show that $\mathbf{x}'\mathbf{A}\mathbf{x}$ is positive definite if $\mathbf{a} \neq \mathbf{0}$.

Some additional results are:

1. The determinant of a positive definite matrix $|\mathbf{A}|$ is positive.

2. If $\mathbf{A}$ is positive definite (semidefinite) then for any nonsingular matrix $\mathbf{P}$, $\mathbf{P}'\mathbf{A}\mathbf{P}$ is positive definite (semidefinite).

3. $\mathbf{A}$ is positive definite if and only if there exists a nonsingular matrix $\mathbf{V}$ such that $\mathbf{A} = \mathbf{V}'\mathbf{V}$.

4. If $\mathbf{A}$ is $(n \times p)$ of rank p then $\mathbf{A}'\mathbf{A}$ is positive definite and $\mathbf{A}\mathbf{A}'$ is positive semidefinite.

5. If $\mathbf{A}$ is positive definite then $\mathbf{A}^{-1}$ is positive definite.

1.5 PARTITIONED MATRICES

A matrix $\mathbf{A}$ can be *partitioned* into submatrices by drawing horizontal and vertical lines between rows and columns of the matrix. Each element of the original matrix is contained in one and only one submatrix. A *partitioned matrix* is usually called a *block matrix*.

$$\mathbf{A} = \left[\begin{array}{ccc|cc} a_{11} & a_{12} & a_{13} & a_{14} & a_{15} \\ a_{21} & a_{22} & a_{23} & a_{24} & a_{25} \\ \hline a_{31} & a_{32} & a_{33} & a_{34} & a_{35} \\ a_{41} & a_{42} & a_{43} & a_{44} & a_{45} \end{array} \right] = \left[\begin{array}{ccc} A_{11} & A_{12} & A_{13} \\ A_{21} & A_{22} & A_{23} \end{array} \right].$$

Product of Partitioned Matrices

The product of two block matrices can be determined if the column partitions of the first matrix correspond to the row partitions of the second matrix. As usual, the total number of columns of the first matrix must be equal to the total number of rows of the second,

$$\mathbf{B} = \left[\begin{array}{c|cc} b_{11} & b_{12} & b_{13} \\ b_{21} & b_{22} & b_{23} \\ \hline b_{31} & b_{32} & b_{33} \\ \hline b_{41} & b_{42} & b_{43} \\ b_{51} & b_{52} & b_{53} \end{array} \right] = \left[\begin{array}{cc} B_{11} & B_{12} \\ B_{21} & B_{22} \\ B_{31} & B_{32} \end{array} \right]$$

and using $\mathbf{A}$ above

$$\mathbf{AB} = \begin{bmatrix} A_{11}B_{11} + A_{12}B_{21} + A_{13}B_{31} & A_{11}B_{12} + A_{12}B_{22} + A_{13}B_{32} \\ A_{21}B_{11} + A_{22}B_{21} + A_{23}B_{31} & A_{21}B_{12} + A_{22}B_{22} + A_{23}B_{32} \end{bmatrix}.$$

Inverse of a Partitioned Matrix

Let the $(n \times p)$ matrix $\mathbf{A}$ be partitioned into four submatrices $\mathbf{A}_{11}(n_1 \times p_1)$, $\mathbf{A}_{12}[n_1 \times (p - p_1)]$, $\mathbf{A}_{21}[(n - n_1) \times p_1]$ and $\mathbf{A}_{22}[(n - n_1) \times (p - p_1)]$ and hence $\mathbf{A} = \begin{bmatrix} \mathbf{A}_{11} & \mathbf{A}_{12} \\ \mathbf{A}_{21} & \mathbf{A}_{22} \end{bmatrix}$. If $\mathbf{B} = \mathbf{A}^{-1}$ exists, then $\mathbf{B} = \begin{bmatrix} \mathbf{B}_{11} & \mathbf{B}_{12} \\ \mathbf{B}_{21} & \mathbf{B}_{22} \end{bmatrix}$ can be related to $\mathbf{A}$ by the following

1. $\mathbf{B}_{11} = [\mathbf{A}_{11} - \mathbf{A}_{12}\mathbf{A}_{22}^{-1}\mathbf{A}_{21}]^{-1} = [\mathbf{A}_{11}^{-1} + \mathbf{A}_{11}^{-1}\mathbf{A}_{12}\mathbf{B}_{22}\mathbf{A}_{21}\mathbf{A}_{11}^{-1}]$.

2. $\mathbf{B}_{12} = -\mathbf{A}_{11}^{-1}\mathbf{A}_{12}[\mathbf{A}_{22} - \mathbf{A}_{21}\mathbf{A}_{11}^{-1}\mathbf{A}_{12}]^{-1} = -\mathbf{A}_{11}^{-1}\mathbf{A}_{12}\mathbf{B}_{22}$.

3. $\mathbf{B}_{21} = -\mathbf{A}_{22}^{-1}\mathbf{A}_{21}[\mathbf{A}_{11} - \mathbf{A}_{12}\mathbf{A}_{22}^{-1}\mathbf{A}_{21}]^{-1} = -\mathbf{A}_{22}^{-1}\mathbf{A}_{21}\mathbf{B}_{11}$.

4. $\mathbf{B}_{22} = [\mathbf{A}_{22} - \mathbf{A}_{21}\mathbf{A}_{11}^{-1}\mathbf{A}_{12}]^{-1} = [\mathbf{A}_{22}^{-1} + \mathbf{A}_{22}^{-1}\mathbf{A}_{21}\mathbf{B}_{11}\mathbf{A}_{12}\mathbf{A}_{22}^{-1}]$.

Exercise

1. (a) Verify the above expressions for $\mathbf{B} = \mathbf{A}^{-1}$ by multiplication to show that $\mathbf{AB} = \mathbf{I}$.

   (b) Verify the above expressions for $\mathbf{B} = \mathbf{A}^{-1}$ by solving for the four submatrices in $\mathbf{B}$ in the equation $\mathbf{AB} = \mathbf{I}$.

2. Use the formulae for the inverse of a partitioned matrix to invert the matrix $\mathbf{A} = \begin{bmatrix} 1 & 6 & 2 \\ 4 & 5 & -3 \\ -7 & 1 & 8 \end{bmatrix}$ by first determining the inverse of $\begin{bmatrix} 5 & -3 \\ 1 & 8 \end{bmatrix}$.

Determinant of a Partitioned Matrix

The determinant of $|\mathbf{A}|$ can be expressed as follows:

1. $|\mathbf{A}| = |\mathbf{A}_{22}||\mathbf{A}_{11} - \mathbf{A}_{12}\mathbf{A}_{22}^{-1}\mathbf{A}_{22}| = |\mathbf{A}_{11}||\mathbf{A}_{22} - \mathbf{A}_{21}\mathbf{A}_{11}^{-1}\mathbf{A}_{12}|$.

2. $|\mathbf{A}| = |\mathbf{A}_{22}|/|\mathbf{B}_{11}| = |\mathbf{A}_{11}|/|\mathbf{B}_{22}|$, where $\mathbf{B}_{11}$ and $\mathbf{B}_{22}$ are defined above with the inverse of a partitioned matrix.

If $\mathbf{A}_{21} = \mathbf{A}_{12} = \mathbf{0}$ then

1. $\mathbf{A}^{-1} = \begin{bmatrix} \mathbf{A}_{11}^{-1} & \mathbf{0} \\ \mathbf{0} & \mathbf{A}_{22}^{-1} \end{bmatrix}$.

2. $|\mathbf{A}| = |\mathbf{A}_{11}||\mathbf{A}_{22}|$.

Exercise

    (a) Use the expression for $|\mathbf{A}|$ in (1) above to determine $|\mathbf{A}|$ in the previous exercise.

    (b) Show that $|\mathbf{A}| = |\mathbf{A}_{11}||\mathbf{A}_{22}|$ using the matrix $\mathbf{A}$ in the previous exercise.

1.6   EXPECTATIONS OF RANDOM MATRICES

An $(n \times p)$ *random matrix* $\mathbf{X}$ is a matrix whose elements x_{ij} $i = 1, 2, \ldots, n$, $j = 1, 2, \ldots, p$, are random variables. The *expected value* of the random matrix $\mathbf{X}$ denoted by $E[\mathbf{X}]$ is the matrix of constants $E[x_{ij}]$ $i = 1, 2, \ldots, n$, $j = 1, 2, \ldots, p$, if these expectations exist.

    Let $\mathbf{Y}$ $(n \times p)$ and $\mathbf{X}$ $(n \times p)$ denote random matrices whose expectations $E[\mathbf{Y}]$ and $E[\mathbf{X}]$ exist. Let $\mathbf{A}$ $(m \times n)$ and $\mathbf{B}$ $(p \times k)$ be matrices of constants. The following properties hold:

    1. $E[\mathbf{AXB}] = \mathbf{A}E[\mathbf{X}]\mathbf{B}$.

    2. $E[\mathbf{X} + \mathbf{Y}] = E[\mathbf{X}] + E[\mathbf{Y}]$.

    Let $\mathbf{z}$ $(n \times 1)$ be a random vector and let $\mathbf{A}$ $(n \times n)$ be a matrix of constants. The expected value of the *quadratic form* $\mathbf{z}'\mathbf{A}\mathbf{z}$ is given by $E[\mathbf{z}'\mathbf{A}\mathbf{z}] = \sum_{i=1}^{n} \sum_{j=1}^{n} a_{ij} E[z_i z_j]$ where $\mathbf{A}$ has elements a_{ij} and $\mathbf{z}$ has elements z_i.

Exercise

    (a) Show that $E[(\mathbf{X} + \mathbf{A})'(\mathbf{X} + \mathbf{A})] = \mathbf{\Sigma} + \mathbf{A}'\mathbf{\Pi} + \mathbf{\Pi}'\mathbf{A} + \mathbf{A}'\mathbf{A}$, where $\mathbf{X}$ is a random matrix $E[\mathbf{X}] = \mathbf{\Pi}$ and $E[\mathbf{X}'\mathbf{X}] = \mathbf{\Sigma}$.

    (b) Show that $E[\mathbf{x}'\mathbf{A}\mathbf{x}] = \sum_{i=1}^{p} \sum_{j=1}^{p} a_{ij}\sigma_{ij} + \sum_{i=1}^{p} \sum_{j=1}^{p} a_{ij}\mu_i\mu_j$, where $E[(x_i - \mu_i)(x_j - \mu_j)] = \sigma_{ij}$, $i, j = 1, 2, \ldots, p$.

1.7   DERIVATIVES OF MATRIX EXPRESSIONS

If the elements of the matrix $\mathbf{A}$ are functions of a random variable x, then the derivative of $\mathbf{A}$ with respect to x is the matrix whose elements are the derivatives of the elements of $\mathbf{A}$. For $\mathbf{A} = (a_{ij})$, $d\mathbf{A}/dx = (da_{ij}/dx)$.

    If $\mathbf{x}$ is a $(p \times 1)$ vector, the derivative of any function of $\mathbf{x}$, say $f(\mathbf{x})$, is the $(p \times 1)$ vector of elements $(\partial f/\partial x_j)$, $j = 1, 2, \ldots, p$.

    Given a vector $\mathbf{x}$ $(p \times 1)$ and a vector $\mathbf{a}$ $(p \times 1)$, the vector derivative of the scalar $\mathbf{x}'\mathbf{a}$ with respect to the vector $\mathbf{x}$ is denoted by $\partial(\mathbf{x}'\mathbf{a})/\partial\mathbf{x} = \mathbf{a}$, which is equivalent to the vector

$$
\begin{bmatrix}
\dfrac{\partial(\mathbf{x}'\mathbf{a})}{\partial x_1} \\[6pt]
\dfrac{\partial(\mathbf{x}'\mathbf{a})}{\partial x_2} \\[6pt]
\vdots \\[6pt]
\dfrac{\partial(\mathbf{x}'\mathbf{a})}{\partial x_p}
\end{bmatrix}.
$$

Given a vector $\mathbf{x}$ $(p \times 1)$ and a matrix $\mathbf{A}$ $(p \times p)$, the vector derivative of $\mathbf{Ax}$ is given by $\partial \mathbf{A}/\partial \mathbf{x} = \mathbf{A}'$ or $\partial \mathbf{A}/\partial \mathbf{x}' = \mathbf{A}$.

Given a vector $\mathbf{x}$ $(p \times 1)$ and a matrix $\mathbf{A}$ $(p \times p)$, the vector derivative of the quadratic form $\mathbf{x}'\mathbf{Ax}$ with respect to the vector $\mathbf{x}$ is denoted by $\partial(\mathbf{x}'\mathbf{Ax})/\partial \mathbf{x} = 2\mathbf{Ax}$, which is equivalent to the vector

$$
\begin{bmatrix}
\dfrac{\partial(\mathbf{x}'\mathbf{Ax})}{\partial x_1} \\[6pt]
\dfrac{\partial(\mathbf{x}'\mathbf{Ax})}{\partial x_2} \\[6pt]
\vdots \\[6pt]
\dfrac{\partial(\mathbf{x}'\mathbf{Ax})}{\partial x_p}
\end{bmatrix}.
$$

Given a matrix $\mathbf{X}$ $(n \times m)$ then the derivative of a function of $\mathbf{X}$, say $f(\mathbf{X})$, with respect to the elements of $\mathbf{X}$ is the $(n \times m)$ matrix of partial derivatives $(\partial f/\partial x_{ij})$. Some useful properties of the *matrix derivative* involving the trace operator are:

1. $\dfrac{\partial}{\partial \mathbf{X}} tr(\mathbf{X}) = \mathbf{I}$.

2. $\dfrac{\partial}{\partial \mathbf{X}} tr(\mathbf{AX}) = \mathbf{A}'$.

3. $\dfrac{\partial}{\partial \mathbf{X}} tr(\mathbf{X}'\mathbf{AX}) = (\mathbf{A} + \mathbf{A}')\mathbf{X}$.

The derivative of a determinant $|\mathbf{X}|$ of a matrix $\mathbf{X}$, with respect to the elements of the matrix, is given by $\partial|\mathbf{X}|/\partial \mathbf{X} = (\mathbf{X}')^* = $ adjoint of $\mathbf{X}'$.

Exercise

(a) Show that if $f = (\mathbf{a} - \mathbf{Bx})'(\mathbf{a} - \mathbf{Bx})$ where $\mathbf{a}$ $(n \times 1)$, $\mathbf{B}$ $(n \times p)$, and $\mathbf{x}$ $(p \times 1)$ then $\partial f/\partial \mathbf{x} = [2\mathbf{B}'\mathbf{Bx} - 2\mathbf{B}'\mathbf{a}]$.

(b) Show that if $\mathbf{f} = tr(\mathbf{A} - \mathbf{BX})'(\mathbf{A} - \mathbf{BX})$ where $\mathbf{A}$ $(n \times p)$, $\mathbf{B}$ $(n \times r)$ and $\mathbf{X}$ $(r \times p)$ then $\partial f/\partial \mathbf{X} = [2\mathbf{B}'\mathbf{BX} - 2\mathbf{B}'\mathbf{A}]$.

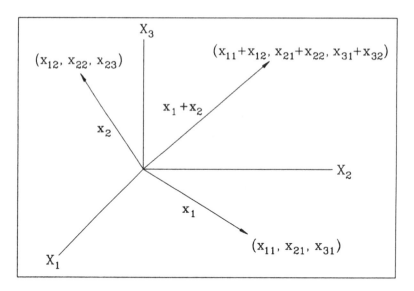

FIGURE A.1. Addition of Two Vectors

2   Linear Algebra

2.1   GEOMETRIC REPRESENTATION FOR VECTORS

A vector $\mathbf{x}' = (x_1, x_2, \ldots, x_n)$ can be represented geometrically in an *n-dimensional space*, as a *directed line segment* from the origin to the point with coordinates $(x_1, x_2, \ldots, x_n)$. The n-dimensional *vector space* is formed by n mutually perpendicular axes $X_1, X_2, \ldots, X_n$. The *coordinates* $(x_1, x_2, \ldots, x_n)$ are values of $X_1, X_2, \ldots, X_n$ respectively. Figure A.1 shows the vectors $\mathbf{x}_1' = (x_{11}, x_{21}, x_{31})$ and $\mathbf{x}_2' = (x_{12}, x_{22}, x_{32})$ in a three-dimensional space. The *addition of the two vectors* $\mathbf{x}_1' + \mathbf{x}_2' = (x_{11} + x_{12}, \; x_{21} + x_{22}, \; x_{31} + x_{32})$ is also shown in Figure A.1. Figure A.2 shows *scalar multiplication* by a scalar k for a two-dimensional vector $\mathbf{x}' = (x_{11}, x_{21})$, where $k\mathbf{x}' = (kx_{11}, kx_{21})$.

The *length of the vector* $\mathbf{x}' = (x_1, x_2, \ldots, x_n)$ is given by the *Euclidean distance* between the origin and the point $(x_1, x_2, \ldots, x_n)$ and is denoted by

$$\|\mathbf{x}\| = \sqrt{(x_1 - 0)^2 + (x_2 - 0)^2 + \ldots + (x_n - 0)^2} = \left[\sum_{i=1}^{n} x_i^2\right]^{1/2}.$$

The *angle θ between* the vector $\mathbf{x}_1 = (x_{11}, x_{21}, \ldots, x_{n1})$ and the vector $\mathbf{x}_2 = (x_{12}, x_{22}, \ldots, x_{n2})$ is given by

$$\cos\theta = \sum_{i=1}^{n} x_{i1} x_{i2} / \left[\sum_{i=1}^{n} x_{i1}^2\right]^{1/2} \left[\sum_{i=1}^{n} x_{i2}^2\right]^{1/2}.$$

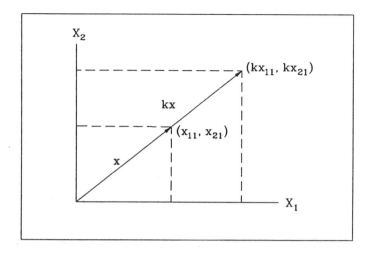

FIGURE A.2. Scalar Multiplication

The matrix product of the $(n \times 1)$ vectors $\mathbf{x}_1$ and $\mathbf{x}_2$ is given by

$$\mathbf{x}_1'\mathbf{x}_2 = \mathbf{x}_2'\mathbf{x}_1 = \sum_{i=1}^{n} x_{i1}x_{i2}$$

and hence

$$\cos\theta = \mathbf{x}_1'\mathbf{x}_2/(\mathbf{x}_1'\mathbf{x}_1)^{1/2}(\mathbf{x}_2'\mathbf{x}_2)^{1/2}.$$

In Figure A.3, the angle θ between $\mathbf{x}_1' = (x_{11}, x_{21})$ and $\mathbf{x}_2' = (x_{12}, x_{22})$ is given by $\cos\theta = [x_{11}x_{12} + x_{21}x_{22}]/[x_{11}^2 + x_{21}^2]^{1/2}[x_{12}^2 + x_{22}^2]^{1/2}$. If $\cos\theta = \pm 1$ the vectors $\mathbf{x}_1$ and $\mathbf{x}_2$ are said to be *orthogonal*.

The distance from the vector point $\mathbf{x}_1' = (x_{11}, x_{21}, \ldots, x_{n1})$ to the vector point $\mathbf{x}_2' = (x_{12}, x_{22}, \ldots, x_{n2})$ is the Euclidean distance between the two points

$$d = \sqrt{(x_{11} - x_{12})^2 + (x_{21} - x_{22})^2 + \ldots + (x_{n1} - x_{n2})^2}.$$

The *projection* of the vector $\mathbf{x}_2$ on the vector $\mathbf{x}_1$ is a vector that is a scalar multiple of $\mathbf{x}_1$ given by

$$\mathbf{x}_2^* = \left[\frac{\|\mathbf{x}_2\|}{\|\mathbf{x}_1\|}\cos\theta\right]\mathbf{x}_1 = k\mathbf{x}_1,$$

where θ is the angle between $\mathbf{x}_2$ and $\mathbf{x}_1$, and k is the scalar given by $\|\mathbf{x}_2\|/\|\mathbf{x}_1\|\cos\theta$.

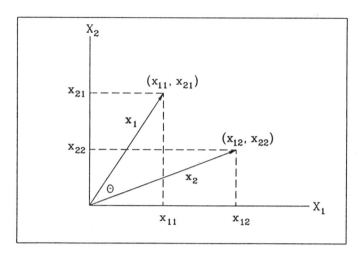

FIGURE A.3. Angle Between Two Vectors

Exercise

(a) Given $x_1 = \begin{bmatrix} 2 \\ 1 \end{bmatrix}$ and $x_2 = \begin{bmatrix} 1 \\ 3 \end{bmatrix}$ plot the vectors x_1 and x_2 in the two-dimensional space formed by the axes X_1 and X_2.

(b) Plot the vector $(x_1 + x_2)$ and the vector $2x_1$ in the two-dimensional space in (a).

(c) Determine the length of the vectors x_1, x_2 and $(x_1 + x_2)$.

(d) Determine the angle θ between the vectors x_1 and x_2.

(e) Determine the distance between the tips of the vectors x_1 and x_2.

(f) Determine the projection $\hat{x}_2$ of the vector x_2 on x_1 and determine the vector $x_2^* = x_2 - \hat{x}_2$. Show that $\hat{x}_2$ and x_2^* are orthogonal.

2.2   LINEAR DEPENDENCE AND LINEAR TRANSFORMATIONS

Linearly Dependent Vectors

If one vector x in n-dimensional space can be written as a linear combination of other vectors in n-dimensional space, then x is said to be *linearly dependent* on the other vectors. If

$$x = a_1 v_1 + a_2 v_2 + \ldots + a_p v_p,$$

then $\mathbf{x}$ is linearly dependent on $\mathbf{v}_1, \ldots, \mathbf{v}_p$.

The set of $(p+1)$ vectors $(\mathbf{x}, \mathbf{v}_1, \ldots, \mathbf{v}_p)$ is also a *linearly dependent set*, if any one of the vectors can be written as a linear combination of the remaining vectors in the set.

Linearly Independent Vectors

A set of vectors is *linearly independent*, if it is not possible to express one vector as a linear combination of the remaining vectors in the set. In an n-dimensional space the maximum number of linearly independent vectors is n. Thus given any set of n linearly independent vectors, all other vectors in the n-dimensional space can be expressed as a linear combination of the linearly independent set.

Basis for an n-Dimensional Space

The set of n linearly independent vectors is said to generate a *basis* for the n-dimensional space.

In two-dimensional space linearly dependent vectors are colinear or lie along the same line. In three-dimensional space linearly dependent vectors are in the same two-dimensional plane and may be colinear. If the set of linearly independent vectors are mutually orthogonal then the basis is said to be orthogonal.

Generation of a Vector Space and Rank of a Matrix

If the matrix $\mathbf{A}$ $(n \times p)$ has rank r, then the vectors formed by the p columns of $\mathbf{A}$ generate an r-dimensional vector space. Similarly the rows of $\mathbf{A}$ generate an r-dimensional vector space. In other words, the rank of a matrix is the maximum number of linearly independent columns, and equivalently the maximum number of linearly independent rows.

Exercise

(a) Given the vectors $\mathbf{x}_1 = \begin{bmatrix} 1 \\ -1 \\ 0 \end{bmatrix}$, $\mathbf{x}_2 = \begin{bmatrix} 0 \\ 1 \\ 2 \end{bmatrix}$ and $\mathbf{x}_3 = \begin{bmatrix} 2 \\ 0 \\ 1 \end{bmatrix}$ show

that they are linearly independent by showing that each cannot be written as a linear combination of the remaining two.

(b) Let $\mathbf{A} = [\mathbf{x}_1, \mathbf{x}_2, \mathbf{x}_3]$ and determine $|\mathbf{A}|$ for the values of $\mathbf{x}_1, \mathbf{x}_2, \mathbf{x}_3$ given in (a).

Linear Transformation

Given a p-dimensional vector $\mathbf{x}$, a *linear transformation* of $\mathbf{x}$, is given by the matrix product

$$\mathbf{y} = \mathbf{A}\mathbf{x},$$

where $\mathbf{A}$ $(n \times p)$ is called a *transformation matrix*. The matrix $\mathbf{A}$ maps the n-dimensional vector $\mathbf{x}$ into a p-dimensional vector $\mathbf{y}$. If $\mathbf{A}$ is a square nonsingular matrix, the transformation is *one to one*, in that $\mathbf{x} = \mathbf{A}^{-1}\mathbf{y}$ and hence each point in $\mathbf{x}$ corresponds to exactly one and only one point in $\mathbf{y}$.

Orthogonal Transformation, Rotation, Orthogonal Matrix

If $\mathbf{A}$ is a square matrix and $\mathbf{A}$ has the property that $\mathbf{A}' = \mathbf{A}^{-1}$, then the equation $\mathbf{y} = \mathbf{A}\mathbf{x}$ is an *orthogonal transformation* or *rotation*. The matrix $\mathbf{A}$ in this case is called an *orthogonal matrix*. If $\mathbf{A}$ is orthogonal, $\mathbf{A}$ also has the property that $|\mathbf{A}| = \pm 1$. The transformation is orthogonal because it can be viewed as a rotation of the coordinate axes through some angle θ. The angle θ between any pair of vectors remains the same after the transformation.

Example

In Figure A.4, the point M has coordinates (x_1, x_2) with respect to the $X_1 - X_2$ axes and has coordinates (z_1, z_2) with respect to the $Z_1 - Z_2$ axes. The angle ϕ between X_1 and Z_1 is the angle of rotation required to rotate the $X_1 - X_2$ axes into the $Z_1 - Z_2$ axes. The coordinates (z_1, z_2) of the point M in $Z_1 - Z_2$ space can be described in terms of the coordinates (x_1, x_2) of M in $X_1 - X_2$ space using the equations

$$
\begin{aligned}
z_1 &= x_1 \cos\phi + x_2 \sin\phi \\
z_2 &= -x_1 \sin\phi + x_2 \cos\phi.
\end{aligned}
$$

In matrix notation the linear transformation to $\begin{bmatrix} z_1 \\ z_2 \end{bmatrix}$ from $\begin{bmatrix} x_1 \\ x_2 \end{bmatrix}$ can be expressed as $\begin{bmatrix} z_1 \\ z_2 \end{bmatrix} = \begin{bmatrix} \cos\phi & \sin\phi \\ -\sin\phi & \cos\phi \end{bmatrix} \begin{bmatrix} x_1 \\ x_2 \end{bmatrix}$. The transformation matrix $\mathbf{A} = \begin{bmatrix} \cos\phi & \sin\phi \\ -\sin\phi & \cos\phi \end{bmatrix}$ has the property that

$$
\mathbf{A}^{-1} = \frac{1}{\cos^2\phi + \sin^2\phi} \begin{bmatrix} \cos\phi & -\sin\phi \\ \sin\phi & \cos\phi \end{bmatrix} = \begin{bmatrix} \cos\phi & -\sin\phi \\ \sin\phi & \cos\phi \end{bmatrix} = \mathbf{A}',
$$

and hence the transformation is orthogonal as required.

Exercise

(a) Let $\mathbf{A} = \begin{bmatrix} \sqrt{2}/2 & -\sqrt{2}/2 \\ \sqrt{2}/2 & \sqrt{2}/2 \end{bmatrix}$ denote a transformation matrix. Show that $\mathbf{A}^{-1} = \mathbf{A}'$ and hence that $\mathbf{A}$ is an orthogonal transformation matrix.

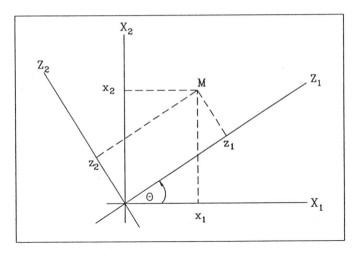

FIGURE A.4. Rotation of Axes

(b) Use the results of the above example to show that the angle of rotation in (a) is 45°.

(c) Give the transformation matrix corresponding to $\phi = 60°$.

2.3 SYSTEMS OF EQUATIONS

Let x_1, x_2, x_3 denote the unknowns in a system of three equations

$$
\begin{aligned}
a_{11}x_1 + a_{12}x_2 + a_{13}x_3 &= b_1 \\
a_{21}x_1 + a_{22}x_2 + a_{23}x_3 &= b_2 \\
a_{31}x_1 + a_{32}x_2 + a_{33}x_3 &= b_3.
\end{aligned}
$$

The *system of equations* can be represented by a matrix equation

$$\mathbf{Ax} = \mathbf{b},$$

where

$$
\mathbf{A} = \begin{bmatrix} a_{11} & a_{12} & a_{12} \\ a_{21} & a_{22} & a_{23} \\ a_{31} & a_{32} & a_{33} \end{bmatrix}, \ \mathbf{x} = \begin{bmatrix} x_1 \\ x_2 \\ x_3 \end{bmatrix}, \ b = \begin{bmatrix} b_1 \\ b_2 \\ b_3 \end{bmatrix}.
$$

Solution Vector for a System of Equations

The *solution vector* $\mathbf{x}$ is given by

$$\mathbf{x} = \mathbf{A}^{-1}\mathbf{b}.$$

Exercise

1. Given the three linearly independent vectors $x_1 = \begin{bmatrix} 1 \\ -1 \\ 0 \end{bmatrix}$, $x_2 = \begin{bmatrix} 0 \\ -1 \\ 2 \end{bmatrix}$ and $x_3 = \begin{bmatrix} 2 \\ 0 \\ 1 \end{bmatrix}$. If $y = \begin{bmatrix} 5 \\ 3 \\ 2 \end{bmatrix}$ is contained in the space generated by x_1, x_2 and x_3 then there exist coefficients a, b and c such that $y = ax_1 + bx_2 + cx_3$. Show that the equation for y can be written as $y = Af$, where $f = \begin{bmatrix} a \\ b \\ c \end{bmatrix}$ and solve for f.

2. Solve the system of equations

$$\begin{aligned} 3x_1 + x_2 - 4x_3 &= 3 \\ 2x_1 + 5x_2 + x_3 &= 10 \\ x_1 - 4x_2 + 3x_3 &= -6 \end{aligned}$$

for x_1, x_2 and x_3 using the expression $x = A^{-1}y$.

Homogeneous Equations — Trivial and Nontrivial Solutions

If the vector b in the system $Ax = b$ is the null vector 0, the system is said to be a system of *homogeneous equations*. The obvious solution $x = 0$ is said to be a *trivial solution*. If A is $(n \times p)$ of rank r, where x and b are $(p \times 1)$, then the system of n homogeneous equations in p unknowns given by $Ax = b$ has $(p - r)$ linearly independent solutions in addition to the trivial solution $x = 0$. In other words, a *nontrivial solution* exists if and only if the rank of A is less than p. If A is a square matrix a nontrivial solution exists if and only if $|A| = 0$.

2.4   COLUMN SPACES, PROJECTION OPERATORS AND LEAST SQUARES

Column Space

Given an $(n \times p)$ matrix X, the columns of X denoted by the p vectors $x_1, x_2, \ldots, x_p$ span a vector space called the *column space* of X. The set of all vectors $\hat{y}$ $(n \times 1)$, defined by $\hat{y} = Xb$ for all vectors b $(p \times 1)$, $b \neq 0$, is the vector space generated by the columns of X.

Orthogonal Complement

The set of all vectors z $(p \times 1)$ such that $Xz = 0$ generates the *orthogonal complement* to the column space of X.

Projection

Given a vector $\mathbf{y}$ $(n \times 1)$ and a p-dimensional column space defined by the matrix $\mathbf{X}$ $(n \times p)$, $p \leq n$, the vector $\hat{\mathbf{y}}$ $(n \times 1)$ is a *projection* of $\mathbf{y}$ onto the column space of $\mathbf{X}$, if there exists a $(p \times 1)$ vector $\mathbf{b}$ such that $\hat{\mathbf{y}} = \mathbf{Xb}$, and a vector $\mathbf{e}$ $(n \times 1)$ such that $\mathbf{e}$ is in the orthogonal complement to the column space of $\mathbf{X}$. If $\hat{\mathbf{y}}$ is the projection of $\mathbf{y}$, then $\mathbf{y} = \hat{\mathbf{y}} + \mathbf{e}$, and the vector $\mathbf{e}$ is the part of $\mathbf{y}$ that is orthogonal to the column space of $\mathbf{X}$.

Ordinary Least Squares Solution Vector

Given an $(n \times p)$ matrix $\mathbf{X}$ of rank p, and a vector $\mathbf{y}$ $(n \times 1)$, $n \geq p$, the projection of $\mathbf{y}$ onto the column space of $\mathbf{X}$ is given by the *ordinary least squares solution vector*

$$\hat{\mathbf{y}} = \mathbf{X}(\mathbf{X}'\mathbf{X})^{-1}\mathbf{X}'\mathbf{y}, \quad \text{where } \mathbf{b} = (\mathbf{X}'\mathbf{X})^{-1}\mathbf{X}'\mathbf{y}.$$

Idempotent Matrix — Projection Operator

The matrix $\mathbf{X}(\mathbf{X}'\mathbf{X})^{-1}\mathbf{X}'$ is an *idempotent matrix* and is called the *projection operator* for the column space of $\mathbf{X}$. The vector $\mathbf{e} = (\mathbf{y} - \hat{\mathbf{y}}) = [\mathbf{I} - \mathbf{X}(\mathbf{X}'\mathbf{X})^{-1}\mathbf{X}']\mathbf{y}$ is orthogonal to the column space of $\mathbf{X}$. The idempotent matrix $[\mathbf{I} - \mathbf{X}(\mathbf{X}'\mathbf{X})^{-1}\mathbf{X}']$ is called the projection operator for the vector space orthogonal to the column space of $\mathbf{X}$.

Exercise

1. Given $\mathbf{X}$ $(n \times p)$ and $\mathbf{y}$ $(n \times 1)$ demonstrate the following:

   (a) $\mathbf{X}(\mathbf{X}'\mathbf{X})^{-1}\mathbf{X}'$ is idempotent.
   (b) $[\mathbf{I} - \mathbf{X}(\mathbf{X}'\mathbf{X})^{-1}\mathbf{X}']$ is idempotent.
   (c) $[\mathbf{I} - \mathbf{X}(\mathbf{X}'\mathbf{X})^{-1}\mathbf{X}'][\mathbf{X}(\mathbf{X}'\mathbf{X})^{-1}\mathbf{X}'] = \mathbf{0}$.
   (d) $\mathbf{y}'(\mathbf{y} - \hat{\mathbf{y}}) = \mathbf{0}$, where $\hat{\mathbf{y}} = \mathbf{X}(\mathbf{X}'\mathbf{X})^{-1}\mathbf{X}'\mathbf{y}$.

2. Given $\mathbf{X} = \begin{bmatrix} 1 & 1 \\ 1 & 2 \\ 1 & 3 \\ 1 & 4 \\ 1 & 5 \end{bmatrix}$ $\mathbf{y} = \begin{bmatrix} 3 \\ 5 \\ 4 \\ 7 \\ 6 \end{bmatrix}$ determine $\hat{\mathbf{b}} = (\mathbf{X}'\mathbf{X})^{-1}\mathbf{X}'\mathbf{y}$, $\hat{\mathbf{y}} = \mathbf{X}(\mathbf{X}'\mathbf{X})^{-1}\mathbf{X}'\mathbf{y}$ and $(\mathbf{y} - \hat{\mathbf{y}})$.

3   Eigenvalue Structure and Singular Value Decomposition

3.1   EIGENVALUE STRUCTURE FOR SQUARE MATRICES

Eigenvalues and Eigenvectors

Given a square matrix $\mathbf{A}$ of order n, the values of the scalars λ and $(n \times 1)$ vectors $\mathbf{v}$, $\mathbf{v} \neq \mathbf{0}$, that satisfy the equation

$$\mathbf{A}\mathbf{v} = \lambda\mathbf{v}, \tag{A.1}$$

are called the *eigenvalues* and *eigenvectors* of the matrix $\mathbf{A}$. The problem of finding λ and $\mathbf{v}$ in (A.1) is commonly referred to as the *eigenvalue problem*. From (A.1) it can be seen that if $\mathbf{v}$ is a solution then $k\mathbf{v}$ is also a solution where k is an arbitrary scalar. The eigenvectors are therefore unique up to a multiplicative constant. It is common to impose the additional constraint that $\mathbf{v}'\mathbf{v} = 1$, and hence that the eigenvector be normalized to have a length of 1.

Characteristic Polynomial, Characteristic Roots, Latent Roots, Eigenvalues

Rewriting (A.1) as

$$(\mathbf{A} - \lambda\mathbf{I})\mathbf{v} = \mathbf{0},$$

we obtain a system of homogeneous equations. A nontrivial solution $\mathbf{v} \neq \mathbf{0}$ requires that

$$|\mathbf{A} - \lambda\mathbf{I}| = 0.$$

This equation yields a polynomial in λ of degree n and is commonly called the *characteristic polynomial*. The solutions of the equation are the roots of the polynomial, and are sometimes referred to as *characteristic roots* or *latent roots*, although the most common term used in statistics is *eigenvalues*.

The characteristic polynomial has n roots or eigenvalues some of which may be equal. For each eigenvalue λ there is a corresponding *eigenvector* $\mathbf{v}$ satisfying (A.1). The matrix $\mathbf{A}$ is singular if and only if at least one eigenvalue is zero.

Some additional properties of eigenvalues and eigenvectors are:

1. The eigenvalues of a diagonal matrix are the diagonal elements.

2. The matrices $\mathbf{A}$ and $\mathbf{A}'$ have the same eigenvalues but not necessarily the same eigenvectors.

3. If λ is an eigenvalue of $\mathbf{A}$ then $1/\lambda$ is an eigenvalue of $\mathbf{A}^{-1}$.

4. If λ is an eigenvalue of $\mathbf{A}$ and $\mathbf{v}$ the corresponding eigenvector, then for the matrix $\mathbf{A}^k$, λ^k is an eigenvalue with corresponding eigenvector $\mathbf{v}$.

5. If the eigenvalues of $\mathbf{A}$ are denoted by $\lambda_1, \lambda_2, \ldots, \lambda_n$, then

$$tr\mathbf{A} = \sum_{j=1}^{n} \lambda_j \quad \text{and} \quad |\mathbf{A}| = \prod_{j=1}^{n} \lambda_j.$$

Eigenvalues and Eigenvectors for Real Symmetric Matrices and Some Properties

In statistics the matrix $\mathbf{A}$ will usually have real elements and be symmetric. In this case the eigenvalues and eigenvectors have additional properties:

1. If the rank of $\mathbf{A}$ is r then $(n-r)$ of the eigenvalues are zero;

2. If k of the eigenvalues are equal the eigenvalue is said to have multiplicity k. In this case there will be k orthogonal eigenvectors corresponding to the common eigenvalue;

3. If two eigenvalues are distinct then the corresponding eigenvectors are orthogonal;

4. An nth order symmetric matrix produces a set of n orthogonal eigenvectors.

5. For idempotent matrices the eigenvalues are zero or one.

6. The maximum value of the quadratic form $\mathbf{v}'\mathbf{A}\mathbf{v}$ subject to $\mathbf{v}'\mathbf{v} = 1$ is given by $\lambda_1 = \mathbf{v}_1'\mathbf{A}\mathbf{v}_1$, where λ_1 is the largest eigenvalue of $\mathbf{A}$ and $\mathbf{v}_1$ is the corresponding eigenvector.

   Similarly the second largest eigenvalue $\lambda_2 = \mathbf{v}_2'\mathbf{A}\mathbf{v}_2$ of $\mathbf{A}$ is the maximum value of $\mathbf{v}'\mathbf{A}\mathbf{v}$ subject to $\mathbf{v}_2'\mathbf{v}_1 = 0$ and $\mathbf{v}_2'\mathbf{v}_2 = 1$, where $\mathbf{v}_2$ is the eigenvector corresponding to λ_2. The kth largest eigenvalue $\lambda_k = \mathbf{v}_k'\mathbf{A}\mathbf{v}_k$ of $\mathbf{A}$ is the maximum value of $\mathbf{v}'\mathbf{A}\mathbf{v}$ subject to $\mathbf{v}_k'\mathbf{v}_1 = \mathbf{v}_k'\mathbf{v}_2 = \ldots = \mathbf{v}_k'\mathbf{v}_{(k-1)} = 0$ and $\mathbf{v}_k'\mathbf{v}_k = 1$, where $\mathbf{v}_k$ is the eigenvector corresponding to λ_k.

   The above properties are essentially summarized by the following statement. If $\mathbf{A}$ is real symmetric of order n, there exists an orthogonal matrix $\mathbf{V}$ such that

$$\mathbf{V}'\mathbf{A}\mathbf{V} = \Lambda \quad \text{or} \quad \mathbf{A} = \mathbf{V}\Lambda\mathbf{V}',$$

   where Λ is a diagonal matrix of eigenvalues of $\mathbf{A}$ with diagonal elements $\lambda_1, \lambda_2, \ldots, \lambda_n$; and $\mathbf{V}$ is the matrix whose columns are the corresponding eigenvectors $\mathbf{v}_1, \mathbf{v}_2, \ldots, \mathbf{v}_n$. The orthogonal matrix $\mathbf{V}$ diagonalizes the matrix $\mathbf{A}$.

7. If $\mathbf{A}$ is nonsingular and symmetric then $\mathbf{A}^n = \mathbf{V}\Lambda^n\mathbf{V}'$, where $\mathbf{A} = \mathbf{V}\Lambda\mathbf{V}'$.

8. If $\mathbf{A} = \mathbf{V}\Lambda\mathbf{V}'$ then the quadratic form $\mathbf{x}'\mathbf{A}\mathbf{x}$ can be written as $\mathbf{y}'\Lambda\mathbf{y}$ where $\mathbf{y} = \mathbf{V}\mathbf{x}$.

Example

Given the matrix

$$
\mathbf{A} = \begin{bmatrix} 6 & \sqrt{\frac{15}{2}} & 0 \\ \sqrt{\frac{15}{2}} & 4 & \sqrt{\frac{3}{2}} \\ 0 & \sqrt{\frac{3}{2}} & 2 \end{bmatrix},
$$

the eigenvalues are obtained by solving the determinantal equation $|\mathbf{A} - \lambda\mathbf{I}| = 0$, which in this case is given by

$$
\begin{vmatrix} (6 - \lambda) & \sqrt{\frac{15}{2}} & 0 \\ \sqrt{\frac{15}{2}} & (4 - \lambda) & \sqrt{\frac{3}{2}} \\ 0 & \sqrt{\frac{3}{2}} & (2 - \lambda) \end{vmatrix} = 0.
$$

The resulting characteristic polynomial is given by

$$
(6 - \lambda)(4 - \lambda)(2 - \lambda) - \frac{3}{2}(6 - \lambda) - \frac{15}{2}(2 - \lambda) = 0.
$$

The characteristic roots or eigenvalues determined from this polynomial are $\lambda = 1$, 3 and 8.

The corresponding eigenvectors are obtained by solving the equation $(\mathbf{A} - \lambda\mathbf{I})\mathbf{v} = 0$ for each of the eigenvalues λ determined above. Corresponding to $\lambda = 3$ the equation becomes

$$
\begin{bmatrix} 3 & \sqrt{\frac{15}{2}} & 0 \\ \sqrt{\frac{15}{2}} & 1 & \sqrt{\frac{3}{2}} \\ 0 & \sqrt{\frac{3}{2}} & -1 \end{bmatrix} \begin{bmatrix} v_1 \\ v_2 \\ v_3 \end{bmatrix} = 0.
$$

Adding the condition that $v_1^2 + v_2^2 + v_3^2 = 1$ yields the eigenvector

$$
\mathbf{v}' = \begin{bmatrix} -\frac{1}{2} & \sqrt{\frac{3}{10}} & \sqrt{\frac{9}{20}} \end{bmatrix},
$$

the remaining two eigenvectors are given by

$$
\begin{bmatrix} -\sqrt{\frac{3}{28}} & \sqrt{\frac{5}{14}} & -\sqrt{\frac{15}{28}} \end{bmatrix} \quad \text{and} \quad \begin{bmatrix} \sqrt{\frac{9}{14}} & \sqrt{\frac{12}{35}} & \sqrt{\frac{1}{70}} \end{bmatrix}
$$

corresponding to $\lambda = 1$ and $\lambda = 8$ respectively. The complete matrix of eigenvectors is given by

$$
\mathbf{V} = \begin{bmatrix} -\sqrt{\frac{3}{28}} & -\frac{1}{2} & \sqrt{\frac{9}{14}} \\ \sqrt{\frac{5}{14}} & \sqrt{\frac{3}{10}} & \sqrt{\frac{12}{35}} \\ -\sqrt{\frac{15}{28}} & \sqrt{\frac{9}{20}} & \sqrt{\frac{1}{70}} \end{bmatrix},
$$

corresponding to $\lambda = 1, 3$ and 8 respectively. The reader should verify that $\mathbf{V'V} = \mathbf{I}$.

Exercise

(a) Show that the eigenvalues of the matrix $\mathbf{A} = \begin{bmatrix} 2 & 1 & 1 \\ 1 & 2 & 1 \\ 1 & 1 & 2 \end{bmatrix}$ are given by 1, 1 and 4.

(b) Show that $\mathbf{v'} = (\frac{1}{\sqrt{3}}, \frac{1}{\sqrt{3}}, \frac{1}{\sqrt{3}})$ is an eigenvector corresponding to $\lambda = 4$.

(c) Show that the remaining two eigenvectors corresponding to the double root $\lambda = 1$ are

$$
\left(0, \ \frac{1}{\sqrt{2}}, \ -\frac{1}{\sqrt{2}}\right) \quad \text{and} \quad \left(-\frac{2}{\sqrt{6}}, \ \frac{1}{\sqrt{6}}, \ \frac{1}{\sqrt{6}}\right).
$$

(d) Show that the 3×3 matrix of eigenvectors

$$
\mathbf{V} = \begin{bmatrix} \frac{1}{\sqrt{3}} & 0 & -\frac{2}{\sqrt{6}} \\ \frac{1}{\sqrt{3}} & \frac{1}{\sqrt{2}} & \frac{1}{\sqrt{6}} \\ \frac{1}{\sqrt{3}} & -\frac{1}{\sqrt{2}} & \frac{1}{\sqrt{6}} \end{bmatrix}
$$

satisfies $\mathbf{V'V} = \mathbf{I}$.

Spectral Decomposition

The equation $\mathbf{A} = \mathbf{V}\mathbf{\Lambda}\mathbf{V'}$ can be written

$$
\mathbf{A} = \lambda_1 \mathbf{v}_1 \mathbf{v}'_1 + \lambda_2 \mathbf{v}_2 \mathbf{v}'_2 + \ldots + \lambda_n \mathbf{v}_n \mathbf{v}'_n,
$$

where $\lambda_1, \lambda_2, \ldots, \lambda_n$ are the eigenvalues of $\mathbf{A}$ and $\mathbf{v}_1, \mathbf{v}_2, \ldots, \mathbf{v}_n$ are the corresponding eigenvectors. This equation gives the *spectral decomposition* of $\mathbf{A}$.

Matrix Approximation

Let $\mathbf{Z}$ $(n \times p)$ be a matrix such that $\mathbf{A} = \mathbf{Z}'\mathbf{Z}$ and let $\mathbf{A}(\ell)$ denote the first ℓ terms of the spectral decomposition $\mathbf{A}(\ell) = \sum_{j=1}^{\ell} \lambda_j \mathbf{v}_j \mathbf{v}_j'$, $\ell < n$. This expression minimizes $tr(\mathbf{Z} - \mathbf{X})(\mathbf{Z} - \mathbf{X})' = \sum_{i=1}^{n} \sum_{j=1}^{p} (a_{ij} - x_{ij})^2$ among all $(n \times p)$ matrices $\mathbf{X}$ of rank ℓ. Thus the first few terms of the spectral decomposition of $\mathbf{A} = \mathbf{Z}'\mathbf{Z}$ can be used to provide a matrix approximation to $\mathbf{A}$.

Example

For the matrix $\mathbf{A}$ of the previous example the matrix of eigenvalues can be approximated in decimal form by

$$\mathbf{V} = \begin{bmatrix} -0.327 & -0.500 & 0.802 \\ 0.598 & 0.547 & 0.586 \\ -0.732 & 0.671 & 0.120 \end{bmatrix}.$$

The spectral decomposition for $\mathbf{A}$ is given by

$$\mathbf{A} = 1 \begin{bmatrix} (0.327)^2 & -(0.327)(0.598) & (0.327)(0.732) \\ -(0.327)(0.598) & (0.598)^2 & -(0.598)(0.732) \\ (0.327)(0.732) & -(0.598)(0.732) & (0.732)^2 \end{bmatrix}$$

$$+3 \begin{bmatrix} (0.500)^2 & -(0.500)(0.547) & -(0.500)(0.671) \\ -(0.500)(0.547) & (0.547)^2 & (0.547)(0.671) \\ -(0.500)(0.671) & (0.547)(0.671) & (0.671)^2 \end{bmatrix}$$

$$+8 \begin{bmatrix} (0.802)^2 & (0.802)(0.586) & (0.802)(0.120) \\ (0.802)(0.586) & (0.586)^2 & (0.586)(0.120) \\ (0.802)(0.120) & (0.586)(0.120) & (0.120)^2 \end{bmatrix}.$$

This simplifies to

$$\mathbf{A} = \begin{bmatrix} 0.107 & -0.196 & 0.239 \\ -0.196 & 0.358 & -0.438 \\ 0.239 & -0.438 & 0.536 \end{bmatrix} + \begin{bmatrix} 0.750 & -0.821 & -1.007 \\ -0.821 & 0.898 & 1.101 \\ -1.007 & 1.101 & 1.351 \end{bmatrix}$$

$$+ \begin{bmatrix} 5.146 & 3.760 & 0.770 \\ 3.760 & 2.747 & 0.563 \\ 0.770 & 0.563 & 0.115 \end{bmatrix}.$$

Except for inaccuracies due to rounding this matrix should be equivalent to the decimal form of

$$\mathbf{A} = \begin{bmatrix} 6.000 & 2.739 & 0.000 \\ 2.739 & 4.000 & 1.225 \\ 0.000 & 1.225 & 1.000 \end{bmatrix}.$$

As a matrix approximation the term of the spectral decomposition corresponding to $\lambda = 8$, say $\widehat{\mathbf{A}}_1$, can be viewed as a first approximation to $\mathbf{A}$.

The difference between the two matrices $\mathbf{A} - \widehat{\mathbf{A}}_1$ is given by

$$\begin{bmatrix} 0.857 & 1.017 & -0.768 \\ 1.017 & 1.256 & 0.663 \\ -0.768 & 0.663 & 1.887 \end{bmatrix}.$$

If two terms ($\lambda = 8$ and $\lambda = 3$) are used, we obtain $\widehat{\mathbf{A}}_2$ with the difference given by $(\mathbf{A} - \widehat{\mathbf{A}}_2)$, which is approximately the matrix term corresponding to $\lambda = 1$ in the spectral decomposition. It would appear that the variation in the magnitudes of the errors is quite small compared to the variation in the magnitudes in the original matrix. In other words the spectral decomposition approximation seems to be weighted toward the larger elements of $\mathbf{A}$.

Exercise

Determine the spectral decomposition for the matrix given in the eigenvalue exercise above and comment on the quality of the approximation based on the largest eigenvalue.

Eigenvalues for Nonnegative Definite Matrices

If the real symmetric matrix $\mathbf{A}$ is *positive definite* then the eigenvalues are all positive. If the matrix is *positive semidefinite* then the eigenvalues are nonnegative with the number of positive eigenvalues equal to the rank of $\mathbf{A}$.

3.2   SINGULAR VALUE DECOMPOSITION

A real $(n \times p)$ matrix $\mathbf{A}$ of rank k can be expressed as the product of three matrices that have a useful interpretation. This decomposition of $\mathbf{A}$ is referred to as a *singular value decomposition* and is given by

$$\mathbf{A} = \mathbf{UDV}',$$

where

1. $\mathbf{D}$ $(k \times k)$ is a diagonal matrix with positive diagonal elements $\alpha_1, \alpha_2, \dots, \alpha_k$, which are called the *singular values* of $\mathbf{A}$, (without loss of generality we assume that the α_j, $j = 1, 2, \dots, k$, are arranged in descending order).

2. The k columns of $\mathbf{U}$ $(n \times k)$, $\mathbf{u}_1, \mathbf{u}_2, \dots, \mathbf{u}_k$ are called the *left singular vectors* of $\mathbf{A}$ and the k columns of $\mathbf{V}$ $(p \times k)$, $\mathbf{v}_1, \mathbf{v}_2, \dots, \mathbf{v}_k$ are called the *right singular vectors* of $\mathbf{A}$.

3. The matrix $\mathbf{A}$ can be written as the sum of k matrices, each with rank 1, $\mathbf{A} = \sum_{j=1}^{k} \alpha_j \mathbf{u}_j \mathbf{v}_j'$. The subtraction of any one of these terms from the sum results in a singular matrix for the remainder of the sum.

4. The matrices $\mathbf{U}$ $(n \times k)$ and $\mathbf{V}$ $(p \times k)$ have the property that $\mathbf{U}'\mathbf{U} = \mathbf{V}'\mathbf{V} = \mathbf{I}$; hence the columns of $\mathbf{U}$ form an orthonormal basis for the columns of $\mathbf{A}$ in n-dimensional space and the columns of $\mathbf{V}$ form an orthonormal basis for the rows of $\mathbf{A}$ in p-dimensional space.

5. Let $\mathbf{A}(\ell)$ denote the first ℓ terms of the singular value decomposition for $\mathbf{A}$; hence $\mathbf{A}(\ell) = \sum_{j=1}^{\ell} \alpha_j \mathbf{u}_j \mathbf{v}_j'$. This expression minimizes $tr[(\mathbf{A} - \mathbf{X})(\mathbf{A} - \mathbf{X})'] = \sum_{i=1}^{n} \sum_{j=1}^{p} (a_{ij} - x_{ij})^2$ among all $(n \times p)$ matrices $\mathbf{X}$ of rank ℓ. Thus the singular value decomposition can be used to provide a *matrix approximation* to $\mathbf{A}$.

Complete Singular Value Decomposition

A *complete singular value decomposition* can be obtained for $\mathbf{A}$ by adding $(n-k)$ orthonormal vectors $\mathbf{u}_{k+1}, \ldots, \mathbf{u}_n$ to the existing set $\mathbf{u}_1, \ldots, \mathbf{u}_k$. Similarly the orthonormal vectors $\mathbf{v}_{k+1}, \ldots, \mathbf{v}_p$ are added to the set $\mathbf{v}_1, \mathbf{v}_2, \ldots,$ $\mathbf{v}_p$. Denoting the n column vectors $\mathbf{u}_j$, $j = 1, 2, \ldots, n$ by $\mathbf{U}^*$ and the p column vectors by $\mathbf{V}^*$ then $\mathbf{U}^{*'}\mathbf{U} = \mathbf{I}$ and $\mathbf{V}^{*'}\mathbf{V}^* = \mathbf{I}$ and $\mathbf{A} = \mathbf{U}^*\mathbf{D}^*\mathbf{V}^*$ is the complete singular value decomposition where $\mathbf{D}^* = \begin{bmatrix} \mathbf{D} & 0 \\ 0 & 0 \end{bmatrix}$.

Generalized Singular Value Decomposition

A *generalized singular value decomposition* permits the left and right singular vectors to be orthonormalized with respect to given diagonal matrices $\boldsymbol{\Omega}$ $(n \times n)$ and $\boldsymbol{\Phi}$ $(p \times p)$ where

$$\mathbf{N}'\boldsymbol{\Omega}\mathbf{N} = \mathbf{M}'\boldsymbol{\Phi}\mathbf{M} = \mathbf{I}.$$

The generalized singular value decomposition of $\mathbf{A}$ $(n \times p)$ is given by

$$\mathbf{A} = \mathbf{N}\mathbf{D}\mathbf{M}' = \sum_{j=1}^{k} \alpha_j \mathbf{n}_j \mathbf{m}_j',$$

where α_j, $j = 1, 2, \ldots, k$, are the diagonal elements of $\mathbf{D}$, $\mathbf{n}_j, j = 1, 2, \ldots, n$, are the columns of $\mathbf{N}$ and $\mathbf{m}_j$, $j = 1, 2, \ldots, p$, are the columns of $\mathbf{M}$. The columns of $\mathbf{N}$ and $\mathbf{M}$ are referred to as the *generalized left and right singular vectors* of $\mathbf{A}$ respectively. The diagonal elements of $\mathbf{D}$ are called the *generalized singular values*.

Let $\mathbf{A}(\ell)$ denote the first ℓ terms of the generalized singular value decomposition for $\mathbf{A}$, hence $\mathbf{A}(\ell) = \sum_{j=1}^{\ell} \alpha_j \mathbf{n}_j \mathbf{m}_j'$. This expression for $\mathbf{A}(\ell)$ minimizes $\sum_{j=1}^{p} \sum_{i=1}^{n} (a_{ij} - x_{ij})^2 \Phi_j \omega_i$ among all matrices $\mathbf{X}$ of rank at most ℓ where the Φ_j and ω_i are the diagonal elements of $\boldsymbol{\Phi}$ and $\boldsymbol{\Omega}$ respectively. The generalized singular value decomposition can be used to provide a *matrix approximation* to $\mathbf{A}$.

The generalized singular value decomposition of $\mathbf{A}$ given by $\mathbf{N}\mathbf{D}\mathbf{M}'$ where $\mathbf{N}'\boldsymbol{\Omega}\mathbf{N} = I$ and $\mathbf{M}'\boldsymbol{\Phi}\mathbf{M} = I$ can be related to the singular value decomposition by writing

$$\Omega^{1/2} A \Phi^{1/2} = \Omega^{1/2} NDM' \Phi^{1/2} = UDV$$

where $U = \Omega^{1/2} N$ and $V = \Phi^{1/2} M$ and $U'U = V'V = I$.

Relationship to Spectral Decomposition and Eigenvalues

If the matrix A is symmetric, the singular value decomposition is equivalent to the spectral decomposition

$$A = UDV' = V\Lambda V'.$$

In this case, the left and right singular vectors are equal to the eigenvectors and the singular values are equal to the eigenvalues.

The singular value decomposition of the $(n \times p)$ matrix A can also be related to the spectral decomposition for the symmetric matrices AA' and $A'A$. For the singular value decomposition

$$A = UDV',$$

the eigenvalues of AA' and $A'A$ are the squares of the singular values of A. The eigenvectors of AA' are the columns of U and the eigenvectors of $A'A$ are the columns of V. Since covariance matrices and correlation matrices can be written in the form $A'A$, the eigenvalues and eigenvectors are often determined using the singular value decomposition.

Data Appendix for Volume II

Introduction

This data appendix contains twenty-two data tables that are used in the chapter exercises throughout Volume II. Outlines to the data sets are given below. In some cases the data tables represent additional observations or variables derived from the data sets used in this text or in Volume I. A listing of the data tables follow the outlines.

Data Set V1   Bus Data

This data set consists of observations on additional variables in the study of bus driver absenteeism used in Chapter 6. The data set is divided into the three contingency tables listed below.

> Part I   — Day by Sex by Attendance
> Part II   — Day by Garage by Attendance
> Part III   — Garage by Sex by Attendance

Data Set V2   Accident Data

This data set consists of observations on some additional variables for the accident data introduced in Chapter 6. The observations are based on the same 86,769 accidents employed in Chapter 6. A four-dimensional contingency table relating INJURY LEVEL, SEATBELT USAGE, POINT OF IMPACT and DRIVER CONDITION is given below. The variable POINT OF IMPACT indicates where on the vehicle the collision occurred. The remaining three variables were used in Chapter 6.

Data Set V3   Accident Data

This data set consists of observations on some additional variables for the accident data introduced in Chapter 6 and in data set V2. The observations are based on the same 86,769 accidents employed in Chapter 6. A four dimensional contingency table relating INJURY LEVEL, SEATBELT USAGE, SPEED LIMIT and DRIVER CONDITION is given below. The variable SPEED LIMIT refers to the posted speed limit for the road on which the accident occurred. The remaining variables were used in Chapter 6.

Data Set V4   Real Estate Data

This data set consists of 116 observations on three bedroom bungalows listed and sold through a real estate multiple listing service in a given year in a particular area of a large city. The variables are LISTP (list price), SELLP (selling price), SQF (square feet), ROOMS (total number of rooms), BEDR (total number of bedrooms), GARAGE (no garage = 0, single garage = 1, double garage = 2), EXTRAS (numerical code 1, 2, or 3 denoting number of extras that come with the house), CHATTELS (numerical code 0, 1, 2, 3 indicating the number of additional items included in the price), AGE (age of the house), BATHR (no. of bathroom pieces i.e. a full bath usually has three pieces while a half bath usually has two pieces), and SELLDAYS (number of days the house was listed before it was sold). Some of the variables from this data set were used in Volume I.

Data Set V5   Automobile Data Part I

This data set contains observations on a sample of 97 automobiles selected from the Fuel Consumption Guide 1985 published by Transport Canada. The table contains observations on ENGSIZE, WEIGHT, AUTOMAT (0 = standard, 1 = automatic transmission), FOR (0 = domestic, 1 = foreign), URBRATE (urban fuel consumption rate), HWRATE (highway fuel consumption rate) and slope shifters; FWFIGHT = FOR $*$ WEIGHT, FENGSIZE = FOR $*$ ENGSIZE, AENGSIZE = AUTOMAT $*$ ENGSIZE, AWEIGHT = AUTOMAT $*$ WEIGHT. An additional sample of observations from this data set was introduced in Volume I.

Data Set V6   Financial Accounting Data

This data set consists of 80 observations collected from the Data Stream data base for a sample of UK companies for the year 1983. An additional sample of observations from this data base was used in Volume I. The variables are RETCAP (return on capital), WCFTDT (ratio of working capital flow to total debt), LOGSALE (log to base 10 of total sales), LOGASST (log to base 10 of total assets), CURRAT (current ratio), QUIKRAT (quick ratio), NFATAST (ratio of net fixed assets to total assets), PAYOUT (payout ratio), WCFTCL (ratio of working capital flow to current liabilities), GEARRAT (gearing ratio or debt-equity ratio), CAPINT (capital intensity or ratio of total sales to total assets), INVTAST (ratio of total inventories to total assets), and FATTOT (gross fixed assets to total assets).

Data Set V7   Air Pollution Data Part I

This data consists of observations on 80 U.S. cities for the year 1960 obtained from Gibbons, Dianne I; Gary C. McDonald and Richard F. Gunst, "The Complementary Use of Regression Diagnostics and Robust Estimators," *Naval Research Logistics* 34, 1, February, 1987. Other cities from this data set were also used in Chapter 9 and 10 and in Volume I, Chapter 4. The variables are defined below.

TMR	– total mortality rate
SMIN	– smallest biweekly sulfate reading ($\mu_g/m^3 \times 10$)
SMEAN	– arithmetic mean of biweekly sulfate readings ($\mu_g/m^3 \times 10$)
SMAX	– largest biweekly sulfate reading ($\mu_g/m^3 \times 10$)
PMIN	– smallest biweekly suspended particulate reading ($\mu_g/m^3 \times 10$)
PMEAN	– arithmetic mean of biweekly suspended particulate reading ($\mu_g/m^3 \times 10$)
PMAX	– largest biweekly suspended particulate reading ($\mu_g/m^3 \times 10$)
PM2	– population density per square mile $\times 0.1$
GE65	– percent of population at least 65×10
PERWH	– percent of whites in population
NONPOOR	– percent of families with income above poverty level
LPOP	– logarithm (base 10) of population $\times 10$

Data Set V8   Shopping Attitude Data

This data set consists of 200 observations obtained from a mail survey designed to obtain attitudes of females pertaining to shopping for clothing. Seven of the items designed to measure shopping orientation are given by

A1:   I like sales people to leave me alone until I find clothes that I want to buy.

A2:   I like to pay cash for clothing purchases.

A3:   Price is a good indicator of the quality of clothes.

A4:   I usually spend more than I planned when shopping for clothes.

A5:   I do not think clothing shops provide enough customer service these days.

A6:   When clothing is sold at a reduced price there is often something wrong with it.

A7:   I like to shop where my friends shop for clothes.

The responses were coded from 1 (strongly agree) to 5 (strongly disagree). In addition to the seven shopping orientation variables two additional variables included in the data set are WORK and AGE. These variables are coded as follows:

	Code			
WORK	0	respondent <u>works</u> outside home		
	1	respondent <u>does not</u> work outside home		

	Code	Age	Code	Age
AGE	1	24 or less	4	45–54
	2	25–34	5	55–64
	3	35–44	6	65 and over

Data Set V9   R.C.M.P. Officer Data

This data set contains observations from Royal Canadian Mounted Police (R.C.M.P.) officers regarding their satisfaction with various aspects of their jobs. The responses were combined into four factors labeled SATF1, SATF2, SATF3 and SATF4. The four factors can be characterized respectively as satisfaction with job characteristics, salary and benefits, commanding officer and co-workers. The observations were obtained from ten different municipal detachments in Alberta, Canada. Observations from this data set were used for examples in Chapters 9 and 10 and also in Volume I.

Data Set V10   Mystery Data

This data set consists of observations on five of the ten variables from the mystery data given in Table 8.11. The five variables are C1, C3, C8, C9 and C10 defined below:

C1:   Importance of more than one murder or crime.
C3:   Importance of powerful opponents.
C8:   Importance of many possible suspects.
C9:   Importance of puzzle being "fair play" by giving clues.
C10:   Importance of suspects appearing as average people.

Data Set V11 and V12   Bank Employee Data

These two data sets represent two different samples of 100 observations each selected from a larger data set. This bank employee data has been used by SPSSX to provide examples for the SPSSX User's Guide. Table V11 is an expansion of Table 7.9. Table V11 includes all the variables in Table 7.9 plus the additional variables SEX, RACE and JOBCAT. Table V12 is identical to Table D3 of Volume I. The variables in these two data sets are listed below:

LCURRENT	– ln (current salary)
LSTART	– ln (starting salary)
SEX	– male = 0, female = 1
JOBCAT	– 1 = clerical, 2 = office trainee, 3 = security officer,
	– 4 = college trainee, 5 = MBA trainee
RACE	– white = 0, nonwhite = 1
SENIOR	– seniority with the bank in months
AGE	– age in years
EXPER	– relevant job experience in years

Data Set V13   Panel Data

This data set is derived from the University of Michigan Panel Study of Income Dynamics and is an expansion of the data summarized in Table 8.25. The purpose of the data in this application is to study the factors that influence married female participation in the labor force. Table V13 contains 200 observations. The last 100 observations in the table are a repetition of the observations in Table 8.25. The variables are defined below.

THISYR	– indicator variable for whether the wife worked outside the home in the year of the survey (1=yes, 0=no)
LASTYR	– indicator variable for whether the wife worked outside the home in the previous year (1=yes, 0=no)
BLACK	– indicator variable for black race (1=black, 0=not black)
EDUC	– education level (years) of the respondent
AGE	– age level of the respondent (years)
CHILD1	– indicator variable for whether there are children in the home under the age of 2 (1=yes, 0=no)
CHILD2	– indicator variable for whether there are children in the home between the ages of 2 and 6 (1=yes, 0=no)
HUBINC	– income of husband in 1000 dollars.

Data Sets V14 and V15   U.S. Divorce Data

Data set V14 provides a summary of the available grounds for divorce in the United States by state (including District of Columbia) in 1982. The data was obtained from *The World Almanac and Book of Facts 1983* published for the Boston Herald American by Newspaper Enterprise Association, Inc., New York. The nine available grounds for divorce are listed below. States that have the ground available are coded 1, if they do not they are coded 0.

BREAK   – marriage breakdown/incompatibility
CRUEL   – cruelty
DESERT   – desertion
NOSUPPOR   – nonsupport
ALCOHOL   – alcohol and/or drug addiction
FELONY   – felony
IMPOTENT   – impotency
INSANE   – insanity
SEPARATE   – living separate and apart for a specified period.

Data set V15 is a subset of V14 consisting of the observations for 20 states.

Data Sets V16 and V17   U.S. Crime Data

Data set V16 summarizes the crime rates by state (per 100,000 population) for nine categories for the United States in 1980. The data set was obtained from *The World Almanac and Book of Facts 1983* published for the Boston Herald American by Newspaper Enterprise Association, Inc. New York. The nine types of crime are violent, property, murder, rape, robbery, assault, burglary, larceny and auto theft.

Data set V17 is a subset of V16 consisting of the observations for 15 states.

Data Set V18   Automobile Data Part II

This data set comes from the same database as V5. This sample consists of observations on 20 different automobiles with respect to COMBRATE (rate of fuel consumption), CYLIND (no. of cylinders), WEIGHT, ENGSIZE and FOR (1=foreign manufacturer, 0=North American manufacturer). The observations on all but FOR are ranks as defined below.

CYLIND   – 4, 6, 8 becomes 1, 2 and 3 respectively
ENGSIZE   – 15–18 = 1, 20–24 = 2, 28–30 = 3, 38–41 = 4, 50 = 5
COMBRATE   – 64–71 = 1, 74–84 = 2, 93–97 = 3, 104–110 = 4
WEIGHT   – (2000,2250) = 1, (2500,2750) = 2, 3000 = 3, 3500 = 4, 4000 = 5

Data Set V19   Cola Similarity Data

This data set consists of a dissimilarity matrix relating ten different brands of cola soft drinks [0 = same, 100 = completely different]. The dissimilarity matrix was derived from an experiment involving ten university students aged 18–21 years. Subjects were blindfolded and then asked to taste ten different colas without swallowing. Subjects rinsed their mouths with distilled water between tastes. Subjects were asked to judge the similarity between all possible pairs (45) of the ten colas over a five-day period. The dissimilarity matrix is an average of the ten dissimilarity matrices of the ten subjects. The data was obtained from Schiffman, Susan S., Lance M. Reynolds and Forest W. Young (1981) *Introduction to Multidimensional Scaling: Theory, Methods and Applications*, New York: Academic Press.

Data Set V20   Car Similarity Data

A number of subjects were asked to compare 11 automobiles with respect to overall similarity. The rank order of the 55 dissimilarities are given in a dissimilarity matrix (1 = most similar pair, 55 = least similar pair). This matrix was obtained from Green, Paul E., Frank J. Carmone and Scott M. Smith (1989) *Multidimensional Scaling: Concepts and Applications*, Boston: Allyn and Bacon.

Data Set V21   Air Pollution Data Part II

This data set consists of a subset of observations from Data Set V7.

Data Set V22   Shopping Attitude Data Part II

This data set consists of 200 observations from a mail survey designed to obtain attitudes of females pertaining to shopping for clothing. Eighteen items designed to measure shopping orientation are contained in the data set. The first seven items were included in Data Set V8 along with variables measuring WORK and AGE. Table V22 contains these items plus the additional 11 items summarized below.

A8:   I always buy my clothes at the same shops.
A9:   I visit several shops before buying clothes for myself.
A10:   I think shops that carry well known makes of clothing are overpriced.
A11:   I like to have someone with me when I shop for clothes.
A12:   I like to make my own buying decisions rather than get advice from others.
A13:   To me, shopping for clothes is fun.
A14:   I feel creative when I go shopping for clothes.
A15:   Buying new clothes gives me a lift.
A16:   I only go shopping for clothes when I really need something.
A17:   Shopping for clothes gives me no satisfaction.
A18:   I like browsing in clothing shops without buying anything.

TABLE V1. Bus Data Part I

Day	Sex	Attend	Frequency
Sun	Male	Present	1676
Mon	Male	Present	4931
Tues	Male	Present	4850
Wed	Male	Present	4835
Thurs	Male	Present	4832
Fri	Male	Present	4857
Sat	Male	Present	2702
Sun	Female	Present	162
Mon	Female	Present	361
Tues	Female	Present	392
Wed	Female	Present	398
Thurs	Female	Present	396
Fri	Female	Present	413
Sat	Female	Present	297
Sun	Male	Absent	132
Mon	Male	Absent	461
Tues	Male	Absent	494
Wed	Male	Absent	501
Thurs	Male	Absent	512
Fri	Male	Absent	479
Sat	Male	Absent	242
Sun	Female	Absent	30
Mon	Female	Absent	79
Tues	Female	Absent	96
Wed	Female	Absent	98
Thurs	Female	Absent	92
Fri	Female	Absent	83
Sat	Female	Absent	63

TABLE V1. Bus Data Part III

Garage	Sex	Attend	Frequency
1	Male	Present	15422
2	Male	Present	10811
3	Male	Present	10023
1	Female	Present	1086
2	Female	Present	801
3	Female	Present	697
1	Male	Absent	1658
2	Male	Absent	1069
3	Male	Absent	977
1	Female	Absent	274
2	Female	Absent	319
3	Female	Absent	223

TABLE V1. Bus Data Part II

Day	Garage	Attend	Frequency
Sun	1	Present	855
Mon	1	Present	2170
Tues	1	Present	2123
Wed	1	Present	2113
Thurs	1	Present	2129
Fri	1	Present	2157
Sat	1	Present	1308
Sun	2	Present	514
Mon	2	Present	1634
Tues	2	Present	1619
Wed	2	Present	1625
Thurs	2	Present	1636
Fri	2	Present	1644
Sat	2	Present	857
Sun	3	Present	460
Mon	3	Present	1475
Tues	3	Present	1482
Wed	3	Present	1469
Thurs	3	Present	1437
Fri	3	Present	1450
Sat	3	Present	813
Sun	1	Absent	81
Mon	1	Absent	222
Tues	1	Absent	269
Wed	1	Absent	279
Thurs	1	Absent	263
Fri	1	Absent	235
Sat	1	Absent	148
Sun	2	Absent	46
Mon	2	Absent	182
Tues	2	Absent	197
Wed	2	Absent	191
Thurs	2	Absent	180
Fri	2	Absent	172
Sat	2	Absent	9
Sun	3	Absent	44
Mon	3	Absent	149
Tues	3	Absent	142
Wed	3	Absent	155
Thurs	3	Absent	187
Fri	3	Absent	174
Sat	3	Absent	83

TABLE V2. Accident Data

Seatbelt	Point of Impact	Injury Level	Driver Condition	Frequency
Yes	Front	None	Normal	7389
Yes	Front	None	Bdrink	199
Yes	Front	Minimal	Normal	308
Yes	Front	Minimal	Bdrink	30
Yes	Front	Minor	Normal	169
Yes	Front	Minor	Bdrink	13
Yes	Front	Majfat	Normal	18
Yes	Front	Majfat	Bdrink	2
Yes	Rear	None	Normal	3509
Yes	Rear	None	Bdrink	79
Yes	Rear	Minimal	Normal	207
Yes	Rear	Minimal	Bdrink	5
Yes	Rear	Minor	Normal	106
Yes	Rear	Minor	Bdrink	1
Yes	Rear	Majfat	Normal	5
Yes	Rear	Majfat	Bdrink	1
Yes	Rside	None	Normal	827
Yes	Rside	None	Bdrink	21
Yes	Rside	Minimal	Normal	36
Yes	Rside	Minimal	Bdrink	5
Yes	Rside	Minor	Normal	27
Yes	Rside	Minor	Bdrink	1
Yes	Rside	Majfat	Normal	8
Yes	Lside	None	Normal	775
Yes	Lside	None	Bdrink	14
Yes	Lside	Minimal	Normal	53
Yes	Lside	Minimal	Bdrink	3
Yes	Lside	Minor	Normal	42
Yes	Lside	Majfat	Normal	7
Yes	Lside	Majfat	Bdrink	1

TABLE V2. Accident Data (continued)

Seatbelt	Point of Impact	Injury Level	Driver Condition	Frequency
No	Front	None	Normal	37492
No	Front	None	Bdrink	2833
No	Front	Minimal	Normal	2025
No	Front	Minimal	Bdrink	384
No	Front	Minor	Normal	1337
No	Front	Minor	Bdrink	278
No	Front	Majfat	Normal	135
No	Front	Majfat	Bdrink	48
No	Rear	None	Normal	16280
No	Rear	None	Bdrink	768
No	Rear	Minimal	Normal	913
No	Rear	Minimal	Bdrink	50
No	Rear	Minor	Normal	491
No	Rear	Minor	Bdrink	42
No	Rear	Majfat	Normal	28
No	Rear	Majfat	Bdrink	6
No	Rside	None	Normal	4165
No	Rside	None	Bdrink	218
No	Rside	Minimal	Normal	397
No	Rside	Minimal	Bdrink	27
No	Rside	Minor	Normal	207
No	Rside	Minor	Bdrink	26
No	Rside	Majfat	Normal	28
No	Rside	Majfat	Bdrink	4
No	Lside	None	Normal	4034
No	Lside	None	Bdrink	173
No	Lside	Minimal	Normal	184
No	Lside	Minimal	Bdrink	20
No	Lside	Minor	Normal	237
No	Lside	Minor	Bdrink	24
No	Lside	Majfat	Normal	46
No	Lside	Majfat	Bdrink	8

TABLE V3. Accident Data

Seatbelt	Speed Limit	Injury Level	Driver Condition	Frequency
Yes	Lt60kph	None	Normal	9838
Yes	Lt60kph	None	Bdrink	234
Yes	Lt60kph	Minimal	Normal	401
Yes	Lt60kph	Minimal	Bdrink	31
Yes	Lt60kph	Minor	Normal	219
Yes	Lt60kph	Minor	Bdrink	10
Yes	Lt60kph	Majfat	Normal	11
Yes	Lt60kph	Majfat	Bdrink	1
Yes	60-89kph	None	Normal	2021
Yes	60-89kph	None	Bdrink	60
Yes	60-89kph	Minimal	Normal	144
Yes	60-89kph	Minimal	Bdrink	11
Yes	60-89kph	Minor	Normal	68
Yes	60-89kph	Minor	Bdrink	2
Yes	60-89kph	Majfat	Normal	6
Yes	60-89kph	Majfat	Bdrink	1
Yes	Gt89kph	None	Normal	641
Yes	Gt89kph	None	Bdrink	19
Yes	Gt89kph	Minimal	Normal	59
Yes	Gt89kph	Minimal	Bdrink	1
Yes	Gt89kph	Minor	Normal	57
Yes	Gt89kph	Minor	Bdrink	3
Yes	Gt89kph	Majfat	Normal	21
Yes	Gt89kph	Majfat	Bdrink	2
No	Lt60kph	None	Normal	52269
No	Lt60kph	None	Bdrink	3242
No	Lt60kph	Minimal	Normal	2531
No	Lt60kph	Minimal	Bdrink	350
No	Lt60kph	Minor	Normal	1609
No	Lt60kph	Minor	Bdrink	228
No	Lt60kph	Majfat	Normal	91
No	Lt60kph	Majfat	Bdrink	26
No	60-89kph	None	Normal	7993
No	60-89kph	None	Bdrink	571
No	60-89kph	Minimal	Normal	790
No	60-89kph	Minimal	Bdrink	87
No	60-89kph	Minor	Normal	452
No	60-89kph	Minor	Bdrink	87
No	60-89kph	Majfat	Normal	58
No	60-89kph	Majfat	Bdrink	16
No	Gt89kph	None	Normal	1709
No	Gt89kph	None	Bdrink	179
No	Gt89kph	Minimal	Normal	198
No	Gt89kph	Minimal	Bdrink	44
No	Gt89kph	Minor	Normal	211
No	Gt89kph	Minor	Bdrink	55
No	Gt89kph	Majfat	Normal	88
No	Gt89kph	Majfat	Bdrink	24

TABLE V4. Real Estate Data

LISTP	SELLP	SQF	ROOMS	BEDR	GARAGE	EXTRAS	CHATTELS	AGE	SELLDAYS	BATHR
79000	75400	1365	7	4	2	3	2	8	31	6
85000	79800	1170	6	3	2	2	1	7	48	4
91000	82000	1160	6	4	2	2	2	6	15	6
107800	98500	1306	6	3	1	3	3	6	18	5
83000	80000	1120	5	3	2	1	2	5	28	6
85000	85000	1040	6	3	2	2	3	7	35	4
85800	84000	1130	6	3	2	2	1	7	57	4
89800	84000	1232	6	3	1	3	2	7	30	6
97600	96800	1364	7	4	2	3	0	7	1	6
99400	95800	1260	6	3	0	2	0	7	30	6
103800	101500	1302	6	3	0	2	3	5	30	6
77800	74000	1040	6	3	0	1	1	8	28	4
103800	101000	1278	6	3	2	1	1	7	33	6
107000	102000	1408	7	3	2	3	3	8	24	6
109800	107000	1225	6	3	2	1	0	6	2	6
93800	90000	1160	6	3	2	3	2	6	44	6
95600	88000	1040	6	3	1	2	1	7	41	4
102000	98000	1424	7	3	2	1	0	6	54	6
111800	104400	1232	6	3	2	3	0	7	36	6
95800	89800	1270	6	3	0	2	0	7	51	6
105800	102000	1298	6	3	2	2	0	7	57	6
135000	128000	1340	6	3	2	3	0	7	38	6
139800	124000	1380	6	3	2	3	2	9	21	6
99800	91600	1200	6	3	2	3	1	10	35	6
109400	106000	1180	6	3	2	3	0	7	63	6
109800	91600	1200	6	3	2	3	1	10	37	6
110800	102000	1260	6	3	2	3	3	7	45	6
97200	94000	1219	6	3	2	1	2	7	43	6
101000	98000	1154	6	3	2	1	0	7	30	6

TABLE V4. Real Estate Data (continued)

LISTP	SELLP	SQF	ROOMS	BEDR	GARAGE	EXTRAS	CHATTELS	AGE	SELLDAYS	BATHR
105800	98600	1221	6	3	2	3	0	7	62	6
111800	100000	1205	6	3	2	2	1	7	112	6
91800	85000	1170	6	3	0	2	3	8	25	4
107800	98000	1200	6	3	2	2	1	7	64	6
95600	93500	1120	6	3	2	1	1	7	16	6
97800	93000	1130	6	3	2	2	0	9	37	4
99800	92000	1270	5	3	0	1	0	5	58	6
103800	98000	1120	6	3	2	2	0	6	67	6
103800	101000	1072	6	3	2	3	1	7	15	7
105000	98000	1130	6	3	2	3	2	7	43	6
86000	84000	1080	6	3	0	1	0	10	59	4
95000	88000	1060	6	3	1	1	1	11	39	4
109800	101000	1232	6	3	2	3	0	8	43	9
123600	114000	1200	6	3	1	3	1	10	26	6
95800	92000	1080	6	3	0	2	0	11	32	4
130000	122000	1460	6	3	2	3	1	7	17	7
91200	87800	1100	6	3	2	3	2	10	61	7
93800	89000	1140	6	3	0	1	1	11	12	4
119600	108000	1248	6	3	2	2	0	12	43	6
89800	87400	1050	6	3	2	1	0	11	26	4
107000	103000	1286	6	3	2	2	0	12	15	6
111800	106000	1250	6	3	2	3	0	9	70	7
77800	77000	1056	5	3	1	3	0	12	8	4
91800	85800	1135	6	3	2	2	0	12	60	4
97800	96000	1280	6	3	2	3	4	10	24	4
100600	96000	1100	5	3	2	2	0	10	73	4
81000	81000	1060	6	3	1	1	0	11	4	3
91800	84000	1080	6	3	0	3	1	10	25	4
95000	90000	1080	6	3	2	1	2	11	20	4

TABLE V4. Real Estate Data (continued)

LISTP	SELLP	SQF	ROOMS	BEDR	GARAGE	EXTRAS	CHATTELS	AGE	SELLDAYS	BATHR
95800	90000	1140	5	3	1	1	1	11	24	8
103800	103000	1160	6	3	2	3	1	9	18	7
105000	101800	1089	6	3	2	2	0	6	24	4
93000	86000	1040	6	3	0	3	1	10	17	4
99800	96000	1200	6	3	2	2	1	9	12	6
107500	104000	1257	6	3	2	2	0	9	7	6
95800	94000	1260	5	3	1	2	1	8	47	6
77000	76000	1106	6	3	2	2	1	5	64	4
77800	76000	1120	6	3	0	1	1	1	10	6
79000	74000	1190	6	3	2	1	0	1	48	6
85000	83000	1264	6	3	1	1	1	3	50	6
87000	83000	1232	6	3	0	1	0	1	36	6
87800	83800	1160	6	3	2	1	0	1	9	6
89800	88000	1190	6	3	2	1	5	1	20	6
93800	91028	1392	6	3	0	2	0	1	44	7
97600	96000	1220	6	3	2	2	3	6	15	6
128400	123000	1450	6	3	2	2	0	1	13	5
87200	87200	1206	6	3	0	1	0	1	11	6
87800	82000	1150	6	3	0	1	1	2	35	4
89800	87800	1204	6	3	0	1	1	1	22	6
95800	92000	1340	6	3	0	1	1	1	12	6
83000	76000	1160	6	3	0	1	2	1	48	6
89800	86000	1200	5	3	0	2	1	1	40	6
91800	89000	1202	6	3	2	1	0	1	57	4
97800	95200	1278	6	3	2	1	0	1	12	6
106150	99000	1260	6	3	1	3	2	6	53	6
119600	110000	1284	6	3	0	3	1	3	2	6
87000	83500	1237	6	3	0	1	2	2	17	4
91800	88000	1050	5	3	0	2	0	7	12	4

TABLE V4. Real Estate Data (continued)

LISTP	SELLP	SQF	ROOMS	BEDR	GARAGE	EXTRAS	CHATTELS	AGE	SELLDAYS	BATHR
101800	99000	1320	6	3	2	2	5	2	14	6
115000	104000	1490	7	3	2	1	2	2	35	6
138400	130000	1491	7	3	2	2	3	1	33	6
89000	83000	1154	6	3	0	1	0	6	38	6
93800	89000	1224	6	3	0	2	0	1	43	6
97800	96000	1220	6	3	2	1	0	6	28	4
103000	99000	1152	6	3	2	1	0	2	13	4
105800	101000	1392	8	4	0	1	2	2	35	6
129800	124000	1480	7	3	2	1	1	0	2	7
95800	92000	1042	5	3	2	3	1	7	20	3
96600	95000	1167	6	3	0	1	0	1	6	6
112000	106000	1246	6	3	2	2	0	1	62	4
114000	100000	1440	6	3	0	2	2	1	34	7
135000	122400	1564	7	3	1	2	3	1	32	7
111000	105000	1224	6	3	2	2	1	7	55	6
115000	114000	1325	6	3	2	2	0	1	63	6
127000	121000	1270	5	3	2	3	1	6	21	7
91800	88000	1340	8	4	0	1	1	2	21	6
105800	104000	1270	6	3	1	1	0	1	48	6
119800	115000	1460	7	3	0	1	0	3	34	6
87800	85000	1001	6	3	2	3	2	7	82	4
101800	97000	1227	6	3	2	1	1	1	67	6
77800	77800	1124	6	3	2	2	0	8	21	4
78000	78000	1070	6	3	1	3	0	12	19	4
78000	73000	1120	5	3	2	2	0	5	21	4
95800	85400	1200	6	3	2	3	0	9	52	6
105800	101000	1308	6	3	0	3	2	7	2	9
97400	96000	1113	6	3	0	1	0	8	37	4
105000	98000	1345	6	3	2	1	1	7	16	6

TABLE V5. Automobile Data Part I

ENGSIZE	WEIGHT	FOR	AUTOMAT	FWEIGHT	FENGSIZE	AENGSIZE	AWEIGHT	URBRATE	HWRATE
16	2500	0	0	0	0	0	0	73	48
18	2500	1	1	2500	18	18	2500	103	73
50	4000	0	1	0	0	50	4000	148	84
23	3000	1	1	3000	23	0	0	121	69
38	4000	0	0	0	0	38	4000	133	78
20	2500	1	1	2500	20	0	0	86	55
18	2500	0	0	0	0	0	0	97	52
12	1750	1	1	1750	12	12	1750	75	50
20	2750	1	1	2750	20	20	2750	87	53
50	4000	0	1	0	0	50	4000	149	86
25	2750	0	1	0	0	25	2750	101	57
34	3500	1	0	3500	34	0	0	145	83
23	3000	1	1	3000	23	23	3000	107	70
28	3000	0	1	0	0	28	3000	120	63
18	2500	1	0	2500	18	0	0	86	56
50	4000	0	1	0	0	50	4000	147	79
18	2750	0	1	0	0	18	2750	117	73
16	2500	0	1	0	0	16	2500	97	63
22	2750	0	0	0	0	0	0	110	59
18	2750	1	1	2750	18	18	2750	101	74
50	3500	0	1	0	0	50	3500	147	80
50	4000	0	0	0	0	50	4000	169	96
20	2500	0	0	0	0	0	0	97	57
23	3000	1	1	3000	23	23	3000	125	83
16	2500	1	0	2500	16	0	0	96	63
23	2750	0	1	0	0	0	0	104	59
22	3000	0	1	0	0	22	3000	117	75
20	2750	0	1	0	0	20	2750	102	60
16	2250	0	0	0	0	0	0	79	51
20	2750	0	0	0	0	0	0	97	57
50	4000	0	1	0	0	50	4000	152	87
50	3500	0	1	0	0	50	3500	149	86
15	2000	1	0	2000	15	0	0	85	55

TABLE V5. Automobile Data Part I (continued)

ENGSIZE	WEIGHT	FOR	AUTOMAT	FWEIGHT	FENGSIZE	AENGSIZE	AWEIGHT	URBRATE	HWRATE
22	2500	0	0	0	0	0	0	87	55
38	3500	0	1	0	0	38	3500	132	81
38	3000	0	1	0	0	38	3000	140	75
50	3500	0	1	0	0	50	3500	147	80
16	2500	0	0	0	0	0	0	74	46
23	3500	0	1	0	0	23	3500	130	86
50	4000	1	1	4000	50	50	4000	172	116
50	3500	0	0	0	0	0	0	172	82
38	3000	0	1	0	0	38	3000	120	81
18	2750	1	1	2750	18	18	2750	104	68
50	3500	0	0	0	0	0	0	141	75
22	3000	0	1	0	0	22	3000	110	72
38	4000	0	1	0	0	38	4000	132	77
34	3500	1	1	3500	34	34	3500	133	77
25	2750	0	0	0	0	0	0	102	54
16	2750	0	1	0	0	16	2750	96	60
16	2500	0	0	0	0	0	0	73	48
32	3000	1	0	3000	32	0	0	134	72
50	3500	0	1	0	0	50	3500	157	77
50	4000	0	1	0	0	50	4000	147	80
25	3500	0	0	0	0	0	0	147	80
58	3000	0	1	0	0	58	3000	107	55
18	4000	1	0	4000	18	0	0	178	95
28	2500	0	1	0	0	28	2500	90	58
22	3000	0	0	0	0	0	0	118	69
38	2750	0	1	0	0	38	2750	93	53
22	3500	0	1	0	0	22	3500	145	77
28	3000	0	0	0	0	0	0	119	79
18	3500	0	1	0	0	18	3500	134	67
20	2500	1	0	2500	20	0	0	97	66
25	2500	0	0	0	0	0	0	97	57
23	3000	0	0	0	0	0	0	99	59
23	3000	1	0	3000	23	0	0	103	64

TABLE V5. Automobile Data Part I (continued)

ENGSIZE	WEIGHT	FOR	AUTOMAT	FWEIGHT	FENGSIZE	AENGSIZE	AWEIGHT	URBRATE	HWRATE
28	3000	0	1	0	0	28	3000	122	71
18	2750	0	1	0	0	18	2750	101	63
28	3000	0	1	0	0	28	3000	118	69
22	2750	0	0	0	0	0	0	112	62
22	3000	0	1	0	0	22	3000	98	64
15	2250	1	0	2250	15	0	0	80	59
20	3000	1	0	3000	20	0	0	113	66
16	2250	0	0	0	0	0	0	88	54
16	2250	1	1	2250	16	16	2250	92	60
15	2000	1	1	2000	15	15	2000	87	59
20	3000	1	1	3000	20	20	3000	130	88
38	3500	0	1	0	0	38	3500	132	81
33	3000	1	0	3000	33	0	0	181	101
20	2700	1	1	2700	20	20	2700	98	67
50	3500	0	0	0	0	0	0	157	77
18	2500	1	0	2500	18	0	0	92	61
16	2500	1	1	2500	16	16	2500	102	62
16	2500	1	1	2500	16	16	2500	91	60
20	2750	0	1	0	0	20	2750	109	70
20	3000	1	1	3000	20	20	3000	98	57
25	3000	0	1	0	0	25	3000	104	58
16	2500	1	0	2500	16	0	0	90	63
30	3000	0	1	0	0	30	3000	118	69
24	2750	1	1	2750	24	24	2750	96	63
25	3000	0	1	0	0	25	3000	92	59
25	2750	0	1	0	0	25	2750	101	57
18	2500	1	0	2500	18	0	0	95	57
16	2250	0	0	0	0	0	0	97	60
41	3500	0	1	0	0	41	3500	134	77
25	3000	1	0	3000	25	0	0	118	67
38	3500	0	1	0	0	38	3500	126	76

TABLE V6. Financial Accounting Data

RETCAP	WCFTCL	WCFTDT	GEARRAT	LOGSALE	LOGASST
0.19	0.16	0.16	0.15	5.2297	4.8375
0.22	0.26	0.16	0.54	4.1495	4.3402
0.17	0.26	0.20	0.49	5.3831	4.8811
0.12	0.08	0.08	0.39	4.1225	3.9333
0.21	0.34	0.34	0.11	4.7795	4.5877
0.12	0.25	0.25	0.19	4.1503	3.9086
0.15	0.25	0.16	0.35	5.6998	5.5577
0.10	0.12	0.09	0.39	4.4162	4.2128
0.08	0.04	0.04	0.50	4.7108	4.5126
0.31	0.12	0.11	0.41	4.4678	4.1928
0.21	0.36	0.33	0.08	4.3899	4.2336
0.22	0.37	0.37	0.16	4.0253	3.8344
0.20	0.48	0.48	0.13	3.8573	3.8764
0.11	0.18	0.15	0.23	3.9068	3.8685
0.38	0.25	0.20	0.27	5.1631	4.6669
0.23	0.24	0.24	0.00	5.7130	4.9772
0.32	0.09	0.09	0.11	4.7114	4.3123
0.13	0.06	0.05	0.55	4.6763	4.4972
0.29	0.60	0.60	0.00	4.5233	4.8709
0.09	0.10	0.09	0.28	4.9876	4.4058
-2.22	-1.28	-1.28	1.78	4.0554	3.5485
0.17	0.12	0.11	0.28	4.2837	3.9679
-0.04	-0.04	-0.04	0.46	4.7616	4.3153
0.26	0.23	0.23	0.00	4.2468	3.8779
0.21	0.40	0.30	0.20	4.4106	4.3829
0.15	0.30	0.21	0.66	4.3984	4.3634
0.23	0.07	0.07	0.11	4.8314	4.4399
0.20	0.33	0.28	0.33	4.2050	4.0364
0.19	0.16	0.14	0.30	4.3139	4.1727
0.08	0.18	0.10	0.35	4.9510	4.8675
0.19	0.15	0.14	0.19	5.5754	5.4405
0.20	0.63	0.35	0.21	4.7722	4.8638
0.14	0.27	0.20	0.30	4.9993	4.8282
0.04	0.07	0.07	0.18	4.1786	3.9151
0.10	0.15	0.12	0.13	5.7613	5.7801
-0.09	-0.46	-0.22	0.68	3.9671	4.0802
0.10	0.18	0.14	0.23	5.6884	5.6334
0.20	0.13	0.12	0.05	4.7908	4.4200
0.13	0.17	0.13	0.22	5.4876	5.3501
0.08	0.14	0.14	0.19	4.0891	3.8737

TABLE V6. Financial Accounting Data (continued)

NFATAST	CAPINT	FATTOT	INVTAST	PAYOUT	QUIKRAT	CURRAT
0.28	2.47	0.36	0.42	0.31	0.54	1.33
0.13	0.64	0.16	0.04	0.45	0.83	0.93
0.43	3.18	0.74	0.13	0.50	0.84	1.09
0.23	1.55	0.50	0.37	0.65	0.50	1.09
0.30	1.56	0.50	0.20	0.25	1.10	1.74
0.34	1.74	0.38	0.31	0.80	1.00	1.89
0.48	1.39	0.62	0.22	0.46	0.73	1.38
0.26	1.60	0.42	0.30	1.03	0.94	1.57
0.25	1.58	0.33	0.31	0.00	0.74	1.28
0.17	1.88	0.25	0.31	0.25	0.66	1.10
0.40	1.43	0.71	0.17	0.61	1.06	1.49
0.42	1.55	0.62	0.17	0.25	0.97	1.38
0.68	0.96	0.97	0.13	0.60	0.61	1.00
0.40	1.09	0.64	0.15	0.80	0.92	1.23
0.21	3.13	0.32	0.38	0.39	0.33	1.39
0.27	5.44	0.38	0.50	0.36	0.24	1.29
0.09	2.51	0.13	0.31	0.53	0.86	1.34
0.24	1.51	0.40	0.42	0.00	0.44	1.14
0.57	0.45	0.58	0.01	0.21	1.18	1.21
0.34	3.82	0.50	0.46	1.52	0.34	1.28
0.16	3.21	0.30	0.37	0.00	0.50	1.06
0.26	2.07	0.32	0.37	0.22	0.67	1.36
0.19	2.79	0.32	0.28	0.00	0.72	1.11
0.21	2.34	0.26	0.27	0.53	1.20	1.83
0.24	1.07	0.36	0.24	0.42	1.77	2.72
0.70	1.08	1.07	0.15	0.00	0.29	0.58
0.17	2.46	0.22	0.00	0.67	0.88	0.88
0.53	1.47	1.16	0.07	0.21	0.77	0.91
0.25	1.38	0.33	0.42	0.52	0.49	1.28
0.31	1.21	0.51	0.27	1.08	1.44	2.36
0.22	1.36	0.36	0.22	0.40	0.96	1.35
0.21	0.81	0.34	0.26	0.51	2.63	3.98
0.72	1.48	0.74	0.09	0.53	0.26	0.54
0.28	1.83	0.54	0.23	4.21	1.08	1.57
0.12	0.96	0.21	0.28	0.43	0.57	1.40
0.62	0.77	0.71	0.19	0.00	0.60	1.45
0.33	1.14	0.52	0.23	0.12	0.83	1.56
0.04	2.35	0.07	0.37	0.33	0.80	1.42
0.26	1.37	0.52	0.41	0.53	0.75	1.73
0.17	1.64	0.27	0.34	0.91	0.74	1.57

TABLE V6. Financial Accounting Data (continued)

RETCAP	WCFTCL	WCFTDT	GEARRAT	LOGSALE	LOGASST
0.14	0.41	0.37	0.21	4.3516	4.3615
0.12	0.30	0.25	0.29	4.2835	4.1800
0.17	0.20	0.20	0.33	4.4271	4.0467
0.27	0.38	0.38	0.00	3.6470	3.9209
0.22	0.45	0.44	0.18	3.8183	3.5728
0.13	0.35	0.35	0.00	4.4159	4.4739
-0.10	-0.70	-0.70	0.00	2.7973	3.8199
0.13	0.18	0.16	0.30	3.7923	3.6307
0.14	0.09	0.07	0.63	6.6360	6.1332
0.07	0.03	0.03	0.49	4.5339	4.5888
0.31	1.57	0.27	0.54	2.0212	3.7313
0.18	0.48	0.41	0.08	4.0326	4.0720
0.26	0.16	0.16	0.72	3.9202	3.8331
0.08	0.13	0.10	0.47	4.5782	4.4681
0.16	0.14	0.14	0.15	4.3202	4.2248
0.09	0.10	0.08	0.45	4.2578	4.2315
0.29	0.12	0.12	0.29	4.6570	4.9039
0.22	0.21	0.19	0.45	3.5813	3.5751
0.12	0.12	0.12	0.39	4.2843	4.0931
0.26	0.78	0.78	0.01	3.7270	3.6617
0.10	2.12	2.12	0.01	0.0000	5.0505
0.15	0.16	0.15	0.44	4.2181	3.9019
0.16	0.20	0.13	0.34	4.2322	4.2077
0.12	0.11	0.11	0.48	4.7735	4.4443
0.16	0.25	0.22	0.11	4.3490	3.9451
0.23	5.46	5.46	0.00	3.7897	4.2260
0.11	0.26	0.20	0.24	3.8366	3.6951
0.15	0.17	0.14	0.28	3.9725	3.8831
-0.02	-0.11	-0.08	0.89	4.3601	4.1623
0.10	0.06	0.05	0.41	4.6635	4.5154
0.08	0.09	0.08	0.05	4.6209	4.4724
0.08	0.08	0.06	0.45	4.0991	3.9227
0.11	0.12	0.11	0.44	4.9139	4.6044
0.10	0.04	0.04	0.47	4.2430	3.8602
0.08	0.04	0.04	0.42	4.8882	5.1535
0.20	0.57	0.57	0.01	3.8650	3.7666
-0.16	-0.54	-0.15	0.54	4.1210	4.8298
0.10	0.37	0.15	0.49	3.5941	4.0571
0.06	0.07	0.06	0.64	4.7757	4.2350
0.26	0.25	0.25	0.46	4.1162	4.3087

TABLE V6. Financial Accounting Data (continued)

NFATAST	CAPINT	FATTOT	INVTAST	PAYOUT	QUIKRAT	CURRAT
0.72	0.98	0.93	0.07	0.47	0.57	0.74
0.31	1.27	0.56	0.33	0.63	0.97	1.91
0.14	2.40	0.23	0.43	0.33	1.00	1.97
0.57	0.53	0.70	0.13	0.57	0.69	0.97
0.13	1.76	0.38	0.45	0.00	1.18	2.43
0.08	0.88	0.41	0.22	0.46	0.58	1.29
0.46	0.09	0.75	0.00	0.00	8.82	8.82
0.24	1.45	0.33	0.52	0.43	0.56	1.82
0.31	3.18	0.37	0.28	0.66	0.60	1.05
0.20	0.88	0.29	0.44	0.81	0.75	1.63
0.06	0.02	0.07	0.00	0.74	2.55	2.55
0.35	0.91	0.53	0.07	0.45	1.91	2.22
0.22	1.22	0.43	0.17	0.00	0.78	1.01
0.51	1.29	0.78	0.18	0.21	0.65	1.03
0.34	1.25	0.58	0.24	0.73	0.72	1.23
0.34	1.06	0.45	0.27	0.83	0.62	1.12
0.27	0.57	0.34	0.02	0.40	1.03	1.06
0.21	1.01	0.29	0.40	0.37	0.63	1.29
0.31	1.55	0.52	0.24	0.64	0.69	1.08
0.45	1.16	0.79	0.07	0.07	1.93	2.23
0.00	0.00	0.00	0.00	0.61	1.97	1.97
0.21	2.07	0.23	0.44	1.40	0.52	1.05
0.45	1.06	0.68	0.16	0.71	0.68	1.09
0.27	2.13	0.43	0.36	0.87	0.65	1.27
0.22	2.53	0.31	0.40	0.20	0.97	1.99
0.26	0.37	0.43	0.00	0.22	6.22	6.22
0.31	1.38	0.59	0.23	0.27	0.99	1.62
0.22	1.23	0.33	0.34	0.54	0.94	1.89
0.47	1.58	0.64	0.22	0.00	0.50	0.85
0.27	1.41	0.55	0.46	0.00	0.52	1.41
0.16	1.41	0.23	0.38	0.65	0.57	1.25
0.40	1.50	0.51	0.34	2.13	0.54	1.25
0.15	2.04	0.25	0.54	0.46	0.49	1.37
0.19	2.41	0.31	0.46	0.00	0.54	1.25
0.22	0.54	0.33	0.00	0.79	1.09	1.09
0.09	1.25	0.18	0.21	0.23	2.10	3.09
0.72	0.20	1.01	0.00	0.00	1.56	1.56
0.27	0.34	0.45	0.59	0.49	0.62	3.33
0.44	3.47	0.54	0.30	0.00	0.46	1.00
0.10	0.64	0.12	0.74	0.07	0.18	1.53

TABLE V7. Air Pollution Data

CITY	TMR	SMIN	SMEAN	SMAX	PMIN	PMEAN	PMAX
Providence	1096	30	163	349	56	119	223
Jackson	789	29	70	161	27	74	124
Johnstown	1072	88	123	245	70	166	452
Jersey City	1199	155	229	340	63	147	253
Huntington	967	60	70	137	56	122	219
Des Moines	950	31	88	188	61	183	329
Denver	841	2	61	188	54	126	229
Reading	1113	50	94	186	34	120	242
Toledo	1031	67	86	309	52	104	193
Fresno	845	18	34	198	45	119	304
Memphis	873	35	48	69	46	102	201
York	957	120	162	488	28	147	408
Milwaukee	921	65	134	236	49	150	299
Savannah	990	49	71	120	46	82	192
Omaha	922	20	74	148	39	107	198
Topeka	904	19	37	91	52	101	158
Columbus	877	94	161	276	74	119	190
Beaumont	728	27	71	144	32	76	190
Winston	802	28	58	128	72	147	306
Detroit	817	52	128	260	59	146	235
El Paso	618	47	87	207	49	150	373
Macon	869	18	27	128	22	122	754
Rockford	842	33	66	210	36	86	143
Jackson	928	41	52	138	39	77	124
Fall River	1157	62	79	136	18	102	254
Boston	1112	42	163	337	55	141	252
Dayton	847	18	106	241	50	132	327
Charlotte	791	43	81	147	62	124	234
Miami	897	44	57	68	33	54	124
Bridgeport	938	137	205	308	32	91	182
Sioux Falls	795	18	55	121	25	108	358
Chicago	1000	75	166	328	88	182	296
South Bend	888	73	77	261	28	90	164
Norfolk	803	49	112	198	39	89	242
Cleveland	969	69	160	282	86	174	336
Austin	689	40	46	58	10	78	157
Knoxville	825	56	77	157	28	135	302
Indianapolis	969	50	139	269	92	178	275
Nashville	919	54	160	362	45	130	310
Seattle	938	1	47	179	32	69	141

TABLE V7. Air Pollution Data (continued)

CITY	PM2	PERWH	NONPOOR	GE65	LPOP
Providemce	116.1	97.9	83.9	109	58.5645
Jackson	21.3	60.0	69.1	64	52.7195
Johnstown	15.8	98.7	73.3	103	54.4829
Jersey City	1357.2	93.1	87.3	103	57.8585
Huntington	18.1	97.0	73.2	93	54.0617
Des Moines	44.8	95.9	87.1	97	54.2540
Denver	25.4	95.8	86.9	82	59.6819
Reading	31.9	98.2	86.1	112	54.3999
Toledo	133.2	90.5	86.1	98	56.5985
Fresno	6.1	92.5	78.5	81	55.6342
Memphis	83.5	63.6	72.5	73	57.9728
York	26.2	97.7	84.8	97	53.7719
Milwaukee	150.2	94.4	90.4	88	60.7711
Savannah	42.7	65.9	72.0	65	52.7485
Omaha	29.9	94.0	86.4	90	56.6075
Topeka	25.9	92.7	84.1	99	51.5010
Columbus	127.2	88.1	86.3	79	58.3440
Beaumont	23.5	79.3	79.9	58	54.8574
Winston	44.7	75.8	79.9	62	52.7744
Detroit	191.5	84.9	86.5	72	65.7546
El Paso	29.8	96.7	77.9	45	54.9703
Macon	28.6	69.0	73.7	62	52.5624
Rockford	40.3	95.8	88.2	85	53.2173
Jackson	18.7	94.3	86.5	90	51.2055
Fall River	71.7	98.7	82.9	116	56.0042
Boston	174.5	97.3	88.5	109	64.9275
Dayton	53.9	89.8	87.1	74	58.4175
Charlotte	50.2	75.4	79.5	57	54.3475
Miami	45.5	85.1	77.2	100	59.7083
Bridgeport	103.3	94.7	90.7	94	58.1530
Sioux Falls	10.6	99.3	82.4	92	49.3739
Chicago	167.5	85.2	89.4	86	67.9385
South Bend	51.1	94.0	88.4	84	53.7770
Norfolk	86.7	73.6	73.1	53	57.6231
Cleveland	261.1	85.5	88.6	89	62.5445
Austin	20.9	87.2	75.2	76	53.2661
Knoxville	25.8	92.5	72.5	74	55.6594
Indianapolis	173.5	85.6	87.2	85	58.4359
Nashville	75.1	80.8	76.5	79	56.0178
Seattle	26.2	95.2	88.8	96	60.4423

TABLE V7. Air Pollution Data (continued)

CITY	TMR	SMIN	SMEAN	SMAX	PMIN	PMEAN	PMAX
Dallas	757	31	69	148	22	96	230
Mobile	823	47	67	248	29	129	284
Phoenix	758	15	86	266	98	247	573
Augusta	823	31	46	158	28	66	142
Youngstown	915	75	145	263	58	148	371
Chattanooga	940	10	105	191	69	186	361
Galveston	873	62	72	86	23	55	125
Fort Worth	789	9	32	73	28	79	152
Flint	747	64	80	229	49	124	468
Charleston	780	15	283	940	55	225	958
New Haven	983	39	124	288	42	88	248
Portland	1146	42	140	287	50	82	147
St. Louis	1004	60	182	299	63	168	295
Atlantic City	1338	54	75	110	25	71	118
New Orleans	1027	49	96	187	62	87	117
Las Vegas	727	31	79	201	50	145	389
Little River	910	19	61	175	29	72	147
San Francisco	925	16	62	202	32	70	183
Raleigh	801	34	49	97	39	80	132
Oklahoma City	812	18	26	63	42	84	173
Worcester	1082	71	90	204	32	99	229
Gary	796	113	190	290	56	170	420
Pittsburgh	1031	55	150	345	43	166	475
Waco	914	20	28	88	10	76	156
Manchester	1102	35	76	129	32	66	156
Terre Haute	1294	63	135	214	53	118	203
Allentown	1059	129	146	305	60	135	261
Richmond	974	49	115	214	26	82	206
Houston	716	18	65	171	26	117	385
Newark	1017	54	131	297	42	113	232
Birmingham	943	55	145	341	38	146	400
Shreveport	891	84	88	272	47	104	197
Columbia	727	11	84	167	55	112	274
Brockton	1139	46	142	332	30	79	165
Tampa	1259	60	105	197	48	94	233
Lansing	799	46	57	226	34	76	160
Kansas City	969	38	141	350	70	142	343
Buffalo	1012	59	114	193	23	131	978
San Bernadino	901	18	131	282	39	140	255
Spokane	1045	14	36	66	19	101	409

TABLE V7. Air Pollution Data (continued)

CITY	PM2	PERWH	NONPOOR	GE65	LPOP
Dallas	29.7	85.4	81.4	71	60.3487
Mobile	25.3	67.7	74.6	57	54.9735
Phoenix	7.2	94.5	80.9	72	58.2185
Augusta	15.2	70.2	67.8	60	53.3574
Youngstown	49.0	90.8	87.1	89	57.0672
Chattanooga	27.7	82.4	74.0	77	54.5205
Galveston	32.7	78.6	76.8	64	51.4726
Fort Worth	35.8	89.3	80.7	73	57.5832
Flint	58.2	90.1	87.8	62	55.7323
Charleston	27.9	94.2	78.6	70	54.0299
New Haven	108.3	94.7	89.9	99	58.1975
Portland	21.4	99.5	83.1	114	52.6186
St. Louis	64.6	85.5	84.9	93	63.1389
Atlantic City	28.0	82.3	77.0	140	52.0650
New Orleans	77.7	69.0	75.7	73	59.3876
Las Vegas	1.6	90.5	88.7	45	51.0386
Little River	31.7	78.5	74.0	86	53.8557
San Francisco	84.0	87.5	88.1	90	64.4457
Raleigh	19.6	73.9	70.9	65	52.2810
Oklahoma City	24.0	90.6	81.4	79	57.0913
Worcester	38.5	99.3	87.2	116	57.6584
Gary	61.1	84.7	89.1	62	57.5857
Pittsburgh	78.8	93.2	85.3	95	63.8119
Waco	14.5	83.9	70.2	98	51.7635
Manchester	20.0	99.7	86.8	109	52.5081
Terre Haute	26.1	95.5	78.1	123	50.3526
Allentown	45.5	99.2	86.4	104	56.9211
Richmond	56.3	73.6	83.2	82	56.1119
Houston	72.7	79.9	81.9	54	60.9453
Newark	242.0	86.6	89.6	94	62.2774
Birmingham	56.8	65.4	74.2	77	58.0268
Shreveport	16.3	65.9	69.6	74	54.4945
Columbia	17.9	71.0	69.2	59	54.1635
Brockton	37.4	98.0	87.7	113	53.9524
Tampa	59.2	88.5	70.0	171	58.8787
Lansing	17.6	97.2	85.6	78	54.7560
Kansas City	63.3	88.6	85.9	92	60.1682
Buffalo	84.6	91.5	93.2	99	61.2872
San Bernadino	4.0	94.3	89.7	109	60.4525
Spokane	16.3	97.7	91.4	111	54.5723

TABLE V8. Shopping Attitude Data Part I

WORK	AGE	A1	A2	A3	A4	A5	A6	A7
1	2	2	5	2	2	4	4	4
1	4	3	2	1	5	1	1	3
0	1	1	3	3	3	5	5	4
0	1	1	3	5	2	5	4	4
1	5	2	1	1	5	1	3	3
1	2	3	1	2	3	4	3	5
0	3	1	1	4	4	3	4	4
1	5	4	1	1	5	3	3	3
1	5	2	1	1	5	3	1	4
1	6	5	2	3	5	3	3	3
0	1	1	3	4	1	5	3	5
1	4	3	1	3	5	2	3	3
0	1	1	2	4	1	5	4	5
0	5	1	1	3	5	1	4	5
1	5	2	1	1	5	3	2	4
1	5	3	1	3	3	3	3	4
0	3	1	1	2	3	3	3	2
1	2	3	2	3	2	3	3	4
1	2	2	4	4	4	3	2	2
0	4	2	1	5	4	5	4	4
0	2	1	4	5	3	3	3	5
0	1	2	2	4	3	4	5	5
0	1	1	3	3	1	5	3	5
0	3	2	1	2	2	2	3	4
0	1	1	4	5	3	4	5	5
0	2	1	5	5	1	5	3	5
1	5	3	1	3	5	1	3	3
1	1	1	4	2	2	3	4	5
1	3	3	1	1	3	3	1	1
0	2	1	5	4	3	5	5	5
1	3	3	1	1	3	1	3	3
1	5	4	2	1	4	2	2	3
0	1	1	4	4	1	5	5	5
0	2	1	3	5	3	4	5	5
0	3	2	2	4	2	3	3	4
1	6	2	1	2	3	1	2	2
1	4	3	1	2	3	3	2	2
0	5	1	2	2	2	3	3	4
1	5	2	1	1	3	1	1	3
1	6	3	1	4	5	3	1	3
1	2	1	3	2	4	3	2	4
1	6	4	2	2	4	2	2	3
1	3	3	1	3	5	3	1	1
0	3	2	4	4	5	2	2	4
0	2	1	3	5	1	4	5	5
0	2	1	3	5	3	3	5	5
0	2	1	5	5	3	5	5	5
0	1	1	4	3	1	5	5	5
1	5	5	1	3	5	1	3	3
0	5	1	3	4	2	3	4	4

TABLE V8. Shopping Attitude Data Part I (continued)

WORK	AGE	A1	A2	A3	A4	A5	A6	A7
1	3	2	1	3	5	1	3	3
0	3	1	1	3	2	2	4	4
1	5	5	4	1	5	2	1	3
1	2	3	1	2	3	3	4	4
0	3	2	3	2	4	2	4	4
0	2	1	5	4	3	5	3	5
1	4	4	1	1	4	3	1	3
1	4	3	1	2	4	3	2	3
1	6	2	4	1	4	1	2	3
0	1	1	5	5	1	3	4	5
0	4	2	2	2	2	1	2	4
1	6	2	3	3	4	1	1	3
0	2	1	3	5	3	3	5	4
0	2	1	2	3	2	3	3	5
0	1	1	4	5	2	4	4	5
1	1	2	3	5	3	5	4	3
0	4	1	4	5	4	4	4	3
1	6	4	2	3	3	1	1	3
0	2	3	5	3	1	5	4	5
1	2	2	2	4	2	4	4	4
0	2	1	2	3	2	4	5	4
0	2	1	2	3	3	5	4	5
0	2	1	3	2	1	2	2	5
0	2	1	5	2	1	2	4	5
1	6	3	1	1	5	3	1	3
0	3	2	2	2	2	3	3	4
0	5	2	2	2	4	4	4	4
1	5	2	1	3	5	1	3	3
1	4	3	2	1	5	1	1	3
0	1	1	4	3	2	5	5	5
1	4	2	1	1	3	1	3	3
1	5	3	1	1	5	1	1	3
1	6	3	1	1	3	3	1	2
1	5	3	1	1	3	3	4	1
1	1	2	2	2	4	4	3	4
0	2	1	3	3	1	4	3	5
1	1	1	4	3	1	4	4	4
1	2	2	3	2	4	2	2	4
1	3	2	1	1	5	1	2	3
0	2	1	4	5	1	4	4	5
1	6	3	1	3	4	1	4	3
1	1	2	2	3	2	2	4	3
0	3	4	2	4	2	2	4	4
1	2	1	2	4	4	2	4	2
1	5	5	1	1	3	2	3	3
1	2	2	2	3	3	3	3	4
0	3	3	4	4	4	3	3	4
1	5	3	1	1	3	3	2	2
1	6	3	3	3	3	1	3	3
0	2	1	5	3	2	3	3	4

TABLE V8. Shopping Attitude Data Part I (continued)

WORK	AGE	A1	A2	A3	A4	A5	A6	A7
1	2	3	3	3	2	3	4	5
1	1	1	5	4	4	4	3	3
1	4	3	1	1	5	3	3	1
0	2	1	3	5	1	3	2	5
1	6	3	2	1	4	1	2	1
0	5	2	5	2	5	3	5	4
0	3	1	2	3	4	3	4	4
0	2	1	5	5	1	5	5	5
1	2	1	3	1	1	5	3	4
1	2	2	2	4	4	4	4	4
1	6	3	1	1	3	1	1	3
1	6	5	1	1	3	3	1	3
0	5	4	1	2	3	2	3	3
1	3	2	1	2	5	3	2	3
1	4	3	1	3	3	3	1	2
1	2	2	4	2	4	4	3	3
1	3	2	1	1	2	3	1	3
1	6	2	2	1	5	1	2	3
0	5	2	2	3	2	2	4	5
1	1	2	3	2	2	3	2	5
1	1	2	4	2	2	4	3	4
0	4	1	4	2	2	2	4	4
1	1	1	4	4	2	3	4	4
1	2	1	2	3	4	4	3	4
0	4	1	2	2	2	4	4	4
1	3	2	1	1	5	3	3	3
0	4	1	5	3	2	3	3	5
0	1	1	5	3	1	5	5	5
1	6	3	1	4	5	2	3	4
1	5	2	1	1	4	1	2	4
1	6	4	1	1	3	1	1	2
1	5	3	1	3	5	3	3	3
1	5	3	3	2	3	3	2	3
1	4	3	1	3	5	1	3	3
0	5	1	1	5	5	1	1	5
0	5	2	1	2	4	4	4	4
0	3	2	2	5	2	4	2	4
0	1	1	3	3	1	5	5	5
1	5	2	1	1	5	2	1	3
1	6	2	1	1	4	1	3	3
1	3	2	2	1	5	2	3	3
1	2	2	2	2	3	3	4	2
1	5	3	1	2	5	3	2	3
0	2	1	5	4	1	5	3	5
0	4	1	3	1	1	4	4	1
0	3	3	3	5	2	2	2	4
0	5	1	2	2	2	3	2	4
1	2	2	2	3	2	4	4	4
1	3	2	1	4	3	4	3	4
0	3	1	1	1	1	4	1	4

TABLE V8. Shopping Attitude Data Part I (continued)

WORK	AGE	A1	A2	A3	A4	A5	A6	A7
1	5	3	1	3	5	3	1	3
1	1	1	2	3	2	2	2	4
0	3	1	3	1	2	3	3	4
0	4	2	2	4	2	1	4	5
0	1	1	5	5	1	3	5	5
0	6	2	2	2	2	4	4	4
0	1	1	5	5	1	5	5	5
0	4	3	2	3	3	4	3	4
1	6	5	3	1	3	1	1	4
0	1	1	5	5	1	3	5	5
1	6	5	1	1	5	3	1	3
0	1	2	5	3	1	5	5	5
1	6	3	1	3	4	1	1	3
0	2	1	2	4	3	4	2	5
0	2	1	3	3	2	4	3	3
0	2	1	3	4	1	3	4	5
0	5	1	1	1	2	3	1	4
1	6	4	1	1	3	2	1	2
1	1	2	3	4	2	4	3	2
0	5	3	1	4	3	2	4	5
0	4	2	2	2	2	2	2	4
0	2	1	3	5	3	4	4	5
1	6	2	1	3	3	2	3	1
1	6	5	1	1	5	1	1	3
0	4	1	1	4	5	3	4	4
1	6	2	1	2	5	1	3	3
1	2	1	3	3	2	2	4	2
0	2	1	5	3	1	5	5	5
0	1	1	3	4	1	4	5	5
0	2	1	3	3	1	4	3	5
0	5	2	5	2	4	2	3	4
1	2	1	2	4	4	3	4	5
1	6	3	1	1	3	2	1	3
0	2	1	5	3	3	5	5	5
0	2	1	3	5	3	3	5	5
0	1	1	3	5	1	5	5	5
1	3	2	2	3	5	3	2	3
0	1	1	3	4	1	5	5	5
0	5	1	1	4	3	4	2	5
0	2	2	5	3	2	3	3	4
1	6	3	1	1	5	1	1	3
1	4	3	1	3	5	3	3	3
0	5	2	2	4	2	3	3	4
0	1	1	4	5	2	5	4	4
0	2	1	5	3	1	5	5	5
1	5	3	3	2	3	1	1	3
0	4	2	2	5	4	2	4	4
1	6	3	1	2	3	2	1	3
0	2	1	5	4	1	4	4	4
1	6	2	1	1	5	1	3	4

TABLE V9. R.C.M.P. Officer Data

DET	SATF1	SATF2	SATF3	SATF4
1	3.8	4.0	4.0	4.0
1	3.0	3.5	5.0	4.0
1	3.6	4.5	3.0	3.0
1	3.2	2.5	4.0	5.0
1	3.8	3.0	5.0	3.0
1	4.4	4.5	5.0	5.0
2	3.8	4.5	5.0	4.0
2	3.2	2.5	4.0	4.0
2	3.8	3.5	5.0	5.0
2	3.8	4.0	4.0	4.0
2	4.4	3.5	4.0	4.0
2	4.2	2.5	5.0	5.0
2	3.6	3.5	4.0	5.0
2	4.4	4.0	5.0	4.0
2	4.6	4.0	5.0	5.0
2	4.2	3.5	5.0	5.0
2	4.0	4.0	5.0	5.0
2	3.6	4.0	4.0	4.0
2	4.2	4.0	5.0	5.0
3	3.2	3.0	4.0	4.0
3	3.6	4.5	5.0	5.0
3	3.6	4.5	5.0	4.0
3	3.8	4.0	5.0	4.0
3	3.8	4.0	5.0	5.0
3	3.4	4.0	4.0	5.0
3	2.8	3.5	5.0	4.0
3	3.6	3.5	5.0	4.0
4	3.0	4.5	5.0	2.0
4	2.8	3.0	4.0	3.0
4	2.8	4.0	3.0	5.0
4	4.0	3.0	5.0	4.0
4	2.4	2.0	5.0	4.0
4	4.4	3.0	5.0	4.0
4	3.2	5.0	5.0	5.0
4	3.4	4.0	3.0	5.0
4	4.6	2.0	3.0	4.0
4	4.0	2.5	4.0	4.0
4	4.6	3.5	5.0	3.0
4	4.6	4.0	4.0	5.0
4	2.4	4.0	3.0	4.0
4	4.0	4.0	5.0	4.0
4	3.2	2.0	3.0	5.0
4	4.4	4.0	4.0	4.0
4	3.6	4.0	5.0	5.0
4	4.2	3.5	4.0	4.0
4	4.0	3.0	5.0	4.0
5	3.8	3.0	4.0	3.0
5	2.8	3.0	1.0	4.0
5	3.6	5.0	4.0	4.0
5	4.0	3.5	2.0	5.0

TABLE V9. R.C.M.P. Officer Data (continued)

DET	SATF1	SATF2	SATF3	SATF4
5	2.2	4.0	1.0	4.0
5	4.0	4.0	4.0	4.0
5	3.4	4.0	2.0	4.0
5	2.6	4.0	1.0	5.0
6	3.8	4.0	2.0	4.0
6	3.2	4.0	3.0	4.0
6	3.6	4.0	2.0	5.0
6	3.8	4.0	3.0	5.0
6	5.0	4.0	2.0	4.0
6	4.2	4.0	3.0	4.0
7	2.2	2.5	5.0	5.0
7	3.8	3.5	5.0	5.0
7	3.6	3.0	5.0	5.0
7	4.0	4.0	5.0	3.0
7	4.2	4.0	5.0	5.0
7	4.4	4.0	5.0	5.0
7	3.4	3.5	4.0	4.0
7	3.4	2.0	3.0	4.0
7	4.0	4.0	2.0	3.0
8	3.0	3.0	4.0	3.0
8	2.2	3.0	1.0	4.0
8	2.8	2.0	2.0	3.0
8	2.8	5.0	3.0	5.0
8	3.6	3.5	4.0	3.0
9	3.8	4.0	4.0	4.0
9	3.8	4.0	5.0	4.0
9	3.4	4.0	4.0	4.0
9	4.4	4.0	4.0	4.0
10	3.6	4.0	4.0	4.0
10	3.8	4.0	4.0	4.0
10	3.2	4.0	3.0	4.0
10	4.0	4.0	5.0	5.0
10	3.2	3.5	5.0	3.0
10	2.2	3.5	2.0	5.0
10	4.8	4.5	5.0	5.0
10	3.6	3.5	4.0	5.0
10	3.6	1.5	4.0	4.0
10	4.2	5.0	2.0	4.0
10	3.4	4.5	5.0	4.0
10	3.4	4.0	5.0	5.0
10	3.8	4.0	4.0	4.0
10	3.8	4.0	4.0	3.0
10	3.6	3.5	4.0	4.0
10	3.8	4.0	4.0	4.0
10	4.0	4.0	5.0	5.0
10	4.6	2.5	5.0	3.0

TABLE V10. Mystery Data

SEXED	C1	C3	C8	C9	C10
7	12	11	16	20	17
5	19	10	11	4	11
3	15	15	11	12	13
8	14	11	20	20	9
4	12	12	12	12	1
7	12	12	11	15	13
6	12	18	18	11	19
8	12	19	14	19	12
8	16	18	15	20	2
1	13	19	18	17	14
8	11	3	16	19	15
8	11	10	10	18	10
7	12	9	17	16	16
5	12	11	11	11	11
6	11	17	17	18	17
7	11	11	11	11	11
8	12	13	20	12	13
6	12	18	18	18	17
7	11	20	20	17	8
8	11	14	13	14	11
6	16	15	17	17	14
6	11	18	20	18	13
4	11	17	14	18	11
8	11	17	17	14	11
5	11	18	18	11	3
6	12	12	12	12	20
6	4	20	11	1	20
5	12	13	12	13	13
8	11	10	17	17	9
7	12	19	19	11	19
8	8	12	18	18	15
7	14	3	18	13	10
6	4	8	15	20	11
8	12	12	13	13	13
7	11	11	11	11	14
3	11	11	8	18	8
7	7	20	12	20	12
4	12	19	17	14	10
8	12	12	14	15	11
7	12	13	17	20	11
7	18	6	19	19	19
8	9	4	8	14	11
8	11	16	18	14	10
3	16	15	16	17	16
3	17	16	11	15	12
6	11	12	14	15	15
1	11	17	10	14	10
5	10	9	13	11	11
4	13	17	15	15	8
5	19	19	4	4	19

TABLE V10. Mystery Data (continued)

SEXED	C1	C3	C8	C9	C10
7	19	12	20	20	20
7	12	15	14	14	14
7	19	12	11	11	11
7	12	15	14	15	9
8	10	11	13	17	10
2	11	20	20	17	1
7	12	19	19	11	11
1	11	7	13	7	11
4	14	18	19	14	6
8	11	15	15	17	16
5	16	15	7	9	5
2	19	11	18	5	12
2	11	18	17	11	10
5	12	10	15	7	10
7	12	14	18	16	14
8	11	7	3	1	3
7	12	17	15	16	18
7	16	11	13	19	16
7	11	17	16	11	18
6	11	12	12	12	12
6	11	19	19	11	11
7	11	20	18	20	20
7	11	19	18	18	19
5	13	8	14	16	14
4	11	18	8	17	8
8	11	11	10	10	10
7	12	15	15	15	12
3	18	16	15	15	10
8	12	11	20	19	11
4	17	16	18	12	13
3	12	19	12	12	13
8	5	0	11	19	18
8	12	11	19	19	10
3	15	16	17	20	16
7	17	16	12	11	15
7	12	19	13	13	13
8	11	3	10	10	15
6	11	17	20	11	17
8	12	17	18	19	10
3	12	19	20	20	8
8	11	16	15	3	7
7	11	20	20	20	10
6	11	16	14	18	10
2	11	10	18	18	20
3	12	9	13	14	15
8	11	15	19	15	18
8	11	13	14	15	13
5	11	6	15	10	11
3	11	15	10	17	12
3	9	17	7	17	16

TABLE V11. Bank Employee Data Part I

LCURRENT	LSTART	SEX	JOBCAT	RACE	EDUC	SENIOR	AGE	EXPER
9.6853	9.0359	0	4	0	16	81	28.50	0.25
10.2524	9.7642	0	5	0	19	83	41.92	13.00
10.0345	9.2873	0	5	0	15	98	41.17	12.00
9.4174	8.6376	0	3	0	12	94	46.25	20.00
9.9282	9.3049	0	5	0	16	83	35.17	5.75
10.2128	9.4575	0	4	0	19	78	30.08	2.92
9.1050	8.7948	0	2	0	12	70	44.50	18.00
9.3038	8.6793	0	2	0	12	70	27.83	3.42
9.6816	9.2301	0	5	0	15	78	35.42	11.08
9.9965	9.3049	0	4	0	16	90	34.33	5.67
10.0858	9.1582	0	4	0	16	94	34.00	4.92
9.6473	9.1582	0	5	0	15	78	38.92	14.67
9.5359	9.3147	0	1	0	16	82	44.42	12.42
10.3735	8.7483	0	2	0	15	82	29.50	2.83
9.3927	8.7483	0	3	0	12	69	63.58	29.00
9.5042	8.7483	0	2	0	12	66	27.42	3.92
9.4556	8.6995	0	3	0	8	85	58.08	36.50
10.0962	9.1049	0	4	0	16	91	33.75	3.67
9.1378	8.6482	0	2	0	12	95	29.33	3.83
10.2921	9.2591	0	4	0	18	94	39.67	4.67
10.1659	9.6158	0	5	0	21	88	56.67	22.00
9.3414	8.7483	0	3	0	8	69	59.42	14.50
9.9035	9.4724	0	5	0	16	77	48.33	22.00
9.4556	8.7483	0	3	0	8	74	63.50	34.00
9.3927	8.7483	0	3	0	12	79	54.17	25.67
9.5539	8.7483	0	3	0	12	78	59.83	32.25
9.8119	9.1377	0	4	0	16	86	32.25	5.00
10.1166	9.5956	0	5	0	19	80	42.58	16.58
10.1266	9.5462	0	5	0	18	75	35.42	6.17
10.5133	9.6800	0	5	0	19	96	44.92	14.58
9.4174	8.5941	0	3	0	12	95	56.00	25.58
10.6334	9.3060	0	5	0	16	97	37.08	5.83
9.4174	8.6995	0	3	0	8	87	64.25	37.58
10.2400	9.5479	0	5	0	19	67	36.92	6.25
10.0266	9.2103	0	5	0	16	93	33.42	2.83
9.5750	8.9226	0	1	0	15	98	37.17	9.50
10.1166	9.3497	0	5	0	19	65	28.42	2.17
9.0572	8.4763	1	1	0	12	98	64.50	31.75
9.0711	8.4118	1	2	0	12	65	23.00	0.00
9.0848	8.4763	1	1	0	12	76	40.17	0.50
9.3927	8.7948	1	1	0	12	95	55.25	19.00
9.1442	8.5941	1	2	0	12	64	45.50	16.50
9.4415	8.9226	1	2	0	12	90	61.67	17.08
9.3673	8.4381	1	2	0	15	93	26.83	0.92
9.1757	8.5370	1	1	0	8	66	46.17	8.00
9.3147	8.9618	1	2	0	15	82	58.75	22.08
8.7578	8.3138	1	1	0	8	74	55.25	3.58
9.2183	8.6586	1	1	0	12	91	33.50	6.92
9.0642	8.5941	1	1	0	8	66	64.25	19.00
10.0301	9.3926	1	5	0	16	73	31.92	1.25

TABLE V11. Bank Employee Data Part I (continued)

LCURRENT	LSTART	SEX	JOBCAT	RACE	EDUC	SENIOR	AGE	EXPER
9.1050	8.3848	1	2	0	8	81	24.33	0.42
9.0070	8.4118	1	2	0	12	69	23.42	0.00
9.2591	8.4381	1	1	0	12	97	53.92	4.00
9.3622	8.5370	1	2	0	17	73	55.58	31.25
9.7527	8.8530	1	4	0	16	79	28.42	1.67
9.6238	8.7483	1	4	0	16	90	29.92	0.58
9.0780	8.2687	1	1	0	12	86	52.00	13.00
8.9847	8.3138	1	1	0	12	72	52.17	4.67
8.8901	8.2687	1	1	0	12	86	62.00	6.00
9.5298	8.9618	1	4	0	16	65	30.75	6.58
9.8389	8.8392	1	4	0	16	86	32.00	1.58
9.0216	8.6482	1	1	0	15	65	59.50	20.08
10.0123	8.9862	1	4	0	16	64	29.00	3.00
9.4970	8.6995	1	2	0	12	90	44.50	0.25
8.9771	8.3138	1	1	0	8	83	59.08	6.25
9.0288	8.5941	1	2	0	12	67	51.50	15.08
9.7642	8.9618	1	4	0	16	68	27.58	0.92
8.8217	8.3138	1	1	0	8	72	56.92	26.58
9.3361	8.4763	0	2	1	8	98	27.83	2.17
9.2301	8.6586	0	2	1	12	86	28.67	3.08
9.5324	8.7483	0	3	1	8	79	48.50	20.50
9.4174	8.6995	0	3	1	8	92	60.67	36.00
9.4026	8.7483	0	3	1	8	68	32.92	12.92
9.3775	8.7483	0	3	1	8	76	35.25	12.00
9.4076	8.6995	0	1	1	12	91	30.33	4.08
10.0758	9.5104	0	5	1	12	74	48.25	22.67
9.4319	8.7483	0	3	1	12	73	38.67	12.92
9.2301	8.4250	0	2	1	12	81	24.75	0.75
10.3546	9.3497	0	4	1	17	70	32.08	5.58
9.4222	8.6995	0	2	1	12	94	34.58	8.50
9.4174	8.1886	0	3	1	12	91	53.50	26.17
10.5966	9.7779	0	5	1	16	66	35.33	10.67
9.3927	8.7483	0	3	1	8	67	47.25	25.42
9.1569	8.5941	0	1	1	15	88	57.50	29.92
9.3201	8.6995	0	2	1	14	79	30.33	3.92
9.1819	8.6995	0	3	1	15	90	42.17	15.92
9.3927	8.6995	0	3	1	8	92	37.83	12.00
9.4174	8.6995	0	3	1	8	83	48.83	25.17
9.1050	8.2687	0	2	1	8	94	29.17	3.00
9.2418	8.5941	1	1	1	15	78	48.67	4.25
9.1881	8.6482	1	1	1	12	74	45.17	9.75
8.8217	8.1886	1	1	1	12	97	60.67	10.33
8.8305	8.3138	1	1	1	8	81	51.50	0.00
9.1313	8.4763	1	1	1	16	84	47.58	17.83
9.3876	8.6995	1	1	1	15	86	40.50	6.58
9.0848	8.3848	1	2	1	12	92	25.50	0.42
9.0143	8.4118	1	2	1	12	73	47.92	12.83
8.9464	8.1886	1	1	1	15	96	60.50	1.92
9.3038	8.3138	1	1	1	12	85	54.17	8.42
9.1881	8.3848	1	2	1	12	75	27.58	2.67

TABLE V12. Bank Employee Data Part II

LCURRENT	LSTART	SEX	JOBCAT	RACE	EDUC	SENIOR	AGE	EXPER
10.6310	10.0858	0	5	0	16	73	40.33	12.50
9.9970	9.2301	0	5	0	15	83	31.08	4.08
9.8627	9.0711	0	4	0	16	93	31.17	1.83
10.2128	9.4724	0	4	0	18	80	29.50	2.42
9.4174	8.7483	0	3	0	12	77	52.92	26.42
9.9988	9.2591	0	4	0	17	93	32.33	2.67
9.8532	9.4880	0	5	0	19	64	31.92	2.25
10.0078	9.5104	0	4	0	19	81	30.75	5.17
10.2219	9.3060	0	4	0	17	89	34.17	3.17
10.2400	9.0711	0	4	0	16	65	28.00	1.58
9.9988	9.5462	0	5	0	19	65	39.75	10.75
10.2036	9.5462	0	5	0	17	83	30.17	0.75
10.3514	9.3049	0	4	0	18	91	30.17	3.92
9.9988	9.4724	0	5	0	18	75	41.17	10.42
10.2887	9.4880	0	4	0	19	78	32.92	3.75
9.4174	8.7483	0	3	0	8	78	63.75	35.75
9.9641	9.0360	0	4	0	16	93	30.67	4.00
9.4174	8.6995	0	3	0	8	84	63.42	31.67
10.1064	9.4880	0	5	0	19	68	29.50	0.75
10.1811	9.8520	0	5	0	16	86	42.42	12.50
10.3498	9.3927	0	4	0	19	93	31.67	0.58
10.0753	9.3927	0	5	0	20	89	35.58	0.50
10.0648	9.0711	0	4	0	16	84	30.25	1.08
10.1659	9.4174	0	4	0	17	78	29.75	2.17
9.4125	8.7483	0	3	0	12	80	61.67	38.33
9.8782	9.0821	0	4	0	16	76	32.67	5.08
9.8201	9.0474	0	4	0	16	93	29.75	2.92
10.1659	9.4566	0	4	0	19	69	28.83	6.17
9.9451	9.3049	0	5	0	19	80	45.67	18.42
9.9321	9.5816	0	5	0	18	78	39.42	12.42
10.5662	9.5471	0	5	0	19	91	34.33	5.67
9.5104	8.6995	0	3	0	12	83	50.25	23.67
9.5324	8.4888	1	2	0	12	77	24.33	0.33
9.5411	8.6691	1	1	0	16	93	31.50	0.67
9.0288	8.4764	1	1	0	8	79	50.17	5.83
9.1695	8.5942	1	1	0	12	98	47.33	20.33
9.1942	8.2990	1	1	0	12	92	44.00	3.67
9.3308	8.5370	1	2	0	15	81	27.17	1.58
9.0711	8.5132	1	1	0	8	74	59.83	26.50
9.3725	8.7193	1	2	0	12	72	25.75	2.50
9.1695	8.4118	1	2	0	12	83	25.83	1.33
9.3725	8.6482	1	2	0	12	63	25.08	0.75
9.1942	8.5942	1	2	0	12	78	27.17	3.92
9.6512	8.6995	1	1	0	16	97	30.58	1.42
9.2648	8.5942	1	2	0	15	94	29.50	0.25
9.1881	8.5717	1	1	0	12	88	54.42	8.92
8.9227	8.3428	1	1	0	15	90	58.00	4.50
9.6037	8.8818	1	2	0	15	75	28.75	0.42
9.6198	8.7483	1	1	0	12	73	54.08	11.00
9.1881	8.4764	1	2	0	12	79	24.33	0.67

TABLE V12. Bank Employee Data Part II (continued)

LCURRENT	LSTART	SEX	JOBCAT	RACE	EDUC	SENIOR	AGE	EXPER
9.3414	8.3428	1	2	0	12	92	25.50	0.50
9.2003	8.3848	1	2	0	12	79	24.67	0.42
9.4742	8.5942	1	2	0	15	90	28.75	1.83
9.9233	8.8818	1	1	0	16	93	32.50	1.83
9.1182	8.4118	1	2	0	12	68	23.42	0.17
9.6434	8.7948	1	2	0	16	90	30.42	0.00
9.1632	8.5370	1	2	0	12	69	24.42	1.67
9.0288	8.4763	1	1	0	12	87	53.92	13.58
8.8901	8.3848	1	2	0	12	80	25.00	0.00
8.7948	8.3138	1	1	0	12	84	62.42	24.00
9.4026	8.7948	1	2	0	16	98	43.92	11.92
9.5148	8.4763	1	1	0	12	82	28.17	0.92
9.2360	8.6482	1	1	0	15	82	30.17	4.25
9.3200	8.5941	1	2	0	12	73	27.33	2.67
9.1819	8.3848	1	1	0	12	93	25.25	0.42
9.3518	8.6269	1	2	0	12	72	27.33	1.50
9.2591	8.3848	1	1	0	12	89	25.83	0.00
9.1182	8.4118	1	2	0	12	69	23.67	0.00
8.8038	8.2687	1	1	0	8	88	62.50	34.33
9.0216	8.4118	1	2	0	12	69	23.67	0.17
9.2183	8.6269	1	2	0	15	64	29.08	4.75
9.0431	8.4763	1	2	0	8	73	60.50	13.25
8.8392	8.3138	1	1	0	12	82	53.92	29.83
9.1313	8.4118	1	2	0	12	81	24.08	1.08
9.2761	8.5941	0	1	1	12	88	29.92	3.17
9.3622	8.5941	0	1	1	12	96	39.50	9.42
9.4970	8.8530	0	1	1	16	67	41.67	10.00
8.9695	8.5370	0	1	1	12	84	44.58	15.00
9.2591	8.7671	0	1	1	8	67	51.42	8.08
9.3254	8.7948	0	1	1	15	66	30.75	7.00
9.0710	8.5370	0	1	1	12	97	53.08	26.25
9.1049	8.6995	0	1	1	15	82	59.75	30.92
9.2873	8.8818	0	1	1	17	70	47.33	16.00
9.2704	8.6995	0	2	1	16	78	33.83	8.75
9.3038	8.6995	0	2	1	12	90	37.50	14.42
9.0982	8.6995	0	1	1	12	80	57.17	22.67
9.4556	8.7483	0	2	1	15	67	29.33	4.83
9.2761	8.7483	0	2	1	12	76	28.33	1.50
9.4366	8.6995	0	2	1	15	96	31.92	4.08
9.4173	8.6995	0	3	1	12	91	45.50	20.00
9.5539	8.7483	0	3	1	12	78	55.33	23.42
9.3147	8.6482	0	2	1	15	98	33.67	2.83
9.4173	8.6995	0	3	1	12	90	43.67	17.42
8.8479	8.3138	1	1	1	12	72	46.50	9.67
9.2928	8.7483	1	1	1	15	84	55.17	19.25
8.7856	8.3138	1	1	1	12	66	60.50	13.58
8.8217	8.3138	1	1	1	12	72	51.50	22.58
9.2534	8.5941	1	1	1	12	72	50.33	14.08
8.9695	8.6269	1	1	1	12	69	50.00	11.08
8.9065	8.3138	1	1	1	12	85	51.00	19.00

TABLE V13. Panel Data

THISYR	LASTYR	HUBINC	AGE	EDUC	BLACK	CHILD1	CHILD2
0	1	7.352	27	10	1	0	0
1	0	6.784	35	12	0	0	0
1	1	6.059	40	12	0	0	0
1	1	6.438	35	12	1	0	0
1	0	4.739	28	10	0	0	1
1	1	6.617	30	10	0	0	1
1	1	6.041	35	10	0	0	1
1	1	4.343	46	7	0	0	0
1	0	9.724	32	10	0	0	0
1	1	1.855	30	10	1	0	0
1	1	11.267	45	12	0	0	1
1	1	15.430	33	16	0	0	0
1	1	8.123	39	12	0	0	0
1	1	5.813	22	12	0	0	1
1	0	6.167	27	12	0	0	0
0	0	4.093	23	10	1	0	1
0	1	2.920	25	12	1	0	1
0	0	8.111	43	12	0	0	1
1	1	11.502	38	12	0	0	0
0	0	11.311	40	12	0	0	1
1	0	6.093	44	12	1	0	0
0	0	7.718	31	12	0	0	1
1	1	9.919	42	12	0	0	0
0	0	0.222	24	12	0	0	0
1	1	7.452	42	12	0	0	0
1	1	6.382	31	12	1	0	0
1	1	7.689	38	12	1	0	0
0	0	6.609	43	12	0	0	0
0	1	7.721	24	12	0	0	1
0	0	36.357	29	12	0	0	0
0	0	3.668	31	12	0	0	0
1	0	8.257	37	12	0	0	0
1	1	11.283	42	12	0	0	0
1	1	5.714	34	12	0	0	0
0	1	9.967	25	10	0	0	1
0	0	7.629	25	10	1	1	0
1	1	10.154	31	12	0	0	0
0	0	10.370	35	12	0	1	0
0	0	15.365	40	12	0	0	1
0	0	11.917	26	12	0	0	1
1	1	11.023	34	12	0	0	0
1	1	17.598	39	12	0	0	0
1	1	7.424	43	12	0	0	0
0	0	8.812	23	12	0	0	1
1	1	4.841	40	12	0	0	0
0	0	10.848	36	12	0	0	0
1	0	9.810	25	12	0	0	1
1	1	15.138	30	12	0	0	1
1	1	9.966	27	12	0	0	0
1	1	8.328	32	12	0	0	1

TABLE V13. Panel Data (continued)

THISYR	LASTYR	HUBINC	AGE	EDUC	BLACK	CHILD1	CHILD2
1	1	8.856	37	12	0	0	0
1	1	7.443	37	10	0	0	0
0	0	14.264	27	12	0	0	1
0	1	0.000	35	10	0	0	1
0	0	3.401	39	12	0	0	1
1	1	0.000	33	16	0	0	0
1	1	9.136	46	12	0	0	0
0	0	4.839	32	7	0	0	0
0	0	14.168	28	12	0	0	1
0	0	7.671	33	12	0	0	1
1	1	1.599	38	12	0	0	0
1	1	3.853	40	12	0	0	0
1	1	5.366	40	12	0	0	1
1	1	7.041	46	12	0	0	1
0	0	12.187	46	12	0	0	0
1	1	12.408	40	12	0	0	0
0	0	5.693	26	12	0	0	1
1	1	3.869	21	12	0	0	0
0	0	3.069	26	12	1	0	1
1	1	7.588	27	16	0	0	0
1	1	7.786	44	12	0	0	0
0	0	12.581	36	12	0	0	0
1	1	5.962	39	10	1	0	0
1	1	4.406	45	10	1	0	0
0	0	6.341	31	12	0	0	0
1	1	7.517	35	12	0	0	1
1	1	7.984	40	12	0	0	0
0	0	7.823	43	12	0	0	0
0	1	6.852	24	12	0	1	0
0	0	8.807	34	12	0	0	1
1	0	9.923	25	12	0	0	1
1	1	9.148	30	12	0	0	1
1	1	4.511	43	7	1	0	1
0	0	15.248	30	12	0	0	1
0	0	7.351	45	12	0	0	0
1	1	20.515	39	12	0	0	0
0	0	8.375	41	12	0	0	0
1	1	7.514	27	12	0	0	1
1	1	9.474	32	12	0	0	0
1	1	10.149	33	18	0	0	0
1	1	4.782	31	12	1	0	0
1	1	10.472	35	12	1	0	1
1	1	6.762	23	12	0	1	0
1	1	9.467	29	12	0	0	1
1	1	10.342	32	12	0	0	1
0	1	1.240	44	12	0	0	0
1	0	13.820	34	12	0	0	0
1	0	9.727	40	12	0	0	0
0	0	8.853	46	12	0	0	0
1	1	8.061	36	12	1	0	0

TABLE V13. Panel Data (continued)

THISYR	LASTYR	HUBINC	AGE	EDUC	BLACK	CHILD1	CHILD2
0	0	4.340	42	12	0	0	0
0	0	13.648	31	12	0	0	1
1	1	4.973	38	10	1	0	1
1	0	8.427	46	12	0	0	0
1	0	18.320	46	18	0	0	0
1	0	7.680	29	10	1	0	1
1	1	5.612	25	12	0	0	1
0	0	13.554	32	12	0	0	1
0	1	5.329	26	12	0	0	0
1	1	10.511	29	12	0	0	0
1	1	10.486	34	12	0	0	0
1	0	14.071	38	16	0	0	0
1	1	9.024	32	12	0	0	0
1	1	14.329	36	12	0	0	1
1	1	5.118	28	18	0	1	0
1	1	3.044	37	12	1	0	0
1	1	2.640	38	7	1	0	0
1	1	2.050	43	7	1	0	0
0	0	6.750	23	12	1	0	1
0	1	3.383	24	12	0	0	0
1	1	6.630	40	12	0	0	0
1	1	7.000	46	12	0	0	0
1	0	8.815	42	12	0	0	0
1	1	3.450	46	12	0	0	0
0	0	12.031	42	12	0	0	0
1	1	6.144	31	12	1	0	0
0	0	11.513	39	12	0	0	1
1	0	12.167	46	12	0	0	1
0	0	9.968	28	16	0	1	0
0	0	5.888	23	12	0	1	0
1	1	10.232	32	12	0	0	0
1	1	8.017	40	12	0	0	0
1	1	11.686	45	12	0	0	0
0	1	28.363	31	12	0	0	1
1	1	4.343	46	7	1	0	0
1	1	10.554	38	12	0	0	0
1	1	2.484	29	10	0	0	1
0	0	5.672	44	12	0	0	0
1	1	13.319	31	18	1	0	0
1	1	7.678	35	18	1	0	0
1	1	7.162	24	12	0	0	0
0	0	7.804	34	12	0	0	0
1	0	13.648	28	16	0	0	1
0	0	9.311	27	12	0	0	1
1	1	27.938	46	12	0	0	0
1	1	6.704	27	12	0	0	1
1	1	7.711	32	12	0	0	0
1	1	8.576	38	16	0	0	0
1	0	7.223	26	16	0	0	1
0	0	11.259	31	16	0	1	0

TABLE V13. Panel Data (continued)

THISYR	LASTYR	HUBINC	AGE	EDUC	BLACK	CHILD1	CHILD2
0	0	26.063	30	12	0	0	1
1	1	11.776	42	12	0	0	0
1	1	12.793	46	18	0	0	0
1	1	11.080	44	12	0	0	0
1	1	7.074	31	12	0	0	0
1	1	6.679	36	12	0	0	1
1	0	15.868	45	12	0	0	0
1	1	7.972	42	16	0	0	0
0	0	0.000	29	12	1	1	0
1	1	3.030	43	10	0	0	0
1	1	2.970	27	16	0	0	0
1	1	9.305	40	12	0	0	0
1	1	8.125	30	12	0	0	0
0	1	13.033	29	10	1	0	1
1	1	0.000	39	12	1	0	0
1	1	2.781	30	12	1	0	1
1	1	3.010	35	12	1	0	0
0	0	26.056	40	12	0	0	0
0	0	5.795	46	12	0	0	0
1	1	0.000	36	12	0	0	1
1	1	2.639	28	12	0	0	1
1	1	9.087	24	12	0	0	0
0	1	12.312	34	12	0	0	0
0	1	7.325	33	12	0	0	0
1	1	3.517	26	10	0	0	0
1	1	17.140	35	12	0	0	0
1	1	24.054	40	12	0	0	0
1	1	6.144	42	12	1	0	0
1	0	13.211	34	12	1	0	0
1	0	9.309	45	12	0	0	0
1	1	3.135	40	10	0	0	0
1	1	2.935	45	10	0	0	0
1	1	9.067	41	12	0	0	0
1	0	10.629	44	12	0	0	0
1	1	8.207	24	12	0	0	0
1	1	9.772	42	12	0	0	0
1	1	8.955	46	12	0	0	0
1	1	6.204	46	10	0	0	0
1	0	9.378	32	12	1	0	0
0	0	54.281	45	12	0	0	0
1	1	7.525	31	12	0	0	1
0	0	11.504	32	12	0	1	0
1	0	5.763	42	12	0	0	0
0	0	5.683	32	12	0	0	1
1	0	10.937	40	12	0	0	0
1	1	9.361	45	12	0	0	0
0	0	6.342	35	12	0	0	1
0	1	7.160	31	10	0	0	0
0	1	7.788	31	12	0	0	1
1	1	2.402	25	10	1	0	0

TABLE V14. U.S. Divorce Data

STATE	BREAK	CRUEL	DESERT	NOSUPPOR	ALCOHOL	FELONY	IMPOTENT	INSANE	SEPERATE
Alabama	1	1	1	1	1	1	1	1	1
Alaska	1	1	1	0	1	1	1	1	0
Arizona	1	0	0	0	0	0	0	0	0
Arkansas	0	1	1	1	1	1	1	1	1
California	1	0	0	0	0	0	0	1	0
Colorado	1	0	0	0	0	0	0	0	0
Connecticut	1	1	1	1	1	1	0	1	1
Delaware	1	0	0	0	0	0	0	0	1
Dist of Columbia	0	0	0	0	0	0	0	1	1
Florida	1	0	1	0	1	1	1	1	0
Georgia	1	0	0	0	0	0	0	0	1
Hawaii	1	1	1	1	1	1	0	1	1
Idaho	1	1	1	1	1	1	0	1	0
Illinois	0	1	1	0	1	1	1	0	0
Indiana	1	1	0	0	0	1	1	1	0
Iowa	1	0	0	0	0	0	0	0	0
Kansas	1	0	0	0	0	0	0	0	0
Kentucky	1	0	1	0	0	0	0	0	1
Louisiana	0	1	1	1	1	1	0	0	1
Maine	1	1	1	1	1	0	1	1	0
Maryland	0	1	1	0	0	1	1	1	1
Massachusetts	1	1	1	1	1	1	1	0	1
Michigan	1	0	0	0	0	0	1	0	0
Minnesota	1	0	0	0	0	0	0	0	0
Mississippi	1	1	1	0	1	1	1	1	0
Missouri	1	0	0	0	0	0	0	0	0

TABLE V14. U.S. Divorce Data (continued)

STATE	BREAK	CRUEL	DESERT	NOSUPPOR	ALCOHOL	FELONY	IMPOTENT	INSANE	SEPERATE
Montana	1	0	0	0	0	0	0	0	0
Nebraska	1	0	0	0	0	0	0	0	0
Nevada	1	0	0	0	0	0	0	1	1
New Hampshire	1	1	1	1	1	1	1	0	1
New Jersey	0	1	1	0	1	1	0	1	1
New Mexico	1	1	1	0	0	0	0	0	0
New York	0	1	1	0	0	1	0	0	1
North Carolina	0	0	0	0	0	0	1	1	1
North Dakota	1	1	1	1	1	1	1	1	0
Ohio	1	1	1	0	1	1	1	1	1
Oklahoma	1	1	1	1	1	1	1	1	0
Oregon	1	0	0	0	0	0	0	0	0
Pennsylvania	1	1	1	0	1	1	1	1	1
Rhode Island	1	1	1	1	1	1	1	0	1
South Carolina	0	1	1	0	1	0	0	0	1
South Dakota	0	1	1	1	1	1	0	0	0
Tennessee	1	1	1	0	1	1	1	0	0
Texas	1	1	1	1	0	1	1	1	1
Utah	0	1	1	1	1	1	1	1	0
Vermont	0	1	1	1	0	1	0	1	1
Virginia	0	1	1	0	0	1	0	0	1
Washington	1	0	0	0	0	0	0	0	0
West Virginia	1	1	1	0	1	1	0	1	1
Wisconsin	1	0	0	0	0	0	0	0	1
Wyoming	1	0	0	0	0	0	0	1	1

TABLE V15. U.S. Divorce Data Subset

STATE	BREAK	CRUEL	DESERT	NOSUPPOR	ALCOHOL	FELONY	IMPOTENT	INSANE	SEPERATE
Alabama	1	1	1	1	1	1	1	1	1
Alaska	1	1	1	0	1	1	1	1	0
Arizona	1	0	0	0	0	0	0	0	0
Arkansas	0	1	1	1	1	1	1	1	1
California	1	0	0	0	0	0	0	1	0
Colorado	1	0	0	0	0	0	0	0	0
Connecticut	1	1	1	1	1	1	0	1	1
Florida	1	0	0	0	0	0	0	1	0
Georgia	1	1	1	0	1	1	1	1	0
Hawaii	1	0	0	0	0	0	0	0	1
Indiana	1	1	0	0	0	1	1	1	0
Iowa	1	0	0	0	0	0	0	0	0
Kansas	1	0	0	0	0	0	0	0	0
Massachusetts	1	1	1	1	1	1	1	0	1
Minnesota	1	0	0	0	0	0	0	0	0
Mississippi	1	1	1	0	1	1	1	1	0
Nebraska	1	0	0	0	0	0	0	0	0
South Carolina	0	1	1	0	1	0	0	0	1
South Dakota	0	1	1	1	1	1	0	0	0
Tennessee	1	1	1	1	1	1	1	0	0

TABLE V16. U.S. Crime Data

STATE	VIOLENT	PROPERTY	MURDER	RAPE	ROBBERY	ASSAULT	BURGLARY	LARCENY	CARTHEFT
Alabama	448	4485	13	30	132	273	1526	2642	316
Alaska	479	5730	9	62	90	317	1385	3727	617
Arizona	650	7519	10	45	193	401	2155	4891	473
Arkansas	335	3475	9	26	80	218	1119	2169	187
California	893	6939	14	58	384	436	2316	3880	742
Colorado	528	6804	6	52	160	309	2030	4325	448
Connecticut	412	5469	4	21	218	168	1700	3089	678
Delaware	474	6301	6	24	137	306	1630	4216	455
Florida	983	7418	14	56	355	556	2506	4434	477
Georgia	555	5048	13	44	197	299	1699	2976	372
Hawaii	299	7182	8	34	190	65	1847	4732	612
Idaho	313	4468	3	22	46	241	1238	2993	236
Illinois	494	4781	10	26	217	239	1242	3039	498
Indiana	377	4552	8	33	141	194	1313	2807	432
Iowa	200	4546	2	14	54	129	1079	3219	247
Kansas	389	4989	6	31	113	237	1521	3196	271
Kentucky	266	3167	8	19	95	143	1040	1875	250
Louisiana	665	4788	15	44	197	407	1523	2888	376
Maine	193	4174	2	12	30	146	1182	2772	219
Maryland	852	5777	9	40	393	410	1698	3629	450
Massachusetts	601	5477	4	27	235	334	1740	2685	1051
Michigan	639	6036	10	46	244	338	1741	3710	584
Minnesota	227	4571	2	23	99	102	1246	3029	295
Mississippi	341	3075	14	24	81	221	1179	1717	178
Missouri	554	4878	11	32	223	287	1668	2795	414

TABLE V16. U.S. Crime Data (continued)

STATE	VIOLENT	PROPERTY	MURDER	RAPE	ROBBERY	ASSAULT	BURGLARY	LARCENY	CARTHEFT
Montana	222	4801	4	21	34	163	950	3529	321
Nebraska	224	4080	4	23	82	114	915	2921	243
Nevada	912	7941	20	67	460	364	2906	4356	678
New Hampshire	179	4499	2	17	42	118	1312	2877	309
New Jersey	604	5797	6	30	303	263	1878	3189	729
New Mexico	615	5364	13	43	127	430	1492	3521	350
New York	1029	5882	12	30	641	344	2061	3064	759
North Carolina	455	4185	10	22	82	339	1422	2546	216
North Dakota	54	2909	1	9	7	35	488	2242	179
Ohio	498	4933	8	34	223	232	1466	3040	426
Oklahoma	419	4633	5	41	104	291	1748	4087	420
Oregon	490	6196	5	41	152	291	1748	4087	360
Pennsylvania	363	3372	6	23	177	156	1038	1915	418
Rhode Island	408	5524	4	17	118	493	1716	2964	843
South Carolina	660	4779	11	37	118	493	1670	2803	306
South Dakota	126	3116	1	12	20	93	692	2255	168
Tennessee	458	4039	10	37	180	229	1501	2175	363
Texas	550	5592	16	47	208	277	1853	3181	558
Utah	303	5577	3	27	80	191	1321	3931	323
Vermont	178	4809	2	29	38	108	1526	2988	294
Virginia	307	4312	8	27	120	151	1202	2882	227
Washington	464	6450	5	52	135	271	1862	4192	395
West Virginia	183	2367	7	15	48	112	738	1429	199
Wisconsin	182	4616	2	14	70	94	1079	3291	245
Wyoming	392	4593	6	28	44	313	903	3344	345

TABLE V17. U.S. Crime Data Subset

STATE	VIOLENT	PROPERTY	MURDER	RAPE	ROBBERY	ASSAULT	BURGLARY	LARCENY	CARTHEFT
Alabama	448	4485	13	30	132	273	1526	2642	316
California	893	6939	14	58	384	436	2316	3880	742
Florida	983	7418	14	56	355	556	2506	4434	477
Massachusetts	601	5477	4	27	235	334	1740	2685	1051
New Hampshire	179	4499	2	17	42	118	1312	2877	309
New Jersey	604	5797	6	30	303	263	1878	3189	729
New Mexico	615	5364	13	43	127	430	1492	3521	350
New York	1029	5882	12	30	641	344	2061	3064	759
North Dakota	54	2909	1	9	7	35	488	2242	179
Oregon	490	6196	5	41	152	291	1748	4087	360
Tennessee	458	4039	10	37	180	229	1501	2175	363
Texas	550	5592	16	47	208	277	1853	3181	558
Vermont	178	4809	2	29	38	108	1526	2988	294
Washington	464	6450	5	52	135	271	1862	4192	395
West Virginia	183	2367	7	15	48	112	738	1429	199

TABLE V18. Automobile Data Part II

TYPE	ENGSIZE	CYLIND	COMBRATE	WEIGHT	FOR
Pontiac Paris	5	3	4	5	0
Honda Civic	1	1	1	1	1
Buick Century	4	2	4	3	0
Subaru GL	1	1	1	2	1
Volvo 740GLE	2	1	2	3	1
Plymouth Caravel	2	1	2	3	0
Honda Accord	1	1	2	2	1
Chev Camaro	3	2	3	4	0
Plymouth Horizon	2	1	2	2	0
Chrysler Daytona	2	1	2	3	0
Cadillac Fleetw	4	3	4	5	0
Ford Mustang	5	3	4	4	0
Toyota Celica	2	1	2	2	1
Ford Escort	1	1	2	2	0
Toyota Tercel	1	1	1	1	1
Toyota Camry	2	1	1	2	1
Mercury Capri	5	3	4	4	0
Toyota Cressida	3	2	3	4	1
Nissan 300ZX	3	2	4	4	1
Nissan Maxima	3	2	4	4	1

TABLE V19. Cola Similarity Data

00										Diet Pepsi
34	00									RC Cola
79	54	00								Yukon
86	56	70	00							Dr. Pepper
76	30	51	66	00						Shasta
63	40	37	90	35	00					Coca-Cola
57	86	77	50	76	77	00				Diet Dr. Pepper
62	80	71	88	67	54	66	00			Tab
65	23	69	66	22	35	76	71	00		Pepsi-Cola
26	60	70	89	63	67	59	33	59	00	Diet-Rite

TABLE V20. Car Similarity Data

00											Mustang SVO
27	00										Cadillac Seville
26	01	00									Lincoln Continental
17	38	39	00								Ford Escort
13	28	29	36	00							Corvette
25	41	42	10	35	00						Chevrolet Corvette
15	32	30	45	09	48	00					Nissan 300ZX
24	40	02	12	43	11	34	00				Renault Alliance
16	33	31	46	08	49	14	44	00			Porshe 944
37	07	05	50	18	52	19	51	21	00		Jaguar XJ6
47	06	04	53	23	55	22	54	20	03	00	Mercedes 500SEL

TABLE V21. Air Pollution Data Subset

CITY	TMR	SMIN	SMEAN	SMAX	PMIN	PMEAN	PMAX	PM2	PERWH	NONPOOR
Jersey City	1199	155	229	340	63	147	253	1357.2	93.1	87.3
Denver	841	2	61	188	54	126	229	25.4	95.8	86.9
Milwaukee	921	65	134	236	49	150	299	150.2	94.4	90.4
Macon	869	18	27	128	22	122	754	28.6	69.0	73.7
Boston	1112	42	163	337	55	141	252	174.5	97.3	88.5
Bridgeport	938	137	205	308	32	91	182	103.3	94.7	90.7
Chicago	1000	75	166	328	88	182	296	167.5	85.2	89.4
Austin	689	40	46	58	10	78	157	20.9	87.2	75.2
Seattle	938	1	47	179	32	69	141	26.2	95.2	88.8
Mobile	823	47	67	248	29	129	284	25.3	67.7	74.6
Augusta	823	31	46	158	28	66	142	15.2	70.2	67.8
Charleston	780	15	283	940	55	225	958	27.9	94.2	78.6

TABLE V22. Shopping Attitude Data Part II

A1	A2	A3	A4	A5	A6	A7	A8	A9	A10	A11	A12	A13	A14	A15	A16	A17	A18	WORK	AGE
2	5	2	2	4	4	4	4	2	2	2	4	4	4	2	2	4	4	2	2
3	2	1	5	1	1	3	4	2	3	2	3	3	4	2	2	4	2	2	4
1	3	3	3	5	5	4	4	2	2	2	4	4	4	4	2	4	2	1	1
1	3	5	2	5	4	4	2	3	2	5	2	2	3	2	4	4	2	1	1
2	1	1	5	1	3	3	4	2	2	4	2	2	4	2	1	5	4	2	5
3	1	2	3	4	3	5	5	3	4	3	2	1	1	1	5	4	4	2	2
1	1	4	4	3	4	4	2	4	2	2	4	4	4	2	2	5	5	1	3
4	1	1	5	3	3	3	2	4	5	4	4	2	2	2	4	4	4	2	5
2	1	3	5	3	1	4	4	2	1	5	1	2	5	2	1	3	4	2	5
5	1	3	5	3	3	3	2	4	4	4	4	4	4	4	1	5	5	2	6
1	2	4	1	5	3	5	4	2	1	5	2	2	3	1	1	4	1	1	1
3	3	3	5	2	3	3	2	4	3	4	2	5	4	2	4	2	5	2	4
1	1	4	1	5	4	5	4	2	1	3	2	4	5	4	2	4	4	1	1
1	2	3	5	1	4	5	1	2	1	1	4	4	5	5	1	2	2	2	5
2	1	1	5	3	2	4	4	4	3	2	2	5	5	2	1	4	5	2	5
3	1	3	3	3	3	4	4	2	4	4	2	5	5	5	1	2	2	2	5
1	2	2	3	3	3	2	2	4	3	2	4	1	1	1	1	2	1	1	3
3	4	3	2	3	3	4	4	2	3	4	2	4	4	2	4	4	4	2	2
1	1	4	4	3	3	2	4	4	3	2	1	3	3	1	4	5	5	1	2
3	2	5	4	5	2	4	4	2	3	4	4	5	5	3	1	3	5	2	4
2	4	5	3	3	4	5	2	2	2	2	4	5	4	5	2	2	4	1	2
2	1	4	1	4	3	5	5	2	2	1	1	1	2	1	5	5	2	1	2
1	4	3	2	5	3	5	4	2	2	4	2	2	2	1	4	4	4	1	4
2	3	2	3	2	3	4	2	2	2	4	1	2	2	1	4	4	2	1	1
1	4	5	3	4	5	5	4	3	3	3	2	2	3	2	2	4	3	1	3
1	1	1	1	1	1	1	1	1	1	1	1	1	1	1	1	1	1	1	1

TABLE V22. Shopping Additude Data Part II (continued)

A1	A2	A3	A4	A5	A6	A7	A8	A9	A10	A11	A12	A13	A14	A15	A16	A17	A18	WORK	AGE
1	5	5	1	5	3	5	5	1	2	3	1	2	1	2	1	5	2	1	2
3	1	3	5	1	3	3	4	3	2	4	1	4	4	2	2	4	4	2	5
1	4	3	2	3	4	5	5	2	3	5	1	2	2	2	1	5	1	2	1
3	1	2	3	3	1	1	4	2	2	4	2	2	2	2	2	2	2	2	3
1	5	1	3	5	5	5	5	1	1	5	1	4	4	2	4	5	2	1	2
3	1	4	3	1	3	3	4	2	3	4	2	2	2	3	4	4	2	2	3
4	2	1	4	2	2	3	5	2	4	5	3	3	4	2	2	3	3	2	5
1	4	1	1	5	5	5	4	1	2	3	2	2	4	4	2	4	4	1	1
1	3	4	3	4	5	5	2	2	4	4	3	2	4	2	2	5	2	1	2
2	2	5	2	3	3	4	4	2	2	4	1	4	4	4	4	4	4	1	3
1	3	5	2	4	3	4	4	2	2	1	3	4	4	2	2	4	2	2	3
2	2	4	3	3	2	2	4	2	3	2	3	2	2	2	2	4	4	2	6
2	1	2	2	1	2	4	4	2	2	2	2	3	4	4	2	4	2	1	4
3	1	2	3	3	3	3	2	2	3	4	2	2	2	2	4	2	4	2	5
1	2	1	2	1	1	3	5	2	2	2	2	4	4	2	3	4	2	2	5
2	1	4	3	3	2	4	3	2	2	2	2	2	4	4	2	1	4	2	6
3	3	2	5	3	2	3	2	4	3	3	2	4	4	2	4	4	4	2	2
1	2	2	4	2	1	4	4	4	2	2	4	5	4	4	4	4	5	2	6
4	1	3	5	3	5	3	4	2	2	2	2	2	4	2	4	2	2	2	3
3	3	4	5	5	5	1	4	2	2	2	2	2	4	2	4	4	2	1	3
2	5	5	1	5	5	4	4	2	2	4	2	4	4	4	4	4	2	1	2
1	4	5	3	5	5	5	5	4	2	4	3	4	4	2	2	2	5	1	2
1	1	3	3	1	3	3	4	4	3	1	1	4	5	5	1	4	4	1	1
5	3	3	5	5	3	5	5	2	3	1	3	3	4	2	2	2	5	2	5
1	3	4	2	3	4	4	4	2	1	4	2	3	4	2	2	5	2	1	5

TABLE V22. Shopping Attitude Data Part II (continued)

A1	A2	A3	A4	A5	A6	A7	A8	A9	A10	A11	A12	A13	A14	A15	A16	A17	A18	WORK	AGE
2	1	3	5	1	3	3	4	2	1	4	1	1	4	1	3	4	4	2	3
1	1	3	2	2	4	4	3	2	3	4	2	4	4	4	2	4	4	1	3
5	4	1	5	2	1	3	4	2	3	3	1	3	3	2	2	4	2	2	5
3	1	2	3	3	4	4	5	2	3	1	5	5	5	5	1	1	5	2	2
2	3	2	4	2	4	4	4	2	4	2	3	4	4	2	2	2	2	1	3
1	5	4	3	5	3	5	2	2	3	4	2	2	2	2	4	5	4	1	2
4	1	1	4	3	1	3	3	2	2	2	4	3	4	2	3	4	2	2	4
3	1	2	4	3	2	3	4	2	3	4	2	5	5	2	3	3	2	2	4
2	4	1	4	1	2	3	4	3	3	5	1	2	2	2	2	5	1	2	6
1	5	5	1	3	4	5	4	2	2	4	2	4	4	2	4	4	2	1	1
2	2	2	2	1	2	4	4	2	1	4	1	2	2	2	4	4	4	1	4
1	3	2	4	1	1	3	4	2	2	5	1	2	4	2	4	4	2	2	6
1	2	5	3	3	5	4	3	2	4	3	3	4	3	2	3	2	2	1	2
1	4	3	2	3	3	5	4	1	2	2	1	4	4	4	2	3	3	1	2
2	3	5	2	4	4	5	4	4	1	5	4	2	4	2	2	4	5	2	1
1	4	5	3	5	4	3	4	2	3	4	2	3	2	2	2	4	2	1	1
4	2	5	4	4	4	3	4	4	2	2	3	3	3	2	2	4	2	2	4
3	5	3	3	1	1	5	2	2	1	4	3	2	3	2	4	5	2	1	6
2	2	3	1	5	4	4	4	4	2	1	1	4	4	2	1	4	2	2	2
1	2	4	2	4	4	4	2	4	2	3	4	4	3	2	2	4	4	1	2
1	2	3	2	4	5	5	5	4	2	5	2	2	5	5	4	4	3	2	2
1	3	3	3	5	4	5	3	1	4	1	1	1	3	3	1	4	5	1	2
1	5	2	1	2	2	5	2	1	1	2	4	2	4	2	3	4	1	1	2
3	1	1	5	3	1	3	2	2	2	2	2	2	4	2	2	4	4	2	6

TABLE V22. Shopping Attitude Data Part II (continued)

A1	A2	A3	A4	A5	A6	A7	A8	A9	A10	A11	A12	A13	A14	A15	A16	A17	A18	WORK	AGE
2	2	2	2	3	3	4	4	2	3	2	4	4	1	3	2	4	4	1	3
2	2	2	4	4	4	4	4	2	2	4	2	2	3	2	2	4	2	1	5
2	1	3	5	1	3	3	4	4	2	5	5	4	2	2	2	5	2	2	5
3	2	1	5	1	3	3	2	4	2	2	2	2	2	1	4	4	4	2	4
1	4	3	2	5	1	5	2	4	2	4	3	3	4	2	4	5	4	1	1
2	1	1	3	5	5	3	4	2	2	4	2	2	2	2	2	5	2	2	4
3	1	1	5	1	3	3	4	4	4	4	4	2	4	2	4	4	2	2	5
3	1	1	3	1	1	2	4	2	3	5	2	4	2	2	2	5	2	2	6
3	1	1	3	3	4	1	3	2	5	2	4	4	2	2	2	4	3	2	5
2	2	2	4	3	3	4	4	2	3	2	3	4	3	2	3	4	4	2	1
1	3	3	1	4	3	5	2	2	3	4	2	2	3	4	4	5	4	1	2
1	4	3	1	4	2	4	5	2	3	2	3	5	3	2	2	4	4	2	1
2	3	2	4	2	2	3	4	2	3	2	3	3	4	2	2	4	2	2	2
2	1	1	5	1	2	5	4	1	2	5	1	4	3	2	2	4	4	2	3
1	4	5	1	4	4	3	4	2	2	2	3	2	3	2	2	4	2	1	2
3	1	3	4	1	4	3	4	2	2	4	2	2	3	2	3	4	2	2	6
2	2	3	2	2	4	3	4	1	2	1	2	2	3	1	5	5	2	2	1
4	2	4	2	2	4	4	4	1	2	4	4	2	2	2	4	4	2	1	3
1	2	1	4	2	3	2	2	2	2	4	2	2	4	2	4	4	4	2	2
5	1	3	3	3	3	3	3	1	3	4	2	4	3	1	3	4	4	2	5
2	2	3	3	3	3	4	3	2	3	2	2	3	3	3	1	2	4	2	2
3	4	4	4	3	2	2	4	4	1	3	1	4	5	1	4	4	4	1	3
3	1	1	3	1	3	3	4	2	1	5	1	1	1	3	2	4	2	2	5
3	3	3	3	3	3	4	4	2	4	2	2	5	4	2	2	4	4	2	6
1	5	3	2	3	3	4	4	2	3	1	2	3	3	2	2	3	4	1	2

TABLE V22. Shopping Attitude Data Part II (continued)

A1	A2	A3	A4	A5	A6	A7	A8	A9	A10	A11	A12	A13	A14	A15	A16	A17	A18	WORK	AGE
3	3	3	2	3	4	5	4	2	3	4	3	4	4	1	2	4	4	2	2
1	5	4	4	4	3	3	4	2	3	4	2	3	4	2	2	4	4	2	1
3	1	1	5	3	3	1	5	2	2	2	4	3	4	1	4	4	2	2	4
1	3	5	1	3	2	5	2	4	4	4	4	1	2	2	4	5	2	1	2
3	2	1	4	1	2	1	4	2	2	2	2	4	5	2	2	4	1	2	6
2	5	2	5	5	5	4	5	5	2	5	5	4	5	1	5	5	1	1	5
1	2	3	4	3	4	4	4	4	2	4	2	4	4	4	2	2	4	1	3
1	5	5	1	3	5	5	3	3	3	4	4	4	4	2	2	3	4	1	2
1	3	1	1	5	3	4	2	2	4	5	1	1	1	1	3	5	1	2	2
2	2	4	4	5	4	4	4	4	4	4	2	4	4	4	2	2	4	2	2
3	1	1	3	4	1	3	4	2	2	4	2	4	4	2	2	4	2	2	6
5	1	1	3	1	3	3	2	4	4	1	3	1	1	1	1	5	3	1	6
4	1	2	2	3	2	3	4	2	4	5	2	2	2	2	4	5	3	2	5
2	1	2	5	3	1	2	4	2	1	4	2	5	5	5	1	2	5	2	3
3	1	3	3	4	3	3	4	2	4	3	4	2	4	2	4	4	2	2	4
2	4	2	2	3	1	3	5	1	3	4	2	2	2	2	2	4	2	2	2
2	1	1	5	4	2	3	4	1	2	4	1	1	1	1	4	4	2	2	3
2	2	3	2	3	4	5	2	2	1	4	2	1	2	1	4	5	1	1	6
2	2	2	2	1	2	5	4	2	4	2	2	4	2	3	2	4	4	2	5
2	3	2	2	2	3	4	2	2	3	2	1	2	1	1	5	2	1	2	1
2	4	4	2	3	4	4	4	4	5	2	4	2	2	2	4	5	1	2	1
1	4	4	2	4	4	4	4	2	2	1	4	2	2	2	4	4	2	1	4
1	2	3	4	3	3	4	2	4	3	4	4	2	2	2	2	5	5	2	1
1	2	2	2	4	4	4	4	2	2	4	4	2	4	2	2	5	4	2	2
1	2	2	4	4	4	4	4	2	2	4	2	2	2	1	2	5	1	1	4

TABLE V22. Shopping Attitude Data Part II (continued)

A1	A2	A3	A4	A5	A6	A7	A8	A9	A10	A11	A12	A13	A14	A15	A16	A17	A18	WORK	AGE
2	1	1	5	3	3	3	2	4	4	1	4	2	3	1	4	4	2	2	3
1	5	3	2	3	3	5	5	5	2	2	3	2	3	2	2	4	4	1	4
1	5	3	1	5	5	5	2	1	2	4	4	4	4	2	4	4	2	1	1
3	1	4	5	2	3	4	5	2	2	4	2	4	4	4	1	4	4	2	6
2	1	1	4	1	2	4	5	2	4	5	1	2	1	1	4	4	2	2	5
4	1	1	3	1	1	2	2	1	2	2	4	5	5	3	2	3	5	2	6
3	1	3	5	3	3	3	4	2	2	4	2	1	2	1	4	5	1	2	5
3	3	2	3	3	2	3	4	2	2	2	3	2	3	2	4	5	2	2	5
3	1	3	5	1	3	3	5	2	2	4	2	3	3	2	2	4	2	2	4
3	1	5	5	1	1	5	1	5	5	5	1	5	5	5	1	1	5	1	5
1	1	2	4	4	4	4	4	2	2	2	2	4	4	4	2	2	4	1	5
2	1	5	2	4	2	4	4	4	1	5	2	5	4	4	2	2	4	1	5
2	2	3	1	5	5	5	4	2	2	4	2	2	2	2	2	4	4	1	3
1	3	1	5	2	1	3	5	3	3	1	2	4	5	4	1	2	4	1	1
2	1	1	4	1	3	3	3	1	2	1	4	3	3	2	2	5	2	2	5
2	1	1	5	2	3	3	4	1	1	1	2	5	4	5	1	3	4	2	6
2	2	2	4	3	3	2	4	3	2	2	2	2	2	1	3	4	1	2	3
3	1	2	5	3	4	3	2	2	4	4	2	3	3	2	3	3	2	2	2
1	5	4	1	5	2	5	4	1	2	3	1	1	2	3	2	4	1	1	5
1	3	1	2	4	3	1	1	4	2	5	3	2	1	1	4	5	3	1	2
3	3	5	2	2	4	4	4	4	2	4	2	4	2	1	4	5	2	1	4
1	2	2	2	3	2	4	2	2	4	3	1	2	4	2	2	3	2	2	3
2	1	3	3	4	4	4	5	1	1	2	4	2	3	2	4	4	1	2	5
2	1	4	3	4	3	4	4	1	1	2	2	2	4	2	1	5	1	2	2
1	1	1	1	4	1	4	4	1	1	4	1	4	4	1	1	4	4	1	3

TABLE V22. Shopping Attitude Data Part II (continued)

A1	A2	A3	A4	A5	A6	A7	A8	A9	A10	A11	A12	A13	A14	A15	A16	A17	A18	WORK	AGE
3	1	3	5	3	1	3	2	2	4	5	4	5	3	2	4	2	3	2	5
1	2	3	2	2	2	4	2	2	1	4	2	3	3	2	3	4	2	2	1
1	3	1	2	3	3	4	5	2	2	3	2	2	3	2	4	4	2	1	3
2	2	4	2	1	4	5	2	1	4	2	4	1	2	2	4	5	1	1	4
1	5	5	1	3	5	5	2	2	2	4	2	2	3	2	5	5	2	1	1
2	2	2	2	4	4	4	4	2	2	2	4	4	4	2	2	2	4	1	6
1	5	5	1	5	5	5	4	4	2	5	2	4	5	3	2	4	5	1	1
3	2	3	3	4	3	4	2	4	3	2	4	4	4	4	2	3	4	1	4
5	3	3	3	5	1	4	4	4	3	4	2	5	5	5	1	2	5	2	6
1	5	1	1	4	1	5	4	2	2	3	3	1	1	1	4	5	2	1	1
5	1	5	5	1	5	3	2	4	4	4	2	4	4	3	2	4	4	2	6
2	5	3	1	3	5	5	4	4	2	2	3	2	3	3	4	5	2	1	1
3	1	3	4	5	1	3	4	2	1	5	2	2	4	2	1	4	2	2	6
1	2	3	3	1	2	5	2	1	2	2	1	2	5	1	3	4	1	1	2
1	3	4	2	4	3	3	4	2	3	4	2	2	3	1	4	4	1	1	2
1	3	3	1	4	4	5	4	2	1	4	2	4	2	1	4	4	2	1	2
4	1	1	2	3	1	4	2	4	3	2	1	2	4	2	1	2	4	2	5
2	1	1	3	3	1	2	4	2	1	1	2	2	4	2	2	4	4	1	6
3	3	4	2	2	3	6	4	4	3	1	5	4	2	1	1	2	2	1	1
2	1	4	3	4	4	5	2	2	3	4	3	2	4	4	2	5	5	1	5
1	3	2	3	2	2	4	2	2	2	4	2	3	2	2	2	4	2	1	4
2	1	5	3	2	4	5	4	2	3	1	4	4	4	1	1	5	2	1	2
5	1	3	3	1	3	1	4	2	2	4	2	2	4	2	2	4	2	2	6
1	1	1	5	1	1	3	4	2	2	4	2	4	4	4	2	5	4	2	6
—	1	4	5	3	4	4	5	2	2	5	2	4	4	4	2	2	4	1	4

TABLE V22. Shopping Attitude Data Part II (continued)

A1	A2	A3	A4	A5	A6	A7	A8	A9	A10	A11	A12	A13	A14	A15	A16	A17	A18	WORK	AGE
2	1	2	5	1	3	3	4	2	2	4	1	3	2	2	4	5	2	2	6
1	3	3	2	2	4	2	4	2	2	1	4	1	4	4	3	5	2	2	2
1	5	3	1	5	5	5	4	2	2	2	2	2	5	1	2	4	2	1	2
1	3	4	1	4	5	5	2	2	4	3	2	2	4	2	4	4	2	1	1
1	3	3	4	4	3	5	3	2	2	4	2	4	2	1	4	5	3	1	2
2	5	2	4	2	3	4	4	2	3	2	2	5	3	2	2	4	2	2	5
1	2	4	4	3	4	5	5	2	4	3	2	2	2	2	5	5	2	2	2
3	1	1	3	2	1	3	4	4	3	2	3	4	4	2	2	2	2	1	6
1	5	3	3	5	5	5	4	2	2	4	2	2	2	4	4	4	2	1	2
1	3	5	3	3	5	5	4	4	1	5	1	2	4	1	2	4	1	2	2
1	3	5	1	5	5	5	5	3	3	4	2	3	4	1	5	5	2	1	1
2	2	3	5	3	2	3	2	1	3	4	4	2	3	1	2	5	5	1	3
1	3	4	1	5	5	5	4	4	1	2	3	1	2	4	4	4	2	2	1
1	1	1	3	4	2	5	5	2	2	1	2	5	5	1	1	4	2	1	5
2	5	3	2	3	3	4	4	2	2	2	2	3	3	2	4	4	4	1	2
3	1	4	5	1	1	3	2	4	2	2	2	4	4	2	2	4	4	1	6
3	1	5	5	3	3	3	4	3	1	4	2	4	2	2	3	4	3	2	4
2	2	3	2	3	3	4	4	2	2	3	2	2	1	2	4	5	2	2	5
1	4	2	2	5	4	4	4	2	2	2	2	2	2	2	2	4	2	1	1
1	5	5	1	5	5	5	4	4	4	2	2	2	4	2	4	5	1	1	2
3	3	3	3	1	1	3	4	2	4	2	4	1	2	2	2	4	4	2	5
2	2	2	4	2	4	4	4	4	4	4	3	5	4	2	2	4	5	1	4
3	1	5	3	2	1	3	2	2	2	2	3	2	3	2	2	4	5	2	6
1	5	2	1	4	4	4	5	4	3	4	4	5	3	2	2	4	5	1	2
2	1	1	5	1	3	4	5	5	3	2	1	2	2	1	4	5	2	2	6

Index

Author Index

A

Agresti, Alan 87, 116
Amemiya, Takeshi 327
Anderberg, M.R. 568, 603
Andersen, Erling B. 46, 87, 116, 306, 327, 453, 465, 467
Anderson, T.W. 169, 190, 191, 241, 327
Arabie, P. 567, 584, 603

B

Ben-Akiva, Moshe 322, 327
Bernstein, Ira H. 467
Bibby, J.M. 144, 157, 169, 190, 191, 327, 468, 568, 604
Bishop, Yvonne M.M. 46, 58, 87, 116

C

Cailliez, Francis 583, 603
Carmone, Frank J. 585, 601, 602, 603
Carroll, J.D. 584, 603
Cattell, R.B. 417, 467
Christensen, Ronald 87, 116
Conway, Delores A. 417, 467
Cooper, Martha C. 564, 565, 566, 567, 604
Coxon, A.P.M. 585, 591, 602, 603

D

Davidson, M.L. 591, 602, 603
de Leeuw, J. 595, 605
Devlin, S.J. 156, 191
Dubes, R.C. 564, 568, 603
Duffy, Dianne E. 87, 116, 306, 328

E

Edelbrock, Craig 567, 603

Everitt, B.S. 46, 116, 467, 568, 603

F

Feldt, L.S. 163, 191
Fienberg, Stephen E. 46, 58, 82, 87, 116, 306, 321, 327
Fomby, Thomas B. 306, 327
Forthoffer, Ron N. 115, 116
Fox, John 306, 327
Freeman, Daniel H. Jr. 46, 115, 116
Freeman, J.L. 115, 116

G

Geisser S. 163, 191
Gibbons, Dianne 347, 467
Gifi, A. 467, 603
Gnanadesikan, R. 157, 191
Gordon, A.D. 568, 603
Gorsuch, R.L. 406, 426, 467
Graef, Jed 592, 605
Green, Paul E. 585, 601, 602, 603
Greenacre, M.J. 453, 457, 461, 462, 468
Greenhouse, S.W. 163, 191
Griffiths, W.E. 286, 306, 327
Grizzle, James E. 115, 116
Gunst, R.F. 347, 467

H

Hakistan, A.R. 417, 467
Harman, H.H. 406, 467
Hartigan, J.A. 603
Hawkins, D.M. 157, 191, 467
Hill, R. Carter 286, 306, 327
Holland, Paul W. 46, 58, 87, 116
Hosmer, David W. 290, 306, 327
Hubert, L. 567, 603
Huynh, H. 163, 191

Subject Index

Springer Texts in Statistics *(continued from page ii)*

Santner and Duffy: The Statistical Analysis of Discrete Data
Saville and Wood: Statistical Methods: The Geometric Approach
Sen and Srivastava: Regression Analysis: Theory, Methods, and Applications
Whittle: Probability via Expectation, Third Edition
Zacks: Introduction to Reliability Analysis: Probability Models and Statistical
   Methods